Encyclopedia of Air

Encyclopedia of Air

David E. Newton

Greenwood Press
Westport, Connecticut • London

Library of Congress Cataloging-in-Publication Data

Newton, David E.
  Encyclopedia of air / David E. Newton.
     p. cm.
  Includes bibliographical references and index.
  ISBN 1–57356–564–4 (alk. paper)
  1. Atmosphere—Encyclopedias. 2. Meteorology—Encyclopedias. I. Title.
  QC854.N48   2003
  551.5′03—dc21   2003044076

British Library Cataloguing in Publication Data is available.

Library of Congress Catalog Card Number: 2003044076
ISBN: 1–57356–564–4

First published in 2003

Greenwood Press, 88 Post Road West, Westport, CT 06881
An imprint of Greenwood Publishing Group, Inc.
www.greenwood.com

Printed in the United States of America

The paper used in this book complies with the
Permanent Paper Standard issued by the National
Information Standards Organization (Z39.48–1984).

10   9   8   7   6   5   4   3   2   1

For Linda Otto and Jeff Blum,
Good and loyal friends,
And how much more can one ask?

Contents

Guide to Selected Topics

Social, Cultural, and Legal Topics

Sports and Recreation

Transportation

Introduction

Many ancient Greek scholars embraced the idea that the world is composed of a few basic materials and that everything we see is merely some combination or another of these fundamental "elements." The elements most commonly identified by the ancient Greeks were earth, air, fire, and water. Today, we know that none of these materials is elemental in the modern sense of the word, that is, incapable of being reduced to some simpler substance. Nonetheless, all four materials are still basic in other, non-chemical senses of the word. For example, both air and water are essential to the survival of all forms of life on Earth. The *Encyclopedia of Air* exams the myriad ways in which one of those materials— air—is fundamental in nature and in the world made by humans.

The book is, in part, a scientific and technical treatise. It outlines some of the basic facts about the composition and properties of air, and it pays special attention to a variety of topics in the science of meteorology, which is based largely on our understanding of the composition and movements of air in the atmosphere. The book also describes many of the ways in which air's properties are put to use in a variety of devices that humans have invented and developed over the years. It also reviews some of the ways in which air is important in the world of living organisms, including issues such as respiration and flight.

But air is important in ways that go beyond physical and biological phenomena. Long before scientists discovered explanations for the properties and behavior of air, humans had tried to understand air in other ways, often by ascribing its properties to supernatural beings. The mythology of nearly every culture contains at least one god or goddess thought to be responsible for the wind, the sky, or other air-related phenomena. Although gods and goddesses are less central to the explanation of natural phenomena today, humans still draw on the properties of air to explain or describe their own thoughts and feelings in works of art, musical compositions, and literary works. The *Encyclopedia of Air* also recalls some of the many ways in which air as a concept has inspired cultural works in many fields.

Although it is not in any sense a thorough and complete review of the topic of air—the subject is much too broad and varied to be covered in a single volume—the *Encyclopedia of Air* does attempt to provide a general introduction to many important themes relating to air. To aid the reader in a further explanation of many of these topics, lists for additional reading and sources for further research are provided at the conclusion of most essays. The *Encyclopedia* also offers an extensive general bibliography, a guide to selected topics to allow the reader to explore broad themes, numerous illustrations, and a detailed subject index.

A

Adiabatic Changes

See Stability of the Air

Aeration

Aeration is the process by which some material is infused with air. The two most common methods of aeration are by bubbling water through the material (usually water) or by spraying the material (again, usually water) into the air. Aeration has many practical applications, the most common of which are soil aeration for improved plant growth, aeration of water to increase the supply of dissolved oxygen for aquatic organisms, and aeration of impure water supplies.

Soil Aeration

Soil aeration is used in circumstances in which the ground has become hard packed, preventing air (oxygen) from reaching the roots of plants in the soil. Roots require oxygen for their growth. The process of aeration is relatively simple. Small plugs of soil are removed, leaving behind narrow cylindrical holes in the ground through which oxygen can have access to plant roots. In addition to increasing the supply of oxygen to roots, aeration may have other benefits also, including an increased uptake of fertilizer

and water, reduction in water runoff and puddling, improved resiliency of soil and plants, and a more rapid decomposition of dead organic material in the soil.

Aeration for Aquatic Organisms

In natural settings, such as lakes, rivers, ponds, and the oceans, aquatic organisms receive all the oxygen they need as a result of the interaction between water and the atmosphere. In non-natural situations, those same conditions may not hold. In home aquaria or commercial fish farms, their may be too many organisms for the amount of oxygen that can be dissolved in a limit supply of water. For example, people who keep fish as pets generally provide a system by which additional air is pumped through their aquarium to ensure that their fish receive an adequate supply of oxygen. Commercial fish farms may also use large pumps to bubble air through their water to ensure that fish obtain enough oxygen not only to stay alive, but also to thrive.

Water Treatment Facilities

Most municipal water treatment facilities include a number of steps designed to remove impurities from raw water to be used for drinking or from wastewater resulting from residential and commercial use. For example, some water treatment plants simply spray water into the air from large pumps, allowing oxygen in the atmosphere to dissolve in water and react with undesirable components.

One purpose of aerating impure water is to increase the amount of oxygen available to zooplankton, single-cell organisms that consume algae in water. Another purpose of aeration is to destroy and remove substances in water that are easily oxidized. If these substances are not removed at this point, they will react with chlorine, generally used at a later stage of water treatment to destroy disease-causing microorganisms. The process of spraying water into the air also increases the rate at which undesirable gases, such as hydrogen sulfide and methane, are released from water. Aeration also increases the availability of oxygen for the oxidation and destruction of a host of organic compounds that may cause unpleasant odors, colors, and taste in water. Finally, oxygen made available during aeration may react with other undesirable components of water. For example, in some parts of the country, natural water contains relatively high concentrations of iron in the form of the iron(II) ion (Fe^{21}). Later stages of water purification that remove many undesirable chemicals are ineffective in removing the iron(II) ion. But during aeration, the iron(II) ion is converted to the iron(III) ion, which precipitates out of water and can be easily removed by filtration.

Water treatment designers must usually choose between two forms of in-water aeration, fine-bubble or coarse-bubble systems. As their name suggests, the two systems differ from each other in the size of air bubbles injected into water. Fine-bubble systems have an important advantage in that the total amount of air exposed to water is much greater than in a coarse-bubble system. At one time, many water treatment plants made use of fine-bubble systems for this reason. The problem is that devices for generating and releasing fine bubbles tend to clog rather easily and systems have to be cleaned frequently. As a result, many treatment systems changed over to coarse-bubble systems over the past few decades. In recent years, research has been conducted on improving systems for releasing fine bubbles, with the result that such systems are once again competitive with coarse-bubble systems for water purification purposes.

Other Functions

Aeration has applications in other industries also. For example, it has been used to some extent in the cooling and drying of grain. Grain that is to be stored for long periods (as in grain elevators) should be kept as cool and dry as possible. These conditions reduce the risk of spontaneous combustion and the growth rate of insects that may live in and off the stored grain. The installation of aeration equipment for the cooling and drying of grain is relatively inexpensive and is becoming increasingly popular in some parts of the world. *See also* Oxygen

Further Reading

Glinski, Jan, and Witold Stepniewski. *Soil Aeration and Its Role for Plants.* Boca Raton, FL: CRC Press, 1985.

Mueller, James A. *Aeration: Principles and Practice.* Boca Raton, FL: CRC Press, 2002. (water purification)

Navarro, Shlomo, and Ronald T. Noyes. *The Mechanics and Physics of Modern Grain Aeration Management.* Boca Raton, FL: CRC Press, 2001.

Aerial Sports

Humans have been intrigued about the possibility of flying through the air as a form of recreation for centuries. In many cases, these aerial sports evolved out of and along with more serious efforts to find ways of traveling through the air. Today's hang gliders,

A Swick-T performs aerobatics. © 2001 EAA/ LeeAnn Abrams.

Formal aerobatic competition began in Germany in 1928 with the development of a mathematical system for judging fliers' performances. As with diving and gymnastics competitions, each contestant was rated on a scale for excellence of performance, and each type of maneuver was assigned a specific degree of difficulty. A contestant's final score was then determined by multiplying his or her average rating times the degree of difficulty for the stunt performed. The system developed more than seven decades ago is still largely in use at national and international competitions.

The first world championship in aerobatics was held in Paris in June 1934 with nine fliers from six countries (Czechoslovakia, France, Germany, Great Britain, Italy, and Portugal). Two contestants were killed in the competition, and one seriously injured. After a lull due to World War II, international aerobatic competitions were

paragliders, aerobatic and competition fliers, skydivers, and target parachutists practice sports that grew out of the development of both lighter-than-air airships (such as balloons and gliders) and heavier-than-air aircraft (such as airplanes).

Aerobatics and Air Racing

In the decade following its invention by the Wright brothers, the airplane rapidly became popular in a variety of recreational sports. One of these sports, acrobatics (short for "aerial acrobatics"), involved the performance of daring feats of flying, such as spins, loops, rolls, and stalls. Aerobatics was also known as stunt flying and was practiced by a group of men and women who traveled around the country performing at county fairs, exhibitions, and other large gatherings. The sport was very dangerous, and many stunt fliers were killed during their acts.

A Skybolt performs aerobatics. © 2001 EAA/ Jim Koepnick.

A Pitts Model 12 performs aerobatics. ©
2001 EAA/ LeeAnn Abrams.

revived in 1960. These competitions have
been held every other year since then, except
for 1974 and 1992.

Aerobatic competitions are held in a box
in space, 1,000 meters (3,300 feet) on a side
and 1,000 meters above the ground. Fliers can
not descend to less than 100 meters (330 feet)
above the ground. Competitions usually consist of three parts: known compulsory,
unknown compulsory, and freestyle. Pilots are
told in advance what stunts they must do in the
first of these, but they are told about stunts for
the second event just before takeoff. As the
name suggests, freestyle events involve stunts
that pilots themselves choose to do.

Air racing also became popular soon
after the Wright brothers's success at Kitty
Hawk. In August 1909, the world's first air
race was held at Rheims, France. Awards
were given for the highest altitude reached,
the longest flight, the most passengers
carried, and the fastest speed over a 10-
kilometer (6-mile) course. The first air race
in the United States was held only a year
later in Los Angeles. The winner of that con-
test was Glenn H. Curtiss, who set a world
record speed of 55 miles per hour.

Air racing reached a peak of popularity
in the United State in the 1920s when con-
tests were held throughout the country. The
largest of these events was the National Air
Races, held in Cleveland, Ohio. More than

500,000 spectators witnessed these races in
1929, after which the Great Depression made
continuation of the events more difficult.

After World War II, air racing once
again became more popular. The National
Air Show was first held in 1951 and continued until 1957. In 1964, the show moved to
Reno, Nevada, which has hosted the event
annually ever since. The Reno show usually
includes competition for five classes of aircraft; a long-distance flight from St. Petersburg, Florida, to Reno; aerobatic events; and
balloon and parachuting competitions.

Further Reading

Handleman, Philip. *Air Racing Today: The
Heavy Iron at Reno.* Osceola, WI: Motorbooks International, 2001.

Matthews, Birch. *Race with the Wind: How Air
Racing Advanced Aviation.* Osceola, WI:
Motorbooks International, 2001.

Szurovy, Geza, and Mike Coulian. *Basic Aerobatics.* New York: McGraw-Hill, 1994.

Bungee Jumping

Bungee jumping is a sport in which a
person is attached to an elastic cord (the
bungee) and then jumps off a tall structure.
The person experiences free fall for a matter of seconds until the bungee cord achieves
maximum extension. As the bungee then
retracts, the person is alternately raised and
lowered until the bungee once more reaches
its greatest extension.

Bungee jumping is said to have been
modeled after an ancient rite of passage
practiced by young males on the Pentecost
Islands in the South Pacific. These young
men tie vines around their ankles and then
launch themselves into the air from tall bamboo towers. Their ability to complete such a
dangerous act signifies their having attained
manhood.

The first bungee jump by Europeans
took place on 1 April 1978, when members
of England's Oxford Dangerous Sports Club

A Panzl performs acrobatics. © 2002 EAA/ LeeAnn Abrams.

jumped from the Clifton Suspension Bridge in Bristol. The sport spread rapidly throughout much of Europe, Australia, New Zealand, and the United States. Today, there are well over a hundred bungee clubs in all parts of Europe, the United States, Australia, New Zealand, Indonesia, the Dominican Republic, Egypt, Iceland, Israel, Japan, South Korea, Nepal, and many other countries.

Bungee jumping has been made possible largely because of the development of very strong cords and reliable harnesses and safety equipment. Bungee cords are of two types: sheathed cords, which consist of a rubber core encased in cotton or nylon, and all-rubber cords. The cord is attached to a chest and waist harness worn by the jumper. Ankle harnesses are also used, usually by more experienced jumpers, allowing a person to do more elaborate stunts during the fall.

The world's record bungee jump is held by Jochen Schwizer, who jumped from a helicopter in September 1997. He wore a bungee cord 284 meters (931 feet) long and fell a maximum distance of 379.8 meters (1,246 feet) from the helicopter.

Further Reading

Glaser, Jason. *Bungee Jumping.* Mankato, MN: Capstone Press, 2000.

Hang Gliding and Paragliding

The sport of hang gliding may be said to have had its birth in the nineteenth century, when a number of individuals attempted to mimic the flight of birds by attaching huge wings to their bodies and jumping off hills or buildings. These early efforts eventually led to the development of large gliders, which remained an important form of air transport until the invention of the heavier-than-air airplane by the Wright brothers in 1903. By that time, however, interest in individual hang gliding had largely disappeared.

That interest was rekindled after World War II with the invention of a new type of glider by the American engineer Francis Rogallo. During the 1940s, Rogallo and his wife built a wind tunnel in their home to test a number of designs for the new flying device. They developed a flexible, delta-wing sail with no rigid supports controlled only by means of a system of ropes, a design for which they received a patent in November 1948. Rogallo's design was significantly improved during the 1960s, when he was an engineer at the National Aeronautic and Space Administration (NASA). NASA officials hoped that the Rogallo wing could be used to bring spaceships back to Earth.

Over the next two decades, a number of designs were tested by inventors. These designs fell into one of two categories: hang gliders and paragliders. The two designs differ from each other in that the hang glider contains a rigid frame that supports the fabric of which the glider is made. By contrast, the paraglider, modeled on Rogallo's original concept, has no rigid support and holds its shape entirely because of air pressure. In the hang glider, the pilot usually lies in a prone position, while in the paraglider, he or

Parasailing in Michigan. © National Parachute Ind., Inc.

she may be either prone or in a sitting position. Of the two aircraft, the hang glider has a more efficient aerodynamic design and is able to travel at higher speeds than can a paraglider.

Both hang gliders and paragliders are launched by having the pilot run across the ground under the wind catches the sail. Once in the air, the pilot controls the forward and sideways motion of the aircraft by moving his or her body forward or backward and to one side or the other, pulling on the control ropes while doing so. When the pilot is ready to return to earth, the front of the glider is pushed upward, the glider drops slowly to the ground, and the pilot lands smoothly.

A number of variations of the hang glider and paraglider have been developed in recent decades. For example, parasailing is a popular sport at many waterside resorts. The rider is placed in a harness attached to a parachute, which is then lifted into the air by a motorboat for a short ride over the water. A paraplane is a parachute that carries its own source of power, such as small automotive engines. After being dropped from an airplane, the paraplane can be maneuvered through the air under its own power. The sport known as paraskiing uses aircraft similar to the paraplane.

Further Reading

Holmes, Len. *Fly the Wing: Hooking into Hang Gliding.* Southern Pines, NC: Karo Hollow Press, 2002.

Pagen, Dennis. *Hang Gliding Training Manual: Learning Hang Gliding Skills for Beginner to Intermediate Pilots.* Black Mountain, NC: Black Mountain Books, 1995.

Whittall, Noel. *Paragliding: The Complete Guide.* Guilford, CT: The Lyons Press, 2000.

Skydiving

Some historians believe that the first use of parachutes for recreational purposes dates to the twelfth century B.C. in China, where people may have used umbrellalike devices to jump off high mounds or small buildings. The development of more effective and reliable parachutes, however, awaited the invention of balloons and airplanes, when emergency descent from a disabled aircraft became a necessity. By the early 1800s, a number of inventors had designed and tried out a variety of designs for parachutes. One of the most famous parachutists of the time was the French inventor Andre Jacques Garnerin, whose most impressive jump was launched from a balloon at an altitude of 2,500 meters (8,000 feet) over London.

With the invention of the airplane by the Wright brothers in 1903, the need for parachutes became a priority, since an escape from a disabled heavier-than-air aircraft was a much more urgent matter than an escape from a lighter glider or balloon. The first decade of the twentieth century, then, saw the invention of many components of the modern parachute, including the parachute pack, ripcord, and pilot chute. In 1912, the first parachute jump from a moving aircraft was made by U.S. Army Captain Albert Berry. Only seven years later, the first intentional jump from an airplane beginning with a period of free fall—the definition of skydiving—occurred when an American by the name of Leslie Irvin jumped from an airplane near Dayton, Ohio.

The formal origin of skydiving is sometimes traced to 1930, when a group of Russian parachutists staged a competition during the annual Sports Festival in which the objective was to land as near as possible to a specific target. Two years later, a similar competition was held during the National Air Races at Roosevelt Field in New York state. In 1935, the first free-drop parachute tower for use in skydiving was constructed at Hightstown, New Jersey. The tower stood some 40 meters (125 feet) high.

Skydiving's popularity spread widely during the 1930s and 1940s and, in 1945, the Federation Aeronautique Internationale, the international body governing sport aviation competition and records, accepted skydiving as a legitimate sport. Today, the governing body of skydiving in the United States is the U.S. Parachute Association. The organization promotes safety and education programs for skydiving and sponsors competitions in six areas: freefall style and accuracy landing; formation skydiving; paraski; freeflying; freestyle skydiving and skysurfing; canopy formation; and collegiates. *See also* Aerobics; Airboat; Balloons; Gliding; Hovercraft; Kite; Parachute

Further Reading

Johnson, Erik. *Understanding the Skydive.* Rome, GA: Lamplighter Press, 2002.

Aerobics

Aerobics is a form of exercise performed strenuously enough to significantly increase a person's rate of respiration and heart beat. For more than two decades, the term *aerobics* has been used primarily to describe classes conducted in schools and colleges and at health clubs and private gyms and conducted by trained or qualified instructors.

The term *aerobic,* in general, refers to any process that takes place only in the presence of air or, more specifically, oxygen. By contrast, the term *anaerobic* is used for processes that take place in the absence of air or oxygen. In some forms of very vigorous exercise, *anaerobic exercise,* the body is not able to extract oxygen fast enough from the atmosphere and its cells switch over to alternative methods of generating energy that do not depend on the presence of oxygen.

Benefits

Proponents of aerobic exercise cite a number of benefits to be gained from the practice. First is a general improvement in one's cardiovascular health. The cardiovascular system includes the heart and blood vessels. Vigorous aerobic exercise is said to improve cardiovascular health by improving the rate and efficiency at which oxygen is taken into the body and moved through the heart and blood vessels. These changes are also said to reduce one's blood pressure and cholesterol level.

Second is a redistribution of body mass, reducing the overall amount of body fat and increasing the amount of muscular tissue. Third is a general improvement in the muscular system, increasing muscular power and endurance. Muscular power is the amount of strength a muscle can exert and endurance is the length of time a muscle can continue to work at a job.

Fourth, aerobics is also believed to improve the health of bones, increasing their strength and density. There is some evidence that such improvements may prevent or reduce the effects of osteoporosis, a condition in which bone density decreases and bones have an increased tendency to break.

Experts point out that such benefits are gained only if exercises are conducted under moderate (that is, aerobic) conditions. If exercise becomes too vigorous, the body may switch over to anaerobic methods of producing energy, obviating the benefits to be gained from aerobic exercise. Charts are available to allow any individual to calculate the correct intensity of an aerobic exercise. For example, according to one recommendation, the maximum heart rate one should attempt to reach in an exercise is equal to 220 minus one's age. For a person 30 years old, then, the maximum heart rate should be $220 - 30$, or 190 beats per minute. This number can then be used to determine the proper exercise rate for various objectives. For example, if fat loss is one's objective, the maximum heart rate to be reached lies between 60 and 75 percent of the maximum heart rate. For the 30-year-old person in the above example, that exercise rate would be between $60\% \times 190$ (114) and $75\% \times 190$ (142) beats per minute.

Finally, aerobics is said to provide psychological and emotional benefits also. It decreases stress, tension, and anxiety, improves one general mood, and reduces sleeping problems.

Types

Aerobics instructors have invented a wide variety of steps that can be performed during an exercise period. These variations provide a more interesting workout as well as one that uses and develops different parts of the body. Some examples of these variations are the basic left and basic right, corner to corner, diagonal, helicopter, jumping jack, reverse turn, T-step, and turnstep moves.

Traditional aerobic exercises are sometimes categorized as *high-impact* and *low-impact* activities. As the name suggests, high-impact aerobics involves more vigorous movements, such as jumping and hopping, in which both feet may be off the floor at least part of the time. High-impact aerobics is especially useful in improving cardiovascular health. By contrast, low-impact aerobics are intended for designed to improve muscular health and are most suitable for elderly people, pregnant women, and people recovering from injuries. In low-impact exercises, one foot is touching the ground at all times.

Step aerobics uses a wooden or plastic box whose height can be adjusted. Exercises involve stepping on and off the box and be carried out in any degree, from moderate to vigorous. The intensity level of an exercise can be changed both by the individual's choice of movement and the height of the box selected.

In *slide aerobics,* a person slides back and forth on a slippery plastic material in stocking feet or special boots. Slide activities are especially effective for development of leg muscles and are popular with athletes preparing for skating, skiing, and tennis.

Aquatic or *water aerobics* is a form of exercise essentially identical to traditional floor aerobics except that it takes place in a swimming pool or a shallow lake. This type of aerobics provides additional benefits because the human body has to work against a much denser medium (water) than in traditional aerobics (air). For example, walking on land uses up about 135 Calories of energy during a 30-minute workout, while the same exercise in water uses up 265 Calories of energy. Aquatic aerobics are especially beneficial in toning and strengthening leg and hip muscles.

Further Reading

Casten, Carole M., and Peg Jordan. *Aerobics Today,* 2nd edition. Belmont, CA: Wadsworth Publishing, 2001.

Mazzeo, Karen S. *Fitness through Aerobics and Step Training,* 3rd edition. Belmont, CA: Wadsworth Publishing, 2001.

Pryor, Esther, and Minda Goodman Kraines. *Keep Moving! Fitness through Aerobics and Step.* New York: Mayfield Publishing, 2000.

Aerobiology

Aerobiology is the study of airborne particles of biological origin, such as pollen, spores, and microorganisms.

Types

A host of organisms are found in the atmosphere, suspended in the air and carried from place to place by winds and air currents. These organisms range in size from one-celled bacteria, fungi, and protozoa to more complex multicellular organisms such as spiders, moths, mites, aphids, and grasshoppers. The general term *aeroplankton* is used to describe the smallest and most common of these floating organisms. The term was introduced to emphasize the parallel between such organisms and marine plankton, free-floating plants and animals found in ponds, lakes, rivers, streams, and the ocean. The term *plankton* itself comes from the Greek word *planktos,* meaning "drifting."

Many airborne organisms carry out all or most life functions, including eating, excreting, and reproducing, as they are carried along on air currents. Some individuals may spend their whole lives aloft, although most are airborne during only a portion of their lives. As is the case on land and in water, complex food webs also occur in the atmosphere, with larger organisms feeding on those smaller than themselves that, in turn, prey on even smaller organisms.

Mechanisms of Transport

Airborne organisms are carried into the atmosphere by a number of mechanisms. For example, many kinds of plants have evolved systems by which their seeds are ejected into the air, after which they may be carried greater or less distances from the mother plant on air currents. Botanists believe that more than 10,000 species of plants reproduce entirely or partially by this method.

For example, all species of orchids use wind propagation as at least one method of seed dispersal. In some species, a single orchid flower can produce more than 4 million tiny seeds, so light that they can easily be carried by the wind or air currents. These seeds are capable of surviving windborne trips of up to 3,000 kilometers (2,000 miles), accounting for the abundance of many types of orchids on remote islands in oceans around the world.

Microorganisms may also be thrown into the air from water as the result of wave action. Wind currents may sweep these organisms off the top of a wave into the atmosphere, where they may remain for hours, days, weeks, or longer. Windborne organisms may, depending to some extent on their size and mass, be carried back to the Earth in downdrafts, in precipitation, or by some other mechanism, or they may remain suspended in the air until they die.

Some organisms have even evolved mechanisms by which the are able to launch themselves into the air and, in some cases and to some extent, control their movement through the atmosphere. Entomologists have now identified certain species of spiders that build simple "sails" from the silk excreted from their spinnerets. They use these primitive sails to lift themselves off a plant or the ground into the air, where they are carried away into the atmosphere. Some evidence suggests that airborne spiders may be able to control the speed and direction of their flight by lengthening or shortening the silk threads they produce.

Reports about floating spiders date back more than a century. For example, Charles Darwin described the appearance of such organisms, perhaps carried by wind currents across the ocean, in the masts of the HMS *Beagle* on which he was serving as naturalist in the mid-eighteenth century. In more recent years, scientists have captured 1,500 spiders from more than 45 species in the atmosphere above the southern United States and northern Mexico ("Living on the Air," http://www.islandnet.com/~see/weather/almanac/arc_1998/98sep02.htm).

Occurrence

Airborne organisms have been found in nearly all parts of the atmosphere around the world. They seem to be most abundant over land where, according to one research, they may reach a concentration of more than a million organisms per square mile of land. (P. A. Glick, "The Distribution of Insects, Spiders and Mites in the Air," *Technical Bulletin USDA No. 673.* Washington, DC: Government Printing Office, 1939). They become less common over open water. One study, for example, found a density of about 50 bacteria per cubic meter of air at a distance of 160 kilometers (100 miles) from shore and a density of only 1 bacterium per cubic meter over the open seas. ("Living on the Air")

The concentration of airborne organisms also tends to decrease with increasing altitude above the Earth's surface. The vast majority of airborne organisms are found within a few hundred meters of Earth's surface. But some have been captured at altitudes of up to 75 kilometers (nearly 50 miles).

Atmospheric conditions affect the vertical distribution of airborne organisms. For example, the water droplets and dust particles that make up clouds provide "resting places" on which organisms can collect, increasing their concentration over that in surrounding, but cloudless, parts of the atmosphere.

Updrafts also affect the distribution of airborne organisms in the atmosphere, sometimes carrying large masses of organisms high into the atmosphere quickly. One researchers reports, for example, that he recovered a spider whose silk "sail" had been caught in an updraft, carrying it to a height of 4,500 meters (15,000 feet). (Glick)

Practical Applications

Research carried out by aerobiologists has some important applications in everyday life. For example, a variety of spores and fungi are responsible for many kinds of allergic reactions experienced by humans. By learning more about the times at which such particles are released into the air, the way they are carried by air currents, the amount of particles that are likely to be released by a plant, the interaction between particles and air pollutants, and the effects of such particles on human biology health scientists can provide more helpful guidance to people with allergies.

International Association for Aerobiology

The International Association for Aerobiology was founded in 1974 at the Hague, the Netherlands. The organization grew out of the International Aerobiology Working Group created as a part of the International Biological Program of 1964. The association sponsors quadrennial congresses, international courses and meetings, and a biannual publication, *International Aerobiology Newsletter. See also* Flying Animals; Seed Propagation (by Wind)

Further Reading

Mandrioli, Paolo, Paul Comtois, and Vincenzo Levizzani, eds. *Methods in Aerobiology.* Bologna, Italy: Pitagora Editrice, 1998.

Mani, M. S. *Ecology and Biogeography of High Altitude Insects.* The Hague: Dr. W. Junk N.V. Publishers, 1968.

Muilenberg, Michael, and Harriet Burge, eds. *Aerobiology.* Boca Raton, FL: Lewis Publishers, 1996.

Aerodynamics

Aerodynamics is the study of the way in which air or other gases travel over or through an object, and the resulting interactions that occur between the air or gas and the surface of the object. The word *aerodynamics* comes from two Greek terms that mean "air" (*aero*) and "powerful" (*dynamikos*). Although the word originally applied to the study of moving air, it is now used to describe the properties and behaviors of other gases, such as carbon dioxide, hydrogen, oxygen, and helium, as they pass through and around solid bodies. Aerodynamics is used to analyze the flow of a fluid in a number of different circumstances, such as over the wings of an airplane, around the body of an automobile, through a pipe or opening in a surface, and around the walls of a building.

Properties of Moving Air

The way air moves through and around an object depends on a number of properties of air, including its velocity, density, temperature, viscosity, and compressibility. In general, the flow of air over or through a body can be described as *laminar* or *turbulent*. Laminar flow is also known as *streamlined* flow. It occurs when all parts of a moving air stream are traveling in the same direction at the same speed. The moving air seems to be traveling in a smooth and regular pattern. By contrast, the air in a turbulent flow is traveling

at different speeds and in different directions in different parts of the airflow. The flow appears to be irregular and variable in various parts of the air. In some cases, the flow of air over or through a material may be laminar in some regions and turbulent in others.

The characteristics of an airflow passing over or through an object change in a very complex way with the speed of the air. For example, one normally thinks of air has having a low viscosity, that is, a low resistance to flow. Its viscosity is, in fact, more than 50 times less than that of water. Still, the viscosity of air is not zero and in every package of moving air, some air molecules tend to stick to a surface over or through which it is moving. As air moves more rapidly over a surface, it becomes compressed and tends to stick more tightly to the surface. An increase in the speed of air over a surface also increases the temperature of the air and the surface. All of these changes affect the way air moves over the surface, changing air flow from laminar to turbulent.

Scientists have devised a number of mathematical measures of the properties of air moving over or through an object. One commonly used measure is called the *Reynolds number,* named for the English engineer Osborne Reynolds (1842–1912). The Reynolds number for any given airflow depends on and can be calculated from four variables: the velocity of the moving air, a characteristic distance over which the air travels, the density of the air, and the air's viscosity. The Reynolds number obtained from such a calculation indicates whether the air flow is likely to be laminar or turbulent.

Aerodynamic behavior depends strongly on the velocity of air, which may also be represented by a measure known as *Mach number,* named after the Austrian physicist Ernst Mach (1838–1916). Mach number is defined as the speed of a moving object or material compared to the speed of sound (331.4 meters per second or 742.3 miles per hour). Thus, a column of air moving at the speed of 1,000 miles per hour has a Mach number of 1.347 (1,000 miles per hour 4 742.3 miles per hour). Air flowing at less than a speed of Mach 1 is known as *subsonic flow*. That flowing at a speed greater than Mach 1 is called *supersonic flow*. Air flowing at a speed greater than about Mach 5 is called *hypersonic flow*. And air in which some regions are traveling at less than Mach 1 and other regions traveling at speeds of more than Mach 1 is said to have *transonic flow*.

Subsonic flow is commonly observed in air passing over and around automobiles, trucks and light, general-aviation airplanes; over and around houses and other buildings; and in slow-moving air and other gases passing through pipes and tubes. Transonic flow occurs in commercial airplanes traveling at speeds somewhat less than that of Mach 1, where air normally flows at speeds less than that of the speed of sound but that, in some regions, may exceed that of the speed of sound. Supersonic airflow is observed with special airplanes that travel at speeds of more than Mach 1 (such as the Concorde and certain fighter planes), and hypersonic flow occurs most commonly during space shuttle flights.

Applications

Some of the best known applications of aerodynamics are in the fields of motor vehicle and aircraft design. Better understandings of airflow over and around airplane wings and bodies and around automobile and truck bodies has led to vast improvement in the design of such structures, allowing them to fly or travel at greater speeds with less wind resistance and improve fuel efficiencies. Architects also make use of aerodynamics to study the flow of air within and around buildings. Such information is often valuable especially for the design of buildings where severe wind problems are likely to occur. Engineers who design heating and air conditioning systems also use data obtained from aerodynamical studies to make those systems more efficient. *See also* Airplane

Further Reading

Ackroyd, J. A. D., B. P. Axcell, and A. I. Ruban. *Early Development of Modern Aerodynamics.* Washington, DC: American Institute of Aeronautics and Astronautics, 2001.

Anderson, John D., Jr. *Fundamentals of Aerodynamics,* 3rd edition. New York: McGraw-Hill, 2001.

Houghton, E. L., and P. W. Carpenter. *Aerodynamics for Engineering,* 5th edition. London: Butterworth-Heinemann, 2001.

Aero-otitis

See Barotrauma

Aeroplankton

See Aerobiology

Aerostat

See Balloons; Dirigible

Aether

The term *aether* is used to describe a substance thought to be present in the natural world with properties unlike those of any other form of matter. The word is though to be derived either from the Greek expression *aei thein* ("to run forever") or *aethein* ("to burn"). A number of important ancient Greek philosophers, including Socrates, Aristotle, and Anaxagoras, postulated the existence of an aether as part of their descriptions of the natural world. The word also appears in the form of *aither* or *ether,* although the latter spelling is avoided because of possible confusion with the family of organic compounds known by that name.

Greek Theories

Among Greek philosophers, the aether was ascribed a location and properties entirely different from those with which humans had any direct contact or experience. Aristotle, for example, argued that the natural world was made up of various combinations of four basic elements: earth, air, fire, and water. Each element had its proper place in the physical world and its proper mode of action. For example, the tendency of objects containing the element fire, Aristotle said, was to move in a straight-line motion upward, while those that contained earth had a tendency to fall toward the ground in a straight-line direction.

The composition of the heavens, however, was different from that of the earth, according to Aristotle. The heavens consisted of a fifth element, or *quintessence,* known as aether. The aether consisted, Aristotle said, of tiny, solid spheres that acted like ball bearings. Each heavenly body—the Sun, Moon, planets, and stars—revolved around the Earth within its own celestial sphere. The celestial spheres, in turn, were supported by and floated on the aether. Unlike motion on the Earth, which was usually in a straight line and unstable, motion in the aether, Aristotle said, was circular and eternal.

Later Theories

The belief in some type of aethereal substance persisted among natural philosophers and scientists for another two thousand years after Aristotle. The existence of some type of aether seemed necessary to explain the behavior of certain kinds of observed phenomena, such as the transmission of light and electrical and magnetic phenomena.

For example, the English physicist Sir Isaac Newton (1642–1727) hypothesized the existence of an aether to explain the properties of reflection, refraction, and diffraction of light. Newton's aether was fundamentally different from that of ordinary matter because the particles of which it was made repelled, rather than attracted, one another. A form of aether invented to explain the properties of light eventually became known as a *luminiferous aether.*

Other types of aethers were hypothesized to explain other natural phenomena. For example, the Dutch physician Hermann Boerhaave (1668–1738) suggested that heat consisted of an aetherlike fluid that could be transferred from one body to another. Boerhaave's theory became popular among physicists, and the substance he hypothesized later became known as *caloric.*

During the early nineteenth century, the study of natural phenomena became more quantitative. Scientists began to derive mathematical equations to describe the behavior of matter and energy. This method of scientific analysis became highly effective in that such equations often described observed phenomena quite precisely and, equally important, predicted new phenomena very accurately.

One problem with the more quantitative approach, however, was that scientists were sometimes uncertain as to what real-world phenomena were represented by various terms in a mathematical equation. In such cases, they sometimes invented a form of aether to match the mathematical terms.

For example, the early nineteenth century saw a growing understanding among physicists of the relationship between electrical and magnetic phenomena. In one theory, designed to explain this relationship, the French physicist André Marie Ampère (1775–1836) hypothesized the existence of an aetherlike substance made of two electrical fluids, one positively charged and one negatively charged. The observed interactions between electricity and magnetism could be explain, Ampère suggested, by the way those two fluids alternately combined and separated from each other.

The concept of an aether was perhaps most sorely needed in theories designed to explain the nature of light. By the early eighteenth century, scientists had discovered that, at least in certain cases, light behaves as if it is propagated by waves, similar in some ways to water waves. But the only waves with which scientists were then familiar were those that traveled by the undulation of some form of matter, such as the up-and-down motion of water in a water wave. What was it, they asked, that was undulating in the case of a light wave?

The most common answer to this question was a luminiferous aether. But the existence of a luminiferous aether posed serious problems for physicists. On the one hand, that aether had to be fairly rigid to permit the undulations necessary to carry a beam of light. On the other hand, the aether had to be thin enough to allow the movement of bodies as large as the planets without any observable effects.

The Michelson-Morley Experiment

In 1887, two American scientists, Albert Michelson (1852–1931) and Edward Morley (1838–1923), conceived of an experiment that might solve the question about the aether once and for all. Their reasoning went as follows: Suppose an aether does exist. And also suppose that light always travels with a constant velocity (a widely held belief at the time). Then what effects would one observe if the velocity of light were measured, first, when light is traveling in the same direction as Earth through the aether and, second, when light is traveling at right angles to the direction of the Earth passing through the aether?

The answer, they said, is that one would measure different velocities of light in these two instances. Michelson and Morley set up just such an experiment and measured the velocity of light traveling in two directions with respect to the path of the Earth through space. The result? They found no differences in the velocity of light in the two cases, and they concluded that an aether does not exist.

High school physics students today are often told that the Michelson-Morley experiment put to rest theories of the aether. That conclusion is not entirely correct. It is true that modern physicists no longer conceive of a space-pervading aethereal substance like that proposed by Aristotle, Newton, Boerhaave, and other earlier scientists. But the equations that describe natural phenomena still contain unknowns whose association with the real world require explanation. Today, those explanations take the form of curved space, a special type of vacuum, and other phenomena that do not easily coincide with phenomena from our everyday lives.

In addition, there are a number of diehards, few of them professional physicists, who continue to believe that something like an Aristotlean or Newtonian aether, or other aetherlike substance, exists. They argue that the Michelson-Morley experiment was improperly designed, conducted, or interpreted and that evidence does not yet exist to support the absence of an aether in the universe. *See also* Air As an Element; Gases and Airs

Further Reading

Cantor, G., and J. Hodge, eds. *Conceptions of Ether: Studies in the History of Ether Theories: 1740–1900.* Cambridge: Cambridge University Press, 1981.

Air

See Physical and Chemical Properties of Air

Air (Musical Term)

In music, the term *air* is used for a melody or simple tune, generally written for the voice or a single solo or accompanied instrument. The term is also spelled as *ayre* and is thought to have evolved from the Italian term *aria,* a selection written for a solo voice, usually quite elaborate in structure and accompanied by a solo instrument or a full orchestra. Some authorities suggest that the air was originally designed to accompany dancing for which other standard musical forms, such as the gavotte or minuet, were not available.

A number of famous airs exist in the classical repertoire, among the most famous of which are Johann Sebastian Bach's "Air from Orchestral Suite Number 3," better known as "Air on a G String" (BWV1068), and "Air with Variations" (for harpsichord; BWV991) and Henry Ghys's "Air par le Roi Louis XIII."

Air Aces

An air ace is a pilot who is credited with damaging or destroying some given number of enemy aircraft. In the United States, an air ace is currently defined as someone who has destroyed five or more enemy aircraft.

Credit for victories in air combat were first given during World War I, when airplanes were first employed in reconnaissance, combat, and other missions. Each country made its own decision as to how the credit for an air "victory" would be awarded. In some cases, a pilot actually had to destroy an enemy aircraft. In other cases, the pilot received credit if an enemy airplane was damaged or forced to land. In the period from the beginning of World War I to the present day, countries have sometimes changed the criteria for awarding victories to a pilot, so that comparisons between countries and over time is virtually impossible.

Credit for a victory was awarded in World War I not only for the destruction of an airplane, but also for shooting down a balloon, aircraft that were relatively widely used in some areas for at least part of the war. The top balloon ace for World War I was Belgian pilot Willy Coppens (1892–1986), who was credited with destroying 35 balloons and two airplanes. The next three most effective pilots against balloons were all French fliers, Leon Bourjade (1889–1924), Michael Coiffard (1892–1918), and Maurice Boyau (1988–1918). The number one balloon ace among American fliers was Frank Luke (1897–1918), credited with the destruction of 14 balloons and four airplanes.

Top Aces of World War I

Records of pilots designated as "aces" are available for 16 nations involved in World War I, nine of which were members of the British Commonwealth. Although comparisons are of limited value because of differing definitions of an "ace," the country with the most pilots given that honor was England, with 607 aces. The next largest lists came from Germany (393 aces), Canada (190 aces), Frances (182 aces), and the United States (120 aces).

The top overall fighter pilot, in terms of "victories" earned, was the German pilot Manfred Albrecht Freiherr von Richthofen, better known as "The Red Baron" in England. Von Richthofen was also called *le petit rouge* by the French and *der rote Kampfflieger* (the Red Battle-Flier) by his fellow Germans. He was credited with a total of 80 victories during the war.

The next most successful pilot, in terms of victories, was the French pilot Rene Fonck (1894–1953), credited with 75 aircraft damaged or destroyed. When the war broke out, Fonck was offered a commission in the French Air Force, but he declined, preferring to serve as an infantryman. He changed his mind in 1915 and began a flying career that was so successful that it earned him the Médaille Militaire, the Légion d'Honneur, and the Officer de la Légion d'Honneur.

The British Commonwealth's top ace during the war was Canadian William Bishop (1894–1956), credited with 72 "victories." His 72 victories eventually earned him the Distinguished Flying Cross and the Victoria Cross, the first Canadian to be honored with that award. The leading ace for the United States and one of the most memorable figures of American aviation was Edward (Eddie) Rickenbacker (1890–1973), who was credited with 26 victories, for which Congress awarded him the Medal of Honor.

Top Aces of World War II

Criteria for aircraft "victories" became somewhat more restrictive during World War II. During the First World War, pilots seldom engaged in the kind of plane-to-plane battles that became routine during the Second World War. Still, given the much greater involvement of airpower in the second war, the number of victories achieved by top aces in all nations increased significantly.

Among the two major axis powers, for example, the top ace among German pilots was Erich Hartmann (1922–1995), credited with 352 victories, all of them against the Russians on the Eastern Front. Hartmann's record is the most number of victories scored by any pilot in history, earning him the honor of "Ace of the Aces" and "Most Successful Fighter Pilot of All Time." His 352 victories were scored over a period of 30 months, an average of more than 11 victories per month.

The most successful Japanese pilot during the war was Hirojoshi Nishizawa (1920–1947), also known as the "Devil of Rabaul" for his role in the terrible air battle over the portion of New Guinea by the same name. Nishizawa was credited with 87 victories, making him the "Ace of Aces" among all Japanese fliers in history.

The top French flying ace during World War II was Marcel Albert (1917–), who was credited with 23 victories while flying for three air forces, the Vichy French air force, the Free French air force, and the Royal Air Force. Great Britain's leading fighter pilot was James Edgar Johnson (1915–2001), credited with 36.91 victories during the war. Johnson served in the Royal Air Force until 1966, after which he retired to hold a number of important civilian posts.

The leading ace for the United States during World War II was Richard Bong (1920–1945), a native of Superior, Wisconsin. Bong became interested in aviation at an early age when President Calvin Coolidge had his mail delivered to him daily by air at his summer retreat near Superior. Bong witnessed the airplane's arrival and departure every day and decided that he wanted a career in flying. He enlisted in the Army Air Corps Aviation Cadet Program in early 1941 and eventually received his pilot wings on 9 January 1942, a month after World War II had begun. He was assigned to the 9th Fighter Squadron of the 49th Fighter Group in Brisbane, Australia, from which he launched his first sorties against Japanese aircraft. Between December 1942 and August 1945, Bong was credited with 40 victories, the largest number credited to any American flier in history. Bong was killed on 6 August 1945, less than a month before the end of World War II, when his airplane crashed on takeoff.

The study of air aces is a topic of considerable interest to a number of professional and amateur historians. For complete lists of aces in both World War I and World War II, as well as many other wars, see the following Web sites:

The Aerodrome: "Aces and Aircraft of World War I," http://www.theaerodrome.com

Air Aces: http://server.mat.fce.vutbr.cz/safarik/ACES

See also Richthofen, Manfred Albrecht Freiherr von; Rickenbacker, Edward

Further Reading

Mason, Francis K. *Aces of the Air.* New York: Mayflower Books, 1981.

Shores, Christopher. *Air Aces.* Novato, CA: Presidio Press, 1983.

Whelan, James R. *Hunters in the Sky: Fighter Aces of WWII.* Washington, DC: Regenery Gateway, 1991.

Air As an Element

In modern terminology, an element is a substance that can not be reduced to any simpler form. Gold, copper, oxygen, nitrogen, lead, and sodium are examples of elements. The concept of some set of simple substances—elements—out of which everything in the world is made originated with the ancient Greeks. These scholars introduced the idea that the natural world cannot, on a fundamental level, possibly be made of the thousands of different materials that exist around us. These materials, they argued, are made of various combinations of a small number of basic substances, the elements.

Scientists today determine which materials are elements by analytical tests conducted in the laboratory. This approach to the study of nature was unfamiliar to the Greeks, who constructed their list of elements by means of logical reasoning. Certain materials in the world around them seemed, for one reason or another, to be more important than others. For example, fire was present everywhere, both on earth and, apparently, in the skies. Fire also had the capacity to bring about remarkable changes, such as converting a solid material into a gas. To some Greek scholars, it seemed obvious that fire was a basic "element" from which the real world was built.

Various Greek scholars chose different materials that they thought to be "elemental." The most common of these materials were earth, air, fire, and water. Each material had its own particular proponent (or proponents), who developed arguments illustrating the special importance of that material. The philosopher who most avidly proposed air as an element was Anaximenes.

Anaximenes was born in Miletus in about 570 B.C. and died about 500 B.C. He was a pupil of Anaximander, one of the greatest of the early astronomers. Little is known about the life of Anaximenes and all that we know of him comes from a few fragments of his works and from writings about him by later scholars. A number of passages attributed to him talk about the essential role of air in the construction of other materials.

For example, Anaximenes was quoted as saying that "air is the first principle of things, for from this all things arise and into this they are all resolved again." He explained that air is of special importance because its properties are intermediary between those of fire and water and, therefore, the basic material from which these two can be formed. Plutarch also quotes Anaximenes as saying that "air is the first principle, and that it is infinite in quantity . . . and all things are generated by a certain condensation or rarefaction of it." In the one fragment that has been attributed to Anaximenes, the scholar writes that "air is the nearest to an immaterial thing; for since we are generated in the flow of air, it is necessary that it should be infinite and abundant, because it is never exhausted." (All quotations are cited in http://history.hanover.edu/texts/presoc/anaximen.htm.)

Lacking any empirical means for determining which substances are truly elemental, the Greeks had no definitive way of knowing which items actually belonged on a list of elements. The substances that were most widely accepted at any one time were adopted on the basic of a philosopher's popularity or the strength of the arguments supporting one material or another. Eventually, however, earth, air, fire, and water became the four substances most commonly accepted as being "elemental." These terms did not, however, mean exactly the same thing to the Greeks as they do to modern scholars. For example, the "water" that was named as an element by the Greeks was not usually thought of as the water one finds in a lake or stream. Instead, it was a waterlike property that was responsible for the fluidity, coldness, low density, and similar properties found in an ordinary material. Similarly, the "air" that Greeks accepted as an element was an "airy" property, such as one might find in light objects, such as clouds and sea foam.

The belief in air as an element survived for nearly two thousand years. It was not until the mid–eighteenth century that modern chemists were finally able to analyze air in such a way as to demonstrate that it is not an element but, in fact, a mixture of other gases (oxygen, nitrogen, carbon dioxide, argon, etc.), some of which themselves are indeed elements in the modern sense of the word. *See also* Gases and Airs

Further Reading

Burnet, John. *Early Greek Philosophy.* New York: Meridian Books, 1957.

Air Bag

An air bag is a safety device that inflates as the result of a collision, protecting the driver and passenger(s) riding in a vehicle. One of the earliest suggestions for an automobile air bag was that proposed by the English inventor Arnold Kent in 1960. Kent outlined the general features of an air bag that would be stowed within the dashboard of a car and that would automatically inflate during a collision.

In the United States, the development of automobile air bag systems followed the lines of Kent's early ideas as well as research on airplane safety systems promoted by the U.S. government. Airplane systems were designed to protect pilots who were forced to make emergency landings in less-than catastrophic circumstances. The first air bag system using electromechanical systems was invented in the United States by Allen K. Breed in 1968. Air bag systems first appeared in commercial automobiles in the late 1980s and, by 1998, had become required equipment in all new cars sold in the United States. A year later, the same requirement was established for all light trucks sold in the country.

Components

An automotive air bag system consists of three essential components: crash sensors, a diagnostic module, and the air bag itself.

Crash Sensors

The purpose of crash sensors is to detect sudden reductions in the speed of a vehicle, as would occur when it collides with a stationary object. Crash sensors are set to operate at different changes in speed, the most common range being between 10 and 15 miles per hour.

Crash sensors are of two general types: electromechanical switches and accelerometers. An electromechanical switch is a device in which the motion of some object is converted into an electrical impulse, which is then passed to the system's diagnostic module. One type of electromechanical switch consists of a ball and magnet. At rest, the ball is held in position by a circular magnet, into which it fits. When a vehicle decelerates (slows down) rapidly, the ball is jarred lose from the magnet. It rolls forward and comes into contact with an electrical circuit, sending a signal to the diagnostic module.

A second type of electromechanical sensor consists of a metal roller attached to a spring. During deceleration, the roller is released from the spring, coming into contact with and activating an electrical circuit. Electromechanical sensors can be calibrated to operate at different rates of deceleration by changing the force of the magnet, as in the first example, or the tension on the spring, as in the second example.

Electromechanical sensors can be mounted in a variety of locations in a vehicle, such as the fender apron, the radiator support, the dashboard, or the diagnostic module.

Accelerometers are miniature devices that sense rapid changes in the speed at which a vehicle is traveling. They are etched on a silicon chip stored inside the diagnostic module. Various types of air bag systems may have only one sensor (a *single-point system*) or a combination of sensors (a *multipoint system*) mounted within the diagnostic module or in two or more different parts of the vehicle.

Diagnostic Module

The diagnostic module is the "brains" of the air bag system, carrying out a number of monitoring functions. Among its many jobs are the monitoring of the electrical parts of the air bag system, providing backup energy in case the system's normal power level falls too low, and activating the air bag itself when necessary.

The diagnostic module can be thought of as the air bag system's "brain" also because it makes the final decision as to whether the air bag should actually be inflated. It receives data from all of the vehicle's crash sensors and compares them with some predetermined standards programmed into it to decide if a dangerous crash has actually occurred.

Air Bag Module

The air bag itself is made of a thin, nylon fabric mounted in the steering wheel, the dashboard, the seat, or the door of the vehicle. The bag is folded to fit into the appropriate space and covered with talc or cornstarch to keep the material from sticking to itself. The bag is filled with three chemicals, sodium azide (NaN_3), sodium nitrate ($NaNO_3$), and silicon dioxide (SiO_2). The bag is attached by electrical wires to the diagnostic module.

When the diagnostic module determines that a dangerous crash has occurred, it sends an electrical signal to the air bag. That electrical signal initiates a chemical reaction in the sodium azide. In the first stage of that reaction, sodium azide decomposes to form sodium (Na) and nitrogen gas (N_2):

$$2NaNO_3 \rightarrow 2Na + 3N_2$$

The nitrogen gas formed in this reaction expands very rapidly, inflating the bag to its full volume.

In the second stage of the reaction, the sodium formed reacts with potassium nitrate to form potassium oxide (K_2O), sodium oxide (Na_2O), and more nitrogen gas:

$$10Na + 2KNO_3 \rightarrow K_2O + 5Na_2O + N_2$$

Since potassium oxide and sodium oxide are both highly caustic materials, they must be neutralized after being formed. The silicon dioxide in the air bag accomplishes this step:

$$K_2O + Na_2O + 2SiO_2 \rightarrow 2NaK(SiO_3)_2$$

The key to successful air bag inflation is speed. Most automobile accidents are completed in about an eighth of a second. To be useful, then, sensors must detect the accident and relay that information to the air bag, and the air bag must inflate in less than that time. With today's technology, air bags inflate in 15–20 msec (milliseconds, or thousandths of a second) at a speed of about 200 miles per hour. They are fully inflated about 30 msec later.

As soon as the air bag is fully inflated, it begins to deflate. Rapid deflation is necessary to protect passengers from colliding with a hard, unmovable object—the inflated air bag—after the accident. Complete deflation usually takes another 20–30 msec. Thus, the total time between impact and complete bag deflation is about 100 msec.

Safety Issues

In general, air bags have been remarkably successful in reducing fatal accidents and injuries as the result of vehicle accidents. In its most recent report (1999), the National Highway Traffic Safety Administration (NHTSA) revealed that the use of air bags saved 4,758 lives. This number represents a 31 percent reduction in deaths for all drivers, 32 percent reduction for passengers, and 36 percent reduction for light truck drivers.

Air bags are significantly more effective as a safety device when used in conjunction with seat belts. The NHTSA estimates that the combination of air bags and seat belts reduces the risk of head injuries in a car accident by 75 percent and of serious chest injuries by 66 percent.

On the other hand, air bags have also been responsible for 146 deaths in vehicle accidents. Slightly more than half of these deaths involved children (84 cases), most of whom (66 cases) were not in rear-facing seats, as is recommended. These injuries typically have occurred when the force with which an individual strikes the air bag is sufficient to cause serious, usually fatal, head trauma. A number of simple safety steps, such as the use of seat belts and rear-facing seats for children and proper driver seat placement, generally prevents the most serious injuries associated with air bag use.

New Developments

Automotive engineers are constantly at work devising more sophisticated and more efficient air bag systems. For example, Automotive Systems Laboratory in Farmington Hills, Michigan, has developed an air bag with two separate sections. When sensors detect a collision, they not only signal the air bag to inflate, but also determine the severity of the accident and decide whether one or both sections of the air bag are to inflate. In more moderate collisions, only one of the two sections is needed to protect a passenger or driver.

A number of technologies have been developed to determine the position of driver and passenger with regard to the air bag. Studies have shown that riders who are closer than 8 inches from the air bag are much more likely to receive serious injuries in a collision. New sensors are now able to locate the position of a driver or passenger's body in relationship to the air bag. The diagnostic module uses this information to decide whether the air bag should be inflated at all during an accident, or whether it should be inflated at less than maximum conditions.

Further Reading

Society of Automotive Engineers. *Automatic Occupant Protection Systems.* Warrendale, PA: Society of Automotive Engineers, 1988.

Information about air bags can be found on the Internet at the home page for the National Highway Traffic Safety Administration, http://www.nhtsa.dot.gov/people/injury/airbags.

Airboat

The term *airboat* is usually applied to a type of boat with a flat bottom and shallow draft that is driven by an airplane or automobile motor and steered by a rudder that operates

AM14 Vanguard Amphibious Marine. © Amphibious Marine.

within the motor's airstream. That is, the operation of an airboat depends completely on the force of moving air and not on any type of "push" against water, as is the case with other types of boats.

Authorities differ as to the date and inventor of the first airboat. One writer claims that the Scottish-American inventor Alexander Graham Bell (1847–1922) built an airboat-type vehicle in Nova Scotia in 1905. The vehicle was used to test aircraft engines and propellers and was commonly known as the "Ugly Duckling" (http://www. pnx.com/gator/ab-FAQ.htm). A report on environmental changes in the Everglades by an intern at the University of Miami Rosenstiel School of Marine and Atmospheric Science claims that airboats were invented and first used in the area in the 1920s (http://www.aoml.noaa.gov/general/lib/ceda

r61.pdf). And *Merriam Webster's Collegiate Dictionary* says that the word *airboat* did not appear in writing until 1946.

The primary advantage of airboats over other types of watercraft is that they can travel over very shallow water, such as that found in swamps and marshes, and even over dry land. Also, airboat enthusiasts claim that such vehicles are, except for the noise they make, more environmentally friendly than other kinds of boats in which the propulsion system, along with its waste products, is submerged.

Probably the major use of airboats is for recreational purposes, in deer, duck, and other types of hunting; fishing; snow and ice travel; and tours. They are also used for weed and insect control since they can be used in areas where traditional land and water vehicles cannot travel. Finally, they have been used in such locations for rescue and survey work.

The term *airboat* is also used for certain types of aircraft. For example, the first scheduled airline using winged aircraft used a Benoit Model 14 airplane in 1914. The airline was called the St. Petersburg–Tampa Airboat Line. It carried passengers and cargo between the two Florida cities, usually operating at no more than about five feet above the water. Some military units also referred to their aircraft as "airboats" in the early days of aviation. For example, the U.S. Navy organized its first aviation service into two division, Air Detachment Seaplanes and Airboat Squadrons.

Further Information
Airboat World Magazine
P.O. Box 5580
Ocala, FL 34478-5580
URL: http://www.airboatworld.com
E-mail: subscriptions@airboatworld.com

Air Brake

An air brake is a device that uses compressed air to slow down or stop the movement of some large mode of transportation, such as a railroad train or a truck.

History
The first railroad cars, developed in Great Britain in the early nineteenth century, used hand brakes to stop. When the driver wanted to stop the car, he or an assistant simply pulled on a lever that tightened a wooden or metal band against the car's wheels. When railroad cars were light and traveled at speeds of no more than a few miles per hour, such braking systems were adequate.

Nonetheless, alternatives to hand braking systems were explored early in the history of railroading. In 1833, for example, English inventor Robert Stephenson designed a braking system that operated on steam. In Stephenson's system, pipes connected the steam chamber in the locomotive of the train to brake cylinders in other cars of the train. When the engineer wanted to stop, he released steam from the steam chamber into the pipes. The steam passed through the pipes and into the brake cylinders in individual cars, forcing the brake shoes against the car wheels.

Braking systems like that invented by Stephenson were generally regarded as a luxury in the early days of railroading. Hand brakes still worked well enough for virtually all purposes in British railroading until the early 1890s. At that point, serious railroad accidents became more common, and trains began to switch to some type of automated braking system.

Railroading in the United States developed more rapidly, largely because of the more extensive land area in this country and the greater need for more, larger, and faster railroad trains. The earliest successful automated braking system in this country, therefore, was invented and adopted three decades earlier than was the case in Great Britain and other European countries. That system was invented by George Westinghouse (1846–1914) in 1869.

The Westinghouse braking system was similar to that of Stephenson's, except that it used compressed air rather than steam to operate brake cylinders. The compressed air was generated by steam in the locomotive's steam chamber and then transported by pipes to air cylinders in each individual train car. As with the Stephenson system, brakes were applied in the original Westinghouse system by releasing compressed air into the pipes, from which it acted on brake cylinders.

The Westinghouse invention encountered two major obstacles, one technological and one sociological. In the first place, a break or leak in the pipes that carried compressed air to the brake cylinders was likely to make the system inoperable. In the second place, thousands of brakemen who depended for their livelihood on traditional methods of hand braking fought this new invention, which, if adopted, would result in their losing their jobs.

By 1872, Westinghouse had found a method for solving the first of these problems, although the second problem was resolved only with the passage of time. In fact, brakemen were retained on most railroad trains many years after automatic braking systems had made their jobs obsolete, protected by union rules that railroads had to adopt to remain in business.

Modern Air Brakes

Westinghouse's resolution of the problems of air loss in his original system was to turn that system on its head. Instead of feeding compressed air into a system of pipes to activate brakes, he designed a system in which those pipes were always kept full of compressed air, which was released when the brakes were to be applied.

Here is how that system worked. Steam from the locomotive's steam chamber is used to drive a piston in a cylinder that compresses air. That compressed air is stored in a reservoir, usually beneath the locomotive. When needed, the compressed air passes out of the reservoir, through one long continuous pipe, into each individual car in the train.

Each car is outfitted with an auxiliary reservoir, in which the air needed to operate brakes on that car is stored. Each car also contains a brake cylinder, whose action it is that actually stops the car's movement. Finally, the various components of the system in each car are attached to each other by means of a triple valve.

To activate the brake, the engineer adjusts the brake handle, releasing air from the pipes. Loss of air pressure is then "recognized" by the triple valve, which changes position, allowing air to flow out of the car's auxiliary reservoir and into the brake cylinder. To release the brakes, the engineer

adjusts the brake handle again, allowing compressed air to flow out of the main reservoir back into the pipes. Again, the triple valve recognizes the increase in pressure in the line, changes position, and allows air to flow out of the brake cylinder into the atmosphere.

Modifications and Improvements

A number of changes have been made over the past century in the original Westinghouse air braking system. One of these improvements was the development of an "emergency" adjunct to the basic braking system. This adjunct makes it possible to apply the brakes more quickly and with greater force than would normally be necessary, as in coming to a slow stop at a station. The heart of the emergency component of the braking system is a second air reservoir in each car. This emergency reservoir is usually combined with the car's auxiliary reservoir within a single large container, separated by a steel plate. The plate has a concave shape, making the emergency side of the reservoir larger than the auxiliary side. Under normal circumstances, an engineer brings a train to a stop by releasing air from the system gradually, slowly increasing the effect of the brakes. Should it be necessary to stop the train very quickly, the engineer can release a large amount of air from the system very quickly. In that case, the triple valve senses the rapid drop of air pressure in the system and releases air from both the auxiliary and emergency reservoirs at once, bringing the train to a very rapid and, often, quite dramatic stop that can even result in the train's derailment.

A second improvement in the basic Westinghouse design was the addition of a device known as *Quick Action*. Quick Action is a method for making sure that the very fast response needed in the braking system during times of emergency will be passed down the train very quickly. Without Quick

Action, the passage of a low-pressure signal from car to car in a very long train can take so long that the pressure change may actually become indistinguishable in the last cars of the train. The Quick Action device is one that, when activated by a drop of air pressure in the main line, itself causes further release of air from the line. Thus, some air is released during an emergency from the pipes in every car, ensuring that the "emergency" signal reaches the last car in the train with undiminished force.

A number of other modifications have been made in Westinghouse's original design, often with the objective of making the entire braking system more responsive with shorter delays in either stopping or starting up. To a large extent, the systems used in freight and passenger trains and many commercial trucks is essentially that of his original design.

A Paasche Model D Compressor and a Paasche Millennium Mil–3 Airbrush.
© Blick Art Materials, www.dickblick.com.

Further Reading

Buckman, Leonard C. *Commercial Vehicle Braking Systems: Air Brakes, ABS and Beyond.* Warrendale, PA: Society of Automotive Engineers, 1998.

"Train Air Brake Description and History," http://www.sdrm.org/faqs/brakes/html.

"The Westinghouse Air-Brake," http://www.railroadextra.com/chapt21.Html.

Williams, Albert Nathaniel. *Air Brakes and Railway Signals: Proud Creations of George Westinghouse's Brilliant Genius.* New York: Newcomen Society in North America, 1949.

Airbrush

An airbrush is a device for applying a liquid to a surface in the form of a fine spray. It is used primarily for the retouching of photographs and artwork.

History

The concept behind an airbrush is nearly as old as humans themselves. Archaeologists believe that some cave paintings dating to more than 30,000 B.C. may have been applied by blowing paint onto a cave wall through a hollow bone or reed. The earliest form of the airbrush, however, was probably conceived by an American inventor, Abner Peeler of Webster City, Iowa. In 1878, Peeler constructed a tool in which paint was sprayed through a wooden pipe by means of compressed air from a pump controlled by a foot pedal. Peeler called his invention a "paint distributor," and it was probably more like a modern paint sprayer than the airbrush used today.

Because of this difference, credit for the invention of the modern airbrush is also given to another American inventor, Charles L. Burdick, who received a British patent for his device in 1893. In Burdick's design, paint or ink was fed into a stream of compressed air and out a nozzle with an opening 0.0007 inch in diameter. Burdick, who called his invention an *aerograph* ("air

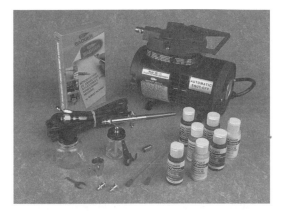

Dick Blick Double-Action Airbrush Kit.
© Blick Art Materials, www.dickblick.com.

drawer"), manufactured and sold it through his London-based company, Fountain Brush, later renamed The Aerograph Company.

Operation

Although there have been many refinements in its design, the modern airbrush has changed remarkably little in its fundamental operation in more than a century. Today's device consists of a long cylindrical tube made of metal with a valve on one side. The valve is connected to a source of compressed air, such as an aerosol can or an electrically driven air compressor. The working liquid, such as paint, ink, varnish, enamel, or lacquer, is stored in a reservoir at the front of the airbrush. A trigger at the top of the cylinder controls the flow of compressed air into the cylinder. When the trigger is depressed, air flows down a narrow tube in the center of the cylinder, and out a thin nozzle at the front of the airbrush. As compressed air moves through the central tube of the cylinder, it creates a partial vacuum at the opening of the liquid reservoir, drawing some liquid into the stream and atomizing it (breaking it into very small particles). The mixture of air and liquid is then directed on the surface being treated, providing a thin, uniform coating.

In the simplest form of an airbrush, the operator can control the flow of air, but not the flow of liquid. In double-action forms of the device, an adjustment screw permits changes in the flow of liquid as well.

Uses

A watercolor artist, Burdick originally invented the airbrush as a way of adding coatings of paint to a picture without blurring or damaging the original layer(s) of paint. Airbrushes soon became more popular, however, for coloring and retouching photographs, which, in the late nineteenth and early twentieth centuries, were available only in black and white or sepia. At one point, around the turn of the century, companies employed large numbers of men and women to airbrush photographs and make them more appealing to consumers.

The use of airbrushes as painting tools grew more slowly, at least partly because some artists thought that using mechanical devices to produce a work of art was not quite legitimate. Nonetheless, by the 1920s, airbrush paintings had gained respectability as a legitimate form of art and had begun to appear widely both in the fine arts and in advertising. Today, airbrush techniques have reached the point where operators can easily control the texture, amount, patterns, thickness, and other properties of not only a piece of fine art, but also of surface coatings for many industrial products.

Further Reading

Misstear, Cecil. *The Advanced Airbrush Book.* New York: Van Nostrand Reinhold, 1984.
Vero, Radu. *Airbrush: The Complete Studio Handbook.* New York: Watson-Guptill Publications, 1983.

Air Cavalry

The term *cavalry* refers to a group of army troops, originally mounted on horseback, whose mission it is to perform certain spe-

Army AH–64 Apache helicopters from the Renegade Troop 4th Squadron, 3rd Armored Calvary Regiment (Fort Carson, Colorado) land at Allen Air Field at Fort Greely (Alaska) to participate in Northern Edge '98, an annual Joint Chiefs of Staff exercise involving air and ground units of all services. © Department of Defense.

cialized military functions. According to one specialist, typical cavalry missions include reconnaissance and counter-reconnaissance, security, raid, deception, attack and counter-attack, exploitation, pursuit, delay, and defense (http://www.uscavalryschool.org/cavalry_history.htm).

Early History

The first mission of cavalry operations is usually traced to the writings of the Greek military leader and theorist Xenophon (ca. 444–357 B.C.). He is credited with writing a long and detailed description of the training, equipping, and operation of cavalry units.

Until the early twentieth century, cavalry operations were conducted almost entirely on horseback and the term itself came to be equated with any operation conducted on horseback. With the development of automotive vehicles, however, military strategists realized that cars, trucks, tanks, and other motorized vehicles could carry out many of the same operations as horses, and usually more quickly and more efficiently. By the

An Air Force Reserve C-5 Galaxy from the 433rd Military Airlift Wing (Kelly AFB, Texas) delivers two AH-64 Apache helicopters to Eielson AFB (Alaska) to take part in Northern Edge '98. © Department of Defense.

early 1920s, therefore, many military cavalry units had replaced horses with such vehicles.

Origin of the Air Cavalry

At nearly the same time that some military leaders were beginning to think in terms of motorized cavalry units, some optimistic students of warfare were even beginning to imagine the role of flying machines for use in cavalry-type operations. In 1909, a young second lieutenant at the Marine Corps Officer Training School at Parris Island, South Carolina, A. A. Vandegrift, wrote a paper on "Aviation, the Cavalry of the Future." Vandegrift's superiors deemed his paper to be "unsatisfactory," but history was to prove the validity of the theory.

World War II definitely demonstrated the essential role of aircraft in modern war-

fare, and Vandegrift's suggestion for a mobile, rapidly dispersable group that could use aircraft to carry out traditional cavalry operations became more clear. The turning point in the development of a modern air cavalry came in the 1950s, at least partly because of an important article written by General James Gavin, "Cavalry, and I Don't Mean Horses." (*Harper's*, April 1954) In his article, Gavin called for the use of helicopters to provide the quick response needed at battlefronts that had not been available during the recently completed Korean War.

By 1956, the U.S. Army Aviation Center had been established at Fort Rucker, Alabama, under the command of Jay D. Vanderpool. Vanderpool began to implement many of General Gavin's ideas with the ultimate objective of providing every army division with the ability to move at least one infantry company (about 60 to 190 soldiers) by air.

Shortly after, the Vietnam War created a new urgency for the development of such airborne units. In April 1962, Secretary of Defense Robert S. McNamara ordered a study that would take a "bold new look at land warfare mobility" that would include an analysis of the role of Army aviation in providing cavalry-type operations. Senior officer on the study committee was Lieutenant General Hamilton H. Howze, and the committee's final report became known as the "Howze Report."

One of the Howze Report's major conclusions was that "adoption by the Army of the airmobile concept is necessary and desirable. In some respects, the transition *is* inevitable just as was that from animal mobility to motor" (http://www.aircav.com/histavn.html). The Howze Board recommended the creation of five air assault divisions and the replacement of 2,339 ground vehicles with 459 aircraft. By 1963, the 11th Air Assault Division had begun to test the principles of air cavalry operations at Fort Benning, Georgia, and other locations across the country. By 1965, the first air mobile division, the 1st Cavalry Division (Airmobile) had been organized and sent to Vietnam.

Air Calvary Operations

Two core operations of the U.S. Army's Air Cavalry Squadron are carried out by Attack Helicopter Troops (ATKHT) and Assault Helicopter Troops (AHT). The mission of each ATKHT includes fixing and preventing enemy penetrations, providing long-range direct antiarmor fire, exploiting successful operations, and carrying out reconnaissance and screening missions. The primary functions of AHTs is to move troops, supplies, and equipment within a combat zone. It may also carry out air

assault operations and assist with medical evacuations. Aircraft currently in use by the Air Cavalry include the AH-1 Cobra, AH-64A and AH-64D Apache, CH-47D Chinook, OH-58C and OH58D Kiowa, and UH-1 and UH-60 Huey helicopters.

Further Reading

Air Cavalry Squadron and Troop Operations. Field Manual No. 1–114. Washington, DC: Headquarters, Department of the Army, 1 February 2000. Online at http://www. adtdl.army.mil/dgi-bin/atdl.dll/fm/1-114/toc.htm.

Dougherty, Kevin J., "The Evolution of Air Assault," *Joint Force Quarterly,* Summer 1999, 51–58.

Air-Conditioning

The term *air-conditioning* is used to describe any system that provides for the treatment of air to change certain of its characteristics, such as temperature, humidity, and concentration of pollutants. In general practice, *air-conditioning* refers to the cooling, rather than the heating, of air. The term *climate control* is now widely used as a synonym for air-conditioning.

History

Throughout recorded history, humans have searched for ways to achieve the most comfortable living conditions in their homes and workplaces, conditions that are not too hot or too cold, not too dry or too moist. In early civilizations, cooling systems often consisted of nothing more complex than damp rags hung over doors and windows, where air currents could cause evaporation of water and thus cooling. Members of the nobility had more alternatives, of course, which included the employment of slaves to provide constant fanning during hot weather. The Roman emperor Varius Avitus is one of many rulers to have ordered that ice and

Home model air conditioner. © Carrier Corporation.

snow be carried from nearby mountains to be placed in parks and gardens, where breezes provided natural air-conditioning.

The beginning of modern air-conditioning systems can be traced to the efforts of a South Carolina doctor, John Gorrie (1803–1855), who was concerned about providing comfort for his patients suffering from the fevers associated with malaria. Gorrie's system was simplicity itself. It consisted of a fan blowing over blocks of ice suspended from the ceiling of his hospital ward.

During the last half of the nineteenth century, a variety of air-cooling systems were invented and patented, most of them consisting of some variation of Gorrie's plan. A system designed by Nathaniel Shaler in 1865, for example, consisted of a series of five ice-filled compartments around which air was blown. In 1880, a similar system was installed at New York's Madison Square Garden. In that system, air was forced through cheesecloth bags and then over huge blocks of ice before being passed into the auditorium itself.

Credit for the design of the modern air-conditioning system is usually given to the American inventor Willis H. Carrier (1876–1950). In 1902, while working for the Buffalo Forge Company, Carrier was asked to solve a problem encountered by a printer in Brooklyn, New York. The printer was having problems producing good color images because temperature and humidity conditions in his workspace were so variable that the paper on which he was printing kept changing dimensions, causing the printed images to blur and lose register.

Carrier's solution to this problem was the first successful mechanical system for controlling the temperature and humidity of air. He described this system in his patent application, "Apparatus for Treating Air," granted in 1906. In Carrier's design, outside air was pulled into the air conditioner by a centrifugal fan through a filter that removed dust, pollen, and other impurities. The air was then passed over a cooling system containing a nontoxic coolant. The cooled and dehumidified air was then blown into the living area or workspace. Although that design has undergone many modifications and improvements, it remains the fundamental pattern on which most air-conditioning systems are based today.

Carrier went on to develop a theoretical basis for his research, developing the fundamentals of a science now known as *psychrometry,* or the measurement of water vapor content in the air or other gases. The mathematical equations he developed to show the relationship of humidity, temperature, and other variables, known as his "rational psychrometric formulas (published in 1911), are still used as the basis for much air-conditioning research and production. Although Carrier is now known as the Father of Modern Air Conditioning, the term itself

Industrial model air conditioner. © Carrier Corporation.

was coined in 1906 by a textile mill architect and engineer, Stuart Cramer.

Operation of an Air-Conditioning System

Air-conditioning systems are used in a wide variety of applications, including residences, office buildings, factories, and other large structures and in all kinds of transportation systems, including automobiles, trucks, trains, and airplanes. The precise design of an air-conditioning system varies according to the specific application, but most operate on a single fundamental design.

This design contains five elements: a compressor, an expansion valve, cooling and evaporation coils, a working fluid (or *refrigerant*), and blowers (fans). The air conditioner is a closed system, which means that the working fluid that produces cooling never leaves or enters the system (except as a result of leaks). The working liquid is a nontoxic chemical that exists in two phases (gaseous and liquid) or in the gaseous stage in a more dense and less dense state. Probably the most common working liquid used today is Freon gas, an organic chemical containing carbon, chloride, and fluorine.

When the working fluid is in the cooling tubes, it exists as a cool gas with a relatively low density. The purpose of the condenser is to convert this cool, low-density gas into a high-density gas, which becomes warm in the process. The high-density gas asses out of the condenser into the evaporation coils, where it gives up its extra heat to the outside atmosphere. In giving up its excess energy, the working fluid changes from a gas to a liquid. After passing through the evaporation coils, the working liquid enters the expansion valve, where it is allowed to increase in volume and change back into a cool gas. The cool gas then passes back into the cooling coils.

The air-conditioning unit is arranged so that the cooling coils are on the inside of the building and the evaporation coils on the outside. When cool working fluid passes through the condensation coils, it picks up heat from the inside of the building. The fluid then carries that heat through the condenser and into the evaporation coils, where it releases the heat to the outside atmosphere. The air-conditioning unit also contains two blowers. One blower is located behind the condensation coils, where it blows cool air from the working fluid into the inside of the building. The second blower is located behind the evaporation coils, where it blows warm air from the working fluid to the outside atmosphere.

In small air-conditioning units, like the window models found in many homes and buildings, all five parts of the system are enclosed in a single metal container. In larger units, the evaporation and condensation coils are separated from each other, with the cooling coils located inside the building and the rest of the air-conditioning unit in a separate container outside the building. The outside container is a familiar sight in many places, with the shape of a large metal barrel that may be wrapped in a metal screen. All of the mechanical changes that take place during the cooling process take place in the outside unit, from which excess heat is also released. Only the cool working fluid is allowed to circulate inside the building.

In some very large buildings, the working part of the air conditioner may be located on top of the building. Ducts carry the cooled working liquid to the inside of the building, where it either cools rooms directly or cools coils of water that, in turn, are used to cool rooms.

Air may also be treated inside the air-conditioning unit to adjust for proper humidity. One way of doing so is to spray a fine mist of water into the cooling tubes to maintain the desired humidity in a room. The actual and desired levels of humidity are controlled by sensors inside the room thermostat. The air-conditioning unit may also contain elements to remove pollutants from air being circulated through a room or building. These elements often consist of one or more filters, with different size pores to remove pollutants of different sizes and chemical composition.

Further Reading

Ackerman, Marsha E. *Cool Comfort: America's Romance with Air-Conditioning.* Washington, DC: Smithsonian Institution Press, 2002.

Cooper, Gail. *Air-Conditioning America: Engineers and the Controlled Environment, 1900–1960.* Baltimore: Johns Hopkins University Press, 1998.

Killinger, Jerry. *Heating and Cooling Essentials.* Tinley Park, IL: Goodheart-Wilcox, 2002.

A series of articles on the history of air-conditioning appeared in the ASHRAE Journal in early 1999. See especially:

Narengast, Bernard, "Early Twentieth-Century Air-Conditioning Engineering," *ASHRAE Journal,* March 1999, 55–62.

Narengast, Bernard, "A History of Comfort Cooling Using Ice, *ASHRAE Journal,* February 1999, 49–57.

Air Force

An air force is a military organization responsible for the air operations necessary for the protection and defense of a nation and its interests. The first airborne devices to be used for military purposes were kites and balloons. In the late seventeenth century, for example, King Petraja of Thailand is said to have tied kegs of gunpowder to a kite and flown it over a rebel army. By the late eighteenth century, balloons were regularly being used for reconnaissance and transportation of soldiers and equipment. In December 1862, for example, the Federal Army used a balloon to carry a squadron of men over the Rappahannock River during an attack on the Confederate Army. By the 1880s, the British army was using both kites and balloons for transportation of men and equipment and, by May 1890, the Balloon Section of the Royal Engineers was formed.

Early History of Air Forces

Heavier-than-air aircraft were employed for military purposes shortly after Wilbur and Orville Wright made their first powered flight at Kitty Hawk, North Carolina, on 17 December 1903. The first use of an aircraft for offensive purposes can be traced to 1 November 1911, when an Italian pilot named Guilio Gevotti dropped several small bombs by hand on Turkish troops at Ain Zaia in Libya. By the outset of World War I in 1914, most major powers had created at least a minimal air force, which they deployed with greater or lesser success during the war.

The British had established the Royal Naval School in 1911 and the Royal Flying Corps a year later. As the war developed and the value of air power became more apparent, the British created the Royal Air Force, independent of the British Army and Royal Navy, in April 1918. Meanwhile, France had formed the Aéronautique Militaire (French Army Air Service) in October 1910, which reached a strength of 132 aircraft in 21 squadrons by 1914. At first, the French used their aircraft for reconnaissance only during the war. But when airman Roland Garros found a way to add deflector plates to his

A 33rd Rescue Squadron HH-60G helicopter takes off from the USS *Juneau* during ship-landing training in the Pacific Ocean. The primary mission of the HH-60G Pave Hawk helicopters is to conduct day or night operations into hostile environments to recover downed aircrew or other isolated personnel during war. Because of its versatility, the HH-60G is also used to perform military operations other than war. These tasks include civil search rescue, emergency aeromedical evacuation, disaster relief, international aid, counter-drug activities, and NASA space shuttle support. © U.S. Navy.

propeller blades, he was able to initiate offensive attacks, making him one of the world's first fighter pilots. After destroying five German airplanes in the early part of April 1915, Garros's plane was itself shot down. The plane was taken to the workshop of inventor Anthony Fokker (1890–1939), who quickly found a way to modify Garros's invention and arm the aircraft with a synchronized machine gun that made air attacks much more successful and lethal.

The development of a military air presence took place more slowly in the United States. On 1 August 1907, the U.S. Army Signal Corps formed an Aeronautical Division to "take charge of all matters pertaining to military ballooning, air machines, and all kindred subjects." At the time, this somewhat grandiose charge referred to a total of eight balloons, some of which dated to the Civil War, three decades earlier. The Division accepted its first heaver-than-air machine, "Airplane No. 1," on 2 August 1909, three months after its first pilots, Frank P. Lahm and Benjamin D. Foulois, had been commissioned.

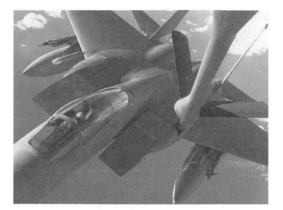

A KC-135R aircraft from the 909th Air Refueling Squadron prepares to refuel an F-15C fighter from the 67th Fighter Squadron on a training mission over the Pacific Ocean. Both aircraft are stationed at Kadena Air Base, Japan. © U.S. Air Force.

The first operational flying unit, the 1st Aero Squadron, was formed in December 1913, and six months later, Congress created the Aviation Section of the Signal Corps to promote the military use of aircraft. In spite of these organizational transformations, U.S. military air power was vastly inferior to that of its European counterparts when the United States entered the war in April 1917. To remedy that situation, President Woodrow Wilson created the Army Air Service as part of the War Department on 24 May 1918. By the time the Armistice was signed six months later, the Air Service had grown to include 19,000 officers, 178,000 enlisted men, and 11,754 aircraft, most of which were trainers.

The U.S. Army Air Force

American political and military leaders, with a few notable exceptions, had not been convinced by the progress of World War I that air power need be an important component of the U.S. military establishment. Perhaps the nation's geographic isolation made air attacks seem unlikely events in any future war. In any case, the Army Air Corps (cre-

ated out of the Army Air Service in 1926) was reduced to a relatively modest peacetime organization. Its roles were reduced from 19,189 officers and 178,149 enlisted men on 11 November 1918 to 919 officers and 8,725 enlisted men in 1926. This small contingent had available 60 pursuit and 169 reconnaissance planes.

By the mid-1930s, the importance of air power in future wars gradually began to dawn on American policy makers. One important factor in this transformation was the striking success of the German Luftwaffe in its conquest of the Sudetenland in 1938 and, a year later, in the Germans' lightning conquest of Poland, Norway, Holland, Belgium, and France. By the end of 1939, leaders of the U.S. Air Corps (renamed in 1926) found themselves, according to one history, "in the novel position of receiving practically anything they requested" ("The Birth of the United States Air Force"). According to pre–World War II plans, this "anything" was to include 84 combat groups with 7,800 aircraft manned by 400,000 airmen by 30 June 1942. When the United States entered the war on 8 December 1941, this plan was quickly accelerated and by war's end, the newly renamed U.S. Army Air Force consisted of 2,253,000 men and women and 63,715 aircraft.

The rapid growth of the Air Force during the war was accompanied by the creation and expansion of auxiliary, administration, and support groups. For example, in December 1940, the various combat divisions within the Air Force were reorganized and designated as the Northeast, Northwest, Southeast, and Southwest Air Districts, later to become the First, Second, Third, and Fourth Air Forces. Simultaneously, the Hawaiian, Panama Canal, and Alaskan Air Forces were redesignated as the Seventh, Sixth, and Eleventh Air Forces, respectively. Noncombat activities were also reorganized into new administration groupings. For example, the Flying Training Command was established in early 1941 and the Air Corps Ferrying Command (later the Air Transport Command) and Air Corps Maintenance Command were formed in mid-1942.

The end of World War II once more brought massive demobilization of personnel. But the lessons of World War I had been learned, and a solid core of personnel and equipment was retained to ensure an effective ongoing air warfare presence in the U.S. military. In 1946, the organizational format of the modern U.S. Air Force was created with (1) the redesignation of the old Continental Air Force as the new Strategic Air Command, and (2) the creation of two new entities, the Air Defense Command and the Tactical Air Command. Along with the existing Air Transport Command, these four entities were responsible for the four major functions of the new U.S. Air Force, strategic, defense, tactical, and airlift. Finally, on 26 July 1947, President Harry S Truman signed the National Security Act of 1947, creating the Department of Defense and a separate and independent Department of the Air Force (USAF). The first secretary of the Air Force, W. Stuart Symington, was appointed to his post on 18 September 1947 and, one week later, Symington appointed General Carl A. Spaatz as the USAF's first chief of staff.

Today's U.S. Air Force

The U.S. Air Force in 2001 consisted of 67,676 officers (55,920 men, 11,754 women, and 2 "unknown" gender) and 280,410 enlisted personnel (225,531 men and 54,879 women). An additional 138,897 civilians (50,226 women and 88,671 men) were employed by the Department of the Air Force. Air Force personnel are arranged in a chain of command that begins, at its lowest level, with the individual airman (referring

to both men and women). Airmen are organized into *flights* of two or more airmen and then into *squadrons,* consisting of two or more flights; *groups,* consisting of two or more squadrons; and *wings,* consisting of two or more groups. During wartime, wings may be designated as a numbered air force, such as the Seventh Air Force. Wings report directly to one of the major air commands, of which there are the following nine:

HQ Air Combat Command (Langley Air Force Base, Virginia)
HQ Air Education and Training Command (Randolph AFB, Texas)
HQ Air Force Materiel Command (Wright-Patterson AFB, Ohio)
HQ Air Force Personnel Center (Randolph AFB, Texas)
HQ Air Force Space Command (Peterson AFB, Colorado)
HQ Air Force Special Operations Command (Hurlburt Field AFB, Florida)
HQ Air Mobility Command (Scott AFB, Illinois)
HQ Pacific Air Forces (Hickam AFB, Hawaii)
HQ U.S. Air Forces in Europe (Ramstein Air Base, Germany)

In addition to the major air commands, there are 34 field operating agencies responsible for a variety of support activities, such as the Air Force Audit Agency, Air Force Center for Environmental Excellence, Air Force Communications Agency, Air Force Historical Research Agency, Air Force Legal Services Agency, Air Force Medical Support Agency, Air Force News Agency, Air Force Personnel Center, Air Force Real Estate Agency, Air Force Weather Bureau, and Air National Guard Readiness Center. Finally, four units report directly to Headquarters, USAF: 11th Wing, Bolling AFB, Washington, DC; Air Force Doctrine Center, Maxwell AFB, Alabama; Air Force Operational Test and Evaluation Center,

Kirtland AFB, Washington; and the U.S. Air Force Academy, Colorado.

Air force groups in other branches of the military are organized somewhat differently. The U.S. Naval Air Force, for example, consists of 26 wings, many of which have mirror organizations for the Atlantic and Pacific theaters, such as a Strike Fighter Wing, Airborne Early Warning Wing, a Helicopter Tactical Wing, a Patrol and Reconnaissance Wing, and a Helicopter Anti-Submarine Wing. There are also separate command wings for the Mediterranean and Western Pacific and six training wings.

By contrast, the U.S. Army aviation mission is divided among a variety of battle groups, such as heavy and light division aviation brigades, theater division brigades, and the Armored Calvary Regiment. The Marine Corps aviation mission is divided into three Marine air wings, whose primary missions are transportation and ground support. Coast Guard flight activities are under the command of individual Coast Guard districts, with three to five aircraft at each location, responsible for transport, surveillance, interdiction, and logistic support.

Aircraft

The U.S. military operates three dozen types of aircraft. Many of these aircraft carry different designations and may have modest modifications for each branch of service that uses them. A complete description of these aircraft can be found on the Internet at http://www.af.mil/news/indexpages/fs_index.shtml. Some examples of these aircraft are the following:

AC-130H Spectre: A heavily armed, side-firing gunship used for reconnaissance, interdiction, and close air support. It is also used for perimeter and point defense, air drops, troop extraction, and combat search and rescue.

B-52 Stratofortress: One of the service's oldest and most dependable aircraft, a long-range heavy bomber capable of flying at altitudes of 15,000 meters (50,000 feet) with a range of 14,000 kilometers (8,800 miles).

C-141B Starlifter: The primary aircraft for delivery of combat forces and equipment, resupply and return of the sick and wounded.

F-15E Strike Eagle: An all-purpose attack fighter capable of flying in all weather, at day or night, and at low altitudes. It is used in both air-to-air battles and in air-to-ground attacks.

UH-1N Huey: A multipurpose helicopter used for transport of security and disaster-response forces, medical evacuation, surveillance of nuclear weapons convoys and test ranges, space shuttle landing support, airborne cable inspection, search and rescue operations, and other functions.

V-22A Osprey: A still-experimental multi-engine vertical takeoff and landing (VTOL) aircraft with tiltable rotors, designed for transport of troops, equipment, and supplies; search and rescue; and special operations functions.

Further Reading

Taylor, Michael John Haddrick. *Encyclopaedia of the World's Air Forces.* Wellingborough, UK: Stephens, 1988.

A good source for books on all aspects of the U.S. and world air forces is bookstores with Web sites. The following are some examples of useful sites:

Aeroplane Books, http://www.aeroplane books.com

Aviation & Aircraft Books, http://www.cahood.com/avian2.htm

Aviation Book Company, http://www.aviation book.com

Aviation Heritage Books, http://www.wind canyon.com

B-29 Bibliography, http://www.pbs.org/shptv/b29/book2.html

"Old Army" Books, http://www.sonic.net/
~bstone/oldarmy/aviation.html

VFW   Mailcall,   http://www.nleditions.com/
AV.html

Fact sheets on many aspects of the U.S. Air Force
can be found at http://www.af.mil/news/ind-
expages/fs_index.shtml.

Air in Literature and Film

Topics related to air, the wind, and the sky
are all common in the fields of literature,
theater, and motion pictures. In some cases,
these works deal with events in the real
world. In other cases, they rely on the symbolic meaning of some aspect of the air for
their theme.

Works on Air

The most popular topics among books
about the air are probably those by and about
pilots and their experiences. Among the best
known of these books are a group by the
French pilot and writer Antoine de Saint-
Exupéry (1900–1944). Saint-Exupéry's
books told about his own experiences in the
air and those of other fliers throughout the
world. They included *Wind Sand and Stars*
(1939), describing his trips above the Pyre-
nees, the Sahara, and the Andes; *Night
Flight* (1931), the story of mail pilots flying
from Chile to Paraguay and Argentina; and
Flight to Arras (1942), the story of a single
night's air raid by French pilot over the city
of Arras.

Other fliers have also written about their
own experiences and those of their colleagues. For example, Charles A. Lindbergh
wrote a number of books on flying, two specifically about his first transatlantic flight in
1927, *We* (1927) and *The Spirit of St. Louis*
(1953). Eddie Rickenbacker wrote two
books about flying, *Fighting: The Flying
Circus* (1919) and his autobiography, *Rickenbacker* (1967).

Other works of nonfiction cover nearly
every aspect of airplane, balloon, glider, and
other types of flight, for recreation and transportation and in warfare. For example,
author Don Dwiggins has written a number
of such books, including *Bailout: The Story
of Parachuting and Skydiving* (1969), *The
Barnstormers: Flying Daredevils of the
Roaring Twenties* (1981), *Famous Flyers
and the Ships They Flew* (1969), and *Into the
Unknown: The Story of Space Shuttles and
Space Stations* (1973). A number of books
are also available on the history of various
aspects of flight. Some examples include
Page Shamburger's *Tracks across the Sky:
The Story of the Pioneers of the U.S. Air
Mail* (1964), Charles Joseph Gross's *Prelude
to the Total Force: The Air National Guard,
1943–1969* (1985), Clive Hart's *The Dream
of Flight: Aeronautics from Classical Times
to the Renaissance* (1972), and Fred E. C.
Culik and Spencer Dunmore's *On Great
White Wings: The Wright Brothers and the
Race for Flight* (2001).

Air warfare and famous air battles have
been especially popular topics for both fic-
tion and nonfiction books. Typical of these
works is the four-volume series *Airwar*, by
Edward Jablonski. The series deals with aer-
ial operations carries out in the period
between 1939 and 1945 of World War II.
Other histories of air warfare include Eric
Melrose Brown's *Duels in the Sky: World
War II Naval Aircraft in Combat* (1988);
Lynn Montross's *Cavalry of the Sky: The
Story of U.S. Marine Combat Helicopters*
(1954); Franklin Noble's *The Bombing
Offensive against Germany* (1965); Alfred
Price's *Luftwaffe Handbook, 1939–1945*
(1977); and Christopher Shores's *Air Aces*
(1983), a biography of famous fighter pilots.
One of the most recent and most successful
of such books is *Air America* (1979), writ-
ten by Christopher Robbins. The book was
later made into a motion picture by the same

name in 1990. It was directed by Roger Spottiswoode and starred Mel Gibson, Robert Downey Jr., and Nancy Travis. The book and film were said to be based on a secret airline operated by the Central Intelligence Agency that was involved in many actions over Southeast Asia from between 1955 and 1974.

The subject of air is an organizing theme in many works of fiction. One such book is Sinclair Lewis's *Free Air* (1919), in which the author tells of a trip taken by Claire Boltwood and her father, as they travel cross country in their Gomez-Dep roadster from Minneapolis to Seattle. The "free air" that they seek and find is the challenging new world of the open spaces of the American West. In his Nobel Prize speech of 1930, Lewis declared the novel "dead before the ink was dry," although it has gained some greater fame in the past few decades.

In one of his early novels, George Orwell wrote about an insurance salesman in the London suburb of West Bletchley, living in a suffocating family life before the start of World War II. In *Coming Up for Air* (1939), the salesman has an opportunity to escape his humdrum life and return to his peaceful childhood home in Lower Binfield. A more recent story of an individual's escape from a suffocating life, this time both real and metaphorical, is Amulya Malladi's 2002 novel *A Breath of Fresh Air.* That novel tells of the road traveled by the protagonist, Anjali, in her effort to escape from a suffocating marriage at the same time she is exposed to a physical threat of suffocation during the Bhopal poison gas leak in India in 1984.

Escapes into the air are also the subject of many science fiction stories and mysteries. In *Chronicles of Air and Dreams: A Novel of Mexico* (1999), for example, author Rosa Martha Villarreal tells of an American archaeologist who survives an earthquake while working inside a Mayan pyramid, but then loses her ability to speak any language other than an ancient Mayan dialect. Her recovery from that experience includes a range of emotions that reveals both ancient Mayan secrets and puzzles of the modern world.

Works on the Wind

As with the subject of air, books that draw on wind as a topic range from the real and concrete to the allegorical. One of the most popular books and films of recent years has been Sebastian Junger's *The Perfect Storm: A True Story of Men Against the Sea* (1997). The book describes the fate of the fishing boat *Andrea Gail* in 1991 when it encountered what has been called "the greatest storm in modern history." The book was made into a motion picture (2000) by the same name, directed by Wolfgang Petersen and starring George Clooney, Mark Wahlberg, Diane Lane, and Mary Elizabeth Mastrantonio.

The notion that the wind has mystical powers and is symbolic of many human emotions reaches back as far as recorded history. The wind gods and goddesses that exist in virtually every known culture represent the notion that a physical force—moving air—can carry with it all manner of events, such as birth, death, cleansing, the arrival of evil, and special blessings. Some of literature's greatest classics include a reference to the wind in their title, suggesting a major theme that occurs in the book itself. Examples of such titles are *Gone with the Wind,* Margaret Mitchell's 1936 novel about Southern society during the last days of the Civil War; Jerome Lawrence and Robert E. Lee's 1955 play, *Inherit the Wind,* a fictionalized account of the famous 1925 Scopes trial; and Kenneth Grahame's children's story, *Wind in the Willows* (1908). All three works were later made into motion pictures, the first two with the same name, in 1939 and 1960, respectively, and the last under the title *Mr. Toad's Wild Ride,* in 1999.

A review of popular fiction titles reveals that the symbolic and mystical attributes of the wind influence the work of modern novelists. These titles include *The Dark Wind* and *The Wailing Wind* (Tony Hillerman, 1990 and 2002); *Ill Wind* (Nevada Barr, 1997); *Restless Wind, Wind of Promise,* and *Wayward Wind* (Dorothy Garlock, 1986, 2002, and 1994); *Riding the East Wind* (Otohiko Kaga, 2002); and *Run before the Wind* (Stuart Woods, 1992).

Works on the Sky

The use of sky images in works of literature and motion pictures draws on concepts of openness, expansive reaches, uplifting, protectiveness, and striving for great spirit values. For example, authors may write about the lives of individuals or communities opening up to new opportunities and new experiences, as if they were opening up to the sky. Or they may represent the tragedies of one's lives as if the sky were falling around them.

Some examples of novels and plays dealing with sky themes are the following:

- C. S. Forester's *The Sky and the Forest* (1948) tells about the invasion of a central African kingdom by European explorers from the standpoint of the natives and not from that of the Europeans.
- The Italian writer Lorenza Mazzetti's *The Sky Falls* (1961) relates her experiences following the murder of her aunt and two cousins in 1944 by German soldiers. The book was made into a highly successful motion picture in 2000.
- "The Sky above Hell" is a short story written by the Russian author Yuri Mamleyev in 1980. It tells of a madman who murders people to see the look in their eyes as their soul leaves their bodies.
- *A Sky So Close* (2001) is a novel by Iraqi writer Betool Kehedairi, who tells of a young woman born to an Iraqi father and English mother torn between two cultures, growing up in a small village outside of Baghdad.
- *Under the Naked Sky* (2001) is a collection of short stories written by 30 Muslim authors about daily life in countries ranging from Morocco to Iraq.
- Paul Bowles's novel, *The Sheltering Sky* (2000), tells of the adventures and musing of three young Americans who set out to walk across the Sahara.

Air Mass

An air mass is a very large body of air with relatively constant temperature and moisture content. Air masses are typically 1,500 kilometers (2,400 miles) wide and a few kilometers thick. They tend to form in certain parts of the world known as *source regions,* out of which they slowly move, carried by prevailing winds.

Source Regions

Source regions have two major characteristics that allow the formation of an air mass. First, they cover a wide area that has a relatively consistent topography. A region like the north central United States is not a source region, for example, because it contains extensive areas covered by both land and water. Since land and water heat at different rates, the air above them is constantly in motion and the uniform heating or cooling needed to produce an air mass is not available. The flat surface of the open ocean, by contrast, provides the consistent topography needed for the formation of an air mass.

A second condition required in a source region is the presence of a relatively calm atmosphere, that is, the absence of strong disturbing winds. The presence of strong air movements allows air of different temperatures and humidity to mix, preventing the formation of an air mass.

Types of Air Masses

Air masses are classified on the basis of two criteria: (1) the nature of the source region (land or water) over which they form, and (2) the latitude of that region. Source regions are classified as continental (c) or maritime (m), while latitude is classified as arctic (A), polar (P), tropical (T), or equatorial (E). These categories and abbreviations can be combined to describe the types of air masses found anywhere in the world.

In North America, weather patterns are influenced by the five types of air masses listed below.

cA air masses form north of the Arctic Circle. They contain very cold and very dry air during the winter when little or no sunlight reaches the Arctic region. Such air masses are uncommon in summer when the region receives significant amounts of sunlight. cP air masses tend to move southward during the winter, bringing very cold weather to central Canada and the north central United States.

cP air masses form in Alaska and central Canada. The air they contain is cold and dry, but to a lesser degree than that found in cA air masses. cP air masses tend to flow in a southeasterly direction over the north central states, bringing cold weather in winter, and cool, pleasant weather in summer. During the winter, cP air masses may flow over the Great Lakes, bringing severe snowstorms common to the eastern shores of those lakes.

mP air masses form on either side of North America in the north Atlantic or north Pacific oceans. They tend to be quite humid, with relatively mild temperatures. Air masses from the Pacific strongly influence weather patterns in the Pacific Northwest, bringing low clouds, rain, and fog to the area. Atlantic air masses are responsible for severe "nor'easter" winter storms in New England and Maritime Canada, but often bring episodes of cool, pleasant weather in summer.

cT air masses originate over hot desert regions of northern Mexico and the southwestern United States. They contain hot, dry air that sweeps upward through the Great Plains, sometimes bringing severe drought. As they move northward, they pick up moisture from the ground, becoming more similar in character to an mT air mass. cT air masses are less common in winter, when temperatures are cooler, than in summer.

mT air masses form over the southern Pacific and southern Atlantic oceans and over the Gulf of Mexico. They contain warm, moist air year-round. Air masses from the latter two regions significantly affect weather patterns in the United States as they move northward across the southeastern states. In summer, they bring hot, humid weather to the eastern half of the country, while they are responsible for heavy precipitation and fog during the winter. Air masses from the south Pacific are less likely to penetrate very far onto land, although they sometimes bring rain and fog to northwest Mexico and southwestern United States in winter and severe storms in summer.

Modifications in Air Masses

The temperature and humidity of an air mass may change as it passes out of its source region. For example, cA and cP air masses usually come into contact with relatively warm land as they move southward. Heat from the land is absorbed by the cold air mass, which becomes warmer. The letters *k* and *w* are sometimes added to an air mass classification to indicate the relative temperature of the air mass and the surface over which it passes. In such cases, *k* indicates that the air mass is cooler, and *w* indicates that the air mass is warmer. Thus, a cAk air mass contains continental Arctic air that is colder than that of the surface over which it passes (as would normally be the case), while a cAw air mass would be an unusual air mass in which temperatures within the mass are higher than those on the underlying surface.

Air masses can also pick up and lose moisture from and to the surface over which they pass. For example, cA and cP air masses typically consist of very dry air. However, when such air masses pass over large bodies of water, they may absorb moisture from the surface water. Such is the case, for example, when a cP air mass passes over the Great Lakes. When that happens, air in the air mass may pick up water from the lakes, becoming more unstable in the process. After passing over the lake, the air mass may release this excess moisture as rain in the summer or as heavy snow during the winter. *See also* Atmosphere; Cyclones and Anticyclones; Global Winds; Wind

Further Reading

Ahrens, C. Donald. *Meteorology Today: An Introduction to Weather, Climate, and the Environment,* 6th edition. Pacific Grove, CA: Brooks/Cole Publishing, 1999.

Allaby, Michael. *Facts on File Weather and Climate Handbook.* New York: Facts on File, 2002.

Lutgens, Frederick K., and Edward J. Tarbuck. *Atmosphere,* 8th edition. Upper Saddle River, NJ: Prentice-Hall, 2000.

Air in Music

See Musical References to Air

Air National Guard

The Air National Guard (ANG) is a reserve component of the U.S. Air Force. In addition to its primary role as a backup source of manpower during times of armed conflict, the Guard has a number of other functions, such as providing air traffic control, parachute rescue operations, airborne firefighting support, and combat communications.

Origin of the Air National Guard

The Air National Guard's history reflects the complex problems of having separate and individual military units organized and administered by the individual states (the "national guard" component) under the larger authority of a federal agency (the U.S. Air Force). Until World War II, the balance was somewhat simpler since ANG units were exclusively state operations with members being called to active service individually as needed.

That tradition originated in November 1915 when the first ANG unit was organized in New York state by Captain Raynal Cawthorne Bolling. That unit, the 1st Aero Company, New York, was mobilized during the 1916 border crisis with Mexico, although it never saw active duty.

Military leaders rejected the idea of bringing ANG units, as units, into active duty during World War I. They did, however, activate individual ANG members, who were then assigned to the U.S. Signal Corps Reserve. Among these air guardsmen was 2nd Lieutenant Erwin R. Bleckley of Kansas, the first Air National Guardsman to win the Medal of Honor (posthumously).

Growth of the Guard

Between the two world wars, interest in and support for the Air National Guard began to grow. By the start of World War II, 29 guard squadrons were in existence. Their primary role was flight training and support for ground services training. Individual units were sometimes employed in dealing with local disasters. The 154th Arkansas Observation Squadron, for example, flew about 20,000 miles in carrying supplies and relief workers during the state's terrible floods of 1927.

World War II proved to be a transitional period for the ANG. As in World War I, most members were activated individually, with little regard for their unit status. As a result, nine ANG units were disbanded entirely during the war. But some units were activated intact and were assigned to specialized missions, such as reconnaissance, liaison, fighter, or bombardier duties. Overall, the most significant contribution of the guard during the war was the training of new enlistees in the Army Air Force itself.

Continuing Problems

The postwar period was difficult for the ANG. For nearly a decade, the Army Air Force (AAF) and the National Guard Bureau had been fighting for control of the organization. AAF officers had little respect for ANG members and had little hope that guardsmen could make a serious contribution to any war effort. In fact, one AAF observer referred to ANG units as "state-sponsored flying clubs" ("Forging the Air National Guard"). The AAF insisted that the Guard could be made into an effective fighting unit only if the Air Force had greater control over the Guard. The National Guard Bureau, on the other hand, argued that control of the Guard should remain in the hands of state governments.

On 18 September 1947, the crisis was formally resolved with the establishment of a separate U.S. Air Force, which included the Air National Guard as one of its components. Unfortunately, the resolution turned out to be largely a paper solution, and the struggle between individual states and the federal government over control of the ANG continued for some time.

The Air National Guard and the Korean War

The Korean War (1950–1953) proved to be a turning point in ANG history. About 80 percent of the nation's air guardsmen were mobilized, about 45,000 men overall, including 66 of the Guard's 99 flying squadrons. These squadrons flew 39,530 combat missions, destroying 39 enemy aircraft but losing 101 U.S. fliers.

As in World War II, many individuals were stripped from their units when they were mobilized. Units that were activated together were assigned randomly to jobs for which they may or may not have been trained. Their overall performance proved that they had been poorly trained with obsolete equipment and were largely unprepared for actual combat assignments. According to one history of the Air National Guard, "It took months and months for them [guard units] to become combat ready. Some units never were" ("Forging the Air National Guard").

Restructuring the Air National Guard

The poor showing of the Air National Guard in the Korean War created demands for a revamping of the organizational structure and operation of the Guard. During the 1950s, therefore, Congress passed a number of laws more clearly defining the Guard's mission and providing a more adequate financial basis for its operation. The new mission of the ANG had four elements. First, the specific role of Guard units in any future combat was to be clearly defined and each Guard unit was to be trained and prepared specifically for that role. Second, the ANG was eventually to provide 100 percent of the nation's air defense interceptor force, a goal that was finally achieved in October 1997.

Third, unit training was to be provided by regular Air Force personnel from the same command under which a unit would serve in wartime. Fourth, the ANG would become part of Secretary of Defense Robert McNamara's "special reserve force pro-

Members of the End of Runway team of the 157th Aircraft Maintenance Unit make final preparations for this F-16CJ to fly another combat sortie on 1 April 2003, during Operation Iraqi Freedom. The 379th AEW is a forward-deploying unit in Southeast Asia. The 157th AMU is a deployed unit from the South Carolina Air National Guard assigned to the 169th Fighter Wing. © U.S. Air Force.

gram," developed in the early 1960s, in which certain designated units were to have additional training and priority access to equipment to make possible rapid and efficient mobilization during wartime.

For all of these efforts, the ANG's problems had still not been resolved by the late 1960s. Their performance during the Berlin Airlift of 1961, for example, was limited and unimpressive, and some regular officers were convinced that Air Guardsmen were still largely "amateurs who had not improved significantly since the Korean War" ("Forging the Air National Guard"). The Vietnam War highlighted another Guard problem when President Lyndon Johnson decided to mobilize only a few ANG units,

confirming a widely held view that the ANG was little more than a "draft haven for relatively affluent young white men" ("Forging the Air National Guard"). President Johnson's view (and that of other observers) has been confirmed by a number of examples over the past decade. For example, President George W. Bush was apparently able to receive an appointment to the Texas Air National Guard in lieu of service with the regular military during the late 1960s. Records indicate that Bush (like many other Air National Guard members) failed to complete even the minimum tasks assigned to guard members. ("George W. Bush's Texas Air National Guard service," http://uggabugga.blogspot.com/2003_01_12_ugga

bugga_archive.html).

The Modern Air National Guard

As the twentieth century drew to a close, the Air National Guard was gaining respect and support for its role in a variety of military actions. During the Persian Gulf crisis of 1991, for example, 10,456 Air Guardsmen were mobilized to carry out tactical reconnaissance missions, broadcast surrender appeals and instructions to Iraqi soldiers, refuel from aerial tankers, and transport personnel and cargo both in the war zone and in Europe.

The authorized strength for the Air National Guard in 2002 was 106,678 personnel organized into more than 140 units in the 50 United States, Puerto Rico, Guam, and the U.S. Virgin Islands. The Guard has a fleet of 10 different aircraft, including the C-130 Hercules, C-5 Galaxy, and C-141 Staflifter cargo and heavy aircraft and the F-15 Eagle and A-10 Thunderbolt fighter jets.

Air National Guardsmen are required to complete a six-week basic training program and then attend regular monthly sessions at a local unit base. They qualify for a number of federal and state benefits, including use of commissary and recreational facilities, federal life insurance and educational benefits, and state educational benefits. *See also* Air Force

Further Information

Air National Guard Bureau
Office of Public Affairs
2500 Army Pentagon
Washington, DC 20310-2500
Telephone: (703) 607-2613
URL: http://www.ang.af.mil
E-mail contacts: see individual state units

Further Reading

"Air National Guard Heritage," http://www.ang.af.mil/history/heritage.asp.

"Forging the Air National Guard," http://www.ang.af.mil/history/forging.asp.
Francillon, Rene J. *The United States Air National Guard.* London: Aerospace Publishing, 1993.

Airplane

An airplane is a type of heavier-than-air aircraft with fixed wings, propelled either by a propeller-driven or jet-powered engine or by some combination of the two. Airplanes are classified with gliders as heavier-than-air methods of transportation, compared to kites and balloons, which are classified as lighter-than-air aircraft, also known as airships or aerostats.

Invention

Orville and Wilbur Wright are generally regarded as the inventors of the modern airplane. They made the first controlled, powered flight at Kitty Hawk, North Carolina, on 17 December 1903. Their work was preceded, however, by a number of earlier inventors who attempted to design and build aircraft that were powered by some type of engine and capable of flight that could be controlled by the pilot. The German inventor Otto Lilienthal (1848–1896) made over two thousand successful glider flights and had plans to construct a glider powered by a steam engine. He was killed when one of his gliders crashed in 1896, before he built such a device, however.

Other inventors were able to build steam-powered aircraft and, in some cases, to get them off the ground. But they were never able to control the flight of an aircraft carrying a pilot. The French inventor Felix Du Temple (1823–1890), for example, built a steam-powered aircraft that remained in the air for two seconds. A contemporary, the Russian engineer Alexander Mozhaiski (1825–1890), constructed an airplane powered by two steam engines, with a similar lack of success. In 1894, the British inventor Hiram Maxim built a huge, 3.5-ton aircraft that remained aloft for a few seconds, but, again, without any mechanism for control. Only in late 1903 were the Wright brothers able to construct an airplane that was able to stay aloft for an extended period of time (12 seconds) over a distance of just over 30 meters (100 feet) completely under Orville's control.

Airplane Aerodynamics

Four forces are at work during the flight of an airplane:

- *Lift* is an upward force, opposite to that of gravity, that raises an airplane into the air.
- *Weight* is a downward force, tending to pull the airplane toward the ground, a result of the Earth's gravitational attraction on the mass of the airplane.
- *Thrust* is a forward force that tends to propel the airplane in a horizontal direction parallel to the ground.
- *Drag* is a backward force that tends to prevent the airplane from traveling in the direction produced by thrust.

When an airplane is in level flight at constant speed, lift and weight forces are equal to each other (the airplane travels neither upward nor downward) and thrust and drag forces are equal to each other (the airplane neither speeds up nor slows down). Weight and drag are forces that tend to prevent an airplane from moving at all, either forward or upward. They are caused by the mass of the airplane itself. To get an airplane to move forward, travel at sufficient speed to get off the ground, and move forward through the air, some means must be provided to produce thrust and lift forces to overcome weight and drag. The two mechanisms used to produce these forces are some type of engine and an airfoil.

Engines

Originally, all airplanes were powered by one or more propellers, driven by an internal combustion engine, similar to that used in automobiles, trucks, and other ground vehicles. Power from the engine is transferred to the propeller by a crankshaft, which causes the propeller to rotate in a vertical plane. The blades of the propeller are shaped like an airfoil (the wing of an airplane), with one edge thicker than the other and twisted at an angle to the vertical. As the propeller blades rotate, they scoop air out from in front of the engine and push it backward across the airplane's wings.

Jet engines operate by generating a stream of air or gas by burning a fuel inside the engine. The ramjet is the simplest form of a jet engine. It is essentially a specially shaped cone open at both ends. Air enters the front of the engine and is used to burn the fuel inside the jet. Hot combustion gases from this reaction exit the rear of the jet, producing a forward thrust on the engine, the wing to which it is attached, and the airplane itself.

A turboprop engine combines the properties of a propeller-driven engine and a ramjet. Some of the exhaust gases produced during combustion in the jet engine are recycled through the engine and used to operate a propeller attached to the jet. In a turboprop, thrust is obtained from two sources: the jet engine itself and the propeller driven by exhaust gases.

Airfoils

The term airfoil, in general, refers to any structure used to provide lift on an aircraft as it moves through the air. Propeller blades are one example of an airfoil. The wings on an airplane are a second example.

Most airplane wings have a classic shape that provides maximum lifting force when the airplane moves through the air. The wing has a flat bottom and a curved top, with a rounded front (leading) edge and a sharp back (trailing) edge. The process by which this shape produces a lifting force is quite complex. Essentially, however, the motion of air over the top of the wing produces a decrease in pressure on the wing. At the same time, air passing over the bottom of the wing produces normal pressure, greater than that on top of the wing. The larger pressure on the bottom of the wing pushes the wing upward, carrying along with the airplane body to which it is attached. The faster the wing moves through the air, the greater the difference in air pressure and the greater the upward force on the wing.

Because airplanes fly at different speeds during a flight, their wing shapes must be flexible. For takeoff and landing, an airplane has to be able to increase or decrease its speed fairly quickly. To make such changes possible, wings have attachments known as *flaps,* movable sections that can be extended or retracted.

When the flaps are extended during takeoff, they increase the area over which air flows and hence the pressure difference above and below the wing, creating more lift. During landing, the flaps are extended and lowered, producing more drag and making it possible to slow the airplane down for landing. After takeoff and before landing, flaps are retracted because they increase the total amount of drag on the airplane and tend to slow it down.

Airplanes also have two airfoils at the back also, horizontal and vertical stabilizers. The horizontal stabilizers look like small wings attached at the very back of the airplane's body. The vertical stabilizer is the tail extending upward between the two halves of the horizontal stabilizer. Both horizontal and vertical stabilizers have flaps attached to them that allow a pilot to change the speed and direction of the airplane.

Controlling Direction

An airplane travels through three-dimensional space and has a tendency to rotate in three directions. Picture a set of three axes with the center of gravity of the airplane at the center of the three axes. One axis, the *longitudinal axis,* runs from the nose of the airplane to its tail. Rotation around this axis is called *roll.* A second axis, the *lateral axis,* runs from one wingtip to the opposite wingtip. Rotation around this axis is known as *pitch.* The third axis, the *vertical axis,* runs from a point directly above the airplane's center of gravity to a second point directly below it. Rotation around this axis is known as *yaw.* An airplane must be fitted with surfaces that allow a pilot to control unwanted rotation around each of these axes. Failure to do so means that the airplane will go out of control and eventually crash.

Rotation around the longitudinal axis is controlled by a pair of *ailerons* (from the French for "little wings") attached to the trailing edge and on the outside of the wings. Ailerons are connected to each other and operate in reverse directions. When a pilot moves the right aileron upward, the left aileron automatically moves downward. By adjusting the ailerons as needed, a pilot can keep an airplane from rolling back and forth around its longitudinal axis.

Rotation around the lateral axis is controlled by the two flaps, or *elevators,* attached to the horizontal stabilizer. When the pilot moves the control wheel in the cockpit forward or backward, the elevators are raised or lowered, causing the nose of the airplane to move upward or downward. For example, during takeoff, the pilot pulls the control wheel toward himself or herself, causing the elevator to go up, which pushes the airplane's nose up and its tail down and makes the airplane move upward.

Rotation around the vertical axis is controlled by the *rudder,* the flap attached to the vertical stabilizer at the back of the airplane. The rudder is controlled by pedals in the cockpit. When the pilot pushes on the right rudder pedal, the tail rudder moves to the left, causing the airplane's nose to rotate to the right. Although the three control mechanisms are structurally independent, they are often used in conjunction with one another to make smooth, efficient turns. *See also* Aerodynamics; Bernoulli's Principle; Wright, Orville and Wilbur

Further Reading

Bristow, Gary V. *Encyclopedia of Technical Aviation.* New York: McGraw-Hill Professional, 2002.

Kermode, Alfred Cotterill. *Mechanics of Flight,* 10th edition. Englewood Cliffs, NJ: Prentice-Hall, 1995.

Air Pollution

Air pollution is a condition that exists when a material is present in the atmosphere at a concentration great enough that is has the potential for harming organisms. For example, smoke occurs naturally in the air as a result of forest fires and other types of combustion. Smoke from forest fires is normally not considered to be a pollutant, however, unless it exists over long periods in concentrations great enough to harm organisms with which it comes into contact. The materials responsible for air pollution are known as *pollutants.*

History

Air pollution has probably existed in some places and to some extent as long as humans have lived on Earth. For example, one can imagine ancient people exposed to very high concentrations of smoke as they cooked meals in their caves. Today, that form of air pollution is known as *indoor air pollution.*

Air pollution became a widespread problem only recently in human civilization, as people began to depend more and more on the combustion of fossil fuels (coal, oil, and natural gas) to heat their homes, cook their meals, drive their vehicles, and carry out other everyday activities. The combus-

Pristine water flowing over rocks in Colorado. © Environmental Protection Agency.

tion of fossil fuels is responsible for most types of air pollutants and for their high concentration in the atmosphere.

Possibly the first regulation attempting to control air pollution was a proclamation issued by England's Queen Elizabeth I in 1307 forbidding the burning of a fuel known as "sea coal," which released large amounts of smoke into the atmosphere.

By the nineteenth century, the widespread use of fossil fuels (especially coal) during by the Industrial Revolution made air pollution a problem in many large cities around the world. In England, for example, residents of most large cities became accustomed to being engulfed in a thick, smoky fog (now known as *smog*) much of the year. Perhaps the most serious smog incident occurred in 1952 when more than 4,000 Londoners died as a result of smog-related health problems. Tragedies such as these prompted governments in many developed countries to begin passing legislation to reduce or eliminate air pollution by the end of the 1950s.

Components

When fossil fuels are burned, they release many products, the two most important of which (in terms of volume) are water and carbon dioxide, both of which are harm-

Discharge from chemical plants in New Jersey. © Environmental Protection Agency.

less to human health. However, many less important (in terms of volume) combustion products pose serious health problems to humans, other animals, plants, and physical structures. The most significant of these products are the following:

Carbon monoxide (CO) is produced when any fuel undergoes incomplete combustion, that is, receives insufficient oxygen to burn completely to carbon dioxide. Carbon monoxide is a colorless, odorless gas that, when inhaled, competes with oxygen for transport through the circulatory system. Since it binds more tightly to hemoglobin in blood, it prevents cells from receiving adequate amounts of oxygen, and they begin to die. At low concentrations (less than 60 parts per million), carbon monoxide can impair the central nervous system, causing headache, fatigue, and drowsiness. At higher concentrations, carbon monoxide can result in respiratory failure, coma, and death.

As with other forms of air pollution, the health effects of a pollutant depends on two factors: the amount of time to which an organism is exposed to the pollutant, and the concentration of the pollutant. In the case of carbon monoxide, for example, exposure to the gas at a concentration of 1,000 parts per million is roughly equivalent to an exposure for five hours at a concentration of about 200 parts per million. In the ordinary course of events, a person is unlikely to be exposed to levels of carbon monoxide that can result in the most serious health consequences, such as respiratory failure or death, although exposures to lower concentrations for long periods may not, under some circumstances, be unusual.

Sulfur dioxide (SO_2) is formed when impurities in fossil fuels react with oxygen

Corn being sprayed with chemicals in Kansas. © Environmental Protection Agency.

during combustion. For example, a variety of sulfur-containing compounds occurs naturally in petroleum. When that petroleum is burned, those compounds are converted in part to sulfur dioxide. Sulfur dioxide released to the atmosphere reacts with oxygen to form sulfur trioxide, which, in turn, dissolves in water droplets in the air to produce sulfuric acid. Sulfuric acid is harmful not only to the respiratory systems of animals that inhale it, but also to plants on which it falls. It also damages certain types of rock, such as limestone, which dissolve when exposed to sulfuric acid.

Oxides of nitrogen (N_2O, NO, NO_2) are formed in the atmosphere when hot waste gases are expelled into the air. Nitrogen is the major component of the atmosphere, but it is a relatively inactive gas and normally does not combine with oxygen, the other major component of the atmosphere. At the high temperatures at which combustion occurs, however, nitrogen becomes more reactive and combines with oxygen to form a variety of oxides. The first of these oxides is nitric oxide (NO), which, depending on conditions, may then be oxidized further to nitrogen dioxide (NO_2). Like sulfur dioxide, nitrogen dioxide dissolves in water droplets in the atmosphere, forming nitric acid, which has effects on living and nonliving matter similar to those of sulfuric acid.

Volatile organic compounds (*VOCs*) are chemical compounds that contain the element carbon and vaporize easily at room temperature. Some examples of VOCs are formaldehyde, benzene, toluene, 1,3-butadiene, acetone, and many types of halogenated hydrocarbons, such as chloromethane, bromomethane, vinyl chloride, methylene chloride, 1,1,1-trichloroethane, and carbon tetrachloride. VOCs are an important factor in indoor air pollution because they are released from a number of common household products and materials used in the construction of buildings. Examples of these products include paints, wood preservatives, aerosol sprays, cleaners and disinfectants, moth repellents, hobby supplies, and dry-cleaned clothing.

The most common health effects of VOCs include irritation of the respiratory system, headaches and loss of coordination, and nausea. Many of the VOCs are thought to have more serious health effects, however, and are classified as carcinogens (cancer-causing agents), teratogens (substances that produce birth defects), and/or mutagens (substances that cause mutations).

Metals make up a small, but sometimes significant, part of air pollution. Perhaps the most troublesome metal found in polluted air is lead. At one time, the major source of lead in the atmosphere was automotive fuels. An additive known as tetraethyl lead was, for many years, added to gasoline to improve its quality. During combustion, that compound tends to break down, releasing free lead into the air. For many decades, the use of tetraethyl lead has been banned from use in gasoline. However, small amounts of the element are still found in air, primarily as a by-product of paint manufacture and use, the operation of lead smelters, and the production of many lead-containing products, such as batteries, bullets, water pipes, and some types of ceramics.

Lead is known to cause severe damage to the central nervous system and can interfere with the digestive system. It is also thought to cause cancer. These effects are observed in both humans and animals.

Particulate is the name given to any solid or liquid material small enough to be suspended in the air for a significant period. Some examples of particulates found in the atmosphere are bacteria (1 to 15 micrometers in diameter), coal dust (10 to 400 micrometers), cement dust (10 to 250 micrometers), fertilizer and ground limestone (30 to 800 micrometers), and foundry sand (200 to 2,000 micrometers). Particulates are produced by a number of human activities, including agriculture, road construction, industrial operations, mining, and transportation, where they are produced as a by-product of the combustion of fossil fuels.

In heavy concentrations, particulates can pose a hazard to visibility and contaminate clothing and other materials exposed to the air. At any level, they may create problems for the health of animals, when they are inhaled and taken into the respiratory system. Over long periods, they may be at least partially responsible for or contribute to respiratory disorders such as allergies, asthma, emphysema, chronic bronchitis, and other forms of lung damage.

Ozone (O_3) is an allotrope (form) of oxygen gas, consisting of three atoms to the molecule rather than two, as in the more common form of the element, diatomic oxygen. Ozone is produced when sufficient energy is available to cause molecules of diatomic oxygen to react with one another to form ozone. Many situations in which fossil fuels are burned create such conditions, making ozone a by-product of the burning of coal, oil, and gas.

Ozone has deleterious effects on both plants and animals, attacking and destroying cells and producing various types of disease and disorders. In humans and other animals, ozone attacks the respiratory system, causing persistent coughs, chest discomfort, shortness of breath, and, in some cases, damage to the lungs.

Point and Nonpoint Emissions

The pollutants present in air can generally be divided into two general categories: those that come from point sources and those produced by nonpoint sources. A point source is a stationary facility that can be specifically identified. Factories and power plants are examples of point sources. Emissions from nonpoint sources cannot readily be associated with a single location or group of specific locations. For example, the vast majority of carbon monoxide in the atmosphere is released from hundreds of millions of automotive vehicles throughout a region, the nation, or around the world. It is impossible to say how much pollution any one car, truck, airplane, or other individual fuel-burning vehicle contributes to the amount of carbon monoxide in any one region.

Point and nonpoint sources present very different problems for efforts to reduce the concentration of pollutants in the atmosphere. With point sources, a regulatory agency can specify that the source develop procedures and install systems to reduce the amount of pollutant(s) it releases to the atmosphere. With nonpoint sources, many more individual units (e.g., cars, trucks, and airplanes) and different types of units (e.g., lawn mowers, golf carts, tractors, chain saws, snowmobiles, and power boats) must be fitted with a variety of pollution control systems to have any effect on the concentration of pollutants in the air.

Air Quality Management

Over the past 50 years, a number of laws have been passed in the United States and other countries in an effort to reduce levels of air pollution. Such laws are necessary because manufacturers, power producers, and others who are responsible for the release of pollutants seldom initiate efforts to improve their own equipment and processes to eliminate pollutants. The cost of such pollution-reduction equipment can reduce a company's profits significantly, so governmental action is usually required to accomplish this goal.

Air pollution control methods generally fall into one of two categories: use of cleaner fuels and installation of pollutant-removing equipment. For example, one of the most successful methods for reducing the amount of sulfur dioxide released into the atmosphere is to use coal, oil, and natural gas with relatively low concentrations of sulfur-containing compounds. When those fuels are burned, then, the amount of sulfur dioxide released to the air is less.

Many kinds of devices have been invented for removing pollutants from air after the combustion of a fossil fuel. For example, particulates can be removed from waste gases by passing those gases into a large centrifuge, which spins the gases at high speeds. Solid and liquid particles in the gases are thrown against the centrifuge's outer walls, where they can be collected and removed.

A variety of other methods are available for removing pollutants from waste gases after combustion. For example, scrubbers are sometimes used to remove oxides of nitrogen and sulfur from such gases. A scrubber is a device that contains a chemical solution that reacts with one or more of the pollutants in the waste gases. For example, a solution of calcium hydroxide may be added to a scrubber to react with sulfur dioxide, producing the solid product calcium sulfite, which can then be removed from the scrubber.

One of the most effective devices for removing pollutants from automotive exhaust gases is the catalytic converter. A catalytic converter is a device inserted into a vehicle's exhaust system, through which waste gases produced in the engine must pass. The catalytic converter contains various chemicals that improve the efficiency with which fuel remaining in the exhaust gases is burned.

In a one-stage converter, engine exhaust enters at the front and passes over a bed of some finely divided metal, usually a precious metal such as platinum or rhodium. The metal increases the efficiency with which oxygen reacts with unburned compounds in the exhaust gases. Carbon monoxide is converted to carbon dioxide, unburned hydrocarbons are converted to water and carbon dioxide, and nitrous oxide is converted to elemental nitrogen.

Most vehicles now use two-stage catalytic converters, with each stage operating at a different temperature. Two-stage converters tend to be more efficient at removing pollutants because the chemical reactions by which pollutant gases are converted to harmless gases may differ, depending on the temperature at which the reaction occurs.

Three approaches are usually recommended for the control of indoor air pollution: source control, improved ventilation, and air cleansing. Source control refers to the storage, covering, or removal of appliances and products that produce pollutants in the first place, sources such as paint cans, gas stoves, and exposed asbestos. Improved ventilation means increasing the flow of air through a home, office, or other building, so that contaminated air is flushed outside and replaced by fresh air. Air cleaners are commercially available devices through which indoor air is passed and filtered. *See also* Air Quality Index; Baghouse; Clean Air Acts; Ozone

Further Information

Cheresmisinoff, Nicholas P. *Handbook of Air Pollution Prevention and Control.* Boston: Butterworth-Heinemann, 2002.

Elsom, Derek M. *Atmospheric Pollution: A Global Problem.* Oxford, UK: Blackwell Publishers, 2002.

Jacobson, Mark Z. *Atmospheric Pollution.* Cambridge, UK: Cambridge University Press, 2002.

Spengler, John D., John F. McCarthy, and Jonathan M. Samet, eds. *Indoor Air Quality Handbook.* New York: McGraw-Hill, 2000.

Air Pressure

Air pressure is the force per unit area exerted by air on a surface. When referring to pressures caused by the atmosphere on a surface, air pressure is called *atmospheric pressure.*

Causes of Air Pressure

Air pressure is caused the collision of air molecules with a surface. Unlike the molecules of liquids and solids, gas molecules feel very little attraction to one another and move very quickly. As a result, each molecule acts essentially as an independent particle, with an energy determined by its mass and its velocity. Each time an air molecule collides with a surface, it produces a force on that surface, producing the effect known as air pressure.

Based on this description, it is clear that air pressure depends largely on two factors: density and temperature. The greater the density of air, the more molecules there are per unit volume and, therefore, the greater the number of collisions per unit area on a surface. The higher the temperature of air, the faster the molecules move and the harder they strike a surface.

Atmospheric Pressure

Air pressure in the atmosphere decreases at a relatively regular rate in traveling upward from Earth's surface. This change in pressure is a result of a corresponding decrease in air density as one moves upward. Imagine a cubic meter of air near Earth's surface. That cubic meter of air is pushed down by air above it in the atmosphere. The closer to Earth's surface the cubic meter of air, the more air there is above it to compress it, and the greater its density. The higher in the atmosphere the cubic meter of air, the less air there is above it to compress it, and the less its density.

The decrease in density (and therefore pressure) of air in the atmosphere is not a linear function. Nearly half of all the air in the atmosphere is found in the lower 5 kilometers (3 miles) of the atmosphere. Almost 90 percent of all the air is found at an altitude of less than 16 kilometers (10 miles).

Above an altitude of 30 kilometers (20 miles), both the density and pressure of air are less than 1 percent of their values at the surface of the Earth.

The actual density and pressure of air in the atmosphere varies due to a number of factors, including temperature and air movements. To compensate for these factors, scientists have developed a measure known as the *standard atmosphere,* which is an idealized model of density and pressure (and temperature) changes at various altitudes in the atmosphere.

Measurement

The pressure exerted by air in the atmosphere is quite significant. Humans and other animals are seldom aware of the intensity of air pressure because, in most cases, air pressure on a surface is balanced by air pressure on opposite sides of the same surface. For example, the pressure exerted by air on Earth's surface is 1.033 kilograms per square centimeter, or 14.7 pounds per square inch. That means that air pushes on the top of a typical school textbook with a force of about 500 kilograms, or 7,350 pounds. The obvious question is why that amount of force does not crush the textbook or the human body, for that matter.

The answer is that air pressure acts in all directions. The textbook is being pushed on from the bottom as well as from the top, preventing it from being crushed. This fact can be illustrated quite simply by removing all the air from a soda can with a vacuum pump. As air is removed from the can, air pressure inside the can decreases until it is not longer large enough to balance the atmospheric pressure of air outside it, and the can is crushed.

Over the years, a variety of units have been used to measure air pressure. In the British system, the most common unit of measure is pounds per square inch (psi) or inches of mercury ("Hg). The latter measurement tells the height of a column of mercury that can be supported by air pressure. In the metric system, the most common units of measure are the kilopascal (kPa) and the millibar (mb). Finally, air pressure is often measure in atmospheres. One atmosphere (atm) is defined as the pressure exerted by air in the atmosphere at sea level. The numerical relationship of these units is as follows:

$$1 \ atm = 101.325 \ kPa = 1 \ 013.25 \ mb = 14.7 \ psi = 29.92 \ "Hg$$

See also Atmosphere, Barometer, Standard Atmosphere; Torricelli, Evangelista

Air Pump

See Vacuum

Air Quality Index

The Air Quality Index (AQI) is a guide developed by the Environmental Protection Agency for describing air quality on any given day in any particular geographic area in the United States. The AQI is determined for five pollutants: ozone, particulate matter, carbon monoxide, sulfur dioxide, and nitrogen dioxide. The index is divided into six categories, as shown in the table on page 54.

The category "unhealthy for certain groups" refers to younger children, senior citizens, individuals with respiratory disorders, and adults who have active outdoor jobs.

A community's AQI is calculated daily based on data collected from thousands of locations around the country. AQIs are then reported by newspapers and local radio and television stations to let citizens know the level of risk resulting from their exposure to air pollution on any given day.

Big Bend National Park on a clear day. © **National Park Service.**

Big Bend National Park on a smoggy day. © **National Park Service.**

Nationwide, most AQIs are regularly less than 100. Seasonal changes, local conditions, and other special factors may result in an unusually high AQI rating for a community during some times of the year. In winter, for example, the AQI for carbon monoxide is likely to increase because of increased burning of wood in fireplaces and because catalytic converters work less effectively in cold weather. *See also* Air Pollution

Further Information

"Air Quality Index: A Guide to Air Quality and Your Health," U.S. Environmental Protection Agency, Publication EPA-454/R-00-005, June 2000.

Airship

See Balloons; Dirigible

Air Quality Index

Air Quality Index Value	Level of Health Concern	Color Code
0 to 50	Good	Green
51 to 100	Moderate	Yellow
101 to 150	Unhealthy for certain groups	Orange
150 to 200	Unhealthy	Red
201 to 300	Very unhealthy	Purple
301 to 500	Hazardous	Maroon

Air Thermometer

An air thermometer is a device for measuring temperature changes by observing changes in a column of air in an enclosed container. The early Greeks were well aware that the volume of air and other gases changes when they are heated or cooled. The Greek physician Galen (130–200), for example, described eight degrees of warmth. One of these was a "neutral temperature," which was halfway between the boiling and freezing points of water.

Philo of Byzantium (born about 300 B.C.) was perhaps the first person to construct a such a device for measuring differences in temperature. He is said to have attached a metal pipe to a hollow lead sphere. When the end of the pipe was submerged in water and the sphere heated, bubbles of air escaped through the end of the tube.

Credit for the invention of the first real air thermometer is usually given to Galileo (1564–1642). According to contemporary writers, he made his thermometer from a small glass flask, about the size of an egg, to which was attached a glass tube about 40 centimeters (16 inches) long. He inverted the device in a container of water so that the open end of the tube was beneath the surface of the water. When he warmed the flask, by placing his hands around it, for example, bubbles of air escaped from the tube, just as with Philo's invention. Galileo also used a caliperlike device to measure the amount of air lost from the heated flask.

Galileo's device is more properly called a *thermoscope,* that is, a device for *observing* changes in temperature, not actually *measuring* such changes. Over the next decade, Galileo and other researchers began to put marks on the glass tube of their thermoscopes, giving them a more specific way to measure changes. The first true thermometer was not constructed, however, until nearly a century later when the French physicist Guillaume Amontons (1663–1705) invented such a device.

A certain amount of controversy surrounded Galileo's claim to be the inventor of the thermoscope. The first printed reference to an air thermometer appeared in a 1612 book by the Italian scientist Santorio Santorio (1561–1636), *Commentaria in artem medicinalem Galeni.* When Galileo heard of Santorio's work, he expressed concern that someone other than himself would receive credit for his invention.

Further Reading

Middleton, W. E. Knowles. *A History of the Thermometer and Its Use in Meteorology.* Baltimore: Johns Hopkins Press, 1966.

Air Traffic Control

The term *air traffic control* refers to all the individuals, equipment, and procedures needed to monitor and direct the flight of airplanes anywhere in the sky. The air traffic control system is essentially the same in all parts of the world.

Structure of the Air Traffic Control System

In the United States, all air traffic control is under the overall supervision and direction of a single agency, the Air Traffic Control System Command Center (ATC-SCC), located in Herndon, Virginia. ATC-SCC is a division of the Federal Aviation Administration (FAA). ATCSCC is usually not involved in the day-to-day operation of flights, concentrating instead on special flight problems, such as those caused by bad weather or unusually busy air traffic in a region.

The management of routine air traffic is handled in a multitiered system. The first of those tiers consists of 21 air route traffic control centers (ARTCC), located in major cities, such as Seattle, Salt Lake City, Chicago, Miami, and New York. Each ARTCC is subdivided into a number of sectors, terminal radar approach control airspaces (TRACONs), and air traffic control tower airspaces (ATCTs).

Each part of the air management system is responsible for some portion of an airplane's flight from one airport to another airport. At the beginning of the flight, the airplane's operation is under the control of personnel assigned to an ATCT. Before the flight even leaves the gate, the planning of its itinerary is under way. The pilot submits

Radar equipment and controls. © Jay Lensch.

a flight plan to a flight data person in the ATCT, who enters that plan in an FAA computer. When the plan has been entered and approved, the flight data person gives the pilot clearance to prepare for departure.

At this point, control of the flight is handed over to a ground controller in the ATCT. That person directs the push-back from the gate and instructs the pilot which taxiways to follow to reach the takeoff point. When the airplane is in position for takeoff, it is handed over to a local controller in the ATCT. The local controller is then responsible for giving permission for takeoff, providing wind and weather information, and announcing the radio frequency at which the pilot will contact air traffic controllers.

Once the flight is in the air, it is handed over again, this time to a departure controller operating out of the nearest TRACON facility. Any one TRACON facility may be responsible for a number of airports within its airspace, which covers a radius of 80 kilometers (50 miles). When the flight has left the TRACON airspace, it is handed over to a center controller operating out of the ARTCC in which the flight began. On long flights, an aircraft may fly through two or more ARTCCs, being handed over each time from one center controller to another center controller in the new ARTCC. Also, each time an airplane leaves the control of one center, a flight progress strip is generated and passed along to the new center.

Throughout an airplane's flight, its progress is monitored by radar. Radar systems within the aircraft send out signals that can be retrieved and monitored in all control centers over which it passes. At the same time, the airplane receives radar signals from control centers to help it locate its current position in the sky.

Toward the end of a flight, the process described above is repeated, but in reverse. The incoming flight is passed off from an ARTCC controller to a TRACON controller, and then to an ATCT controller, who directs and monitors the flight's final descent into the airport. When the plane is on the ground, its progress becomes the responsibility of a ground controller, who directs the pilot to the proper gate for parking and disembarkation.

Airspaces

In general, the term *airspace* refers to a portion of the atmosphere above a certain part of the Earth. For example, French airspace is all of the atmosphere located above the land and water areas that make up the nation of France. Every nation has legal control over its own airspace and may permit or deny passage of aircraft from other nations through its airspace. In some instances, a nation may decline to allow aircraft from unfriendly nations passage through its airspace. An American airplane flying from Alaska to India, as an example, might at one time (and still, in some cases) have had to fly around airspaces controlled by Russia, North Korea, China, and other nations with whom our relations were not friendly.

Air traffic controllers use more refined and more complex definitions of the term airspace to describe flying conditions that must be observed within those spaces. For example, the FAA's Class A airspace refers to the airspace between 18,000 and 60,000 feet above the 48 contiguous states and Alaska and within 12 nautical miles of the coastline of those states. In general, all flight within this airspace must be operated under instrument flight rules (IFR), which are different from visual flight rules (VFR).

Class B airspace includes the space from ground level to an altitude of 10,000 feet around the nation's busiest airports. Within a Class B airspace, clearance from an air traffic controller is required for operation, and IFR rules are in effect unless the sky is completely clear of clouds. Class C, D, E, and G airspaces have been designated for other, more limited flying areas where other specific rules of operation are in effect.

Further Reading

Field, Arnold. *International Air Traffic Control: Management of the World's Airspace.* Oxford, UK: Pergamon Press, 1985.

Illman, Paul E. *The Pilot's Air Traffic Control Handbook.* New York: McGraw-Hill Professional Publishing, 1999.

Noland, Michael S. *Fundamentals of Air Traffic Control.* Belmont, CA: Wadsworth Publishing Company, 1999.

Air University

Air University (AU) is the primary educational institution of the U.S. Air Force, providing training for military and civilian personnel from the United States and other countries of the world, from preflight basic training to graduate education. The university was formed in 1946 as plans were being made to establish the U.S. Air Force as a separate branch of the U.S. armed forces. It was designed to serve as a single central institution with a variety of functions and responsibilities.

The mission statement of the Air University reads as follows:

Air University conducts professional military education, graduate education, and professional continuing education for officers, enlisted personnel and civilians to prepare them for command, staff, leadership and management responsibilities. Specialized and degree-granting programs provide education to meet Air Force requirements in scientific, technological managerial, and other professional areas. In addition, Air University is responsible for research in designated fields of aerospace education, leadership and management, provides pre-commissioning training, and offers selected courses for enlisted personnel leading to the awarding of select Air Force specialty credentials. Air University also contributes to the development and testing of Air Force doctrine, concepts and strategy.

Air University programs designed to carry out this mission fall into five major categories: Air Force Officer Accession and Training Schools, professional military education programs, academic education programs, continuing professional education programs, and other related programs.

Air Force Officer Accession and Training Schools (AFOATS)

AFOATS consists of two previously independent programs, the Air Force Reserve Officer Training Corps (AFROTC) and the Officer Training School (OTS). The AFROTC is the largest and oldest source of commissioned officers for the U.S. Air Force. It has programs at more than 140 colleges and universities, many of which serve other institutions in their geographic area. Overall, nearly 2,000 students from nearly 1,000 institutions graduate each year from AFROTC programs and receive their commissions as second lieutenants. A related Junior AFROTC program operates on over 600 high school campuses nationwide, enrolling about 91,000 participants.

The Officer Training Program consists of two major division, Basic Officer Training (BOT), a 13.5-week program that pro-

vides overall basic education in technical, professional, and physical aspects of commissioned service, and Commissioned Officer Training (COT), a program that provides specialized education for judge advocates, chaplains, and medical officers. About 7,000 new lieutenants graduate from the BOT program each year, and 2,700 officers graduate from the COT program annually.

Professional Military Education Program (PMEP)

The Air University's professional military education program consists of five major units: the Squadron Officer School, Air Command and Staff College, Air War College, College for Enlisted Professional Military Education, and U.S. Air Force Senior Noncommissioned Officer Academy.

The *Squadron Officer School* is a seven-week course that constitutes the first phase of an Air Force officer's professional education. It focuses on four curriculum areas: officer duties and responsibilities, air and space power, leadership tools, and applications. Each year, nearly 4,000 individuals, including about 100 civilians and some international officers, are graduated from the program.

Air Command and Staff College is a 40-week program that constitutes the second phase of an officer's professional military education. The program is organized around nine major courses of study dealing with the profession of armed warfare, the requisites of command, the nature of war, and the application of air and space power to specific regions of battle. Each year, about 600 resident and 7,000 nonresident officers and Department of Defense civilians complete this program.

Air War College (AWC) is a 44-week program enrolling about 250 officers from all branches of the U.S. military, international officers, and certain senior civilian employees annually. The program is designed to prepare participants for positions of command and management of air and space power. The highlight of the program is a one-week National Security Forum, where business, civic, and professional leaders from around the United States meet and interact with AWC students.

College for Enlisted Professional Military Education (CEPME) is responsible for designing curricula taught at three programs for noncommissioned officers (NCOs): the Airman Leadership Schools, the Air Force Senior Noncommissioned Officer Academy, and other noncommissioned officer academies. It is also directly responsible for the operation of the latter two operations. Airman Leadership Schools are operated by individual Air Force major commands, using curricula developed by the CEPME.

U.S. Air Force Senior Noncommissioned Officer Academy is the highest level of professional education available to enlisted Air Force personnel. It is designed to produce skilled professional managers and focuses on courses in leadership and management, communication skills, and military studies. Five seven-week sessions are held each year, with an annual enrollment of about 1,800 students.

Academic Education Program

The Air Force's Academic Education Program consists of two major components, the Community College of the Air Force (CCAF) and the Air Force Institute of Technology (AFIT). The Air Force claims that the CCAF is "the only degree-granting institution of higher learning in the world dedicated exclusively to enlisted people." It is open to active-duty personnel and members of the Air National Guard and the Air Force Reserve. It is accredited by the Southern Association of Colleges and Schools and has awarded more than 125,000 associate in applied science degrees since it opened in 1972.

The AFIT consists of two division, the Graduate School of Engineering and the Graduate School of Logistics and Acquisition Management, both located at Wright-Patterson Air Force Base in Dayton, Ohio. The Institute offers both masters and doctoral degrees and supervises students enrolled in the separate Civilian Institute Program, located at nonmilitary universities, hospitals, research centers, and industrial organizations throughout the United States. More than 30,000 students graduate each year from these and other AFIT-administered programs.

Continuing Professional Education Program

Air University's Continuing Professional Education Program consists of two institutions, the Ira C. Eaker College for Professional Development and the College of Aerospace Doctrine, Research and Education (CADRE). Eaker College is AU's largest residential college, consisting of six schools that offer courses for chaplains and chaplain assistants, comptrollers, judge advocates, historians, first sergeants, commanders, family support center managers, paralegal specialists, and personnel and manpower managers.

CADRE offers courses in Contingency Wartime Planning, Joint Doctrine Air Campaign, Joint Flag Officer Warfighting, and Joint Forces Air Component Command. It also carries out research in Air Force doctrine, concepts and strategies, and conducts the Air for Wargaming Institute.

Other AU Organizations

Air University is also home to the Civil Air Patrol and the Office of Academic Support, which includes the AU Library, the Academic Instructor School, the International Officer School, AU Television, and Extension Course Institute, AU Press, and Educational Technology. *See also* Air Force

Further Information

Air University
42nd Air Base Wing
Public Affairs Office
50 Le May Plaza South
Maxwell AFB, AL 36112-6334
Telephone: (205) 953-2014
URL: http://www.au.af.mil
E-mail: au.pressorder@maxwell.af.mil

American Wind Energy Association

The American Wind Energy Association (AWEA) was founded in 1974 to promote the use of wind energy as a clean source of electricity for consumers around the world. The association consists of more than 1,000 members representing wind power plant developers, wind turbine manufacturers, utilities, consultants, insurers, financiers, researchers, and wind energy advocates from all parts of the world. AWEA is a clearinghouse for information on wind energy projects going on around the world, companies working in the wind energy field, research and development in wind technology, and policy developments involving the use of wind energy and other forms of renewable energy.

An important area of AWEA activity is cooperation with legislative bodies to ensure that wind industry interests are represented in making of new legislation, rules, and regulations. The association is especially interested in promoting policies that boost wind energy programs, such as the Production Tax Credit and the Renewables Portfolio Standard. The association also maintains an extensive educational outreach, providing statistics and information on the development of domestic and international wind energy markets. An important component of that effort is a weekly wind energy newsletter.

Each year, AWEA sponsors WIND-POWER, a conference providing presenta-tions on latest industry trends, technological developments, and information about political issues relating to wind energy. The association also publishes a number of books, reports, wind energy fact sheets, and other publications. Among the free publications available from AWEA are "Wind Energy Applications Guide," "Global Market Reports," "Public Attitudes on Wind Energy, Renewables, and the Environment," "Wind Energy Information Guide," "Wind Energy in the U.S.: A State-by-State Guide," and "Wind Energy OUTLOOK." Examples of databases available from the association in print and online are "Current and Planned Wind Power Projects Database," "Green Power Database," and "Emissions Spreadsheet."

Further Information

American Wind Energy Association
122 C Street, NW, Suite 380
Washington, DC 20001
Telephone: (202) 383-2500
Fax: (202) 383-2505
URL: http://www.awea.org
E-mail: windmail@awea.org

Amontons, Guillaume (1663–1705)

Guillaume Amontons made a number of important discoveries about the properties of gases during his short life of 42 years. In 1695, he invented a barometer that could be used at sea. This invention was significant because traditional barometers were fragile and difficult to use on ships. The narrow tube containing mercury in Amontons's barometer was shaped like a funnel, narrower at the closed end. The tube was narrow enough that mercury did not run out, making it more stable in unsettled weather. Many years later, in 1688, he suggested another type of "compound barometer" consisting of a very long tube with three bends, making an instrument with three parallel tubes. The outer tubes contained mercury columns, as in traditional barometers, while the inner tube contained air.

Amontons was born in Paris on 31 August 1663. He became deaf at an early age, a situation that might have discouraged some people. But Amontons is said to have made no complaint about his disability since it allowed him to concentrate more closely on his scientific studies. He was largely self-taught and received no formal university education. He was interested in a vast array of subjects, ranging from the physical sciences and mathematics to architecture, surveying, and drawing.

His special interest was meteorological instruments, especially thermometers and barometers. In addition to the two barometers described above, he invented a type of hygrometer, for measuring the amount of moisture in the air, and an improved version of Galileo's air thermometer. In his research on gases, he anticipated Charles's gas law by discovering that the volume of any gas always changes by the same amount for any given change in temperature.

Amontons is also said to have invented a type of optical telegraph in which messages where sent from one windmill to another (the highest points in the landscape in most areas). An observer at one windmill would use a telescope to read the message at the preceding windmill and then post that message on his own windmill for further transmission. Amontons is said to have demonstrated this concept for the royal family sometime between 1688 and 1695, but no further details are available, nor is it known whether the system was ever put to use. Amontons also invented a thermic motor in 1699 that used hot air to produce mechanical motion.

Among Amontons's surviving works are a single book dealing with meteorological instruments, *Remarques et expériences physiques sur la construction d'une nouvelle clepsydre, sur les barometres, thermometeres, et hydrometers* (1695). He died in Paris on 11 October 1705. *See also* Barometer

Further Reading

Middleton, W. E. Knowles. *The History of the Barometer.* Baltimore: Johns Hopkins Press, 1964.

Taton, René, ed. *History of Science.* New York: Basic Books, 1966–1966, passim.

Anaximenes

See Air As an Element

Anemometer

See Wind Measurement

Animals That Fly

See Flying Animals

Anticyclone

See Cyclones and Anticyclones

Aphorisms and Sayings

Air is the subject of a number of common aphorisms and sayings. Some examples of these expressions are as follows:

Beating the air can be traced to the Apostle Paul's First Epistle to the Corinthians (chapter 9, verse 26), in which he explains, "Well, I do not run aimlessly, I do not box as one beating the air" (Revised Standard Edition). Paul suggests that he does not waste his time trying to do useless work or to take on tasks that have not specific purpose, a meaning the phrase continues to have today.

Castles in the air probably was first used in the English language by the poet Sir Philip Sidney (1554–1586) in an essay, "In Defence of Poesy," first published in 1595.

Sidney argues that one contribution poets can make is to fantasize about wonderful visions that still have some substance to them, or, as he says:

And that poet hath that idea is manifest by delivering them forth in such excellency as he had imagined them, which delivering forth, also, is not wholly imaginative, as we are wont to say by them that build castles in the air. . . ." (1891 version)

Etymologist Robert Hendrickson argues, however, that the phrase may have even older roots, dating back to the eleventh century, when the idea appears in a poem that includes the lines

Thou shalt make castels thanne in Spayne,
And dream of Ioye [joy] all but in vayne. (Hendrickson, 1997, p. 132)

Over time, Hendrickson suggests, the idea of "castles in Spain" as beautiful flights of fancy eventually evolved into the more familiar "castles in the air" used today.

On cloud nine first appeared in 1959, according to the Merriam Webster Collegiate Dictionary. However, one can find commentators who claim an older history for the phrase, claiming that it may have arisen from the special charm of the number 9, often regarded as being lucky because it is the product of two other lucky numbers, 3 times 3. Wilton's Word & Phrase Origins Web site (http://www.wordorigins.org) also points out that the phrase "on cloud eight" dates to the 1930s, referring to a state of drunkenness. In any case, the phrase usually refers to a state of elation or well-being, that is, being "way out there."

Sow the wind and reap the whirlwind is another phrase that derives from the Bible. In the book of Hosea (chapter 8, verse 7), the prophet berates the people of Israel for forsaking their God, pointing out that they have carried out evil practices ("sown the wind") that will result in punishment far worse than their own deeds ("reap the whirlwind").

Take the wind out of one's sails is an aphorism that comes directly from a maneuver once used commonly by pirates in their attacks on merchant vessels. By placing their own ship between their target and the direction from which the wind was blowing, the pirates were able to deprive their victim of the wind they needed to escape attack. The modern sense of the phrase is that "taking the wind out of one's sails" is to slow someone down or bring them to a stop.

Three sheets to the wind is a nautical expression that often refers to a person who is very drunk or otherwise out of control. The "sheets" mentioned in the saying are not sails, but ropes that control the position and movement of sails. If a single sheet (rope) is loose, the attached sail is likely to flap in the wind. With two sheets loose, the sail will be under even less control, and with three sheets loose, one is likely to lose control of the boat entirely.

Tilting at windmills is a phrase whose origin can be determined rather easily and specifically. It comes from the famous novel *Don Quixote,* by the Spanish writer Miguel de Cervantes (1547–1616). In that book, a middle-aged, mentally unbalanced gentlemen, Don Quixote de la Mancha, has lost the ability to differentiate between the real world and the world of chivalry, about which he constantly reads. He sets out in search of his "fair maiden," Dulcinea. Along the way, he encounters a group of 40 windmills, which he takes to be giants whom he must conquer. He undertakes the hopeless task, only to have his spear caught in the windmills and find himself unhorsed. The experience leads to the modern interpretation of the phrase, referring to any person who takes on an impossible task.

Windfall is a fifteenth-century expression that refers to an unexpected stroke of good luck. The phrase appears to have arisen from the custom that peasants were allowed to cut trees only under certain circumstances and with special permission from their rulers. They were, however, allowed to pick up branches and trees that may have been blown over in a storm, an act, for them, of unexpected good fortune.

Further Reading

Ewart, Neil. *Everyday Phrases: Their Origins and Meanings.* Poole, Dorset, Blandford Press, 1983.

Hendrickson, Robert. *The Facts on File Encyclopedia of Word and Phrase Origins,* Revised and Expanded Edition. New York: Facts on File, 1997.

Lurie, Charles N. *Everyday Sayings: Their Meanings Explained, Their Origins Given.* New York: G. P. Putnam's Sons, 1928.

Archimedes (287–212 B.C.)

Archimedes was a Greek mathematician, natural philosopher, and engineer. One of his most lasting discoveries was the law of floating bodies, now known as *Archimedes's Principle.* In addition, he is reputed to have invented a number of other devices, including one for raising water from a lower level to a higher level, a device we now call the *Archimedes's screw.*

Archimedes was born in Syracuse in 287 B.C., the son of Phidias, an astronomer. For his education, he traveled to Alexandria, where he studied with the Greek mathematician Conon, who had been a pupil of the great Euclid. The only biography of Archimedes of which we know is one written during his life by a friend named Heracleides. Still, a great deal is known about Archimedes because of the frequent mention of his work in the writings of his contemporaries. In addition, a number of Archimedes's own works remain, including *The Sandreckoner, On Floating Bodies, On the Sphere and the Cylinder, Measurement of a Circle, On Spirals, On Concoids and Spheroids,* and *The Method.*

One of the many stories about Archimedes's life found in contemporary works tells of how he was killed. It is said that he was working on a mathematical problem at the time a Roman army landed in Syracuse. To aid in the solution of this problem, Archimedes had drawn some geometrical figures in the sand. When a Roman soldier seemed about to walk across those figures, Archimedes is said to have warned him off with a cry of "Don't disturb my circles!" Unwilling to be ordered about by an old man, the solider drew his sword and killed Archimedes on the spot.

Archimedes's Principle

Many middle and high school students today have heard about the discovery of Archimedes's Principle. The philosopher had been asked by the king of Syracuse, Hieron II, to find out whether a crown he had been given was actually made of pure gold, as the king had requested. Archimedes was faced with solving this problem, of course, without taking the crown apart or otherwise damaging it.

For some time, the philosopher was unable to think of any way to meet Hieron's request. Then, the answer came to him "in a flash." He is reputed to have been sitting in his bathtub one day when the solution to the problem came to him. As he lowered himself into the tub, a certain amount of water overflowed and spilled out of the tub. The amount of water his body displaced, Archimedes realized, was a function of his own weight and volume.

If that were true, he could use the same process to test the purity of gold in the king's crown. The first step was to place the crown in a container of water filled to the brim and to measure the amount of water displaced. The second step was to place an equal weight of gold in the same container of water and measure the amount of water displaced in this instance. If the two volumes of displaced water were the same, the crown was really made of pure gold. If not, the gold had been mixed with a less valuable metal such as, as it turned out to be the case, silver.

Archimedes was so excited about his discovery that he is said to have jumped from the bathtub and run through the streets of Syracuse naked shouting "Eureka, eureka," that is, "I have it, I have it." His solution to King Hieron's puzzle brought him everlasting fame but, unfortunately, it also brought death to the dishonest goldsmith who had made the crown.

Today, Archimedes's Principle is stated in a more formal way, although its meaning is the same as in the original story. A modern statement of the principle is that an object immersed in a fluid is pushed up with a force that is equal to the weight of the displaced fluid. *See also* Buoyancy

Further Reading

"Archimedes Home Page," http://www.mcs.drexel.edu/~corres/Archimedes

Dijksterhuis, E. J. *Archimedes.* Translated by C. Dikshoorn. Princeton, NJ: Princeton University Press, 1987.

Heath, Sir Thomas Little. *Archimedes.* New York: Macmillan, 1920.

Stein, Sherman K. *Archimedes: What Did He Do Besides Cry Eureka?* Washington, DC: Mathematical Association of America, 1999.

Argon

Argon is one of the elements that makes up Group 18 in the periodic table of the elements, the group sometimes known as the *rare gases* or the *noble gases*. Argon is the third most abundant gas in the Earth's atmosphere, after nitrogen and oxygen. It is present to the extent of about 0.93 percent. It also occurs rarely in the Earth's crust, with an abundance of about 4 parts per million.

Physical and Chemical Properties

Argon is a colorless, odorless, tasteless gas with a density of 1.784 grams per liter,

about 40 percent more dense than air itself (1.29 grams per liter). It has a boiling point of −185.86°C (−302.55°F) and a melting point of −189.3°C (−308.7°F). Its atomic number is 18, its atomic mass is 39.948, and its chemical symbol is Ar.

Like the other elements in Group 18 of the periodic table, argon is chemically inactive. Research chemists have made a few compoundlike substances of argon in the laboratory, but no such compounds are known to exist in the natural world.

Discovery and Naming

Argon was discovered in 1894 by the English chemist John William Strutt, Lord Rayleigh (1842–1919) and the Scottish chemist William Ramsay (1852–1916). Its existence had been suspected at least a century earlier by the English scientist Henry Cavendish (1731–1810). After removing both nitrogen and oxygen from a sample of air, Cavendish found that a small bubble of an unknown gas remained. He suspected the presence of a new element, but he was unable to learn enough about the gas to claim discovery. When Rayleigh and Ramsay repeated Cavendish's experiment in the 1890s, they were able to use the newly discovered technique of spectroscopy to show that the bubble was a new gas, which they named *argon,* after the Greek word *argos* for "lazy." The term seemed appropriate since they were unable to make the gas react with any other substance.

Production and Uses

Most argon is produced commercially by the fractional distillation of liquid air. When liquid air is allowed to warm, the gases of which it is composed evaporate, one at a time. Nitrogen evaporates first because it has the lowest boiling point (−195.79°C), followed by argon (−185.86°C), and then by oxygen (−182.96°C). A small amount of argon occurs in wells with natural gas and can be obtained from that source also.

Most of argon's major uses depend on its chemical inertness. For example, it is often used in incandescent lightbulbs to prevent the hot metal wire inside the bulb from reacting with oxygen or other gases. Argon is also used in welding for the same reason. If welding were done in air, the hot metals formed in the process would react with oxygen and, perhaps, nitrogen, making a weld impossible. If welding is done in a confined space filled with argon, however, the hot metals do not react with any other element or compound in the region.

Argon is also used in lasers for some procedures. For example, argon-dye lasers are used in eye surgery because their frequencies of operation can be adjusted with high precision. Other types of argon and argon-dye lasers are used to treat skin conditions that cause unsightly spots or growths. *See also* Gases and Airs

Further Reading

Cook, Gerhard A. *Argon, Helium, and the Rare Gases: The Elements of the Helium Group.* New York: Interscience Publishers, 1961.

Atmosphere

The atmosphere is the envelope of air that surrounds the Earth. The atmosphere, lithosphere (solid portion of the Earth), and hydrosphere (liquid portion) make up the three basic physical components of the planet.

Composition

The atmosphere consists primarily of four gases: nitrogen (78.08%), oxygen (20.95%), argon (0.93%), and carbon dioxide (0.0369%). The atmosphere also contains other gases, such as neon (0.0018%), helium (0.000 52%), methane (0.000 15%), krypton (0.000 11%), and hydrogen (0.000 05%). A number of other gases, including xenon, nitrous oxide, carbon monoxide, ammonia, nitrogen dioxide, sulfur dioxide, hydrogen sulfide, and ozone, exist in trace amounts.

The atmosphere's composition is variable, depending upon vertical distance above the Earth's surface, geographical location, and time. For example, ozone (O_3), the triatomic form of oxygen, is very rare in the lower parts of the Earth's atmosphere (except in the air just above urban areas), with a concentration of less than one part per million. At an altitude between 10 and 50 kilometers (6–30 miles), however, its concentration is more than ten times as great, or about one part per hundred thousand.

The composition of the atmosphere at lower altitudes also varies depending on geography and topography. Over the oceans, for example, the concentration of water vapor is relatively high (but still very low in absolute terms), and substances otherwise absent from the atmosphere, such as salt (sodium chloride; NaCl), are also present. In regions above and around volcanic areas, still other gases, such as sulfur dioxide (SO_2) and hydrogen sulfide (H_2S) are present in small amounts. In spite of these localized variations, the Earth's atmosphere remains largely constant, primarily because of vertical and horizontal wind systems that constantly move and mix air from all parts of the planet.

Finally, the composition of the atmosphere changes over long periods of time because of both natural and anthropogenic factors. Carbon dioxide is perhaps the best example. According to one highly regarded study of the concentration of carbon dioxide in the atmosphere, conducted at the Mauna Loa Astronomical Observatory on the Island of Hawaii, the concentration of carbon dioxide in the atmospherc has increased from 316 parts per million in 1958, when the study began, to 369 parts per million in 2000, the last year for which data are available. Scientists attribute this increase largely to the combustion of fossil fuels by humans (of which carbon dioxide is a major product) before and during that time period.

Vertical Structure

Scientists divide the atmosphere into four major layers, largely on the basis of temperature and pressure changes found at increasing altitudes. The lowest layer is the *troposphere,* extending from sea level to a height of about 10 kilometers (6 miles). Both temperature and pressure decrease at a relatively constant rate in the troposphere. The temperature drops from a planet-wide annual average of about 12°C (54°F) at sea level to about −60°C (−76°F) at the top of the troposphere. The pressure drops from about 1,000 millibars at sea level to about 300 millibars at a height of 10 kilometers.

The temperature change in the troposphere occurs in a linear fashion, with a decrease of about 6.5°C per kilometer (3.3°F per 1,000 feet) of altitude, a change known as the *environmental lapse rate.* By contrast, the pressure decrease occurs geometrically. That is, the greater the height above the Earth's surface, the more rapid is the *rate* of pressure decrease. In fact, at an altitude of 16 kilometers (10 miles) above the Earth's surface, the pressure has dropped to 10 percent of its value at sea level, and at an altitude of 32 kilometers (20 miles), it has decreased to 1 percent of its surface value. Each additional increase in altitude of 16 kilometers is accompanied by a drop in pressure by a factor of 10.

This decrease in air pressure is a function of the gravitational force attracting air molecules to the Earth. That force decreases as the square of the distance from the Earth's center. As a consequence, the density of atmospheric gases (and, hence, the pressure they exert) drops off rapidly as one ascends in the atmosphere.

At an altitude of about 10 kilometers, the temperature ceases to decrease and,

through a height of another 10 kilometers, remains about constant at about −56°C (−69°F). The region of the atmosphere in which this change begins, at a height of about 10 kilometers, is known as the *tropopause.* The height of the tropopause differs to some extent over various latitudes on Earth. Above the equator, for example, the tropopause occurs at an altitude of more than 16 kilometers, while it is less than 9 kilometers in height above the poles.

The tropopause marks the beginning of the second layer of the atmosphere, the *stratosphere.* As one travels upward through the stratosphere, atmospheric pressure continues to drop at a steady, geometric rate. It continues to do so, in fact, to the outer reaches of the atmosphere. The temperature, however, behaves in a somewhat erratic fashion, beginning to rise slowly at an altitude of about 20 kilometers, and then more rapidly above a height of 30 kilometers. At an altitude of about 50 kilometers, the upper boundary of the stratosphere, or *stratopause,* it reaches a maximum of about 0°C (32°F). This increase in temperature is due largely to the presence of ozone, which absorbs ultraviolet radiation from sunlight.

The stratopause is the lower boundary of the third layer of the atmosphere, the *mesosphere,* which stretches to a height of about 80 kilometers (50 miles). Within the mesosphere, the temperature once more begins to decrease, reaching a minimum of about −90°C (−130°F) at its upper limit. That point, the *mesopause,* marks the lower boundary of the uppermost part of the atmosphere, the *thermosphere.* The thermosphere has no discrete upper boundary. The density of air molecules gradually becomes lower and lower the farther one travels from Earth's surface until it is no longer possible to distinguish between "the atmosphere" and "outer space."

Temperatures in the thermosphere are relatively constant at about −90°C to an altitude of about 90 kilometers, at which point they begin a constant increase through the rest of the atmosphere. Temperatures increase in this part of the thermosphere because of the absorption of solar energy by atoms of oxygen and nitrogen. Although temperatures may reach more than 1,000°C in the upper parts of the thermosphere, these numbers do not have the same meaning they do on Earth's surface. Temperature is defined as the average kinetic energy of a mass of atoms and molecules, which, in the upper thermosphere, is very high. The number of particles with this amount of energy, however, is so low that the amount of heat they contain is very low.

The region between altitudes of about 80 and 400 kilometers (50 and 245 miles) is also known as the *ionosphere*. The name comes from the fact that solar energy at these altitudes is able to ionize molecules of oxygen and nitrogen, producing positively charged ions of the elements and free electrons:

$$N_2 + solar\ energy \rightarrow N_2^+ + e^-$$
$$O_2 + solar\ energy \rightarrow O_2^+ + e^-$$

The production of these positively charged ions and electrons is responsible for certain natural phenomena, such as the aurora borealis (northern lights) and aurora australis (southern lights). Those dramatic displays in the skies are produced when ions and electrons produced during period of high solar activity are trapped in the Earth's magnetic field and carried to its magnetic poles.

The ionosphere is also responsible for the fact that AM radio signals can be transmitted over long distances during the night, but not during the day. At night, AM signals transmitted from Earth's surface reflect off the upper layer (or "F" layer) of the ionosphere and return to Earth hundreds of kilometers from their source. The F layer of the ionosphere is always present, both day and night. During the day, however, those radio signals are absorbed by lower layers of the ionosphere, the "D" and "E" layers, which disappear when the sun sets.

Evolution of the Atmosphere

The Earth was formed, by a process still not entirely understood, about 4.6 billion years ago. During its early stages, the Earth probably consisted of two or more layers, the outer of which (the primitive atmosphere) consisted primarily of hydrogen and helium gases. The Earth's gravitational attraction was not strong enough, however, to hold these gases very long and they rather quickly escaped into outer space. Today's atmosphere contains only trace amounts of both gases.

As soon as the solid Earth began to cool, gases began to escape from its interior, much as gases are lost today during volcanic eruptions and other types of *outgassing*. Outgassing probably released water vapor, carbon dioxide, carbon monoxide, sulfur dioxide, nitrogen, ammonia, methane, hydrogen sulfide, and chlorine. Oxygen was *not* among the gases released during outgassing. Of the two main gases produced by outgassing, water vapor condensed and fell back to the Earth's surface to form oceans, lakes, and other forms of surface water. Carbon dioxide was left behind as the primary constituent of the primitive atmosphere.

Over time, however, carbon dioxide began to react with water and a variety of other compounds on the Earth's surface to form a variety of rocks and minerals. Scientists believe that the Earth's atmosphere changed from one containing nearly 100 percent carbon dioxide shortly after its birth to one containing no more than about 10 percent carbon dioxide a billion years later. During that time, the concentration of methane gas, and, to a lesser extent, nitrogen gas, began to increase. By a billion years after the Earth's formation, the atmosphere probably contained about 80 percent methane and related carbon compounds and less than 10 percent nitrogen.

About 3.5 billion years ago, an important event began to make a dramatic effect in the composition of the Earth's atmosphere. That event was the appearance of photosynthetic bacteria, phytoplankton, and algae, which are organisms with the ability to take carbon dioxide from the atmosphere and convert it to carbohydrates. An important by-product of that reaction is oxygen. With the appearance of these organisms, then, a mechanism was introduced for the production of oxygen in the atmosphere. As long as one-celled organisms were the only organisms capable of producing oxygen, the concentration of the gas remained small. Much of the oxygen produced was probably used up quite quickly in reactions with other elements and compounds on the Earth.

After the appearance of photosynthetic organisms, the concentration of methane and related gases peaked (shortly after the 1 billion year mark), while the concentration of nitrogen began to increase. By the 2.5 billion year mark, the Earth's atmosphere consisted almost entirely of nitrogen (over 95 percent), with traces of carbon dioxide, methane, and oxygen. These levels of atmospheric gases survived for 1.5 billion years. Then, another change in the Earth's biological structure occurred: the first eukaryotic organisms appeared. These organisms were far more efficient at converting carbon dioxide and water into carbohydrates, with the resultant formation of free oxygen gas. As the number of eukaryotes grew, larger and larger quantities of oxygen were released into the atmosphere. The concentration of oxygen began to increase rapidly at about the 4 billion year mark, increasing from nearly zero percent to its present level of about 20 percent over the next 600 million years. *See also* Air Mass; Atmospheric Optical Phenomena; Atmospheric Scattering and Absorption; Global Winds; Stability of Air; Standard Atmosphere; Wind

Further Reading

Allen, Oliver E., and the editors of Time-Life Books. *Atmosphere.* Alexandria, VA: Time-Life Books, 1983.

Walker, James C. G. *Evolution of the Atmosphere.* New York: Macmillan, 1977.

Atmospheric Optical Phenomena

Atmospheric optical phenomena are visible events that occur in the atmosphere because of the way sunlight is reflected, refracted, or diffracted by atoms, molecules, solid particles, and liquid droplets present in the air. Some examples of such phenomena include mirages, halos, sun dogs, solar pillars, rainbows and fogbows, glories, coronas, and crepuscular rays.

Reflection, Refraction, and Diffraction

Reflection is the bending of light that occurs when it strikes a surface. During reflection, light rays bounce off the surface at an angle equal to that at which the light strikes the surface. For example, if a light ray strikes a surface at an angle of 30° for the vertical, it will be reflected back into the air at the same angle of 30° from the vertical opposite from the source.

When light strikes a smooth surface, such as that of a mirror, all of the reflected light travels in the same direction, a fact that explains why the image of an object seen in a mirror is identical to the object itself. If light strikes a very small object, such as an atom or a molecule, or if it strikes an uneven surface, it may be reflected in many possible directions, a process known as *scattering.* The process of scattering explains some optical phenomena. For example, the sky is blue because molecules of oxygen and nitro-

Partial halo with parhelia (sun dogs) on both sides of the halo. © NOAA.

gen in the atmosphere scatter light of shorter wavelengths, such as those of blue and violet, more efficiently than they do light of longer wavelengths, such as red and orange.

Reflection also explains the color of objects. Certain materials absorb light of some wavelengths, but not of others. A material that absorbs all wavelengths but green, for example, will appear to be green because that is the only color reflected.

Refraction is the process by which light is bent as it travels at an oblique angle between two transparent media of different densities. For example, imagine a beam of light travel traveling through air and striking the surface of water at an angle of 30° from the vertical. When the light enters the water, it begins to travel more slowly than it did in air. As the light beam slows down, its path changes. Some of the most common examples of refraction occur when one attempts to observe an object under water. The object

appears to be displaced from its actual location, a fact that one appreciates in trying to take hold of the object.

Diffraction is the process by which light rays are bent as they pass the sharp edge of an object. Perhaps the most familiar example of diffraction is a shadow. A shadow is formed when light rays are blocked out by an object. A shadow never has sharp edges, however, since the light passing around the edge of the object is bent to some extent, giving the shadow its typical "fuzzy" edges.

Atmospheric Phenomena

Some of the atmospheric phenomena that can be explained by reflection, refraction, and diffraction include mirages, the twinkling of stars, the green flash, and rainbows and fogbows.

Mirages

A mirage is an optical phenomenon in which an object appears to be displaced

A diffused rainbow due primarily to variable drop sizes, North Carolina. © NOAA.

from its actual position because two layers of the atmosphere are heated to different temperatures. Mirages occur most commonly over the oceans and deserts, or above smooth highways, parking lots, or other man-made spaces. One type of mirage, the *inferior mirage,* occurs on very hot days when air near the ground is significantly warmer than air just above it. A familiar example of an inferior mirage is the apparent displacement of an oncoming automobile while driving on a very hot highway. In such cases, light rays reflected off the object (the oncoming car) pass first through the upper, cooler, more dense layer of air and then through the lower, warmer, less dense layer. Rays of light from the object are refracted during this process. They are

refracted again after being reflected off the road and traveling through the warmer, and then the cooler, layers of air to the observer. As a result of these refractions, the light from the object appears to be coming from a point below its actual position. The oncoming car seems to be lower in elevation than it actually is.

Another example of an inferior mirage is the appearance of a body of water as one travels across the very hot surface of a desert. In this case, light from the sky is refracted in passing through two layers of air, and the sky appears to be lower in space than it actually is . . . as a body of water in the middle of the desert.

A *superior mirage* occurs when conditions are just the opposite of the above situation, with a cooler lower layer underlying a warmer upper layer. In this case, light is also refracted in passing from the warmer, upper, less dense layer into the cooler, lower, more dense layer. The light rays reaching an observer's eyes would then appear to be coming from a point higher in space than the object's actual position. An example of a superior mirage is the situation sometimes observed on the ocean, where the layer of air just above the water is much cooler than air just above it. In such cases, a ship seen in the distance may appear to be farther above the horizon that it actually is.

In some instances, the vertical displacement of an image by refraction is so great that an object may seem to be floating in the sky. This extreme case of a superior mirage is also known as *looming.*

A *lateral mirage* is produced when two vertical columns of air of different temperature and density are adjacent to each other. A light beam that passes through two such columns will be refracted in such a way as to produce an image of the real object that is displaced to the left or right of the object itself.

Under some conditions, the atmosphere contains many cells of air at different temperatures with different densities. When light travels through two or more of these cells, more complex visual images can be produced. One example is the phenomenon known as *towering,* in which the size of an object is magnified, as would be the case with certain types of shaving mirrors. A particularly interesting form of towering is called *fata morgana* after the fictional sister of King Arthur. Morgana was reputed to have the ability to construct castles in the air, so the modern term refers to unusually large images produced as a result of refraction. Towering and fata morganas are most likely to be seen from a ship as it approaches land, making rather ordinary cliffs look like huge mountains.

Twinkling of Stars and the Green Flash

Two other optical phenomenon that occur because of refraction are the twinkling of stars and the green flash. The twinkling of stars, or *stellar scintillation* (or *astronomical scintillation*), is a phenomenon familiar to most people. When one looks into the sky on a clear night, one sees untold numbers of stars, all shining brightly, but seldom with a clear, consistent light. Instead, all stars in the night sky appear to blink on and off. The reason for this effect is that stars are so far away that the light from them that reaches Earth is essentially no more than a single point of light. As that point of light passes through Earth's atmosphere, it encounters regions of the atmosphere with differing temperatures and densities. It is refracted back and forth in a horizontal direction as it travels to Earth's surface. An observer's experience of looking at the star's light first head-on, then from the side, then head-on, then from the side, and so on, results in the twinkling effect. That effect often becomes more pronounced before sunrise and just after sunset since the star's light travels a greater distance through the atmosphere in both cases, encounters more air, and undergoes a greater number of refractions.

One of the most dramatic optical phenomena caused by refraction is the so-called green flash that occurs in the last fraction of a second before the sun sets. The green flash was first described on 17 January 1837 by a Captain Back of the British sailing ship HMS *Terror.* He wrote that at 9:45 P.M., he observed "the upper limb of the sun, as it filled a triangular cleft on the ridge of the headland, of the most brilliant emerald colour, a phenomenon which I had not witnessed before" ("The Green Flash" http://www.reefnet.on.ca/gearbag/wwwgfl.html).

The green flash occurs because different colors of sunlight are refracted by different amounts. As the sun sets, the colors that are least refracted—yellow, orange, and red—are the first to disappear. As the sun continues to drop lower on the horizon, eventually only the colors that are refracted the most—blue, green, and violet—are still visible. In the last fraction of a second, it is only these colors that can be seen, the yellows, oranges, and reds having disappeared beneath the horizon.

From this description, one would expect a "blue flash" to be seen, rather than a green flash, since blue light is refracted more than green light. Indeed, some observers claim to have seen such a "blue flash." The reason a green flash is more commonly observed, scientists think, is that the retina of the human eye is more sensitive to light in the midrange of the solar spectrum—the range that includes greens—than that of any other color.

Rainbows and Fogbows

Arguably the most spectacular optical phenomenon observed in Earth's atmosphere is the rainbow. A rainbow is an arc or a circle of colored light formed in the sky when rain droplets are present. The rainbow always contains all seven primary colors, with red on the outside, violet on the inside and orange, yellow, green, blue, and indigo between the two.

Rainbows are formed when light is refracted while passing into and out of water droplets in the atmosphere. As a single light ray enters a water droplet, it is refracted. Each color of light contained within the light ray is refracted to a slightly different extent, a maximum difference of about 2° for red and violet components of the sunlight. After passing through the middle of the water droplet, the refracted light rays strike the back inside of the droplet and are reflected back toward the front of the droplet. As they pass out the front of the droplet, they are refracted again, making the angle of separation between colors even greater. The process by which a water droplet separates white light into its constituent colors is called *dispersion*. Dispersion is also observed in other situations, as when white light is passed through a glass or plastic prism.

If only one droplet of water were present in the atmosphere, an observer would see only one color of light in the rainbow, the color that would be directed at his or her eyes by the angle at which the light ray left the droplet. But, of course, the atmosphere consists of billions upon billions of water droplets, each reflecting an array of colored rays across a wide region of the sky. To an observer, this array of colors is seen as the rainbow with the colors arranged in sequence from red to violet.

The first mathematical description of a rainbow was developed by the French mathematician and physicist René Descartes (1596–1650) in 1637. Descartes calculated the path that a light ray would follow passing into and out of a water droplet with reflection from its inner back surface. He found that the least amount of scattering and greatest focus of light occurred at an angle of 42° from the path of the sun's rays. In honor of this discovery, the colored rays of light that emerge from a water droplet are sometimes referred to as *Descartes rays.*

Under suitable conditions, a *secondary rainbow,* in addition to the *primary rainbow* described above, may be formed. The secondary rainbow occurs in an arc about 8° above the primary rainbow. It is produced when a light ray enters the bottom of a water droplet and is reflected twice, increasing the amount of dispersion produced.

Fogbows, dewbows, and moon bows are similar to rainbows, but each has its own unique properties. *Fogbows* form in misty weather, when water droplets are very small, less than about 0.05 millimeter. Droplets of this size are not large enough to disperse white light into its components effectively, and the refraction and reflection processes simply yield a rainbow-type event that is almost completely white.

Dewbows may also form under the same condition as fogbows except, with dew drops of larger size, they may form a hyperbolic rainbow with distinct colors like those of a solar rainbow that lies on top of the ground. Finally, *moon bows* form in exactly the same way as solar rainbows, except the intensity of moon light is so much weaker that the colors formed are much lighter and less distinct.

Halos, Sun Dogs, Sun Pillars, and Other Phenomena

When sunlight passes through a cirrus cloud, a variety of optical phenomena may be observed. One such phenomenon is a *halo,* a ring of light surrounding the sun (or, in the evening, the moon). Halos are usually white, but they may have some pale color under certain conditions. Colored halos have a red-orange-yellow inner fringe and a green-blue-violet outer fringe.

Halos are caused when light passes through ice crystals and is refracted twice, once upon entering the crystal and once upon leaving it. Ice crystals must have certain geometric shapes to produce refractions of this kind. They must all be hexagonal (six-sided), although they may look like flat hexagonal plates, tall hexagonal cylinders, bolt-shaped columns, or bullet-shaped hexagons. Such crystals must have diameters of less than 20.5 m (micrometers) for refraction to occur in such a way as to produce a halo. Refraction through a hexagonal ice crystal produces a ray that is displaced by 22° from the original path of the light ray. Such halos are called, therefore, *22° halos.*

Larger halos, known as *46° halos,* also occur. Such halos are produced when light enters one side of a hexagonal crystal and exits through the bottom of the crystal. By contrast, a 22° halo is produced when light enters one side of a crystal and exits the opposite side.

Sun dogs are formed by almost the same process as are halos and are usually seen in conjunction with them. The difference is that the ice crystals through which light passes are somewhat larger, 30 m, rather than 20.5 m or less, and the crystals must be oriented so that their flat surface is parallel to the Earth's surface. Under these conditions, light refracted through the ice crystals to an observer's eye produces a pair of brightly colored spots, the sun dogs, to the left and right of the sun. Sun dogs are also known as *mock suns* or *perihelia.* They are most commonly seen when the sun is near the horizon, at sunrise or sunset.

Another optical phenomenon related to halos is a *sun pillar,* a vertical column of bright light extending upward from the rising or setting sun. Sun pillars are formed by reflection when sunlight bounces off ice crystals that are floating gently to the Earth's surface.

Coronas are the only atmospheric optical phenomena observed more commonly with the moon than with the sun. Coronas are formed when sunlight or moonlight passes through a cloud whose droplets are very close to each other, about the same as the wavelength of light itself. In such a case, the light is diffracted and dispersed as it passes through the cloud, producing a whitish disk with the sun or moon at its center. When the water droplets within the cloud are all of nearly the same size, the diffraction may produce a separation of colors, like that which occurs in the formation of a rainbow, with the outer fringe of the corona having a reddish tint and the inner fringe, a bluish tint.

A *glory* is an optical phenomenon usually observed from an airplane. As one looks downward out of the airplane window, he or she may see a shadow of the plane itself, surrounded by one or more rainbow bands. These bands are produced in essentially the same way as a rainbow except that the water droplets through which refraction occurs are much smaller in size than those responsible for rainbow formation.

A particularly interesting form of a glory is the *heiligenschein,* or "holy light." The phenomenon is said to have been first observed by the Italian sculptor and goldsmith Benvenuto Cellini (1500–1571). While walking with a friend on the dewy grass one morning, Cellini observed what appeared to be a whitish halo surrounding his shadow on the ground. Since he did not see a similar halo around his friend's head, he assumed that the vision confirmed his belief that he was, indeed, a very special person. For some time after this event, later observed by many other people, the phenomenon was known as *Cellini's halo.*

The heiligenschein is usually observed in the fog or mist, or the dewy ground, or on dusty land. The effect is produced when the tiny particles of which fog, mist or dust is formed reflect light back to the observer, forming a bright halo around the person's otherwise dark shadow. The effect occurs as a halo because light reflected by particles very far from the shadow are not reflected to the observer, but to some more distant point to the left or right. It is for this reason that Cellini saw his only halo, but not that of his friends, being too far from the friend to observe the reflected light around *his* head. Indeed, the friend probably saw a halo around his own head, but not around Cellini's. A variation of the heiligenschein is known as the *specter of the Brocken,* named after a mountain in Germany frequently covered with fog. Hikers on the mountain often see halos around their shadows in the fog, just as Cellini first saw a halo around his own shadow.

Further Reading

Greenler, Robert. *Rainbows, Halos, and Glories.* New York: Cambridge University Press, 1980.

Lynch, David K., and William Livingston. *Color and Light in Nature.* Cambridge, UK. Cambridge University Press, 2001.

Minnaert, Marcel G. J. *The Nature of Light and Colour in the Open Air.* New York: Dover Publications, 1954.

Tape, Watler. *Atmospheric Halos.* Washington, DC: American Geophysical Union, 1994.

Atmospheric Scattering and Absorption

Solar radiation that enters the Earth's atmosphere experiences one of three fates. It may be scattered or absorbed by particles in the atmosphere, or it may pass through the atmosphere and strike the Earth's surface. In all three cases, characteristics of the radiation are changed as a result of its interaction with particles in the atmosphere or the Earth's surface.

Scattering

Scattering occurs when radiation collides with particles and changes direction as a result of this interaction. Three types of scattering are possible, depending on the relative size of the particle(s) involved and the wavelength of the radiation. The first type of scattering is known as *Rayleigh scattering,* after the English physicist John William Strutt, Lord Rayleigh (1842–1919). Rayleigh scattering occurs when the wavelength of radiation is much larger than the size of particles with which it comes into contact. It occurs more frequently with radiation of short wavelengths and is responsible for the fact that the sky appears to be blue. Since blue light has shorter wavelengths than that of most other colors contained within sunlight, it is most completely scattered, producing a blue coloration for the sky.

Mie scattering, named after the German physicist Gustav Mie (1868–1957), occurs when the wavelength of radiation is about the same size as the particles with which it interacts. Mie scattering tends to occur more frequently with radiation of longer wavelengths and is observed most commonly in the lower parts of the atmosphere.

Nonselective scattering occurs when the wavelength of radiation is much smaller than that of the particles with which it comes into contact. The term comes from the fact that all wavelengths of radiation are affected about equally. Nonselective scattering is responsible for the fact that clouds appear to be white because all wavelengths of sunlight are scattered about equally.

Absorption

Absorption is the process by which radiant energy comes into contact with a particle and converted into another form of energy with a different wavelength. Absorption in the atmosphere is a complex process because solar energy consists of radiation of many different wavelengths, and molecules and particles in the atmosphere are each capable of absorbing energy of unique wavelengths.

For example, oxygen and ozone tend to absorb radiation with wavelengths less than about 0.3m (microns, or micrometers), while carbon dioxide and water vapor tend to absorb radiation greater than about 3m. By contrast, nitrogen absorbs little or no incoming radiation at any wavelength.

These statements are broad generalizations. The absorption pattern for any one element or compound differs considerably for various wavelengths. In the case of carbon dioxide, for example, the gas absorbs virtually no radiation at wavelengths of less than 2m, then demonstrates high absorptivity at wavelengths of about 2.5m and 4m. It is then transparent (does not absorb radiation) again between wavelengths of about 4m and 10m, after which its absorptivity once again increases to a maximum.

From the above discussion, it is clear that none of the most abundant gases in the atmosphere (nitrogen, oxygen, or carbon dioxide) absorb incoming radiation in the range between about 0.3m and 0.8m, the portion of the electromagnetic spectrum that makes up visible light. A portion of the electromagnetic spectrum that is able to pass effectively through the atmosphere without absorption is called an *atmospheric window,* as an analogy to the window in a building that allows light to pass through.

When molecules absorb solar radiation, they begin to vibrate more rapidly. That is, their thermal energy increases. At some point, molecules re-radiate the absorbed energy to the surrounding atmosphere. When they does so, however, the re-radiated energy has a longer wavelength than that of the energy originally absorbed. The absorption and re-radiation of solar energy by oxygen, ozone, and water vapor result in a warming of the atmosphere.

Overall, about 19 percent of the solar radiation that reaches our atmosphere is absorbed (primarily by oxygen and ozone) and converted to heat energy. About 32 percent of solar radiation is scattered, a fifth back into space and four-fifths to the Earth's surface. An additional 29 percent of all solar radiation passes through the atmospheric window and reaches Earth's surface directly. Most of that (86 percent) is absorbed by the Earth's surface and the remainder (14 percent) is reflected back into space. The remaining 20 percent of solar energy that passes into the Earth's atmosphere is reflected off clouds, back into space.

The Greenhouse Effect

Just over half of all solar radiation reaching Earth's atmosphere (25 percent through an atmospheric window and 26 percent through scattering) strikes its surface. Even though much of that radiation is absorbed by the surface, it is then later re-radiated. The re-radiated energy has much longer wavelengths (in the range between 1m and 30m) than that of solar energy. Between them, carbon dioxide and water vapor absorb radiation within this range efficiently *except* for one small region, that between 8m and 11m. This part of the spectrum, then, represents another atmospheric window that is largely transparent to radiation. It is through this window that terrestrial radiation (radiation from the Earth's surface) is able to escape back into space.

The vast majority of terrestrial radiation, however, is absorbed by carbon dioxide and water vapor, the molecules of which then become excited and, therefore, "warmer." At some point, these molecules re-radiate the absorb energy, some of which returns to the Earth's surface and some of which is radiated back into space. The re-radiated energy returning to Earth's surface is again absorbed and re-radiated into space, resulting in a Ping-Pong effect in which the atmosphere is gradually heated to a relatively constant temperature with a relatively constant variation throughout the troposphere. This constant variation, known as the *environmental lapse rate,* amounts to an average of about 6.5°C per 1,000 meters (5.5°F per 1,000 feet) of altitude.

Further Reading

"Earth's Energy Budget," http://okfirst.ocs. ou.edu/train/meteorology/EnergyBudget.htm.

Atmospheric scattering and absorption are discussed in most introductory texts on meteorology. See, for example, Frederick K. Lutgens and Edward J. Tarbuck, *The Atmosphere: An Introduction to Meteorology.* Upper Saddle River, NJ: Prentice-Hall, 2001, Chapter 2.

B

Baghouse

A baghouse is a facility for the removal of very small particles from waste gases produced in a variety of industrial processes, such as papermaking, the manufacture of cement, lumbering operations, smelting and other metal-processing procedures, and the modification and purification of many natural products. The waste gases from such operations typically consist of a complex mixture of gases, liquids, and solid particles of many sizes, some of which may pose environmental hazards. For example, finely divided solid particles can, if vented into the atmosphere, be inhaled by humans and other animals and lodge in the respiratory passage, leading to a variety of respiratory disorders.

At one time, exhaust gases from most industrial operations were released directly to the atmosphere through smokestacks. Today, environmental regulations prohibit the direct release of most gases into the air. In most air pollution control schemes, therefore, exhaust gases pass through a series of treatment stages before they are allowed to escape into the atmosphere. In the first step of a treatment procedure, gases are carried into a *knockout chamber*, where heavier particles in the exhaust gases fall out as a result

of gravitational forces. These particles are carried away on a conveyor belt on the floor of the chamber.

Exhaust gases then pass into the baghouse, a tall structure that may reach a height of 10 meters (30 feet). The baghouse is filled with hundreds of long, cylindrical bags, that were once made of nylon or cotton but that are now available in dozens of natural and synthetic fibers. The material chosen for such bags must be tough, resistant to high temperatures, long-lasting, and capable of filtering out solid particles of some desired size. Baghouse filters are typically 15–30 centimeters (6–12 inches) in diameter and as long as the baghouse's height. Some filters can withstand temperatures of up to 550°C (1,000°F) and can remain in service for two years or more. In some facilities, the filter bags have been replaced by *candles,* long tubes of porous ceramic that have properties similar to those of traditional cloth bags

Exhaust gases flow into one side of the baghouse and out the opposite side. As they pass through the baghouse, small particles in the gases are trapped in the filters. Periodically, a mechanical shaker agitates the filters, causing solid particles to be released. These particles fall onto a conveyor belt, which carries them out of the facility. Because of their very fine composition, baghouse wastes often receive special treatment before permanent disposal. They may, for example, be compacted and then sealed in airtight bags before being buried in a landfill. *See also* Air Pollution

Balloons

A balloon is a bag made of a tough, nonporous material and filled with hot air or any gas with a density less than that of air. Balloons were invented during the late eighteenth century and are now used for a variety of purposes, including scientific research and recreation.

Balloons belong to a category of aircraft known as *aerostats,* or *lighter-than-air* machines. The name comes from the fact that all such aircraft have a total mass less than that of the air they displace. Other lighter-than-air machines include blimps and dirigibles. Although these terms are sometimes used somewhat interchangeably, balloons and blimps are usually thought of as nonrigid aircraft; that is, their shape depends on the gas with which they are filled. Dirigibles, by contrast, have a rigid frame made of wood or metal. Hybrid aircraft that contain limited amounts of rigid support have also been invented.

Early History

The first balloons were built by the French papermaker Joseph Montgolfier (1745–1799) and his brother Jacques. The story is told that Joseph got the idea for a balloon by watching partially burned paper being lifted by the hot air in the fireplace of his home. He tested his idea by making a small balloon of silk and filling it with hot air. The balloon floated to the ceiling of the room, confirming his theory, at least in principle.

Over a number of months, Joseph and Jacques tested a variety of balloon designs, using paper, linen, and other materials for

GRIS balloon launch in 1995. GRIS stands for gamma-ray imaging spectrometer. © NASA.

the balloon *envelope* (the bag itself) and various lifting materials, including hydrogen, smoke, and hot air. On 5 June 1783, the Montgolfier brothers conducted the first public demonstration of their invention in the town square of Annonay, in the Rhône-Alpes region of France between Lyon and Marseilles. They used a balloon made of linen, lined with paper, about 30 meters (100 feet) in diameter. Lifting power for the balloon was provided by a fire of chopped wood and wet straw in a basket hung beneath the envelope. The balloon lifted off, attained a height of about 2,000 meters (6,500 feet), stayed aloft for ten minutes, and came back to earth about two kilometers (just over a mile) from its launch point.

Three months later, on 19 September, the Montgolfiers repeated their demonstration, this time with three passengers: a duck, a sheep, and a rooster. The balloon was launched from the gardens at the Palace of Versailles, stayed aloft for eight minutes, and landed with all three animals in good condition. Finally, on 21 November of the same year, the Montgolfiers launched their first balloon with two human passengers, Pilâtre de

Balloon racing in Indianapolis, 1909. © Library of Congress.

Rozier and the Marquis d'Arlandes. The passengers carried with them a bucket of water in case the balloon's source of power got out of control. But they experienced no problems of any kind, floating more than 150 meters (500 feet) into the air above Paris and remaining airborne for nearly half an hour before landing safely. (A firsthand account of this trip can be found at http://www.newlink.com/~lwnelson/roz.htm.)

At about the same time of the Montgolfier experiments, the French physicist Jacques Charles (1746–1823) was studying the use of hydrogen gas to lift a balloon. Hydrogen had been discovered in 1766 by the English chemist and physicist Henry Cavendish (1731–1810). The low density of the gas, about one-fourteenth that of air, made it an obvious choice to use in balloons. Charles encountered two major problems, however. First, hydrogen was rather difficult to make. The only process then known was the reaction between certain metals and an acid. This process was slow and relatively expensive. In addition, hydrogen is flammable and sometimes explosive, requiring great care in its use.

Nonetheless, Charles was able to launch a hydrogen-filled balloon at about the same time as Montgolfier's successes, on 27 August 1783. The balloon landed in a field outside Paris, where it met with an unfortunate end. Farmers attacked the airship, believing that it was some type of monster descending on earth from the skies. In spite of its fate, Charles's balloon contained many of the elements present in modern-day balloons, including a vent through which waste gas could be expelled, a wicker gondola suspended beneath the envelope, and sandbags as ballast for raising the device.

Charles did not give up on his balloon experiments because of the fate of his first attempt. Indeed, a month later he launched a second balloon, made of rubberized silk and filled with hydrogen, from the Tuileries Gardens in Paris. By this time, interest in balloons had become so great that a crowd of 400,000 people were present to witness the great demonstration. The balloon was 8.5 meters (28 meters) in diameter and remained in flight for two hours, during which time it covered a distance of more than 43 kilometers (27 miles).

Over the next few years, balloonists continued to experiment with different construction materials and different systems for powering their aircraft. One successful design involved the use of a hydrogen balloon and a hot-air balloon tied together. Rozier attempted to use this combination to cross the English Channel in 1785, but the hydrogen balloon exploded only a half-hour after takeoff and Rozier was killed. Within a matter of months, however, the cross-channel trip was attempted again, this time by French balloonist Pierre Blanchard (1753–1809) and one of his financial supporters, Dr. John Jeffries, an American living in England. This trip was successful, although it ended somewhat farcically. Blanchard and Jeffries had to jettison everything in their gondola, eventually including all of their clothing, to stay aloft long enough to land on the French shore, which they did with no extra time left to them.

Uses of Balloons

Balloons were put to use for two purposes almost immediately after their invention. One use was for entertainment and sport. A number of individuals, Blanchard among them, found that they could make a very nice living by demonstrating balloon travel at fairs, exhibitions, and other community gatherings. The second use was in warfare.

The first use of balloons by the military dates to 26 June 1794 in the Battle of Fleurus between France and Austria. The French raised balloons from which to observe the battlefield and to direct artillery fire on the enemy. In most cases, these balloons were *captive balloons,* that is, they were tethered to the ground by means of a long rope or metal cable. An observer sat in a gondola suspended beneath the balloon to make the necessary observations.

The success of this early venture prompted the French government to organize a balloon corps, the world's first air force, and to begin construction of observational balloons. French hopes proved to be unfounded, however; enthusiasm for the use of balloon rapidly dissipated, and they disappeared from military arsenals for nearly a century. One of the major problems with military balloons was that they were entirely at the mercy of the winds. Careful observations, the purpose for which the balloons were designed, could often not be made because they were so buffeted above by the wind.

For more than seven decades, balloons were used almost entirely for sport and recreation. One memorable exception occurred during the siege of Paris in 1870 by the Prussian army. For a period of five months, balloons were essentially the only way by which matériel and messages could be sent into and out of the city to other parts of France.

One of the most extensive uses of balloons occurred during the American Civil War between 1861 and 1865. Probably the most important figure in the use of balloons during this period was a New Hampshire aviator named Thaddeus Sobieski Constantine Lowe (1832–1923). Lowe had become fascinated with balloon travel at an early age and staged demonstrations of ballooning around the country. Shortly after the beginning of the war, Lowe convinced the U.S. Army Department to make use of balloons for observation of battlefield activity. Lowe developed envelopes made of silk coated with linseed oil and powered by hydrogen. He found it necessary to invent a portable hydrogen generator so that balloons could be inflated and launched at the point where they were needed, rather than being carried back to some permanent gas works. Lowe's enthusiasm for the use of balloons proved to be overly optimistic, however, as winds and uncertain weather made their contributions to observation largely worthless. Finally, after the Battle of Fredericksburg in 1862, General Joseph Hooker concluded that "observation on the moon would disclose as much as to the movements of the enemy" as would balloon observations, Lowe's pay was cut, his staff reduced, and, in essence, the experiment on observational balloons was ended (http://www.fredericksburg.com/CivilWar/Battle/0707cw).

The use of balloons for military observation did not completely come to an end with the failure of Lowe's experiment. Indeed, they have seen at least some use in most major wars conducted over the last century. During World War I, for example, the Americans used an blimp-shaped observation balloon called the Caquot, named after its French inventor, Lieutenant Albert Caquot. The Caquot balloon was 28 meters (92 feet) long and nearly 10 meters (32 feet) in diameter. It held just over 900,000 liters (32,200 cubic feet) of hydrogen and could ascend to about 1,200 meters (4,000 feet). In clear weather, observers could see a distance of up to 65 kilometers (40 miles). Nearly a thousand such balloons were used during the war.

The Caquot was used again during World War II, this time by British forces. But it saw much less use than it did during the earlier conflict. Even in relatively recent times, balloons have been used for military observation. During the Cold War between the United States and the Soviet Union, for example, the Central Intelligence Agency (CIA) launched balloons from Western Europe carrying cameras capable of recording activity on the ground in the Soviet Union. Under the best of circumstances, these balloons were carried by prevailing wins across the Soviet Union and into the Pacific Ocean, where their camera packages were recovered. As in previous instances, however, the unpredictable nature of the weather and wind patterns made this type of observation relatively ineffective.

One of the most effective military uses of balloons is as defensive devices, in which case they are often known as *barrage balloons*. Barrage balloons are captive balloons that are lifted into the air above a city or some other area to be protected from air raids. The theory is that aircraft will not be able to fly over the area without coming into contact with and being destroyed by the barrage balloons. Barrage balloons were used extensively by the British during World War II. These balloons were erected and maintained by members of the Women's Auxiliary Air Force and rose as far as 1,500 meters (5,000 feet) into the air. By the end of 1944, about 3,000 barrage balloons had been built and used. Some of their most effective work came against the German V-1 "flying bomb" attacks on southern England. They destroyed more than a hundred of these bombs before they reached their targets.

Blimps

Although the terms *balloon* and *blimp* are sometimes used interchangeably, the later word is more properly used for balloons that have a sausage or cigar shape, like that of the well-known Goodyear blimp. The first such aircraft to be successfully flown was built by the French engineer Henri Gifford (or Giffard; 1825–1882). Gifford's blimp was filled with hydrogen and powered by a three-horsepower steam engine suspended in a gondola below the envelope of the blimp. On 24 September 1852, Gifford flew his blimp over Paris at a speed of about 9 kilometers per hour (5 miles per hour) for a distance of about 27 kilometers (16 miles). His machine used a three-bladded propeller and a boat rudder to drive and steer the airship.

Three decades later, the Tissandier brothers, Albert (1836–1906) and Gaston (1843–1899), built a 1,000 cubic meter (37,000 cubic foot) airship powered by a battery-operated motor. During the same period, one of the most active researchers on blimp design was the Brazilian aviation pioneer Alberto Santos-Dumont (1873–1932), who built and tested a number of blimps in France, all powered by gasoline engines.

At first, there were some hopes that blimps might be used in the same way as balloons, for observation and transportation, for example. But even with a new source of propulsion, they proved to be of little practical value, because they flew slowly, were subject to wind and weather, and could carry limited amounts of cargo. Nonetheless, a limited market for blimps has existed since the mid-1920s, when the Goodyear Tire & Rubber Company began building lighter-than-air machines. During World War II, the U.S. Navy employed more than 150 blimps for minesweeping and submarine patrol missions on both the Atlantic and Pacific coasts and in the Gulf of Mexico.

Goodyear built its first commercial blimp in 1925, the *Pilgrim,* and continued producing similar airships for than four decades. After a period of time during which they no longer built blimps, the company renewed construction of blimps in September 2002. A number of companies around the world also produce blimps, primarily for commercial and military purposes. The largest airship in existence is the *Sentinel 1000,* built by Westinghouse Airships in 1991. The *Sentinel 1000* is 68 meters (220

feet) long, 20.16 meters (64.88 feet) high, and 16.67 meters (54.7 feet) wide, with a volume of 10,000 cubic meters (350,000 cubic feet). It is powered by twin Porsche turbocharged 300-horsepower engines capable of driving the airship at a maximum speed of 111 kilometers per hour (69 miles per hour). Plans were made to build an even larger model of the airship, the *Sentinel 5000,* with an envelope seven times the size of *Sentinel 1000*'s. A combination of design and finance problems made that project unfeasible, however.

The blimps in use today find one of three major applications: military, police and civil aid, and advertising and tourism. Military applications are largely surveillance activities, such as patrolling important installations, borders and the oceans; carrying out missions to counter narcotics and smuggling operations; and conducting communications and signal intelligence activities. Blimps are also used as alternatives to police patrol helicopters for reconnaissance and search and rescue activities, at least partly because they are quieter, require less frequent refueling, and are less polluting than helicopters.

Finally, blimps are popular as flying advertisements for companies and events. Companies as diverse as Budweiser, Whitman's, Reebok, MasterCard, EuroDisney, and the World Wrestling Federation have used blimps to advertise their products and services.

Weather Balloons

One of the most common and most important uses of balloons has long been in the field of meteorology. On 17 July 1862, the British meteorologist James Glaisher (1809–1903) and his balloonist colleague Henry Coxwell (1819–1900) made the first ascent into the atmosphere to measure weather conditions, such as temperature, pressure, and humidity. Their balloon, *Mammoth,* used methane gas and rose to a height of 7,977 meters (26,177 feet). At that altitude, the density of oxygen is so low that Glaisher and Coxwell began to feel faint, and were just able to return to Earth before falling unconscious. The two men later made 27 more ascents between 1862 and 1866. (Glaisher's account of their most dangerous ascent can be found on the Internet at http://members.lycos.co.uk/Vigilant/balloons/glaisher.html.)

Over the next half-century, manned balloons became a very popular form of studying atmospheric conditions. The U.S. Weather Bureau had largely replaced weather kites with balloons by the early 1910s, and observers routinely traveled to heights of about 20 kilometers (12 miles) to register their observations. An important breakthrough in balloon meteorology techniques came in 1972 when the Russian meteorologist Pavel A. Molchanov (1893–1941) invented the radiosonde. A radiosonde is a small expendable package containing a variety of meteorological instruments suspended from a hydrogen- or helium-filled balloon. Measurements of temperature, pressure, humidity, and other atmospheric characteristics are transmitted from the radiosonde to receiving stations on the ground. After completing its measurements, the radiosonde is ejected from the balloon and carried back to earth by parachute.

The National Weather Service releases about 75,000 radiosondes each year. Data are collected at 92 stations in the United States twice a day, at 00:00 and 12:00 UTC (Coordinated Universal Time). These data are used not only for weather predictions, but also for weather and climate change research, air pollution studies, severe storm forecasts, and satellite confirmation studies. *See also* Aerial Sports; Dirigible

Further Reading

Denniston, George C. *The Joy of Ballooning.* Philadelphia: Courage Books, 1999.

Edita, S.A. *The Romance of Ballooning: The Story of the Early Aeronauts.* New York: Viking Press, 1971.

Evans, Charles M. *War of the Aeronauts: The History of Ballooning in the Civil War.* Mechanicsburg, PA: Stackpole Books, 2002.

Hall, George, and Baron Wolman. *Blimp!* New York: Van Nostrand Reinhold, 1981.

Wirth, Dick. *Ballooning: The Complete Guide to Riding the Winds.* New York: Random House, 1980.

Current balloon research is described on NASA's Balloon Program home page on the Internet, http://www.wff.nasa.gov/~code820.

Barometer

A barometer is an instrument for measuring atmospheric pressure. It is an essential device in measuring current weather conditions and predicting weather patterns. The barometer was given its name by the English chemist and physicist Robert Boyle (1627–1691) in 1665. Boyle chose the name from two Greek words meaning *pressure* ("baro-") and *measurement* ("-metro").

Invention

The barometer was invented by the Italian physicist Evangelista Torricelli (1608–1647) in 1643. Torricelli's invention came about as the result of his interest in a practical problem that had long troubled engineers involving the operation of a water pump. Water pumps were used for a variety of purposes, such as lifting water out of a mine. But they were able to lift water only to a certain height, about 10 meters (34 feet).

At the time, the explanation for the operation of a water pump was based on a widely accepted scientific principle, namely that "nature abhors a vacuum." When the piston in a water pump is raised, a vacuum is produced inside the pump. Since nature will not "allow" a vacuum to exist, scholars believed, water rushes in to fill the vacuum. The only problem with this explanation was that no one could explain why nature abhorred a vacuum only to a height of 10 meters.

Torricelli's teacher and friend, Galileo Galilei (1564–1642) suggested that Torricelli study this problem. Torricelli began with a discovery that Galileo himself had made, namely that air has weight. Prior to Galileo's research, scientists believed that air had no weight and, in fact, possessed a property of "levity" that produced a lifting effect.

But if air has weight, Torricelli reasoned, then it must push down on water, the ground, and any other object on or above the Earth's surface. Perhaps air pressure accounted for the inability of water pumps to lift beyond a height of 10 meters.

To test this idea, Torricelli sealed a long glass tube at one end and filled the tube with mercury. He then placed his thumb over the open end and inverted the tube into a bowl of mercury. He found that a small amount of mercury flowed out of the tube into the bowl. But a column of mercury about 76 centimeters (30 inches) high remained standing in the tube.

To explain this result, Torricelli reasoned that the empty space about the mercury column in the tube was entirely empty; it contained a vacuum. The column of mercury in the tube was supported by the pressure of air in the atmosphere pushing down on the mercury in the bowl, that pressure then being transmitted to the mercury in the tube.

Early Developments

In theory, Torricelli could have built his barometer of any liquid at all. A water barometer, however, would have to be tall enough to hold a column of water 10 meters high because water is only one-fourteenth as dense as mercury. In fact, the German physicist Otto von Guericke (1602–1686) did make a water barometer that contained a glass ball at the top. Inside the ball was the figure of a man, whose rise and fall von Guericke used to predict the weather. This clever, but very impractical, device was one of the first barometers used for weather prediction.

Another possible application of the barometer was studied in the late 1640s by the French physicist and mathematician Blaise Pascal (1623–1662). Pascal reasoned that, if air has weight, its effect would be less noticeable on a mountain, where the air is thinner, than at sea level. In 1648, he convinced his brother-in-law to carry two mercury barometers more than a mile up the Puy-de-Dôme to test his theory. He found that the level of mercury in the barometers fell more than three inches, confirming his suspicions. Pascal's discovery suggested that it might be possible to use some modification of a simple barometer to measure one's altitude above sea level.

Later Developments

The earliest barometers were very simple instruments consisting of a mercury-filled tube on which lines were drawn to indicate the height of the mercury. These barometers were generally poorly made and quite inaccurate. Besides, no one had found any important practical application for air pressure readings, so accuracy was not very important.

As the value of barometers in making weather predictions became more obvious, however, a number of changes were made to render them more accurate. For example, thermometers were attached to the side of a barometer to permit measurement of the temperature of the mercury in the tube. Mercury, like all other substances, changes density with changing temperature. As the temperature rises, mercury expands; as the temperature drops, mercury contracts.

Later barometers were also fitted with verniers to improve the accuracy of readings. A vernier is a short scale attached to the side of a barometer and fitted so that it can slide up and down against the barometer. By comparing the divisions marked on a vernier with the divisions marked on the barometer, it is possible to make an air pressure reading with an accuracy one decimal place greater than without the vernier.

For even more accurate measurements, additional modifications must be made in a barometer. For example, suppose that atmospheric pressure decreases. Air pressure is able to support a somewhat smaller amount of mercury in the barometer tube, and some mercury runs out into the reservoir in which the tube sits. But one can no longer simply read the level of the mercury in the tube, however, since the level of mercury in the reservoir has increased slightly.

Two ways have been found to deal with this problem. In one case, the so-called *Kew barometer,* the graduations on the mercury tube are not the same distance apart all the way up the tube. Instead, they are made slightly smaller the farther up one goes on the tube. This contraction of scale adjusts for the change in mercury level in the reservoir itself. Kew barometers also tend to be somewhat smaller in diameter than other barometers because they are frequently used by mariners. A narrower tube makes it easier to read the height of the mercury in a barometer as the ship rocks back and forth on the seas.

Another barometer modification was invented by the French physicist Jean Fortin (1750–1831). In the *Fortin barometer* a flexible bag with an extra supply of mercury is attached to the bottom of the mercury reservoir. The flow of mercury between bag and reservoir is controlled by an adjustable screw whose point is set so as just to touch the surface of the mercury in the reservoir. As atmospheric pressure and mercury levels change, modifications of the adjustable screw keep the mercury level in the reservoir at a constant height.

Modern barometers have even more modifications to obtain precise readings of atmospheric pressure. The United Kingdom's National Standard Barometer, for example, measures changes in mercury level by directing a laser beam at the top of the mercury column, permitting the measurement of very small changes in column height and, hence, atmospheric pressure.

Aneroid Barometers

Even in its most technologically advanced form, the mercury barometer has some significant disadvantages. The mercury-filled glass tube must always be about a meter high, making it awkward to move from one place to another. The glass is breakable, and mercury is a somewhat dangerous substance with which to work. It is hardly surprising, then, that inventors began exploring possible alternatives to the mercury barometer not long after Torricelli made his first such instrument.

The first successful alternative to the mercury barometer was invented in the early 1840s by the French scientist Lucius Vidie. Today's *aneroid* ("without liquid") *barometer* operates on essentially the same principle devised by Vidie. The barometer consists of a flexible metal container from which most of the air has been removed. The container is able to change size and shape as atmospheric pressure increases or decreases. An increase in air pressure pushes down on the metal container, causing it to decrease in size, while a decrease in air pressure reduces the pressure on the container, causing it to increase in size.

The movement of the metal container is reflected in the motion of a pointer attached to one end of the barometer. The pointer rides up and down with changes in atmospheric pressure. One way to observe the motion of the pointer is to attach it to the hand on a dial that moves around a circular scale, from low pressure to high pressure.

The simple clocklike aneroid barometer present in many homes operates on this basis.

Another way to observe the movement of the point is to have it rest on the side of a rotating cylinder wrapped with graph paper. As the cylinder rotates on its own axis, the pen-equipped pointer makes a tracing on the paper that reflects increases and decreases in pressure. A recording barometer of this design is known as a *barograph.*

Altimeters

An *altimeter* is an instrument that measures one's height above sea level. The simplest type of altimeter is simply an aneroid barometer that is carried aloft in an airplane, a balloon, or some other device that travels up and down in the atmosphere. The only modification needed is to convert the actual atmospheric pressure shown on the face of the aneroid barometer to its comparable height in meters, feet, or some other unit. *See also* Torricelli, Evangelista

Further Reading

Banfield, Edwin. *Barometers: Aneroid and Barographs.* Trowbridge, Wiltshire, UK: Baros Books, 1985.

Brombacher, W. G. *Mercury Barometers and Manometers.* Washington, DC: U.S. Department of Commerce, National Bureau of Standards, 1960.

Middleton, W. E. Knowles. *The History of the Barometer.* Baltimore: Johns Hopkins Press, 1964.

Barotrauma

Barotrauma is any injury to the body caused by an increase in air pressure. The two most common situations in which barotrauma may occur are during an airplane flight and during scuba diving. Barotrauma may affect the whole body or only one portion of it, most commonly the ear. Barotrauma that occurs during descents into deep water is also known as *decompression sickness,* while that which is experienced during the descent of an airplane is also known as *aero-otitis.*

The human ear consists of three parts, the outer, middle, and inner ears. The outer and middle ears are separated by a thin membrane known as the *tympanum* or *eardrum.* Pressure on the front and back sides of the tympanum is kept constant by means of a narrow tube, the *Eustachian tube,* that leads from the middle ear to the back of the nose. Each time one swallows, a tiny bubble of air passes upward from the back of the mouth, through the Eustachian tube, into the middle ear. The movement of that bubble is responsible for the "click" or popping sound that occurs when one swallows.

The normal function of the Eustachian tube may be disturbed by a variety of causes. For example, a viral infection (such as a cold or the flu) may cause the inner linings of the tube to swell, interrupting the flow of air into the middle ear. During the ascent or descent of an airplane, air may not flow into the middle ear rapidly enough.

In such cases, a partial vacuum may form within the inner ear. The tympanum is pushed into the middle ear by normal air pressure from the outer ear. Increased pressure on the tympanum may cause feelings of pain and a need to "pop" one's ear. Distortion of the tympanum may also affect its ability to vibrate properly, muffling or altering the sound waves transmitted from the outer ear to the middle ear.

In more severe cases, fluids will seep into the middle ear from membranes on its inner lining. These fluids are not in and of themselves harmful, but they can cause pain and further affect a person's hearing. In the most serious cases, the tympanum may actually break, causing bleeding and additional loss of fluid into the middle ear. Such an occurrence may result in the loss of hearing.

Mild cases of aero-otitis are very common and generally disappear shortly after one has disembarked from an airplane. The condition is relatively easy to avoid by chewing gum, "popping" one's ears during descent, or swallowing or yawning frequently. More serious cases of the condition that involve bleeding or drainage of fluid into the ear require medical attention.

Further Reading

"Ears and Altitude," http://www.entnet.org/healthinfo/ears/altitude.cfm.

Barrage Balloon

See Balloons

Beaufort, Sir Francis (1774–1857)

Sir Francis Beaufort is probably best known for a wind force scale and wind speed scale that he developed in 1831. At the time, relatively few navigational and meteorological instruments were available to mariners. As a result, plotting voyages across open water required an almost intuitive knowledge of weather systems and considerable experience. Beaufort based his chart on his own experience with the effects of winds of various force and speed on British sailing ships. The chart was adopted in 1838 by the British Admiralty for use on all its ships. The scale has undergone constant revision over the years and is now available for both land and sea conditions.

Francis Beaufort was born in County Meath, Ireland, on 7 May 1774. He entered the navy when he was only 13 years old and spent the next 68 years associated with the navy in one form or another. In 1796, he was promoted to the rank of lieutenant on the sailing ship *Phaeton*. Four years later, he received 19 musket and sword wounds on his head, arms, and body in the battle of Malaga. He received command of his own vessel, HMS *Woolwich*, in 1805. He was assigned the task of making a complete hydrographic survey of the Rio de la Plata region of South America. It was during this voyage that Beaufort first began collecting data for and making initial sketches of his wind charts.

In 1829 Beaufort was appointed chief hydrographer of the Royal Navy, a post he

Beaufort's Original Wind Force Scale

0	Calm		
1	Light air	Or just sufficient to give steerage way	
2	Light breeze	Or that in which a man-of-war with all sail set, and clean full would go in smooth water from	1 to 2 knots
3	Gentle breeze		3 to 4 knots
4	Moderate breeze		5 to 6 knots
5	Fresh breeze	Or that to which a well-conditioned man-of-war could just carry in chase, full and by	Royals, etc.
6	Strong breeze		Single-reefed topsails and top-galley sail
7	Moderate gale		Double-reefed topsails, jib, etc.
8	Fresh gale		Treble-reefed topsails, etc.
9	Strong gale		Close-reefed topsails and courses
10	Whole gale	Or that with which she could scarcely bear close-reefed main-topsail and reefed fore-sail	
11	Storm	Or that which would reduce her to storm staysails	
12	Hurricane	Or that which no canvas could withstand	

held for 26 years. He retired from active duty at the age of 81 in 1855. He died two years later in London on 17 December 1857. Beaufort's original wind force scale is shown here. The strength of the wind is indicated in each category by the number and types of sails that would be deployed for the condition. *See also* Wind Measurement

Beaufort Wind Scale

See Beaufort, Sir Francis; Wind Measurement

Bernoulli's Principle

Bernoulli's principle states that the pressure exerted by a moving fluid, such as water or air, is a function of the speed with which the fluid is moving. This principle was discovered in 1738 by the Swiss mathematician and physicist Daniel Bernoulli (1700–1782).

Imagine that air is flowing through a pipe of varying dimensions. When air passes into a portion of the pipe with small diameter, its velocity increases; when it passes into a portion of the pipe with a larger diameter, its velocity decreases. Or imagine that air is flowing over the surface of an object. The faster the air moves over the surface, the less the pressure it exerts on the object.

Bernoulli's principle explains many everyday phenomena. For example, a shower curtain often appears to be "sucked in" against a person's body while he or she is taking a shower. The explanation for this phenomenon is that the flow of water from the showerhead results in a decrease of pressure on the inside of the shower curtain. Normal air pressure on the outside of the curtain pushes it inward against the person's body.

The operation of an atomizer also depends on Bernoulli's principle. An atomizer consists of a bottle containing a liquid, a vertical tube inserted into the liquid, and a horizontal tube attached to the vertical tube. A rubber bulb is attached to one end of the horizontal tube, which has a small opening at the opposite end. When one squeezes the rubber bulb, air is forced out of the bulb, through the horizontal tube, and out the open end. The fast moving air creates a region of low pressure at the top of the vertical tube. Normal air pressure inside the bottle pushes down on the liquid in the bottle, forcing it upward in the tube and into the horizontal spray of air. The liquid is "atomized," that is, broken up into small particles and carried out the front end of the tube.

The carburetor in a internal combustion engine operates on the same principle. A stream of air passing over a reservoir of gasoline creates a region of reduced pressure. Normal air pressure pushes liquid gasoline upward through a vertical tube into the flow of air, where it is atomized into small droplets that burn in an attached cylinder.

In baseball, the movement of a ball that occurs when a pitcher throws a "curve ball" also depends on Bernoulli's principle. The pitcher imparts a spin on the ball so that air on one side of the ball is moving more rapidly than air on the opposite side of the ball. Normal air pressure pushes the ball in the direction of lower pressure, causing its path to curve.

The lift provided by airplane wings is also explained by Bernoulli's principle. Airplane wings are shaped so that their lower surface is straight and flat, while their upper surface is curved. As the airplane moves forward, air travels backward over the wing. It travels more rapidly over the top of the wing than it does over the bottom, producing an area of low pressure on top of the wing. Normal air pressure at the bottom of the wing pushes it upward, lifting the wing and the rest of the airplane upward. *See also* Aerodynamics; Airplane

Blimp

See Balloons

Boyle, Robert (1627–1691)

Robert Boyle made a number of important discoveries about the properties of gases. He is probably best known for his discovery about the relationship of the volume of a confined gas and the pressure exerted on the gas. Mathematically, that relationship can be expressed as $pV = k$, where p is the pressure of the gas, V is its volume, and k is some constant. On the European Continent, the same law is known as *Mariotte's Law* after Edmé Mariotte (1620–1684), who discovered the relationship at about the same time as Boyle. Mariotte did not announce his discovery in writing, however, until 1676, sixteen years after Boyle did.

Boyle was born at Lismore Castle, County Waterford, Ireland, on 25 January 1627. His father, the earl of Cork, was reputed to have been the richest man in England at one time. Young Boyle demonstrated extraordinary intellectual talent at an early age. He entered Eton at the age of 8, by which time he had already mastered Greek and Latin, both of which he spoke fluently. By the age of 14, he had also mastered Italian in preparation for a visit to that country. During that visit, he became familiar with the writings of Galileo, which came to have enormous impact on his intellectual growth. He became interested in scientific experimentation, a pursuit to which he spent the rest of his life.

Boyle returned home in 1645 to find his father dead and himself independently wealthy. He took up residence in Oxford, where he became part of an group of scholars informally known as the Invisible College. In 1663, King Charles II formally recognized the group as the Royal Society, with Boyle one of the charter members.

Boyle was one of the rare scientists of the time interested in experimentation. The strong influence of Aristotle continued to reign over most scientists, who still believed that the exercise of reason was the surest way to scientific knowledge. One of Boyle's earliest experiments involved the construction of an air pump, invented by the German physicist Otto von Guericke (1602–1686) in 1650. Working with his assistant, Robert Hooke (1635–1703), Boyle carried out a variety of experiments with the air pump over three years. Boyle first announced his law in the appendix to his report on those experiments, *New Experiments Physio-Mechanicall, Touching the Spring of Air and Its Effects* (1660).

Boyle's Law was significant for a number of reasons. For one thing, it was one of the earliest quantitative formulations of a physical relationship. Also, it demonstrated that air (and presumably other gases) consisted of particles widely separated from each other by empty space. This conclusion was the only one possible based on the discovery that gases could be forced into smaller and smaller volumes with increasing pressure. Boyle's experiments were also important because he kept meticulous records that allowed other scientists to repeat them to confirm or deny his results.

Boyle made a number of other important contributions to science. In his 1661 book *The Sceptical Chemist,* for example, he provided the first modern definition of a chemical element, that is, a substance that cannot be broken down into any simpler substance. He was also the first person to distinguish among acids, bases, and neutral substances.

As Boyle grew older, he became more interested in religious studies and learned both Hebrew and Aramaic to carry out his research. He was elected president of the Royal Society in 1680, but he declined to take the oath of office for religious reasons. He died in London on 30 December 1691. In his will, he left funds for a series of lectures, not on scientific topics, as one might imagine, but on the defense of the Christian religion against unbelievers.

Further Reading

Conant, James Bryant. *Robert Boyle's Experiments in Pneumatics.* Cambridge, MA: Harvard University Press, 1965.

Hunter, Michael. *Robert Boyle: 1627–91.* Woodbridge, UK: Boydell Press, 2000.

Sargent, Rose-Mary. *The Diffident Naturalist: Robert Boyle and the Philosophy of Experiment.* Chicago: University of Chicago Press, 1995.

Shapin, Steven. *Leviathan and the Air-Pump: Hobbes, Boyle, and the Experimental Life.* Princeton, NJ: Princeton University Press, 1989.

Brush, Charles F. (1849–1929)

Charles F. Brush is one of the pioneers of the electric industry in the United States. In 1887, he built a very large wind power turbine, 17 meters (50 feet) in diameter with 144 wooden rotor blades. It was the largest windmill in the world and the first to be used for the large-scale generation of electrical energy. The turbine continued to operate for two decades, during which time Brush used the electrical power generated to recharge batteries in his mansion in Cleveland, Ohio. The turbine had limited commercial potential, however, because of mild and unpredictable winds in the Cleveland area.

Brush was born in Euclid, Ohio, on 17 March 1849. He attended public school in Cleveland, and went on to earn a degree in mining engineering from the University of Michigan in 1869. He then returned to Cleveland, where he worked as an analytical chemist for four years. He also went into partnership with Charles E. Bingham as a dealer in iron and iron products.

By 1877, Brush had begun devoting most of his time to inventing, a lifelong passion. In addition to the giant windmill, he invented an electric dynamo and an electric arc light. In 1879, he demonstrated his arc light at a public ceremony in Cleveland's Public Square. A year later, he founded the Brush Electric Company. The company was soon successful enough to allow Brush to devote as much time as he wanted to inventing.

In 1889, he sold the Brush Electric Company to the Thompson-Houston Electric Company. Two years later, this company merged with the Edison General Electric Company to form General Electric, now one of the world's largest corporations.

In addition to his involvement in the electrical industry, Brush served as president of the Air Linde Products Company and was cofounder of the Sandusky Portland Cement Company. He was also one of the incorporators of the Case School of Applied Science, which later became the Case Institute of Technology and, after merging with Western Reserve University, Case Western Reserve University.

Brush was awarded the French Legion of Honor, the Rumford Medal, the Edison Medal, the Franklin Medal, and three honorary doctorates. He spent much of his later years in philanthropic activities, supporting a number of charitable, religious, and scientific organizations. He died in Cleveland on 15 June 1929. *See also* Windmill

Buoyancy

Buoyancy is the tendency of a body to rise when submerged in a fluid. The principle of buoyancy is also known as Archimedes's Principle, after the Greek mathematician who discovered it in the third century B.C. It applies to all types of fluids, both liquids and gasses.

Archimedes's Principle can be expressed as follows: The buoyant force on an object immersed in a fluid is equal to the weight of the fluid displaced by the object. For example, suppose that a cork with a volume of 10 milliliters is immersed in water. The density of cork is about 0.2 grams per milliliter, making its total weight 2 grams. The density of water is 1.0 gram per milliliter. The weight of the water displaced by the cork is 10 milliliters 3 1.0 grams per milliliter, or 10 grams. The buoyant force exerted on the cork, then, is 10 grams. Since the cork itself weighs only 2 grams, it will be pushed upward by the water.

Cause of Buoyancy

Buoyancy can be explained as the net effect of two forces: the pressure of the fluid in which the object is immersed and the gravitational force exerted on the object itself, its weight. The pressure exerted by a fluid increased with depth and is exerted equally in all directions, a fact discovered by the mathematician Blaise Pascal (1623–1662). Thus, any object immersed in a fluid feels a pressure on its bottom that is greater than the pressure on its top. This difference in pressures results in a net upward force, tending to push the object upward.

The second force acting on the object is the force of gravity, tending to push it downward. If the gravitational force on the object is greater than the net upward force of the fluid, the object will sink. If the gravitational force is less than the net upward force of the fluid, the object will be pushed upward.

Buoyancy in Air

The behavior of a cork immersed in air is different from one immersed in water. The density of air is about 1.29 grams per liter, or 0.001 29 grams per milliliter. The weight of air displaced by a cork 10 milliliters in volume would be 10 milliliters 3 0.001 29 grams per milliliter, or 0.012 9 grams. The upward force on a cork exerted by air if, of course, far less than that needed to push the cork upward.

It is clear that the only objects that can float in air are those with densities less than that of air. For example, the least dense of all gases, hydrogen, has a density of about 0.09 grams per liter, or 0.000 009 grams per milliliter. Imagine a thin balloon containing 10 milliliters of hydrogen immersed in air. The total mass of the hydrogen is 10 milliliters 3 0.000 09 grams per milliliters, or 0.000 9 grams. The total mass of the air displaced by the hydrogen is (from above) 0.0129 grams. The surrounding air is thus able to exert enough upward force on the hydrogen to cause it to rise.

The buoyancy of air is used in the operation of all lighter-than-air machines, such as balloons, blimps, and dirigibles. In all such machines, some gas is chosen with a density less than that of air. Although hydrogen would be desirable for such uses (and, at one time, was widely used in such machines), it has the disadvantage of being flammable and sometimes explosive. Methane also has a sufficiently low density for use in lighter-than-air machines, and was also used in early balloons and blimps. It is also flammable, however, so it is no longer used for such purposes. By far the most popular gas used in lighter-than-air aircraft is helium. Helium is nearly twice as dense as hydrogen (0.1785 grams per liter), but it is neither flammable nor explosive. *See also* Archimedes; Balloons; Dirigibles

Further Information

Herbert, Don. *Buoyancy and Displacement.* Columbus, OH: Merrill Publishers, 1988 (videorecording).
Buoyancy is discussed in most introductory physics and physical science textbooks.

C

Carbon Dioxide

Carbon dioxide is the fourth most abundant gas in the atmosphere, after nitrogen, oxygen, and argon. Unlike those gases, its composition tends to be variable. According to measurements made at the Mauna Loa Observatory in Hawaii, its abundance has been steadily increasing since 1958, from 315 parts per million to 369 parts per million in 2000. Most scientists believe that this increase is caused by the release of carbon dioxide from anthropogenic sources, primarily the combustion of fossil fuels.

Many scientists believe that increasing concentrations of carbon dioxide in the atmosphere pose a serious environmental threat to the planet, described as *global warming* or *global climate change*. Since carbon dioxide in the atmosphere has a tendency to trap heat reflected from the Earth's surface, increasing concentrations of the gas may result in a higher annual average temperature on the Earth. That change, in turn, may result in an increase in the rate at which glaciers and ice caps melt and an increase in the temperature of the Earth's oceans, both factors resulting in a rise in sea levels worldwide.

Physical and Chemical Properties

Carbon dioxide is a colorless, odorless gas with a density of 1.97 grams per liter, about 50 percent greater than that of air (1.29 grams per liter). Under a pressure of about 60 kilograms per square centimeter (870 pounds per square foot), carbon dioxide is converted directly to a solid, called *dry ice,* without passing through the liquid phase. Dry ice is a snowlike white powder with a density of 1.56 grams per milliliter that sublimes at a temperature of $-78.5°C$ ($-109°F$). Carbon dioxide is readily soluble in water, forming a solution (carbonic acid) present in all "soda pop" formulas.

Carbon dioxide is not flammable and does not react readily with other elements and compounds. In fact, it is commonly used as a fire extinguisher because it prevents a burning fuel from having access to oxygen.

Discovery and Naming

Carbon dioxide was first recognized as a gas different from air itself by the Flemish scientist Johannes Baptista van Helmont (1579–1644) in about 1630. Van Helmont collected the gas produced by burning wood, recognized that it differed from atmospheric air, and gave it the name *gas sylvestre* ("gas from wood"). More than a hundred years later, the gas was rediscovered by the Scottish physician Joseph Black (1728–1799), who produced carbon dioxide by heating magnesium carbonate (then called *magnesia alba,* "white magnesia").

Black also obtained the gas from exhaled breath, determined its density, and studied its other physical and chemical properties. He called the gas *fixed air,* a name by which it was known until 1873, when it was given its modern name of carbon dioxide.

Biological Function

Plants absorb carbon dioxide from the atmosphere and, through photosynthesis, convert it to carbohydrates. The general form of the reaction is as follows:

$$CO_2 + H_2O \rightarrow C_xH_{2y}O_y,$$

where $C_xH_{2y}O_y$ represents sugars, starches, and other carbohydrates synthesized in plant cells. When plants are eaten by animals, those carbohydrates are converted into other basic biochemical compounds, such as lipids (fats and oils), proteins, and nucleic acids.

In animal cells, these biochemical compounds are metabolized to produce the energy needed for all life functions. During metabolism, carbon dioxide is a by-product in a reaction that is basically the reverse of photosynthesis:

$$C_xH_{2y}O_y \text{ (carbohydrates, lipids, proteins)} + O_2 \rightarrow CO_2 + H_2O$$

Production and Uses

Carbon dioxide is obtained as the by-product of a number of industrial operations, such as the synthesis of ammonia and its related products. It is also produced during the cracking of hydrocarbons in the petroleum refinery industry. Some carbon dioxide is also obtained from natural springs and wells. In small quantities, the gas can be produced by the action of an acid on a carbonate, such as calcium carbonate:

$$2HCl + CaCO_3 \rightarrow CaCl_2 + CO_2 + H_2O$$

About half of the carbon dioxide consumed in the United States is used for refrigeration systems. The next largest consumer of the gas is the beverage industry. Smaller amounts of carbon dioxide are also used in the fabrication of metals and in chemical manufacturing, municipal water treatment systems, mining and petroleum industries, and as a fire extinguisher. *See also* Atmosphere; Respiration

Further Reading

Most introductory textbooks in chemistry and the high school and college level contain a discussion of carbon dioxide.

Center for Clean Air Policy

The Center for Clean Air Policy (CCAP) was created in 1985 by a bipartisan group of state governors. Its purpose is to promote and implement innovative solutions for major environmental and energy problems related to issues of air pollution. The center's underlying philosophy is that most such problems can be solved or ameliorated by using market-based approaches in which solutions to pollution problems are reached through economic incentives and disincentives. The center is concerned with problems both in the United States and in other parts of the world. It maintains two central offices, one in Washington, DC, and the other in Prague, the Czech Republic.

The center sponsors and promotes a number of programs, presentations, events, and other activities, as well as providing a broad range of publications dealing with air pollution issues. Its efforts are divided into four major program areas: air quality, climate, transportation, and education and exchange. Some examples of projects currently being conducted under the auspices of the Air Quality Program are "Exploring New Directions in Clean Air Policy,"

"Encouraging Multiple-Pollutant Reduction Strategies," "Air Quality Benefits from Clean Power Developments on Brownfields," and "Reducing Aviation Emissions."

As part of its educational and exchange program, the center sponsors a number of fellowships, exchange of personnel, study tours, and other opportunities for elected officials and other involved individuals to meet with counterparts in other parts of the United States and from other parts of the world. These programs are conducted under the sponsorship of the German Marshall Fund Environmental Exchange and the Transatlantic Exchange on Environmental Sustainability.

CCAP publications are available in both print and electronic format. Some examples of those publications include "Acid Rain: Road to Middleground Solution," "The Untold Story: The Silver Lining for West Virginian in Acid Rain Control," "Disclosure in the Electricity Marketplace: A Policy Handbook for States," and "The Potential use for Biomass in Hungary." *See also* Clean Air Acts

Further Information
Center for Clean Air Policy
750 First Street, NE, Suite 940
Washington, DC 20002
Telephone: (202) 408-9260
Fax: (202) 408-8896
URL: http://www.ccap.org
E-mail: general@ccap.org

Chemical Properties of Air

See Physical and Chemical Properties of Air

Civil Air Patrol

The Civil Air Patrol (CAP), headquartered at Maxwell Air Force Base in Alabama, is a division of the U.S. Air Force. The CAP was established on 1 December 1941, less than a week before the United States entered World War II. It was formed to provide a way by which private fliers and aviation enthusiasts could make a special contribution to the nation's civil defense program. In 1943, the organization was incorporated into the U.S. Army Air Force and, on 1 July 1946, the CAP was made a permanent peacetime institution under Public Law 476. Two years later, in May 1948, the CAP was made a part of the U.S. Air Force, and the secretary of the Air Force was authorized to provide staff liaison personnel at all levels of the Patrol.

There are currently 35,000 adult members of the CAP flying more than 3,700 privately owned and 530 government aircraft. The organization also enrolls more than 26,000 cadets in its Cadet Program. The organization is divided into 52 wings, one for each state, the District of Columbia, and Puerto Rico. Each wing is divided into groups, squadrons, and, in some cases, flights. Overall, CAP is made up of about 1,700 individual flying units.

The Civil Air patrol has three main functions: emergency services, aerospace education, and cadet training. The core of its emergency services assignment includes search and rescue efforts, disaster relief, and civilian support activities necessitated by natural disasters, such as tornadoes and earthquakes. As of 2002, the CAP was flying more than 85 percent of all federally operated search and rescue operations, saving an average of more than a hundred lives each year.

Over the years, the CAP has assumed additional responsibilities in the nation's war against drugs. In November 1985, it began flying reconnaissance flights along U.S. borders to assist the U.S. Customs Service in its effort of drug interdiction and control. In 1989, it was assumed similar responsibilities for the U.S. Drug Enforcement Agency and the U.S. Forest Service. In 1999, CAP pilots flew more than 6,500 missions as part of these agreements. Pilots engage in patrol and communication during those flights, but not in actual enforcement activities.

The CAP's aerospace education program centers on approximately 200 workshops for teachers held each year at about 100 colleges and universities. These workshops are designed to improve teachers' understanding of various aerospace topics and to prepare them for teaching about the subject. The CAP sponsors an annual National Congress on Aviation and Space Education for aerospace teachers. It also develops and publishes curriculum materials on aerospace topics for use in the nation's schools and colleges.

The CAP Cadet Program is open to all U.S. citizens and legal residents between the ages of 12 and 21 with at least a fifth-grade education. The program consists of classes in aerospace science and technology, leadership, moral education, physical fitness, and special topics. Upon completion of the program, cadets are granted the General Billy Mitchell Award, which entitles graduates to enter the U.S. Air Force, should they choose to do so, as an airman first class. *See also* Air Force

Further Information
Civil Air Patrol
Air University
Public Affairs Office
Attn: CAP-USAF
55 Le May Plaza South
Maxwell AFB, AL 36112-6332
Telephone (334) 953-4241
URL: http://www.capnhq.gov
E-mail: hyla.person@maxwell.af.mil

Clean Air Acts

The Clean Air Acts are a series of legislative bills passed by the U.S. Congress in an effort to control air pollution and improve air quality. The first of these acts was the Air Pollution Control Act of 1955, providing funds for research on polluted air by the U.S. Public Health Service (USPHS). The legislation required that the USPHS provide individual states with the results of this research.

Clean Air Acts of the 1960s
A series of acts were passed during the 1960s attempting to control air pollutants emitted by a variety of mobile and stationary sources. In many cases, these acts were based on air quality legislation developed and adopted by the state of California, long the governmental entity in the United States with the most advanced air pollution control laws. The first of these pieces of legislation was the Clean Air Act of 1963. That act set emission standards for pollutants released by stationary sources, such as power plants and steel mills. The law also provided grants to individual states to assist them in carrying out whatever air quality programs they developed on their own. The law did not deal, however, with the most important source of air pollutants at the time: cars, trucks, other motor vehicles, trains, and airplanes.

That deficiency and other air pollution problems were dealt with in amendments to the Clean Air Act of 1963 passed in 1965, 1966, 1967, and 1969. The 1965 amendments, for example, authorized the secretary of Health, Education and Welfare to establish nationwide standards for automobile exhaust emissions. This legislation and later amendments also authorized the surgeon general to study the effects of air pollutants on human health, expanded local air quality programs, set compliance deadlines for meeting new air quality standards, established air quality control regions (AQCRs), and authorized research on low-emission fuels and more fuel-efficient automobiles.

The Clean Air Act of 1960 and its amendments contained the important basic principles needed to improve the nation's air quality. However, enforcement of the acts was left largely to the state and, in some cases, the financial support needed to meet the acts' goals was insufficient. As a result, little improvement in the nation's air quality was seen by the end of the decade.

Clean Air Act of 1970

By 1970, the nation's environmental consciousness had developed considerably, and both legislators and the general public had become aware of the continuing problem of air pollution and the inadequacy of existing laws and regulations. As a result, Congress passed and President Richard Nixon signed the Clean Air Act of 1970. Although passed as an amendment to the original Clean Air Act of 1960, the 1970 legislation was actually a more extensive and more aggressive approach to the problems of air pollution in the United States.

The 1970 act established National Ambient Air Quality Standards (NAAQSs) in 247 AQCRs and set New Source Performance Standards (NSPSs) that regulated the amount of emissions to be permitted from a new source in an area. It also required individual states to create their own state implementation plans (SIPs), to be approved by the Environmental Protection Agency (EPA) and to enforce air quality standards for stationary sources. Enforcement of mobile source regulations was retained in the hands of federal agencies. The amendment for the first time also allowed individual citizens to take legal action against any organization, including any governmental body, in violation of emission standards. Finally, the amendments provided $30 million for new and continuing research on noise pollution in urban areas.

Probably the most significant features of the 1970 act were the provision for stiff fines for violation of clean air regulations and the establishment of a clear and specific schedule for compliance with the new legislation.

Additional amendments to the Clean Air Act, passed in 1977, dealt primarily with motor vehicle emission standards. Since adoption of the 1970 act, representatives of industry, particularly the automotive industry, had expressed concern about meeting the deadlines set in that legislation. They requested an extension for compliance, a request that was granted in the 1977 amendments. The 1977 act also included the first legislation dealing with the destruction of stratospheric ozone.

Clean Air Act of 1990

Dissatisfied with the progress of air quality improvement over the previous three decades, President George Bush and Congress attacked the problems of air pollution once again in June 1989. The result of this effort was the Clean Air Act Amendments of 1990, a piece of legislation that covered previous ground, established even more rigorous standards and methods of achieving clean air, and attacked new issues. Title I of the act focuses on problems of urban air pollution, ozone, carbon monoxide, and particulate matter. It allows the EPA to designate areas that have not met minimum air quality standards and to impose stringent programs for improving air quality in these areas. It also requires that areas make use of the best available control technologies (BACT) in areas with severe problems of air pollution.

Title II increased standards for exhaust emissions from cars and trucks and from evaporation at gasoline pumps during refueling. Reduction in the sulfur content of fuels was also mandated, and a "clean fuel car pilot research program" was established for the development of more fuel efficient automobiles. Title III introduced a new list of 189 "air toxics" not covered by previous

Clean Air Act legislation with standards to be established for each toxic material.

Titles IV and VI dealt with two atmospheric problems, acid deposition (acid rain), and ozone depletion and global warming, respectively. To deal with the first problem, a 10 million ton reduction in sulfur dioxide from 1980 levels by 1995 was mandated, with additional reductions in future years. A ban on the production of substances that deplete the ozone layer was the main feature of Title VI.

Title V was described by an EPA press release as "in many ways the most important procedural reform contained in the new law" ("Overview: The Clean Air Act Amendments of 1990," http://www.epa.gov/oar/caa/overview.txt). The title requires that states develop programs for issuing permits to all entities that release pollutants to the atmosphere.

Other titles in the act dealt with additional air quality issues, such as support for the federal acid rain research effort, the development of new methods for air monitoring, provisions for the improvement of visibility near national parks, and the creation of unemployment benefits for workers laid off as a result of a company's compliance with the provisions of the Clean Air Act. *See also* Air Pollution; Center for Clean Air Policy; Clean Air Trust; Clear the Air; Foundation for Clean Air Progress

Further Reading

Bryner, Gary C. *Blue Skies, Green Politics: The Clean Air Act of 1990 and Its Implementation.* Washington, DC: Congressional Quarterly Books, 1995.

Legislation: A Look at U.S. Air Pollution Laws and Their Amendments, http://www.ametsoc.org/AMS/sloan/cleanair/cleanairlegisl.html.

U.S. Environmental Protection Agency, Office of Air and Radiation. *The Benefits and Costs of the Clean Air Act, 1970 to 1990: EPA Report to Congress.* Washington, DC: U.S. Environmental Protection Agency, Office of Air and Radiation, Office of Policy, [1997].

———. *The Benefits and Costs of the Clean Air Act, 1990 to 2010: EPA Report to Congress.* Washington, DC: U.S. Environmental Protection Agency, Office of Air and Radiation, Office of Policy, [1999].

Wark, Kenneth, Cecil F. Warner, and Wayne T. Davis. *Air Pollution: Its Origin and Control,* 3rd edition. Menlo Park, CA: Addison-Wesley, 1998.

The home page for the EPA's Clean Air Act is http://www.epa.gov/oar/oaq_caa.html.

Clean Air Trust

The Clean Air Trust (CAT) was established as a nonprofit organization in 1995 by former Senators Edmund Muskie of Maine and Robert Stafford of Vermont. The purpose of the organization is to educate the general public and policymakers about the provisions and value of the Clean Air Act. Among its accomplishments over the past decade have been the development of a program to clean up dirty diesel trucks, initiated efforts to promote the use of cleaner gasoline, coordinated efforts of health and environmental groups to press for stricter standards for particulate pollution, helped defeat efforts to weaken the Clean Air Act, and commissioned and published public opinion surveys on public attitudes about pollution cleanup. The trust publishes a monthly list and description of "clean air villains" who contribute to the nation's problems of air pollution, a variety of bulletins on topics of special interest, and publications on certain topics of special concern related to air pollution. *See also* Air Pollution; Clean Air Acts

Further Information

Clean Air Trust
1625 K Street, #790
Washington, DC 20006

Telephone: (202) 785-9625
URL: http://www.cleanairtrust.org
E-mail: frank@cleanairtrust.org

Clear the Air

Clear the Air (CTA) is a national program of education aimed at improving air quality, especially by focusing on the role of coal-burning plants in producing air pollution. The organization is a joint project of the Clean Air Task Force, National Environmental Trust, and U.S. PIRG Education Fund. It was established under a grant from the Pew Charitable Trusts through a grant to Pace University.

The four major foci of the organization's efforts are:

- seeking Congressional and state legislative action on issues of air pollution;
- pushing for strong regulatory action by the U.S. Environmental Protection Agency;
- supporting legal action to protect existing air quality standards; and
- using a major media campaign to inform the public about air quality issues.

Clean the Air acts as a coordinating network for five dozen environmental groups in 30 states. *See also* Air Pollution; Clean Air Acts

Further Information
Clear the Air
1200 18th Street, N.W., 5th Floor
Washington, DC 20036
Telephone: (202) 887-1715
Fax: (202) 887-8880
URL: http://cta.policy.net
E-mail: info@cleartheair.org

Coriolis, Gaspard Gustave de (1792–1843)

Gaspard Gustave de Coriolis is best known for his experimental research and mathematical theory dealing with the path tra-versed by a freely moving object in relationship to a rotating surface. That theory explains the curved path taken by any such object and can be applied to a number of natural phenomena, such as the movement of ocean currents and of masses of air in the Earth's atmosphere.

Coriolis was born in Paris on 21 May 1792. He enrolled in France's premier scientific school, the École Polytechnique, in 1808, from which he eventually earned his degree in highway engineering. Coriolis accepted a job with the corps of engineers, but he was forced to resign in 1816 because of his father's death and his own poor health. In fact, Coriolis was plagued with poor health his whole life, a condition that did not, however, prevent him for carrying on an active and productive career.

After leaving the corps of engineers, Coriolis accepted an assignment as assistant professor of Analysis and Mechanics at the École Polytechnique. In 1829, he left that position to become chair in mechanics at the École Centrale des Arts et Manufactures. In 1836, he assumed the same post at the École des Ponts et Chaussées. Two years later, he returned to the École Polytechnique as director of studies, a post he held until his death in Paris on 19 September 1843.

In addition to his study of the effect named for him, Coriolis was very much interested in formalizing the study of physics and linking the work of scientists and technicians. He argued that it was necessary to make very clear definitions of scientific terms that could be understood and used by both theoreticians and workers in the field. In that direction, he was successful in replacing the somewhat vague expression "force-displacement" than in use by the modern term *work*. He also defined mathematically the previously ambiguous term "kinetic energy" with a clear mathematical definition. Kinetic energy, he said, is equal to one half the product of the mass of a body times the square of its velocity, or, K.E. = $1/2mv^2$. *See also* Coriolis Effect; Wind

Further Reading

Freiman, L. S. *Gaspard Gustave Coriolis.* Moscow: Nauka, 1961.

Coriolis Effect

The Coriolis effect is the change observed in freely moving objects that travel through the Earth's atmosphere or oceans as a result of the Earth's rotation. The effect was named after its discoverer, Gustave Gaspard de Coriolis (1792–1843).

To understand the Coriolis effect, imagine that a bullet is fired from a gun in a horizontal direction at some point on the Earth's surface. To an observer on the Earth, the bullet has only one force acting on it, the propulsive force from the gun. The observer will estimate the direction of the bullet by the direction of that propulsive force.

However, a second force also acts on the bullet, one that is not readily apparent to the Earth-bound observer. That force is the Earth's west-to-east rotation. The bullet's path is, therefore, determined by some combination of two forces, the propulsive force provided by the gun and the force of inertia provided by the Earth's rotation.

Motion to the South and North

The effect of the Earth's rotation on the path of a moving bullet depends on the latitude at which it is fired. The reason for this fact is that the Earth moves at different linear speeds at the equator, at the poles, and at all latitudes in between. An object at the equator, for example, travels faster than at any other point on the globe. By comparison, an object at the pole moves very slowly and, at the exact poles themselves, has zero velocity.

Now consider a bullet fired southward from a point near the North Pole. The two components of the bullet's path are the southward component provided by the propulsive force of the gun and an eastward component provided by the Earth's rotation. The eastward component, in this example, is relatively small. As the bullet travels in a southerly direction, it passes over the surface of the Earth, all points of which are moving more rapidly to the east than is the original starting point from which the bullet was fired. When the bullet finally lands, it will do so at a point farther to the west than would have been expected. In other words, the bullet's path would appear to be curved to the right from its original starting position.

Now consider a bullet fired from a gun in a northerly direction from the equator. In this case, the bullet still has two components, the northerly component resulting from the propulsive force of the gun and an easterly component provided by the rotation of the Earth at the equator. In this case, the eastward component is a substantial one since the Earth is rotating very rapidly at the equator. As the bullet travels to the north, its eastward component is greater than the land or water over which it is traveling. When it comes to rest on the ground, it will have landed farther to the east than would have been expected. Again, it would appear to have traveled to the right from the expected straight path.

The behavior of objects fired to the north or south at latitudes between the equator and the North Pole can be determined by a similar argument. In all cases, however, it happens that the moving object will be deflected to the right from its expected straight-line path by some "mysterious" force, the Coriolis force. A similar argument can be made for objects traveling to the north or south in the Southern Hemisphere. In this case, however, the moving object would follow a curved path bending to the left from the expected straight-line path.

Motion to the East and West

The Coriolis effect is also observed with objects moving in an easterly or westerly direction on the planet. Suppose, for example, that a bullet is fired from a gun in an easterly direction at some time, t = 0. At that moment, "east" for the bullet and for the gun from which it was fired means the same thing. But a few moments later, that situation has changed. The bullet continues to travel in a straight-line "easterly" direction that is exactly the same as it was at time t = 0. But the gun is following a different path. Since it is "attached" to the Earth's surface, its path is curved and its sense as to what constitutes "east" is continually changing. To an observer on the Earth's surface, the bullet would appear to fall to Earth at a point to the right (in the Northern Hemisphere) or the left (in the Southern Hemisphere) of its intended straight-line direction.

Practical Effects

The Coriolis effect can be observed in a variety of natural phenomena. For example, the movement of sunspots across the face of the sun can partially be understood as an expression of the Coriolis effect. Also, the movement of ocean currents is influenced by the force. And all wind systems that blow across the Earth's atmosphere, from massive prevailing winds to local wind storms, are produced and controlled by the Coriolis force. *See also* Coriolis, Gaspard Gustave de; Global Winds

Further Reading

Stommel, Henry M., and Dennis W. Moore. *An Introduction to the Coriolis Force.* New York: Columbia University Press, 1989.

Ctesibius (Second Century B.C.)

Very little is known about Ctesibius (also Ktesibius). His personal life is largely a mystery, and all that we know of his research comes from the writings of later historians. We are told that he was born the son of a barber in Alexandria and that his first invention was made when he was still a boy. He devised a balancing mirror that his father could use in working with customers of different heights. The device consisted of a horizontal pole balanced on a vertical pole. The mirror hung from one end of the horizontal pole and a lead weight as heavy as the mirror was suspended from the opposite end. The lead weight hung inside a pipe and sometimes made a whistling noise as it dropped through the tube.

Ctesibius was intrigued by this discovery and decided to make a simple musical instrument that operated on the principle. At first, his primitive organ made a sound only when a weight dropped through a pipe. No sound could then be made again until the weight was raised and dropped a second time. To solve this problem, Ctesibius used running water to force air constantly through a series of pipes, producing a continuous series of sounds that the organ player could control. This water-driven air organ, a *hydraulis,* was the forerunner of many other instruments built in later centuries.

The invention for which Ctesibius is probably most famous is the *clepsydra,* or water clock. The device consisted essentially of a container with a small hole in the bottom through which water was released at a regular and predictable rate. Although very primitive, it was one of the earliest reasonably accurate devices available for measuring time. It survived as the timekeeping instrument of choice until Christiaan Huygens (1629–1695) invented the pendulum clock in 1656.

Further Reading

Drachmann, A. G. *Ktesibios, Philon and Heron: A Study in Ancient Pneumatics.* Copenhagen: E. Munksgaard, 1948.

Cyclones and Anticyclones

A cyclone is an area of low atmospheric pressure, while an anticyclone is an area of high atmospheric pressure. The wind patterns that develop around each area are known, respectively, as a *cyclonic* or *anticyclonic* wind. Cyclonic winds blow in a counterclockwise direction in the Northern Hemisphere and in a clockwise direction in the Southern Hemisphere. The reverse is true for anticyclonic winds.

Cyclones and anticyclones are easily recognized on weather maps as closed or nearly closed circular or nearly circular isobars surrounded by wind flags indicating the direction and speed of winds around the high or low pressure region. On satellite photographs, cyclones and anticyclones can be identified by the characteristic comma-shaped clouds that form around them.

Cyclogenesis

The theory of cyclogenesis (cyclone formation) was first developed during World War I by a group of Norwegian meteorologists that included Jacob Bjerknes, Vilhelm Bjerknes, Halvors Solberg, and Tor Bergeron. This theory was first published in 1918 in an article titled "On the Structure of Moving Cyclones." The theory was developed to explain the formation and development of cyclones in northern Europe and comparable regions around the world. Such cyclones are now known as *middle-latitude* or *wave cyclones*. The Norwegian theory, also known as the *polar front theory,* was based entirely on observations of winds at and just above the Earth's surface. More recent data about high-level winds have made possible modifications of the original theory. To a large extent, however, the original 1918 theory is largely accurate.

According to this theory, a cyclone begins to form when a polar front moves southward toward the equator and comes into contact with a warm air mass. A polar front is defined, in fact, as the boundary between a cold polar air mass and a warm subtropical air mass. The boundary between warm and cold air masses is seldom perfectly smooth but, instead, consists of discontinuities produced by topographical irregularities, temperature differences, ocean currents, or other surface phenomena, or, more commonly, by variations in the polar jet stream.

The discontinuities produced may cause the polar front to take on a wavy appearance, not unlike that of a water wave. The troughs and crests of the wave front may tend to dissipate and disappear, or they may tend to reinforce each other and grow larger. In some cases, the waves may extend several hundred kilometers across the Earth's surface. They tend to travel west to east, following the direction of the polar air mass, and they may last anywhere from a few days to a few weeks.

Air Movement around a Cyclone

Since is a cyclone is an area of low pressure, one might expect air at the ground to rush in, equalizing pressure within and around the cyclone and causing the cyclone to dissipate rather quickly. This sometimes happens, but the reason that it does so only rarely is explained by wind patterns in the upper atmosphere above the cyclone. In the midlatitude regions, the polar jet stream blows from west to east at an altitude of between 7,500 and 12,000 meters (25,000–40,000 feet). It sometimes blows along a relatively straight path, roughly west to east. When it does so, conditions for cyclone development are unfavorable.

More often, the jet stream follows a meandering path, traveling sometimes to the south and sometimes to the north of a direct west-to-east path. When it does so, air tends to converge or "bunch up" in some parts of the jet stream and to diverge or "spread out" in other parts. When convergence and divergence occur (and they often occur in parts of the atmosphere close to each other) over areas of low or high pressure, conditions for the formation of cyclones and anticyclones are ideal.

In such cases, the divergence of air in one part of the jet stream over a region of low pressure causes air to be "sucked up," as in a vacuum cleaner, from the ground into the atmosphere, where it spills out into the jet stream. (Actually, the air is pushed up by surrounding areas of high pressure, although the "sucking" image may be easier to visualize.) As the air rises around the low pressure area, the Coriolis effect causes it to start spinning in a counterclockwise direction.

Essentially the same argument can be used to describe the formation of an anticyclone. In this case, however, air is pushed downward from a convergence in the jet stream over a region of high pressure and then out into surrounding areas of low pressure. Again, the Coriolis effect sets these winds to spinning in a clockwise direction.

Cyclone Movement

Cyclones tend to move across the land or water over a period of days or weeks. As they do so, the cold and warm air masses from which they were originally produced gradually become more thoroughly mixed until temperature gradients within the boundary disappear or become insignificant. Cyclonic and anticyclonic winds then tend to dissipate.

As they travel across the land, cyclonic and anticyclonic winds tend to bring with them certain characteristic types of weather.

Those weather patterns differ somewhat depending on the part of the cyclonic system one studies. The northern or poleward part of the system tends to receive the worst, longest-lasting effects of cyclogenesis as warm air from the south and east is pushed

northward and westward, where it piles up over the incoming cold front. These conditions product wet, cloudy weather that may last for days.

Areas farther south tend to experience a greater variety of weather patterns that change over time. The first appearance of a developing cyclone may be high cirrus clouds in an otherwise clear blue sky ("mare's tails") that precede a cold front by as much as 1,000 kilometers (600 miles). For a short period (12–24 hours), light rains may fall and winds gradually change to a more southerly direction as the cold front moves in, pushing warm air ahead of it. The area is briefly under the influence of the warm air mass, which produces clear skies, with cumulus clouds and warm temperatures. Eventually, however, the cold air mass overtakes the area, bringing with it heavy, dark storm clouds (such as nimbostratus and altostratus clouds) that drop heavy precipitation and may produce thunderstorms and tornadoes. As the cold front moves through the area, temperatures drop and the weather changes, producing clear blue skies with high cirrus clouds.

Under some circumstances, cyclonic and anticyclonic winds may intensify, contract, and become far more destructive than usual. The most common examples of such storms are tornadoes and hurricanes. Tornadoes usually form from cumulonimbus clouds over land, while hurricanes are comparable storms that develop over bodies of water. Both tornadoes and hurricanes are themselves sometimes known simply as "cyclones."

Tropical and Midlatitude Cyclones

When the term cyclone (or anticyclone) is used to describe a storm system, rather than simply an area of low pressure, it may be further designated as a *tropical cyclone* or a *midlatitude cyclone*. The process of formation for both storms is similar, but the source of energy from which each develops is different. In the case of a midlatitude cyclone, this source of energy is almost entirely a difference in temperature between two adjacent air masses. In the case of a tropical cyclone, the source of energy is the evaporation of water from the ocean and the condensation of water vapor in upper parts of the storm system. *See also* Coriolis Effect; Global Winds; Hurricane; Tornado; Wind

Further Information

Landsea, Christopher W., "FAQ: Hurricanes, Typhoons, and Tropical Cyclones," http://www.aoml.noaa.gov/hrd/tcfaq/tcfaqA.html

"Midlatitude Cyclones," http://ww2010.atmos.uiuc.edu/(Gh)/guides/mtr/cyc/home.rxml.

Shapiro, Melvyn A., and Sigbjørn Grønås, eds. *The Life Cycles of Extratropical Cyclones.* Boston: American Meteorological Society, 1999.

Sharkov, Eugene A. *Global Tropical Cyclogenesis.* Chichester, UK: Springer, 2000.

D

Dirigible

A dirigible is a lighter-than-air flying machine belonging to a class of aircraft known as *aerostats*. The other machines included in this class are blimps. Dirigibles are also known as *airships* because, unlike balloons, they have propulsion and steering systems that allow them to operate under the control of a pilot. The term *dirigible* originally arose from the French expression *ballon dirigeable*, or "steerable balloon."

Dirigibles differ from blimps in that they have a rigid internal structure, usually made of aluminum, that supports the shape of the ship. By contrast, blimps contain no such structure and keep their shape because of the pressure exerted by the gas contained within it. Both dirigibles and blimps are filled either with hydrogen or, more commonly today, with helium. Both types of aerostat also have engine-driven propellers, rudders to control lateral motion, and elevators to control vertical motion.

Invention

The dirigible was invented by retired a German cavalry officer, Ferdinand Adolf August Heinrich Graf von Zeppelin (1838–1917). His first airship was launched on 2 July 1900 near Lake Constance in southern Germany. The airship was 126 meters (419 feet) long and 11.4 meters (38 feet) in diameter, with a volume of 9,580 cubic meters (338,400 cubic feet) of hydrogen. The dirigible envelope was divided into 16 cells and was powered by two 16-horsepower engines. Almost from the onset, the dirigible was commonly known by the name of its inventor, *Zeppelin.* To be strictly correct, however, the term *Zeppelin* is a proprietary name that should not be used for airships built by any company other than von Zeppelin's own Luftshiffbau Zeppelin G.m.b.H.

For a short time, dirigibles achieved a small measure of popularity for civil and military air transportation. Between 1900 and 1938, Zeppelin built 119 dirigibles, of which the vast majority (103) were delivered to the German Army during World War I. Some of these airships were used for bombing raids over England. Probably the most famous of the dirigibles was the *Graf Zeppelin,* launched in 1928. For a decade, it made flights across Europe and to North America, South America, the Middle East, and even the Arctic. It carried 20 passengers in its 66-hour transatlantic trip on one of the most elegant voyages available at the time.

Dirigibles were also built in Great Britain (a total of 14) and the United States (a total of five), but interest in the aircraft dwindled rapidly in the late 1930s. One reason for the decline in interest was the development of heavier-than-air aircraft, which were soon able to offer faster, more dependable, and nearly as luxurious travel as did dirigibles. Most authorities believe, however, that the fate of dirigibles was sealed on 6 May 1937 when the dirigible *Hindenburg* burned and crashed while landing at Lakehurst, New Jersey, with the loss of 35 lives.

No dirigibles have been built since the fateful crash of the *Hindenburg.* The only airships still operating are a number of blimps used for surveillance work by the military and civilian law enforcement agencies and as advertising platforms for important sporting and other events. *See also* Zeppelin, Ferdinand Adolf August Heinrich, Count von

Further Reading

Clarke, Basil. *The History of Airships.* London: H. Jenkins, 1961.

Dick, Harold G., and Douglas Hill Robinson. *The Golden Age of the Great Passenger Airships, Graf Zeppelin & Hindenburg.* Washington, D.C.: Smithsonian Institution Press, 1985.

Hannah, Barry. *Airships.* New York: Vintage Books, 1985.

Topping, Dale. *When Giants Roamed the Sky: Karl Arnstein and the Rise of Airships from Zeppelin to Goodyear.* Akron, OH: University of Akron Press, 2000.

Downburst Phenomena

A downburst is a strong downward flow of air through a thunderstorm cloud that results in an outward flow of air on or just above the Earth's surface. The term was first suggested in 1978 by Ted Fujita, one of the world's most famous severe storm researchers. Fujita classified downbursts into two categories, *macrobursts,* with a diameter of more than 4 kilometers (2.5 miles) and *microbursts,* with diameters of less than 4

kilometers. The ground-level winds that result from downbursts can be very destructive and, in some cases, even more dangerous than those of a tornado.

Formation

Downbursts form after thunderstorm clouds have reached a mature stage of development. Early in the formation of a thunderstorm cloud, the movement of air is primarily upward, as warm surface-level air is pushed upward by cool air on the ground around it. Eventually, this rising air reaches an altitude in which moisture in the air begins to condense and form water droplets and/or ice crystals. At that point, precipitation forms, and snow or rain begins to fall through the cloud. The downward flow of water droplets and/or ice crystals carries with it surrounding air in the middle of the thunderstorm cloud. At the same time, cool (and sometimes dry) air around the top of the cloud is drawn into its center, and the downward flow of air is accelerated.

This downward flow of air produces downbursts, whose lateral surface winds often precede by minutes or hours the release of precipitation from the cloud. The size of the area over which these winds are felt determines whether the event is classified as a macroburst or a microburst. In either case, the downward and outward flow of air from the storm cloud and the precipitation released results in the rapid dissipation of the cloud.

Microbursts

The special problems for aircraft posed by microbursts were studied by Fujita in the 1970s, largely motivated by the crash of Eastern Airlines flight 66 at Kennedy International Airport in June 1975. Fujita found that the wind patterns around the center of a microburst were responsible for the crash and showed that such winds pose a hazard for aircraft.

When the downdrafts from a microburst strike the ground, Fujita found, they spread outward and rapidly reach a maximum lateral velocity. As they spread outward, they curl around, flowing upward and then back toward the central downdraft. Shortly thereafter, a cushion of outflowing air at the base of the downdraft develops, pushing more air outward and increasing the speed at which it moves through the curls.

These conditions pose a severe hazard for pilots who fly though one of these curls. At first, the pilot experiences a headwind as he or she encounters the outer part of the curl. In attempting to compensate for this headwind, the pilot may then point the aircraft's nose downward. Within seconds, however, the aircraft has flow through the curl and has encountered winds in the opposite direction. Unless the pilot can compensate very quickly for this change in wind direction, the tailwinds now being experience will drive the aircraft into the ground. Crashes involving aircraft operated by Continental, Delta, Pan Am, United, and U.S. Airways airlines have all been attributed to microburst phenomena.

As a result of Fujita's research, meteorologists have a much better understanding of the atmospheric conditions that may cause microbursts, the visual signs of their impending arrival, and the warnings meteorologists can provide to pilots. *See also* Fujita, Tetsuya Theodore; Wind Shear

Further Reading

Caracena, Fernando, Ronald L. Holle, and Charles A. Doswell III, *Microbursts: A Handbook for Visual Identification,* 2nd edition. Boulder, CO: U.S. Department of Commerce, National Oceanic and Atmospheric Administration, Environmental Research Laboratories, National Severe Storms Laboratory, 1990. Also available online at http://www.nssl.noaa.gov/~doswell/microbursts/Handbook.html.

Dust Bowl

See Sandstorms and Dust Storms

Dust Devil

A dust devil is a small whirlwind with superficial characteristics similar to those of a tornado. Dust devils differ from tornadoes, however, in that they tend to be much smaller and less intense than their larger counterparts. A typical dust devil is only a few meters in diameter and no more than about 100 meters in height. Wind speeds in a dust devil seldom exceed about 100 kilometers per hour (60 miles per hour). They also tend to be short-lived, often exhausting themselves in a matter of minutes. By contrast, tornadoes are usually much larger, with diameters of up to 600 meters (2,000 feet) and wind speeds approaching 500 kilometers per hour (300 miles per hour). They may last more than three hours before dying out. Finally, dust devils are formed from the ground up, while tornadoes build from the sky down.

Dust devils are produced by the differential heating of two adjacent areas of ground, such as at the boundary between cultivated and uncultivated land. Ground that is darker and that, therefore, absorbs heat more readily will transfer heat to the air above it more quickly than will lighter ground. As the hot air rises, cool air around it flows in and under the warm air, pushing it upward. As the warm air rises, it begins to rotate because of the Coriolis effect. As more air flows inward, the rising warm air spins faster, much as a spinning skater rotates faster when his or her arms are drawn inward. Dust and other debris stirred up by the spinning air makes the dust devil visible. Most dust devils tend to form on the afternoons of calm, sunny days when the sun's heat is most intense.

Most dust devils are harmless, although the most severe forms have been known to damage tents, other light structures, and poorly made buildings. They may injure humans and other animals as the result of loose boards and other heavy objects carried into the air.

Dust devils have been observed on Mars since the first Viking orbiter missions of 1976–1980. Although their mode of formation appears to be the same as for those on Earth, Martian dust devils tend to be larger and longer lasting than those on Earth. The largest of a group observed by the Mars Global Surveyor in 1999, for example, was nearly 8 kilometers (5 miles) high, much taller than a typical tornado cloud on Earth. *See also* Tornado

Dust Storm

See Sandstorms and Dust Storms

E

Earhart, Amelia (1897–1937)

Amelia Earhart was one of the world's most famous women fliers, having set a number of records for airplane flights for women.

Earhart was born in Atchison, Kansas, on 24 July 1897. Her family moved frequently, and she eventually graduated from Hyde Park High School in Chicago in 1916, at the height of World War I. Rather than going directly to college, Earhart served as a volunteer nurse at Spadina Military Convalescent Hospital in Toronto from 1917 to the end of the war. She then enrolled in a premedical program at Columbia University, which she left after one semester. She then joined her parents in Los Angeles, where she took her first flying lesson with Frank Hawks, a famous exhibition flier. After just one lesson, she knew she wanted to become a pilot and began saving money for the $1,000 cost of flying lessons. Earhart completed those lessons and soloed in 1921. A year later, she had saved enough money to buy her own airplane, a Kinner Airster.

In 1924, Earhart moved to Boston with her mother and sister and took a job teaching at Denison House, where she taught English to immigrant children. During her stay in Boston, she gradually became more involved in flying. In 1928, for example, she was invited to be the first woman to cross the Atlantic (as a passenger). Four years later, she repeated that journey, this time piloting her own plane in a record crossing time of 13 hours, 30 minutes. She was only the second person (and first woman) to record this achievement. Over the years, Earhart set a number of records for women fliers, including the highest altitude reached (4,250 meters, or 14,000 feet) in 1922, the fastest speed, 291.85 kilometers per hour (181.18 miles per hour) in 1930, and the fastest nonstop transcontinental flight (Los Angeles to Newark, New Jersey, in 19 hours 5 minutes) in 1932. She later improved on some of these records.

In June 1937, Earhart and navigator Fred Noonan set out on what was to be their last trip, an attempt to fly around the world in a twin-engine Lockheed Electra. The first part of the flight went very well. The team traveled from Miami to South America, across the Atlantic Ocean and over central Africa, and then on to Thailand, Singapore, and Australia. On 2 July, however, Earhart sent out her last message. Apparently some problem had developed with her airplane and she was concerned about not being able to make her next stop, at Howland Island. The U.S. Navy began a search for Earhart's plane, but it was never discovered, nor was any further word from the two fliers ever received. The cause of her disappearance is one of the great mysteries in the history of aviation.

Further Reading

Butler, Susan. *East to the Dawn: The Life of Amelia Earhart.* Reading, MA: Addison-Wesley, 1997.

Goldstein, Donald M. *Amelia: The Centennial Biography of an Aviation Pioneer.* Washington, DC: Brassey's, 1997.

Rich, Doris L. *Amelia Earhart: A Biography.* Washington, DC: Smithsonian Institution, 1989.

Energy Budget (Earth's)

See Atmospheric Scattering and Absorption

Erosion by Wind

Wind erosion is caused by the movement of air across the ground, resulting in the breakup and removal of soil particles, which is generally followed by the deposition of those particles in other locations.

The extent to which wind erosion occurs depends on two factors: the speed with which the wind blows and the character of the ground over which it passes. In most cases, the wind has little or no effect on land that is covered with vegetation, bare rock, or earthy particles with diameters of about 4 millimeters or more. Exceptions to this general statement may occur in the case of severe windstorms, such as hurricanes and tornadoes.

Because vegetation and soil moisture are highly effective in preventing the loss of soil particles during windstorms, wind erosion tends to occur only in specialized geographic areas, such as deserts and other arid regions and in places where human activity has resulted in the loss of natural vegetation. For example, land that has been continually cultivated, but on which no crops are currently being grown, is a likely candidate for erosion by wind.

Wind erosion on Kaibito Plateau, near Tuba, Arizona. Coconino County, Echo Cliffs quadrangle, July 1913. © U.S. Geological Society.

Wind erosion, Kaibito Plateau, Coconino County, Echo Cliffs quadrangle, August 1914. © U.S. Geological Society.

Mechanisms of Wind Erosion

Moving air can erode the ground in one of two ways: by degrading rocky material and by carrying away loose soil. In theory, the force of the wind against rocky material may be sufficient to break apart the material or to tear off loose pieces of rock. In practice, the wind seldom blows strongly enough to produce such an effect. More commonly, small pieces of rock picked up and carried by the wind may collide with larger pieces of rock and abrade off or split apart the rock. Generally speaking, however, wind is a very ineffective force in the actual degradation of rocky material.

Wind is a more efficient mechanism for transporting small pieces of rock that have already been abraded by other mechanisms, such as the action of running water. Transport by moving air may carry away particles of rock by one of three mechanisms: suspension, saltation, or surface creep. *Suspen-sion* occurs when winds pick up particles smaller than about 0.2 millimeters and carry them upward into the atmosphere. These particles commonly do not settle back to the Earth's surface very readily and may be carried tens, hundreds, or thousands of kilometers before they are once more deposited. In some cases, they may remain as a part of the atmosphere for years. About 20 percent of all soil removed by winds is carried away by the process of suspension.

Saltation occurs when the wind is strong enough to pick up small particles with diameters of about 0.2 to 2 millimeters. These particles are typically raised a short distance into the air, after which they settle back to Earth's surface along a low, sloping path that seldom exceeds about 2 meters (6 feet) in height. During saltation, particles are seldom transported more than a few meters from their original position. The process is thought to be responsible for up to 80 percent of all wind erosion that occurs.

Rocky Mountain National Park, Colorado. Timberline (limber pine) tree on Longs Peak, July 21, 1916. Its shape is a result of prevailing westerly wind, which sweeps snow from the highland of the eastern slopes. © U.S. Geological Society.

Surface creep occurs when the wind pushes against particles larger than about 2 millimeters, causing them to slide or roll across the ground. Again, the process moves particles no more than a few centimeters at a time from their original position.

Deposition of Particles

When the speed with which air moves decreases, it loses its ability to carry particles and begins to deposit them back on the Earth's surface. This process of deposition may result in the formation of distinctive topographical features.

In a perfectly homogeneous and uniform landscape, the only effect of wind erosion would be to move solid particles from one place to another, determined by the direction from which the wind was blowing. But no surface is entirely homogeneous or uniform. Vegetation, ridges, grooves, buildings, and other surface irregularities interrupt the flow of air, increasing or decreasing the speed and distance with which it transports particles. This effect can be seen with special clarity on deserts or along beaches, where the particles deposited as a result of reduced wind speed form ripples or dunes.

Both ripples and dunes are formed when some surface irregularity causes particles to be deposited abruptly, forming small mounds on the Earth's surface. These mounds, in turn, become addition irregularities that cause further deposition, causing the ripple or dune to begin building up. Sand dunes along beaches can be particularly dramatic examples of this process, often reaching to the height of a few tens of meters. Over time, the action of winds pick up sand from the windward side of the dune and drop it on the lee side, producing an effect of the dune's "walking" across the beach over time.

The Wind Erosion Equation

Scientists have understood the general principles of wind erosion for over 70 years. But the earliest attempt to quantify the effects of moving air on rocky particles date only to the mid-1950s. During that period, William S. Chepil published the earliest version of the wind erosion equation (WEE), derived while he was working at the Swift Current Research Station in Swift Current, Saskatchewan. The equation has been revised a number of times and is now a complex mathematical function that describes the amount of erosion that will occur over a period of time as a function of the soil erodibility index (that is, the tendency of the soil to be eroded), the roughness of the ground's surface, the climatic characteristics of the area, the distance across which the wind blows, and the amount of vegetation in the area. The equation does not take wind velocity directly into effect since, except for severe storms, the amount of erosion that occurs in a given region is influenced to a much greater degree by the above factors. Today, computer programs are available for calculating the value of the amount of erosion likely to occur, given any specific set of environmental factors.

Effects and Prevention

Wind erosion is a very severe problem in some parts of the world and some parts of the United States, and a negligible problem in other parts. Northwest China, most parts of Australia, the Siberian plains, southern portions of South America, and selected regions in North America are some of the regions most likely to be affected by wind erosion. In the United States, the Great Plains and the eastern seaboard, from New York to the northern part of Florida, are especially sensitive to wind erosion.

Estimates of the amount of soil lost to wind erosion worldwide are difficult to come by. Conditions often change from year to year and region to region, even within a particular geographical area. By one estimate, wind erosion may be responsible for the loss of 28 million hectares (69 million acres) of land annually in the United States ("The Problem of Wind Erosion," http://www.weru.ksu.edu/problem.html).

Soil and agricultural scientists recommend a number of steps landowners can take to reduce the effects of wind erosion. These steps include the following:

- retain as much vegetative cover on land as possible;
- spread crop or animal residues;
- till the land in such a way as to provide irregular surfaces;
- provide irrigation to increase soil moisture;
- plant or construct wind barriers; and
- use additives to increase soil compactness or spray the soil with materials that increase compactness.

Further Reading

Pye, K., and N. Lancaster. *Aeolian Sediments: Ancient and Modern.* Oxford, UK: Blackwell Scientific Publications, 1993.

Shao, Yaping. *Physics and Modeling of Wind Erosion.* Dordrecht, the Netherlands: Kluwer Academic Publishers, 2000.

F

Ferrel, William (1817–1891)

William Ferrel is best known for his contributions to the field of geophysical fluid dynamics, the behavior of gases and liquids on and above the Earth's surface. He first proposed the three-cell model of the global circulation of air in the mid-1850s. In recognition of this accomplishment, the central of those three cells in today's theory of global circulation is named the Ferrel cell. Ferrel also made important contributions to the understanding of tidal patterns.

Ferrel was born on 29 January 1817 in rural Fulton County, Pennsylvania, the eldest of six boys and two girls. His early education was limited to a few years of schooling during the winter, when there were no crops to tend. He was, however, intensely interested in natural and scientific phenomena and read every book on the subjects that came to his attention.

By 1839, Ferrel had saved enough money to enroll in Marshall College (now Franklin and Marshall College) in Mercersburg, Pennsylvania. He later transferred to the newly established Bethany College in Bethany, West Virginia, from which he graduated in the college's first graduating class in 1844.

Ferrel took a teaching job in Liberty, Missouri, where he remained for six years. He then accepted a similar position in Allensville, Kentucky, where he remained until 1854. In that year, he moved to Nashville, Tennessee, and opened his own school.

While still a teenager, Ferrel had become interested in celestial phenomena and tried to work out for himself the methods by which eclipses could be predicted. As he moved to larger cities, he gained access to basic texts in physics, such as Isaac Newton's *Principia* and Pierre-Simon Laplace's *Mécanique céleste*. In 1853, he published his first scientific paper, an analysis of Laplace's explanation of tidal behavior. Ferrel pointed out an error that he found in the great French physicist's work in ignoring second-order effects.

Two years later, Ferrel published the first in a series of papers for which he is probably most famous, "Essay on the Winds and Currents of the Ocean," in the *Nashville Journal of Medicine and Surgery.* In this and succeeding papers, Ferrel outlined the mathematical basis for what has since become known as Ferrel's Law: "That if a body is moving in any direction, there is a force arising from the earth's rotation, which always deflects it to the right in the northern hemisphere, and to the left in the southern."

Ferrel's work came to the attention of Benjamin Apthorp Gould, editor of the

Astronomical Journal, in which an early Ferrel paper had appeared. Gould arranged an appointment for Ferrel on the staff of the *American Ephemeris and Nautical Almanac* in Cambridge, Massachusetts, where he worked from 1858 to 1867. He was then asked by Benjamin Peirce to join him at the U.S. Coast Survey, of which Peirce had just been appointed superintendent. Ferrel's assignment at the Coast Survey was to work on a general theory of the tides, a subject of considerable interest to ships of all kinds.

In 1880, Ferrel submitted a report to the superintendent of the Coast and Geodetic Survey on a tide-predicting machine that he had invented. The machine was finally built and put into operation in late 1882. It was first used to produce tide tables for 1885, a task that had formerly required the work of 40 computing devices working together. In recognition of this achievement, the National Oceanic and Atmospheric Administration later named one of its research ships in his honor.

In 1882, Ferrel left the Coast and Geodetic Survey to join the U.S. Army's Signal Service (predecessor of the Weather Bureau). He remained with the Signal Service until his retirement in 1886. He died on 18 September 1891, in Maywood, Kansas. *See also* Global Winds

Flying Animals

Many members of the animal kingdom have evolved the ability to fly. In general, flying animals use one of two general methods of transport: gliding or parachuting on the one hand, and true, powered flight, on the other. Perhaps the best-known flying animals are the members of the class *Aves* (birds), of whom almost all are true fliers. But certain members of other animals classes are also capable of some form of flight. These include the bat, the flying squirrel, and the flying lemur (all members of the class *Mam-*

malia); the flying fish (of the order *Beloniformes*); the flying frog (of the class *Lissamphibia*); the gliding lizard and gliding snake (of the class *Reptilia*); and many members of the class *Insecta*.

The Fossil Record

Perhaps the best represented class of flying animals preserved in the fossil record is the insects. Evidence of a very large number of flying insects of many varieties has been found. In some cases, flying insects are found complete. In other cases, only wings or other body parts remain. In still other cases, other types of evidence, such as nests, have been discovered.

Some of the most interesting fossil remains of flying animals are those of the first vertebrates to achieve true flight, the *Pterosauria* (pterosaurs). The pterosaurs are thought to have evolved from running dinosaurs about 225 million years ago. They are, therefore, members of the class *Reptilia*, consisting of two suborders, the more primitive *Rhamophorhynchoidea*, who survived for about 100 million years, and the more evolved *Pterodactyloidea*, who died out only about 65 million years ago.

A primary difference between *Rhamophorhynchoidea* and *Pterodactyloidea* was that the former had long tails and short wings, while the latter had short tails and long wings. *Rhamophorhynchoidea* was also a much smaller animal, usually no more than about 60 centimeters (2 feet) long, while *Pterodactyloidea* grew as large as about 15 meters (50 feet).

Pterosaurs had many of the same morphological characteristics found in birds today, contributing to their ability to fly. For example, they had a special body structure known as a keeled sternum to which were attached flight muscles and a short, strong arm bone, to which the wings themselves were attached. Both skull and limb bones were hollow, as are those of birds, reducing the weight that has to be supported during flight. Pterosaurs did not, however, had feathers, until recently and important defining characteristic of birds.

The pterosaur's wing consisted of a very thin membrane attached to a very long fourth finger on each front foot. Its first three fingers were very small digits attached to the base of the wing. The few complete skeletons available in the fossil record suggest that pterosaurs may have been able to walk, although scientists believe that they may have spent most of their time in the air or on water.

The earliest member of the class *Aves* (birds) to appear in the fossil record is archaeopteryx. Of the seven specimens of archaeopteryx that have been found, the earliest dates to about 150 millions. Biologists believe that the animal may have evolved from bipedal dinosaurs that lived on the land. Unlike modern birds, however, archaeopteryx had a long tail, a full set of teeth, a relatively flat sternum, and three claws on each wing.

Gliding Flight

A number of animal species have evolved body structures that allow them to glide or parachute from one tree or bush to another or to glide to earth from a tree. One of the most successful animal groups in this regard consists of about a dozen genera of the family *Sciuridae* (consisting of the squirrels, chipmunks, woodchucks, and prairie dogs). The members of these genera have a flap of loose skin connecting their front and hind legs. To get from one place to another, the squirrel launches itself into the air and spreads out its legs. The flap of skin then acts as a parachute, or the wings of a glider, allowing it to float safely to its destination. Two New World representatives of the flying squirrel include the species, *Glaucomys volans* and *Glaucomys sabrinus.* These animals are usually about 20 cen-

timeters (8 inches) in length (not including a tail of almost equal length), capable of flying up to about 50 meters (150 feet). Other genera of flying squirrels are found throughout the world, from northern Europe, across Asia, to some South Pacific Islands.

Another mechanism that is used for flying can be observed in Wallace's flying frog (*Rhacophorus nigropalmatus*), found in Malaysia and Borneo. This frog has evolved very long webbed toes with specialized pads that allow them to cling to trees. When they jump from a tree, they spread out their toes, and the webs act as a parachute, allowing the frog to glide to the ground.

Flying fish are members of the family *Exocoetidae.* They have enlarged pectoral fins and elongated tail fins that act like wings and allow them to float in the air when they thrust themselves out of the water. They tend to be found in tropical waters, although one species, *Cypselurus heterurus,* is found in the Atlantic Ocean, especially in the Gulf Stream.

One of the most unusual forms of gliding observed in the animal world is used by five species of snakes who are members of the genus *Chrysopelea,* whose habitat is Southeast Asia and Sri Lanka. These animals have no appendages or other adaptations that allow them to float through the air. Instead, they "fly" by flattening their bodies and undulating back and forth as they fall to the ground. Scientists are able to describe in detail how the snake moves its body to remain in air, but they have as yet been unable to explain the aerodynamics by which this process works.

True Flight

The only mammals to have evolved the ability to fly under their own power are the bats, members of the order *Chiroptera.* Excepting the rodents, the bats make up the largest and most diverse group of mammals. More than 1,000 species of bats exist in virtually every part of the world. The ability of bats to fly results from an evolutionary adaptation in which their second, third, fourth, and fifth fingers have become greatly elongated and attached to each other by means of a thin membrane. The second and third fingers act as the leading edge of the bat's wing, while the fourth and fifth fingers provide points to which the membrane is attached at the rear of the wing. The membrane acts as a wing that provides lift as the bat sails through the air.

When bats fly, they use strong muscles attached to their wings to push downward into the air, forcing it back over their wings. The action is a bit like the rowing stroke than one uses to propel a boat through the water. The lift provided by air flying over a bat's wings raises the animal into the air, in a manner similar to that by which airplane wings help an aircraft fly. Unlike airplane wings, however, a bat's wings can be moved and adjusted to obtain maximum lift from the air that flows over them. In many cases, bats fold their wings inward during the upstroke of their flight, reducing air friction when power is not being produced.

The classic fliers in the animal world are the birds, members of the class *Aves*. According to ornithologists, there are about 9,500 species of birds in the world today. The fossil record indicates, however, that as many as 150,000 species may have existed at one time or another in Earth history.

The anatomical structure and flight pattern of various bird species differ in many respects. However, all birds have some common structures in common that make it possible for them to fly with greater or lesser ease. For example, they are covered with feathers of different types, each with its own function for flight efficiency. Many birds also have hollow bones, which gives them a rigid structure with minimal weight. Bird bodies are also aerodynamically efficient,

with a pointed nose or beak and slim, smooth body lines and tails that can be used for maneuvering.

When a bird takes off, it propels itself forward and upward by pushing air downward with its wings. To achieve this effect, most birds use primary feathers that can be spread out, providing a wide surface area with which to push against the air. In this configuration, the bird's wings have a shape much like that of an airplane wing. As the bird pushes downward, its feathers act like the propeller of an airplane, pushing air backward across its wings and its body.

On the return upward stroke of its wings, a bird rotates the position of its primary feathers, orienting them at right angles to the wind. In this position, air is able to flow between the feathers, offering relatively little resistance to the wind. When the wings have returned to their original position, feathers return once more to their flattened shape, allowing the next round of push against the air.

Once in flight, the bird obtains lift in much the way that an airplane does. Air flows over the curved upper surface of its wings more rapidly than it does over the flatter lower surface. According to Bernoulli's principle, the more rapid flow of air over the upper surface results in a decreased pressure there, allowing normal air pressure on the bottom of the wing to push the bird upward.

A major difference between bird wings and airplane wings is that the former are much more flexible than the latter. They are able not only to push downward against the air, but they are also able to change the shape of their wings to obtain the best possible position to collect and move air over the bird's body. For example, the angle of the wing compared to its direction of flight is often much greater at the wing tip than closer to the bird's body. This kind of wing flexibility increases the efficiency of the bird's flight and allows the animal to make sharp turns and rapid upward and downward flight to pursue prey or to avoid being taken by a predator.

Wing shapes differ according to the type of flight required by a bird. Some ocean birds, who spend most of their time gliding, have long, narrow wings. Birds that require rapid takeoffs tend to have shorter, thicker, more powerful wings. The rate at which a bird moves its wings also differs widely among various species. Gulls and herons, for example, may have wing beats of no more than two or three per second during flight, while some hummingbirds achieve wing beats of up to 80 per second.

Further Reading

Burton, Robert. *Birdflight: An Illustrated Study of Birds' Aerial Mastery.* New York: Facts on File, 1990.

Leister, Mary. *Flying Fur, Fin and Scale: Strange Animals That Swoop and Soar.* Owings Mills, MD: Stemmer House, 1977.

Foundation for Clean Air Progress

The Foundation for Clean Air Progress was formed in 1995 for the purpose of informing and educating the general public about air quality issues. It is a nonprofit, nonpartisan organizations whose members represent various sectors of business and industry, including energy, transportation, farming, tourism, and manufacturing. The organization does not engage in lobbying or other advocacy activities, but concentrates on the development of reports on clean air issues. Some of these reports are "Breathing Easier about Energy: A Healthy Economy and Healthier Air," "Profiles of Local Clean Air Innovation Empowering Communities to Meet the Air Quality Challenges of the 21st Century," "Progress in Reducing Ozone Excedance Days in Ten Major U.S. Cities

(1987–1999)," and "Meteorologists Views on Air Quality." *See also* Air Pollution; Clean Air Acts

Further Information

Foundation for Clean Air Progress
1801 K Street, N.W., Suite 1000L
Washington, DC 20036
Telephone: (800) 272-1604
URL: http://www.cleanairprogress.org
E-mail: info@cleanairprogress.org

Fujita, Tetsuya Theodore (1920–1998)

Tetsuya Theodore ("Ted") Fujita was a meteorologist who discovered the phenomenon known as *microbursts* and who developed a standard for measuring the strength of tornadoes. Microbursts are sudden severe downdrafts that occur over relatively limited areas during thunderstorms. They may produce wind speeds in excess of 250 kilometers per hour (150 miles per hour) for relatively brief periods. Fujita developed his theory about microbursts in 1975 while studying an otherwise unexplained crash of Eastern Airlines flight 66 at New York's Kennedy airport. Fujita's work made it possible for meteorologists to develop more sophisticated methods for monitoring the appearance of microbursts near airports, preventing future such disasters. Because of his research on severe storms, Fujita was often known as "Mr. Tornado."

Fujita was born in Sone town, now part of Kitakyushu City on the island of Kyushu, Japan, on 23 October 1920. He attended Meiji College of Technology (now the Kyushu Institute of Technology) in Kyushu, from which he received his bachelor's degree in mechanical engineering in 1943. He was then appointed assistant professor of Physics at Meiji.

At the conclusion of World War II, Fujita began a study of damage caused by

Fujita Tornado Scale

F-Scale Number	Intensity Phrase	Wind Speed	Type of Damage Done
F0	Gale tornado	64–116 kph 40–72 mph 35–62 knots	Some damage to chimneys; branches broken off trees; shallow-rooted trees uprooted; sign boards damaged.
F1	Moderate tornado	117–180 kph 73–112 mph 63–97 knots	Lower limit of hurricane winds; peels surface off roofs; mobile homes pushed off foundations or overturned; moving cars pushed off roads; attached garages may be destroyed.
F2	Significant tornado	181–253 kph 113–157 mph 98–136 knots	Considerable damage; roofs torn off frame houses; mobile homes demolished; boxcars turned over; large trees broken and uprooted; light objects turned into missiles.
F3	Severe tornado	254–332 kph 158–206 mph 137–179 knots	Roofs and some walls torn off well-built houses; trains overturned; most trees in forests uprooted.
F4	Devastating tornado	333–412 kph 207–260 mph 180–226 knots	Well-built houses leveled; structures with weak foundations blown some distance; cars thrown; heavy objects turned into missiles.
F5	Incredible tornado	420–512 kph 261–318 mph 227–276 knots	Strong frame houses lifted off foundations and carried considerable distances and destroyed; car-size objects turned into missiles and thrown through the air in excess of 100 meters; bark torn off trees; steel-reinforced concrete structures badly damaged.
F6	Inconceivable tornado	513–610 kph 319–379 mph 277–329 knots	Winds of this intensity highly unlikely.*

*The small area of damage that winds might produce would probably not be recognizable among the mess produced by F4 and F5 winds surrounding the F6 winds. Missiles such as cars and refrigerators would do serious secondary damage that could not be directly identified as F6 damage. If this level is ever achieved, evidence for it might be found only in some manner of ground swirl pattern, for it may never be identifiable through engineering studies.

American atomic bombs in Hiroshima and Nagasaki. He then moved on to research on volcanic eruptions and the effects of severe thunderstorms. In 1951, Fujita enrolled in a doctoral program at the University of Tokyo and, two years later, received his D.Sc. degree for a study on typhoons. In the same year, he joined the University of Chicago as a research associate in meteorology. In 1956, he became a research professor and senior meteorologist at Chicago, where he also established the Severe Local Storms Projects (SLSP). The primary focus of this project was the study of tornadoes and other types of severe storms. SLSP was later reorganized as the Mesometeorology Research Project (MRP) in 1961, and again as the Satellite and Mesometeorology Research Project (SMRP) in 1964, and finally as the Wind Research Laboratory (WRL) in 1988.

Fujita's health began to deteriorate in 1995, but he continued to work out of his home until late 1998. He died in his sleep in Chicago on 19 November 1998. The WRL was closed on 1 October 1999, nearly a year after Fujita's death. *See also* Downburst Phenomena; Fujita Tornado Scale

Further Reading

"Tetsuya Theodore Fujita (1920–1998)," http://www.msu.edu/~fujita/tornado/ttfu-jita/biography.html.

Fujita Tornado Scale

The Fujita Tornado Scale is a system for designating the intensity of tornadoes developed by University of Chicago meteorologist Tetsuya Theodore ("Ted") Fujita and his wife, Sumiko, in 1971. Each tornado is assigned an "F-Scale Number," according to the system shown in the table.

Tornadoes rated F0 and F1 are generally referred to as "weak" tornadoes. Those rated F2 and F3 are called "strong," while those rated F4 and F5 are called "violent." About 74 percent of all tornadoes fall into the "weak" category, about 25 percent are "strong," and only about 1 percent are "violent." Even though they constitute the smallest fraction of tornadoes, violent tornadoes cause the worst damage. For example, between 1950 and 1999, violent tornadoes caused 67 percent of all deaths from tornadoes (2,988 deaths), while 1,293 died in strong tornadoes, and 179 died in weak tornadoes. *See also* Fujita, Tetsuya Theodore; Tornado; Wind

Further Information

"Fujita Tornado Damage Scale," http://www.spc.noaa.gov/faq/tornado/f-scale.html.

G

Gases and Airs

Modern scientists recognize that air is a mixture of many gases and that the many discrete gases that occur in air and elsewhere are distinctly different from air itself. That understanding evolved slowly over a period of two centuries between the early 1600s and the late 1700s.

Early Beliefs

Early scientists saw no difference between "air" and "gas" for a number of reasons. First, as late as the seventeenth century, most scientists, like their predecessors from ancient Greece, still believed air to be an element. Differences that they observed in different samples of air or different kinds of air they attributed to impurities in the air. Second, studies by the English scientist Robert Boyle (1627–1691) showed that all kinds of "air" followed a common pattern when subjected to changes in pressure. Boyle's Law, as it is now known, appeared to confirm that all forms of air were, in fact, fundamentally alike.

Third, methods for collecting and studying gaseous substances were now well developed until the mid-1700s. The term *gas* itself was not even introduced until the early 1600s by Johann Baptista van Helmont (1577–1644), who suggested the term

in frustration over the wild behavior of gases he worked with. Those gases were sometime so uncontrollable that they damaged or even destroyed the equipment with which he was working, suggesting to van Helmont the term *gas,* from the Greek word for "chaos."

Defining Gases

Van Helmont was among the first scientists to recognize the differences between atmospheric air and individual gases. This step was made possible at least in part because van Helmont had rejected the traditional view that nature was created basically from four elements: earth, air, fire, and water. In his research, van Helmont found that a certain type of gas (which he called *gas sylvestre*, "gas from wood"), was produced during fermentation, by the action of acids on seashells, and as the result of natural changes that take place within caves. He also discovered a second type of gas, which he named *gas pingue* (now known as methane), produced when organic matter was heated and during the decay of dead plants and animals.

Van Helmont's work was hampered by the fact that both gas sylvestre and gas pingue were, inn fact, mixtures of gases, rather than pure gases themselves. Far more important, however, was his recognition that gaseous substances distinct from air can be produced in a variety of ways. The first cracks in the long tradition of "air as an element" had begun to appear.

A second breakthrough in the study of gases resulted from the work of the English biologist Stephen Hales (1677–1761). Hales was interested in quantitative changes that take place in living organisms. For example, he attempted to measure the amount of "air" that plants take on and give off through their leaves. To conduct such studies, he invented one of the first (perhaps *the* first) pneumatic troughs, a device for collecting gas over water.

Hale's pneumatic trough consisted of a bent gun barrel sealed at one end. The substance to be heated or otherwise treated was placed in the closed end of the gun barrel, and the open end was placed beneath a jar filled with water supported in a larger tub of water. Gases produced within the gun barrel escaped from the open end and were captured in the jar, after which they could be studied. The pneumatic trough so familiar to high school and college students of chemistry has essentially the same design as that of Hales's original equipment.

Discovery of Gases

As valuable as was the invention of the pneumatic trough, Hale contributed little to the understanding of gases themselves. He was interested in the quantities of gases involved in his research, not in their chemical identity.

Progress in that direction was not to occur until a quarter century later with the work of the Scottish scientist Joseph Black (1728–1799). As a graduate student at the University of Edinburgh, Black became interested in the reaction between acids and the mineral known as *magnesia alba* (magnesium carbonate: $MgCO_3$). Black found that this reaction resulted in the formation of a type of "air" which he called *fixed air.* He adopted that name because the gas is,

under some circumstances, "fixed" or "contained within" some other substance (*magnesia alba*) before being released by a chemical reaction.

Black went on to study the properties of fixed air (which we now called carbon dioxide, CO_2) and its preparation from other rocks and minerals, such as limestone (calcium carbonate; $CaCO_3$). He eventually worked out the process by which carbonates and their oxides are interconvertible, as with limestone, for example:

$$CaCO_3 \rightarrow CaO + CO_2$$
limestone   lime   fixed air (carbon dioxide)

Finally, and of special importance, Black showed that fixed air is present in, but is different and distinct from, atmospheric air. As part of that demonstration, he weighed a given sample of air and showed that its density was significantly different from that of atmospheric air.

During the last quarter of the eighteenth century, the distinction between "air" and "gas" was finally becoming clear as additional discrete gases with unique characteristic properties were identified: hydrogen by Henry Cavendish (1731–1810) in 1766, oxygen independently by Karl Wilhelm Scheele (1742–1786) and Joseph Priestley (1733–1804) in 1771 and 1774, respectively, and nitrogen by Daniel Rutherford (1749–1819) in 1772. Finally, the last important step in distinguishing between "air" and "gas" was taken by the French chemist, Antoine Laurent Lavoisier (1743–1794), who recognized that gases made up the third state of matter (along with solids and liquids). *See also* Air as an Element

Further Reading

Ihde, Aaron J. *The Development of Modern Chemistry,* 3rd edition. New York: Harper & Row, 1970, Chapter Two, "Pneumatic Chemistry."

Lilienthal glider in flight, 1895. © Library of Congress.

Geostrophic Wind

See Wind

Gliding

Gliding is a method of flying a heavier-than-air aircraft that operates without an engine. Gliding is also called *soaring.*

Early History

Humans have probably always been intrigued with the process by which birds and other flying animals have traveled through the air and have tried to imagine ways in which they could replicate this behavior. Many stories exist of attempts by humans to construct winglike appendages with which they might be able to fly, always without success. Some of the earliest designs for gliders were those of the great Italian inventor Leonardo da Vinci (1452–1519), whose sketchbooks contain many designs for such devices. All of them show some type of birdlike wing attached to a human body, with the ability to control the movement of the wings or a tail. Although there is still some dispute about the question, it seems unlikely that da Vinci ever built or tested any of his designs.

The machines designed by da Vinci are known today as *hang gliders;* that is, they consist of a winglike structure from which a person is suspended in a horizontal position. The person maneuvers the glider by moving his or her body and pulling on ropes that control the orientation of the glider wings. Hang gliding has become a popular sport in many parts of the world.

Cayley and Lilienthal

Virtually no progress was made in the design and construction of gliders for nearly 400 years after da Vinci. Then, in the early 1800s, English inventor George Cayley (1773–1857) designed and built the first working glider of which we know. His first design, completed in 1799, consisted of a clam-shaped biplane made of cloth sails with a horizontal tail and two lateral fins at the base of the tail. The pilot sat inside the sails, from which he maneuvered the aircraft. Over the next 50 years, Cayley tried out a number of designs, concluding with a triplane design that carried a man about 275 meters (900 feet).

In spite of Cayley's successes, glider development came to a halt for another half century. Then, in the late nineteenth century, the German engineer Otto Lilienthal (1848–1896) made a number of breakthroughs in the design of gliding machines. Unlike most of his predecessors, Lilienthal made a careful study of the flight of birds and the principles of aerodynamics, determined to base his glider designs on the best information available on the way objects fly through the air. His 1889 book, *Bird-flight as the Basis of Aviation,* eventually became required reading for other aeronautical pioneers. In his book, he maintained that the most successful approach to human flight was to find a way of duplicating the flapping action of a bird's wings, a concept that he incorporated in all of his glider designs.

Over a ten-year period, Lilienthal designed and built 18 types of gliders, all of which he tested himself. In 1894, he constructed an artificial hill 50 meters (150 feet) high from which to launch his glider flights. He also used natural hills in the region around Berlin for his tests. He was eventually successful in flying his gliders distances of more than 200 meters (750 feet). His career came to a tragic end in August 1896 when a glider he was testing was caught in a gust of wind, dropping him to the ground and killing him.

Among the many aviation researchers influenced by Lilienthal's work was the French-American engineer Octave Chanute (1832–1910), who was consulted by the Wright brothers about their plans for a heavier-than-air machine. The Wrights' earliest successful flights, in fact, were made in glider-type machines that owed their design to contributions of Lilienthal, Chanute, and other aeronautical inventors from many countries. Ironically, the success of the Wright brothers led to a massive interest in the development of heavier-than-air engine-powered aircraft and an end, at least for a while, of interest in glider research.

Following World War I, glider flying became a popular recreational activity in many countries. The 1930s saw the formation of a number of glider flying clubs in Great Britain and other countries of Western Europe. Some research was conducted in various countries on the use of gliders for military purposes, but such research was limited and not very fruitful. The only nation that pursued this research with much vigor was Germany, which was prevented by the Treaty of Versailles from developing a military air force. The Germans built a number of gliders in the postwar period through civilian air clubs and used the fly-ing machines for training pilots and developing methods of moving troops. When World War II began, the Germans made limited use of gliders in delivering troops to designated battlefronts. On 20 May 1941, gliders carried 23,000 troops to Crete for an assault on that island. So many loses were sustained, however, that the Germans essentially curtailed glider operations for the remainder of the war.

After World War II, there was a renewed interest in gliding in most developed nations of the world. By the 1950s, gliding clubs had grown up around the world and national and international competitions were being held. World soaring championships, first held in 1937 and 1948, had become annual affairs by the early 1950s. Today, world championships are held in 11 gliding categories.

Principles of Glider Flight

Since gliders have no engines, they must rely on natural forces to power their flight. These natural forces are of three general types: thermal lift, ridge lift, and wave lift. *Thermal lift* is provided when a mass of air is warmed and rises upward in the atmosphere. The objective of the glider pilot is to find such thermals and use them to lift the glider upward. The challenge is for a pilot to recognize where thermals are located and to guide the aircraft in such a way as to stay within them.

Ridge lift is produced when winds blow against the side of a mountain or mountain range and are deflected upward into the atmosphere. The most important factor in determining the amount of ridge lift is the slope of the mountain. The greater the slope, the greater the deflection of the winds and the faster the air moves upward.

A comparable lifting force sometimes occurs on the lee side of a mountain when air flows over the top of the mountain and comes into contact with a stable air mass on its opposite side. In such a case, the moving air will be deflected upward, producing a force known as a *wave lift*. The name comes from the fact that the air deflected upward flows backward toward the mountain, down the leeward side of the mountain, and back again into the stable air mass. The air is, therefore, converted into a standing wave that may exist for many cycles before it dissipates.

Launching a Glider

Lacking an engine, gliders need some external mechanism for getting them into the air, where they can find a thermal, ridge, or wave lift to continue their flight. These mechanisms can be classified as a ground launch or aerotow. Two types of ground launch are commonly used. In one, a large winch is built into one end of a runway. The winch holds several thousand feet of cable or rubberized rope, the end of which is attached to the glider. At a signal, the winch begins reeling in the cable or rope at speeds of up to 150 kilometers per hour (90 miles per hour). Wind passing over the wings of the glider provides the lifting force needed to get it off the ground and more than 100 meters into the air, at which point the glider begins to fly on its own.

Ground launches can also be carried out by having a car or truck tow the glider down the runway until it reaches sufficient speed to take off. In either form of ground launch, a release mechanism allows the pilot to drop the cable or rope when takeoff speed has been attained. Aerotowing involves the use of a conventional airplane to tow the glider until it gets airborne. When the pilot has reached the desired altitude, the tow rope is dropped and the glider flies off on its own.

Modern gliders have achieved some remarkable records. Some have been piloted to heights of nearly 15,000 meters (50,000 feet), and the long distance record for glider flight is 2,463 kilometers (1,530 miles), set by two German glider pilots in November 2000. World records speeds for gliders are in the 100 kilometers per hour (60 miles per hour) range, although those speeds have been exceeded for brief periods in a run. *See also* Aerial Sports

Further Information

The Soaring Society of America, Inc.
P.O. Box E
Hobbs, NM 88241-7504
Telephone: (505) 392-1177
Fax: (505) 392-8154
URL: http://www.ssa.org
E-mail: dennis@ssa.org

Further Reading

Piggott, Derek. *Understanding Gliding: The Principles of Soaring Flight.* London: A. and C. Black, 1977.
Schweizer, Paul A. *Wings Like an Eagle: The Story of Soaring in the United States.* Washington, DC: Smithsonian Institution, 1988.
Welch, Ann Courtenay Edmonds. *The Story of Gliding,* 2nd edition. London: J. Murray, 1993.
Whelan, Robert F. *Cloud Dancing: Your Introduction to Gliding and Motorless Flight.* Highland City, FL: Rainbow Books, 1996.

Global Winds

Global winds are winds that blow over large parts of the planet for extended periods of time, such as many months or years. The term *global circulation* is often used to describe these winds and the patterns they follow.

History

The earliest widespread interest in global circulation patterns in the Western World can be traced to the late fifteenth century when explorers such as Christopher Columbus (1451–1506) learned how to use prevailing winds to take them back and forth across the Atlantic Ocean. Columbus's success in reaching the New World in slightly over a month, in fact, was due to his guess that prevailing wind patterns blowing off the European coast would carry him to his desired destination, China. Although he failed to reach China, he made perhaps an even more important contribution to history. He proved that large-scale wind patterns did exist across the Atlantic and that those wind patterns could be used to take explorers back and forth between Europe and the New World.

The Halley Model

It was nearly two centuries more before anyone attempted a scientific explanation for such large-scale winds. In 1686, the English astronomer Edmond Halley (1656–1742) proposed the first such theory about the global circulation of air. Halley began with the assumption that the air above the equator receives more direct sunlight for a longer time than at any other place on Earth. As a result, this air has a tendency to expand and rise into the upper atmosphere. By contrast, air over the poles receives relatively little sunlight for much shorter periods of time. Its tendency is to contract and sink to the Earth's surface.

With air rising above the equator and settling at the poles, a giant cell of moving air forms. Air rises at the equator, Halley said, migrates northward and southward to the two poles in the upper atmosphere, sinks to Earth at the poles, and travels back to the equator along the Earth's surface.

Halley's theory also attempted to explain the best-known characteristic of global air circulation: prevailing winds. In certain parts of the world, winds tend to blow strongly in one direction for long periods. Probably the best known of these prevailing winds at the time were the *trade winds,* which blow from east to west in the regions of about 30° both above and below the equator. The trade winds are caused, Halley said, because air in the massive planetary cell is pulled toward the west by the sun's rays.

The Hadley Model

Halley's theory was correct only in its most general form. A more accurate explanation of global circulation patterns was offered a half century later in 1735 by the English lawyer and amateur meteorologist George Hadley (1685–1786). Hadley agreed in general with Halley's theory of a single massive cell consisting of air flowing from equator to pole and back again. However, he had a different explanation for the east-west pattern of the prevailing winds.

The key factor in producing this effect, Hadley said, was the Earth's rotation on its axis. As the Earth rotates, so does the air above its surface. Both Earth and atmosphere rotate at different speeds, however, depending on latitude. The Earth's rotational speed at the equator is the greatest and at the poles, its least. Air traveling through the massive equator-to-pole-to-equator cell would be affected by this difference in speed. Air returning to the equator from the poles along the Earth's surface from would be moving more slowly than the air into which it flows near the equator. To an observer on the Earth's surface, this difference in air movements would appear as an east-to-west wind.

The Ferrel Model

The next important missing piece to the puzzle of global circulation was discovered in 1835 by the French physicist Gaspard Gustave de Coriolis (1792–1843). At the time, Coriolis was studying the behavior of moving objects on a spinning surface. He was able to show mathematically that any such object would follow a curved path relative to any fixed point on the surface. If one were to roll a marble across the surface of a spinning record, for example, the marble would not travel in a straight line, but in a curved path.

One of the first scholars to find a practical application of Coriolis's discovery was the American naval officer and meteorologist William Ferrel (1817–1891). Ferrel recognized that a mass of air flowing over the (rotating) Earth's surface was a situation to which Coriolis's law would apply. In applying Coriolis's discovery to global circulation, Ferrel discovered and used huge amounts of meteorological data commissioned and compiled by American naval officer Matthew Fontaine Maury (1806–1873). Maury's data showed, among other things, that there are certain regions of the globe with relatively constant air pressure, where prevailing winds are essentially absent. These areas include areas of low pressure at the equator and 30° from the poles and areas of high pressure at the poles themselves and at latitudes 30° north and south of the equator.

Ferrel incorporated Maury's data into a revised theory of global circulation. Our modern theory of global circulation is very similar to that of Ferrel's. According to this theory, air flowing through the Earth's atmosphere forms not one large cell (as envisioned by Halley and Hadley), but three smaller, although still very large, cells. The cells are formed in the following manner.

First, air above the equator rises into the upper atmosphere because of intense heating by the sun's radiation. As that air rises, it cools and moves outward away from the equator. The region of the Earth characterized by rising air currents and consistent low pressure is called the *intertropical convergence zone* (ITCZ). As Maury's data confirmed, winds are generally calm in this region of the globe. Sailors had long been aware of the risks of becoming trapped in these windless regions and had given them the name *doldrums.*

At this point, Ferrel's model differed from that of Halley and Hadley because he recognized that air from the equator does not travel all the way to the poles before descending back to Earth's surface. Instead, it cools sufficiently so that some begins to settle back to Earth's surface after traveling a distance of only about 30° of latitude north and south of the equator. At that point, some air reaches the ground and flows in two directions, back toward the equator and outward toward the poles.

The air that flows downward to the Earth then travels back toward the equator along the Earth's surface. This air flow completes a cell, now known as a *Hadley cell,* similar to the single massive cell envisioned by Halley and Hadley. It covers only a third of the globe above and below the equator, however.

Also, Ferrel recognized that air returning to the Earth's surface at about 30°N and 30°S latitude would be affected by the Coriolis effect. The descending air would be diverted into a westerly direction in both regions, accounting for the existence of the trade winds. Descending winds in the regions around 30°N and 30°S latitude produce, Ferrel said, areas of consistent high pressure, explaining the observations in Maury's data about these regions. These regions of constant high pressure are also characterized by calm weather and had been given the name *horse latitudes* in the northern hemisphere. The reason for that name is said to be that ships transporting horses were often becalmed for such long periods that they had to throw their cargo (horses) overboard because they had run out of food and water for them.

Refinements on the Three-Cell Model

Meteorologists now have a better understanding as to the flow of air in the areas around 30°N and 30°S latitude. Air traveling outward from the equator in the upper atmosphere also experiences the Coriolis effect. By the time this air reaches the 30°N and 30°S latitude points, it has been diverted by the Coriolis effect from a south-to-north direction to a west-to-east direction now known as the *subtropical jet stream.* Air begins to "pile up" in the upper atmosphere, forcing some of it back towards Earth's surface, producing the downward flow of air hypothesized by Ferrel. Some of the air in the upper atmosphere, however, continues to flow outward toward the north and south poles.

Air flow in the regions between 30°N and 60°N and between 30°S and 60°S is somewhat more complex than that in the Hadley cells. In these regions, now known as *Ferrel cells,* surface winds are deflected by the Coriolis effect from a south-to-north direction to a west-to-east direction. These *prevailing westerlies* had long been known to land dwellers in North America, Europe, and Asia as being responsible for prevailing weather patterns in their areas.

Upper air movements within Ferrel cells are largely a continuation of the equator-to-pole air flow in Hadley cells. These air movements are also influenced by the Coriolis effect and gradually change from south-to-north patterns to a west-to-east flow, producing the *polar jet streams* at latitudes of about 60°N and 60°S.

At ground levels beneath the polar jet streams, the northeasterly flow of air (the prevailing westerlies) comes into contact with cold air flowing away from the poles. This air is part of the third major cell that makes up the global circulation pattern, the *polar cell.* The polar cells consist of air moving between the poles themselves and regions around 60°N and 60°S latitude.

One portion of these cells consist of downward drafts of very cold air above the poles themselves. Some of this air does, in fact, reach the polar regions from the equa-

tor, has Halley and Hadley had originally hypothesized. The downward flow of cold air at the poles produces areas of high pressure at the poles. Air near the ground is pushed outward by these downward drafts, resulting in air movements outward and away from the poles. These air movements are affected by the Coriolis effect, producing the winds known as the *polar easterlies.* In the region where the polar easterlies collide with the prevailing westerlies (about 60°N and 60°S latitude), air is pushed upward into the atmosphere, producing areas of low pressure known as the *subpolar lows* or *midlatitude cyclones.*

Theory versus Reality

Much of the information about global circulation patterns outlined above is based on somewhat simplified models of the Earth and its atmosphere. Simplified models are used because so many factors are involved in determining these patterns, and data on these factors are often incomplete or missing. Moving from scientific theories and mathematical models about global circulation is even more difficult, however, for one reason: The Earth is not a homogeneous, smooth sphere, as most theories and models assume. Instead, the Earth's surface is covered by different kinds of materials (such as rock, sand, ice, and water) that is distributed irregularly with differing elevations. Surface irregularities such as these divert air movements out of the paths predicted for them by theory into more complex patterns.

For example, the intertropical convergence zone does not form a regular band around the equator, as predicted by theory. Instead, it consists of a wavy zone that tends to extend farther to the south of the equator over land masses, such as South America, Africa, and Australia, and farther to the north of the equator over open water, such as the Pacific and Atlantic Oceans. In addition, the ITCZ is not stationary throughout the year. Instead, it tends to move farther north of the equator in the middle of the year (June–August) and farther south at year-end (November–February).

In fact, about the only place where global circulation patterns approach those predicted by theory is over the Pacific Ocean, where a relatively smooth water surface covers a huge expanse. In this region, there are very few disruptions of land or elevation to alter the flow of air that would be expected from the description above.

By contrast, the flow of air in the northern hemisphere is disturbed, sometimes quite dramatically, by extensive land areas that include mountain ranges, deserts, plains, and a variety of other landforms. These topographical features can alter air movement within the three major cells to produce smaller cells, each with its own distinctive wind pattern. These patterns also are subject to change throughout the year.

Ultimately, meteorologists and climatologists still find it necessary to rely on actual weather observations to understand global circulation patterns while they work to develop more sophisticated models. On the basis of observations made over many decades, for example, that cells of high pressure tend to form each year in eastern Asia, off the west coast of North America, and in the middle of the north and south Atlantic Oceans at the end of the year (around December). By contrast, cells of high pressure area are more likely to be found over Australia and off the coast of Alaska during the middle of the year (around July).

The location and movement of these local air cells is sometimes responsible for distinctive weather patterns in a region. For example, the appearance of a high-pressure cell over eastern Asia in the winter and a low-pressure cell over the Indian subcontinent in the summer is a major factor in the weather system known as *monsoon circulation. See also* Atmosphere; Wind

Further Reading

Most introductory texts on meteorology contain a description of global wind patterns. See, for example, Frederick K. Lutgens and Edward J. Tarbuck, *The Atmosphere: An Introduction to Meteorology,* 6th edition, Englewood Cliffs, NJ: Prentice-Hall, 1995, Chapter Seven, "Circulation of the Atmosphere."

Gods and Goddesses of Air

Gods and goddesses of the air appear in every human culture of which we know, dating from the earliest human history to the modern day. The role of these air deities, as for all gods and goddesses, is to help humans understand phenomena and events for which there was (or is) no other explanation, events such as the creation of the world, regular astronomical and terrestrial phenomena (such as the rising and setting of the sun and stars), natural disasters, and all manner of human experiences, such as birth, death, and disease.

Mythic systems vary considerably from culture to culture and from time to time within a given culture. Such systems, furthermore, are often complex and internally inconsistent. Within any one culture, it is not unusual for a god or goddess to have more than one name, set of parents and/or offspring, and collection of powers and responsibilities.

Nonetheless, some striking similarities exist across cultures and across time about the nature and role of air deities. In the first place, most such deities can be roughly divided into three groups: those of the sky, of the winds, and of air itself. Sky deities are most commonly associated with creation myths. Those gods and goddesses usually appear out of the distant unknown region beyond the thin layer of air above Earth's surface, known as "the sky," to create the Earth itself; the oceans, planets, and animals; and human life. Wind gods and god-

In the act of creation, the air god Shu holds aloft the sky goddess Nut, separating her from Geb, god of the earth. Detail from the painted sarcophagus of Butehamun. © Werner Forman/ Art Resource, New York.

desses have a wide range of functions from the practical (carrying ships across the sea and bringing spring to the world) to the spiritual (breathing life into the newborn and carrying away spirits of the dead). By contrast, the number of purely air gods is relatively small and insignificant.

Evolution of Air Deities

It seems likely that cultural conceptions of air deities has evolved over many centuries. Consider the case of wind gods and goddesses. At first, humans seem to have taken a relatively simplistic view of the winds as a natural force that could be appeased and controlled by certain formalized acts. In his classic study of animistic practices among early peoples, *The Golden Bough,* Sir James Frazer provides many examples of this belief system. For example, the Yakut Indian from Siberia who sets out on a journey on a hot day attempts to arouse the wind by looking for a stone that has been swallowed by an animal or fish. That stone is then wrapped with horsehair and attached to a stick. Finally, the stick is waved in the air until the wind begins to blow. If the Yakut wants the wind to last for more than a week, the stone must first be dipped in the blood of a bird or some other animal. The traveler then turns three times opposite the direction of the journey.

Rites for reducing or stopping the wind were also known. The Payague Indians of South America, for example, attempted to calm a storm by lighting firebrands and running at the wind, trying to frighten it away. The Kayans of Borneo, similarly, were known to threaten an approaching thunder-storm by drawing their swords and waving them at the winds.

Over time, control of the winds evolved into a somewhat more complex and philo-sophical process. In the first place, humans began to personify the winds, imbuing them with recognizably human traits but super-human powers. Also, the rites needed to assuage and encourage the newly invented wind spirits were often assigned to mem-bers of a community with special powers—shaman, magicians, wizards, and priests—rather than remaining as acts that any individual could perform. In many cul-tures, the new gods and goddesses came to represent the four cardinal directions from which the winds blow, north, east, south, and west. This custom may have arisen, according to Sir Edward Burnett Tylor (*The Origins of Culture*), because "Man natu-rally divides his horizon into four quarters, before and behind, right and left, and thus comes to fancy the world a square, and to refer the winds to its four corners." In any case, it is of some interest that the identifi-cation of spatial directions on the basis of wind directions developed far earlier than that based on any other geographical factor, such as mapping.

Frazer cites a number of examples of the way wind control was assigned to specific individuals. In Finland, for example, certain wizards sold ropes containing three knots to mariners about to set out on a journey. Unty-ing one knot was supposed to release a mild wind; untying the second knot produced a half-gale; and untying the third knot would unleash a hurricane. The staying power of such beliefs is reflected in Frazer's note that, at the time of his writing (1922), Estonian peasants who lived across the Baltic Sea from Finland still attributed unusually severe winds to the machinations of Finnish wiz-ards.

Air Deities in Early Civilizations

Myths about air gods and goddesses already appear in relatively complex and sophisticated forms in records of the earli-est human civilizations. In Egypt, for exam-ple, two creations myth exist, one that developed in the Upper Kingdom and one in the Lower Kingdom. In the former, the world arose out of Nun, a vast ocean of chaos. Out of Nun came the creator god, known as Amun (also Amon, Amun-Ra, Amon-Ra, Atum, Atum-Ra, and so on). (In some early traditions, Amun was himself revered as a god of the air and wind.) Amun then gave birth to two children, Shu and Tefnut. Shu was given authority over the air, or, more specifically, dry air, while Tefnut was made goddess of moist air and water. In a more abstract sense, Shu came to represent preservation and the absence of change, while Tefnut came to stand for decay and change.

Shu and Tefnut, brother and sister, then had two children of their own, Geb, god of the earth, and Nut, goddess of the sky. The two children snuggled up so closely to each other that the sun was unable to travel across the sky between them. So Amun ordered Shu to come between them, creating enough space for the sun's journey. At the end of each day, however, Shu moved away and let Nut once more come to rest on top of Geb. A by-product of this story was that Geb and Nut eventually had two children of their own, the first two humans and the founders of the Egyptian civilization, Isis and Osiris.

Probably the next oldest myth of which we know arose in the Fertile Crescent civi-lizations of Sumeria and Babylonia. Accord-ing to this complex and fascinating story, the world began when Apsu, the god of fresh water, and Tiamat, the goddess of the seas, gave birth to two children, Anshar and Kishar, representing the horizon between earth and sky. Anshar and Kishar in turn had a child of their own, Anu (or An), eventually to become the most important god in the Sumerian pantheon of deities, the "Father God," "Sky God," and the "King of Gods."

To Anu and his consort, Ki, was born Enlil, a name that means Lord Air. Enlil was associated not only with the air and winds, but more specifically with the spring winds, which bring birth and renewal. He is credited with creating agricultural devices, such as the pickax, hoe, and plow, and giving these devices to humans. Beyond these practical gifts, he was believed to be the source of inspiration, imagination, and psychic powers.

Enlil was regarded as the second most important god after Anu. Rulers of Sumeria and Babylonia believed that they had received their power and authority from Enlil, given to him by his own father, Anu. On a more profound level, Enlil also had the ability to create gods and promote older gods to higher positions.

Enlil's wife is Ninlil, whom he first rapes, an act that begets the moon goddess Nanna. The story of Enlil and Ninlil is one of the most interesting to be found in all ancient creation myths (see "Enlil: Lord Air/Wind, Master of the Divine Word, Inspirer and Empowerer" at http://www.gate waystobabylon.com/gods/lords/lordenlil. html).

Gods of the sky, wind, and air were also common in Far Eastern myths of the earliest ages. A prominent Chinese god from early mythic stories is Fei Lian (or Fei Lien), depicted as a winged dragon with the head of a stag and the tail of a snake. He carries the winds about with him in a bag. In his human form, he is known as Feng Po (or Feng Bo), the "Earl of the Wind." In later myths, Feng Po is replaced by Feng Po Po, or "Lady Wind," a wrinkled old woman who rides across the sky with the winds trapped in a goatskin bag on her back.

As in other cultures, Hinduism has a long and complex tradition of creation myths that evolved over time. In one such story, Prajapati, god of the sky, had intercourse with his daughter, out of which arose the world as humans know it, consisting of three parts: the solid earth, the atmosphere, and the distant heavens.

Two wind gods have special prominence in the Vedic tradition of Hinduism: Rudra and Vayu. Both were powerful, ferocious gods, often associated with strong and potentially destructive winds. Rudra's name means "Lord of Teras" and "the Howler." He was sometimes associated with Agni, the god of fire, and was largely regarded as a destructive force in the world. Vayu was the father of two sons also associated with the wind, Hanuman and Bhima, and of the storm gods, the Maruts. According to legend, the wise man Narada asked Vayu to blow the top off Mount Meru, the center of the universe. When Vayu did so, the mountaintop landed in the ocean, forming the present-day island of Sri Lanka.

Later Civilizations

The mythology of sky, wind, and air gods and goddesses in ancient Greece is well documented in classic tales, such as the *Iliad* and *Odyssey,* and physical clues remain in statues, monuments, friezes, and structures, such as the Tower of Winds in the Roman *agora* in Athens. The primary wind deity among the Greeks was Aeolus, a son of Poseidon, according to some, or of Hippotes, according to others. Aeolus was also known as Astraeus and, in some legends, was said to have been married to Eos, goddess of the dawn. Aeolus is generally acknowledged to be the father of Boreas, Eurus, Notus, and Zephyrus, gods of the north, east, south, and west winds, respectively. Aeolus was said to have lived on the mythical island of Aeolia, which some scholars have identified as one of the Lipara islands off the coast of Sicily.

Aeolus dispersed the winds at the bidding of other gods, or as he chose to do so on his own. For example, he gave Odysseus a bag of winds to provide his ship a sure and safe journey back to Ithaca. Shortly before reaching their destination, however, Odysseus's crew opened the bag of winds to see what treasures it might hold. The winds released turned into a storm that blew the ship back to Aeolus, where the god now refused to welcome Odysseus and blew his ship back to sea.

Each of the four subsidiary winds, Boreas, Eurus, Notus, and Zephyrus, had his own personality and powers. Boreas, for example, is credited with destroying the Persian navy under Xerxes when it attacked Athens. Eurus and Notus were often worshipped for the warmth and rain they brought in spring. But Eurus was also feared for his violent side, expressed in terrible storms that destroyed ships at sea and threatened crops. The story most commonly associated with Zephyrus involves his competition with Apollo for the love of the boy Hyakinthos. As Apollo and Hyakinthos were playing one day with a discus, it is said, Zephyrus blew the discus back into the face of Hyakinthus. The boy died instantly and from his blood sprang the hyacinth flower, whose petals still carry the boy's initial.

As with other Greek deities, the wind gods were largely adopted by the later Roman civilization with only changes of names and minor modifications of their personalities and powers. Those transformations are as follows:

Aeolus: (no Roman equivalent)
Boreas: Aquilo
Eurus: Vulturnus
Notus: Auster
Zephyrus: Favonius

Remnants of Greek wind myths are retained today in the eastern Mediterranean, where certain local winds are known by classical names. North winds blowing across the Aegean Sea, for example, are called boreas, south winds are known as notios, and west winds are referred to as zephyros.

The Greek pantheon of air deities included other wind gods and goddesses. One of these was Typhon (also Typhaeon and Typhoeus), the son of Gaia (goddess of the earth) and Tartaros (god of the underworld). Typhon was usually described as having the body of a dragon covered with 100 burning snake heads. He breathed out fire and made terrible noises to frighten his enemies. Before Zeus became supreme ruler of the gods, Typhon wounded him and imprisoned him in a cave. When saved by Pan and Hermes, Zeus fought Typhon once more, defeated him, and buried him either in Hades or under Mount Aetna. Typhon was thought to be responsible for terrible wind storms and earthquakes. His name is remarkably similar to the Chinese term *ta* ("great") *feng* ("wind"), from which the modern term *typhoon* is thought to be derived.

Another group of wind deities were the harpies, flying females described in a variety of ways by different authors. In their earliest manifestations, their role appears to have been the removal of souls from dead bodies to Hades. Over time, their description and functions became more complex and more diverse. In some stories, there were only four harpies, born to Typhon and his sister Echinda (or, some say, to Electra and Thaumas). They were Aello (rainsquall), Celaeno (storm-dark), Okypete (swift-flying), and Podarge (swift-foot). Other writers mention only Aello and Okypete, while Homer refers only to Celaeno. In some later myths, large groups of harpies, numbering more than two dozen, are described.

Most authors agree that the harpies are ugly creatures with perhaps beautiful hair and faces, but the bodies of vultures. Their claws were long and sharp, sometimes said to be made of brass. They were said to be constantly hungry and afflicted with diarrhea, accounting for the terrible stench of their bodies. Their only fear was the sound of a brass trumpet. In addition to transporting the human soul to Hades, harpies were thought to be responsible for severe storms and hurricanes.

Deities of Northern Europe

Cultures throughout northern Europe also had creation myths and pantheons of gods and goddesses, some of whom were associated with the sky, the air, and/or the winds. Two examples of these deities are Jumala and Njord. Jumala was the supreme Finnish god, ruler of the sky and thunder. He and his wife, Akka, created the world and all living beings. Over time, Jumala appears to have evolved into the god Ukko, the supreme deity referred to in *The Kalevala* (the Finnish classical epic) as "the God of Breezes," "the Silver Ruler of the Air," and "the Golden King." He controls the weather and often expresses his anger by hurling thunderbolts at the Earth.

Njord (also Niord) was the Norse god of the wind and sea. He was also known as the "stiller of storms" because of his ability to control the wind and put an end to storms. The story for which he is most famous involves his betrothal to the Giantess Skadi, who was allowed to choose a god for her husband. The only condition was that she had to choose by looking only at the god's feet. She chose Njord, thinking that he was a more beautiful god named Balder, because of Njord's beautiful feet. The story accounts for Njord's other name, "he of the beautiful feet."

Deities of the New World

Virtually every New World culture that has been studied, from the Caribbean to Alaska, appears to have had one or more deities devoted to the sky, air, and/or winds. The presence of such deities permeates the stories of creation throughout pre-Hispanic Mesoamerica. For example, in one of the few written histories of creation adopted by the Mayans, the world is created by a single creature with seven manifestations (somewhat similar to the Father, Son, and Holy Ghost of Christian theology). One of the seven forms in which this God Seven appears is Alom, the god of wind. Like his six brother manifestations, he tries to create life with which to populate the new world and, like them, he fails. Only when the seven brothers learn how to work together are they successful in creating Ixpiyacoo and Ixmucane, the grandparents of all Mayans and all other humans on Earth. Later, the Mayans also adopted four specific wind gods for each of the four cardinal directions, Cauac, Ix, Kan, and Mulac.

The most important god of the interlocked Tolmec-Mixteca-Mayan-Aztec tradition, Kukulcan (also Kukumatz), is also associated with the wind. He was thought to be the supreme god of creation, resurrection, and reincarnation. He was thought to have emerged from the ocean, bringing all knowledge with him to the Toltecs, and then returning to the ocean. His representations usually include one feature for each of the four elements: a vulture for air, an ear of corn (maize) for earth, a fish for water, and a lizard for fire.

After the fall of the Mayan civilization, the Aztecs adapted Kukulcan and integrated him into their own supreme deity, Quetzalcoatl. Quetzalcoatl was depicted as a feathered serpent that could manifest himself in many ways, one of which was the wind god Ehecatl. The second day of each month in the Aztec calendar is devoted to Ehecatl, who is shown wearing a billed mask. In the form of Ehecatl, Quetzalcoatl is said to have breathed life into humans and all other life on Earth and to provide the winds that bring rain for crops. In some parts of Mexico, local windstorms are still referred to as ehecacoatls ("wind serpents"), and a company that manufactures wind-power generators is named Ehecatl Mexicana.

Native American tribes almost universally believe in a sky deity responsible for the world's creation. A typical myth is that told by the Iroquois people in the northeastern United States. According to this legend, Atahensic was the daughter of the Evening Star, who said that her husband-to-be was to be found on Earth. To find her mate, Atahensic left the heavens and came to Earth, gaining the title of "the woman who fell from the sky." She was regarded as the First Woman, as both the creator and destroyer of life, as the source of disease and trouble, and as the transporter of dead souls. Atahensic's first daughter was Gusts of Wind, who returned to Atahensic's womb after her birth when her mother fell (or was pushed) over a cliff into a deep lake. The First Mother also appears in the mythology of other North American tribes, including that of the Huron (where she is known by the same name), the Seneca (where she was called Eagentci), and the Navajo of the southwestern United States (as First Woman).

Wind gods also play an important role in the creation stories of the Dakotas of the Central Plains. According to this mythology, two evil gods, the Old Man Waziya and his wife, Wakauka, had a beautiful daughter, Ite. Ite was a subordinate of Skan, the sky god, and was responsible for carrying away the souls of the dead to the Spirit Trail (the Milky Way). The wind god Tate fell in love with Ite and married her. Together, Tate and Ite had four quadruplet sons, the north, south, east,

and west winds. Tate was later criticized for marrying a nongod and was banished to a home "beyond the pines," where he lived with his four wind-sons and Yumni, the whirlwind, a fifth son. Each day, Tate sent one or more of his sons out into the world to carry out some specific task, such as blowing rains to a field. On one occasion, the brothers returned to find that their father had taken a beautiful woman named Whope into their home. They soon learned that Whope was the daughter of the Sun and the Moon. Each brother wanted Whope as his own wife, but she was eventually given to Okega, the south wind. In celebration of their marriage, the gods presented to Tate and his sons the whole world and everything in it.

Further Reading

Andrews, Tamara. *Legends of the Earth, Sea, and Sky: An Encyclopedia of Nature Myths.* Santa Barbara, CA: ABC CLIO, 1998.

Cavendish, Richard, and Brian Innes, eds. *Man, Myth, & Magic: The Illustrated Encyclopedia of Mythology, Religion, and the Unknown,* 21 vols. New York: M. Cavendish, 1995.

Coulter, Charles Russell. *Encyclopedia of Ancient Deities.* Jefferson, NC: Mc Farland, 2000.

"Encyclopedia Mythica," http://www.pantheon.org.

Graves, Robert. *Greek Myths.* Garden City, NY: Doubleday, 1981.

"North American Indian Gods," http://nikki.sitenation.com/namerican/northgods.html.

Read, Kay Almere, and Jason J. González. *Handbook of Mesoamerican Mythology.* Santa Barbara, CA: ABC-CLIO, 2000.

Gradient Wind

See Wind

Guericke, Otto von (1602–1686)

Otto von Guericke invented the air pump in about 1650, inspired by the question of whether or not a vacuum could exist. He used the air pump in a number of dramatic public demonstrations to illustrate the force of air pressure. Perhaps his most famous demonstration involved two copper hemispheres, called the *Magdeburg hemispheres,* after his home town of Magdeburg. Von Guericke attached the two hemispheres to each other and pumped out the air from inside the sphere so formed. He then tried to have the two hemispheres pulled apart by attaching a team of horses to each. His failure to do so prompted great amazement and interest among the general public and among scientists throughout Europe.

Otto von Guericke was born in Magdeburg, Germany, on 20 November 1602 into a long-established wealthy and politically powerful family. He attended the University of Leipzig from 1617 to 1620 and then transferred to the University of Helmstedt for one more year of study. He studied law at the University of Jena from 1621 to 1622 and then moved to the University of Leiden, where he attended lectures on mathematics and engineering as well as continuing his law studies.

In 1626, von Guericke returned to Magdeburg and entered politics. Germany was then engaged in the Thirty Years' War, and in 1631, Magdeburg was largely destroyed by French troops. Von Guericke and his family managed to escape from the city, and he joined the army of the Swedish king Gustavus Adolphus II. When the tide turned and Magdeburg was retaken, von Guericke returned to the city and was instrumental in helping rebuild it. Partly in recognition for this work, he was elected burgomaster (mayor) in 1646. He held that post for 35 years before retiring to Hamburg, where he spent the rest of his life until dying there on 11 May 1686.

The question as to whether a vacuum could exist had intrigued scholars for more than 2000 years. The great Greek philosopher Aristotle (384–322 B.C.) had argued in the negative, teaching that "nature abhors a vac-

uum" and that one could not, therefore, exist. Von Guericke decided to build his air pump to see if that proposition could be supported experimentally. In doing so, he was one of the first scientists to test Aristotle's position with experiments rather than reason. Like his counterpart in England, Robert Boyle (1627–1691), von Guericke helped establish a new scientific tradition in which questions were answered on the basis of empirical evidence rather than by logical deduction.

Von Guericke's air pump was simple in design. It consisted of a cylinder and piston with two flap valves. The piston was moved in and out of the cylinder by a man's pushing and pulling on it. The opposite end of the cylinder was attached to a wooden keg. When the piston pulled outward from the cylinder, air rushed out of the wooden keg into the cylinder through one flap valve. When the piston's movement stopped, that flap valve closed, preventing air from returning to the keg. At that moment, the second flap valve opened, allowing air trapped in the cylinder to flow outward into the atmosphere. The process was repeated many times, with more and more air removed from the keg each time. The fact that a vacuum was being created was clear to von Guericke because he could hear the whistling sound of air rushing into the keg.

Since the wooden keg was too porous for a vacuum to form, von Guericke began experimenting with metal casks. He found that he was successful in removing a very large fraction of the air in the metal casks with his simple air pump. As this experiment was repeated many times over, von Guericke made a number of improvements in his air pump and evacuation vessel. For example, he found that any deviations from perfect sphericity in the vessel increased the probability of its collapse from outside atmospheric pressure.

In another of his dramatic experiments, von Guericke attached a rope to the piston of his air pump and had 50 men hold the rope. He then attached the air pump to a spherical copper vessel and attached a second air pump to the opposite side of the copper vessel. When the second air pump began working, the 50 men were unable to keep the rope from being drawn into the first cylinder by the force of atmospheric pressure on the vacuum formed inside the vessel.

Von Guericke was also interested in a number of other scientific subjects. He developed astronomical theories, for example, based on the existence of countless numbers of stars, each of which had its own solar system and was independent of each other. To encompass this theory, he proposed the existence of an infinite space, a concept contrary to almost all existing astronomical concepts at the time, but one that later became more popular.

Von Guericke was intrigued by magnetism and conducted a number of studies to learn more about the phenomenon. In one of them, he invented a machine modeled on the supposed composition of the Earth in which magnetic substances rubbed against each other. In the process, an electrical charge was created, a device that might be thought of as the first electrical machine. *See also* Vacuum

Further Reading

Guericke, Otto von. *The New (So-Called) Magdeburg Experiments of Otto von Guericke,* translated by Margaret Glover Foley Ames. Dordrecht and Boston: Kluwer Academic, 1994.

Gusts and Squalls

Gusts are sudden, rapid, and brief changes in wind speed. Squalls are gusts of higher speed and longer duration.

Even when the wind appears to be blowing steadily, there are likely to be frequent

Cold front squall line. © NOAA.

Gust probe. © NOAA.

fluctuations in wind speed and direction. High points in wind speed are known as *peaks* and low points, *lulls*. Wind fluctuations are caused by a number of factors, one of which is the *roughness* of the surface over which air moves. Roughness can be expressed numerically by designating a *roughness class,* ranging from 0 (very smooth) to 4 (very rough). The open sea often has a roughness class approaching 0, where gusts may not exceed about 30 percent of the average wind speed. By contrast, regions with higher roughness classes may experience gusts twice that of the average wind speed.

The appearance of gusts is also affected by precipitation factors. The availability of water vapor in the air may increase the amount of energy released during precipitation, also increasing the intensity of wind gusts. Squalls are, therefore, a common occurrence over lakes and the oceans.

According to the World Meteorological Organization and the U.S. National Weather Service, a gust is defined as any wind speed of at least 16 knots (18 miles per hour) that involves a change in wind velocity with a difference between peak and lull of at least 10 knots (12 miles per hour) lasting for less than 20 seconds. By contrast, a squall is defined as having a wind speed of at least 16 knots (18 miles per hour) that is sustained at 22 knots (25 miles per hour) or more for at least two minutes (in the United States) or one minute (in other parts of the world).

Groups of gusts and squalls sometimes occur in the form of a *gust front* or a *squall line.* Both formations tend to occur at the leading edge of a group of thunderstorms, formed by the downdraft of air pushed outward by the advancing thunderstorms. *See also* Wind

Further Reading

Federal Meteorological Handbook No. 1: Surface Weather Observations and Reports. Washington, DC: National Weather Service, Department of Commerce, 1996.

Huschke, Ralph E., ed. *Glossary of Meteorology.* Boston: American Meteorological Society, 1989.

H

Hadley, George (1685–1768)

George Hadley was an English meteorologist best known for his theory about the global circulation of air, published in 1735. One cell of moving air in current theories about global circulation, the Hadley cell, has been named in his honor.

George Hadley was born in London on 12 February 1685, the younger brother of a well-known inventor, John Hadley. He entered Pembroke College in 1700 to study law. Although he was admitted to the bar nine years later, he appears not to have been especially interested in the law and spent most of his time carrying out experiments and developing theories in mechanics and the physical sciences.

On 20 February 1735, Hadley was elected a fellow of the Royal Society and, just three months later, presented his paper, "Concerning the Cause of the General Trade Winds," on which his fame rests. He summarizes his theory by suggesting that air "as it moves from the tropicks towards the æquator, having a less velocity of diurnal rotation than the parts of the earth it arrives at, will have a relative motion contrary to that of the earth in those parts, which being combined with the motion towards the æquator, a N.E. wind will be produced on this side of the æquator, and a S.E. on the other side."

Hadley was later placed in charge of meteorological observations for the Royal Society, a post he held for seven years. During that time, he published two exhaustive reports on his studies during this period. Hadley spent the last years of his life with his nephew, Hadley Cox, in the village of Flitton, where he died on 28 June 1768. *See also* Global Winds; Wind

Hampson, William (1854?–1926)

William Hampson developed a method for liquefying air in the late 1890s, about the same time that a similar discovery was being made in Germany by Karl von Linde (1852–1934). The two men obtained patents for their inventions within two weeks of each other in 1895. Hampson's process was the simpler of the two. He compressed air under a pressure of 200 atmospheres and then suddenly reduced the pressure to one atmosphere. Rapid expansion of the air caused it to cool. The cool air was then used to cool incoming gas in the next cycle of the operation. Although the process was simple, it was not very efficient. The best Hampson's equipment was able to do was produce about a liter of liquid air per hour at the expenditure of 3.7 kilowatts of power.

Little is known about the details of Hampson's life. He is believed to have been born in 1854 in Bebington, Cheshire, England. He attended Manchester Grammar School and Trinity College, Oxford, from which he received his M.A. in 1881. He then joined the Inner Temple (a group of lawyers), although there is no evidence that he ever practiced law. In fact, his life is essentially a blank page to historians until 1895, when he received his patent for a liquid air machine.

Hampson soon sold the rights to his machine to Brin's Oxygen Company (later British Oxygen Company), and he continued to serve as a consultant to the company. He also joined forces a research team headed by William Ramsay at University College, London. Ramsay's team was engaged in the process of separating and identifying a group of previously unknown elemental gases known as the *inert gases, noble gases,* or *rare earth gases.* Since these gases were all obtained from liquid air, Ramsay's work was aided immeasurably by Hampson's understanding of this topic.

After a period working with Ramsay, Hampson became embroiled with another researcher studying the liquefaction of gases, James Dewar (1842–1923). Hampson apparently retreated from the dispute with Dewar and largely disappeared from the public scene. He eventually became qualified as a medical practitioner and worked in various London hospitals on the medical applications of x rays and electricity. He wrote two books on popular science, *Paradoxes of Science* (1904) and *The Explanation of Random* (1906) and one political tract, *Modern Thraldom* (1907), in which he claimed that the practice of lending money was at the basis of all modern social prob-

lems. *See also* Linde, Karl Paul Gottfried von; Liquid Air

Heron of Alexandria (about A.D. 10–75)

Heron, also known as Hero, of Alexandria was an important mathematician and engineer who lived in Alexandria after Egypt had been conquered by the Romans. There is a fair amount of confusion about the details of his life. Many references in later writings refer to both "Heron" and "Hero," and it is often unclear as to whether these references are to the same person or to different individuals. His birth date was established even approximately only in the last century when an eclipse that occurred over Alexandria in A.D. 62 to which he alludes makes it possible to guess at this information.

Authorities believe that Heron taught at the Museum of Alexandria, and most of what we know about him appears to have been prepared in the form of lecture notes on topics in mathematics, physics, pneumatics, and mechanics. One of his most fascinating books on pneumatics describes a number of his inventions that operate on steam or air. That book is available in translation online at http://www.history.rochester.edu/steam/hero.

Heron made some important discoveries about the properties of air that were largely ignored by his contemporaries and immediate successors. For example, he demonstrated that air is a form of matter that takes up space by plunging the open end of a bottle into a container of water. Water was not able to flow into the bottle because the bottle was already full of air. He also showed that air can be compressed and concluded from this fact that air is made of tiny particles separated by large spaces, a concept that was not to reappear in scientific thought until the time of Robert Boyle (1627–1691) more than 1,500 years later.

Further Reading

Drachmann, A. G. *Ktesibios, Philon and Heron: A Study in Ancient Pneumatics.* Copenhagen: E. Munksgaard, 1948.

Hovercraft

A hovercraft is a vehicle that travels over land or water on a cushion of air. It can be used on almost any type of surface—including water, grass, sand, or ice—provided that the surface is relatively smooth. Since hovercrafts typically travel on a "bubble" of water generated by their internal fans, they are often known as *air cushion vehicles* (or ACVs).

Invention

The first attempt to build a vehicle able to travel on air is thought to have been the Swedish philosopher, theologian, and inventor Emmanuel Swedenborg (1688–1772). In 1712, Swedenborg designed a vehicle that looked something like an upside-down rowboat. The vehicle was designed to travel on a cushion of air that was pumped into the space beneath the hull by means of oarlike scoops. Swedenborg never actually built such a vehicle because he knew of no way to force air beneath the hull fast enough.

The first operating hovercraft was built in the 1950s by two engineers operating independently, Jean Bertin (1917–1975) in France and Christopher Cockerell (later Sir Christopher Cockerell; 1910–1999) in England. In a now-famous experiment, Cockerell forced a stream of air from a vacuum cleaner under an empty cat food can inside a coffee can. He adjusted the air flow to allow the cat food can to just barely float above the bottom of the coffee can. By 1956, he had produced the first working model of a hovercraft and, three years later, on 29 July 1959, the first commercial hovercraft traveled across the English Channel.

Bertin's research took him in three related but somewhat different directions. His naviplane, terraplane, and hovertrain were designed to travel, respectively, over water and land, the last of these designed to carry both passengers and cargo. In 1961, he obtained a patent for an "air cushion" system on which to operate these vehicles.

Specifications, Operation, and Uses

Most hovercraft today range in length from about 10 feet to about 20 feet, with widths of about 6 to 8 feet. They are powered by 60- to 100-horsepower engines and travel at speeds of up to about 60 miles per hour, depending partly on the terrain over which they travel.

The hovercraft's engine is used to drive a fan, usually about 36 inches in diameter, which generates a stream of air aimed in two direction. One stream of air is forced downward, beneath the hull of the hovercraft. It produces the cushion of air on which the vehicle rides. The second stream of air is forced out the back of the vehicle, provid-

Racers in Michigan pit their Hovercrafts against each other. © Universal Hovercraft.

ing the driving force that moves the vehicle forward.

Hovercrafts are now used in a large variety of applications, including personal recreation, fire and rescue operations (including ambulance services), military reconnaissance and assault operations, patrol, pest control, surveying and drilling, scientific research, and personnel and cargo transportation.

Hydrofoils

Another vehicle similar in design to the hovercraft is the hydrofoil. A hydrofoil is a hybrid vehicle that incorporates characteristics of both an airplane (upon whose principle it operates) and a boat or ship (which it looks like). Like the hovercraft, the hydrofoil has a long history. It was invented by the Italian inventor Enrico Forlanini (1848–1930) in 1905. Forlanini's vehicle consisted of a boat to which were attached a horizontal series of foils built like airplane wings. (A airfoil is, in fact, another name for an airplane wing.) As the vehicle was powered forward, air flowing over the foils produced the Bernoulli effect, in which normal air pressure below the foils pushed them upward against the reduced pressure on top of the foils.

Alexander Graham Bell also built an early form of the hydrofoil in 1911, the HD-1. Bell's original model attained a speed of 50 miles per hour. The Father of the Modern Hydrofoil, however, is generally thought to be the German inventor Hanns von Schertel (1902–1985). One of von Schertel's most important achievements was the solution of problems relating to the hydrofoil's propulsion system and its stability in rough water.

Modern hydrofoils are of two general types: surface piercing and submerged. In the surface piercing design, the boat's foils travel just below the surface of the water. In the submerged design, they are always fully under water. Surface-piercing hydrofoils are able to travel at very high speeds on calm waters, but they become unstable in rough water because the foils tend to follow closely the water's irregular surface. Submerged hydrofoils require additional components to maneuver through turns and to control irregular motions, such as yawing and pitching. However, they operate more reliably in rough water.

Further Reading

Amyot, Joseph R. *Hovercraft Technology, Economics, and Applications.* Amsterdam: Elsevier, 1989.

Hurricane

A hurricane is a cyclonic storm that forms over tropical oceans with rotating winds in excess of 119 kilometers per hour (74 miles per hour). They range in size from 100 to 1,500 kilometers (60 to 1,000 miles) in diameter with pressure differentials as great as 60 millibars between the inside of the storm and the surrounding areas. The worst such storms may cause dozens or hundreds of deaths and billions of dollars in property damage. The word *hurricane* comes from the name of a Taino (a Caribbean people) god of evil, Huracan. In parts of the Pacific Ocean, hurricanes are known as typhoons, and in the Indian Ocean, they are usually called cyclones.

Hurricane Formation

Hurricane formation can be understood in terms of one simple physical phenomenon: the condensation of water vapor. A unit volume of water vapor contains more heat energy than does a comparable mass of liquid water. When water vapor condenses to form liquid water, that excess heat (known as *latent heat*) is released to the surrounding air.

That air becomes warm and less dense and begins to rise in the atmosphere. (Actu-

Eye wall of a hurricane. © NOAA.

and south of the equator. Hurricanes do not form in the latitudes between 5° north and south of the equator because Coriolis forces are too weak to produce strongly rotating winds. They are also rare or absent in most parts of the eastern South Pacific and all of the South Atlantic.

The pressure differentials that develop during hurricane formation can be quite dramatic. They show up on weather maps as closely spaced isobars around cyclones, areas of low atmospheric pressure. The pressure at the center of such storms are among the lowest ever measured on Earth. The pressure at the center of Hurricane Gilbert as it passed over the Caribbean on route to the Yucatan peninsula on 13 September 1988, for example, was 888 millibars, the lowest such measurement ever made. It broke the previous low pressure record of 892 millibars at the center of a hurricane that struck the Florida Keys on Labor Day in 1935.

Hurricane Development

Hurricanes develop out of tropical thunderstorms, where vast amounts of water vapor are condensing to form liquid water. Regions where such thunderstorms are developing are called *tropical disturbances.* Hurricanes grow from tropical disturbances through two preliminary stages before developing into full-blown hurricanes. The first stage of development is called a *tropical depression,* an area where a group of thunderstorms has come together around a central cyclone. Winds begin to rotate around the cyclone at speeds of about 37–63 kilometers per hour (23–39 miles per hour). A tropical depression can be recognized on weather maps by the presence of a closed isobar and from satellite photographs as a large undifferentiated cloud mass with slowly rotating winds.

Over time, a tropical depression may become more organized, taking on a more

ally, the warm air is pushed upward by cooler air surrounding it.) As air rises, it begins to rotate because of the Coriolis effect, in a counterclockwise direction in the Northern Hemisphere and clockwise in the Southern Hemisphere. The process of hurricane formation is similar to that for any cyclonic or anticyclonic wind or for a tornado except that much larger amounts of heat are available in hurricane formation because of the latent heat released during the condensation of water vapor.

It is apparent that hurricane formation can take place only in regions where there is a large supply of warm water. Scientists have learned that water must have a temperature of at least 26.5°C (81°F) before hurricanes can begin to form. Such conditions are found in certain limited parts of the Earth's oceans, between 5° and 20° north circular form with winds increasing to speeds of up to 119 kilometers per hour (74 miles per hour). At this point, the storm is called a *tropical storm*. Each year, 80 to 100 tropical storms develop over the oceans. About half of these never develop any further. They may, for example, come into contact with land, where the driving source for their development—warm water—is no longer available. They then turn into storms that, because of their heavy winds and rains, may still be destructive, but not of the severity of a hurricane.

Tropical storms that remain over warm water long enough may eventually evolve into hurricanes with clearly defined rotating winds with speeds greater than 119 kilometers per hour (74 miles per hour). The most recognizable feature of a fully developed hurricane is its central eye, a vertical column of air directly over the center of the cyclone. Air at the center of the eye flows downward and has relatively little rotational component. It is the calmest part of the storm, seeming to offer respite as it passes over an area.

Hurricane Movement

Hurricanes typically travel across water at the rate of up to 500 kilometers (300 miles) per day. The path they take is largely determined by the prevailing wind patterns in the region where they are formed. Hurricanes formed over the Atlantic, for example, tend to be carried by the tropical easterly tradewinds toward the Caribbean Sea, the Gulf of Mexico, and the southeastern United States. Those formed over the Indian Ocean tend to be carried both northward and northwesterly over the Indian subcontinent and southwesterly along the east coast of Africa. Hurricanes formed just north and south of the equator in the western Pacific are carried to the northwest and southwest, respectively. They continue to move until (1)

they reach land, (2) they pass over cooler water, or (3) upper atmospheric conditions prevent their survival.

The severity of a hurricane is measured on the Saffir-Simpson Hurricane Damage Potential Scale. On this scale, hurricanes are given a rating of 1 to 5, based on wind speeds and, hence, on the potential damage they may cause. A medium-strength hurricane with a rating of 3, for example, has wind speeds of 178–209 kilometers per hour (111–130 mph) and is expected to destroy mobile homes, blow down large trees, and damage small buildings.

Hurricane Watches and Warnings

Meteorologists use an array of data collection devices and techniques to track the development of possible hurricanes and to warn the public of the possible arrival of such storms. These data come from satellite photographs, aircraft reconnaissance, Doppler radar devices, and fixed-location buoys. When a hurricane appears to pose a threat to an area, the Tropical Prediction Center in Miami, Florida, releases a hurricane watch or a hurricane warning.

A *hurricane watch* warns that a hurricane or tropical storm may pose a threat to a specific coastline region within a 36-hour period. A *hurricane warning* is issued when hurricane-force winds of 119 kilometers per hour (74 miles per hour) or more are expected to pass a specific coastal region within 24 hours or less. Watches and warnings provide a basis for governmental agencies and individuals to decide what actions they should take to avoid or reduce the risk of personal injuries and property damage. Many coastal communities in the United States, for example, have evacuation plans that provide for escape routes and temporary housing for individuals in the projected path of a hurricane. The hurricane watch/warning system has greatly reduced the number of personal injuries and deaths associated with hurricane landfalls, although there is a limit to their effectiveness in reducing property damage and loss. *See also* Coriolis Effect; Tornado

Further Reading

Davis, Pete. *Inside the Hurricane: Face to Face with Nature's Deadliest Storms.* New York: Henry Holt, 2000.

Fitzpatrick, Patrick J. *Natural Disasters: Hurricanes: A Reference Handbook.* Santa Barbara, CA: ABC-CLIO, 1999.

"Hurricanes," http://ww2010.atmos.uiuc.edu/(Gh)/guides/mtr/hurr/home.rxml.

Sheets, Bob, and Jack Williams. *Hurricane Watch: Forecasting the Deadliest Storms on Earth.* New York: Vintage Books, 2001.

I

Icarus

Icarus was the son of Daedalus, a renowned architect, sculptor, craftsman and inventor, and a descendant of Erechtheus (also Erichthonius), the mythical first king of Athens. Daedalus was said to have been given his skills by the goddess Athena. According to some legends, Daedalus was banished from Athens when he killed his nephew and pupil, Talus, who had begun to show greater skill than his master. He traveled to Crete, where he was installed at the court of King Minos II. He built a hollow wooden cow for Minos's wife, Pasiphae, with which she became pregnant and bore the monster known as the Minotaur. Daedalus then constructed the famous labyrinth, a series of twisting passageways at the center of which the Minotaur was installed. Each year, seven young men and women were fed to the Minotaur as sacrifices.

When Theseus, the king of Athens, arrived in Crete to kill the Minotaur, Daedalus agreed to provide him with the secret by which he could find his way into and back out of the labyrinth. When Minos was told of Daedalus's treachery, he had him and his son, Icarus, confined to a prison, which was either a tall tower, according to some stories, or the labyrinth itself, according to others.

**Charles Paul Landon's *Daedalus and Icarus*
© Erich Lessing/ Art Resource, New York.**

In planning a method of escape, Daedalus collected as many bird feathers as he could and glued them to his arms and those of Icarus. Then, by flapping their arms like birds, they both flew out of their prison toward their freedom. Surprised and delighted with his newfound ability to fly, Icarus soared high into the sky. Daedalus flew more conservatively close to the ground and warned his son not to rise too high. Icarus was so enamored of his new skill, however, that he ignored his father's warnings and eventually flew high enough that the glue holding his feathers to his body melted. The feathers flew off, and Icarus plunged into the sea.

According to some stories, Daedalus flew down to the sea, trying to save his son, but saw that the boy had drowned, surrounded by the feathers that had come loose from his body. Other legends say that Icarus's body was found by sailors and carried to a nearby island, while still other tales say that his body was washed ashore on the island of Sicily. In any case, Daedalus eventually retrieved his son's body and built a temple in his honor on an island that now bears his name, Ikaria.

Further Reading

Schnabel, Ernst. *Story for Icarus: Projects, Incidents, and Conclusions from the Life of Daedalus, Engineer.* New York: Harcourt Brace, 1961.

International Aerobiology Newsletter

See Aerobiology

International Association for Aerobiology

See Aerobiology

Isobar

Isobars are lines drawn on a weather map connecting places of equal sea-level pressure. The data required for the construction of a weather map come from more than 10,000 weather stations located in virtually every part of the United States and at land and ocean locations around the world. Four times each day, these data are transmitted to three national meteorological centers at Washington, DC, Moscow, and Melbourne, Australia, where they are compiled, analyzed, and then retransmitted to local meteorological stations. Each station then uses these data to make its own site-specific forecast.

The isobars on weather maps are usually separated by some common pressure difference, such as every 3 millibars or every 4 millibars. One isobar, for example, might represent all locations in an area with a surface pressure of 1,000 mbar; the next isobar, one with a pressure of 1,004 mbar; the next, with a pressure of 1,008 mbar; and so on. The spacing between any two adjacent isobars is an indicator of the pressure difference between these two locations. If the two isobars are close to each other, pressure drops off rapidly in the region, and strong winds are to be expected. If two adjacent isobars are widely separated, the pressure differences occurs over a greater geographical distance and only mild winds are expected. The pressure difference between two adjacent isobars is called the *pressure gradient* and can be represented mathematically as $\Delta p/d$, that is, the change in pressure divided by the distance between the two isobars.

Wind arrows (or *wind barbs* or *wind flags*) attached to isobars show the direction and intensity of the winds blowing in the region. Each wind arrow contains up to five vertical arrows of various lengths indicating the wind intensity. A wind arrow with two full vertical lines and one half line, for example, represents a wind speed of 26–31 miles per hour.

Isobars may be drawn for any level of the atmosphere, from sea level to altitudes of 10 kilometers or more. Upper atmosphere isobars differ from their surface counterparts, however, in that they usually show the height (in meters) at which some given pressure (such as 500 mbar) is found.

Isobars are seldom straight lines. They are usually curved because of surface features, such as the proximity of water and dry land or the presence of mountain ranges, that cause unequal heating of the air and, thus, unequal pressures. In some cases, isobars close in on themselves, forming closed loops (or nearly closed loops) that represent a "low" or "high," that is, a center of low pressure or high pressure. *See also* Air Pressure; Cyclone and Anticyclone; Wind

K

Kite

A kite is a flying object usually consisting of a light frame covered with a thin material, such as paper or linen cloth. Kites appear to have been used in many cultures throughout the world for both utilitarian and recreational purposes almost since the beginning of recorded time. The earliest hints of kite flying may be found on stone carvings found in Egypt dating to 500 B.C. or earlier that show people holding strings attached to objects in the sky that may have been some type of kite. Tradition has it that kites had also been invented in Greece by about 500 B.C., although there is no concrete evidence for this assertion.

Early History

The earliest records of kite flying that can be accepted with some certainty are those from China dating to about the fifth century B.C. Some historians believe that credit for the invention of the kite should go to the philosopher Mo Zi (478–392 B.C.), who made a wooden kite in the shape of a hawk that flew for one day. Certainly by the second century B.C., kite flying had become very popular throughout China, described in detail in numerous histories of the time. A number of legends had by then grown up explaining how the practice had originally

A teen in Ohio files her kite. © Premier Kites.

developed. According to one such legend, a Chinese farmer working in the fields had his bamboo hat blown off one day by a stiff wind. It happened that the tie attached to his hat was long enough that the hat began to act like a kite. The farmer realized the sport that could be made of hat/kite flying, and the practice was born.

Ancient legends about kite flying still exist throughout the Far East. In some parts of Polynesia, for example, people re-enact with kite-flying ceremonies a mythic battle between two wind gods, Tane and Ragne, in which the former is defeated when his kite tail becomes entangled in a tree. In Malaysia, a legend that predates written history tells how people released hundreds of small kites made of palm leaves to appease the monsoon gods.

From their earliest history, kites were used for a variety of purposes. According to one legend, the Chinese general Han Hsin (died 196 B.C.) had his troops build large kites equipped with whistles to fly over the enemy's camp. The purpose was to convince enemy troops that the gods were warning them that disaster was about to strike, driving them from the field of battle. Han Hsin is also said to have used kites to carry messages and conduct surveillance of areas he was about to attack.

The latter activity suggests that early kites were large enough to carry humans, and such appears to have been the case. A kabuki play by the Japanese writer Bakin Takizawa (1767–1848), for example, is based on a twelfth-century legend about the Shogun Minamoto-no-Tametomo and his son, who had been banished to Oshima island in Tokyo Bay. The son is strapped to a large kite and flown to the mainland, where he organizes an army to restore his father to his lands. Marco Polo writes about a century later of a practice among Chinese sailors of building kites that lift seamen into the air. The success or failure of this practice, he writes, is taken as an omen for any planned voyages.

Kite flying was commonly used for more mundane purposes also. In Polynesia, for example, kites have long been used to carry fishing lines far out beyond the shore. The practice can be a more effective way of fishing, since there is not boat to warn or disturb fish. In Japan, kites were sometimes used to lift materials to the top of buildings being constructed.

Kite flying took on both religious and ceremonial significance early on in its history. People quite logically believed that flying a kite might be a way of carrying away unpleasant, evil, or harmful aspects of their lives. During the Sung Dynasty in China (A.D. 950–1126), for example, a Kite Day was specified annually on which a person could load a whole year of bad luck onto a kite and send it into the sky. In both Japan and Korea, kites were often inscribed with the name and birth date of male children and then allowed to float away into the sky, bearing with them any evil spirits that may have associated themselves with the child. As a modern remnant of this practice, Boys Festival is still celebrated in Japan on 5 May each year with the flying of a carp-shaped windsock and kites.

Kite Flying in Europe

Knowledge about kite building appears to have traveled from Asia to Europe by three routes. First, Roman troops fighting in the Middle East seem to have learned about the windsock-shaped kite from their Muslim enemies. They adopted the style for their own banners and called it a *draco* (for "dragon"). In addition to its ceremonial function, the draco was used as an indicator of wind direction for use by Roman archers. The art of kite making was also reported on to Europeans both Marco Polo in the thirteenth century and, later, by Dutch travelers returning from the East Indies in the mid–sixteenth century.

The first concrete evidence for kite flying in Europe is an illustration in a fourteenth-century book showing three soldiers using a kite to drop a bomb on a castle. Until the mid–eighteenth century, however, kite flying in Europe was regarded largely as an idle pastime of interest primarily to children. Then, a few scientific investigators began to appreciate their potential for research into the atmosphere. One of the first such efforts was that of the Scottish astronomer Alexander Wilson (1766–1813). In 1794, Wilson attached a set of meteorological instruments to a train of six kites, with which he measured the temperature, humidity, and wind strength at various altitudes. The system was designed to release each instrument at a pre-designated altitude.

Three years later, Benjamin Franklin (1706–1790) carried out one of the most famous kite-flying experiments of all time. Although perhaps best known in the United States as a statesman, Franklin also made important contributions in a number of scientific fields, especially, the theory of electricity. On 10 June 1752, Franklin released a kite containing a metal antenna into a thunderstorm. He found that an electrical discharge released by a bolt of lightning was identical to the electrical discharges produced in his own laboratory, clarifying for the first time the fundamental nature of atmospheric lightning. Franklin also used kites for less scholarly pursuits as, for example, when he used a kite to pull him across the water while he was swimming.

Throughout the nineteenth century, kites were sometimes used for atmospheric research in the United States and Europe. In 1893, for example, the Australian inventor Lawrence Hargrave (1850–1915) invented a folding kite for use by the U.S. Weather Bureau in collecting atmospheric data, such as temperature, humidity, and wind patterns, and by the U.S. Army for observational flights. Eventually Hargrave's kites were put into use at 17 weather stations around the United States, where they survived until 1933.

Hargrave's folding kite was only one of his many inventions aimed at finding the most aerodynamically stable and efficient lighter-than-air flying device. His most successful design was a simple box-shaped kite which he called his "cellular kite." Hargrave investigated the effects of adding horizontal panels on the front, back and sides of the cellular kite's basic design, as well as various combinations of the basic kite, such as a double-box biplane pattern. He published the results of his investigations in the *Journal of the Royal Society of New South Wales,* gave lectures throughout England, and sent copies of his models to other aviation pioneers in many countries of the world, thus spreading his knowledge throughout the fledgling industry. When the Wrights began their own research on heavier-than-air flight, they started with large gliders that were little more than very large box kites of the design originally suggested by Hargrave. Their first actual flight in such a device took place in 1901 on an oversized box kite 18 feet long and 5 feet wide.

Research on the use of kites for transportation was also well advanced in England. As early as 1804, the English inventor Sir George Cayley (1773–1857) had invented a peg-top kite (a kite with a curved, airplane-wing-style shape) capable of carrying a single person into the air. One of the most successful kite-based transportation systems in England was the *char volant,* invented by an English schoolteacher, George Pocock (1789–1845). Pocock's earliest design consisted of a normal horse-drawn carriage capable of holding four passengers towed by a pair of arch-top kites. With maximum winds, the device traveled at about 30 kilometers per hour (20 miles per hour) over flat ground. In later designs, the char volant could tow carriages holding up to 16 people or a half-ton of cargo under the control of a four-line tow rope system that made the device competitive with horse-drawn coaches over long distances. Pocock's invention benefited from the vast amount of knowledge about wind control obtained from decades of large, efficient sailing ships. Pocock's char volant was doomed to commercial failure, coming as close as it did to the invention of the steam engine. However, his work served as an example for later studies on the aerodynamics of flight that led to the Wright brothers' great breakthrough.

Other Applications

Kites have long had a number of limited and specialized uses. During the Spanish-American War, for example, an American photograph named William Eddy used kites to take pictures of battles in Cuba. He then advertised his photographs for sale on kites carrying flags and lanterns flown over cities. A short time later, a New Yorker by the name of Gilbert Totter Woglom raised an American flag 3 meters (10 feet) long to a height of 300 meters (1,000 feet) to celebrate the opening of the Washington Memorial Arch in New York City.

Kites have had limited application in modern warfare too. In World War II, for example, they were used to deploy barrage cables above merchant ships to protect them from dive bombers. Kites were sometimes used by the German Navy to carry observers from the top of surfaced submarines to locate enemy ships. And kites with radios were used by the U.S. Navy as a way of locating downed airmen.

Kite Types

Over the centuries, kites in almost every imaginable shape and design have been constructed. One of the most common designs consists of a thin sheet of light material (paper, silk, plastic, or cotton) held in place with a sturdy frame of a light material (bamboo, balsa, or wire). The sheet can be almost any geometric shape, such as square, rectangular, triangular, hexagonal, or octagonal. The diamond-shaped kite is perhaps one of the most familiar. Other types of kites are named for their distinctive features, such as the arch-top, box, peg-top, and lozenge kite.

Another class of kites consists of those with three-dimensional shapes. The most common of these is the box kite, usually consisting of a rigid cube or rectangle, at least two sides of which are covered with a thin sheet of light material. Both flat and three-dimensional kites may have a variety of appendages, such as one or more tails and/or one or more horizontal wings for decorative or navigational purposes.

Kite Fights

Kite fighting and other types of kite-flying competitions have very long histories and are especially popular in certain parts of the world, such as Japan, Korea, Malaysia, and other parts of the Far East. According to one tradition, kite flying in Japan began in the early eighteenth century when one Lord Mizoguski convinced two neighboring communities to settle a disagreement by fighting each other with kites rather than in hand-to-hand battle. That tradition continues in the same region of Japan even today with residents battling each other annually with more than 200 giant kites and 1,500 smaller ones.

In 1921, kite fighting was declared the national sport of Thailand, where battles are still conducted between large *chula* (male) kites and smaller *pakpao* (female) kites. The object of such battles is to capture a kite of the opposite sex and return it to one's own territory.

Further Reading

Hart, Clive. *Kites: An Historical Survey.* New York: Praeger, 1967.

Moulton, Ron, and Pat Lloyd. *Kites: A Practical Handbook.* Los Angeles: Transatlantic Publishers, 1997.

Newman, Lee Scott, and Jay Hartley Newman. *Kite Craft: The History and Processes of Kitemaking throughout the World.* New York: Crown Publishers, 1974.

Pelham, David. *The Penguin Book of Kites.* Harmondsworth, UK: Penguin Books, 1976.

Further Information
American Kitefliers Association
P.O. Box 1614
Walla Walla, WA 99362
URL: http://www.aka.kite.org

L

Lighter-Than-Air Machines

See Balloons; Dirigible

Lindbergh, Charles A. (1902–1974)

Lindbergh made aviation history in 1927 when he made the first nonstop transatlantic flight between New York and Paris. He flew an airplane named the *Spirit of St. Louis,* which he had helped design and which was built by Ryan Airlines, Inc., of San Diego. Lindbergh attempted the historic flight in an effort to win a $25,000 prize offered by New York businessman Raymond Orteig in 1919.

Charles Augustus Lindbergh was born at his grandmother's house in Detroit, Michigan, on 4 February 1902. A month later, his family returned to their home in Little Falls, Minnesota, where his father was a lawyer and, later, a representative to Congress from the state's Sixth Congressional District. Lindbergh attended Little Falls High School and the University of Wisconsin at Madison, where he majored in mechanical engineering. In 1922, he left the university and entered a flying school in Lincoln, Nebraska. After earning his pilot's license, he began a career of barnstorming throughout the Midwest.

Charles Lindbergh. © **Library of Congress.**

In 1924, Lindbergh enlisted in the Army Air Service and, after completing his flight training, was commissioned a second lieutenant. His career with the Army was short-lived, however, as there was little demand for pilots at the time. He returned to his barnstorming career and also took work as a test pilot and a mail pilot. In the fall of 1926, he decided to attempt the transatlantic flight that would earn him the Orteig Prize.

Lindbergh's flight was an extraordinary accomplishment, especially when compared to the efforts made by others who had attempted the transatlantic crossing with multi-engine airplanes and crews of two or more. From the moment he landed at Paris' Le Bourget Airfield, he became an instantaneous international hero. He was awarded the Medal of Honor and the first Distinguished Flying Cross ever awarded by the U.S. government. After his return to the United States, he made a tour of more than 80 cities in the *Spirit of St. Louis,* providing a significant impetus to the nation's growing interest in aviation.

In 1932, Lindbergh and his wife, Anne Morrow Lindbergh, suffered a terrible tragedy when their first child, Charles Jr., was kidnapped and later killed. In response to that tragedy and the resulting publicity, the Lindberghs moved to England, where they lived for many years. For much of his adult life, Lindbergh was involved in a number of social and political controversies. During the 1930s, for example, he spoke out against the involvement of the United States in the war then developing in Europe, later to become World War II. His involvement in an isolationist group known as America First eventually prompted his resignation from the Army Air Corps Reserve.

Once the United States did enter the conflict, however, Lindbergh became actively involved in the nation's war effort. He served as an advisor on the production of bombers to the Ford Motor Company and as technical advisor and test pilot for United Aircraft (now United Technologies).

During the early 1960s, Lindbergh also became increasingly concerned about the impact of human activities on the environment. In July 1964, he wrote an important article on the topic for the *Reader's Digest,* "Is Civilization Progress." He later worked on efforts to protect endangered species, such as the humpback and blue whales, and to promote the development of national parks in the United States.

Lindbergh was a prolific author, with seven books to his credit, including his autobiography, *Autobiography of Values,* published posthumously in 1978, and *The Spirit of St. Louis,* for which he won the 1954 Pulitzer Prize.

Lindbergh died of cancer on 26 August 1974 in Maui, Hawaii, his home for the preceding five years.

Further Reading

Berg, A. Scott. *Lindbergh.* New York: G. P. Putnam's, 1998.

Gray, Susan M. *Charles A. Lindbergh and the American Dilemma: The Conflict of Technology and Human Values.* Bowling Green, OH: Bowling Green State University Popular Press, 1988.

Lindbergh, Charles A. *Autobiography of Values.* New York: Harcourt Brace Jovanovich, 1978.

———. *"We."* New York: G. P. Putnam's Sons, 1927.

Linde, Karl Paul Gottfried von (1842–1934)

Karl von Linde was one of the most successful inventors of methods for liquefying air. In 1876, while professor of Theoretical Engineering at the Munich Polytechnic, Linde invented the first reliable refrigerator, a device that operated on liquid ammonia. More than two decades later, he turned his attention to the problem of liquefying air and, in 1895, invented a method by which the product could be manufactured at the rate of nearly a thousand cubic feet per hour. This accomplishment made oxygen, which is obtained from liquid air, cheap and abundant. The element had long been very popular for oxygen therapy in hospitals, as an oxidizer in a variety of industrial processes, and, later, for use in the oxyacetylene torch (invented in 1904), and as part of a propellant mixture for rockets.

Linde was born in Berndorf, Bavaria, on 11 June 1842. From 1861 to 1864, he studied science and engineering under the great Rudolf Clausius (1822–1888) at the Eidgenössische Polytechnikum in Zürich. He then spent two years in the study of locomotives and machines, primarily at the factory of A. Borsig in Tegel, Germany. In 1866, he was appointed head of the technical department of the newly founded Krauss and Company, locomotive manufacturers in Munich.

In 1868, Linde was offered a position as extraordinary professor at the newly founded Munich Polytechnische Schule, later the Technische Hochschule. Four years later, he was appointed full professor of theoretical engineering. During his tenure in Munich, Linde lectured on a number of theoretical and practical topics, including the theory of steam engines, turbines, water wheels, and railroad locomotives.

Linde began his research on refrigeration in 1870. By 1879, his new invention had become so popular that he left his teaching position and founded his own company to produce and sell refrigeration machines, the Gesellschaft für Linde's Eismaschinen (Linde's Ice Machines) in 1875. His company was successful, selling more than a thousand machines over the next decade. But by 1891, he had tired of the business world and decided to return to teaching and research. He accepted an offer to return to Munich, where he concentrated his research and teaching on the theory of refrigeration. In 1902, he was instrumental in creating the institute's first laboratory for applied physics.

In 1907, Linde and his associates formed Linde Air Products in Cleveland to make and sell his refrigeration equipment.

That company later joined with four other companies to produce Union Carbide and Carbon Corporation. In the 1990s, the refrigeration and air liquefaction part of that corporation was again separated to become a new company, Praxair. Linde died in Munich on 16 November 1934. *See also* Hampson, William; Liquid Air

Liquid Air

Liquid air is air that has been cooled below the boiling point of its major components, a temperature of about −195.5°C. At this temperature, nitrogen (b.p. = −195.5°C), oxygen (b.p. = −183°C), argon (b.p. = −189.3°C), and carbon dioxide (m.p. = −78.5°C) have all either liquefied or turned to a solid (in the case of carbon dioxide).

Early Attempts to Liquefy Gases

The first person to liquefy a gas was the English chemist and physicist Michael Faraday (1791–1867). In 1823, Faraday prepared liquid chlorine by heating a hydrate of chlorine (a compound of chlorine and water) at one end of a bent glass tube sealed at both ends. The opposite end of the bent tube was immersed in ice water. As gaseous chlorine was released from the warm end of the tube, it traveled to and condensed in the cooler end.

Faraday and his mentor, Sir Humphry Davy (1772–1829) used this technique to liquefy a number of other gases, including ammonia, hydrogen chloride, hydrogen sulfide, nitrous oxide, and sulfur dioxide. By about 1850, all but six of the known gases had been liquefied. These six gases (oxygen, nitrogen, nitrogen dioxide, carbon monoxide, methane, and hydrogen) were thought by some to be incapable of liquefaction and were called *permanent gases*. (Helium was later added to this list when it was discovered in 1895.)

The method used by Faraday and Davy illustrate two techniques by which a gas can be converted to a liquid: (1) by cooling the gas below its boiling point, and (2) by compressing the gas under high pressure, usually of the order of 50 atmospheres pressure or more. In either of these cases, the molecules of which the gas is composed are compressed until they are close enough together to take on the liquid state.

In 1877, the French physicist Louis Paul Cailletet (1832–1913) liquefied three of the six permanent gases, carbon monoxide, nitrogen, and oxygen. Cailletet's work depended on an important discovery made by the Irish chemist Thomas Andrews in 1869. Andrews found that there is some temperature, called the *critical temperature,* above which no amount of pressure alone can liquefy a gas. Andrews's discovery showed that simply applying more and more pressure to a gas was not sufficient by itself to liquefy the gas if it were above its critical temperature. It would be necessary to use some combination of cooling and pressure to bring about liquefaction.

Cailletet used a method based on the Joule-Thompson effect, a physical phenomenon originally discovered by the English physicist James Prescott Joule (1818–1889) and the Scottish physicist William Thomson, Lord Kelvin (1824–1907). Joule and Thompson had discovered that a gas that expands from an area of high pressure to one of lower pressure undergoes a change in temperature. The temperature change depends on the gas and the temperature, but for most gases at or near room temperature, expansion results in a decrease in temperature. For example, air at room temperature decreases in temperature by about 0.25K (degree kelvin) for each one atmosphere decrease in pressure.

Cailletet's approach, then, was to begin by cooling a gas as much as possible by means of ice, frozen carbon dioxide, or other very cold substances. He then allowed the cold gas to expand, causing it to cool even further. In the case of carbon monoxide, nitrogen, and oxygen, this approach was adequate to produce small amounts of liquid forms of each gas.

Results similar to those of Cailletet's were being obtained at about the same time by other scientists. In Switzerland, for example, the chemist Raoul Pierre Pictet (1846–1929) used somewhat more sophisticated equipment and was able to produce larger amounts of the permanent gases in their liquefied states. In Poland, Zygmunt von Wroblewski (1845–1888), professor of physics, and his colleague Karol Olszewski (1846–1915) were also successful in liquefying relatively large amounts of carbon monoxide, nitrogen, and oxygen. They also were probably the first scientists to liquefy hydrogen, as they were able to observe a fine mist when they subjected hydrogen gas to their compression and cooling technique.

Regenerative Cooling

By the late nineteenth century, scientists had developed yet another technique for cooling gases, a process known as *regenerative cooling.* In this process, a gas that has been cooled and condensed is then used itself as a coolant. The principle behind regenerative cooling had first been introduced in 1857 by the German-British inventor Sir William Siemens (1823–1883). Siemens first proposed the concept of *regenerative heating,* in which hot waste gases are used to heat a steam engine, as a way to increase the engine's efficiency. The same principle proved to be an effective way of cooling gases to even lower temperatures than had previously been obtained.

Patents for methods of regenerative cooling were issued to two individuals in 1895 within two weeks of each other. The patents went to the English inventor William Hampson (1854–1926) and the German chemist Karl von Linde (1842–1934). Hampson's method was the simpler of the two. He compressed air to a pressure of 200 atmospheres and then expanded it to a pressure of 1 atmosphere. The very cold air thus formed was then used to cool incoming air in a second stage of that process. Eventually the air became cool enough to liquefy. Hampson's technique, although simple, was not very efficient and soon lost out to the approach developed by von Linde.

In Linde's process, incoming air was first cooled in an ammonia refrigeration system, then cooled twice by a Hampson-type compression and expansion process. The crucial parts of Linde's apparatus were well insulated with a wool-filled wooden case. With Linde's method, liquid air could be produced in volume for commercial use, rather than in small samples of only research interest.

The methods of Linde, Hampson and their predecessors was brought to an even higher point of development with the work of the Scottish chemist and physicist Sir James Dewar (1842–1923). Dewar was able in 1898 to liquefy the last of the permanent gases, hydrogen. He made about 20 cubic centimeters of the product by using a regenerative cooling procedure that began with pressures of 180 atmospheres on a sample of hydrogen gas.

In 1893, Dewar also took the final step in condensing air by producing a sample of solid air from liquid air. A few years later, in 1902, he exhibited his command of liquid and solid air by constructing and exhibiting for the prince and princess of Wales a 6-foot-tall fountain of liquid air issuing from a block of solid air.

Uses of Liquid Air

The most important commercial use of liquid air is as a source of elements normally found in air, such as oxygen, nitrogen, and argon. When a sample of liquid air is allowed to warm up under controlled conditions, the gases of which it is composed begin to boil off, one at a time. The first fraction to escape consists largely of nitrogen, with small amounts of helium and neon. The liquid air that remains consists largely of oxygen and argon. Very small amounts of krypton and xenon can also be removed from the last fraction.

Liquid air is becoming popular for other uses as well. For example, refrigeration systems have been designed that operate on liquid air, rather than ammonia or other traditional heat exchange mediums. Also, some liquid air generators have been designed for use where limited amounts of oxygen (for which liquid air is the major commercial source) can be generated easily and inexpensively. These units are sold to hospitals and emergency rescue services. Liquid air also has limited applications in cryogenic (low-temperature) research and production. *See also* Hampson, William; Linde, Karl Paul Gottfried von

Further Reading

Liquid Air, IGC Document 82/01/E. Brussels: European Industrial Gases Association, 2001. Available online at the association's Web site, http://www.eiga.org.

Literary References to Air

See Air in Literature and Film

M

Montgolfier, Joseph Michel (1740–1810), and Jacques Étienne (1745–1799)

Joseph Michel and Jacques Étienne Montgolfier, brothers, conducted the first successful experiments in the construction and operation of balloons. They were both born in Vidalon-lez-Annonay, France, on 26 August 1740 and 6 January 1745, respectively, the sons of a wealthy paper manufacturer. Legend has it that the Montgolfier family learned the process of papermaking from an ancestor who fought in the Crusades and was introduced to the procedure while a captive of the Muslims.

The Montgolfier brothers first began to think about balloons as they watched bits of unburned paper carried upward by the heat of a fire. They first built small paper bags, which they filled with warm air from a fire, and watched them ascend to the ceiling of a room. Eventually, they constructed larger and larger balloons of sturdier materials. On 5 June 1783, they assembled a linen bag about 10 meters (35 feet) in diameter weighing about 225 kilograms (500 pounds) in the marketplace of their hometown. They ignited a fire of paper and wood beneath the balloon, which rose to a height of about 2,000 meters (6,500 feet). The balloon trav-

French balloonist Joseph Montgolfier holding on to balloon ropes. © Library of Congress.

eled a distance of about two kilometers (1.5 miles) before coming to rest.

The Montgolfiers continued to improve their balloon designs, building larger and larger containers that could travel higher and farther. But their work was soon overshadowed by the suggestion by countryman Jacques Charles (1746–1823) that hydrogen be used rather than hot air as a lifting mechanism for balloons. In recognition of their work, however, the Montgolfiers were elected correspondents to the French Académie des Sciences in 1783, which recognized them "as scientists to whom we are indebted for a new art that will make an epoch in the history of human science."

Jacques Étienne died at Serriéres on 1 August 1799. His brother then discontinued his research on balloons and turned his attention to other scientific topics. He invented a parachute, a calorimeter, a hydraulic ram, and a hydraulic press. He was noted as a great humanitarian offering refuge to people condemned during the French Revolution. He died in Balaruc-les-Bains on 26 June 1810. *See also* Balloons

Motion Pictures, Air Themes

See Air in Literature and Film

Musical References to Air

Throughout the ages, musical composers have used air, in one manifestation or another, as the subject of their compositions. Although air itself appears rather infrequently as the focus of a musical composition, both wind and sky are relatively common subjects.

Classical Music

Classical music can be classified as *programmatic* or *absolute.* Programmatic music has some connection with the real world. It may attempt to tell a story or produce sounds similar to those found in everyday life. The Third Movement of Beethoven's *Sixth Symphony,* for example, is a easily recognizable telling of a summer day in the country, replete with thunder, lightning, storms, and bright sunshine.

Air, wind, and the sky seldom appear in classical music in the form of programmatic music. One can not read the title of a work and expect that it will tell anything about the physical nature of the sky, the wind, or some other aspect of air. The forces that motivated a composer to include these words in the title of a work are usually unknown to listeners. Some examples of classical works with air-related themes include the following:

- "Clear Sky—Sunshine," by Danish composer Niels Wilhelm Gade (1817–1890, a piano selection in A flat major.
- "The Sky above the Roof," by English composer Ralph Vaughan Williams (1872–1958), a song set to a poem written by the French poet Paul Verlaine (1844–1896).
- "At the Corner of the Sky," written by American composer Leo Smit (1921–), a selection for flute, oboe, speaker, and chorus.
- "Small Sky," by Japanese composer Toru Takemitsu (1930–1996), a song for a cappella choir, based in part on the traditional Irish air, "Danny Boy."
- "Starlit Sky," written by American composer Barry Rose (1934–), as a religious anthem for choir.
- "The Sky Is Low, the Clouds Are Mean," by American composer Anthony Iannaccone (1943–), written for chorus.

Popular Music

The choice of an air-related themes in popular music is much easier to understand. Lyrics of popular songs frequently call upon some real physical aspect of the air, wind, or sky for their theme or, more frequently, use an aspect of air to represent some specific thought or emotion expressed in the song's words. Some examples of the use of air-related ideas in popular music are the following:

- "Slow Hot Wind," by Norman Gimbel and Henry Mancini (1963), compares a hot summer breeze with the stares of someone watching the singer with "slow fire in his eyes."
- "Summer Wind," by Henry Mayer and Johnny Mercer (1965), tells of two lovers who share the summer with the summer wind.
- "Black-Throated Wind" is one of the Grateful Dead's less commonly performed works, written in 1972 by Bob Weir and John Barlow. It tells of the unhappy journeys of the singer for whom the wind "speaks of a life that passes like dew."
- "Goodbye Blue Sky," written by Roger Waters and David Gilmour in 1979, is a lament about the bright promises of a future filled with peace that have been defeated by increased militarism around the world

- "Against the Wind," by Bob Seger (1980) is a recounting of the composer's life-long battle against various types of adversity, battles "against the wind."
- "Castles in the Air," written in 1981 by Don McLean, tells of the singer's unhappy break-up with a "city woman" and his search for a simpler, more natural life.
- "The Wind beneath My Wings," written by Jeff Silbar and Larry Henley in 1982, offers praise to "my hero," who lifts the singer into the sky and acts as his/her "wind beneath my wings."
- "Candle in the Wind," by Elton John and Bernie Taupin, written in 1997 for the funeral of Princess Diana, praising her for standing up for so many worthwhile causes, "like a candle in the wind."

N

National Oceanic and Atmospheric Administration (NOAA)

The National Oceanic and Atmospheric Administration is the lead U.S. agency for monitoring, educational, and research programs on the Earth's atmosphere. It was established in 1970 in Congress's Reorganization Plan Number 4, bringing together the previously independent National Weather Service, National Ocean Survey, and National Marine Fisheries Service. The agency's scope and responsibilities have been further broadened in later legislation, including the Coastal Zone Management Act of 1972, the Marine Mammal Protection Act of 1972, the Sanctuaries Act of 1972, the Weather Modification Reporting Act of 1972, the Endangered Species Act of 1973, the Offshore Shrimp Fisheries Act of 1973, the Fishery Conservation and Management Act of 1976, and the Land Remote Sensing Act of 1984.

The agency primarily responsible for atmosphere-related issues is the National Weather Service (NWS). NWS was created in 1870 as an arm of the Department of War with the mission of "taking meteorological observations at the military stations in the interior of the continent and at other points in the States and Territories . . . and for giv-ing notice on the northern (Great) Lakes and on the seacoast by magnetic telegraph and marine signals, of the approach and force of storms." In 1890, the NWS was transferred to the Department of Agriculture, and its responsibilities were broadened to include many nonmilitary functions, such as providing weather information, forecasts, and warnings to the general public.

Today, NOAA's work is divided into eight general areas: weather, ocean, satellites, fisheries, climate, coasts, charting and navigation, and research. Activities of the National Weather Service fall into two general categories: (1) information about normal and severe weather patterns for the aviation industry, firefighting organizations, mariners, and the general public, and (2) research on weather and climate problems.

An example of the type of work sponsored by the NWS is the Wind Profiler Network, established in 1986. The Wind Profiler Network consists of groups of Doppler radar units that measure wind speed and direction up to an altitude of 16 kilometers in a region. The network is intended to supplement and eventually replace wind measurements currently being made by manually operated weather balloons. Network data are critically important in general weather forecasting and are of special value to airports, where they can be used to detect and forecast wind shear and downbursts that create serious hazards for commercial and private aircraft.

Another important NWS function is NOAA Weather Radio (NWR), a nationwide network of radio stations broadcasting weather information 24 hours a day. The network carries general weather information as well as forecasts, warnings, and post-event information on all types of weather emergencies, both natural and human-made. NWR, sometimes called the "Voice of the National Weather Service," has nearly 500 transmitters in all 50 states and adjacent coastal waters, Puerto Rico, the U.S. Virgin Islands, and the U.S. Pacific territories. NWR broadcasts on seven frequencies between 162.400 MHz and 162.550 MHz. It is now part of the Federal Communication Commission's Emergency Alert System, providing a single source for all types of emergency information for the general public.

An example of the atmospheric work conducted by NOAA is the Climate Modeling and Diagnostic Laboratory in Boulder, Colorado. The function of the Laboratory is to conduct global samples of carbon dioxide, halocarbons, and other compounds that may have an effect on global climate change. It was created to meet one of NOAA's primary challenges, to monitor, model, and study worldwide environmental changes that may affect the lives of all Americans.

Further Information
National Oceanic and Atmospheric
Administration
14th Street & Constitution Avenue, NW
Room 6013
Washington, DC 20230
Telephone: (202) 482-6090
Fax: (202) 482-3154
URL: http://www.noaa.gov
E-mail: answers@noaa.gov

National Weather Service
1325 East West Highway
Silver Spring, MD 20910
URL: http://www.nws.noaa.gov
E-mail: w-news.webmaster@noaa.gov

National Weather Service

See National Oceanic and Atmospheric Administration (NOAA)

National Wind Coordinating Committee

The National Wind Coordinating Committee (NWCC) was founded in 1994 for the purpose of providing a forum for "identifying and discussing issues that impact the use of wind power, to catalyze actions addressing key issues, and to build consensus among varied stakeholder groups." Members include representatives of the wind industry (such as the Atlantic Renewable Energy Corporation, the American Wind Energy Association, and Vermont Environmental Research Associates, Inc.); state legislators (include state senators and representatives from Iowa, Kansas, and North Dakota); environmental organizations (such as Environmental & Energy Study Institute, Land & Water Fund of the Rockies, and the Union of Concerned Scientists); green power marketers (such as Competitive Utility Strategies, Ed Holt & Associates, and Green Marketer); utility regulators (such as the Montana and Wyoming Public Service Commissions and the Pennsylvania and South Dakota Public Utilities Commissions); local, regional, tribal, state, and federal agencies (such as the California Energy Commission, Intertribal Council on Utility Policy, and the U.S. Department of Energy Wind Energy Program); agricultural and economic development agencies (such as the economic development director of the

City of Lake Benton, Minnesota, and the U.S. Department of Agriculture); and other utilities and support organizations (such as the Alliance of Energy Suppliers, American Electric Power, and Bonneville Power Administration).

The NWCC has produced more than a dozen white papers on wind energy topics, including

- "The Benefits of Wind Energy;"
- "Wind Energy Environmental Issues;"
- "Wind in a Restructured Electric Industry;"
- "Wind Energy Costs;" and
- "Distributed Model for Delivery of Electricity."

The committee has also produced a number of reports on specialized issues related to the development of wind energy. These include "Transmitting Wind Energy: Issues and Options in Competitive Electric Markets," "Wind Power Transmission Case Studies Project," "Permitting of Wind Energy Facilities: A Handbook," "Studying Wind Energy/Bird Interaction: A Guidance Document," "Understanding Consumer Demand for Green Power," "Credit Trading: Issues and Opportunities," and "NWCC Guidelines for Assessing the Economic Development Impacts of Wind Power."

Further Information
National Wind Coordinating Committee
1255 23rd Street, NW, Suite 275
Washington, DC 20037
Telephone: (202) 965-6398 or (888) 764-WIND
Fax: (202) 338-1264
URL: http://www.nationalwind.org
E-mail: nwcc@resolv.org

National Wind Technology Center

The National Wind Technology Center is one of two major research laboratories (the other being the Sandia National Laboratory) at which research and development in the area of wind power is conducted. The center is located in Golden, Colorado, where it is operated by the U.S. Department of Energy's National Renewable Energy Laboratory.

The primary focus of the center is to work with the U.S. wind industry to design and develop improved forms of wind technology to broaden the nation's range of energy resources and to reduce the amount of greenhouse gases emitted to the atmosphere.

As one way of meeting this objective, the center provides expert assistance to private wind power companies about new technologies that have been developed and that are available for transfer to private industry. The center also has an extensive collection of data and information of value to private industry. For example, its Wind Resource Database contains a copy of the *Wind Energy Resource Atlas of the United States;* a collection of U.S., state, and international wind resource maps; a wind resource bibliography; and an extensive collection of weather data from the center, the Weather Channel, and *USA Today* weather site.

The center also maintains an online bibliography of relevant reports containing more than 100 titles. Some examples of the topics covered include "Avian Risk and Fatality Protocol," "Certification Testing for Small Turbines," "Economics of Grid-Connected Small Wind Turbines in the Domestic Marketplace," "IEA Wind Energy Annual Reports," "Mongolia Wind Resource Assessment Project," and "Renewable Energy for Rural Health Clinics."

Further Information
National Wind Technology Center
National Renewable energy Laboratory
1617 Cole Boulevard
Golden, CO 80401

Telephone: (303) 384-6922
URL: http://www.nrel.gov/wind

Nitrogen

Nitrogen is a colorless, odorless, tasteless gas that makes up about 78 percent of the Earth's atmosphere. It also occurs in a number of rocks and minerals in the Earth's crust, the most important of which are potassium nitrate (saltpeter or niter) and sodium nitrate (soda nitre or Chilean saltpeter).

Physical and Chemical Properties

Nitrogen has an atomic number of 7, an atomic mass of 14.0067, and a chemical symbol of N. Its boiling point is $-195.79°C$ ($-320.42°F$) and its freezing point, $-210.01°C$ ($-346.02°F$). Frozen, it forms a white solid that looks like snow. Nitrogen has a density of 1.25046 grams per liter, slightly less than that of air (1.29 grams per liter). The gas is slightly soluble in water. About 2 liters of nitrogen can be dissolved in 100 liters of water.

Nitrogen is a relatively inactive element at room temperature. It becomes somewhat more active at elevated temperature and reacts readily with oxygen in the atmosphere in the presence of lightning. The product of that reaction is nitric oxide (NO):

$$N_2 + O_2 \rightarrow 2NO$$

Nitric oxide is considerably more reactive than is nitrogen gas itself. It combines readily with oxygen and water in the atmosphere to form nitric acid (HNO_3). Nitric acid is carried to the Earth's surface when it rains or snows. There it reacts with metals, minerals, organic material, and other naturally occurring substances to form nitrates and nitrites.

The process by which elemental nitrogen is converted to nitrogen compounds is called *nitrogen fixation*. Nitrogen fixation also occurs in the soil, where certain types of bacteria have evolved the ability to convert nitrogen gas dissolved in the ground into nitrates. Those nitrates are used by plants to make proteins and other essential biochemical substances.

Discovery and Naming

Nitrogen was discovered in 1772 by the Scottish physician and chemist Daniel Rutherford (1749–1819). Rutherford collected a sample of air under a bell jar. He then placed a mouse and a burning candle in the air. Eventually, the candle burned out, and the mouse died. Rutherford concluded that the gas remaining after the oxygen had been used up by the burning candle was incapable of supporting life.

Nitrogen was discovered almost simultaneously and independently by the English chemist Joseph Priestley (1733–1804) and the Swedish chemist Carl Wilhelm Scheele (1742–1786). The element was first given the name of *azote* by the French chemist Antoine Laurent Lavoisier (1743–1794). The name comes from the French expression meaning "without life," reflecting the fact that the element does not support breathing. In 1790, the modern name of nitrogen was suggested by the French chemist Jean Antoine Claude Chaptal (1756–1832). Chaptal chose the name because the gas is found in both nitrates and nitric acid. The name *nitrogen,* then, means "nitrate and nitric acid" (*nitro-*) and "origin of" (*-gen*).

Production and Uses

Nitrogen is produced commercially by the fractional distillation of liquid air. When liquid air is allowed to warm up, its component gases evaporate, one at a time. The first gas to boil off is nitrogen, the gas with the lowest boiling point. About 9 million short tons of nitrogen gas are produced by this method in the United States each year.

By far the most important use of nitrogen gas is in the preparation of ammonia

(NH_3), a compound widely used in the production of synthetic fertilizers. About 90 percent of all nitrogen produced is converted to ammonia. The ammonia, in turn, is converted to nitric acid, ammonium, sulfate, ammonium phosphate, and other compounds used in fertilizers. Smaller amounts of nitrogen are in liquid form, to freeze other substances, and as a protective environment in the preservation of valuable documents. *See also* Atmosphere

Further Reading

Information on nitrogen is available in most introductory high school and college chemistry textbooks.

Nitrogen Fixation

See Nitrogen

O

Optical Phenomena

See Atmospheric Optical Phenomena

Oxygen

Oxygen is a colorless, odorless, tasteless gas that makes up nearly 21 percent of the Earth's atmosphere. It also occurs as a component of water in oceans, lakes, rivers, and ice caps. Nearly 89 percent of the weight of water is oxygen. Oxygen is also the most abundant element in the Earth's crust. Its abundance is estimated at about 45 percent by weight, making it almost twice as abundant as the next most common element, silicon. Some common oxygen-containing minerals are oxides (such as silicon dioxide, or sand), carbonates (such as calcium carbonate, the main ingredient in chalk, marble, limestone, and calcite), nitrates (such as sodium nitrate, or Chilean saltpetre), and phosphates (such as calcium phosphate, a component of "phosphate rock").

Physical and Chemical Properties

Oxygen has an atomic number of 8, an atomic mass of 15.9994, and a chemical symbol of O. It changes from a gas to a bluish liquid at a temperature of $-182.96°C$ ($-297.33°F$). It can then be solidified at a temperature of $-218.4°C$ ($-361.2°F$). Oxygen has a density of 1.429 grams per liter, slightly greater than the density of air (1.29 grams per liter).

Oxygen is not very active at room temperature, but its activity increases significantly with increases in temperature. For example, oxygen reacts with metals only quite slowly at room temperature, a process known as *rusting*. It also reacts slowly with organic materials in the process known as *decay*. But at higher temperatures, it reacts very rapidly with a much greater variety of materials in the process known as *combustion*.

Oxygen exists in three allotropic forms. Allotropes are forms of an element with differing chemical and physical properties. The three allotropes of oxygen are normal oxygen, also known as dioxygen and diatomic oxygen (chemical formula: (O_2); nascent, atomic, or monatomic oxygen (chemical formula: O); and ozone, or triatomic oxygen (O_3).

Diatomic oxygen is the most common form of oxygen because both nascent oxygen and ozone react to form diatomic oxygen. In the case of nascent oxygen, two oxygen atoms combine to form an oxygen molecule:

$$O + O \rightarrow O_2$$

Ozone molecules are relatively unstable and react or break down to form diatomic oxygen:

$$O_3 + O_3 \rightarrow 3O_2.$$

Oxygen in the atmosphere is constantly undergoing changes from one allotrope to another. For example, in the presence of energy, such as the energy of sunlight, oxygen molecules may break apart and recombine to form ozone:

$$O_2 \xrightarrow{Energy} O + O$$
$$O + O_2 \rightarrow O_3$$

Discovery and Naming

Oxygen was discovered independently in 1774 by the Swedish chemist Carl Wilhelm Scheele (1742–1786) and the English chemist Joseph Priestley (1733–1804). The element was named by the French chemist Antoine Laurent Lavoisier (1743–1794), sometimes called "The Father of Modern Chemistry." The word means "acid" (*oxy-*) "former" (*-gen*) because Lavoisier thought that oxygen was a component of all acids, a point about which he was wrong.

Biological Function

Oxygen is a component of the four major families of biochemical compounds that make up all living things: carbohydrates, proteins, lipids, and nucleic acids. It also appears in virtually every other compound found in plants and animals. Plants obtain the oxygen they need from three sources: atmospheric oxygen (O_2), carbon dioxide in the air (CO_2), and water. Most animals obtain the oxygen they need for respiration from the atmosphere or from air dissolved in water.

Organisms use oxygen to generate the energy they need for normal life functions, such as growth, reproduction, and maintenance of biological systems. This energy is produced in the process of metabolism, by which cellular compounds are converted to simpler forms with the release of energy.

Production and Uses

Oxygen is produced commercially by the fractional distillation of liquid air. Liquid air is allowed to evaporate under controlled conditions. The first portion obtained during this process is nitrogen since it has the lowest boiling point of all major components of air. When nitrogen has boiled off from liquid air, the next portion to evaporate is oxygen.

One of the best known uses of oxygen is in artificial respiratory systems for people with breathing problems. Oxygen may be delivered through breathing masks or, in more serious conditions, through oxygen tents. Oxygen, in the liquid form, is also used in rocket propulsion systems.

By far the largest amount of oxygen, however, goes to the preparation and treatment of metals. For example, oxygen is used to burn off carbon and other impurities in the preparation of iron and steel. Oxygen is also used to extract metals from their ores. As an example, copper, lead, and zinc are produced first by heating their ores (the sulfides) in the presence of oxygen. For example:

$$2ZnS + 3O_2 \rightarrow 2ZnO + 2SO_2$$

The oxide formed in this process (ZnO in this example) is then treated with carbon to obtain the pure metal. For example:

$$2ZnO + C \rightarrow 2Zn + CO_2$$

See also Atmosphere; Liquid Air; Ozone; Respiration

Further Reading

Sawyer, Donald T. *Oxygen Chemistry.* New York: Oxford University Press, 1991.

Information on oxygen is available in most introductory high school and college chemistry textbooks.

Ozone

Ozone is an allotrope of oxygen with the chemical formula O_3. Allotropes are forms of an element with differing chemical and physical properties. Ozone differs from "normal" (diatomic) oxygen in that its molecules contain three atoms of oxygen each (thus, O_3), while the form of oxygen normally encountered contains two atoms of oxygen per molecule (O_2).

Physical and Chemical Properties

Ozone is pale-blue gas with a distinctive pungent odor. Its boiling point is about $-112°C$ ($-170°F$) and its freezing point, about $-192.5°C$ ($-314.5°F$). Its density is 2.143 grams per liter, about 50 percent greater than that of diatomic oxygen. Ozone is nearly twice as soluble in water as is diatomic oxygen. Liquid ozone is a deep blue color and highly reactive. Solid ozone is deep purple, nearly black.

Ozone is highly reactive because it is unstable. It decomposes to release an atom of nascent oxygen (O) and a molecule of diatomic oxygen (O_2):

$$O_3 \rightarrow O + O_2$$

Nascent oxygen, in turn, is also highly reactive, attacking metals, nonmetals, organic materials, and other substances with which it comes into contact. This property of ozone explains one of its most important uses, in the purification of drinking water, swimming pools, sewage, and other systems.

Discovery and Naming

Ozone was first observed in 1785 by a Dutch chemist Martin van Marum (1750–1837). He noticed the characteristically strong and pungent smell of the gas around electrical machinery. Ozone is commonly produced in such cases because the energy of an electrical charge can cause the formation of ozone from diatomic oxygen molecules:

$$3O_2 \rightarrow 2O_3$$

Van Marum did not isolate or study the gas, however, so credit for its discovery goes instead to a German chemist Christian Schönbein (1799–1868) who isolated, studied, and named the gas in 1840. He chose the name *ozone* because of its distinctive odor. In Greek, *ozo* means "smell."

It was not until 1886, however, that the first use of ozone as a germicide was investigated by a French scientist by the name of de Meritens. Within the decade, the utility of ozone as a purification agent had been established and drinking water treatment plants had been established in Oudshoorn, The Netherlands, and Paderborn and Weisbaden, Germany.

Production and Uses

Ozone's instability makes it both dangerous and expensive to transport, so the gas is usually produced at the site where it is to be used. The simplest procedure is to irradiate air electronically, producing a mixture containing about 2 percent ozone. The primary use of ozone is in purification systems, such as those used in swimming pools and water and sewage treatment plants. Ozone is also used commercially as a bleaching agent, as in the bleaching of textiles, paper, and waxes and oils. Smaller ozone units are also used in homes and offices to eliminate odors and provide germicide action. Ozone is also used in industrial operations as a source of oxygen.

Environmental Issues

Ozone exists in every part of the atmosphere. But it is considerably more abundant in the stratosphere, the region of the atmosphere between about 10 to 16 kilometers (about 6 to 10 miles) to a distance of about 50 kilometers (about 30 miles) above the Earth's surface. The largest concentration of ozone occurs in a relatively thin layer located between 20 and 30 kilometers (12 and 18 miles) high. In the ozone layer, the concentration of the gas is at least six times as great as it is just above Earth's surface and about 300 times as great as the concentration of ozone in the middle of the troposphere (about 5 kilometers, or 3 miles, above Earth's surface).

The ozone layer is of considerable environmental significance because it shields living organisms on Earth from the harmful effects of certain parts of solar radiation, known as ultraviolet (UV) radiation. For example, UV radiation is known to kill microscopic plants that form the basis of food chains in the open seas. It has also been implicated in higher levels of skin cancer in humans and suppressed immune systems.

Over the past half century, there has been growing concern about damage to the Earth's ozone layer from certain synthetic chemical products known as chlorofluorocarbons (CFCs). These chemicals react in the stratosphere with ozone molecules, breaking them down into ordinary oxygen. The consequence of this change for the health of humans, other animals, and plants has been sufficient to bring out a worldwide ban on the production and use of CFCs. *See also* Air Pollution; Atmosphere; Oxygen

Further Reading

Horváth, M., L. Bilitzky, and J. Huttner. *Ozone.* Amsterdam: Elsevier, 1985.

P

Parachute

A parachute is a device for slowing a person's descent through the air, usually made of a hemispherical piece of fabric with cords to which the person is attached.

Early History

As with so many other inventions, the idea of a parachute in its modern form was thought to have been invented by Leonardo da Vinci (1452–1519). In one of his sketch books, he shows a man suspended from a pyramidal piece of cloth about 7 meters (36 feet) across at its base. One Fauste Veranzio is said to have constructed a parachute along the lines of da Vinci's invention and used it to descend from a tower in Venice in 1617.

In fact, it was not until three centuries later that the first parachute descent of which we know took place. In 1783, the French physicist Louis-Sebastien Lenormand (1757–1839) launched himself from the tower of the Montpelier Observatory and drifted safely to the ground. Within a few years, other inventors were testing their own parachute designs. In 1785, the French balloonist Jean Pierre Blanchard (1753–1809) attached a parachute to his dog and dropped it from a height of a few hundred meters. The dog landed safely but is said to have run away and never been seen again.

Action parachute. © National Parachute Ind., Inc.

The first parachutist to attempt making a living from the activity was the Frenchman Andre-Jacques Garnerin (1769–1832). Garnerin's first descent took place on the outskirts of Paris on 22 October 1979 when he fell a distance of about 700 meters (2,400 feet) from a balloon. Garnerin's accomplishment was significant for three reasons. First, his parachute was one of the first to have the now-familiar hemispherical shape, rather than the pyramidal form recommended by da Vinci and used by nearly all Garnerin's predecessors. Second, the parachute had a hole in the middle of the parachute, eliminating the unpredictable and sometimes dangerous descents caused by air trapped in the parachute. Third, the parachute lacked a rigid frame that had been part of all previous parachutes. Garnerin went on to make numerous exhibition jumps over France and England, his most impressive being a 2,500-meter (8,000-foot) jump over England in 1802.

Development of the Modern Parachute

During the nineteenth century, parachuting remained an activity limited to amateur enthusiasts. Except for those who used balloons for some recreational or utilitarian purpose, the parachute had no practical value. It was the efforts of these amateur enthusiasts, however, that led to a number of important inventions in parachute design. For example, a French inventor named Bourget in 1804 developed the first parachute that could be folded up. Nearly all previous devices had been made of canvas, cotton, silk, or some other material attached to a wooden support system that kept them open.

Late in the century, U.S. Army Captain Thomas Scott Baldwin (1854–1923) and his brothers invented a parachute made out of silk with no stiffeners of any kind, essentially like parachutes used today. Baldwin offered to jump from a balloon with his parachute for a dollar a foot. The first exhibition of the new Baldwin parachute took place on 30 January 1885 at San Francisco's Golden Gate Park when Thomas jumped from a height of 5,000 feet. He became so successful and so well known that he earned a number of honors and trophies, including induction to the National Aviation Hall of Fame in 1964. Some historians have called him the Father of the Modern Parachute.

The invention of the airplane by the Wright brothers in 1903 provided an important role for the parachute. It provided a mechanism by which fliers would have a way of escaping from a damaged aircraft and saving their own lives. The early twentieth century saw additional refinements in the construction of parachutes, giving them their contemporary appearance. In 1911, for example, an Italian by the name of Pino invented the drogue parachute. In Pino's design, the parachute system consisted of two parts, the major parachute to carry a person to the ground and a smaller parachute, the drogue. When a person jumps from an airplane, he or she first pulls a cord that releases the smaller drogue parachute, slowing the rate of descent. The drogue parachute then pulls a cord on the larger parachute, causing it to open and inflate to full size. This system is essentially the one on which all modern parachutes are based.

Parachutes in Wartime

During World War I, parachutes were used almost exclusively by the German army for escape from damaged balloons and airplanes. The French, English, and Americans used parachutes rarely, primarily for men assigned to balloon duty. Staff officers were concerned that pilots might abandon a damaged aircraft too quickly if they had parachutes at their disposal. Also, there was a feeling among some fliers that parachutes were "unmanly" and they were not willing to wear them during flight. Finally, the technology of parachute design was not well developed, and army officials worried that fliers would be more at risk from jumping out of their planes with a parachute than trying to land a stricken aircraft.

One of the earliest proponents of parachute use in the United States military was Colonel Billy Mitchell, who proposed outfitting all fliers with parachutes. Mitchell's suggestion came within a month of the end of the war, however, and it was quickly dropped. Mitchell was also apparently the first military official to suggest the possibility of using parachutes to drop soldiers behind enemy lines in a war, a suggestion that was largely ignored because of the war's termination.

During the late 1930s, Russia, Germany, and France began to train military personnel, called *paratroopers,* to jump from aircraft into a battle zone using parachutes. The first nation to use paratroopers in actual combat was the Soviet Union (now the Russian Federation) in its campaign against Finland in the winter of 1939–1940. In the spring of 1940, the Germans also used paratroopers extensively in their assaults on Norway, Belgium, the Netherlands, and Luxembourg. The success of the Russians and Germans in the use of paratroopers convinced the U.S. Army to organize its first paratroop unit, the 501st Regiment, stationed at Fort Benning, Georgia, in April 1941. The first combat jump made by American paratroopers occurred on 8 November 1942 in Tunisia.

Principle of Operation

The first step in deploying a parachute is to pull on a ring attached to a *rip cord,* which releases the parachute from the bag in which it is contained. In most cases, a drogue (or *pilot*) parachute is released first, which, in turn, releases the main parachute from its container. At this point, the main body of the parachute, its *canopy,* is free to open and assume its shape. While most parachutes are still hemispherical in shape, they are also made in other configuration, such as hemicylindrical or square.

Once the parachute is open, it begins to exert a drag on the downward motion of the person to whom it is attached. A *drag* is defined as any resistance to the motion of an object through a fluid. In this case, the parachute acts in opposition to the free fall of a person caused by the Earth's gravitational attraction. Without a parachute, a person falls through the air at a maximum speed (*terminal velocity*) of about 54 meters per second (120 miles per hour). The effect of the parachute is reduce the person's fall to a speed of about 7 meters of second (15 miles per hour).

The speed of descent depends on a number of parachute characteristics, such as its shape, surface area, permeability of the material of which the parachute is made, length of the suspension (or *shroud*) lines, and total weight (including the payload). A formula that can be used for calculating descent speed for a round parachute, for example, is:

$$v = \sqrt{\frac{2W}{\rho CS}}$$

where

v is the descent velocity in meters per second;
W is the total mass of the parachute and load in newtons;
ρ is a constant with a value of 1.225 kilograms per cubic meter;
C is the drag coefficient (about 0.8 for a hemispherical parachute); and
S is the surface area of the parachute, in square meters.

Today, parachutes are an essential component of emergency escape systems in all types of aircraft, both military and civilian. They are also used by military units for the delivery of troops and supplies. During peacetime, parachutes are also employed to drop personnel, supplies, and equipment in cases of disasters, such as forest fires or shipwrecks. Parachutes are used to slow down fast-moving vehicles, such as race cars, airplanes, and the space shuttle. Finally, parachuting has become a widely popular sport in a variety of different forms. *See also* Aerial Sports

Further Information
The Parachute Industry Association
3833 West Oakton Street
Skokie, IL 60076
Telephone: (847) 674-9742
Fax: (847) 674-9743
URL: http://www.pia.com
E-mail: hq@pia.com

Further Reading
Parachute History, http://parachutehistory.com
Poynter, Dan. *The Parachute Manual: A Technical Treatise on Aerodynamic Decelerators,* 3rd edition. Santa Barbara, CA: Para Publishers, 1984.

Philon (about 280 B.C.–?)

Philon was one of the earliest thinkers to carry out experiments on air and to consider its basic nature and properties. Only one of his writings has survived, a work called *Mechanics*. This work, divided into nine parts, includes discussions of lever, catapults, pneumatics, automatic theaters, fortresses, and the besieging and defending of towns.

Philon was born in Byzantium about 280 B.C., although almost nothing is known about his life, including the date and place of his death. He is referred to by some important later writers, including Heron, Eutocius, and Vitruvius. These authors describe Philon's contributions to architecture, the study of mechanics, and warfare. Perhaps his most significant contribution in method his the method he developed for duplicating a cube. Among his contemporaries, he appears to have been best known for his research on developing the catapult and other types of weapons used in warfare.

In the field of pneumatics, Philon discovered that air expands with heat, and he may have tried to use this concept in a very early form of the air thermometer, later developed by Galileo (1564–1642) in 1593. He also found that a torch ignited in a closed volume of air is eventually snuffed out and that a certain portion of the air in the container is consumed. This fact was later rediscovered and interpreted in the eighteenth century by the English chemist Joseph Priestley (1733–1804) and the French chemist Antoine Laurent Lavoisier (1743–1794).

Further Reading
Drachmann, A. G. *Ktesibios, Philon and Heron: A Study in Ancient Pneumatics.* Copenhagen: E. Munksgaard, 1948.
"Philon of Byzantium," http://www.groups.dcs. st-and.ac.uk/~history/Mathematicians/ Philon.html.

Physical and Chemical Properties of Air

Air is a mixture of gases, consisting primarily of nitrogen, oxygen, argon, and carbon dioxide. A mixture is a heterogeneous (nonuniform) blend of elements and compounds with no specific, fixed composition. For example, the precise composition of air varies in the Earth's atmosphere with altitude and various other conditions. When one speaks of the physical and chemical properties of air, therefore, one refers to some average or range of the properties of the individual gases of which air consists.

Properties of air are often given at certain "standard conditions," for which standard pressure is 1 atmosphere (or its equivalents, 1,103.25 hectopascals or 1,103.25 millibars or 760 millimeters or 29.92 inches of mercury) and standard temperature is 0°C (or 273K or 32°F). Under these conditions, the density of dry air (air

Composition of Dry Air at Sea Level

Substance	Percent by Volume	Percent by Weight
Nitrogen	78.00	75.53
Oxygen	20.95	23.26
Argon	0.93	0.12
Carbon dioxide	0.036*	0.005
Neon	0.0018	<0.0001
Helium	0.0005	<0.0001
Methane	0.0002	<0.0001
Krypton	0.0001	<0.0001
Hydrogen	0.000 05	<0.000 01
Nitrous oxide	0.000 05	<0.000 01
Xenon	0.000 008	<0.000 01
Ozone	0.000 001	<0.000 001

that contains no water) is 1.29 grams per liter or 0.080 pounds per cubic foot. Air can be liquefied at about the boiling point of its component gas having the lowest boiling point, nitrogen, $-195.5°C$. *See also* Argon; Atmosphere; Carbon Dioxide; Nitrogen; Oxygen

Plant Propagation (by Wind)

See Seed Propagation (by Wind)

Prandtl, Ludwig (1875–1953)

Prandtl is widely regarded as the Father of Modern Theoretical Aerodynamics. During his lifetime, he studied a number of fundamental problems arising out of the flow of air and other fluids, such as turbulence, supersonic flow, and airflow over wing surfaces.

Ludwig Prandtl was born in Freising, Germany, on 4 February 1875. He obtained his degree in mechanical engineering from the Technische Hochschule in Munich in 1898 and his doctorate from the same institution in 1900. His first academic appointment was as professor at the Technische Hochschule of Hanover in 1901, after which he was appointed professor of Applied Mechanics at the University of Göttingen in 1904, where he remained for the rest of his life. He died at Göttingen on 15 August 1953.

Prandtl became interested in problems of fluid dynamics in 1900, when he worked briefly at the Maschinenfabrik Augsburg-Nürenberg. His first assignment there was to find a way to remove metal shavings from machines. He found that existing theoretical formulations dealing with fluid dynamics, while elegant, did not fit conditions in the real world very well. As a result of his research, he made an important fundamental discovery, now known as the theory of boundary layers. In his paper on this subject, published in 1904, he showed that any time a fluid flows through a pipe, a thin layer of the fluid remains attached to the pipe itself. Neighboring layers gradually move more rapidly the farther they are from the surface of the pipe.

Prandtl later demonstrated how this thin static layer of fluid affected the lift and drag on airfoils for aircraft being developed. He and his colleague, Theodore von Kármán (1881–1963), discovered the effect of turbulent flow on the surfaces of airplanes, resulting in dramatic improvements in the design of wings and airframes. It can be said that, to a large extent, the general design of most modern aircraft and space ships is a result of the research carried out by Prandtl and von Kármán. *See also* Aerodynamics

R

Red Baron

See Richthofen, Manfred Albrecht Freiherr von (1892–1918)

Respiration

The term *respiration* has two distinct, but related, meanings. First, respiration refers to the exchange of gases between an organism and its surrounding environment, as when an animal breathes in (inhales) oxygen and breathes out (exhales) carbon dioxide. All organisms respire in one way or another. A second type of respiration, known as *cellular respiration,* consists of all of those chemical reactions that occur within a cell by means of which carbohydrates are converted into energy. The two forms of respiration are related to each other in that the oxygen taken in by an organism is used by cells during cellular respiration, one product of which is carbon dioxide, expelled by the organism to the surrounding environment.

"Whole Body" Respiration

Most organisms have evolved some mechanism by which they can exchange gases with the environment. This statement is true for nearly all living organisms, including bacteria, fungi, protists, plants, and animals.

Respiration Requirements

The mechanisms used by different organisms for respiration vary considerably. But they all have certain characteristics in common. First, they must all occur within a moist environment. The gases that move into and out of an organism are carried in solution, so sufficient water must be available to permit transport of the dissolved gases. Second, the surface across which respiration occurs must be quite large so that adequate amounts of inhaled and exhaled gases may enter and leave the organism. Third, some system for moving gases from the organism's surface to its interior must exist. It is essential not only that gases be able to enter and leave the surface of the organism, but that they also be able to pass between the organism's interior and exterior. Finally, some structure must exist to protect the fragile surface at which respiration occurs.

Whatever the respiratory system used by an organism, the force that drives gas exchange is always the same, the difference in gas concentration between the interior and exterior of the organism's body. For example, when the concentration of oxygen is greater outside a cell wall than it is inside the cell wall, oxygen diffuses across (passes through) the cell wall and into the cell from the surrounding environment. Similarly, when the concentration of carbon dioxide inside the cell is greater than in the surrounding environment, the gas diffuses across the cell wall, out of the cell and into the surrounding environment.

Plant Respiration

Gas exchange in plants takes place in green leaves, stems, and roots. The epidermis (outer layer) of a green leaf, for example, contains a number of openings, known as *stomata* (singular: *stoma*), through which gases may flow into and out of the interior of the leaf. Once inside the epidermis, gases flow easily through the porous interior of the spongelike mesophyll, the middle layer of the leaf. Because of the extensive empty spaces within the mesophyll, a very large fraction of the surface areas of cells is exposed to gases, increasing the efficiency of gas diffusion into and out of the cells.

The flow of gases through stomata is controlled by specialized cells located on either side of each stoma, cells known as *guard cells*. When guard cells are turgid (filled with water), they push away from each other and the stomata between them are open. When they are flaccid (relatively dry), they collapse and close the stomata between them. The opening and closing of guard cells occurs in response to environmental conditions and controls the amount of gases that flow into and out of the leaf.

Animal Respiration

Unicellular and multicellular organisms use a variety of mechanisms for respiration. In one-celled organisms, gas exchange takes place directly across the cell membrane. No special structures are needed to provide an adequate movement of gases into and out of the organisms. In some of the smaller and less complex multicellular organisms—such as hydra, planaria, sponges, jellyfish, and terrestrial flatworms—gas exchange may

also take place directly across the organism's outermost cells. In some cases, these organisms may have quite large surface areas, increasing the efficiency of respiration. Sponges, for example, have very large surface area compared to their total volume, making respiration highly efficient.

The efficiency of respiration is increased in some organisms because of the fact that blood vessels lie very close to the epidermis, permitting gases to pass almost directly into and out of the organism's circulatory system from and into the surrounding environment. Amphibians and annelids (segmented worms) have this type of respiratory system. Although it contains no specialized organs, this type of respiratory system is somewhat more complex than the simple, direct exchange of gases used by the simplest organisms.

A third type of respiratory system uses openings in an organism's bodies. The outer portion of these openings is called a *spiracle,* which leads to a tube that reaches deeper into the animal's body, the *trachea.* Inside the organism's body, the trachea subdivides into a number of smaller branches that are in contact with its organs and tissues. Gases flow into and out of the trachea, directly to interior cells, where they are used and produced. Insects and terrestrial arthropods have tracheal respiratory systems.

A fourth type of respiratory system uses gills, which are special types of tissue that extend from the organism's body. Gills are thin, narrow strips of tissue that contain blood vessels lying close to the surface. Gases move across the epidermis of the gill, directly into or out of the circulatory system.

Most air-breathing animals have evolved a respiratory system based on lungs. In humans, for example, the respiratory system consists of the nose, through which air enters the body, and the pharynx, larynx, trachea, and bronchi, a series of tubes through which air moves before reaching the lungs. There are two bronchi, each serving one of the two lobes (parts) of the lungs. Within the lungs, each bronchus branches off into smaller tubes, known as *bronchioles,* which end in tiny air sacs called *alveoli.* The normal human lung is thought to contain about 300 million alveoli. Gases pass into and out of the lungs by diffusing across the membrane surrounding each alveolus. They then enter or leave tiny capillaries with which each alveolus is filled. From there, they make their way through the circulatory system to cells or out of the body.

Cellular Respiration

All organisms constantly need to produce energy simply to maintain life, to grow, to fight off disease, and to reproduce. This energy is produced within individual cells by means of a complex series of chemical reactions known as cellular respiration. Before cellular respiration can occur, food eaten by the organism must be digested and changed into relatively simple chemical compounds, the most important of which are the carbohydrates. During cellular respiration, carbohydrates react with oxygen to produce carbon dioxide, water, and energy. A very simple representation of that series of reaction is as follows:

$$C_6H_{12}O_6 + 6O_2 \rightarrow 6CO_2 + 6H_2O + energy$$

The energy produced in this series of reactions is then used to carry out all of the other chemical reactions needed by the body for maintenance, growth, and reproduction. The carbon dioxide and water are waste products that are excreted from the body through the respiratory system (for carbon dioxide) and the excretory system (for water).

Under certain circumstances, organisms are able to convert carbohydrates to energy without the use of oxygen. During vigorous exercise, for example, the respiratory system may not be able to supply oxygen to the body as fast as cells need it. In such cases, alternative reactions may take over and use substances other than oxygen to obtain energy from carbohydrates. The reactions that make up that process are know *anaerobic* (without oxygen) *respiration.*

Further Reading

Hlastala, Michael P., and Albert J. Berger. *Physiology of Respiration.* 2nd edition, Oxford: Oxford University Press, 2001.

Sebel, Peter, et al. *Respiration: The Breath of Life.* New York: Torstar Books, 1985.

West, John B. *Respiratory Physiology,* 5th edition. Baltimore: Williams & Wilkins, 1995.

Richthofen, Manfred Albrecht Freiherr von (1892–1918)

Manfred Albrecht Freiherr von Richthofen was the most successful fighter pilot during World War I in terms of the number of enemy aircraft destroyed. He is credited with 80 "victories" over his opponents. He was highly regarded not only by his fellow officers and countrymen, but also by many of his opponents in air warfare. He was given the name *der rote Kampfflieger* (The Red Battle-Flier) by the Germans, *le petit rouge* (The Little Red One) by the French, and *the Red Baron* by the English and Americans. These names were derived from the blood-red color of the Fokker Dr.1 triplane he flew.

Von Richthofen was born on 2 May 1892 in Breslau, Germany (now Wroclaw, Poland), the son of a Prussian nobleman. He attended the military school at Wahlstatt and the Royal Military Academy at Lichterfelde. He was commissioned as a cavalry officer in the Prussian army in April 1911 and served as a scout on both the eastern and western fronts on the German nation. When war broke out in 1914, von Richthofen decided that the cavalry did not provide enough battle opportunity for him, and he asked to be transferred to the fledgling German air service (the *Fliegertruppe*).

Von Richthofen's first official confirmed air victory came on 17 September 1916 when he shot down a French airplane. Over the next 18 months, von Richthofen was credited with an additional 79 victories over enemy aircraft. Finally, on 21 April 1918, he was himself shot down over Vaux sur Sommes, France, by an enemy aviator whose identify has never been confirmed.

Eddie Rickenbacker. © Library of Congress.

Further Reading

Gibbons, Floyd Phillips. *The Red Knight of Germany.* New York: Arno Press, 1980.

Kilduff, Peter. *Richthofen: Beyond the Legend of the Red Baron.* New York: John Wiley & Sons, 1993.

Rickenbacker, Edward (1890–1973)

Edward ("Eddie") Rickenbacker was America's leading aviation ace during World War I, receiving credit for 26 "victories" over enemy aircraft. A "victory" was described as the destruction of an enemy airplane.

Rickenbacker was born on 8 October 1890 in Columbus, Ohio, the third of eight children born to William and Elizabeth (Basler) Rickenbacher. The Rickenbachers were Swiss emigrants who met and married in Columbus. Rickenbacker later changed the spelling of his last name and added a middle name, Vernon. He grew up in poverty and, when his father died in 1904, left school to help support the family. Over the next decade, he held a variety of jobs at a beer factory, a bowling alley, a glass factory, and a cemetery monument yard, as well as working with the Pennsylvania Railroad.

In 1905, he was given a job at the Frayer-Miller automobile manufacturing plant in Columbus and, for the next 12 years, he was involved in the manufacture, sales, and racing of automobiles.

Shortly after the United States entered the war, in May 1917, Rickenbacker enlisted in the U.S. Army. He was sent to France, where he was at first assigned as a staff driver for General John Pershing. Dissatisfied with the assignment, he sought and received a transfer to the U.S. Army Air Corps. During his first month as a fighter pilot, he shot down five enemy planes, earning him the title of "ace" and the French *Croix de Guerre.* By the end of the war, Rickenbacker had scored 21 more victories, making him the leading ace among U.S. pilots.

After the war, Rickenbacker returned to his first love, automobiles, and founded the Rickenbacker Automobile Company. When the company failed in 1927, he raised enough money to buy the Indianapolis Speedway, where Rickenbacker had himself first raced in 1911. He held the track for another twenty years before selling it in 1947.

Rickenbacker also maintained his interest in aviation and in 1925 founded Florida Airways, later to become Eastern Airlines. The venture suffered severe financial hardships as a result of a hurricane the following year, and Rickenbacker moved on to work with General Motors and to American Airlines for more than a decade. In 1938, he was able to purchase and restore Eastern, a company he served as chief executive officer until 1963.

One of the best-known events in Rickenbacker's life took place in October 1942 when an airplane on which he was traveling from Hawaii to New Guinea had to land in the Pacific Ocean. Rickenbacker and six other passengers were stranded at sea for 24 days, during which they floated more than 500 miles in their life raft.

Rickenbacker died in Zurich, Switzerland, on 27 July 1973. He is buried in Columbus.

Further Reading

Farr, Finis. *Rickenbacker's Luck: An American Life.* Boston: Houghton Mifflin, 1979.

Rickenbacker, Eddie. *Rickenbacker.* Englewood Cliffs, NJ: Prentice-Hall, 1967.

S

Saffir-Simpson Hurricane Scale

The Saffir-Simpson Hurricane Scale, shown on page 192, was developed by and named for Herbert Saffir, a consulting engineer, and Bob Simpson, director of the National Hurricane Center, in 1969. The purpose of the scale is to make comparisons among various hurricanes easier and to help emergency disaster workers prepare for oncoming hurricanes. *See also* Hurricane; Wind

Further Information

"Saffir-Simpson Hurricane Scale," http://www. nhc.noaa.gov/aboutsshs.html.

Sandia National Laboratories

Sandia National Laboratories is one of the two organizations that carry out wind research under the auspices of the U.S. Wind Energy Program. The other organization is the National Wind Technology Center in Golden, Colorado.

Sandia National Laboratories was created in 1945 at Sandia Base, near Albuquerque, New Mexico, as part of the U.S. Manhattan Project, under which the first nuclear bomb was constructed. Four years later, operational control of the laboratories

Sandia National Laboratories researcher Richard Griffith studies blueprints of a building to better understand airflow. He has led a team that developed modeling and simulation tools for assessing the threat and vulnerability of buildings to biological and chemical assaults. This includes looking at how these agents move and are deposited inside a building. © Sandia National Laboratories/ Randy Montoya.

was turned over to the American Telephone & Telegraph Company (AT&T), which managed the facilities until 1993. In that year, management of the laboratories was taken over by Martin Marietta Corporation, now Lockheed Martin.

Since the end of World War II, Sandia has expanded its mission to include basic and

Saffir-Simpson Hurricane Scale

Category	Wind Speeds And Pressure	Effects
One	119–153 kph 74–95 mph 65–83 knots >980 mb	Storm surge generally 4–5 feet (1.2–1.5 meters) above normal. No real damage to building structures. Damage primarily to unanchored mobile homes, shrubbery, and trees. Some damage to poorly constructed signs. Some coastal road flooding and minor pier damage. Examples: Hurricane Allison (1995), Hurricane Danny (1997).
Two	154–177 kph 96–110 mph 84–95 knots 980–965 mb	Considerable damage to mobile homes, poorly constructed signs, and piers. Coastal and low-lying escape routes flood 2–4 hours before arrival of hurricane center. Small craft in unprotected anchorages break moorings. Examples: Hurricane Bertha (1996), Hurricane Marilyn (1995)
Three	178–209 kph 111–130 mph 96–113 knots 964–945 mb	Storm surge generally 9–12 feet (2.7–3.6 meters) above normal. Some structural damage to small residences and utility buildings with a minor amount of curtainwall failure. Damage to shrubbery and trees with foliage blown off trees and large trees blown down. Mobile homes and poorly constructed signs are destroyed. Low-lying escape routes are cut by rising water 3–5 hours before arrival of hurricane center. Flooding near the coast destroys smaller structures with larger structures damaged by battering of floating debris. Terrain continuously lower than 5 feet (1.5 meters) above mean sea level may be flooded inland 8 miles (13 km) or more. Evacuation of low-lying residences with several blocks on the shoreline may be required. Examples: Hurricane Roxanne (1995), Hurricane Fran (1996).
Four	210–249 kph 131–155 mph 114–134 knots 944–920 mb	Storm surge generally 13-18 feet (4.0–5.5 meters) above normal. More extensive curtainwall failures with some complete roof structure failures on small residences. Shrubs, trees, and all signs are blown down. Complete destruction of mobile homes. Extensive damage to doors and windows. Low-lying escape routes may be cut by rising water 3–5 hours before arrival of hurricane center. Major damage to lower floors of structures near the shore. Terrain lower than 10 feet above sea level may be flooded, requiring massive evacuation of residential areas as far inland as 6 miles (10 km). Examples: Hurricane Luis (1995); Hurricane Felix (1995), Hurricane Opal (1995).
Five	249+ kph 155+ mph 135+ knots <920 mb	Storm surge generally greater than 18 feet (5.5 meters) above normal. Complete roof failure on many residences and industrial buildings. Some complete building failures with small utility buildings blown over or away. All shrubs, trees, and signs blown down. Complete destruction of mobile homes. Severe and extensive window and door damage. Low-lying escape routes are cut by rising water 3–5 hours before arrival of hurricane center. Major damage to lower floors of all structures located less than 15 feet (4.5 meters) above sea level and within 500 yards of the shoreline. Massive evacuation of residential areas on low ground within 5–10 miles (8–16 km) of the shoreline may be required. Example: Hurricane Gilbert (1988), the strongest Atlantic tropical cyclone on record.

applied research in nuclear weapons, nuclear power, and nuclear waste; electronics; space; health and medicine; cryptography and other aspects of national security; and geothermal energy and wind energy development.

Today, Sandia's wind research takes three forms. First, the laboratory conducts applied research on aerodynamics, structural dynamics, fatigue, materials, manufacturing, controls, and system integration of wind power systems. Second, it helps private organizations to solve specific technical problems in their wind power devices and systems. Third, it explores the potential effect on wind power systems that might result from rare atmospheric events and equipment modifications that might protect such systems.

Further Information

The Wind Group
Sandia National Laboratories
P.O. Box 5800
Albuquerque, NM 87185—Mail Stop 0165
Telephone: (505) 844–8066
URL: http://www.sandia.gov/wind/
E-mail: ashanse@sandia.gov

Sandstorms and Dust Storms

Sandstorms and dust storms are atmospheric phenomena in which solid particles are stirred up by the wind and carried into the air. No clear distinction exists between sandstorms and dust storms except that the particles carried away in the former tend to be larger and heavier than those stirred up by the latter. In geology, the term *dust* is often reserved for particles with diameters of less than 1/16 millimeter.

Sandstorms are generally more severe than are dust storms. The worst such storms may carry solid particles more than 1,500 meters (5,000 feet) into the air, travel up to 80 kilometers per hour (50 miles per hour), and cover very large land areas. It is not uncommon for such storms to stir up so much solid material that the sun is completely blocked out. Sandstorms in North Africa and the Mideast (known as *simooms* or *haboobs*) produce fallouts as far west as Western Europe. Scientists estimate that sandstorms and dust storms are the most potent of all erosional agents, carrying away 60–200 million metric tons of soil every year from the Sahara alone. Some of this soil is carried as far away as the Atlantic Ocean and North America.

Sandstorms and dust storms are serious environmental problems because they remove valuable top soil, greatly reducing the agricultural value of land. They are also a threat to humans and other animals. A three-day sandstorm in the Ningxia Hui

A dust storm approaching Stratford, Texas, during the Dust Bowl days. © NOAA.

Autonomous Region of China in May 1993, for example, resulted in the death of more than 120,000 animals and 85 people. In addition, 4,412 houses were destroyed or covered by the storm. Winds reached a speed of 37.9 meters per second (84.8 miles per hour) with a visibility of less than 50 meters.

Sandstorms and dust storms are natural phenomena that occur in arid areas where the soil is loose and easily disturbed by the wind. Most commonly, a storm begins when the ground becomes heated, air begins to rise, and the winds created are of sufficient velocity to break loose and carry away particles of soil. Factors that increase the likelihood of sandstorm and dust storm formation include topography and human activity. In areas where mountainous regions are adjacent to plateaus or plains, large pressure gradients may form resulting in strong winds that lead to the formation of sandstorms and dust storms. Arid regions where large bodies of water are in contact with flat, dry areas also produce large temperature differences and, hence, large pressure gradients that lead to strong winds.

An increasingly important factor in the formation of sandstorms and dust storms is human disruption the land. Overgrazing, deforestation, urban development, and improper use of agricultural and water resources are major factors in the loss of plant cover on lands, converting once productive soil to arid or desert conditions. Studies have shown that the number and severity of sandstorms and dust storms increased dramatically over the last half of the twentieth century, largely as a result of improper land-use practices.

In contrast, the development of wise land-use practices has resulted in a reduction in the number and severity of sandstorms and dust storms. For example, a severe sandstorm hit Turpan County in the Xinjiang Uygur Autonomous Region of China in 1961, destroying 85 percent of the area's crops. Shortly thereafter, the county planted large shelterbelts of trees, sowed extensive areas to grass, and put into practice more conservative water use practices. Later sandstorms that struck the area caused damage to no more than 8 percent of the crops.

Probably the most famous period of sandstorms and dust storms in the United States was the mid-1930s over an area known as the *Dust Bowl*. The term refers to an area of 400,000 square kilometers (150,000 square miles) in Oklahoma, Texas, Kansas, Colorado, and New Mexico. A series of unusually hot summers along with disastrous agricultural practice by settlers made the region ripe for a series of devastating sandstorms and dust storms from 1934 through 1937. Storm clouds visible from hundreds of miles away swept across the land, burying homes, crops, and farm animals and displacing countless thousands of families. In spite of aggressive replanting projects and the introduction of more conservative agricultural practices, the area has continued to be plagued by similar, albeit less severe, sandstorms and dust storms.

Dust storms have also been observed on Mars. As far back as the 1970s, Viking and Mariner spacecraft took pictures of such storms, some as large as 1,000 kilometers (600 miles) in diameter. In the following decades, dust storms were recorded in some years, but not all. One of the largest and most recent was a storm observed at the Martian north polar cap between 18 September and 15 October 1996. Scientists do not yet know the cause of such storms on Mars, although they suspect that the storms arise because of the same factors operating on Earth.

Further Reading

Cooke, Ronald, Andrew Warren, and Andrew Goudie. *Desert Geomorphology.* New York: Chapman & Hall, 1993.

Hurt, R. Douglas. *The Dust Bowl: An Agricultural and Social History.* Chicago: Nelson-Hall, 1981.

Seed Propagation (by Wind)

An essential function of all plants is the release and dispersal of seeds from which new plants will grow. Plants have evolved three major mechanisms for seed dispersal. In one process, seeds are contained within structures that readily stick to passing animals, which carry the seeds to locations distant from the parent plant. In another process, seeds are dropped into rivers, streams, lakes, the oceans, or other bodies of water, which transport them to new locations. In a third process, seeds are carried away from parent plants in gusts of air or on wind currents.

Seed dispersal can be classified into one of six major categories: ground transport; light, cottonlike structure; gliders; parachutes; helicopters; and spinners. The most common type of ground transport is the tumbleweed, a member of the genus *Salsola* introduced into the United States from eastern Russia in the late nineteenth century. After it finishes blooming, the tumbleweed dries and breaks off at its base. The dry plant is blown across the ground, dropping individual seeds as it goes. A single tumbleweed plant may produce as many as 50,000 seeds.

The seeds of a number of plants are contained within or attached to a light, fluffy material that is easily blown away by the wind. As the mother plant releases these seeds, they are carried away by the slightest air movement. Some plants that propagate by this method are members of the willow family (*Salicaceae*), the cattail family (*Typhaceae*), the sycamore family (*Platanaceae*), and the kapok tree (*Ceiba pentandra*).

Structures that disperse seeds by means of a gliding process generally look much like an airplane wing. In such structures, the seed is generally located at the center of the two wings. When the wind blows over such structures, they are carried away by the same aerodynamic principles that explain the flight of an airplane. One of the most dramatic examples of seed dispersal by this method is the tropical vine *Alsomitra macrocarpa,* with a wingspread of up to 15 centimeters (6 inches).

Seed dispersal by parachute usually takes one of two general forms. In some cases, a single seed is attached to its own parachute-like structure, consisting of dozens of long, fine hairs attached to the seed. In other cases, dozens or hundreds of such seeds are packed together in a spherical ball that is easily carried away by the wind. Perhaps the best known example of plants that use this type of seed dispersal is the common dandelion (*Taraxicum officinale*). Many members of the sunflower family (*Asteraceae*), artichoke family (*Cynara*), and milkweed family (*Asclepiadaceae*) also use this method of seed dispersal.

Seeds dispersed by a helicopter-like mechanism have one or two wings, thicker along one edge than the other, with the seed at the base of the wing. As the structure, called a *samara,* falls to the ground, it spins around with a motion similar to that of a helicopter. Plants may produce single or twinned samaras. Twin samaras may be carried along by the air intact, or they may break apart as they fall to the ground. Many members of the maple family (*Aceraceae*), for example, propagate by this method, as do some members of the pine (*Pinus*), fir (*Abies*), and spruce (*Picea*) families.

Seeds that travel by spinning are often enclosed in structures similar to those that use helicopterlike travel for dispersion. With spinners, however, the seed is more likely to be surrounded by a flat, circular membrane shaped like a fan. As the structure is carried off by the wind, it tends to spin around the seed itself, or it may be buffeted back and forth, producing a fluttering motion. Some plants that propagate by this method include members of the elm family (*Ulmaceae*), figwort family (*Scrophulariaceae*), bignonia family (*Bignoniaceae*), and the catalpa tree

(*Catalpa speciosa*), hop seed plant (*Dodonea viscosa*), and jacaranda (*Jacaranda mimosifolia*). *See also* Aerobiology

Further Reading

Murray, David E. *Seed Dispersal.* Orlando, FL: Academic Press, 1986.

Pijl, L. van der. *Principles of Dispersal in Higher Plants,* 3rd edition. Berlin: Springer-Verlag, 1982.

Soaring

See Gliding

Squalls

See Gusts and Squalls

Stability of Air

The stability of air is a measure of its tendency to rise or sink in the atmosphere. To understand the concept of ability, it is first necessary to review two other concepts: environmental lapse rate and adiabatic temperature change.

Environmental Lapse Rate and Adiabatic Temperature Change

Environmental lapse rate is defined as the rate at which temperature decreases in the troposphere. The average, or *normal lapse rate,* is 6.5°C per kilometer, but can vary widely depending on local conditions. For example, a cold front moving into an area may result in temperatures near the ground cooler than those at higher altitudes. In such cases, temperatures actually increase in moving from sea level to higher altitudes, a condition known as a *temperature inversion.*

The term *adiabatic temperature change* refers to temperature changes that take place in air not because heat is added or taken from the air, but because of expansion or compression of the air. When air expands, it gives up energy and becomes cooler. Conversely, when air is compressed, it gains energy and becomes warmer.

A parcel of air at sea level that is allowed to rise passes through areas of successively lower pressure. (Atmospheric pressure drops at a regular rate in moving from sea level to higher altitudes.) As a result, that parcel of air expands and becomes cooler because of adiabatic temperature changes. Conversely, a parcel of air that sinks encounters higher pressures surrounding it, is compressed, and becomes warmer. The rate of cooling or warming that occurs as air moves upward or downward in the atmosphere as a result adiabatic changes is known as the *adiabatic lapse rate.*

The adiabatic lapse rate differs depending on whether the parcel of air is dry or moist. When water vapor is present, the adiabatic lapse rate is affected because, at some point, the water vapor begins to condense, releasing additional heat to the parcel of air. The dry adiabatic lapse rate is 10°C per 1,000 meters (5.4°F per 1,000 feet), while the wet adiabatic lapse rate varies depending on the amount of moisture present. Its value tends to vary from about 5°C per 1,000 meters (3°F per 1,000 feet) for very moist air to 10°C per 1,000 meters (6°F per 1,000 feet) for less moist air.

Stable and Unstable Air

The stability of a parcel of air can be determined by comparing the environmental lapse rate with the adiabatic lapse rate. That is, a parcel of air has:

Absolute stability when the environmental lapse rate is less than the wet adiabatic lapse rate. For example, consider the case in which the environmental lapse rate is 5°C per 1,000 meters and the dry adiabatic lapse rate is 10°C per 1,000 meters. In such a case, as the parcel of air rises in the atmosphere, it is always cooler than the surrounding air and

U.S. Standard Atmosphere, 1976

Altitude (m)	Temperature (°C)	Pressure (mb)	Density (kg/g/m³)
0	15	1013.2	1.225
500	12	954.6	1.167
1,000	9	898.8	1.111
1,500	5	845.6	1.058
2,000	2	795.0	1.006
2,500	-1	746.8	0.957
3,000	-4	701.1	0.909
3,500	-8	657.6	0.863
4,000	-11	616.4	0.819
4,500	-14	577.3	0.777
5,000	-17	540.2	0.736
6,000	-24	471.8	0.660
7,000	-30	410.6	0.589
8,000	-37	356.0	0.525
9,000	-43	307.4	0.466
10,000	-50	264.4	0.413
12,000	-56	193.3	0.311
14,000	-56	146.0	0.227
16,000	-56	102.9	0.0165
18,000	-56	75.0	0.121
20,000	-56	54.7	0.088
30,000	-46	11.7	0.018
40,000	-22	2.7	0.0038
50,000	-2	0.76	0.000 97
60,000	-28	0.028	0.000 28
70,000	-56	0.000 07	0.000 07

has a tendency to settle back toward the ground. Under such conditions, there is little tendency for clouds to form.

Absolute instability when the environmental lapse rate is greater than the dry adiabatic lapse rate. Consider the case in which the environmental lapse rate is 12°C per 1,000 meters and the dry adiabatic lapse rate is 10°C per 1,000 meters. In this case, the parcel of air is always warmer and less dense than the surrounding air, and it will always continue to rise in the atmosphere.

Conditional instability when the environmental lapse rate is *less* than the dry adiabatic lapse rate and *greater than* the moist adiabatic lapse rate. For example, consider the case in which the environmental lapse rate is 8°C per 1,000 meters and the dry and wet adiabatic lapse rates are 10°C and 6°C per 1,000 meters respectively. In this instance, the parcel of air rises as long as the surrounding atmosphere is relatively dry, but then reaches a point at which it becomes stable or unstable depending on the moisture content of the surrounding atmosphere. If it reaches a point where the environmental lapse rate is greater than the moist adiabatic lapse rate, the parcel of air continues to rise in the atmosphere.

Atmosphere Conditions for Stability and Instability

Any condition that produces cooler temperatures near the ground than aloft, resulting in a temperature inversion, causes stable atmospheric conditions. For example, the ground may cool rapidly on a clear night, producing cooler temperatures near the ground than at higher altitudes. Or an approaching cold mass may force cooler air beneath a warm air mass or, conversely, an approaching warm air mass may flow over an existing cold air mass. In both cases, a temperature inversion may occur as cooler temperatures develop nearer the ground and higher temperatures at higher elevations.

Instability occurs under the reverse of these conditions. For example, heating of the Earth's surface causes the air above it to become warmer and less dense, allowing cooler surrounding air to flow in and push the warmer air upward. Or a warm air mass may move inward along the Earth's surface, heating the ground and pushing warm air upward.

Atmospheric stability and instability is an important factor in determining weather patterns. In general, the clouds formed when unstable air moves upward tend to become filled with moisture, resulting in relatively heavy precipitation. The upward flow of stable air by outside forces, by contrast, tends to form light vertical clouds with relatively little moisture content and little precipitation. The upward flow of air, or lack of it, also determines whether pollutants produced by human activities are carried away (by upwardly moving unstable air) or retained (by stable air trapped by a temperature inversion). *See also* Atmosphere

Further Reading
"Atmospheric Stability & Clouds," http://psb. usu.edu/courses/bmet2000/stability.html.

Standard Atmosphere

The standard atmosphere is an idealized model of the distribution of temperature, pressure, density, and other physical properties of air at various altitudes in the atmosphere, from sea level to heights of 50 kilometers or more.

The properties of air in the atmosphere change in a relatively predictable and well-known way in moving from sea level to the upper atmosphere. For example, pressure drops geometrically at a relatively constant rate. But actual readings of physical proper-

ties such as pressure, density, and temperature vary somewhat around the planet depending on a variety of factors, such as surface features (land versus water and mountains versus plains, for example) and on latitude (equator versus poles).

The standard atmosphere is a theoretical construct obtained by averaging out these variations and developing an ideal model that shows the broad, overall trends in physical properties of air at various altitudes in the atmosphere. Two versions of the standard atmosphere exist, an international version and a U.S. version. The U.S. Standard Atmosphere was first developed in 1958 by the U.S. Committee on Extension to the Standard Atmosphere. It was revised in 1962, 1966, and 1976. The 1976 version is now available as a joint publication of the National Oceanic and Atmospheric Administration, the National Aeronautics and Space Administration, and the U.S. Air Force. Standard Atmosphere tables can be posted in a variety of formats. The table shown on page 197 is an abbreviated version of one widely used form of the table. Notice in this table that pressures drop off rapidly near the Earth's surface, but much less rapidly at higher altitudes. By contrast, temperatures decrease quite regularly from sea level to an altitude of 12–20 kilometers, after which they begin to increase to the maximum height shown in the table.

Further Reading
(International Civil Aviation Organization) Atmosphere, http://142.26.194.131/aerodynamics/Appendix/ICAO.html
U.S. Standard Atmosphere 1976, http://nssdc. gsfc.nasa.gov/space/model/atmos/us_standard.html.

T

Thermoscope

See Air Thermometer

Tornado

A tornado is a violently rotating column of air in contact with the ground and suspended from a cumulonimbus cloud. Tornadoes are also known as cyclones or twisters in some parts of the world.

Types of Tornadoes

Tornadoes are commonly categorized as "weak," "strong," or "violent" depending on the speed of the winds associated with them, their general shape and appearance, and their destructive potential. "Weak" tornadoes are the mildest and most common types of tornadoes, accounting for about 70 percent of such storms. The winds associated with a weak tornado do not exceed 180 kilometers per hour (110 miles per hour). They can often be recognized by their narrow, ropelike shape.

About 30 percent of all tornadoes are classified as "strong," that is, having winds of 180–320 kilometers per hour (110–200 miles per hour). They are usually identifiable from their funnel-shaped clouds, sometimes attributed to all types of tornadoes. About 1 percent of all tornadoes are classi-

fied as "violent," with winds in excess of 320 kilometers per hour (200 miles per hour). Although relatively few in number, violent tornadoes cause by far the greatest amount of destruction. They have been known to carry railway cars, automobiles, tractors, and other heavy objects through the air and deposit them dozens or hundreds of meters away from their original location. Violent tornadoes may also tear apart solidly built structures that are safe from "weak" and "strong" tornadoes.

The weakest tornadoes have relatively short lifetimes, often only a few minutes, during which they travel less than a kilometer over a path 100 meters or less in width. By contrast, strong and violent tornadoes may travel up to 30 kilometers (20 miles) at a rate of 45 kilometers per hour (30 miles per hour) over a path up to 600 meters (2,000 feet) wide.

Tornado Formation

Tornadoes form out of severe thunderstorms that may themselves cause moderate destruction to property and injury to humans and other animals. Even today, after extensive research, the detailed steps involved in tornado formation are only partially understood, at least partly because of the problems of approaching, measuring, and otherwise studying phenomena of such violent properties.

Still, the general outline of tornado formation is clear. Tornadoes form in areas with large temperature differences and, hence, large surface pressure differences. For example, very cold polar air from Canada may sweep down across the Great Plains in the spring, coming into contact with warmer tropical air from the Gulf of Mexico. Along the cold front that develops between these two air masses, dramatic temperature differences may produce pressure differentials of up to 10 percent between the exterior and interior of a storm cloud. These conditions are ideal for the formation of a tornado.

At such points, air from high pressure areas outside the storm cloud will rush inward and then be pushed upward into the center of the cloud. As the air rises, it begins to rotate because of the Coriolis effect, spinning faster and faster the higher it rises. The rising air may also cool adiabatically, causing moisture to condense inside the cloud and giving it a grayish appearance. In-rushing air may also pick up dirt and debris, darkening the cloud even further and giving it the ominous shape and color associated with a tornado.

Tornado Patterns

Tornadoes tend to develop in regions where there are few or no natural topographic barriers to the collision between cold and warm air masses. The central United States, along the Rocky and Appalachian Mountains, is such a region, reflected in the fact that Oklahoma averages more tornadoes annually than any other state. Indiana, Kansas, Nebraska, and Texas are not far behind in tornado frequency. In fact, the likelihood of tornado formation is greater along the North Dakota–Texas axis than anywhere else in the world.

Tornadoes also tend to form during seasons when temperature differentials are the greatest, especially in spring. In one 27-year survey of tornado frequency in the United States, the greatest number of storms occurred between April and June with a peak average of five per day in May. By contrast, only one tornado every other day was reported between November and February.

Tornado Prediction

Tornado prediction is difficult primarily for two reasons. First, tornadoes are localized phenomena, covering an area of less than 40 kilometers (25 miles) in width, and they last only a few hours. Most weather stations are at least 150 kilometers (100 miles) apart, making it difficult for a tornado to be detected unless it develops close to a station. Secondly, meteorologists still do not know how to tell which thunderstorms are most likely to develop into a tornado or how severe the storm is likely to become.

As a result, the National Severe Storms Forecast Center (a part of the National Weather Service) has developed a two-tier system for alerting the public as to the possible appearance of a tornado. At the first tier, a *tornado watch* indicates that conditions for tornado formation are favorable and that the public should listen for further information about the appearance of a tornado. Tornado watches are usually issued for a six-hour period over an area of 65,000 square kilometers (25,000 square miles).

A *tornado warning* is issued when a tornado has been sighted and there is reason to believe that it will travel over a certain area. Tornado warnings are short-term alerts, usually valid for an hour or less, and are directed at the fairly limited area through the storm is expected to pass. *See also* Air Mass; Coriolis Effect; Cyclones and Anticyclones; Fujita Tornado Scale

Further Reading

Bluestein, Howard B. *Tornado Alley: Monster Storms of the Great Plains.* New York: Oxford University Press, 1999.

Grazulis, T. P. *The Tornado: Nature's Ultimate Windstorm.* Norman: University of Oklahoma Press, 2001.

Weems, John Edward. *The Tornado.* College Station: Texas A&M Press, 1991.

WWW2010, http://ww2010.atmos.uiuc.edu/(Gh)/guides/mtr/svr/torn/home.rxml.

Torricelli, Evangelista (1608–1647)

Evangelista Torricelli's most important contribution to science was his invention of the barometer in 1643. He was inspired to carry out this research by Galileo Galilei (1564–1642), whose secretary he was at the time. Along with a number of other scientists, Galileo was puzzled as to why it was that pumps could lift water up to heights of about 10 meters (34 feet), but no higher. To investigate this problem, Torricelli filled a glass tube with mercury, which he then inverted in a bowl of mercury. He found that a small amount of mercury ran out of the tube into the bowl, leaving a vacuum at the top of the tube. Torricelli concluded that atmospheric pressure pushing on the surface of the mercury was sufficient to support a column of mercury 76 centimeters (about 30 inches) high. Since a water column of comparable high would be about 34 feet high, Torricelli had solved the problem given him by Galileo.

Torricelli was born in Faenza, Italy, on 15 October 1608. Both parents died shortly after his birth. Through the efforts of his uncle, he was educated at the Sapienza College in Rome, primarily in mathematics. There he met a man by the name of Castelli, Galileo's favorite pupil. Castelli taught Torricelli the fundamentals of Galileo's work and encouraged Torrecelli to carry out similar research on his own. In 1641, he wrote his own book on physical movement, *De Motu,* which came to Galileo's attention.

Galileo was so impressed with Torricelli's work that he invited him to come to Rome and work as his secretary. Galileo lived only three months after Torricelli's arrival, however.

After Galileo's death, Torricelli was appointed professor of Mathematics at the University of Florence. He held that post until he died of typhoid fever in Florence on 25 October 1647. *See also* Barometer

Tower of the Winds

The Tower of the Winds is an octagonal building made of white marble that stands at the edge of the Roman *agora,* or market, in Athens. It was designed by the architect Andronikos of Kyrrhos (a town in Macedonia) and is, therefore, sometimes called the Horologion of Andronikos. Andronikos is thought to have lived from the late second century to the mid–first century B.C. The date of the tower's construction is uncertain, the earliest estimate being the late second century B.C. Since the structure was mentioned by the Roman historian Varro in 37 B.C., it can not be any newer than that date.

The tower is about 12 meters (40 feet) high, and each side is 3.2 meters (10.5 feet) long. It gets its name from eight friezes at the top of each side, each frieze representing a Greek wind deity, as follows:

North side:	Boreas blowing through a conch shell
Northeast side:	Kaikias dumping hailstones from his shield
East side:	Apeliotes carrying fruit and grain
Southeast side:	Eurus wearing a cloak
South side:	Notos emptying a pitcher of rain on Earth
Southwest side:	Lips holding a ship's stern decoration

| West side: | Zephyros scattering flowers |
| Northwest side: | Skiron pouring out ashes from a bronze urn |

The Tower of Winds is often referred to as a *horologion* because of rods projecting from its sides with which the time of day and season of the year could be determined. The direction of the rods' shadows gave the time of day and their length, the time of year. A bronze wind vane shaped like a Triton (a son of the sea god Poseidon) placed on top of the tower showed the wind's direction. Finally, the tower contained an elaborate water clock, powered by water pumped from a spring on the Acropolis.

After the rise of Christianity, the tower was used for other purposes, usually a church, chapel, or baptistery. In the eighteenth century, it was used as a place of lodging by the sect known as the Whirling Dervishes. An engraving shows members of the sect conducting their spinning ceremony within the tower.

Over the centuries, the tower was buried to about half its height by wind-blown soil. It was excavated and restored between 1837 and 1845, between 1916 and 1919, and again in 1976. Today it exists largely in its original state except for the wind vane and water clock, both of which have been destroyed. *See also* Gods and Goddesses

Turbulence

See Aerodynamics

U

U.S. Wind Energy Program

The U.S. Wind Energy Program is a function of the U.S. Department of Energy's Office of Wind and Geothermal Technologies. The purpose of the program is to guide and assist the development of wind power so that the nation will have a more diverse variety of energy sources and be able to reduce emissions of greenhouse gases that contribute to climate change. The program has a number of facets, including basic research on wind energy; research and development on wind turbines; and support for utilities, industry, and international wind energy projects.

The program has adopted three short- and long-term goals. They are as follows:

- By 2002, the development of wind turbine technologies that will reduce the cost of wind energy to 2.5 cents per kilowatt-hour (in 15 mile per hour winds);
- By 2005, the establishment of the U.S. wind industry as the international leader in wind technology, with the goal of capturing 25% of world markets; and
- By 2010, the realization of 10,000 megawatts of installed wind-powered electrical generating facilities in the United States.

Examples of the types of projects supported by the program include:

- Development of computer models for improved wind turbines;
- Design of better control systems to maximize energy production and minimize wear in turbines;
- Production of more efficient turbine blades;
- Improvements in wind forecasting techniques;
- Advanced studies on integration of wind energy with other methods of energy production, transmission, and distribution;
- Certification testing of new wind turbines in the United States;
- Development of standards for safety, power performance and blade testing that will be applicable in the United States and around the world; and
- Provision of technical assistance in the areas of wind resource evaluation and technology development in other nations of the world.

Research and development studies sponsored by the program are carried out primarily at two locations: the National Wind Technology Center in Golden, Colorado, and Sandia National Laboratories in Albuquerque, New Mexico.

Further Information
Office of Wind & Geothermal Technologies
Office of Power Technologies
U.S. Department of Energy
Forrestal Building, 5H-021
1000 Independence Avenue, SW
Washington, DC 20585
Telephone: (202) 586–5348
URL: http://www.eren.doe.gov/wind

V

Vacuum

Vacuum is described most simply as any space that contains no matter. In the field of physics, the term has a somewhat more complex meaning, referring to any space that contains neither matter nor energy. Both definitions are of value primarily as theoretical constructs since scientists believe that there is no place in the universe where one can find a space that meets either of these definitions, that is, a space devoid of both matter and energy. For example, even in the most distant regions of space, at least some small number of atoms and/or molecules are to be found everywhere. A cubic kilometer of space may contain less than a hundred atoms and molecules, but it is not completely empty of matter.

From a more practical standpoint, a vacuum is usually defined as a space that contains significantly fewer atoms and molecules than are to be found in normal air. Any vacuum less complete than a perfect vacuum is sometimes called a *partial vacuum*. The concentration of matter in a given volume, and therefore a measure of its "emptiness," is usually expressed as the pressure exerted by the atoms and molecules that make up the volume. For example, vacuums used for many industrial and research purposes have pressures of about 10^{-4} to 10^{-9} torr, where 1 torr is equal to atmospheric pressure at Earth's surface. The best vacuums ever produced by humans are of the order of 10^{-15} torr.

History of the Concept

The concept of vacuum was understood by Greek natural philosophers as early as the fifth century B.C. Indeed, to some scholars, the concept was essential to their explanation of the nature of the world. Among these scholars were the atomists, who believed that the world consisted of countless numbers of very tiny particles that combined in various ways to form all known forms of matter. Vacuum was essential to atomistic theories because it was necessary for there to be an empty space in which these tiny particles could be located and move about.

The concept of vacuum was not embraced by all Greek philosophers, however, most notably the most famous of them all, Aristotle (384–322 B.C.). Aristotle claimed that a vacuum could not exist because Nature abhorred the concept of an empty space. If humans should attempt to create such a space, Nature would immediately force something into the momentarily empty space to prevent the creation of a vacuum.

Aristotle's influence was such that his view of vacuum persisted for nearly two thousand years. It was not until the early seventeenth century that scholars began to question his "Nature abhors a vacuum" teachings. Probably the most important factor in initiating this debate was the operation of water pumps.

By the mid–seventeenth century, water pumps were widely used for a number of purposes, one of which was the removal of water from underground mines. In the most common type of water pump, a piston is raised within a cylinder, leaving an empty space behind. Water then rushes in from some outside source (such as the floor of a mine) to fill the empty space. When the piston is pushed back down into the cylinder, the water is pushed out of the pump into a pipe from which it is discharged. Engineers discovered early on, however, that water pumps had a distinct limitation. They could not raise water from a depth of more than about 10 meters (30 feet). It appeared that the Aristotelian theory about vacuum had a limitation: Apparently Nature abhorred a vacuum only to a depth of about 10 meters.

That conclusion, however, is clearly absurd. One scientist who become involved in attempts to solve the problem of water pump limitations was Galileo Galilei (1564–1642). Galileo suggested to his pupil, Evangelista Torricelli (1608–1647), that he find an explanation for the inability of pumps to lift water more than about 10 meters. Torricelli discovered that the reason pumps could move water from a lower elevation to a higher elevation was not that something pulled the water up (Nature's abhorrence of a vacuum), but that something pushed it up (atmospheric pressure).

In carrying out his research, Torricelli actually created one of the first clearly demonstrable vacuums. He filled a glass tube, sealed at one end, with mercury and then inverted the tube, placing its open end into a bowl of mercury. He found that a small amount of mercury ran out of the tube and into the bowl, leaving behind a column about 76 centimeters (30 inches) high. The empty space at the top of the tube, above the column of mercury was, indeed, empty, or nearly so. (It may have contained a very small amount of air trapped in the mercury and some mercury vapor.) The upper end of the tube, then, contained a vacuum. In demonstrating the existence of a vacuum and explaining the operation of water pumps, Torricelli also created as a result of his research one of the first and most important of all meteorological instruments, the barometer.

Over the next century, scientists became increasingly interested in the properties of a vacuum. In 1660, for example, the English chemist Robert Boyle (1627–1691) showed that forces of electrical and magnetic attraction could be detected in a vacuum, just as they could in normal air. He also showed that heat could be produced by friction between two bodies in a vacuum.

Somewhat later, scientists began to appreciate the value of vacuums in carrying out certain types of experiments. Such experiments required that individual atoms, molecules, and ions be able to travel relatively great distances without colliding with other particles. By carrying out such experiments in a vacuum, collisions could largely be avoided. For example, in 1855, the German physicist Johann Heinrich Wilhelm Geissler (1815–1879) invented a method for making glass tubes that would hold very good vacuums. He used a mercury pump, based on Torricelli's original concept, to remove air from glass tubes and then sealed the tubes to retain the vacuum thus produced. These tubes, widely known by the inventor's name as *Geissler tubes,* later became popular for the study of the individual particles that make up the structure of atoms.

Vacuum Applications and Devices

Although greatly modified and improved, the principle of the Geissler tube continues to be used widely in many fields of scientific research. For example, particle accelerators are machines used to accelerate the motion of electrons, positrons, and other subatomic particles to very high speed. They are used in studies on the composition of matter and for a number of practical applications. It is essential that the particles accelerated in such machines not encounter other particles as they traverse the dimensions of the machine. The interior of the machine must, therefore, be maintained at a high vacuum.

Specialized microscopes for viewing of very small objects also require the use of a vacuum. In an electron microscope, for example, electrons are released from a gun and aimed at the object to be viewed. Those electrons bounce off the object, back to the collector, where they are used to form a picture of the object. The interior of the microscope must be kept at a high vacuum to prevent electrons from colliding with other subatomic particles and being diverted.

Vacuums are used for other research purposes also. For example, equipment and methods to be used in space travel and research are tested in very large chambers with pressures similar to those of interstellar space, about 10^{-6} torr or less.

Another major use of vacuums is in cleaning processes. Perhaps the best known of such devices is the vacuum cleaner. In a vacuum cleaner, a fan at the base of the machine rotates in such a way as to push air away from the front of the machine and upward into the bag of the machine. The air then travels out of the bag through the pores in the material of which it is made. As air is pushed out of the base of the machine, a partial vacuum is created. Air from outside the base rushes in to replace the air pushed upward into the bag. As the machine is pushed across the floor, a brush rotates against the floor or rug, stirring up dust and dirt. The dust and dirt are carried into the base of the machine by the in-rushing air. It travels up into the bag, but is prevented from leaving the bag by the small size of the pores in the bag.

The principle of a vacuum cleaner is used in many industrial operations. In the manufacture of silicon chips for use in computing devices, for example, care must be taken that no foreign material of any kind is deposited on the chips being produced. To attain a "clean room" in which chips can be assembled, then, an oversize vacuum-cleaner-type device is installed in the ceiling of the room. A fan in the device pulls air out of the room, allowing air from the room to flow in and replace it. Any dust or other contamination in the room is carried out with the out-flowing air and is trapped in large fabric bags, similar to those found in a vacuum cleaner.

The first major use for vacuums developed as the result of the invention of the incandescent light bulb by Thomas A. Edison (1847–1931) in 1879. In Edison's invention, an electric current passed through a thin wire enclosed in a glass bulb heated the wire and caused it to glow. If the bulb contained air, the hot wire would react with oxygen in the air, changing its chemical nature and preventing electricity from flowing through the bulb. One solution to this problem was to remove the air from the bulb, that is, to create a vacuum inside the bulb. In this way, no atoms or molecules would be present to react with the wire, and it could continue to conduct an electric current for a long time.

A vacuum is also used in the manufacture of specialized containers used to keep objects hot or cold. Such containers are called *vacuum bottles* or *thermos bottles.* They consist of an interior container, which holds the material being kept hot or cold. The interior container is surrounded by a second container filled with a vacuum. The vacuum prevents the transfer of heat into or out of the interior container.

Vacuum Pumps

Many devices have been invented for the production of a vacuum. The oldest and simplest device was developed by the German physicist Otto von Guericke (1602–1686). It consisted of a cylinder fitted with a piston and two valves. When the piston was pulled out of the cylinder, one valve opened and the other closed. When the piston was pushed into the cylinder, the reverse operation occurred. To evacuate a container, von Guericke attached the open end of the brass cylinder to a hole in the container and then pushed and pulled the piston into and out of the cylinder. When the piston was pulled out, air flowed out of the container into the pump. When the piston was pushed in, air flowed out of the cylinder into the atmosphere. With enough repetitions, sufficient quantities of air were removed from the container to produce a reasonably satisfactory vacuum.

Perhaps the closest modern analogue to von Guericke's device is the rotary oil-sealed vacuum pump. This device consists of a short metal cylinder with a second eccentric ("off-center") cylinder inside it. The outer cylinder contains two ports, one for the entry of air and one for its discharge. When the interior eccentric cylinder rotates, it is always in contact with the inner wall of the outer cylinder at some point and away from the rest of the inner wall. During the cylinder's rotation, the inlet valve is first open, allowing air to flow into the pump. As the cylinder continues to rotate, that air is carried around inside the pump until it reaches the discharge valve, at which point it is released to the atmosphere. Every time the eccentric cylinder rotates it carries another parcel of air from the inlet valve to the discharge valve, until the desired amount of evacuation is achieved.

Another widely used type of vacuum pump is the ejector pump. An ejector pump consists of a long cylinder, open at both ends, with an inlet port near the top of the pump. Its operation depends on a principle discovered by the Swiss mathematician and physicist Daniel Bernoulli (1700–1782) in 1738. Bernoulli found that the pressure exerted by a liquid on the walls of a container was in inverse proportion to its speed. That is, the faster a liquid flows through a container, the less pressure it exerts on the walls of the container. In an ejector pump, some type of fluid, such as water or steam, is injected at high speed at the upper end of the pump. Because of its high speed, the fluid exerts a relatively low pressure on the walls of the cylinder. As the fluid rushes past the inlet port, air is pushed out of the container being evacuated, through the inlet port, and into the downward-flowing fluid. The air is ejected from the bottom of the cylinder along with the working fluid. *See also* Guericke, Otto von

Further Reading

Chambers, A., R. K. Fitch, and B. S. Halliday. *Basic Vacuum Technology.* Bristol, UK: A. Hilger, 1989.

Hoffman, Dorothy M., Bawa Singh, and John H. Thomas, eds. *Handbook of Vacuum Science and Technology.* San Diego: Academic Press, 1997.

O'Hanlong, John F. *A User's Guide to Vacuum Technology,* 2nd edition. New York: Wiley Interscience, 1989.

W

Weather Balloons

See Balloons

Wind

A *wind* is any flow of air. The term most commonly refers to a flow of air across the Earth's surface. Nearly all winds are three-dimensional in nature. That is, they tend to move both horizontally and vertically with respect to the Earth's surface, although the horizontal component tends to be much larger than the vertical component. A typical mild wind, for example, has a horizontal velocity of a few tens of kilometers per hour, while its vertical component is likely to be less than a kilometer per hour.

Types of Wind

Winds can be classified in a variety of ways. According to one system of classification, the winds that blow over the largest areas of the Earth's surface for the longest periods are known as *macroscale* or *planetary* winds. Such winds tend to cover thousands of kilometers of land or ocean and blow for weeks, months, or even years at a time. Prevailing winds, such as the trade winds or prevailing westerlies are examples of such macroscale winds.

Mesoscale winds tend to blow over less extensive areas, usually a few hundreds or thousands of kilometers, and for shorter periods, usually a few days or weeks. Land and sea breezes, mountain and valley breezes, chinook winds, katabatic winds, and hurricanes and other severe storms are examples of such winds.

Microscale winds are those that cover only very small distances, usually less than a kilometer, and for generally no more than a few seconds or minutes. Some examples of microscale winds are dust devils and wind gusts.

Macroscale winds are discussed in the entry **Global Circulation,** and two types of microscale winds are discussed in the entries entitled **Dust Devils** and **Turbulence.** Some major types of mesoscale winds are discussed later in this entry.

Causes

Winds of all kinds are caused by some combination of three major factors: (1) differences in air pressure; (2) the Coriolis effect; and (3) frictional forces.

Air Pressure Effects

The fundamental cause of wind is the unequal heating of the Earth's surface by solar energy. Some surface features, such as newly turned soil, absorb solar energy well and become warmed. Other features, such as ice fields, tend to reflect sunlight, absorbing relatively little solar energy and remaining relatively cool.

This unequal heating results in different air pressure. The air over a warm region of the Earth's surface tends to expand and rise, generating a region of low air pressure. By contrast, the air over a cool region of the Earth's surface tends to contract and settle, resulting in a region of high air pressure.

Regions of similar and differing atmospheric air pressure are represented on weather maps by means of isobars. *Isobars* are lines on a weather map joining points of equal atmospheric pressure. The distance between two adjacent isobars indicates the difference in air pressure between the two regions. If two adjacent isobars on a map are labeled 1,016 mb (millibars) and 1,020 mb, for example, the difference in air pressure over that distance is 1,020 mb − 1,016 mb, or 4 mb. The pressure difference between two adjacent isobars is known as the *pressure gradient* between the two isobars. Mathematically, the pressure gradient between two isobars can be expressed as the pressure difference divided between the distance between the two isobars, or:

$$PG = \frac{\Delta\,\widehat{pressure}}{d}$$

The pressure gradient over an area determines the force with which winds will blow. If two isobars on a weather map are close together, the pressure gradient per unit distance is relatively large, and strong winds will blow. If two isobars are widely separated on a map, the pressure gradient per unit distance is relatively small, and winds will be less severe.

Coriolis Effect

In principle, the flow of air across a pressure gradient from a region of higher pressure to one of lower pressure would always follow a path perpendicular to the isobars. In fact, this type of air flow is never observed. The most important force responsible for the deflection of air flow across a pressure gradient is the Coriolis effect. The *Coriolis effect* is a phenomenon observed with any freely moving object traveling on or above the Earth's surface. It was first described in detail by the French physicist Gustave Gaspard de Coriolis (1792–1843) in 1835.

The Coriolis effect is caused by the Earth's rotation on its own axis. Imagine a bullet fired from a gun from a point on the Earth in a southerly direction. One might expect that bullet to follow a straight path toward the south. In fact, while the bullet is traveling through the air, the Earth itself is rotating beneath it. If one tracks the actual track of the bullet, it becomes clear that the path is not a straight line, but a curved line that is deflected to the right of its original path in the Northern Hemisphere and to the left of its original path in the Southern Hemisphere.

The Coriolis effect can be observed in a host of phenomena involving moving objects. For example, the erosion along a river that is apparently flowing in a straight line is actually greater on one bank (the right bank in the northern hemisphere) than on the other. The Coriolis effect also affects the path followed by air movements (winds) as they move across the pressure gradient between two isobars. The path followed by the wind is not a straight line perpendicular to the isobars, but a curved line that follows a counterclockwise direction in the Northern Hemisphere and a clockwise direction in the Southern Hemisphere.

Frictional Forces

The third force that affects the speed and direction of wind is friction between the moving air that makes up the wind and surface features, such as hills, rocks, and bodies of water. Frictional forces significantly affect winds that develop at altitudes of less than about a kilometer about the Earth's surface. Above that height, moving air comes into relatively little contact with surface features, and frictional forces are insignificant.

Geostrophic Winds. Consider a parcel of air momentarily at rest at an altitude of about 5 kilometers above the Earth's surface. Because of gradient pressure forces, that parcel of air will eventually begin to move to a region of lower air pressure. It will, at least for a fraction of a moment, move perpendicularly from the region of higher pressure to that of lower pressure. As soon as the air moves, however, it is subject to Coriolis forces and will begin to move in a curved direction. The faster the air flows, the stronger the Coriolis force acting on it.

Eventually, the parcel of air will be flowing in a direction parallel to and between two adjacent isobars. Winds of this type are known as *geostrophic winds.* The strength of a geostrophic wind depends on the pressure gradient between adjacent isobars. The greater the pressure gradient, the stronger the winds will flow.

The picture of geostrophic winds outlined here is oversimplified. Other factors usually alter the path of a geostrophic wind. The picture is, however, sufficiently correct to form a useful tool for meteorologists in understanding and predicting weather patterns.

Gradient Winds. The flow of air in lower portions of the air is generally affected by surface features and the frictional forces they cause. This portion of the atmosphere, about a kilometer or two in depth, is known as the *boundary layer.* Geostrophic winds like those described above are seldom encountered in the boundary layer because of frictional effects. Exceptions do occur, however. Winds blowing over the smooth

surface of a lake or sea, for example, may encounter so little friction that their flow approximates that of a geostrophic wind.

The effect of surface features, such as hills and valleys and bodies of water, can be seen in the shape of isobars on a weather map. These isobars are seldom straight. Instead, they tend to be curved, sometimes to an extent that they form closed circles surrounding areas of high or low pressure. The wind flow through such systems is affected not only by pressure gradient forces and the Coriolis force, but also by a third force, centrifugal force.

Consider a parcel of air flowing into the space between two isobars around a low-pressure center, a *trough.* As air flows into this space, it is acted on by the pressure gradient force, which pulls it inward toward the low-pressure center, and by the Coriolis effect, which pulls it outward, away from the center. But since the air parcel is traveling in a curved path, it also experiences a third force, centrifugal force, which also pushes it outward.

The path followed by this parcel of air, called a *gradient wind,* represents a balance among these three forces. That is, the total inward force (the pressure gradient force; f_g) must equal the total outward force (the sum of the Coriolis force, f_C, and the centrifugal force, f_c):

$$f_g = f_C + f_c$$

This situation differs from that in a geostrophic wind, where:

$$f_g = f_C$$

In the case of a gradient wind, balance can be achieved only if the Coriolis force decreases as the centrifugal force increases. And the only way for the Coriolis force to decrease is for the speed of the gradient wind to decrease. From this argument, it is clear that gradient winds blow through a low-pressure trough parallel to and between two isobars, but at a speed less than that of geostrophic winds above them. Gradient winds that travel around a low-pressure center are also known as *cyclones* and the movement of air is called a *cyclonic flow.*

A similar argument can be made for winds blowing around a high-pressure area, or high-pressure *ridge.* In this case, however, centrifugal forces act in the same direction as the pressure gradient. To keep the gradient winds in a path between two isobars, then, their speed must increase; that is, the Coriolis effect must increase. In the case of a high-pressure ridge, then, gradient winds blow with greater velocity than do geostrophic winds above them. Such winds are known as *anticyclones* and the flow of air around the high pressure center, an *anticyclonic flow.*

Once more, the description of air movements presented here is overly simplified because it does not take into account the effect of surface features. Hills, valleys, bodies of water, and other surface features change both the direction of winds and the speed with which they blow. In mountainous areas, for example, winds may be reduced in speed by as much as half and may be diverted from a flow parallel to that of isobars by as much as 45°. Even smooth surfaces, such as those of a desert or smooth lake, may produce enough friction to reduce winds speeds by a third of geostrophic winds overhead and alter their path by up to 20° from parallel paths between isobars.

Mesoscale Winds

Mesoscale winds, also known as *local winds,* occur in every part of the world. Some of the hundreds of different names that have been given to local winds can be found in Chapter 35 of *The American Prac-tical Navigator,* originally prepared by Nathaniel Bowditch and now published by the National Ocean Service. Some Web sites have expanded on this list. One, "Whirling Winds of the World" (http://freespace.virgin.net/mike.ryding/index.htm), lists more than 400 specialized names given to local winds. These names range from the *abroholos,* a violent squall that blows across parts of Brazil from May to August each year, to the *kloof,* a damp, cool wind common in the Simon's Bay region of South Africa, to the *zonda,* a flow of moist tropical air that blows down from northern Argentina into Uruguay.

Local winds are generally produced by the irregular heating of the Earth's surface because of differing topographical features, especially land and water areas that are adjacent to each other, and mountainous areas.

For example, *land breezes* and *sea breezes* are generated because water and land absorb solar energy at different rates. Consider what happens to a coastal area during a bright, sunny day. Land absorbs heat and becomes warm faster than does the adjacent water. Air above the land expands and rises, creating an area of low pressure. Cooler air over the adjacent water flows into the low-pressure area, producing a sea breeze. These breezes may produce a drop in temperature of as much as 10°C compared to areas only a few kilometers inland.

At night, conditions are reversed. Dry land gives up heat more rapidly than does water. Air above the land condenses and becomes more dense, resulting in a high-pressure area over the land. Air from this high-pressure region flows outward across the water, resulting in a land breeze.

Similar changes occur in mountainous regions. During the day, the sides of a mountain are exposed to sunlight for a longer period than is any adjacent valley. As the mountain slopes are warmed, air above them becomes less dense, producing an area of low pressure on the mountain slopes. Air from the valley below flows upward into the low-pressure region, producing *valley breezes.*

At night, these conditions are reversed. Exposed mountain slopes give up their heat and the air above them becomes cooler and more dense. This cool air settles to the ground and forms a region of high pressure. Air from the high-pressure area flows outward and down the mountain slope into the valley, resulting in a *mountain breeze.*

Mountain breezes are one form of *gravity wind,* also known as a *drainage wind* or *katabatic wind,* caused when air that is relatively cool and dense flows down the side of a mountain. In some parts of the world and at some times of the year, this flow of air may become quite intense, resulting in very strong winds that have been given characteristic names. For example, the *mistral* is a gravity wind that flows out of the French Alps toward the Mediterranean Sea. The *bora* is a wind that originates in the eastern Alps and flows downward to the Adriatic Sea.

Another type of wind that forms on the leeward side of a mountain is the *chinook* or *föhn* (or *foehn*). These winds develop when air from the windward side of the mountain flows up and over the mountain tops into a low-pressure area on the leeward side of the mountain. This air is often considerably warmer than the air it replaces on the leeward slopes, resulting in a distinct warming trend there. In fact, the word *chinook* derives from the term "snoweater" used by Pacific Northwest American Indians.

One of the best-known forms of a chinook wind in the western United States is a *Santa Ana wind.* Santa Ana winds form when hot desert air from Nevada and Arizona flows over and through the San Gabriel and San Bernadino Mountains on its way to the Pacific Ocean. As the air passes through canyons and other gaps in the mountain, it picks up speed, is compressed, and becomes hotter. On an unfortunately regular basis, these winds carry fire to trees, brush, and houses that stand in their way, causing massive financial damage and, in some cases, loss of human life.

Record Wind Speeds

The highest wind speeds occur in severe storms, such as tornadoes and hurricanes. It is generally difficult or impossible to measure those wind speeds, however, since they may exceed the speed that most instruments can withstand. It appears that the strongest wind ever measured occurred during a tornado that struck Oklahoma City on 13 May 1999. A truck-mounted Doppler radar device clocked that wind at 509 kilometers per hour (318 miles per hour). The previous record for wind speed—460 kilometers per hour (286 miles per hour)—was also measured by Doppler radar on 26 April 1991 in a tornado that struck near Red Rock, Oklahoma.

The highest wind speed on record for conventional (nonhurricane, nontornado) events was observed on top of Mount Washington, New Hampshire, on 12 April 1934. That speed was measured to be 372 kilometers per hour (231 miles per hour). *See also* Dust Devil; Wind Measurement

Further Reading

DeBlieu, Jan. *Wind: How the Flow of Air Has Shaped Life, Myth, and Land.* Boston: Houghton Mifflin, 1998.

Posey, Carl A. *The Living Earth Book of Wind & Weather.* Pleasantville, NY: Reader's Digest Association, 1994.

Wind Chill

Wind chill is a measure of the combined effect of atmospheric temperature and wind

Wind Chill Chart

Wind (mph) \ Temperature (°F)	40	35	30	25	20	15	10	5	0	-5	-10	-15	-20	-25	-30	-35	-40	-45
Calm																		
5	36	31	25	19	13	7	1	-5	-11	-16	-22	-28	-34	-40	-46	-52	-57	-63
10	34	27	21	15	9	3	-4	-10	-16	-22	-28	-35	-41	-47	-53	-59	-66	-72
15	32	25	19	13	6	0	-7	-13	-19	-26	-32	-39	-45	-51	-58	-64	-71	-77
20	30	24	17	11	4	-2	-9	-15	-22	-29	-35	-42	-48	-55	-61	-68	-74	-81
25	29	23	16	9	3	-4	-11	-17	-24	-31	-37	-44	-51	-58	-64	-71	-78	-84
30	28	22	15	8	1	-5	-12	-19	-26	-33	-39	-46	-53	-60	-67	-73	-80	-87
35	28	21	14	7	0	-7	-14	-21	-27	-34	-41	-48	-55	-62	-69	-76	-82	-89
40	27	20	13	6	-1	-8	-15	-22	-29	-36	-43	-50	-57	-64	-71	-78	-84	-91
45	26	19	12	5	-2	-9	-16	-23	-30	-37	-44	-51	-58	-65	-72	-79	-86	-93
50	26	19	12	4	-3	-10	-17	-24	-31	-38	-45	-52	-60	-67	-74	-81	-88	-95
55	25	18	11	4	-3	-11	-18	-25	-32	-39	-46	-54	-61	-68	-75	-82	-89	-97
60	25	17	10	3	-4	-11	-19	-26	-33	-40	-48	-55	-62	-69	-76	-84	-91	-98

Frostbite Times   ☐ 30 minutes   ☐ 10 minutes   ☐ 5 minutes

$$\text{Wind Chill (°F)} = 35.74 + 0.6215T - 35.75(V^{0.16}) + 0.4275T(V^{0.16})$$

Where, T = Air Temperature (°F)   V = Wind Speed (mph)   Effective 11/01/01

NWS Wind Chill Temperature Index. Courtesy, National Weather Service, Office of Climate, Water and Weather Services.

speed on the rate at which the human body cools. The wind chill concept was developed during the late 1930s and early 1940s by the American geographer Paul Siple and geologist Charles Passel during their research in the Antarctic. Wind chill is more accurately called the *windchill equivalent temperature,* or *WET.*

WET is simply a precise scientific formulation of the everyday observation that a person feels colder on a day when the wind is blowing than when it is not, even if the atmospheric temperature is the same in both cases. Two factors account for this effect. First, on a windy day, the movement of air carries away body heat and replaces it with cool air, a phenomenon that does not occur in calm weather. Second, the movement of air over the skin causes perspiration to evap-

orate more rapidly than in calm weather. The body loses heat when perspiration evaporates, adding to the wind's cooling effect.

The formula originally developed by Siple and Passel for determining WET at any given temperature and wind velocity is as follows:

$$WET(°F) = 0.0817(3.71V^{0.5} + 5.81 - 0.25V)(T—9.14) + 9.14$$

where T is the temperature in degrees Fahrenheit and V is the velocity in miles per hour. In 2001, the weather services of the United States and Canada decided to replace the original formula with a new one that matches conditions in the real world somewhat more accurately than did the original formula. That formula is as follows:

$$WET(°F) = 35.74 + 0.6215T—35.75V^{0.16}$$
$$+ 0.4275TV^{0.16}$$

These formulas are of relatively modest interest to ordinary citizens, and weather organizations usually publish charts that provide the necessary information in a form that is easy to read and understand. One such chart is shown here. Notice that the chart shows three areas that indicate the time within which frostbite is likely to occur.

Further Information

"National Weather Service Implements a New Wind Chill Temperature Index," http://www.nws.noaa.gov/om/windchill/index.shtml.

Wind Energy

See Wind Energy Systems

Wind Energy Systems

A wind energy system is any device or group of devices that captures the kinetic energy of moving air and converts it into mechanical or electrical energy to be used for some practical purpose, such as lifting water or operating electric appliances. Two of the oldest wind energy systems are sails and windmills, both first used by human more than 5,000 years ago. Sails are now used in wind energy systems, and this entry focuses on the use of windmills to generate electricity. When used in this way, windmills are more commonly known as *wind turbines.*

History

The first windmill built to generate electrical power was constructed by Charles F. Brush (1849–1929) in 1888 in Cleveland, Ohio. Brush's wind turbine had a 17 meter (50 foot) diameter rotor made of 144 cedar "picket-fence" style blades. It operated for 20 years, providing a relatively modest 12 kilowatts of power, which Brush used to charge a bank of batteries in the basement of his home.

At about the same time, the Danish meteorologist and inventor Poul la Cour (1846–1908) designed the first wind turbine based on aerodynamic principles, including primitive airfoil shapes for its blades. La Cour's turbines were more aerodynamically efficient than Brush's machine, making them the first machines from which electrical power could be generated economically on a large scale.

Little progress in the development of wind power systems took place until after World War I. Then, research on wind energy began to become popular among a relatively small number of inventors in parts of Europe and the United States. In the United States, that research was directed at the development of small machines capable of generating 1 to 3 kilowatts that could be used on farms for lighting, radio operation, and other electric uses. These efforts fell by the wayside in the 1930s, however, at least partly because President Franklin D. Roosevelt's rural electrification program brought power lines to all but the most remote rural areas of the country.

The first large-scale wind turbine designed for commercial use was built in Russia in 1931 on the shore of the Caspian Sea. The turbine generated 100 kilowatts of electricity and operated for two years. A decade later, a 1.25 megawatt turbine with a rotor consisting of two 53 meter (175 foot) stainless steel blades was built in Vermont. The turbine operated for four years before one of the blades broke off. Still later, one of la Cour's students, Johannes Juul, designed the world's first alternating current (AC) wind turbine. The first machine of this design was built in 1956 at Gedser in southern Denmark. It operated for 11 years with essentially no maintenance, generating 200 kilowatts of electrical power.

The development of wind energy systems for the first two-thirds of the twentieth century was seriously hampered by growing dependence on fossil fuel energy systems. Electrical power plants powered by coal, oil, and natural gas were dramatically less expensive than similar facilities based on alternative sources of energy, including wind systems. The oil crisis of 1973 somewhat changed that scenario, however, as oil prices suddenly increased dramatically and many governments began to realize that alternative energy systems would have to be developed, if not for immediate purposes then at least for long-term implementation. That realization has continued to motivate research on alternative energy systems over the past three decades, even as fossil fuel prices have returned to more modest levels.

An important breakthrough in wind turbine technology was the Nordtank 55 kilowatt turbine developed in Denmark in the early 1980s. The three-blade, propeller-style rotor had a 60 meter (200 foot) diameter that reduced the cost of wind energy by about 50 percent. Nordtank turbines were soon being installed in a number of locations in Denmark and other countries. Thousands were sold in the United States, where they become the components of massive "wind farms," primarily in California. The rush to develop wind energy in the United States dropped off quickly, however, as the cost of conventional fossil fuels returned to their pre-1973 levels. Today, Germany has the largest wind power capacity (2,874 MW [megawatts], as of 1998), followed by the United States (1,884 MW), Denmark (1,450 MW), India (968 MW), Spain (834 MW), and the Netherlands (363 MW).

Turbine Design

Wind turbines are of two basic designs, the horizontal-axis and vertical-axis styles. A horizontal-axis turbine consists of a vertical tower at the top of which is located a horizontal tube, called the *nacelle,* that contains the gearbox and generator. Attached to one end of the nacelle is the rotor, usually consisting of two or three blades. As the rotor spins, the linear kinetic energy of moving air is converted to the rotational energy of a central shaft in the gearbox; the shaft's energy is converted to electrical energy in the generator.

The vertical-axis, or "egg-beater," turbine was invented by the French engineer Georges Darrieus (1888–1979) in the 1920s. It consists of a vertical pole to which are attached two or more looped blades at the top and bottom, giving the machine its characteristic "egg beater" appearance. The gearbox and generator are in a box at the bottom of the vertical pole.

Another type of vertical-axis turbine is the S-shaped Savonius machine, designed by the Finnish inventor Sigurd Savonius (1884–1931). One important advantage of a vertical-axis turbine is that it does not have to be turned to face the wind. No matter what direction the wind strikes it, it will cause the rotor to start spinning.

Wind turbines operate on either one of two aerodynamic principles: lift or drag. In a lift machine, rotor blades are shaped like airplane wings, wider at the front and narrower at the back. As wind passes over the blades, it causes them to lift and begin rotating. All horizontal axis turbines operate on lift forces. No external source of power is needed to set such turbines in motion.

Vertical axis turbines may use lift forces (as in the Darrieus rotor) or drag forces (as in the Savonius rotor). In aerodynamic terms, drag is equivalent to air resistance, such as the drag (resistance) an object experiences as it falls toward the earth. In drag machines, rotor blades are put into motion as the result of the wind pushing against them.

Turbine Characteristics

The power output of a wind turbine depends primarily on two factors: the area of the rotor and the wind speed. In general, the larger a rotor, the more wind it can capture and the greater power in can generate. For example, the minimum rotor diameter for a 10 kilowatt turbine designed for private use is about 7 meters (23 feet), while a utility-size 750 kilowatt turbine requires a rotor diameter of at least 24 meters (79 feet). The tower needed to support a rotor of given size depends on a variety of factors, not the least of which being the length of the rotor blades (They can't come into contact with the ground, of course!). Most utility turbines are at least 60 meters (200 feet) tall.

Wind speed is a critical factor in turbine operation because the power generated by a turbine is proportional to the cube of the wind speed. If the wind speed doubles, therefore, the amount of power generated increases by a factor of eight (2^3). To achieve the power output expected from a turbine designed for private use, the average annual wind speed must be at least 4 meters per second (9 miles per hour), while utility size turbines require minimum wind speeds of 6 meters per second (13 miles per hour).

The importance of wind speed in power production means that the topography around a wind turbine is very important. The presence of hills, trees, buildings, and other objects that can deflect the flow of air will decrease wind speeds and make the operation of a wind turbine less efficient. Wind engineers, who refer to this factor as *roughness,* classify various areas according to a *roughness class* from 0 (very smooth) to 4 (very rough). As an example, the typical runway of a modern airport has a roughness class of about 0.5.

Roughness considerations have made off-shore wind turbines an attractive alternative to comparable land-based machines.

Of course, the ocean is sometimes rough, but, under many conditions, its roughness class may approach 0 for long periods. In countries with large coastline-to-area ratio, offshore wind turbines are becoming more popular. The Danish government's "Energy 21" long-term energy plan, for example, calls for the production of 4,000 MW of offshore wind power by 2030, which, along with 1,500 MW of land-based wind power, will provide more than half of the nation's annual electrical needs.

Wind Power Economics

The cost of wind energy in the United States has dropped dramatically from nearly 40 cents per kilowatt hour in 1980 to about 4 cents in 2001, a price that is competitive with traditional fossil-fuel-generated electricity. These reductions have resulted from significant improvements in wind turbine technology at research centers such as the Sandia National Laboratory near Albuquerque, New Mexico, and the Risoe National Laboratory in Roskilde, Denmark.

The great advantage of wind power is, of course, that the raw material needed to run wind turbines—air—is free. The primary cost of wind energy systems is the construction of the wind turbines themselves. Yet, once these machines are constructed, they are remarkably long lasting and efficient. The average wind turbine built today has a lifetime of about 120,000 operating hours (roughly 20 years) with an availability factor (percentage of time the turbine is actually available for operation) of about 98 percent.

A major disadvantage of wind energy systems is that the wind does not blow constantly in most places and so wind turbines can operate for only a certain number of hours per year. That number of hours can be increased by a judicious choice of siting, but some "down time" is inevitable with almost any wind energy system.

These electricity-generating turbines are spinning in the wind in the Texas Panhandle as part of the long-term experiments by Sandia Laboratories' wind energy technology department. © Sandia National Laboratories.

One solution to that problem has been the development of a program known as *net metering,* or *net billing.* In this system, a private residence generates its own electric power with wind turbines when conditions permit (that is, when the wind blows), but draws from an existing electrical network operated with fossil-fueled, nuclear, or other power plants when winds are calm. The private residence often generates more electricity than it can use at some times, for example, at night when no appliances are in use. It is allowed to sell this excess electricity back to the electrical network during these periods. The private customer is billed only for the *net* amount of electricity used, that is, the amount of electricity taken from the network less any electricity returned to the network. In the United States by 2001, 29 states had enacted laws permitting some form of net metering, with indications that the practice was like to expand.

Potential for Wind Energy

Proponents of wind power development see a rosy future for the industry. They point to studies by the Batelle Pacific Northwest Laboratory, a federal research facility, showing that 20 percent of the U.S. need for electricity could be met by wind power. Conditions in any one of five states—North and South Dakota, Texas, Kansas, and Montana—would be sufficient, the study showed, to meet a third of the nation's demand for electricity. That potential does not seem unreasonable when compared to the fact that some parts of Spain and Denmark currently receive 20 to 25 percent of their electricity from wind turbines.

In 1999, the U.S. Department of Energy announced a new program, Wind Powering America, with the goal of increasing the share of wind-powered electricity in the United States to 5 percent by 2020. The plan also calls for an increase in the number of states with more than 20 MW of wind energy to 16 by 2005 and to 24 by 2010.

Similar, or even more rapid progress may be expected in other parts of the world. According to the Danish Wind Industry Association, annual growth rates of 20 percent per year in the use of wind turbines are expected over the next decade. Wind power is expected to be especially popular in developing countries, both because of its low operational cost and its lack of damage to the environment.

Environmental Effects

Wind turbines are among the most environmentally benign sources of energy available. They produce no pollutants of any kind. By comparison, traditional fossil-fueled power plants produce a host of waste products, such as sulfur dioxide, oxides of nitrogen, carbon monoxide, carbon dioxide, and particulate matter, that harm the biotic and abiotic environment. The American Wind Energy Association estimates that a more extensive use of wind for the generation of electricity in the United States could eliminate up to one-third of all harmful emissions from fossil-fueled plants. Even the relatively modest step of developing 10 percent of the available wind power in the ten windiest states would essentially eliminate the waste products that cause acid rain and reduce carbon dioxide emissions in the United States by one-third.

A few relatively modest environmental problems have, in the past, been associated with the generation of wind energy, although these problems have now been largely resolved. At one time, the mechanical operation of the rotor and gearbox had a tendency to be noisy. Technological breakthroughs have essentially solved this problem, and the only noise noticeable at most wind farms is a quiet "whooshing" sound like that of the wind itself.

At one point, there was also some concern about the number of birds killed by collision with wind turbines at the Altamont Pass wind farm in California. That problem seems to be a local issues that has not occurred to any great extent at other wind power sites. Some people are also concerned about the visual impact of hundreds of wind turbines on the landscape. Studies have shown, however, that this concern is not widespread and that people sometimes become more supportive of wind energy when turbines are actually built in their area.

Further Reading

American Wind Energy Association, "The Most Frequently Asked Questions about Wind Energy," Washington, DC, 1999. Available online at http://www.awea.org.

Burton, Tony, et al., eds. *Wind Energy Handbook.* New York: John Wiley, 2001.

Guided Tour on Wind Energy, http://www.windpower.dk/tour/index.htm.

Hau, Erich. *Windturbines: Fundamentals, Technologies, Application and Economics.* New York: Springer Verlag, 2000.

Pasqualetti, Martin J., Paul Gipe, and Robert W. Righter. *Wind Power in View: Energy Landscapes in a Crowded World.* San Diego: Academic Press, 2002.

Patel, Mukund R. *Wind and Solar Power Systems.* Boca Raton, FL: CRC Press, 1999.

Perez, Karen. *Wind Energy Basic: A Guide to Small and Micro Wind Systems.* White River Junction, VT: Chelsea Green Publishers, 1999.

Wind Erosion

See Erosion by Wind

Wind Farm

See Wind Energy Systems

Wind Instruments

A *wind instrument* is any device that produces music by causing the vibration of a column of air in a pipe or tube. Musical instruments can be classified in a number of ways, one of the most useful of which was devised by the German musician Eric Sachs and the Austrian musician Eric M. von Hornbostel in 1914. According to the Sachs-von Hornbostel system, all musical instruments can be divided into four categories: chordophones, or stringed instruments; aerophones, or wind instruments; idiophones, or percussion instruments made of wood or metal; and membranophones, or percussion instruments made of a stretched skin or membrane.

All musical instruments produce a sound when a string, a column of air, a stretched membrane, or some other object is caused to vibrate. The vibration of that object causes air in the vicinity to vibrate, producing air waves that travel to the ear and are interpreted as sound or music. Music is a special type of sound (in contrast to noise, for example) that consists of regular air waves and often consists of harmonious patterns known as overtones. The formation of

Various wind instruments. © C.G. Conn division of Conn-Selmer, Inc.

musical sounds by an instrument is generally complex and consists of the vibration not only of a single form of matter (such as a string or a piece of metal), but also of other materials to which it is attached and of the air and other materials surrounding the instrument.

Organs and the human voice can be classified as aerophones because they make sounds by causing columns of air to vibrate. Both instruments are often considered separately, however, from other types of wind instruments.

History

Aerophones are among the oldest—if not the oldest—musical instruments used by humans. A flute made of bird bone found at Brassenpouy, France, has been given an age of about 25,000 years by carbon dating. Representations and remains of flutes and other wind instruments appeared as early as 2500 B.C. in Egypt. Indian gods and goddesses dating to the first millennium B.C. also show primitive aerophones. In one case, for example, the god Vishnu is shown holding a conch shell horn.

Over the centuries, virtually every human society has developed one or more forms of an aerophone. Some examples include the shofar (a trumpet made out of a ram's horn) in Israel; bagpipes among the Celts; the Laotian mouth organ (khaen), also found in parts of Thailand; the didgeridoo, used by Australian aborigines since the dawn of their civilization; Incan pipes that date back to at least 1250 B.C. and are still popular in Peru; and wooden and reed flutes that have always been part of Native American tribal ceremonies.

Types

The pipe organ differs from other forms of aerophones in that a column of air is caused to vibrate by some mechanical means, such as by releasing forced air into or out of a pipe. By contrast, other wind instruments produce sounds when a human player blows into the device. Aerophones of this type are often classified into one of two categories: woodwinds or brass winds (or, more simply, brasses). Some examples of woodwinds are the flutes (such as the flute, fife, panpipes, piccolo, and recorder), clarinet family, and the oboe family (including the English horn, bassoon and contrabassoon). Among the most familiar brasses are the bugle, cornet, trombone, trumpet, tuba, and French horn.

Their names alone would suggest that woodwinds and brasses differ essentially on the basis of the materials of which they are made: wood in the one case and metal in the other. In fact, the primary difference between the two types of aerophone is the way in which sound is produced, with or without reeds in one case (woodwinds) and with some type of mouthpiece in the other (brasses). Thus, flutes made of metal are still classified as woodwinds, in spite of the material from which they are made, because of the way in which sound is produced.

The three major families of woodwinds differ from one another primarily in terms of the way in which air is caused to vibrate within them. In the flutes, the player blows over an opening in the instrument (as in the flute itself) or into the end of the device (such as some simple wooden whistles). Members of the clarinet family, by contrast, all have a reed made of cane or plastic that vibrates when air is blown across it. Members of the oboe family have a double reed, the two parts of which vibrate in unison, so that the tube is sometimes closed (when the two reeds are in contact with each other) and sometimes open (when they are apart).

In brass winds, sounds are produced by blowing into a cup-shaped mouthpiece (bugle, cornet, trombone, tube, trumpet) or a funnel-shaped mouthpiece (French horn). In brasses, the player's lips play the role of a reed. By changing the tension of his or her lip muscles, the player can produce an extended range of tones.

The tone produced by vibrating air in a pipe or tube depends on the length of the pipe or tube. A pipe organ, for example, consists of many pipes, each of which produces a single tone, determined by the pipe's length. Some types of aerophones operate on a similar principle. Panpipes, for example, are a set of reed, wooden, or plastic pipes, each of a different length. The panpipes can play only a given number of tones, those produced by the single pipes of which the instrument is made.

In both woodwinds and brasses, the number of tones produced is controlled by a series of holes drilled into the instrument's pipe. A player can open or close these holes with his or her fingers or by means of mechanical valves, keys, and pistons, which effectively change the length of the vibrating column of air within the instrument. Brass wind makers have also invented a variety of ways of changing the effective length of a vibrating column of air within an instrument, such as by using slides (as in a slide trombone) or by adding extra lengths of tubes (called crooks and shanks) to the instruments that can be opened and closed by the player.

Further Reading

Carse, Adam. *Musical Wind Instruments.* New York: Dover, 2002. Reprint of 1965 edition.

Daubney, Ulric. *Orchestral Wind Instruments, Ancient and Modern.* Freeport, NY: Books for Libraries Press, 1970. Reprint of 1920 edition.

Whitwell, David. *A New History of Wind Music.* Evanston, IL: Instrumentalist, 1972.

Wind Measurement

For the purpose of studying weather and predicting future weather patterns, meteorologists make measurements on many properties of air, most important of which are wind direction, speed, moisture content, and temperature. This entry deals with instruments and methods for measuring the first two of these properties.

Weather Vanes

One of the oldest weather measuring devices is a weather vane. The first such device of which we have any record was built by the Greek astronomer Andronicus in 48 B.C. The weather vane is said to have been about 4 to 8 feet (1.2 to 2.4 meters) long with the head and body of a man and the tail of a fish. The weather vane was erected atop the Tower of the Winds in Athens in honor of the god Triton.

The key feature of any weather vane is its long, flat shape, allowing it to rotate in such a way as to face into an oncoming wind. Many vanes have sidearms that carry the directions north, south, east, and west so that the wind's direction can be easily noted. More advanced weather vanes are erected on a flat circular plate with specific angular directions ranging from 0° to 360°.

Over time, weather vanes assumed functions other than weather prediction. For example, a ninth-century pope decreed that every church in Europe have a cock-shaped weather vane on top of its highest point to reminder parishioners of Jesus' prophecy that the cock would not crow on the morning after the Last Supper until Peter had denied him three times (Luke 22:34). Today, weather vanes are still popular ornamental devices on homes, farms, and other buildings even though they no longer serve much of a meteorological function.

One exception to that statement is at small airports, rowing clubs, and other locations where individuals want to have an estimate of the direction and speed with which the wind is blowing. One modern improvement on the traditional weather (wind) vane is the *aerovane*. An aerovane consists of a long cylindrical tube with a propeller at the front end and a fin at the back end. The fin keeps the aerovane faced into the wind (indicating its direction), while the rate at which the propeller rotates indicates the speed with which the wind is blowing. Both direction and wind speed are typically detected, transmitted and recorded electronically, allowing pilots, boaters, and others to access these data.

Windsock

Another device for indicating the direction and speed of the wind is a windsock. The most popular use of windsocks is as a landing guide for small airplanes at airports. Pilots can estimate the direction and intensity of the wind by noticing the direction in which the windsock is blowing and how far it is extended from the vertical. Windsocks are also used on the open water to permit sailors to estimate the speed and direction of the wind.

The simplest form of a windsock consists of a piece of thin material, such as nylon, cut and sewn in the shape of a truncated cone (a cone with its point cut off). The wider opening of the windsock is attached to a tall pole by means of strings or ropes, so that the device can hang loosely and swing freely in the wind. As the wind blows, it enters the mouth of the sock. The stronger the wind, the more the sock is extended in a horizontal direction.

The windsock is not a precise measuring device. It shows only the approximate direction and speed of the wind. This information, however, is usually sufficient for the needs of small-aircraft pilots, rowers, and small-boat enthusiasts.

Most light windsocks can be used to estimate wind speeds of about 15 to 25 knots (17 to 30 miles per hour). Windsocks made of heavier material can be used to indicate stronger winds. Pilots and others who use windsocks learn to estimate wind speed from the shape of the sock. For example, the Boulder Rowing Club informs its members that a horizontal windsock indicates a speed of about 15 knots, while a sock hanging at an angle of 45° indicates a speed of about 10 knots.

Many variations and improvements of the basic windsock are now available. For example, one company (Action Flags Manufacturing Company) offers the Wind Dancer System, which consists of a long, rectangular flag hanging from a tall pole. The angle at which the flag projects from the pole indicates the wind speed.

Beaufort Wind Scale

During the days of sailing ships, it was crucial for ship captains to know the speed and direction of the wind so they could plot their courses on the open seas. The first widely successful and reasonably precise system for making such measurements was developed in 1805 by Sir Francis Beaufort (1774–1857), a member of the British Navy for 68 years. Beaufort based his system on easily observed phenomena that could be correlated with known wind speeds. A modern version of his scale, adapted for use on

Beaufort Wind Scale

Beaufort Number	Description	Wind Speed (in mph)	Visual Clues and Damage Effects
0	calm	0	Calm wind. Smoke rises vertically with little if any drift.
1	Light air	1–3	Direction of wind shown by smoke drift, not by wind vanes. Little if any movement of flags. Wind barely moves tree leaves.
2	Light breeze	4–7	Wind felt on face. Leaves rustle and small twigs move. Ordinary wind vanes move.
3	Gentle breeze	8–12	Leaves and small twigs in constant motion. Wind blows up dry leaves from the ground. Flags are extended out.
4	Moderate breeze	13–18	Wind moves small branches. Wind raises dust and loose paper from the ground and drives them along.
5	Fresh breeze	19–24	Large branches and small trees in leaf begin to sway. Crested wavelets form on inland lakes and large rivers.
6	Strong breeze	25–31	Large branches in continuous motion. Whistling sounds heard in overhead or nearby power and telephone lines. Umbrellas used with difficulty.
7	Near gale	32–38	Whole trees in motion. Inconvenience felt when walking against the wind.
8	Gale	39–46	Wind breaks twigs and small branches. Wind generally impedes walking.
9	Strong gale	47–54	Structural damage occurs, such as chimney covers, roofing tiles blow off, and television antennas damaged. Ground is littered with many small twigs and broken branches.
10	Whole gale	55–63	Considerable structural damage occurs, especially on roofs. Small trees may be blown over and uprooted.
11	Storm force	64–75	Widespread damage occurs. Larger trees blown over and uprooted.
12	Hurricane force	over 75	Severe and extensive damage. Roofs can be peeled off. Windows broken. Trees uprooted. RVs and small mobile homes overturned. Moving automobiles can be pushed off the roadways.

Adapted from "Storm Spotter Scales and Conversions Information," National Weather Service, Portland, Oregon, online at http://www.wrh.noaa.gov/Portland/scale.html.

land as well as water, is widely used today. The chart shows one form of the Beaufort Scale used on land. A comparable chart is available for use on the water.

Anemometers

An anemometer is yet another device for measuring the direction and speed of the wind. The first anemometer of which we have written evidence was constructed in 1450 by the Italian artist and architect Leone Battista Alberti (1404–1472). Alberti's anemometer consisted of a flat disk attached to a vertical rod around which it could rotate. When placed in the wind, the disk rotated to indicate the direction from which the wind

was blowing. The speed with which the disk rotated also gave an inexact measure as to the wind's speed. Similar devices were invented independently by a number of other men, including the English physicist Robert Hooke (1635–1703) and English scholar William Whewell (1794–1866). All were similar in that they used some type of flat disk that would catch the wind and rotate at a speed proportional to the speed of the wind.

An important improvement on Alberti's design was produced in 1864 by the Irish astronomer Thomas Romney Robinson (1792–1882). Robinson's device consisted of a vertical rod to which were attached, at the top, four horizontal arms. A copper cup, similar to a coffee cup without its handle, was attached at the end of each arm. This "spinning cup" anemometer caught the wind more effectively than did a flat disk and was able to provide more precise measurements of the wind's speed and direction.

Modern anemometers operate on essentially the same principle as that of Robinson's spinning-cup anemometer. The rotation of the anemometer is, however, now measured more precisely by translating the rotational movement of the cups into an electrical signal that can be read or recorded. For example, some anemometers have a dial on which the wind speed can be read much as one reads an automobile speedometer. Other anemometers are attached to recording graphs that show changes in wind speed and direction continuously. Many anemometers also have a weather-vane arrow attached to the top of the central rod to indicate the direction in which the wind is blowing.

Another variation of the anemometer is the *propeller anemometer.* Two horizontal bars are attached to a central vertical rod at right angles to each other, and a small propeller is attached to the end of each horizontal bar. Unlike the cup anemometer, the horizontal bars do not rotate. Instead, the rate of rotation of each propeller is measured and then combined by means of vector analysis to determine the precise direction and speed of the wind. Some propeller anemometers have a third propeller attached at the top of the central rod to measure vertical wind movements.

The above explanations describe only the most basic forms of anemometers. Today, scientists and engineers have developed a number of variations of these instruments, some that measure very low wind movements or very high wind movements, some that measure the volume of wind as well as its speed and direction, and some that are very sensitive and others that provide only rough approximations of speed and direction.

New Developments

Scientific developments have produced a number of new devices and methods for measuring wind speeds. For example, radar can be used for tracking air movements because particles carried along in the wind reflect radar waves. The movement of these particles can be measured in essentially the same way that a police radar is used to measure the speed of an automobile.

The most widely used type of radar today, especially for severe storms, is *Doppler radar.* Doppler radar operates on the basis of the *Doppler effect,* a phenomenon discovered in 1842 by the Austrian physicist Christian Johann Doppler (1803–1853). Doppler found that the frequency of waves emitted by a moving source changes if the source is moving toward or away from an observer. The amount by which the frequency changes indicates the speed with which the object is moving as well. A common example of the Doppler effect is the change in pitch of the sound one hears as a railroad locomotive approaches or recedes from an observer. Doppler radar is used to determine wind speed by directing a beam of microwaves into the air, measuring the change in frequency of reflected waves, and then comparing the reflected frequency with the original microwave frequency.

Rawinsondes are also used to measure wind velocity and direction at upper altitudes in the Earth's atmosphere. A rawinsonde is a type of radiosonde, a package of weather instruments carried aloft by balloons. These instruments measure such properties as the temperature, humidity, and pressure at various altitudes. The data they collect is transmitted back to Earth by radio signals. In a rawinsonde, the movement of the balloon and instrument package are tracked by radar or a radio detection finder, allowing a measurement of the wind's speed and direction at the balloon's altitude.

Satellite tracking can also be used to measure atmospheric wind speeds from above. In one approach, continuous photographs of clouds or other atmospheric phenomena make it possible to determine the direction and speed of air movements at various altitudes. Until recently, this technology was relatively ineffective in producing data on large-scale air movements at all altitudes in the Earth's atmosphere. This deficiency prevented meteorologists and climatologists from obtaining some of the most important data they need in analyzing and predicting weather patterns.

In 1997, NASA announced a new technology designed to overcome this handicap, the SPAce Readiness Coherent Lidar Experiment (SPARCLE). In this experiment, laser beams are fired from a space shuttle downward through the atmosphere. The time between initial pulse and echo, along with changes in the color of the laser beam line, can be analyzed to determine with a high degree of precision the direction and speed of air movements in every part of the atmosphere, from Earth's surface to a height of 20 km (12 miles). *See also* Beaufort, Sir Francis

Further Reading

U.S. Weather Bureau. *History of Weather Bureau Wind Measurements.* Washington, DC: U.S. Government Printing Office, 1963.

Windmill

A windmill is a machine that converts wind energy into some form of mechanical work, such as lifting water or grinding grain. When a windmill is used to generate electrical energy, it is more commonly called a *wind turbine generator.* Windmills have been, and sometimes still are, used to run a saw mill; process various types of foodstuffs, such as spices, cocoa, paints, dyes, and tobacco; make paper; and press oil from seeds.

Origins

Some authorities believe that the first windmills were constructed in China more than 2,000 years ago. Little concrete evidence exists for such machines, however. The first windmills of which we have certain evidence were built in ancient Persia (now Iran) about A.D. 500–900. Our knowledge of these devices is based on written descriptions since no drawings, paintings, or other visual depictions remain.

The first drawings of Persian windmills date to about 950, at which time the machines were apparently in wide use for pumping water and grinding grain. Persian windmills were of a *vertical design* in which rectangular sails hang from horizontal bars that are, in turn, attached to a central vertical pole. These machines had the general appearance of a modern circular clothesline in which clothes are left to dry hanging from four crossbars attached to the top of a vertical pole. The sails in Persian windmills were originally made of reeds or thin sheets of wood. Windmills of this design are also known as *panemone.*

When the wind blows on the sails of a panemone, it causes them to revolve around the central pole. The pole also rotates on its own axis A grinding stone attached to the base of the central pole also rotates, grinding grain placed beneath it. By devising a set of gears, the motion of the rotating central pole can also be used to lift water from a well, out of a swamp, or into an irrigation ditch.

Introduction to Europe

Windmills first appeared in Europe in the twelfth century. Some authorities believe that the concept behind their design was brought to Europe from the Mideast by returning Crusaders. Others think that windmills were invented independently in Europe. One piece of evidence in support of the latter view is that European windmills had a design that was fundamentally different from those used in Persia, China, and other parts of Asia. European windmills were (and are) almost exclusively of a *horizontal design.*

A windmill of this style consisted of a vertical pole at the center of the mill's main structure, or *body.* In its earliest form, the central pole was able to rotate on its own axis, carrying with it the rest of the body. That is, the windmill as a whole rotated on its own axis in the wind. Projecting from the central pole was a single horizontal arm, to which was attached a wheel containing a set of sails. The sails were made out of thin wood, cloth, or some other light material and oriented in the wheel to face the wind. The horizontal windmill design is the one with which most people in the Western world are probably still familiar.

In its earliest design, the European horizontal mill was built so that it could be rotated to face the wind. Operators pushed on a horizontal beam, known as the *tailpole* or *tiller beam,* to face the whole mill building in the correct direction. Then, when the wind blew, it caught the sails and caused them to revolve around the horizontal axis. The visual effect of the spinning windmill is similar to that of the revolving propellers on an airplane or air-conditioning fan. The spinning of the horizontal axis was transmitted to the central pole by a series of gears, and the pole rotated on its own axis. The energy of the rotating central pole could be used to grind grain, lift water, or perform other work.

Some historians believe that the distinctly European horizontal windmills were originally designed after waterwheels, which had long been used in many parts of Europe to grind grain and perform other functions.

The first type of windmill used in Europe, described above, is generally known as a *post mill,* after the central post around which it is built. Later modifications in this design resulted in the *smock mill,* so named because of its general appearance to protective garment worn by workers in many occupations at the time. In a smock mill, the body of the mill was fixed to the ground. Only a cap at the top of the body could be rotated so as to face the sails to the wind. The upper cap was rotated manually at first and, in later designs, by the addition of a *fantail,* extending from the back of the central pole. The fantail was designed to catch the wind and face the sails in the direction from which the wind was coming. Both post mills and stock mills were made of wood.

The greatest benefit of the smock mill over its predecessors was that the mill building itself could be made much larger (since the body of the mill no longer had to be rotated) and taller. The taller the body, in turn, the greater the amount of wind captured by the sails.

The next development in windmill design was an extension of the smock mill concept. Stone, brick, and other sturdy building materials, rather than wood, were used in the construction of the body, making it possible to construct the building to even larger dimensions. This design of windmill was called the *tower mill* because of its similarity to watchtowers and similar tall structures.

Improvements in Design

Over the years, inventors and engineers made a number of improvements in the design of windmills. These improvements took advantage of a growing understanding of aerodynamic principles to produce windmills that made more efficient use of the wind, operated at a relatively constant speed, and lasted longer. For example, a British blacksmith, Edmund Lee (dates unknown) received a patent in 1745 for a fantail. The fantail was and eight-bladed device that looked like a small windmill itself. It was attached to the back of the windmill cap at right angles to the main sails of the windmill. When the wind is blowing against the main sails, it blows at right angles to the fantail, having no effect on its blades. When the wind changes direction, it blows into the fantail blade, setting it to spinning and causing it, and the cap to which it is attached, to turn so as to orient the main sails once more into the wind.

A second improvement was the *spring sail,* invented by the Scottish engineer Andrew Meikle (1719–1811) in 1772. The spring sail consists of wooden shutters that can be opened and closed, either manually or automatically, to change the amount of wind they catch. Sails of this type adjust the rate at which the windmill wheel turns by changing shape as the wind's speed changes.

Use of Windmills

By the fifteenth century in Europe, windmills were rapidly becoming an attractive alternative to animal power and water power in areas where winds blew often enough and strongly enough. In some countries, they were major sources of power well into the nineteenth century, when steam power and electricity gradually made their use uneconomical. In the mid–eighteenth century, for example, there were said to be more than 10,000 windmills in the Netherlands, probably the greatest number of such devices per square mile anywhere in the world. About the same time, in 1846, the tiny island of Barbados listed 506 windmills, built to take advantage of regular sea breezes.

Windmills were also an important part of the American colonies founded in the seventeenth century. The first such devices were built in the Jamestown colony in the 1620s, where they were used for grinding flour. But the real breakthrough in American windmills did not come for more than two centuries. Then, in 1854, a machinist from Vermont, Daniel Halladay, invented a self-regulating windmill that quickly became widely popular. The wooden blades in Halladay's windmill changed their pitch automatically when the wind speed changed. This design prevented the wheel from tearing apart at high wind speeds. Within a year, the Illinois Railroad Company had placed orders for dozens of the Halladay windmill, to be used to pump water into the company's steam engines at regular locations along its lines.

Windmills also became popular on farms and in small communities in the growing western states, where wind was abundant but water and other sources of power were in short supply. In the century after Halladay's invention, more than six million small windmills with energy output of less than one horsepower were installed in the country. Many of these windmills remained in service well into the twentieth century because they were often the only inexpensive source of energy available in less populated areas.

By the beginning of the twentieth century, however, the limitations of wind power in its battle against steam power and electricity was already becoming evident. In fact, the first windmill designed to generate electricity—rather than lift water, grind flour, or perform some other traditional function—was built by a mechanical engineer, Charles F. Brush (1849–1929), in Cleveland in 1888. Brush's machine consisted of a multiblade, "picket-fence" rotor 17 meters (56 feet) in diameter. It developed

12 kilowatts of power at an operating speed of 500 rpm (revolutions per minute).

Only three years later, a Danish meteorologist, Poul la Cour (1846–1908), designed and built the prototype of what was to become the world's first wind turbine generators. La Cour incorporated the most up-to-date aerodynamic principles in his sleek four-bladed, airfoil-shaped machines. He also constructed a wind tunnel to test his machines, gave a number of courses on the new wind technology, and established the world's first journal on wind power, *Journal of Wind Electricity. See also* Brush, Charles F.; Wind Power

Further Reading

Baker, T. Lindsay. *A Field Guide to American Windmills.* Norman: University of Oklahoma Press, 1985.

Brooks, Laura. *Windmills.* New York: Metro Books, 1999.

Larkin, David. *Mill: The History and Future of Naturally Powered Buildings.* New York: Universe Publishing, 2000.

Wind Power

See Wind Energy Systems

Wind Power Monthly

Wind Power Monthly states that its aim is to "extend your knowledge of the role being played by an important new technology in the battle for today's electricity markets—and of its potential to play a far greater role in meeting a fundamental need of tomorrow: the clean and safe supply of electricity to all."

Each issue of the magazine includes breaking news about wind power from all parts of the world; special-focus articles on topics such as research and development, market economics, and social and political issues; editorial comments; and The Windicator, a section on market indicators in the field of wind energy.

Further Information

(U.S. address)
Windpower Monthly
PMB #217
P.O. Box 496007
Redding, CA 96049-6007
URL: http://www.wpm.co.nz
E-mail: mail@windpower-monthly.com

Wind Propagation (of Seeds)

See Seed Propagation (by Wind)

Wind Rose

The term *wind rose* has two meanings. It refers, first of all, to a type of icon used on very old maps to show the directions from which winds blow. This icon had the general appearance of a *compass rose* now used on maps to show the directions north, south, east, and west. The wind rose was used before the compass was invented in the thirteenth century. It served the same purpose as a compass rose since the geographical designations north, south, east, and west were taken to be identical with the directions from which winds blow.

Today, a wind rose is a diagram that shows the relative frequency with which winds blow from each direction. It may also show the speed and power with which those winds occur. A wind rose has the everyday function of telling mariners the direction from which winds are likely to appear and how strongly they are likely to blow. It can also be used in the siting and design of windmills and wind power systems.

The wind rose shown here is constructed from data collected by the state of Oklahoma's Mesonet project. That project maintains a network of monitoring stations that collects a variety of environmental data,

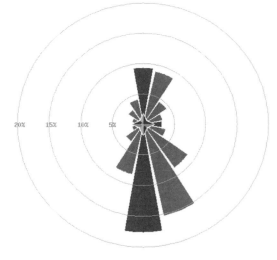

Sample Mesonet wind rose product. Oklahoma Department of Commerce.

including wind speeds. This wind rose summarizes wind patterns at one of those monitoring stations for one season of the year.

In this diagram, the outlying circle represents the horizon surrounding the monitoring station. It has been divided into 16 segments, representing the cardinal directions north, south, east, and west, and certain intercardinal points between them. Wind roses may consist of any number of segments, although eight and 12 are also common choices.

The radius of each of the 12 sectors of the circle represents the relative frequency with which winds blow from the direction represented by the sector. For example, the sectors facing toward the south and southeast have the longest lengths, indicating that the wind blows from these directions most often at this time of the year. The dotted circles inside the horizon circle show the relative frequency of the winds from each direction. In this example, about 17 percent of the time the wind was blowing from the

south and about 15 percent of the time, from the southeast. You can tell that the wind never blew from the west and only rarely from the east.

Some wind roses are more complex than the one shown here. They may also contain segments within the segments shown here. The internal segments may show the average wind speed from each direction. The average wind speed is calculated by the total average wind speed overall at the station multiplied by the fraction blowing from each direction. Also, the wind power available from each direction can be found by a similar procedure. The total wind power (the cube of the wind speed) is multiplied by the fraction blowing from each direction to find the wind power available from each direction. This piece of information is especially useful in determining locations at which wind power generators can profitably be situated.

Finally, another more complex type of wind rose may contain a set of radial arrows drawn in each direction from the center of the circle. Each set of arrows in any one particular direction show the frequencies of winds with different speeds. If the south winds shown in the wind rose reproduced here were divided by wind speed, for example, the single wedge shown in the drawing might consist of five or more separate wedges. The thickness of each wedge would indicate how frequently winds of each class of speed were observed from the southerly direction.

Wind roses vary in appearance for different locations on the Earth's surface and for different months of the year. They tend to remain relatively constant, however, from year to year.

Wind Shear

Wind shear is a variation in the speed or direction of wind speed over a short distance. Wind shear can occur in either a horizontal or vertical direction and at virtually any altitude from sea level to the upper stratosphere. An example of wind shear in the upper atmosphere is the region surrounding the jet stream, where one parcel of air (the jet stream) is moving very rapidly in comparison to the air surrounding it.

A number of factors may be responsible for the development of wind shear, most commonly conditions involving significant temperature differences in two adjacent parcels of air. For example, imagine that a cold air mass is traveling across the land, coming into contact with a warm air mass. At the boundary between these two air masses (the *cold front,* in this case), temperatures may differ quite dramatically over only a few tens of kilometers. These temperature at the base of the cold front, where the cold air mass has worked its way under the warm air mass, is likely to be cool. Warm air remains in place several hundreds or thousands of meters above the ground at the same spot, however, resulting in a significant temperature differential and, hence, different wind patterns at various altitudes.

The approach of a thunderstorm may also produce wind shear. Cold air flowing downward in the center of the storm may strike the ground and spread out laterally in all directions. Winds at or just about ground level may become much stronger and oriented differently from those only a few hundred meters above the ground. The most dramatic example of such wind shears is a *downburst,* a short-lived, very severe wind pattern that may involve winds of very high speed.

Topographical features may also cause the formation of wind shear. A dominant wind pattern may come into contact with a mountain range, a group of trees, buildings, or some other obstruction, changing its speed and direction from adjacent, unobstructed winds.

Wind shear may be responsible for or related to other atmospheric phenomena. For example, friction may develop between a stream of rapidly moving air and a relatively quiet part of the atmosphere surrounding it. As friction develops, the flow of air may begin to break apart along its boundaries, resulting in the formation of turbulent eddies. Wind shear can be involved in both the formation and dissipation of large storms, such as thunderstorms, tornadoes, and hurricanes. A rapid flow of air in the upper part of the troposphere, for example, may actually shear off the top of a hurricane cloud, interrupt the process of hurricane formation, and convert the system into a harmless group of rain clouds.

Wind shear is a problem of special concern to pilots. Once in flight, a pilot can usually adjust to the bumpiness that may be encountered when an airplane flies through a wind shear. But the problem may be more serious during takeoff and landing. During these periods, a pilot adjusts the aircraft's controls to lift off and land against certain known (or expected) wind patterns. If those wind patterns change suddenly, either in direction or speed, as happens in a wind shear, the pilot must make very rapid adjustments to prevent the aircraft from crashing into the ground. *See also* Downburst Phenomena

Further Reading

Aeronautical Information Manual. Washington, DC: Federal Aviation Administration, 2002, Chapter 7. Also online at http://www.faa.gov/ATPubs/AIM/Chap7.

Committee on Low-Altitude Wind Shear and Its Hazard to Aviation. *Low-Altitude Wind Shear and Its Hazard to Aviation.* Washington, DC: National Academy Press, 1983.

Wind Storm

See Cyclones and Anticyclones; Hurricane

For the first time ever, engineers at NASA Ames Research Center, located in California's Silicon Valley, will test a research wind turbine in the world's largest wind tunnel, to learn how to design and operate wind turbines more efficiently, April 17, 2000. © NASA.

Wind Tunnel

A wind tunnel is a chamber in which the effects of air flowing past objects can be observed, measured, and studied. The basic principle behind wind tunnel design is that the characteristics of an object at rest in a steady flow of air are the same as the characteristics of that body moving through air at rest.

Wind tunnels are used to study many phenomena, the largest majority of which relate to the design of aircraft and spaceships. Such tunnels are used to study properties such as the way and rate at which heat is transferred in various parts of an airframe, the drag experienced by an aircraft, the lifting capacity for wings of various design, and the performance of propellers. Some other applications of wind tunnel research involve research on the flow of air through and around buildings, bridges, and other large structures; the behavior of bombs and missiles dropped from aircraft; flow patterns of smoke released by factories; and wind patterns in and around natural structures, such as hills and valleys.

History

Researchers as far back as the eighteenth century were interested in the effects of moving air on the characteristics of a body. The English engineer Benjamin Robins (1707–1751) is thought to have built the first device for measuring drag on an object. The device consisted of a central spool to which was attached a horizontal arm that could spin around the spool. Robins hung various objects from the end of the arm and then made it spin at various rates. Robins's "whirling arm" device was later used by a number of other scientists to study drag.

The first wind tunnel was built in 1871 by the English inventor Francis Herbert Wenham (1824–1908). Wenham's tunnel consisted of a horizontal box about 12 feet long and 18 inches square. He blew air through the tunnel with a fan and suspended objects from the ceiling of the tunnel to observe the way they were affected by the flow of air.

Perhaps the most famous early wind tunnel was built by Wilbur and Orville Wright in 1902. The Wrights had spent much of the 1901 flying season testing various glider designs based on mathematical formulas developed by the German aeronautical pioneer Otto Lilienthal (1848–1896). When none of these designs worked as well as the Wrights had expected, they decided to build a wind tunnel in which to test new wind designs. Their wind tunnel was simple, consisting of a wooden box with a glass window in one side. They blew air through the box with a fan powered by a one-horsepower gasoline engine. Their newly found ability to observe the effects of airflow over various types of wings was an important factor in the successful first heavier-than-air flight they made on 17 December 1903 at Kitty Hawk, North Carolina.

Types

Wind tunnels today are usually classified into one of four major groups: subsonic (or low-speed), transonic, supersonic, and hypersonic. These four terms are derived from the speed of the airflow that passes through each type of tunnel. In subsonic wind tunnels, the highest speed of airflow is about 500 kilometers per hour (300 miles per hour). Transonic wind tunnels operate with air speeds on either side of the speed of sound, anywhere from Mach 0.7 to Mach 1.4. A Mach number of 1.0 is equal to the speed of sound, 331 meters per second or 1,220 kilometers per hour (758 miles per hour). Supersonic wind tunnels operate at Mach numbers of about 1.0 to 5, while hypersonic wind tunnels use airflow rates with Mach numbers of greater than 5.

Wind tunnels of all four types may be configured as either *open-* or *closed-circuit* systems. Both open- and closed-circuit systems have a roughly similar design. Air is blown into one end of the tunnel by a powerful fan. The air moves through a long tunnel with a gradually increasing diameter. At a certain distance within the tunnel, there is a contraction area, a point at which the diameter of the tunnel decreases significantly. As air reaches the contraction area, it is compressed and then forced through the narrow end of the area into a test area. It leaves the contraction area with greatly increased speed and flows over the object being tested in the test area. After the air passes through the test area, it enters a diffuser, a region where the tunnel diameter again increases. The purpose of the diffuser is to reduce the speed and turbulence of air leaving the test section.

The primary difference between open- and closed-circuit systems is that, in the former, air that leaves the diffuser is allowed to escape into the atmosphere or the laboratory. It is "lost" to the experiment and is not re-used. In a closed-circuit system, air that leaves the diffuser is then returned to the front of the tunnel, where it is forced once more by fans back into the tunnel. That is, air used in the experiment does not escape from the system (except through leaks), but is used over and over again in the tunnel.

Special adaptations are necessary in wind tunnels to be used at transonic speeds. The reason for these adaptations is that air traveling near the speed of sound generates shock waves which interfere with measurements of the object being studied. To deal with these shock waves, the test section is surrounded by a larger chamber and vents are cut into the walls between the test section and outer chamber. These vents absorb most of the shock waves produced at transonic speeds and permit accurate measurements of an object in the test section.

The primary problem in constructing supersonic and hypersonic wind tunnels is generating a flow of air fast enough to meet the needs of such devices. A common approach is to install a high-pressure air tank at the entrance to the wind tunnel. Air released from such tanks then moves through the tunnel with speeds great enough to meet Mach 1.0 and greater requirements.

Another way of producing high speed air flows is to install a vacuum chamber at the end of the wind tunnel. In this case, the pressure difference between air entering the tunnel and exiting it into the vacuum chamber can be sufficient to produce the wind speeds necessary in the test section. In the case of a hypersonic wind tunnel, some combination of high-pressure air tank at the entrance and vacuum chamber at the exit can be used to produce Mach 5 air flow speeds.

Vertical Wind Tunnels

Vertical wind tunnels have also been built for both research and recreational purposes. Fire researchers, for example, are interested in studying the way smoke, toxic particles, biological materials, and other substances travel upward through large buildings and "urban canyons" between buildings. The movement of such particles can be studied in wind tunnels oriented in a vertical rather than horizontal direction. Vertical wind tunnels are also used for recreational purposes. Fast streams of air are forced upward from the base of the tunnel and out of the open upper end. Individuals jump downward into the upflowing air, producing a force similar to that of skydiving. About two dozen such vertical wind tunnels have been built in the United States, Great Britain, Israel, France, Germany, and other countries around the world.

Further Reading

Barlow, Jewel B. *Low-Speed Wind Tunnel Testing.* New York: Wiley, 1999.

Pope, Alan. *Wind-Tunnel Testing,* 2nd edition. New York: Wiley, 1954.

Wright, Orville (1871–1948), and Wilbur (1867–1912)

The Wright brothers are generally given credit for inventing the successful heavier-than-air, engine-powered aircraft. They completed their first flight in this machine, later known as Flyer I, on 17 December 1903 at Kitty Hawk, North Carolina.

Wilbur Wright was born near Millvale, Indiana, on 16 April 1867, the third of five children to Bishop Milton Wright and his wife, Susan Catharine Koerner Wright. His brother Orville was born 19 August 1871 in Dayton, Ohio, where his family had since moved. The Wright children grew up in a conservative family, reflected in the fact that Wilbur and Orville never smoked, drank, or danced. Nor did either marry. Wilbur was seriously injured at the age of 18 when a hockey puck smashed into his face. As a result of the injury, he suffered from poor health for much of his life and barely finished high school. Orville lost interest in school during his senior year of high school, stopped attending school, and never graduated.

Aviation Experiments

From an early age, the Wright brothers were fascinated with mechanical devices. After leaving school, they decided to open a bicycle shop, which was very successful. The designed, built, and sold their own bicycle, the Wright Special, which they sold for $18. Before long, they had also become interested in aviation, and they began to attack the theoretical and practical problems of flying with enthusiasm. They read everything they could find on the subject and wrote to a number of pioneers, including the French-American inventor Octave Chanute.

The Wright brothers understood that before they could attempt to develop a heavier-than-air powered aircraft, they would have to learn everything they could about nonpowered flying. They traveled to Kitty Hawk, recommended by the U.S. Weather Bureau as having nearly ideal weather for aviation experiments, where they built and tested their first gliders. They carried out their first tests in October 1900, but failed to achieve much success. The returned to Dayton, where they designed and tested a variety of new glider designs. A year later, they returned to Kitty Hawk to test their new designs, with somewhat more success than they had experienced the prior year.

Wilbur and Orville Wright with their second powered machine, Dayton, Ohio, 1904.
© **Library of Congress.**

The Wright brothers realized, however, that there were still some fundamental flaws in their designs. Eventually they came to the conclusion that some of the data they had been using from the experiments of the German glider pilot Otto Lilienthal were faulty, and they decided to create new and more accurate tables on lift, wing size and shape, and other variables. To accomplish this task, they not only designed dozens of wing shapes, but they also built one of the first wind tunnels to test each new shape.

In August 1902, the Wright brothers returned to Kitty Hawk a third time. They were finally able to fly their glider successfully, keeping it aloft, directing its path, and making controlled turns. They decided that they were ready to attempt the design and test of a heavier-than-air craft.

Over the next year, they continued their research in Dayton on the construction of a gasoline engine for the aircraft, on a variety of wing shapes and sizes, and on control systems for the operation of the wing and tail flaps. By November 1903, they were ready to begin tests on their first heavier-than-air machine. For over two months after they arrived in Kitty Hawk in September, they had experienced failure after failure. Finally, however, on 17 December, Orville launched the machine from a low hill and flew it for a distance of about 30 meters (100 feet) at a height of about 3 meters (10 feet) above the ground. Buoyed by their success, they flew again a number of times before the end of the day. Unfortunately, the day came to a disappointing conclusion when the wind caught the aircraft as it was being towed to the hangar, tossed it upside down, and destroyed it beyond repair. When the Wright brothers published a brief statement of their success on 5 January 1904, the world took almost no notice of their success.

Later Lives

The Wright brothers were not discouraged by the loss of Flyer I nor the lack of interest in their announcement of their accomplishments. They continued to work on aircraft design, producing Flyer II in mid-1904. Wilbur flew the new airplane on a circular course at Huffman Farm, near Dayton, on 20 September 1904. The flight lasted for 1 minute 35 seconds and covered a distance of just over a kilometer (three-quarters of a mile). A year later, Flyer III was ready for testing, and it was flown for a distance of 38.9 kilometers (24.2 miles) in a time of about 38 minutes.

After their successes with Flyer II and Flyer III, the Wright brothers attempted to interest the War Department in signing a contract for the construction of airplanes for the army. Although the War Department showed virtually no interest in the project, private inventors and businessmen in the United States and Europe quickly began to "borrow" and adapt many of the Wrights' ideas for airplane design. Over the next decade, the Wright brothers fought a number of lawsuits with individuals and companies who attempted to use their ideas. The Wrights won most of those battles, but so much energy was expended in lawsuits that progress in design and construction lagged. In Europe, by contrast, aircraft development went forward rapidly and for many years vastly outstripped work in the United States.

Wilbur did not live to see the conclusion of the many legal battles surrounding his inventions or the ultimate triumph of modern aviation. He developed typhoid fever and died in Dayton on 30 May 1912. His brother outlived him by 36 years, dying of a heart attack in Dayton on 30 January 1948. *See also* Airplane

Further Reading

Brouch, Tom D. *The Bishop's Boys: A Life of Wilbur and Orville Wright.* New York: W. W. Norton, 1989.

Culick, F. *On Great White Wings: The Wright Brothers and the Race for Flight.* New York: Hyperion, 2001.

Freedman, Russell. *The Wright Brothers: How They Invented the Airplane.* Orlando, FL: Harcourt Brace & Company, 1994.

Howard, Fred. *Wilbur and Orville: A Biography of the Wright Brothers.* New York: Knopf, 1987.

Z

Zeppelin

See Dirigible

Zeppelin, Ferdinand Adolf August Heinrich, Count von (1838–1917)

Ferdinand Adolf August Heinrich von Zeppelin was a member of the aristocracy of the state of Würtemburg and the inventor of the first rigid airship, the dirigible. He was born in Konstanz, Baden, on 8 July 1838. His father was a member of the court of the Duke of Hohenzollern-Sigmaringen and his mother was the daughter of a successful textile manufacturer. As with most upper-class children, he received his early education at the hand of personal tutors before entering the Ludwigsburg Military Academy and the University of Tübingen.

In 1858, Zeppelin became a cavalry officer in the army of Würtemburg and, five years later, took a leave of absence to visit the United States and observe the military strategy of the Union Army during the Civil War. During his visit to the United States, he took his first balloon ride in St. Paul, Minnesota, inspiring a passion for air travel that was to remain with him the rest of his life.

Upon his return to Europe, Zeppelin took part in the Seven Weeks' War between Austria and Prussia and, in 1870, fought with the Prussian Army against France in the War of 1870. In 1891, he retired from active service with the rank of lieutenant general. He then turned his attention and his considerable financial resources to the study of lighter-than-air machines. At the time, the only devices available for travel through the air were kites and balloons. Both worked well enough for some purposes, but they suffered from one serious handicap: Their movement was completely subject to air currents and wind patterns. Air travel visionaries knew that they had to develop a machine that could travel under its own power and that could be steered.

In 1900, Zeppelin completed the design of an airship that would meet these criteria. On 2 July of that year, he launched the first lighter-than-air machine, which was given the name *dirigible,* from the French expression *ballon dirigeable,* or "steerable balloon." That machine was 126 meters (419 feet) long and 11.4 meters (38 feet) in diameter, with a volume of 9,580 cubic meters (338,400 cubic feet) of hydrogen. The dirigible envelope was divided into 16 cells and was powered by two 16-horsepower engines.

Zeppelin's design had been made possible by the discovery in the 1880s of an inexpensive method for producing aluminum metal by the American chemist Charles Martin Hall (1863–1914). The ready availability of inexpensive aluminum made it possible for Zeppelin to use strips of the light but strong metal for the internal framework around which the airship was constructed. In its first trip, Zeppelin's dirigible, the LZ-1, carried five passengers to an altitude of 400 meters (1,300 feet) on a ride that covered about 6 kilometers (4 miles) and lasted just over 15 minutes. The dirigible was slightly damaged upon landing, but Zeppelin had proved that his ideas were sound.

In 1898, Zeppelin had founded a new company, the Aktiengesselschaft zur Förderung der Luftschiffahrt, to build and market his dirigibles. Within five years, however, Zeppelin had exhausted his own financial resources, liquidated the company, and appealed to Kaiser Wilhelm II and the general public for support to continue his research. He received enough response to continue with his work and to continually improve his airship.

Unfortunately for Zeppelin, dirigibles never attained the popularity he had hoped for them. More than a hundred were built between 1900 and 1938, all but 16 of which were purchased by the German army for use during World War I. They were used to a limited extent for bombing raids over London, but those raids demonstrated their serious weaknesses. They were so large and so dangerous (since they were filled with combustible hydrogen) that they were easy targets for enemy guns. At least 40 were shot down and destroyed in the London raids.

The use of dirigibles for luxury transportation was somewhat more successful and longer lived. German dirigibles were used for flights throughout Europe, to North and South America and the Mideast, and, in one case, for a grand around-the-world tour. The explosion and destruction of one of the grandest dirigibles, the *Hindenburg,* on 6 May 1937 destroyed all hopes for the use of these airships for commercial travel. By the time of that disaster, Zeppelin himself had long been dead. He died on 8 March 1917 after surgery in Berlin. *See also* Dirigible

Further Reading

De Syon, Guillaume. *Zeppelin! Germany and the Airship, 1900–1939.* Baltimore: Johns Hopkins University Press, 2002.

Eckener, Hugo. *Count Zeppelin, the Man and His Work.* Translated by Leigh Farnell. London: Massie Publishing Company, 1938.

Goldsmith, Margaret L. *Zeppelin: A Biography.* New York: W. Morrow, 1931.

Zone of Aeration

The zone of aeration is the layer of Earth's surface just beneath the ground. It is given that name because it is not completely saturated with water and, therefore, is sometimes called the *unsaturated zone.* During most parts of the year, the spaces between solid particles that make up the zone of aeration are filled with air, although in times of heavy rain, even these spaces may also fill with water.

The zone of aeration may be divided into three parts. The upper layer is called the *soil water zone* or *soil moisture zone.* Water in this region is readily available to plant roots. The next deeper layer, the *intermediate zone,* separates the soil water zone from the deepest layer of the zone of aeration, the *capillary fringe.* The capillary fringe contains many fine cracks through which water can seep upward, against the force of gravity, into the upper zone of aeration. The upward seeping water derives from the lowest layer of soil, the region of *groundwater,* where air is entirely absent and the pores between solid particles is always filled with water.

Further Reading

Tindall, James A. *Unsaturated Zone Hydrology for Scientists and Engineers.* Upper Saddle River, NJ: Prentice-Hall, 1999.

Bibliography

Ackerman, Marsha E. *Cool Comfort: America's Romance with Air-Conditioning.* Washington, DC: Smithsonian Institution Press, 2002.

Ackroyd, J. A. D., B. P. Axcell, and A. I. Ruban. *Early Development of Modern Aerodynamics.* Washington, DC: American Institute of Aeronautics and Astronautics, 2001.

Ahrens, C. Donald. *Meteorology Today: An Introduction to Weather, Climate, and the Environment,* 6th edition. Pacific Grove, CA: Brooks/Cole Publishing, 1999.

Allaby, Michael. *Facts on File Weather and Climate Handbook.* New York: Facts on File, 2002.

Allen, Oliver E., and the editors of Time-Life Books. *Atmosphere.* Alexandria, VA: Time-Life Books, 1983.

Amyot, Joseph R. *Hovercraft Technology, Economics, and Applications.* Amsterdam: Elsevier, 1989.

Anderson, John D., Jr. *Fundamentals of Aerodynamics,* 3rd edition. New York: McGraw-Hill, 2001.

Baker, T. Lindsay. *A Field Guide to American Windmills.* Norman: University of Oklahoma Press, 1985.

Banfield, Edwin. *Barometers: Aneroid and Barographs.* Trowbridge, Wiltshire, UK: Baros Books, 1985.

Barlow, Jewel B. *Low-Speed Wind Tunnel Testing.* New York: Wiley, 1999.

Berg, A. Scott. *Lindbergh.* New York: G. P. Putnam's, 1998.

Bluestein, Howard B. *Tornado Alley: Monster Storms of the Great Plains.* New York: Oxford University Press, 1999.

Bristow, Gary V. *Encyclopedia of Technical Aviation.* New York: McGraw-Hill Professional, 2002.

Brombacher, W. G. *Mercury Barometers and Manometers.* Washington, DC: U.S. Department of Commerce, National Bureau of Standards, 1960.

Brooks, Laura. *Windmills.* New York: Metro Books, 1999.

Brouch, Tom D. *The Bishop's Boys: A Life of Wilbur and Orville Wright.* New York: W. W. Norton, 1989.

Bryner, Gary C. *Blue Skies, Green Politics: The Clean Air Act of 1990 and Its Implementation.* Washington, DC: Congressional Quarterly Books, 1995.

Buckman, Leonard C. *Commercial Vehicle Braking Systems: Air Brakes, ABS and Beyond.* Warrendale, PA: Society of Automotive Engineers, 1998.

Burnet, John. *Early Greek Philosophy.* New York: Meridian Books, 1957.

Burton, Robert. *Birdflight: An Illustrated Study of Birds' Aerial Mastery.* New York: Facts on File, 1990.

Burton, Tony, et al., eds. *Wind Energy Handbook.* New York: John Wiley, 2001.

Butler, Susan. *East to the Dawn: The Life of Amelia Earhart.* Reading, MA: Addison-Wesley, 1997.

Cantor, G., and J. Hodge, eds. *Conceptions of Ether: Studies in the History of Ether Theories: 1740–1900.* Cambridge, U. K.: Cambridge University Press, 1981.

Caracena, Fernando, Ronald L. Holle, and Charles A. Doswell III, *Microbursts: A Handbook for Visual Identification,* 2nd edition. Boulder, CO: U.S. Department of Commerce, National Oceanic and Atmospheric Administration, Environmental Research Laboratories, National Severe Storms Laboratory, 1990.

Carse, Adam. *Musical Wind Instruments.* New York: Dover, 2002. Reprint of 1965 edition.

Casten, Carole M., and Peg Jordan. *Aerobics Today,* 2nd edition. Belmont, CA: Wadsworth Publishing, 2001.

Chambers, A., R. K. Fitch, and B. S. Halliday. *Basic Vacuum Technology.* Bristol, UK: A. Hilger, 1989.

Cheresmisinoff, Nicholas P. Handbook of Air Pollution Prevention and Control. Boston: Butterworth-Heinemann, 2002.

Clarke, Basil. *The History of Airships.* London: H. Jenkins, 1961.

Committee on Low-Altitude Wind Shear and Its Hazard to Aviation. *Low-Altitude Wind Shear and Its Hazard to Aviation.* Washington, DC: National Academy Press, 1983.

Conant, James Bryant. *Robert Boyle's Experiments in Pneumatics.* Cambridge, MA: Harvard University Press, 1965.

Cook, Gerhard A. *Argon, Helium, and the Rare Gases: The Elements of the Helium Group.* New York: Interscience Publishers, 1961.

Cooke, Ronald, Andrew Warren, and Andrew Goudie. *Desert Geomorphology.* New York: Chapman & Hall, 1993.

Cooper, Gail. *Air-Conditioning America: Engineers and the Controlled Environment, 1900–1960.* Baltimore: Johns Hopkins University Press, 1998.

Culick, F. *On Great White Wings: The Wright Brothers and the Race for Flight.* New York: Hyperion, 2001.

Daubney, Ulric. *Orchestral Wind Instruments, Ancient and Modern.* Freeport, NY: Books for Libraries Press, 1970. Reprint of 1920 edition.

Davis, Pete. *Inside the Hurricane: Face to Face with Nature's Deadliest Storms.* New York: Henry Holt, 2000.

DeBlieu, Jan. *Wind: How the Flow of Air Has Shaped Life, Myth, and Land.* Boston: Houghton Mifflin, 1998.

Denniston, George C. *The Joy of Ballooning.* Philadelphia: Courage Books, 1999.

De Syon, Guillaume. *Zeppelin! Germany and the Airship, 1900–1939.* Baltimore: Johns Hopkins University Press, 2002.

Dick, Harold G., and Douglas Hill Robinson. *The Golden Age of the Great Passenger Airships, Graf Zeppelin & Hindenburg.* Washington, DC: Smithsonian Institution Press, 1985.

Dijksterhuis, E. J. *Archimedes.* Translated by C. Dikshoorn. Princeton, NJ: Princeton University Press, 1987.

Drachmann, A. G. *Ktesibios, Philon and Heron: A Study in Ancient Pneumatics.* Copenhagen: E. Munksgaard, 1948.

Eckener, Hugo. *Count Zeppelin, the Man and His Work.* Translated by Leigh Farnell. London: Massie Publishing Company, 1938.

Edita, S. A. [firm]. *The Romance of Ballooning: The Story of the Early Aeronauts.* New York: Viking Press, 1971.

Evans, Charles M. *War of the Aeronauts: The History of Ballooning in the Civil War.* Mechanicsburg, PA: Stackpole Books, 2002.

Farr, Finis. *Rickenbacker's Luck: An American Life.* Boston: Houghton Mifflin, 1979.

Federal Meteorological Handbook No. 1: Surface Weather Observations and Reports. Washington, DC: National Weather Service, Department of Commerce, 1996.

Field, Arnold. *International Air Traffic Control: Management of the World's Airspace.* Oxford: Pergamon Press, 1985.

Fitzpatrick, Patrick J. *Natural Disasters: Hurricanes: A Reference Handbook.* Santa Barbara, CA: ABC-CLIO, 1999.

Francillon, Rene J. *The United States Air National Guard.* London: Aerospace Publishing, 1993.

Freedman, Russell. *The Wright Brothers: How They Invented the Airplane.* Orlando, FL: Harcourt Brace & Company, 1994.

Freiman, L. S. *Gaspard Gustave Coriolis.* Moscow: Nauka, 1961.

Gibbons, Floyd Phillips. *The Red Knight of Germany.* New York: Arno Press, 1980.

Glaser, Jason. *Bungee Jumping.* Mankato, MN: Capstone Press, 2000.

Glinski, Jan, and Witold Stepniewski. *Soil Aeration and Its Role for Plants.* Boca Raton, FL: CRC Press, 1985.

Goldsmith, Margaret L. *Zeppelin: A Biography.* New York: W. Morrow, 1931.

Goldstein, Donald M. *Amelia: The Centennial Biography of an Aviation Pioneer.* Washington, DC: Brassey's, 1997.

Gray, Susan M. *Charles A. Lindbergh and the American Dilemma: The Conflict of Technology and Human Values.* Bowling Green, OH: Bowling Green State University Popular Press, 1988.

Grazulis, T. P. *The Tornado: Nature's Ultimate Windstorm.* Norman: University of Oklahoma Press, 2001.

Greenler, Robert. *Rainbows, Halos, and Glories.* New York: Cambridge University Press, 1980.

Guericke, Otto von. *The New (So-Called) Magdeburg Experiments of Otto von Guericke,* translated by Margaret Glover Foley Ames. Dordrecht and Boston: Kluwer Academic, 1994.

Hall, George, and Baron Wolman. *Blimp!* New York: Van Nostrand Reinhold, 1981.

Handleman, Philip. *Air Racing Today: The Heavy Iron at Reno.* Osceola, WI: Motorbooks International, 2001.

Hannah, Barry. *Airships.* New York: Vintage Books, 1985.

Hart, Clive. *Kites: An Historical Survey.* New York: Praeger, 1967.

Hau, Erich. *Windturbines: Fundamentals, Technologies, Application and Economics.* New York: Springer Verlag, 2000.

Heath, Sir Thomas Little. *Archimedes.* New York: Macmillan, 1920.

Hlastala, Michael P., and Albert J. Berger. *Physiology of Respiration,* 2nd edition. Oxford, UK: Oxford University Press, 2001.

Hoffman, Dorothy M., Bawa Singh, and John H. Thomas, eds. *Handbook of Vacuum Science and Technology.* San Diego: Academic Press, 1997.

Holmes, Len. *Fly the Wing: Hooking into Hang Gliding.* Southern Pines, NC: Karo Hollow Press, 2002.

Horváth, M., L. Bilitzky, and J. Huttner. *Ozone.* Amsterdam: Elsevier, 1985.

Houghton, E. L., and P. W. Carpenter. *Aerodynamics for Engineering,* 5th edition. London: Butterworth-Heinemann, 2001.

Howard, Fred. *Wilbur and Orville: A Biography of the Wright Brothers.* New York: Knopf, 1987.

Hunter, Michael. *Robert Boyle: 1627–91.* Woodbridge, UK: Boydell Press, 2000.

Hurt, R. Douglas. *The Dust Bowl: An Agricultural and Social History.* Chicago: Nelson-Hall, 1981.

Huschke, Ralph E., ed. *Glossary of Meteorology.* Boston: American Meteorological Society, 1989.

Illman, Paul E. *The Pilot's Air Traffic Control Handbook.* New York: McGraw-Hill Professional Publishing, 1999.

Johnson, Erik. *Understanding the Skydive.* Rome, GA: Lamplighter Press, 2002.

Kermode, Alfred Cotterill. *Mechanics of Flight,* 10th edition. Englewood Cliffs, NJ: Prentice-Hall, 1995.

Kilduff, Peter. *Richthofen: Beyond the Legend of the Red Baron.* New York: John Wiley & Sons, 1993.

Killinger, Jerry. *Heating and Cooling Essentials.* Tinley Park, IL: Goodheart-Wilcox, 2002.

Larkin, David. *Mill: The History and Future of Naturally Powered Buildings.* New York: Universe Publishing, 2000.

Leister, Mary. *Flying Fur, Fin and Scale: Strange Animals That Swoop and Soar.* Owings Mills, MD: Stemmer House, 1977.

Lindbergh, Charles A. *Autobiography of Values.* New York: Harcourt Brace Jovanovich, 1978.

———. *"We."* New York: G. P. Putnam's Sons, 1927.

Liquid Air, IGC Document 82/01/E. Brussels: European Industrial Gases Association, 2001. Available online at the association's Web site, http://www.eiga.org.

Lutgens, Frederick K., and Edward J. Tarbuck. *Atmosphere,* 8th edition. Upper Saddle River, NJ: Prentice-Hall, 2000.

Lynch, David K., and William Livingston. *Color and Light in Nature.*
Cambridge, U.K.: Cambridge University Press, 2001.

Mandrioli, Paolo, Paul Comtois, and Vincenzo Levizzani, eds. *Methods in Aerobiology.* Bologna, Italy: Pitagora Editrice, 1998.

Mani, M. S. *Ecology and Biogeography of High Altitude Insects.* The Hague: Dr. W. Junk N.V. Publishers, 1968.

Mason, Francis K. *Aces of the Air.* New York: Mayflower Books, 1981.

Matthews, Birch. *Race with the Wind: How Air Racing Advanced Aviation.* Osceola, WI: Motorbooks International, 2001.

Mazzeo, Karen S. *Fitness through Aerobics and Step Training,* 3rd edition. Belmont, CA: Wadsworth Publishing, 2001.

Middleton, W. E. Knowles. *The History of the Barometer.* Baltimore: Johns Hopkins Press, 1964.

———. *A History of the Thermometer and Its Use in Meteorology.* Baltimore: Johns Hopkins Press, 1966.

Minnaert, Marcel G. J. *The Nature of Light and Colour in the Open Air.* New York: Dover Publications, 1954.

Misstear, Cecil. *The Advanced Airbrush Book.* New York: Van Nostrand Reinhold, 1984.

Moulton, Ron, and Pat Lloyd. *Kites: A Practical Handbook.* Los Angeles: Transatlantic Publishers, 1997.

Mueller, James A. *Aeration: Principles and Practice.* Boca Raton, FL: CRC Press, 2002.

Muilenberg, Michael, and Harriet Burge, eds. *Aerobiology.* Boca Raton, FL: Lewis Publishers, 1996.

Murray, David E. *Seed Dispersal.* Orlando, FL: Academic Press, 1986.

Navarro, Shlomo, and Ronald T. Noyes. *The Mechanics and Physics of Modern Grain Aeration Management.* Boca Raton, FL: CRC Press, 2001.

Newman, Lee Scott, and Jay Hartley Newman. *Kite Craft: The History and Processes of Kitemaking throughout the World.* New York: Crown Publishers, 1974.

Noland, Michael S. *Fundamentals of Air Traffic Control.* Belmont, CA: Wadsworth Publishing Company, 1999.

O'Hanlong, John F. *A User's Guide to Vacuum Technology,* 2nd edition. New York: Wiley Interscience, 1989.

Pagen, Dennis. *Hang Gliding Training Manual: Learning Hang Gliding Skills for Beginner to Intermediate Pilots.* Black Mountain, NC: Black Mountain Books, 1995.

Pasqualetti, Martin J., Paul Gipe, and Robert W. Righter. *Wind Power in View: Energy Landscapes in a Crowded World.* San Diego: Academic Press, 2002.

Patel, Mukund R. *Wind and Solar Power Systems.* Boca Raton, FL: CRC Press, 1999.

Pelham, David. *The Penguin Book of Kites.* Harmondsworth, UK: Penguin Books, 1976.

Perez, Karen. *Wind Energy Basic: A Guide to Small and Micro Wind Systems.* White River Junction, VT: Chelsea Green Publishers, 1999.

Piggott, Derek. *Understanding Gliding: The Principles of Soaring Flight.* London: A. and C. Black, 1977.

Pijl, L. van der. *Principles of Dispersal in Higher Plants,* 3rd edition. Berlin: Springer-Verlag, 1982.

Pope, Alan. *Wind-Tunnel Testing,* 2nd edition. New York: Wiley, 1954.

Posey, Carl A. *The Living Earth Book of Wind & Weather.* Pleasantville, NY: Reader's Digest Association, 1994.

Poynter, Dan. *The Parachute Manual: A Technical Treatise on Aerodynamic Decelerators,* 3rd edition. Santa Barbara, CA: Para Publishers, 1984.

Pryor, Esther, and Minda Goodman Kraines. *Keep Moving! Fitness through Aerobics and Step.* New York: Mayfield Publishing, 2000.

Pye, K., and N. Lancaster. *Aeolian Sediments: Ancient and Modern.* Oxford: Blackwell Scientific Publications, 1993.

Rich, Doris L. *Amelia Earhart: A Biography.* Washington, DC: Smithsonian Institution, 1989.

Rickenbacker, Eddie. *Rickenbacker.* Englewood Cliffs, NJ: Prentice-Hall, 1967.

Sargent, Rose-Mary. *The Diffident Naturalist: Robert Boyle and the Philosophy of Experiment.* Chicago: University of Chicago Press, 1995.

Sawyer, Donald T. *Oxygen Chemistry.* New York: Oxford University Press, 1991.

Schnabel, Ernst. *Story for Icarus: Projects, Incidents, and Conclusions from the Life of Daedalus, Engineer.* New York: Harcourt Brace, 1961.

Schweizer, Paul A. *Wings Like an Eagle: The Story of Soaring in the United States.* Washington, DC: Smithsonian Institution, 1988.

Sebel, Peter, et al. *Respiration: The Breath of Life.* New York: Torstar Books, 1985.

Shao, Yaping. *Physics and Modeling of Wind Erosion.* Dordrecht, the Netherlands: Kluwer Academic Publishers, 2000.

Shapin, Steven. *Leviathan and the Air-Pump: Hobbes, Boyle, and the Experimental Life.* Princeton, NJ: Princeton University Press, 1989.

Sheets, Bob, and Jack Williams. *Hurricane Watch: Forecasting the Deadliest Storms on Earth.* New York: Vintage Books, 2001.

Shores, Christopher. *Air Aces.* Novato, CA: Presidio Press, 1983.

Society of Automotive Engineers. *Automatic Occupant Protection Systems.* Warrendale, PA: Society of Automotive Engineers, 1988.

Spengler, John D., John F. McCarthy, and Jonathan M. Samet, eds. *Indoor Air Quality Handbook.* New York: McGraw-Hill, 2000.

Stein, Sherman K. *Archimedes: What Did He Do Besides Cry Eureka?* Washington, DC: Mathematical Association of America, 1999.

Stommel, Henry M., and Dennis W. Moore. *An Introduction to the Coriolis Force.* New York: Columbia University Press, 1989.

Szurovy, Geza, and Mike Coulian. *Basic Aerobatics.* New York: McGraw-Hill, 1994.

Tape, Watler. *Atmospheric Halos.* Washington, DC: American Geophysical Union, 1994.

Taylor, Michael John Haddrick. *Encyclopaedia of the World's Air Forces.* Wellingborough, UK: Stephens, 1988.

Tindall, James A. *Unsaturated Zone Hydrology for Scientists and Engineers.* Upper Saddle River, NJ: Prentice-Hall, 1999.

Topping, Dale. *When Giants Roamed the Sky: Karl Arnstein and the Rise of Airships from Zeppelin to Goodyear.* Akron, OH: University of Akron Press, 2000.

U.S. Environmental Protection Agency, Office of Air and Radiation. *The Benefits and Costs of the Clean Air Act, 1970 to 1990: EPA Report to Congress.* Washington, DC: U.S. Environmental Protection Agency, Office of Air and Radiation, Office of Policy, [1997].

———. *The Benefits and Costs of the Clean Air Act, 1990 to 2010: EPA Report to Congress.* Washington, DC: U.S. Environmental Protection Agency, Office of Air and Radiation, Office of Policy, [1999].

U.S. Weather Bureau. *History of Weather Bureau Wind Measurements.* Washington, DC: U.S. Government Printing Office, 1963.

Vero, Radu. *Airbrush: The Complete Studio Handbook.* New York: Watson-Guptill Publications, 1983.

Walker, James C. G. *Evolution of the Atmosphere.* New York: Macmillan, 1977.

Wark, Kenneth, Cecil F. Warner, and Wayne T. Davis. *Air Pollution: Its Origin and Control,* 3rd edition. Menlo Park, CA: Addison-Wesley, 1998.

Weems, John Edward. *The Tornado.* College Station: Texas A&M Press, 1991.

Welch, Ann Courtenay Edmonds. *The Story of Gliding,* 2nd edition. London: J. Murray, 1993.

West, John B. *Respiratory Physiology,* 5th edition. Baltimore: Williams & Wilkins, 1995.

Whelan, Robert F. *Cloud Dancing: Your Introduction to Gliding and Motorless Flight.* Highland City, FL: Rainbow Books, 1996.

Whelan, James R. *Hunters in the Sky: Fighter Aces of WWII.* Washington, DC: Regenery Gateway, 1991.

Whittall, Noel. *Paragliding: The Complete Guide.* Guilford, CT: The Lyons Press, 2000.

Whitwell, David. *A New History of Wind Music.* Evanston, IL: Instrumentalist, 1972.

Williams, Albert Nathaniel. *Air Brakes and Railway Signals: Proud Creations of George Westinghouse's Brilliant Genius.* New York: Newcomen Society in North America, 1949.

Wirth, Dick. *Ballooning: The Complete Guide to Riding the Winds.* New York: Random House, 1980.

INDEX

About the Author

DAVID E. NEWTON is the author of more than 400 textbooks, encyclopedias, resource books, research manuals, and other educational materials. He taught math and physical sciences in Grand Rapids, Michigan, for 13 years; was professor of chemistry and physics at Salem (Massachusetts) State College for 15 years; and was adjunct professor in the College of Professional Studies at the University of San Francisco for 10 years.

MUMMIES, DISEASE, AND ANCIENT CULTURES

Mummies, Disease, and Ancient Cultures

Edited by AIDAN *and* EVE COCKBURN

CAMBRIDGE UNIVERSITY PRESS

CAMBRIDGE

LONDON   NEW YORK   NEW ROCHELLE

MELBOURNE   SYDNEY

Published by the Press Syndicate of the University of Cambridge
The Pitt Building, Trumpington Street, Cambridge CB2 1RP
32 East 57th Street, New York, NY 10022, USA
296 Beaconsfield Parade, Middle Park, Melbourne 3206, Australia

First published 1980

Printed in the United States of America
Typeset by The Composing Room of Michigan, Inc., Grand Rapids, Michigan
Printed and bound by Halliday Lithograph Corporation, West Hanover, Massachusetts

Library of Congress Cataloging in Publication Data

Mummies, disease, and ancient cultures.

Includes index.

1. Mummies. 2. Paleopathology. I. Cockburn,
Aidan. II. Cockburn, Eve.
GN293.M85 616′.00932 79–25682
ISBN 0 521 23020 9

To everyone who made this book possible:

First, to our patient authors, who have responded nobly to our demands for four years, from every corner of the globe, giving unstintingly the fruits of their firsthand, original research and adapting with understanding to the format we required.

Second, to the many scientists who processed specimens, identified organisms, and generously gave of their knowledge.

Third, to the backstage workers who cannot be mentioned individually in the text – typists, technicians, and support staff of all kinds – without whose willing help a book of this complexity could never get off the drawing board.

Contents

Preface

Why mummies? That is the question we are often asked. How did an otherwise respectable physician and a senior member of the University of Oxford, whose field is modern language and literature, find themselves regarded as "the mummy experts"?

The story begins in Aidan's early medical years. He is cursed with the 'satiable curiosity of the elephant's child and always wants to know "why?" Why are diseases the way they are? Were they always like this? Where did they come from? Under the influence of the nineteenth-century ideas of Darwin and Huxley, he worked out a series of theories that would explain how disease organisms evolved, how they changed during the different epochs of the development of human society, and how the interaction of these two streams of evolution resulted in our current infectious disease patterns. Eve, with her nonscientific background, found herself looking at these ideas with the cold and critical eye of an outsider, then helping with the sorting and organizing of theories – and so a partnership was born.

After Aidan's first two books on the evolution of infectious diseases (1963 and 1967), there was a hiatus of several years. Then came two casual conversations, which led to the present line of research and the present book. At a meeting of the American Association of Physical Anthropologists in Boston in 1971, Lucile St. Hoyme of the Smithsonian Institution remarked: "Aidan, why don't you apply for a grant from the Smithsonian to study in some area where you could find facts to back up your theories? We have local currency funds available in at least seventeen countries." She listed them, and the obvious one that would provide a fertile field for research was Egypt. Aidan applied for and received a grant to go to Egypt on a reconnaissance trip to investigate the possibility of organizing a project for the autopsy of large numbers of mummies, thus obtaining facts to back up his, until then, largely speculative ideas.

Then came casual conversation number two. Eve was talking to William H. Peck, Curator of Ancient Art at the Detroit Institute of Arts, about the projected trip. Bill asked whether Aidan had ever autopsied or in any way examined a mummy before, and when the answer was no, he suggested that Aidan might like to practice on one of those in storage in the institute's basement. The story of this first, primitive autopsy has already been fully described (*Smithsonian,* November 1973); its importance lies in the idea of examining mummies in American museums rather than those in Egypt.

While in Egypt, Aidan met David O'Connor, who became a major contributor to the final program. At the Pennsylvania University Museum, where he was Egyptian curator, there were several mummies, and David invited Aidan to examine these if he needed to. The first autopsy (PUM I), conducted in Philadelphia at the university, was an unmitigated disaster. No one really knew what to do, and readers of this book will find only passing reference to the project – but it *was* a valuable learning experience. The media had been invited and turned up in full force, so the examination became a three-ring circus, with photographers and cameramen taking over the autopsy room; at one stage there was

even a class of visiting third-graders, complete with teachers, wandering through and getting underfoot. Not an atmosphere conducive to serious scientific work!

However, three more mummies were provided by David and successfully autopsied in Detroit, with conditions strictly controlled. The Smithsonian Institution, the Detroit Institute of Arts, and Wayne State University School of Medicine collaborated in the sponsorship of these studies. The first study, of PUM II, became the basis of the Paleopathology Association. Papers presented at the symposium held in conjunction with the autopsy, which had been given the somewhat fanciful name of "Death and Disease in Ancient Egypt," were printed with a covering letter of information under the grandiose title, *Paleopathology Newsletter*, Number 1. At that time, it really was debatable whether there would ever be an issue number two! However, the publication found an immediate audience, and so the Paleopathology Association was born. There are no association dues, no formal organization, no by-laws. The *Newsletter* is now a viable entity, with more than 300 subscribers in 25 countries, and it is from these that contributors to the present book are drawn. During the past 5 years, a great deal of major scientific work has been performed, all on a strictly voluntary basis. People work because they are interested, consumed by that same 'satiable curiosity that started Aidan off in the first place. We are grateful for what their enthusiasm and energy has produced–and we hope readers of this book will feel the same.

E. G. C.

Introduction

AIDAN COCKBURN
President, Paleopathology Association
Detroit, Michigan, U.S.A.

What is a mummy? For most people, the word immediately brings to mind visions of Egypt and, in particular, pictures of a body wrapped in swaddling bands of cloth. This was the original idea of the term, and indeed from the earliest days of antiquity, the preserved bodies of ancient Egypt have gripped the imagination of all who knew about them, whether rich or poor, educated or not. This was so much the case that when the Romans took over Egypt and found the art of preservation to be badly degenerated, they tried to revive the old ways. But it was too late. The ability to read hieroglyphics and ancient writings had been lost when the Greeks under the Ptolemies conquered the country and introduced their much superior Greek script. However, some form of body preservation was continued up to the eighth century A.D. At that time, the invading Arabs swept all before them in Egypt. To them, the practice of embalming the dead was abhorrent, and they put a stop to it.

Today, the term *mummy* has been extended to cover all well-preserved dead bodies. The majority of these are found in dry places such as the sands of deserts or dry caves, where desiccation has taken place rapidly, doing naturally what Egyptians did by artifice. The basic procedure in either process is the same: Water is extracted rapidly from the tissues. There is no mystery in this, for people since antiquity have been preserving fish and meat in the same basic ways, either by drying in the sun or by packing in salt. The Egyptian embalmers used a naturally occurring salt called *natron* instead of common table salt and supplemented this with oils, resins, and bitumen. The word *mummy* is derived from the Persian *mumeia* or *mum*, meaning "pitch" or "asphalt." This substance had been used in classical times in medical prescriptions, but medieval physicians introduced a refinement with preparations of pitch from Egyptian mummies. These "exudations" of mummies became very popular and remained so up to the nineteenth century. The first use of the word referring to medicine dates back to the early fifteenth century (*Encyclopaedia Britannica* 1911). As applied to a preserved body, however, the earliest record is 1615 (*Oxford English Dictionary*).

Occasionally, bodies are found preserved in other ways. Most of these are frozen, like an Inca boy who had been sacrificed on a high mountain in Chile. Apparently, he had been drugged and left to freeze. In Siberia, mammoths and extinct horses have been found in the permafrost. In the Altai mountains of Russia, Scythian bodies from about 400 B.C. have been recovered, encased in ice, from their tombs. The first use of *mummy* applied to a body frozen in ice was in 1727 (*Oxford English Dictionary*).

More baffling is the wonderfully preserved corpse of a Chinese princess of 2,000 years ago. The coffin in the tomb was hermetically sealed and still contained the preserving fluid, a weak mercurial solution. The tissues were still elastic and the joints could be bent. Whether this survival was attributable to the exclusion of oxygen, as suggested by the Chinese scientists, or to the mercurial solution, or to a combination of both is uncertain.

Serious scientific studies of mummies on

an organized basis began in Egypt shortly after the turn of the century (Dawson 1938). This coincided with the period when Egypt was dominated by the British and with the foundation of a school of medicine and the creation of the first Aswan Dam. The two events were interrelated. The first great dam at Aswan was completed in 1902, and the reservoir behind it was filled in the spring of 1903. By this action, the First Cataract on the Nile was obliterated, Philae was inundated, and much of the valley of the Nile was flooded. Many antiquities were ruined and many ancient burials destroyed by the inundation and the seepage.

Much public resentment had been expressed at this destruction of historical records, and the pathetic sight of the Temple of Philae, standing half drowned in the muddy water, had appealed to innumerable tourists as a sacrifice of the beautiful and historic on the altar of modern utilitarianism. In 1907, the Egyptian Government proposed to increase the height of the dam by another seven meters. Such a project entailed the flooding of a very large area. The Government wisely decided that before the inundation took place, the area should be thoroughly surveyed and examined. All antiquities were to be recorded; all burials were to be examined, described, photographed, and rescued before the raised Nile could reach them. [Wood Jones in Dawson 1938]

Coincidental with this event was the founding of the English-language Government School of Medicine in Cairo on the ruins of the former French school, which had become defunct 10 years earlier. The professors at the school were excellent, and three of them shaped the future of the study of mummies for decades. They were Grafton Elliot Smith (anatomy), Armand Ruffer (bacteriology), and Alfred Lucas (chemistry). Elliot Smith left Egypt after 7 years to return to England, but for the rest of his life he continued to develop the ideas he had conceived in Egypt. Ruffer explored means of examining soft tissues, and his techniques are still used today. The world lost a distinguished scientist when he was killed during World War I,

while serving in a hospital ship that was sunk. Lucas continued in Egypt; when Tutankhamun's tomb was found, he was called in as a consultant. Elliot Smith made the greatest impact at the time, but today it is Armand Ruffer who reigns supreme in the field of paleopathology.

Elliot Smith began his studies of Egyptian bodies in the Thebaid in 1901, and in 1905 he made the first of his detailed examinations of the technique of mummification. In early 1903, the tomb of Pharaoh Tuthmosis IV was found, and Maspero, the director of the Service des Antiquités, ordered that the mummy of the king be unwrapped and examined. The result was a public spectacle for the elite of Cairo, but the examination had no scientific value. However, Elliot Smith was able to make a later, private examination that included roentgenography. At that time, there was only one X-ray machine in Cairo, so Elliot Smith and Howard Carter took the rigid pharaoh in a cab to the nursing home to have it X-rayed. This was a historic first. Following this, Elliot Smith made a study of all the royal mummies found in the two great caches of Deir el-Bahri (1881) and the tomb of Amenophis (1898). Later he investigated a series of mummies from different periods in order to determine how the embalming process had changed over the centuries. On a visit to his native Australia, he found two mummies of Papuans from the Torres Strait in a museum in Adelaide. These had so many features in common with the mummies of Egypt that he developed his concept of cultural diffusion, claiming that the idea of mummification had spread from Egypt to the Torres Strait. Pretty and Calder discuss this theory in Chapter 11.

The investigations that resulted from the raising of the Aswan Dam proved beyond the capacity of the staff of the Government Medical School, so additional assistance, notably that of W. R. Dawson and Frederic Wood Jones, was obtained from England. The task was so great that in one month in one small area, the archeologists uncovered the tombs

of 2,000 persons. The workers did their best, but by modern standards the whole operation was a very crude and rushed affair. Under similar circumstances, it is doubtful whether we could have done any better today. Regardless of this, the pathologies of ancient Egypt were revealed for the first time. In all, autopsies were performed on about 8,000 mummies.

After that initial period of excitement, work on ancient bodies dwindled to almost nothing until recent times, when an even bigger Aswan Dam was built and the whole business of saving ancient relics began anew.

Sir Armand Ruffer's work was not in the blistering heat of the exposed desert, but within the four walls of his laboratory. Whereas Elliot Smith measured bones and studied mummification, Ruffer explored the possibility of restoring ancient tissues to something approximating their condition before death. He was so successful that Ruffer's fluid is still in use today. He was the first to show *Schistosoma* ova in kidneys, bacteria in tissues, and organized structures in organs dead for two or three millennia. We must all salute his memory.

Lucas, a painstaking chemist, analyzed the materials used by ancient Egyptians. By modern standards, his techniques seem old-fashioned, but his results have never been surpassed. His work (1962) is a classic that even today is indispensable. He also experimented with mummification. According to Herodotus, the body was immersed in a large vat that contained a solution of natron, although apparently this interpretation depends on the translation of a single word. If the word is taken in its other meaning, the interpretation is different. Lucas doubted that immersion in a solution of natron would produce mummies as we know them today, so he took pigeons, soaked them as described, and found that the flesh became soft and separated from the bones. Typical mummies could be produced by packing dry natron inside the birds and covering them with the salt externally. Treated this way, they dried out

quickly. The process was so successful that decades later his mummified pigeons still sit in the Department of Antiquities just as he left them, though kept at room temperature without any special care for so many years.

How many mummies are there in the world? This question is often asked, especially in criticism of the sometimes destructive methods used in an autopsy. The implication is that something irreplaceable is being destroyed, but this is not always correct, for two of the six mummies studied by our group have been enhanced in value and are now on exhibition in major museums instead of being hidden in basements.

A head count of mummies is not possible, but some idea of the total number can be gained by a review of their history. In desert areas of North Africa that have been dry for thousands of years, large numbers of bodies must still remain preserved in the sand. In Peru the same applies, with the additional factors of careful burial and wrapping in cotton. So over a period of 2,000 or 3,000 years, many millions of bodies must have been interred. To start with, artificial mummification in Egypt was probably reserved for the pharaoh, his family, and the nobles; but eventually, as everyone wanted to live for eternity, the practice spread. Even if the population had been no more than 1 million people (and surely it was much more), with an average life expectancy of 40 years, about 1 million mummies would have been laid in the ground every 40 years. In the course of 2,000 years this would amount to more than the present population of Egypt.

The question of numbers was discussed in Egypt in 1972 with officials from the Department of Antiquities. They say that tombs containing mummies are discovered almost every time a new road or airfield is constructed. There are so many mummies that those that appear to be of no special interest are reburied in the sand.

On the other hand, Egypt has been exporting or using mummies for centuries, so vast

numbers must have been destroyed or dispersed. The process started in the fifteenth century, when it was claimed that ground-up mummy had medicinal properties, and this became an expensive and valued remedy for many diseases. How many hundreds of tons of mummy tissues were swallowed by credulous sufferers, before the practice died out early in the nineteenth century, is anyone's guess.

Another drain on the mummy population in Egypt was caused by lively interest in the Western world, possibly sparked by reports from the savants who accompanied Napoleon to that country. By the end of the nineteenth century, it was de rigueur for every museum to have at least one mummy on exhibition. Even today many small towns have a specimen dating from this period, although the larger museums frequently relegate the bodies to the basement.

In Canada during the nineteenth century, mummy cloth was used in the manufacture of paper. Because the supply of rags for paper making proved inadequate, Canadian paper manufacturers imported thousands of mummies just for their wrappings. What happened to the bodies is not known.

In *Innocents Abroad*, Mark Twain tells of another way in which mummies were destroyed. They were used instead of coal in the engines of the newly constructed railway! Some mummies were destroyed in areas where irrigation was extended and the level of the subsoil water rose. This was especially true in the delta region, where silt from the Nile pushed the land increasingly into the Mediterranean and the land sank slightly as a result of the extra weight.

The biggest destruction of mummies came in the period of dam building that began about 1900 and continues to the present. More and more land was covered by water, and any bodies interred there were damaged. Rescue operations touched only the fringe of the problem, and in any event, most of the bodies found were simply reinterred after a preliminary examination, then left to the mercy of the rising waters. However, Nubia and the areas above the Aswan dams were not noted for the practice of artificial mummification, and so far the prime areas have been left undisturbed by this particular form of cultural development.

When all these factors are taken into account, it seems probable that many hundreds of thousands of mummies have been lost; even so, millions still remain in the sands and tombs of Egypt. Add to this figure the millions in Peru and other dry areas of South America, and it becomes clear that a huge store remains for future generations to study.

Of course, mummification was not confined to humans: Animals were treated the same way. The Egyptians embalmed specimens of almost every animal in their ecosystem – ranging from bulls through birds, cats, fish, and bats down to shrews – and the numbers were enormous. Their sacred bird, the ibis, may have become extinct simply because every one found was killed and stuffed.

It is in the cold regions, however, that future studies on animals look most promising, especially for the field of biochemistry. The frozen mammoths of Siberia are well known, and a baby mammoth in good condition was recently exposed by a bulldozer in Siberia (Figure 1). Rhinoceroses, horses, and small mammals have also been identified. The permafrost of the Arctic must be a well-stocked refrigerator whose contents are virtually unexplored. At the other pole, conditions are different, but the potential is also great. There, the extreme cold and dryness of the air freeze-dry any animals that die on ice or on land. Thousands of dead seals have been found, some thousands of years old, and there are even reports of bacteria remaining alive in the soil after 10,000 years. Possibilities like this make the future of mummy research most exciting.

Among the most satisfying discoveries associated with mummies are objects that have been included by chance. Quite apart from the religious ritual and ceremony that apper-

Figure 1. *Baby mammoth found frozen in Siberia, 1977. (Courtesy of Professor N. K. Vereshchagin, Academy of Science, USSR)*

tain to the processing of a body for preservation, objects were occasionally included that were insignificant to the embalmers but shed bright beams of light in corners that would otherwise remain dark for us.

Our own group experienced one such example of serendipity with the finding of a ball of cotton in the wrappings of PUM II. This is the earliest cotton recorded in Western civilization, although the textile was used in both India and America perhaps 2,000 years earlier. How it reached Egypt and what it was doing in a mummy's wrappings are matters for conjecture. Meryl Johnson, who was the first to notice it, is inclined to believe that perhaps the ball of cotton was regarded as a valuable object and was included for that reason (Chapter 4).

Another adventitious find was the only

example of an Etruscan text that has come down to us (Wellard 1973). It was half a century before anyone realized the writing on the shroud was Etruscan, and in that time some 80 percent of the wrappings disappeared. The mummy was bought in 1848 by a Croatian, Michael Barie, who was employed by the Hungarian Chancellery in Alexandria. He took it home with him, and on his death his brother gave it to the museum of Agram (now Zagreb). The museum noted that the wrappings were "covered with writing in an unknown and hitherto undeciphered language."

Dr. Heinrich Brugsch, an Egyptologist, viewed the writing and could make nothing of it; he mentioned it to Richard Burton, the famous explorer and linguist, who was at that time British Consul at Trieste. The vice-

consul copied part of the text for him, and this was published in the *Transactions of the Royal Society of Literature* in 1882. The previous year Burton had published a book, *Etruscan Bologna*, but in spite of these studies he did not suspect the nature of the writing. In the end, the bandages were sent to the University of Vienna, and there Professor Jakob Krall made the proper identification. How this inscribed linen came to be wrapped around a dead Egyptian girl is not known. It has been suggested that the linen had nothing to do with the girl, but that the embalmers simply bought a sheet of second-hand linen and tore it roughly into strips for their purposes. As will be seen in Chapter 6, a similar method was used on PUM III. We still do not know what the writing says, though it appears to deal with the religious code of the Etruscan people. Perhaps some day an interpretation will be made that will help us to understand more about these mysterious people.

Why were bodies preserved in this way, to last long periods of time? Naturally preserved bodies have come down to us simply because of accidents or environmental conditions, without deliberate human thought. The general climate or microclimate at the time of death produced a situation in which the tissues were dehydrated or frozen, so that the usual biochemical changes of degradation in dead bodies were inhibited. This occurred in hot dry areas, at high altitudes, or in arctic surroundings.

When, however, deliberate efforts were made to ensure that the body would continue to exist in a form somewhat resembling that of the living person, the questions of why this was done and how it was done become matters of considerable interest, going as they do to the very basis of man's attitude to death.

Death is a fearsome thing. Within a brief period – 24 hours in the tropics and a few days elsewhere – a close friend, a relative, or a colleague becomes a bloated, horrifying caricature of the living human being; the body melts and eventually turns into a heap of bones. This has happened since the origin of life and is a basic fact: Everything that lives must die.

The human race has constantly rebelled against this idea, producing various concepts to show that death is not the end and that life in one form or another continues after the physical disintegration of the body. This is the essence of most religions. There are many variations on the theme – the Valhalla of the Vikings, Paradise for Muslims, Resurrection among Christians, the reincarnation beliefs of the Hindus – but the central premise is that death is not final: Something better comes afterward.

In most cases where mummies have been preserved deliberately, the objective seems to have been to keep the body intact and recognizable for this afterlife, even to the point of burying with it clothing, food, and utensils for the future. The Egyptians believed that the spirit of a person could not continue to exist if the physical body disappeared; therefore, to attain immortality, the body had to be mummified. Eventually, almost all Egyptian bodies were so treated, to universal satisfaction: The people believed they had conquered death.

Not all peoples are afraid of death. One has only to think of Spartans at Thermopylae, Scythian youths going willingly to their deaths at the funeral ceremonies of their kings, sacrificial maidens of the Incas treated with the greatest respect during their lives and dying in full expectation of a magnificent thereafter, Christian martyrs in Roman times welcoming death in a state of ecstasy, or even, in modern times, the kamikaze pilots of Japan. To people like these, death is not a disaster, but the gateway to a better existence in another world.

Looking at the matter on a global scale, we can say that most people believe in some form of life after death and often come to the conclusion that the body of the deceased should

be prepared for this continuing existence. This can be achieved either by burying the person along with objects for use in the next life or, better still, by preserving the body itself. It can be noted in passing that almost all dead bodies in the United States today are embalmed. This can have no other purpose but to give survivors reassurance that life continues in some way: From the purely public health point of view, it is meaningless.

Throughout the world, nations and tribes have striven to preserve the bodies of their leaders and great men. One unsuccessful effort was that of the Chinese ruler who had a marvellous suit made of small pieces of jade stitched together with gold wires and shaped to enclose his whole body. Jade was believed to preserve bodies, but in this case it was ineffective, and the body inside the suit changed into dust. In Vienna, according to Ekkehard Kleiss (personal communication 1977), there are mummies of the kings and princes of Austria. A peculiar feature here is that in some instances the bodies are kept in one cathedral and the internal organs in another. To this day, the bodies of the Jewish patriarchs and their families are kept in tombs in Israel. It is probable that all of them were mummified as described in the Bible.

Among mummies that have been lost are those of the Ptolemies in Egypt. These Greek pharaohs adopted the customs of their conquered country, and that must have included preserving the body after death. The greatest of all the Greek leaders was Alexander, and the site of his tomb in Alexandria has also been forgotten. An active search is even now in progress for the tomb of the world's first known genius, Imhotep, near the pyramid he built for his pharaoh. This step pyramid at Saqqara was the first one, and archeologists feel certain that Imhotep was buried near his master.

Preservation of the body was undertaken not only for the future good of the dead, but often out of fear by the living of the spirits of the dead. If there is life after death, as so many

people have believed, then it follows that the ghosts may return to haunt and hurt the living. Therefore, the dead must be placated by adequate care during burial and kept in a friendly state by presents of food and other gifts long after death. The ancestor worship of the Chinese is probably based on such beliefs. The mummies of the Inca rulers were displayed at intervals in the square in Cuzco, as though they were still living, and were offered food and drink.

In many parts of the world, bodies resting in churches, cathedrals, and temples were preserved by accidents of climate or circumstance. I saw one example in the making in 1959 in East Pakistan (now Bangladesh) while fighting a smallpox epidemic. One night in Cox's Bazaar, the district commissioner came to see me and asked me if I would visit a monastery 60 km away on a river that formed the boundary with Burma. Some 2,000 years ago, there had been 1,500 monks there, but now only one was left and he was very ill. On arrival, I was escorted to the temple, but had to enter alone.

The last surviving monk was lying before the statue of Buddha, almost unconscious, scarcely breathing, and with a very feeble heartbeat. His legs were gangrenous, and because this was the hot and dry season, they had dried up and were in fact mummified. He could not drink the water I gave him, and it was obvious that he was near death: In fact, he died the next day. In time, he would have dried up completely. But in that climate the body could not have stayed mummified for long; with the arrival of the monsoon rains, humidity, insects, rats, and fungi would soon have reduced the soft tissues to dust so that only a skeleton remained.

In many countries, however, bodies like his have survived. A spectacular example is in the Capuchin Catacombs of Palermo, Sicily, where 8,000 mummies of men, women, and children, dressed in their best clothes, line the walls, their flesh preserved by the dryness of the air. These mummies represent a

cross section of the populace of the nineteenth century, coming from all sections of society. The practice was abandoned in the early twentieth century, but the mummies remain as dressed a hundred or more years ago.

In Venzone, Italy, there are mummies displayed in white sheets in erect positions (one wonders if they survived the recent earthquake that destroyed the town). In Vienna, kings and princes were preserved and their organs distributed among the cathedrals and churches of that city. In Corfu, Greece, the mummy of Saint Spiridion, patron saint of the island, is paraded with full ceremony around town during his festival. The list is endless.

An advance in embalming technology in which the tissues are infiltrated with paraffin wax first appeared in the Argentine at the beginning of this century. The most superb example of this method is said to be the body of Eva Perón, who appears to be merely asleep in spite of the peregrinations of her remains to Italy and back home again. Many observers have commented on the waxy face of Lenin, and some on that of Stalin, so perhaps they also had the paraffin treatment.

This longing for everlasting life has taken a new twist in recent years as a result of advancements in technology. It has long been known that freezing can, to some extent, suspend animation; for example, fish frozen in lake ice sometimes swim away when they thaw out. For decades microbiologists have preserved living viruses, bacteria, and simple organisms by keeping them at very low temperatures. Recently, it was reported from the Antarctic that living bacteria, up to 10,000 years old, have been found deep down in the frozen soil.

It is, therefore, not surprising that a new "science" of cryogenics has arisen in which the bodies of the dying or recently dead – the time at which the change from life to death takes place is a debatable legal point of considerable importance – are placed in deep freezes for revival in the future.

The reason this procedure is deemed worthwhile is the belief that a disease that kills today is likely to be curable tomorrow. It is hoped that, at some future time, life can be restored and the disease treated. In this way life can be continued, with occasional pauses for renewal, through eternity. Already, numerous bodies have been prepared in this way and stored in lockers at low temperatures. This technique may or may not work, but it will surely supply excellent specimens for research by paleopathologists in the future!

REFERENCES

Dawson, W. R. 1938. *Sir Grafton Elliot Smith.* London: Cape.

Lucas, A. 1962. *Ancient Egyptian materials and industries,* 4th ed., revised and enlarged by J. R. Harris. London: Edward Arnold.

Wellard, J. 1973. *The search for the Etruscans.* New York: Saturday Review Press.

PART I
Mummies of Egypt

1

Mummies of ancient Egypt

WILLIAM H. PECK
Curator of Ancient Art
Detroit Institute of Arts
Detroit, Michigan, U.S.A.

In the modern mind no single type of artifact from the ancient world excites more interest than the Egyptian mummy, and no other kind of object is considered more typically Egyptian. The very word *mummy* brings to mind a host of associated ideas – the Egyptian belief in life after death, the seemingly pervasive concern with the notion of death, and the elaborate preparations that were made for it. It is well to state at the outset that religious beliefs made it necessary to preserve the dead, and what seems a preoccupation with death was actually the outgrowth of a love of life and an attempt to prepare for a continuation in the next world of life as it is known in this.

A considerable literature, much of it of a speculative nature, has grown up around the modern interest in the process of mummification. In recent decades the progress of science has done much to dispel earlier misconceptions, but many of these have become firmly fixed and die hard. The process of mummification is still considered to be a "lost art" by many who would rather remain content with an intriguing mystery than be disappointed with a simple explanation. The process was the result of a continuous development based on trial and error and observable results. The details of technique can now be discussed with some confidence and accuracy.

Modern literature on the subject of mummification is extensive; the historic cornerstone of the study in English is Thomas Pettigrew's *History of Egyptian Mummies*, published in 1834. For the time at which it appeared, the work was a monumental undertaking. Based on scholarly research and practical experience, Pettigrew's work was a summation of almost all that was known concerning Egyptian funerary practices. He compiled all the ancient sources and commented on them, as well as discussing many examples of mummified remains investigated by or known to him (Figure 1.1). This work is illustrated by engraved plates by George Cruikshank (better known for his satirical drawings) that are the product of careful observation.

It was not until 1924 that another work of comparable stature appeared. *Egyptian Mummies* by G. Elliot Smith and Warren R. Dawson is still a standard text to which the interested reader can turn with confidence. Alfred Lucas was another twentieth-century pioneer in the scientific study of the process of mummification. In addition to many articles on various aspects of the subject, he devoted a chapter to his findings in *Ancient Egyptian Materials and Industries* (Lucas 1962). His work has been carried on by the Egyptian, Zaki Iskander (Mokhtar, et al. 1973). Recently, James Harris and Kent Weeks (1973) have published a popular report of the work done on the royal mummies in the Cairo Museum in *X-Raying the Pharaohs*.

The purpose of mummification in ancient Egypt was twofold. The body of the deceased, it was believed, had to be treated to render it incorruptible. At the same time the physical appearance had to be maintained as nearly as possible to what it had been in life. The Egyptian conception of life after death developed early, as is evidenced by the burials of the predynastic (prehistoric) age. It would seem that the notion of preserving human and ani-

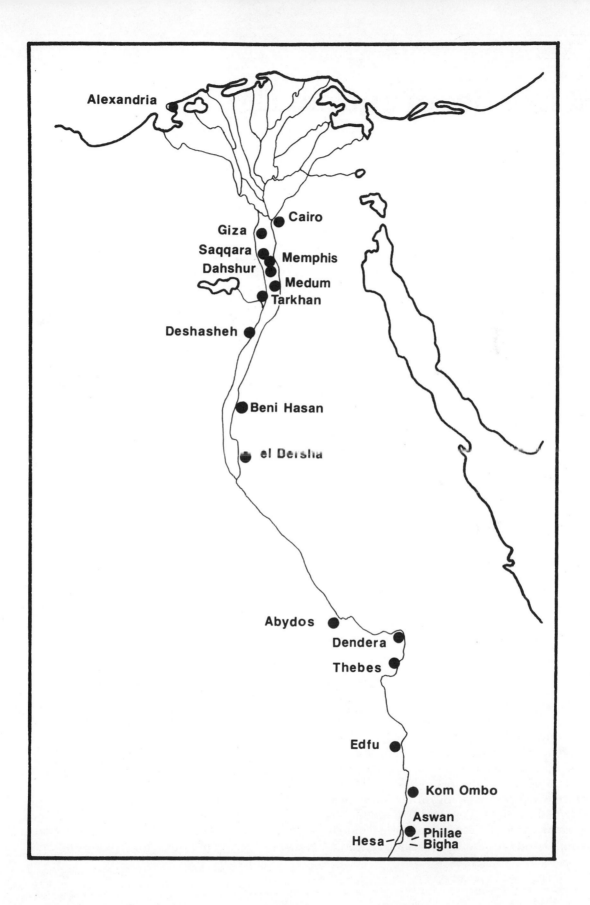

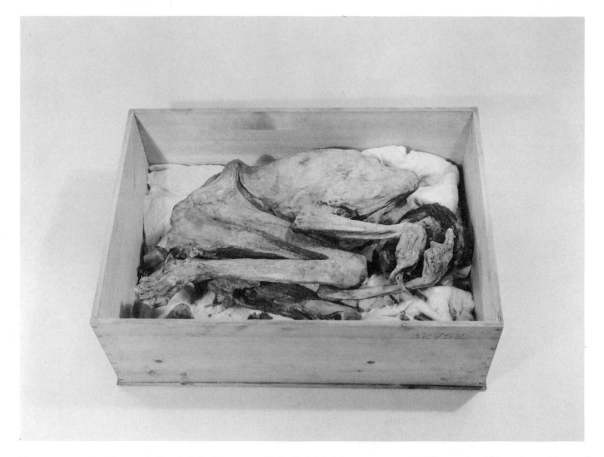

Figure 1.2. Predynastic burial, before 3000 B.C. British Museum 32752. The typical flexed position of most predynastic burials is illustrated here, as is the high state of preservation resulting from the dry sand rather than a complicated embalming process. The subject is an adult female.

mal dead came about naturally in the dry climate of Egypt. Predynastic burials were simple and practical (Figure 1.2). The corpse was placed in a hole in the sand, usually in a contracted position, accompanied by such grave goods as pottery and other useful objects. No embalming process was carried out; in no way was the body prepared (mummified) for the burial, but it was often wrapped in linen, reed matting, or hide. The pit was sometimes lined with matting, boards, or bricks, but the cavity that received the body was more grave than tomb. A small tumulus was erected over the grave, never large enough to interfere

with the warming effect of the sun. It was the hot, dry sand that served to desiccate the tissue. The result, to be observed in countless examples, is a well-preserved corpse. From the simple fact that objects were included with the burial, we can deduce that they were meant to serve the spirit of the dead in some fashion in the next life.

At the beginning of the dynastic age (around 3000 B.C.), the religious beliefs and accompanying funerary ritual appear to have been already well developed. The tomb structures of early dynastic kings were designed as imitation palaces and fortresses that must

Figure 1.1. Major sites in ancient Egypt. (Map by Timothy Motz, Detroit Institute of Arts)

have reflected the style of living architecture of the time. The great quantity of funerary offerings in these burials indicates that a king or member of the royal family expected to be able to use such material in a continued existence in the next life. The subsidiary interment of retainers nearby suggests the ability to confer immortality on them, if for no other purpose than to serve their master in the spirit world. As tomb structures became more complex, the position of the body in relation to the surface layers of warm sand was altered; the body was placed lower in the earth, at the bottom of a tomb shaft.

The notion of a "home" for the spirit continued throughout pharaonic history; the form of the structure underwent many changes, but the fundamental purpose remained the same. In the Old Kingdom the private tomb superstructure became more houselike, providing a protection for the burial, which was placed deep in the earth beneath it. The superstructure also provided the necessary rooms for the conduct of ritual at the time of the burial and after, as well as a storage area for ritual objects and offerings. Because the development of the tomb resulted in the removal of the body from the surface area of warm sand, it became necessary to invent a technology that would accomplish the preservation of the physical remains, a process that had occurred naturally in more simple times. In all cases where the body of the deceased had received "proper" burial, we can assume that some effort was made to treat the corpse and render it resistant to decay. The key factor in the preservation of the human body, as it was practiced by the Egyptians, is the removal of all body fluids. It is difficult to imagine the impetus for the initial steps in the development of the craft of mummification, but it has been suggested that some accidental knowledge of predynastic burials must have become available to the people of the early dynasties. By simple reasoning, it could have been determined that the removal of body fluid was the most important factor in the preparation of

the dead. The inspiration may even have come from observation of the processes of drying meat and fish. In any case, during the first 400 years of pharaonic history, the essential details of Egyptian mummification were evolved.

According to the *Oxford English Dictionary*, the word *mummy* is recorded in the English language as early as the fourteenth century. It existed in medieval Latin as *mumia* and was ultimately derived from the Arabic and Persian designations for an embalmed body by way of those for wax or bitumen. In modern usage, the word *mummy* is taken to mean the body of a human or animal that has been embalmed by the ancient Egyptian or some similar method as a preparation for burial. By analogy, many corpses are called "mummies" even if they have nothing to do with ancient Egypt. As a result, it is common to speak of Peruvian mummies, Aleutian mummies, the mummified Capuchins of Palermo, and the like. In English, the word *mummy* has been used to designate medicinal materials prepared from the mummified bodies of Egypt, a brown pigment from the same source used in oil painting, and, in a somewhat more specialized use, as a slang term for Egyptian issues on the English stock exchange. Current in the sixteenth century was the use of the word *mummy* for any dead flesh: "The water swells a man; and what a thing should I have been, when I had been swel'd? I should have been a mountain of Mummie" (William Shakespeare, *The Merry Wives of Windsor*).

It has been traditional in the past to base any study of mummification on the accounts given by a few classical authors. The few Egyptian texts that can be used to supplement these are tantalizing in the extreme. It is to be hoped that a complete account of the embalming process may be found for some period in Egyptian history, but until now, this has not happened. By accident of preservation, a composition entitled *The Ritual of Embalming* exists in a fragmentary state in two versions (Goyon 1972). The information

it contains is more of a ritual nature than a step-by-step handbook on the technique of mummification. Its material consists of three parts: ceremonial acts to be performed on the mummy, prayers and incantations to be said during the process, and the methods of applying ointments and bandages to some parts of the body (arms, hands, legs, feet, back, and head). If either of the two late Papyri that contain this partial text were more complete, at least a sequence of wrapping the body might be explained. The most important lack is the absence of any information on the earlier stages of the mummification process, including evisceration and desiccation. From a number of minor Egyptian sources on stelae, ostraca, and papyri, the total length of time necessary for the total mummification process is established as 70 days. This includes the long period in which the body was allowed to dry.

The two classical authors who have given the best and most complete account of the process of mummification as they understood it are Herodotus and Diodorus Siculus. The account of Herodotus is by far the better known of the two, probably because the *Persian Wars,* of which it is a part, makes such interesting reading and because his account of Egypt has received such widespread publication. It must always be remembered that he was a Greek from Halicarnassus writing in the fifth century B.C. and that he is often accused by modern historians of reporting a considerable amount of what may be termed hearsay evidence. The reliability of Herodotus in regard to his descriptions of Egyptian customs and daily life has been the subject of much contemporary criticism; what he had to say about mummification must be weighed against the physical evidence of the mummies themselves.

Herodotus' account in the Rawlinson translation is as follows:

There are a set of men in Egypt who practice the art of embalming, and make it their proper business. When a body is brought to them these persons show the bearers various models of corpses, made in wood, and painted so as to resemble nature. The most perfect is said to be after the manner of him whom I do not think it religious to name in connection with such a matter; the second sort is inferior to the first, and less costly; the third is the cheapest of all. All this the embalmers explain, and then ask in which way it is wished that the corpse should be prepared. The bearers tell them, and having concluded their bargain, take their departure, while the embalmers, left to themselves, proceed to their task. The mode of embalming, according to the most perfect process, is the following: they take first a crooked piece of iron, and with it draw out the brain through the nostrils, thus getting rid of a portion, while the skull is cleared of the rest by rinsing with drugs; next they make a cut along the flank with a sharp Ethiopian stone, and take out the whole contents of the abdomen, which they then cleanse, washing it thoroughly with palm-wine, and again frequently with an infusion of pounded aromatics. After this, they fill the cavity with the purest bruised myrrh, with cassia, and every other sort of spicery except frankincense, and sew up the opening. Then, the body is placed in natrum for seventy days, and covered entirely over. After the expiration of that space of time, which must not be exceeded, the body is washed, and wrapped round, from head to foot, with bandages of fine linen cloth, smeared over with gum, which is used generally by the Egyptians in the place of glue, and in this state it is given back to the relations, who enclose it in a wooden case which they have had made for the purpose, shaped into the figure of a man. Then fastening the case, they place it in a sepulchral chamber, upright against the wall. Such is the most costly way of embalming the dead. [Herodotus, *History,* Book II:86]

If persons wish to avoid expense, and choose the second process, the following is the method pursued: Syringes are filled with oil made from the cedar-tree, which is then, without any incision or disembowelling injected into the bowel. The passage is stopped, and the body laid in natrum the prescribed number of days. At the end of the time the cedar-oil is allowed to make its escape; and such is its power that it brings with it the whole stomach and intestines in a liquid state. The natrum meanwhile has dissolved the flesh, and so nothing is left of the dead body but the skin and the bones. It is returned in this condition to the relatives without any further trouble being bestowed upon it. [Herodotus, *History,* Book II:87]

The third method of embalming, which is practiced in the case of the poorer classes, is to clear out the intestines with a purge, and let the body lie in natrum the seventy days, after which it is at once given to those who come to fetch it away. [Herodotus, *History*, Book II:88]

Herodotus' account may well give a description of the mummification process as it existed in the fifth century and as it may have been related to him, but it must be remembered that over 2,000 years of development cannot always be measured by a description of so late a stage. It will be helpful to suggest what can be learned from his account. That the embalmers were a special class of workers we are reasonably certain. The presentation of models of the various classes of mummification to the family cannot be verified, but contained in the description is a reference to "him whom I do not think it religious to name" (the god Osiris), which may be a reference to small mummiform statues, and an attempted explanation of their purpose. The brain was often removed through the nose, but evidence exists for its removal through the base of the skull and other openings. The incision in the abdomen usually was made on the left side and is seldom found to have been stitched up. Herodotus states emphatically that the body was covered with natron. This part of his description holds the most important key to the process. In earlier translations this was misinterpreted as natron in solution involving a prolonged soaking. It has been proved very satisfactorily that dry natron was used for the important step of desiccating the body and that the 70 days assigned to this stage actually refer to the entire mummification process. If the mummy was given back to the relatives for placement in a coffin, it was probably for an inspection of the embalmer's work. There is little evidence that the mummy was placed standing in the tomb, but mummy cases from the Ptolemaic period do have a baselike section at the foot. Good evidence of the second method exists, for mummies have been found with no abdominal incision, yet with internal organs missing and the anus plugged with linen packing.

Herodotus' account of mummification continues with some details that need not be quoted in full. He says that the bodies of women of high rank or great beauty are not delivered to the embalmers immediately, but after 3 or 4 days, to prevent the possibility of intercourse with the dead body. He also adds that the bodies of those who have fallen in the Nile, or who have been attacked by crocodiles, must be embalmed by the inhabitants of the nearest city and buried by them. Of the delay in embalming important women there is some evidence, but the second assertion is difficult to prove.

The second important classical source for the process of mummification is the account of Diodorus Siculus. A native of Sicily, as his name implies, he drew heavily on Herodotus, added a few details, and has left us some additional information. Because he was writing in the first century A.D., and because he agrees so much with Herodotus, it is hard to believe that his account contains an accurate description of mummification in his own time. To Herodotus' statement that embalmers were of a special class, he adds that the occupation was hereditary. This is likely, considering the number of other trades of ancient Egypt that were passed on in families. Diodorus adds designatory titles for the specialists who performed the different stages of the process. He identifies the heart and kidneys as having been left in place in the body when the other organs were removed, which seems to be accurate considering the number of times the heart has been found in mummified remains. He says that the cleansed corpse was treated with cedar oil and other substances, but he omits mention of the removal of the brain.

After listing the three grades of mummification, Diodorus describes only the most expensive, but gives prices for all three. Two interesting details are added in his account. According to him, the embalmer who made

the incision in the side, even though it was necessary to the embalming process, had to flee to escape the wrath of his fellow workmen. The explanation given is that any injury to the body of the deceased had to be punished, and this suggests that the ritual of protecting the corpse was taken seriously by the practitioners of the embalming craft. The second addition supplied by Diodorus is a description of the embalmed body, which, according to him, was preserved in every detail. So lifelike was the state of preservation that the body could be kept as a sort of display piece for the edification of the living. This agrees with Petrie's theory that some mummies of the Roman period must have been on view in the home for a considerable time before they were interred. Petrie was referring to mummies that had painted face coverings; the implication in Diodorus is that the face was still visible. The latter would be particularly curious and is not supported by the evidence of existing mummies.

In addition to Herodotus and Diodorus, Elliot Smith and Dawson refer to several papyri that give additional information, such as the prices of the various materials used in the mummification process. They also cite several late references to Egyptian embalming from Plutarch, Porphyry, Augustine, and others. For a somewhat distant source that contains mention of mummification, the Book of Genesis in the Old Testament should be quoted: "And Joseph commanded his servants, the physicians, to embalm his father; and the physicians embalmed Israel (Jacob)" (Gen. 50:2); "And forty days were fulfilled for him; and so are fulfilled the days of those which are embalmed; and the Egyptians mourned for him, three score and ten days" (Gen. 50:3); "So Joseph died, being an hundred and ten years old; and they embalmed him, and he was put into a coffin in Egypt" (Gen. 50:22). These short statements from the Old Testament add little to the Egyptian or classical sources beyond suggesting that the author had some familiarity with, or

access to, a tradition concerning Egyptian mummification.

No ancient Egyptian illustrations of the mummification process exist as such. The tomb paintings that have been preserved depict stages in the ritual and the offerings of prayers, but none of the physical treatment of the body itself. There are numerous instances in which the mummy is shown on the funerary bier, while it is being transported to the tomb, and before the tomb entrance at the time of the Opening of the Mouth ceremony, but the physical mummification seems not to have been an appropriate subject for tomb decoration. Any modern study of the process of mummification is dependent, then, on the physical remains of mummified bodies supplemented by the Egyptian, classical, and other references that deal with mummification and that have to be tested against the examples of mummification preserved.

The preservation of buried bodies in the predynastic age has already been commented on. The state of the development of the art of mummification in the early dynastic period is difficult to determine and must be inferred from the evidence of examples of later times. From Petrie's excavation of the Royal Tombs at Abydos came the bones of an arm that was wrapped in linen and still decorated with jewelry, but this indicates only the use of wrapping and gives no indication concerning the other preparations of the body. Quibell found at Saqqara the remains of a female of the Second Dynasty (2780–2635 B.C.). A contracted burial contained in a wooden coffin, the corpse was wrapped in over 16 layers of linen, but again, the condition of preservation made it impossible to determine exactly how the body had been treated. One important burial from the beginning of the Fourth Dynasty does a great deal more to suggest the state to which the science had progressed. In the early years of this century, the Boston–Harvard expedition, working at Giza under the direction of G. A. Reisner, discovered a tomb of Queen Hetepheres, the wife of Snef-

eru and the mother of Cheops (Ca. 2500 B.C.). Apparently, the remains of the queen had been entombed first at some other location, perhaps Dahshur, for the simple shaft tomb at Giza gave every appearance of a reburial. When the stone sarcophagus was opened, the body was missing, but the compartmented chest that contained the queen's viscera was found undisturbed. The packages of internal organs were still preserved in a solution of natron. The fact that the solution was still liquid after 4,500 years was incredible, but the real value of the discovery is the evidence it gives for the developed practice at this early time of the removal of the viscera and their inclusion in the burial in a special container of their own. The course of events to be inferred is obvious. At some point between the end of the predynastic period and the time of Hetepheres' burial, the technique developed of removing from the body those organs that were most likely to decay. It is not surprising that, in the 500 years this historical period covers, methods of embalming should have developed to such an advanced state.

The outward form of vessels used in the preservation of the viscera varied during the dynastic period. In the case of Hetepheres, the container used was a compartmented chest of alabaster. Individual canopic jars of the Fourth Dynasty exist. The idea of sets of four containers (or one divided into four parts) continues throughout much of dynastic history. The form changes – a miniature coffin occasionally replaced the jars – but the central idea was that the parts removed from the body were still a part of it and had to be treated as the body was, as well as being buried with it. The tradition was so strong that in the late period, when the organs were returned to the body cavity after being treated, dummy or imitation jars were still included in the burial.

By the Fourth Dynasty, the techniques of mummification had advanced so far that they may be studied in detail. That the process took a considerable time is suggested by the lack of well-preserved evidence from the ear-

lier dynasties. A single foot found in the burial chamber of the Step Pyramid at Saqqara, tentatively identified as once a part of the mummy of Zoser, offers no evidence about the techniques of mummification, but it does suggest, from the layers of linen wrapping, that the corpse was padded out in some semblance of a lifelike form.

The number of preserved bodies from the Fourth Dynasty on (and, of course, the number of tombs that have been identified) makes it clear that the preservation process was used for the nobility as well as for royalty. One well-known example found at Medum by W. M. F. Petrie in 1891 was provisionally dated by Elliot Smith and Dawson to the Fifth Dynasty, but they also stated that "the exact age of this mummy is uncertain. On archaeological evidence, it may be as early as the IIIrd Dynasty, but the extended position and the great advance in technique which it displays would seem to indicate a somewhat later date, probably Vth Dynasty" [2450–2290 B.C.]. Petrie's description, made at the time of the mummy's discovery, is worth quoting:

The mode of embalming was very singular. The body was shrunk, wrapped in a linen cloth, then modelled all over with resin, into the natural form and plumpness of the living figure, completely restoring all the fullness of the form, and this was wrapped around in a few turns of the finest gauze. The eyes and eyebrows were painted on the outer wrapping with green." [Petrie 1892]

Elliot Smith and Dawson examined this specimen at the Royal College of Surgeons in London, where it had been deposited by Petrie, and they described it as being wrapped in large quantites of linen with the outer layers soaked in resin and modeled to resemble the fine details of the body. Even the genitals were so treated, with such care as to allow the investigators to determine that circumcision had been practiced. They also observed that the body cavity had been packed with resin-impregnated linen. This mummy, found by Petrie at Medum and identified by him as a man named Ranofer, is now dated to the

Fourth Dynasty. The descriptions of it give a concrete idea of the attempt that had been made to create a lifelike appearance by modeling and wrapping. The mummy is no longer available for further examination, as it was destroyed in an air raid in World War II (Lucas 1962).

The fully developed intention at the height of the Old Kingdom seems to have been to effect the most lifelike suggestion possible of the original appearance of the body before death. It is a pity that there are not more and better preserved examples of mummification that span the time between the beginning of the dynastic period and the pyramid age. Nevertheless, the conclusion is the same: The practice of preservation was accompanied by a desire to create of the dead body a resemblance to the deceased as he had been in life. One additional example from the Fifth Dynasty adds support to this assertion. The mummy of a man named Nefer, found at Saqqara, is described in *X-Raying the Pharaohs* as looking like a man asleep. The wrappings of this specimen were soaked in an adhesive and molded to suggest the shape of the body, "the genitalia were particularly well modeled. Eyes, eyebrows and mustache were carefully drawn in ink on the moulded linen" (Harris and Weeks 1973). From the description, this mummy seems to provide an almost exact parallel to the one from Medum, now destroyed.

The desire to preserve the outward appearance of the deceased took another direction, in addition to the coating of the body with resin and the modeling of the features in that material. Masks of plaster applied over the face are known from as early as the Fourth Dynasty. By the time of the Middle Kingdom, the facial features were often modeled in cartonnage, a combination either of cloth and glue or of papyrus and plaster (Figure 1.3). In the Middle Kingdom, the entire body was sometimes covered with cartonnage as a final layer of the wrapping process. The face seems naturally to have been the most important part of the body for realistic treatment. The

technique varied from the plaster masks of the Old Kingdom to the flat, painted portraits of the Roman period (Figure 1.7) and includes such notable examples as the solid gold face mask of Tutankhamun. Like the coverings of the face, the style of wrapping in linen bandages varied. At its best-developed stage in the New Kingdom, every individual part, including each finger and toe, was wrapped separately. After this, each larger unit was covered, and finally the total mummy was enwrapped. Single sheets as long as 13 or 17 m have been recorded.

Changes in the technique of embalming were progressive and continuous, but not until the time of the New Kingdom can the complete sequence of steps in the process be detailed. Working on the basic studies carried out by Lucas, and with continued experimentation, Zaki Iskander (Mokhtar et al. 1973) outlined what he considered the complete method employed at its fully developed stage:

1. Putting the corpse on the operating table
2. Extraction of the brain
3. Extraction of the viscera
4. Sterilization of the body cavities and viscera
5. Embalming the viscera
6. Temporary stuffing of the thoracic and abdominal cavities
7. Dehydration of the body
8. Removal of the temporary stuffing material
9. Packing the body cavities with permanent stuffing material
10. Anointing the body
11. Packing the face openings
12. Smearing the skin with molten resin
13. Adorning and bandaging the mummy

The following is a commentary on the 13 steps.

1. The body of the deceased was taken to the place of mummification soon after death. The clothing was removed, and the body was placed on a work table

Figure 1.3.   *Body of an adult female, Twenty-first Dynasty, 1080–946 B.C. British Museum 48971. The well-preserved outer wrapping of this mummy is in the typical arrangement for its period. The large outer cloths have been tied in place with a simple but decorative strapping, which is also well preserved.*

for the succeeding processes. Embalmers' tables have been found and recognized as such. Little other comment is needed for what would have been the most obvious first stage of the working procedure.

2. The brain, one of the organs most subject to rapid putrefaction, was probably removed first; this is verified by Herodotus' account and seems practical. A passage was opened, through the nose usually, and the cerebral matter taken out with a hooked metal rod. Implements have been identified as possibly the types used for this operation. The brain was apparently not preserved. There is no mention on any container connected with the mummification process or on burial material of their use for the brain, but there are certainly many mummies in which it can be demonstrated that the brain is absent.

3. An incision was made in the abdomen, usually on the left side, but it may be found in other locations as well. The abdominal organs were extracted except for the kidneys, but this exception was not consistently observed; the kidneys are also missing sometimes. Next, the diaphragm was cut out and the contents of the thorax, except the heart and usually the aorta, were removed. This is the usual condition of well-prepared mummies. In mummification of the less expensive types, there is no incision and the removal of the organs is less consistent. The necessity for this stage in the preservation of the body needs little comment, for the extraction of the soft organs, particularly the intestines, would greatly aid the preservation of the body. The heart was usually left in place because it was considered the "seat of the mind." This may help to explain why the brain seems not to have been preserved. The heart, in

any case, is muscle and would not be as apt to decay as the organs that were removed.

4. After the removal of the internal organs, the thorax and the abdomen were cleansed, probably with palm wine, which would have had some sterilizing effect. Diodorus mentions the cleansing of the viscera with palm wine, and it is natural to assume that this would have also been done to the body cavities. The operation would have left no detectable trace, but it is a natural assumption that some sort of internal cleansing was carried out.

5. The viscera were separated, emptied, cleansed, and dried. They were then treated with molten resin, wrapped in separate linen packages, and placed in containers. The so-called canopic jars or chests took different forms at different times, as will be discussed elsewhere. This stage could have been carried out concurrently with the next.

6. After the cleaning mentioned in step 4, the thorax and abdomen were stuffed with temporary packing material to ensure the complete desiccation of the body. There is evidence from refuse material examined by Iskander that such temporary packing existed. Although this step has been doubted by some authorities, it would have been practical and useful, not only for the drying effect, but also for the maintenance of the shape of the corpse. Herodotus mentioned packing the cavities before the complete desiccation, but he did not specifically indicate that it was of a temporary nature.

7. The complete desiccation of the body could now be accomplished. Exactly how this was done has been the subject of considerable debate, but the general conclusion reached in modern scholarship indicates that the process

involved the use of dry natron. The body, probably on a slanting bed, was completely covered with natron. This had the effect of removing any remaining body liquid and consequently ensuring against any further putrefaction. The drying-out process lasted 40 days; the total embalming process, 70 days. There is a good deal of evidence to support this time scheme.

8. After the drying-out process, the temporary packing material was removed. It was not discarded, because it had come into contact with the body. Caches of embalmers' material have been found in sufficient quantity to make it clear that the temporary packing was buried near the tomb.

9. Resin or resin-impregnated cloth was put into the cranium. This is often evident in X rays of the skull, in which the resin can be seen as having reached its own level while liquid and then solidified. The body cavities were stuffed with linen cloth and bags of other materials including natron, sawdust, earth, and occasionally a few onions. The incision was closed with resin, wax, or linen and covered with a plate of metal or wax. The embalming incision is seldom found actually to have been stitched together.

10. The body was presumably anointed at this point with fragrant materials.

11. The orifices of the head were packed with wax or resin-soaked linen, and pads of linen were placed over the eyeballs.

12. Liquefied resin was smeared over the whole body. This acted as a preventative against the reentry of moisture and tended to strengthen the skin.

13. Amulets and other jewelry were placed on the mummy. The amount and quality of such materials depended on the wealth and position of the deceased. In a royal mummy, such as that of Tutankhamun, the quantity of objects was very large, including bracelets, rings, necklaces, pectorals, finger and toe coverings, as well as amuletic devices prepared especially for the burial. The mummy was then wrapped in linen bandages, sometimes with resin between the layers. The most complex wrapping began with the individual fingers and toes, proceeded to the limbs, and ultimately encased the total corpse in many layers of linen.

These 13 steps outline the most complete manner of preparing the body for the tomb in the New Kingdom. A mummy such as this would have stood a good chance of being completely preserved (Figures 1.4 through 1.7). There were, undoubtedly, many variations on this model program, but the principal operations of desiccation, washing with palm wine (which contained alcohol), and anointing with resin all resulted in the desired end of rendering the physical remains incorruptible.

One of the most complex arguments about the mummification process has centered around the chemical material used in the desiccation of the body. The principal ingredients that have been considered are salt and natron. To the earlier investigators, salt seemed the most likely desiccating agent employed, probably from actual experience with salted meat and fish. Considerable investigation and study have proved that salt appears only as an adulterant and was not the principal means of preserving the body. For a considerable time, natron was believed to have been used in a solution, mainly because of faulty translations of the description of the mummification process in Herodotus. Lucas and others have proved satisfactorily that the body was packed in dry natron, not soaked in a solution. Lucas made a series of experiments in which he treated the bodies of pigeons by four possible methods: salt in solution; dry salt packing; natron in solution; and dry natron packing. The conclusion he

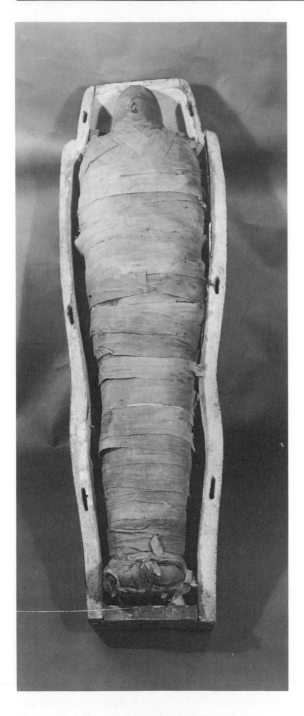

reached was that the mummification process depended for its success on the use of dry natron. Zaki Iskander treated a group of ducks using the principle suggested by Lucas. After 30 years, in 1973, he stated that they were "still in a very good state." Iskander had kept the mummified ducks in a laboratory under normal conditions of atmosphere and humidity. It can be generally assumed that the packing of the mummy in dry natron, a material abundant in ancient Egypt, was the simplest and most practical method of desiccation. The physical problem of obtaining containers of sufficient size to hold a human body should have long ago ruled out the possibility of a bath, whereas the dry method could have been employed by arranging the remains on a table or matting and simply heaping the natron around it.

Three other arguments for the use of a "pickling" solution have been propounded. The fact that mummies have been found with separate limbs that did not belong to them suggested that parts had become detached in a soaking process and had been incorrectly reassembled. In fact, it is more likely that such cases were the result of a later rewrapping of vandalized bodies. The purpose of finger and toe stalls was, at one time, explained as being to keep the fingernails and toenails in place during soaking. In actuality, these objects were probably more decorative than useful. The third argument for the use of a soaking bath was the apparent lack of epidermis in examples inspected; the theory was that the epidermis had soaked away. The lack of epidermis can usually be explained by its having come away with the wrapping or actually having been present but not recognized as such. The physical state of well-preserved mummies, taken together with accurate translations of the ancient texts, makes the use of a liquid bath and prolonged soak-

Figure 1.4. Body of an old man, Ptolemaic period, 332–30 B.C. British Museum 20650. The somewhat rough and wide outer wrapping strips suggest that some further decoration is missing, such as additional bandings, a shroud, or other

elements. In the catalog, Mummies and Human Remains, *of the British Museum there is mention of beads found in the wrapping; this may indicate that the missing element is a bead netting.*

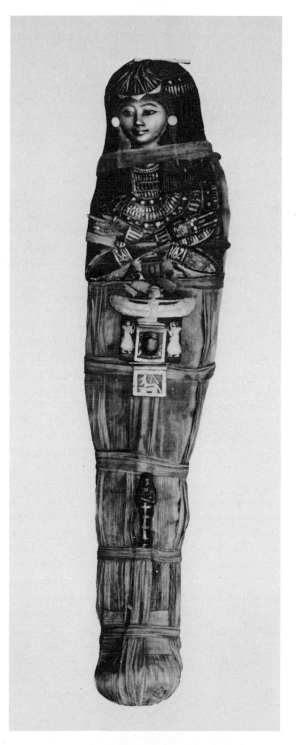

ing seem unlikely. The use of dry natron for complete packing of the body seems, at present, to offer the most understandable explanation of how desiccation was carried out.

From the earliest time for which there is evidence, the process of preservation was accompanied by a decided attempt to create a lifelike appearance of the mummy. In the Old Kingdom, this was accomplished mainly by wrapping with sufficient linen to restore the natural contours of the body. Special mention should be made of the practice followed in the Twenty-first Dynasty. Modern taxidermists would find the techniques of the embalmers in this period similar to their own work, for the basic steps outlined above as typical of the New Kingdom were augmented by a process designed to produce an even more natural effect. Stuffings of various materials, principally linen cloth, were inserted under the skin through incisions made for that purpose. The body cavity was filled and the arms and legs rounded out either from inside the trunk or through minor openings made in the limbs. The loss of body mass was made up wherever needed, and the face was stuffed from inside the mouth. Elliot Smith and Dawson suggest that this elaborate treatment of the body occurred at a time of, and may have been responsible for, less emphasis on substitute images of the deceased as alternate dwelling places for the spirit. It is true that during this period there are fewer examples of the *ka* statues made in the likeness of the dead. In any case, mummies from this time exhibit a definite attempt to restore an appearance of life and can be easily recognized as belonging to this period.

The evolution of the mummification process was accompanied by a development of the necessary funerary "furniture," such as containers for the body and the viscera. The early pit burial of the predynastic age required no container except for the simple wrapping

Figure 1.5.   Body of an old woman, Ptolemaic to Roman period, date uncertain. British Museum 6665. This example illustrates the use, even at a late date, of a modeled face mask on the mummy. The other objects that embellish the front of the body may not belong, and as a consequence, the actual date of the mummy is difficult to determine.

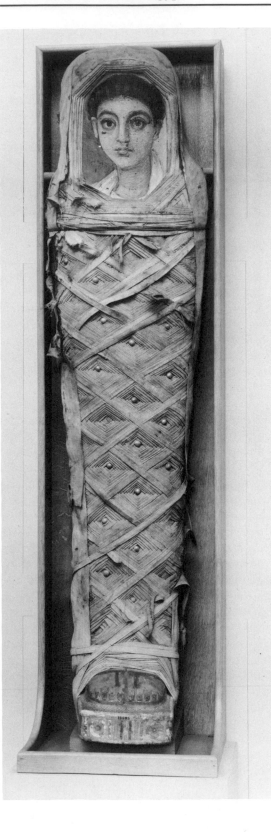

of the body in cloth, reed matting, or hide. Because the pit or hole was sometimes lined with brick, boards, or matting, it may be considered that this acted as a substitute for a portable container. In the early dynastic period, the notion of a miniature house for the deceased was suggested by the use of small paneled chests or boxes, suitable only for contracted burials. In form, these were imitative of lower Egyptian houses with paneled walls and arched roofs. As religious ritual developed, these small containers gave way to larger rectangular sarcophagi of wood or stone (limestone, granite, or alabaster). As the burial of the remains of Queen Hetepheres has already received some comment above, her sarcophagus can serve as an example of the beginning of Dynasty Four. Made of fine alabaster with a close-fitting lid, Hetepheres' sarcophagus was of a proportion that indicates it was made to accommodate an extended rather than a contracted body. When found by excavators in the twentieth century, it was, unfortunately, empty.

A rectangular container for the body was used throughout the Old Kingdom, but it is in the painted wooden coffins of the Middle Kingdom that the form reached its height as a decorated object. A typical coffin of the Eleventh and Twelfth dynasties was decorated with painted architectural motifs on the outside, which resemble the so-called palace facade design. The left, or east, side was usually decorated with some hieroglyphic texts and with a false door and a pair of eyes to allow the spirit the means by which he could communicate with the outside world. The interior of the coffin was decorated with paintings in registers or horizontal bands.

Figure 1.6.   Body of an adolescent boy, Roman period, probably first to second century A.D. British Museum 13595. The typical crossed outer wrapping of narrow strips with gilded metal buttons common on mummies of the Roman period is shown here. The use of a flat, painted portrait is a Greco-Roman contribution to the manner of decorating the mummy.

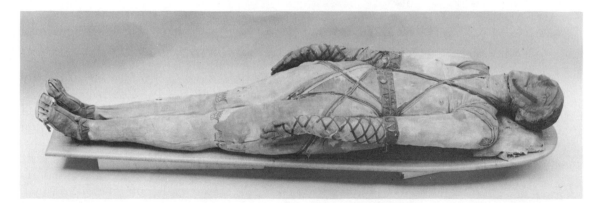

Figure 1.7.   Body of an adult male, Roman period. British Museum 6704. An unusual (for its time) treatment of the final wrapping is illustrated here. The limbs are wrapped separately, and the face has had the features painted, a technique going back to the Old Kingdom but no longer common in later Egyptian history.

The uppermost of these contained a carefully painted invocation on behalf of the occupant; the middle register, a series of depictions of funerary offerings and necessities for the next life; and the lower areas, a series of finely written prayers, spells, and amuletic sayings drawn from a larger body of religious literature known as the Coffin Texts.

The anthropoid coffin makes its first appearance in the Middle Kingdom. It is probable that the notion of designing the complete container for the body in such a way that it resembles a human figure developed directly from the Old Kingdom tradition of covering the face of the deceased with a lifelike mask. Although the long-hallowed use of a rectangular coffin persisted into the New Kingdom, particularly for royal burials, the body at that time was first enclosed in an anthropoid coffin or a series of nested coffins of such shape. The notion, probably, was that the coffin helped to ensure the preservation of the shape of the body, while the stone sarcophagus served as a house or a shrine within which to contain it. Throughout the history of the anthropoid coffin the face was nearly always modeled in relief (Figure 1.5). The hands, arms, and breasts also were sometimes treated as relief decoration. Wooden examples could be elaborately painted inside and out with a combination of religious texts and vignettes illustrating the funerary ritual, protective divinities, and the progress of the spirit in the next life. Wooden coffins could be items kept in stock, with the name and title of the individual added at the time of use.

The elaborate protection for the mummified body was extended to the internal organs that had been removed from it. They were considered with the same care because they were, in effect, part of the human remains and had to be treated with the same degree of respect. As said above, one of the earliest pieces of solid evidence for the removal of the viscera as a part of mummification was the canopic box of Hetepheres. The alabaster box was divided into four compartments, was sealed with a tight-fitting lid, and still contained the embalmed viscera when found.

The division of the viscera, as it was later standardized, was into four parts: the liver, the stomach, the lungs, and the intestines. At most times in the history of mummification, four containers, or symbolic substitutes for them, were provided. These containers are usually called canopic jars because their form, with the lid or stopper in the shape of a head, was thought to resemble the burial of Canopus, the priest of Menelaus, who was revered in the form of a bulging jar with a human head.

During the Old Kingdom, the canopic jar was typically a rough-hewn, slightly bulging limestone jar with a convex lid. There is preserved in the Metropolitan Museum a set of jars of this description found in the burial chamber of the tomb of Pery-neb of the late Fourth Dynasty. Although these four containers are identifiable as to their intended purpose, they were clean inside and were never used. By the beginning of the Middle Kingdom, the identification of the viscera as part of the body was reinforced by the shape of the jar lid or stopper, which was modeled in the shape of a human head. That the four jars had human heads as late as the end of Dynasty Eighteen is attested to by the four stoppers from the canopic chest of Tutankhamun. As a complete example of the treatment of the viscera in a royal burial, the canopic chest of this king should be described. Inside a gilt wooden chest was found an alabaster chest of similar shape. This was divided into four interior compartments, each of which was plugged with a stopper fashioned in a likeness of the king's head. In the four compartments were four miniature inlaid gold coffins that actually contained the visceral packets.

In the Eighteenth Dynasty, a second treatment of the four canopic lids developed. One continued to be made in the form of a human head, but the other three were fashioned as the heads of a jackal, a baboon, and a falcon. The four heads identified the viscera as being protected by four genii (called the four sons of Horus). When it became customary to package the organs that had been removed and return them to the body cavities, the tradition of the canopic containers was continued, but the jars were false or imitation, often solid, the body and lid carved from the same piece of stone.

The materials from which the canopic containers were made were extremely varied. Clay, limestone, alabaster, wood, faience, and cartonnage were all employed. The canopic jar could be a carefully designed work of art or a crude, roughly fashioned receptacle. As is the case with all aspects of the preparation for burial, the social status and ability to pay of the deceased or his family were the key factors in the choice of the methods and materials employed. This has resulted in a wide variety of tomb goods and, naturally, the method of embalming and wrapping during any one period. What had begun as a necessary precaution during the early dynastic times for the protection and preservation of the king's body had gradually become available to any who could pay the price. The number of objects – coffins, papyrus scrolls of the *Book of the Dead*, heart amulets – that have been found with the name of the deceased left blank attest to the common practice of producing standard funerary objects to be sold at a price. The place in the text appropriate for the name was intended to be filled in at the time of purchase. In some instances, this was not done, for what reason we shall probably never know.

The religious basis of the mummification process was rooted in the necessity of preserving the physical remains as a resting place for the spirit. What had been accomplished accidentally in the predynastic period was done by a gradually developed process over the centuries. The techniques and procedures, which became more and more complicated through the course of Egyptian history, served this single purpose: the preservation of the human form, and particularly the features of the face, from decay. With the embalming process there grew up attendant aids for the protection of the spirit. The ritual decoration of the mummy, coffin, and sarcophagus, and the amuletic devices placed on and around the body served this end. The mummy has become a symbolic touchstone that conjures up the mysteries of ancient Egypt for modern man. What the physical remains of the ancient Egyptians can tell us through scientific techniques is only now becoming evident.

REFERENCES

Diodorus Siculus. 1935. *History*. Translated by
    C. H. Oldfather. Cambridge, Mass.: Harvard
    University Press.

Elliot Smith, G. and Dawson, W. R. 1924. *Egyptian mummies*. New York: Dial Press.

Goyon, J. C. 1972. *Rituels funéraires de l'ancienne Egypte*. Paris: Cerf.

Harris, J. E., and Weeks, K. R. 1973. *X-raying the pharaohs*. New York: Scribner

Herodotus. *History*. Translated by G. Rawlinson. 1910. London: Dent.

Lucas, A. 1962. *Ancient Egyptian materials and industries*, 4th ed., revised and enlarged by J. R. Harris. London: Edward Arnold.

Mokhtar, G.; Riad, H.; and Iskander, Z. 1973. *Mummification in ancient Egypt*. Cairo: Cairo Museum.

Petrie, W. M. F. 1892. *Medum*. London: David Nutt.

Pettigrew, T. J. 1834. *A history of Egyptian mummies*. London: Longmans.

Reisner, G. A. 1927–32. Articles on Queen Hetepheres. *Bulletin of the Museum of Fine Arts* (Boston). 25 (1927), 26 (1928), 27 (1929), 30 (1932).

2

Diseases in ancient Egypt

A. T. SANDISON
Department of Pathology
Western Infirmary
Glasgow, Scotland

All men and women share certain experiences. All are born (and during birth may injure their mothers); all suffer illnesses during their lives; and all must sooner or later die, whether from disease, degenerative process, accident, or violence. The historian's overall view of ancient peoples is incomplete if he fails to take into account these phenomena of states of health or disease.

The major lines of study of ancient diseases comprise examination of literary sources by scholars in collaboration with physicians, study of artistic representations in sculpture and painting, and study of skeletal remains and mummies by macroscopic examination, supplemented by radiography and by histological examination using light, polarizing, and electron microscopes.

The major literary sources for our knowledge of disease processes in Egypt are the Ebers, Edwin Smith, and Kahun papyri (Dawson 1953). The first deals with medical diseases and includes, among many others, descriptions of urinary disorders, virtually certainly including schistosomiasis and parasitic gut infestations. The second is surgical and contains accurate prognostic comments on traumatic and certain inflammatory diseases. The third concerns obstetrical and gynecological disorders. Precise diagnoses are, in many instances, difficult to make from the symptoms listed. Nevertheless, the papyri will continue to engage scholars and medical historians for many years. From these, as well as other literary sources, specialist scholars have adduced evidence of trachoma, seasonal ophthalmia, skin diseases, hernia, hemorrhoids, and so on.

Studies of artistic representations have yielded clear evidence of achondroplastic dwarfism and highly probable diagnoses of such states as bilateral dislocation of the hips, postpoliomyelitic limb atrophy, and pituitary disorder (Aldred and Sandison 1962; Wells 1964). Macroscopic examination of ancient Egyptian skeletons has revealed a wealth of pathological changes. These have been reviewed by Brothwell and Powers (1968) and Sandison (1968) with regard, respectively, to congenital abnormalities and acquired disease. Among the congenital lesions are included achondroplasia, acrocephaly, talipes equinovarus, Klippel-Feil syndrome, hipjoint dysplasia, hydrocephaly, and cleft palate. Bilateral parietal bone thinning has also been observed and must be differentiated from osteoporosis symmetrica. This has been noted in the mummies of Meritamen and Tuthmosis III as well as in Khety of the Twelfth Dynasty.

Acquired diseases include such common conditions as osteoarthritis, nonspecific osteitis, middle ear infection, and rarer lesions such as tuberculous and leprous osteitis, osteoma, osteochondroma, possible chondrosarcoma of the pelvis, cranial changes attributable to meningioma and possible nasopharyngeal cancer, gout, and osteoporosis. Satinoff (1968) described examples of osteoarthritis, otorhinolaryngologic disease, osteoma, possible metastatic tumors of the skull, and most interestingly, the recently recognized basal cell nevus syndrome in the Egyptian skeletal collection at Torino. Salib (1967) has considered fractures in ancient Egyptians.

In the macroscopic examination of bones, artifacts are troublesome in producing pseudopathological changes; these are fully discussed by Wells (1967), and I have summarized them elsewhere (Sandison 1968). Such pseudopathological changes may be caused by depredation by insects and rodents, effects of plant roots, high winds, pressure of overlying soil or matrix, and impregnation by chemical substances.

The study of actual mummies by dissection (Ruffer 1921; Elliot Smith and Dawson 1924), band-saw cutting and slab radiography (Sandison 1968), and histological examination (Ruffer 1921; Sandison 1955, 1970) has yielded much interesting information. If dissection is not possible macroscopic examination of mummies may be supplemented by radiography. Gray has carried out extensive surveys of mummies in several European museums and has discovered evidence of arthritis, arterial calcification, cholelithiasis, and possible bone infarction. It is now clear that artifacts must be carefully excluded; earlier radiographic diagnoses of alkaptonuric arthropathy are now known to have been erroneous (Gray 1967). However, Lee and Stenn (1978) have shown a homogentisic acid polymer in material from an Egyptian mummy of 1500 B.C.; this appears to be a proved case of ochronosis. In spite of this, the majority opinion today is against the presence of ochronosis and in favor of postmortem artifact.

If tibial bones are subjected to X-ray examination, transverse (so-called Harris's) lines may be seen. They have been thought to indicate episodes of intermittent disease or malnutrition. There is no way of telling what condition was causal, and lines may become absorbed. Nevertheless, the fact that about 30 percent of Egyptian mummies show Harris's lines suggests a generally poor state of health in childhood and adolescence in ancient Egypt.

Organisms are seen in abundance in many microscopic sections of mummy tissues, but they are putrefactive in type and have multiplied in the tissues during the period between death and effective dehydration of the corpse. That putrefaction was a problem in Egyptian mummification is indicated by the fact that "stench" is implied in some of the names for the place of embalming.

After this preamble, we shall look at the evidence of disease in ancient Egypt.

INFECTIVE DISEASES

Ruffer's (1910b) identification of calcified ova of Schistosoma (Bilharzia) haematobium in the kidneys of two mummies of the Twentieth Dynasty is unquestioned. Hematuria was probably common in ancient Egypt. Schistosomiasis has certainly been very common in Egypt in the twentieth century. Ferguson (1910) reported that 40 percent of over 1,000 Egyptian males between the ages of 5 and 60 years who came to necropsy at the Kasr Aini Hospital in Cairo showed evidence of schistosomiasis. Larrey (1812–1817) reported frequent examples of hematuria, presumably of bilharzial origin, in French troops during the campaign of Napoleon Bonaparte in Egypt in 1799–1801. The parasite is small and was not recognized until 1851 by Bilharz, so it is unlikely that the ancient Egyptians ever identified it. Recently, in Toronto, mummy ROM I has been shown to harbor not only a tapeworm (Taenia) with eggs but also ova of Schistosoma haematobium and to display changes in the liver that may have resulted from the schistosomal infestation. It may be useful to mention also at this point that Ascaris eggs have been identified in mummy PUM II. We thus have positive identification of Schistosoma haematobium, Ascaris, and Taenia in Egyptian mummies. Ruffer tentatively diagnosed malaria in some Coptic bodies with splenomegaly, but this is not good evidence in a warm country.

The case of leprosy in a Coptic Christian body discovered by Elliot Smith and Derry (1910) at El Bigha in Nubia and redescribed by Elliot Smith and Dawson (1924) is unquestioned. This case was reviewed macroscopi-

cally by Rowling (1960) and macroscopically and radiologically by Møller-Christensen (1967); both accepted the diagnosis. I have examined skin, subcutaneous tissue, and nervous tissue from this specimen. Using modifications of the Ziehl-Neelsen method, it was impossible to demonstrate acid-fast bacilli. This is not surprising, as it is known that, even in paraffin block, tissues containing numerous *Mycobacterium leprae* may, after a few years, appear to be free of bacilli.

Møller-Christensen (1967) has also described a female skull from the same cemetery as the El Bigha body that shows the facies leprosa. So far these are the only acceptable cases of leprosy found in all the thousands of carefully examined mummies, dried bodies, and skeletons from ancient Egypt and Nubia. Leprosy is thus unlikely to have been a common disease.

If we turn now to the other important mycobacterial disease – tuberculosis – we find that Morse (1967) and Morse et al. (1964) have discussed the evidence for its presence in ancient Egypt. Their conclusions, which were derived from a critical survey of all the available evidence, both artistic and pathological, are completely acceptable. The famous mummy of the priest Nesperehan of the Twenty-first Dynasty has been unreservedly accepted as a case of typical spinal tuberculosis with characteristic psoas abscess. Morse et al. conclude that in all there are 31 acceptable cases of skeletal and mummy tuberculosis. Of these, 16 were culled from the literature and 15 were reported for the first time. The dates for these 31 cases are not all certain, nor is the provenance of the mummies absolutely clear, but probably the dates range from 3,700 to 1,000 B.C. Morse et al. accept that the fibrous adhesions and collapsed left lung found in a Byzantine Nubian female body from the Island of Hesa (Elliot Smith and Wood Jones 1908) are good evidence of pulmonary tuberculosis. This diagnosis is perfectly tenable, but other conditions might lead to similar appearances. I believe that these 31 acceptable cases probably represent only a part of those found in Egyptian sites. Doubtless many have been discarded unrecognized and others may languish in museums unrecognized and undescribed. Probably some form of granulomatous tuberculosis has infected man since Neolithic times and may have resulted from closer contact with livestock, presumably bovine, following the Neolithic Revolution. The contemporary domestic dog may suffer from pulmonary or, less commonly, renal tuberculosis, but this is attributable to a human form of bacillus and is probably derived from contact with the owner. The dog is therefore *not* likely to have been the source of origin of Neolithic human tuberculosis.

Another important granulomatous disease is syphilis caused by *Treponema pallidum*. Other treponemal diseases include yaws, pinta, and bejel, and some workers have concluded that all these disease processes may be caused by variants of one organism. This may explain the present obscurity of the origin of syphilis, and it seems to me there is no likelihood of a convincing early resolution of the problem. There is, however, no clear evidence of syphilis in the vast amount of material examined in Egypt and Nubia from the ancient period, although venereal syphilis is not uncommon in Egypt at the present time.

The interpretation of changes in the temporal bone of the skull is difficult. Although the ancient medical papyri give prescriptions for what must be middle ear infection, there is no literary evidence of surgical intervention. An early dynastic skull from Tarkhan shows the appearance of mastoiditis, but a fairly well healed trephined opening is situated some distance directly above the mastoid region and may have been of magicomedical significance. Some authors have claimed that mastoid disease was very common in ancient Egypt and Nubia, but only 6 acceptable cases appear to have been noted in a total of at least 10,000 individual remains examined. Apart from the Tarkhan skull, there is a predynastic case of probable acute mastoiditis, another case of predynastic or early dynastic date,

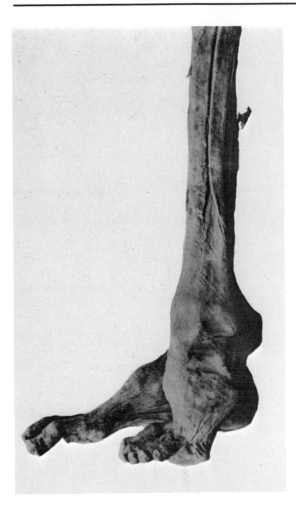

Figure 2.1.   Probable clubfoot of Pharaoh Siptah. (Courtesy of Service of Antiquities)

another in a New Kingdom female, another in a Coptic female, a further Nubian case dated to the first millennium A.D., and one of Meroitic date. Schultz (1939) reported 3 cases of maxillary sinus infection in 35 adult chimpanzees. Maxillary sinus infection is common in contemporary man, and it is probable that cases must have occurred in ancient Egypt.

Polymyelitis is a virus affection of the anterior horn cells of the spinal cord, and its presence can be deduced only by a deformity in persons who have survived the acute stage. Mitchell (1900) noted in an early Egyptian body from Deshasheh shortening of the left leg, which he interpreted as evidence of poliomyelitis. The clubfoot of the Pharaoh Siptah (Elliot Smith 1912) (Figure 2.1) is more probably attributable to congenital abnormality than to poliomyelitis, so also is the deformity of Khnumu-Nekht of the Twelfth Dynasty (Cameron 1910). Some authorities, however, believe that the deformities in the last two individuals may be postmortem artifacts.

I think it possible that smallpox may have existed in ancient Egypt and that unrecorded epidemics occurred. The changes described by Ruffer and Ferguson (1911) in the skin of a male mummy may well have been those of variola, despite criticism of this diagnosis by Unna. So also the Pharaoh Rameses V (Figure 2.2) may have been the victim of lethal smallpox (Elliot Smith 1912). It would be of interest to submit to electron microscopy material from these and any other possible cases that come to light in the future. Other examples of skin conditions are hyperkeratosis of the skin and senile acne (Figure 2.3).

I have discussed elsewhere (Sandison 1967c) the evidence for infective ocular disease in ancient Egypt. Seasonal ophthalmia may have occurred. The evidence for ocular disease is, however, purely literary, and no paleopathological evidence is available. Blindness certainly is portrayed in ancient Egyptian art.

Rowling (1967) has surveyed the evidence for respiratory disease and accepts Ruffer's (1910a) cases of pneumonia (two of which are Twentieth Dynasty and one Greco-Egyptian), as well as that of Shaw (1938). However, I believe these should be considered with some reservation because of the possibility of artifacts of fungal nature that may simulate a cellular reaction. Rowling believes from literary evidence that bronchiectasis and pulmonary tuberculosis may have occurred; both are possible, but not certain.

Gallstones were illustrated in a Twenty-first-Dynasty mummy by Elliot Smith and Dawson (1924), and Gray has noted

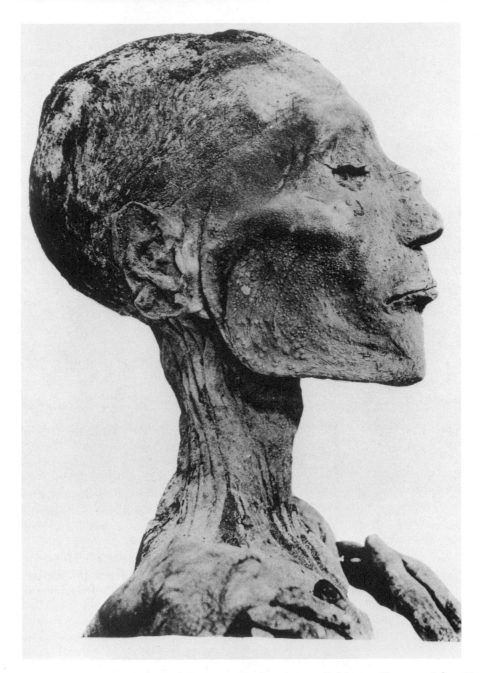

Figure 2.2.   *Head of Pharaoh Rameses V, showing probable smallpox vesicles. (Courtesy of Service of Antiquities)*

gallstones in radiography of a female mummy. However, cholelithiasis may well be the result of metabolic abnormality or stagnation of bile, and chronic cholecystitis need not necessarily be associated with stone formation, although it sometimes is. Renal stones also have been rarely discovered in association with ancient remains in Egypt and

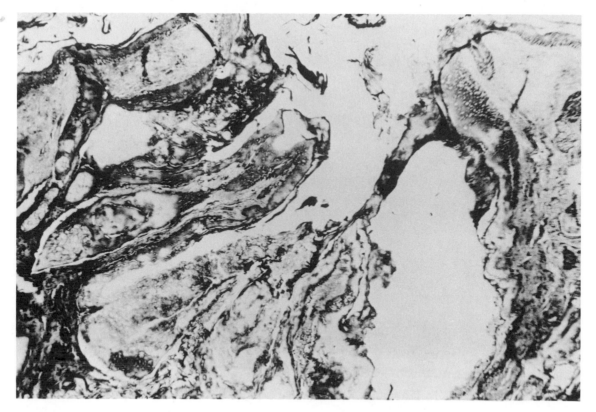

Figure 2.3.   Senile acne of the face in an old New Kingdom mummy. Phosphotungstic acid–hematoxylin.

Nubia. Renal lithiasis is not necessarily associated with infection, although this may occasionally be the case, either as a primary or as a secondary phenomenon.

More convincing as evidence of abdominal inflammation are the appendicular adhesions noted in a Byzantine body from Hesa by Elliot Smith and Wood Jones (1910). These probably followed resolving acute appendicitis, although the topography of the adhesion, which crosses the pelvis, is somewhat unusual.

BONE DISORDERS

Nonspecific bone inflammation is not rare in human remains from older societies. The evidence for this is overwhelming, and the phenomenon is of interest in view of the opinions formerly held by some morbid anatomists that the inflammatory reaction may have been a late evolutionary acquisition. Similar inflammatory processes also are seen in fossil animals as far back as the Mesozoic period, but their causes are likely to remain obscure. It has been suggested that many specific infections were probably rare or absent before the gregariousness permitted by the Neolithic Revolution. Nonspecific infections might, however, have been possible without close personal contiguity and might have been caused by a wide range of organisms. Such nonspecific changes are of significant frequency in early cemeteries, but are difficult to interpret.

It is customary to divide bone inflammation into periostitis, osteitis, and osteomyelitis, depending on whether perios-

teum, bone proper, or bone marrow is most obviously involved. This is to some extent artificial, because bone is a biological unit and not a series of distinct tissue entities. These phenomena may lead one to another; all three may be present together. When confronted with evidence of periosteal reaction in an ancient bone, it is difficult to say with any confidence whether the cause was infection or trauma. It must, however, be conceded that site is important, particularly the tibia, which has a large subcutaneous area. Even here we are on uncertain ground; the subcutaneous tibia is vulnerable to trauma, but also to extension of infection from the skin to the periosteum. Wells (1964) noted that one in six of a group of Saxon agriculturists had tibial periosteal reactions. Compared with Anglo-Saxons, ancient Egyptians show a lower fracture rate for the leg, and yet periosteal reaction is as common in ancient Egyptians. Wells (personal communication, 1970) found that among 92 ancient Egyptian tibiae (dating from predynastic to Coptic times) 14 (15.2 percent) had "well-marked" periostitic changes. Wells, therefore, suspects that something other than trauma may be operating, and suggests that in ancient Egypt infection of insect bites or simple abrasions may have been causal.

There are no recent incidence figures for ancient Egypt other than those of Wells, but there are isolated reports. Derry (1940–1941) noted only dental infection and osteoarthritis in the bones of Pharaoh Psusennes I of the Twenty-first Dynasty; Derry (1942) only arthritis in Pharaoh Amenenopet and no abnormality in Har-Nakht; Derry (1947) no disease in Prince Ptah-Shepses of the Fifth and Sixth dynasties; Batrawi (1947) some evidence of dental and maxillary sinus infection in a middle-aged male and possible postfracture sepsis of the radius in a middle-aged female from Shawaf, but no disease in Pharaoh Djed-Ka-Re of the Fifth Dynasty; Batrawi (1948) possible dental and air sinus infection of Akhet-Hetep, but only arthritis in

his wife; and Batrawi (1951) no evidence of infection in remains from the Northern Pyramid of Sneferu of the Fourth Dynasty. Apart from such isolated reports, we must fall back upon the *Archaeological Survey of Nubia* (Elliot Smith and Wood Jones 1910).

Among this enormous mass of material, although occasional bodies gave clear evidence of the cause of death, a very large proportion did not. Elliot Smith and Wood Jones (neither of whom was a pathologist) were therefore left in ignorance of the factors causing death in the vast majority of persons buried in ancient Nubia. They reached some important conclusions, in that they believed that examination of 6,000 bodies revealed no evidence of tuberculosis, syphilis, or rickets. As we have seen, however, this view had to be revised with regard to tuberculosis. They also believed correctly that malignant disease must have been exceptionally rare. Where the actual cause of death was clear to them, the death had usually resulted from violence. Although some remains showed evidence of disease, it was impossible even to guess at the precise cause of death. Elliot Smith and Wood Jones concluded that inflammatory diseases of bone were rarely seen in ancient Egyptian skeletons. Even when fractures had been severe and necessarily compound, sepsis rarely seems to have followed. Well-healed fractures are also commonly seen in wild apes. They postulate, in contrast to contemporary experience in surgical practice, that there must have been a remarkable resistance to infection in ancient Nubia.

An alternative explanation might, of course, be that organisms were of lower virulence. Elliot Smith and Wood Jones (1910) found that neglected dental disease accounted for practically all septic conditions of the facial bones, but in one female skull there were traces of chronic inflammation around the margin of the nares and destruction of the turbinates and part of the nasal septum. There was a direct communication through the hard palate posteriorly be-

tween nasal and oral cavities. The remainder of the skeleton was unexceptional, and a diagnosis of long-standing nasal infection was made. The cause of this must remain obscure.

Elliot Smith and Wood Jones (1910) also illustrated two examples of cranial disease, one in the parietal bone and the other in the frontal bone; these showed peripheral reaction and central necrosis, which were attributed to extension of scalp infection. The appearances are certainly not those of syphilis. An isolated humerus of a child of New Kingdom date showed necrosis of the lower half of the shaft, with a line of heaped-up reactive bone running down the middle of the anterior surface and a more superficial reaction over the lateral area of this line. The internal bone was necrotic and constituted a sequestrum. This appears to have been severe osteomyelitis. Sepsis following fracture appears to have been rare, even in apparently compound fractures. This is especially notable in skull wounds in which the scalp must surely have been severely affected. Nevertheless, in two New Kingdom scapulae there was evidence of reaction in transverse fractures, where there must have been severe soft tissue injury, but death occurred before healing took place.

Of 65 fractures of the upper limb, only one showed evidence of sepsis. This was in a woman of the Christian period, both of whose forearms were fractured; the right ulna had united, but the left showed periostitis. Another skeleton of this period from Hesa showed inflammation and necrosis following fracture of the clavicles, and there were two other ununited clavicular fractures. Of 38 fractures of the lower limb, only 2 showed septic changes. One early dynastic male with a right-sided fracture of tibia and fibula showed malunion and much inflammatory bone extending as periostitis over the bone shafts below the area of injury. The right femur of a male of the same period had an inflammatory reaction around the fractured lower portion.

Peter Ker Gray (1967) has discovered what appears almost certainly to be an old infarct of bone in a mummy from the Horniman Museum, and Campbell Golding (1960) a further example in a Ptolemaic female mummy from the same museum. Caution is necessary in accepting these findings. Very similar radiographs from clinical cases have been published and in these the lesions have been attributed to arteriosclerosis. I have shown that Ruffer's conclusion that arterial disease occurred in ancient Egypt is tenable and have published photographs of arteriosclerotic change (Sandison 1962); thus, a possible etiological factor is not lacking.

Healed fractures are not rare, and splints have been described. The mummy of Pharaoh Seknenre of the Seventeenth Dynasty shows that he was attacked by at least two men armed with ax and spear, possibly while asleep, and suffered severe wounds (Elliot Smith 1912). Wood Jones described judicial hanging and decapitation in Roman Egyptian skeletons.

One of the most common diseases found in ancient bones is arthritis. There are numerous accounts from many prehistoric and historical periods by many authors. Arthritis in animal fossils is discussed at length by Moodie (1923). Ankylosed vertebrae have been noted in animal remains over a wide span of time, from fossil reptiles to cave bears and even to domesticated animals from ancient Egypt. Arthritic changes have also been noted in human remains during the whole of Egyptian history from predynastic to Coptic times (Elliot Smith and Wood Jones 1908, 1910; Ruffer 1912; Bourke 1967). Zorab (1961) examined by radiography eight British Museum bodies of the New Kingdom and Roman period from Egypt. No cases of ankylosing spondylitis were found, but osteoarthritic lipping of the vertebrae was noted in four instances.

Rowling (1960) drew attention to specimen number 178A in the Nubian Collection. Hyperplastic new bone on the femora represented the adductor muscles and was more

marked on the right side. Some new bone was also formed on the pelvis. There was ossification of the interosseous ligament between radius and ulna. The humeri were normal. There was a small spina bifida with a defect in the first and second sacral vertebrae. The appearances are not those of osteomyelitis or myositis ossificans. Rowling interprets these strange findings as ossification attributable to partial paraplegia consequent on the spina bifida. The forearm changes are, however, difficult to explain, and Rowling suggests that the patient supported himself on his hands. Brothwell believes this is an example of osteogenesis imperfecta and points out the similarity to cases illustrated by Fairbank (1951).

A metabolic disorder that produces recognizable changes is gout. There is a classic case reported by Elliot Smith and Dawson (1924): an old Coptic male from Philae on the Nile. The radiographs are characteristic, and analysis of tophi by W. A. Schmidt showed uric acid and urates. Rowling (1960) reviewed this specimen and entirely agreed with the diagnosis. Because of the intense sunlight, rickets is never seen in Egyptian material.

VASCULAR DISEASE

With regard to vascular disease, we are on firm ground and have direct evidence. Blood vessels are often well preserved in Egyptian mummies and dried bodies. Czermack (1852) described aortic calcification, and Shattock (1909) made sections of the calcified aorta of Pharaoh Merneptah. Elliot Smith (1912) noted this change in his macroscopic description of the royal body, and also described calcification of the temporal arteries in Rameses II. Ruffer (1910a, 1912) described histological changes in Egyptian mummy vessels from the New Kingdom to the Coptic period. Long (1931), examining the mummy of Lady Teye of the Twenty-first Dynasty described degenerative disease of the aorta and coronary arteries, with arteriosclerosis of the kidney

and myocardial fibrosis. Moodie (1923) described radiological evidence of calcification of superficial vessels in a predynastic body. It is often difficult to assess the older descriptions that are unaccompanied by photographs; Sandison (1962, 1967a) examined and photographed mummy arteries (Figures 2.4 to 2.6) using modern histological methods. Arteries were tapelike in mummy tissues, but could readily be dissected. Arteriosclerosis, atheroma with lipid depositions, reduplications of the internal elastic lamina, and medial calcification could readily be seen. Atheromatous lesions in mummy arteries tend to form sectoral clefts: This should not be interpreted as dissecting aneurysm. It is evident that the stresses of highly civilized life are not, at any rate, the sole causes of degenerative vascular disease.

TUMORS

With regard to tumors there is a marked paucity of evidence, possibly because the expectation of life in earlier times was short. Moreover, most examples are to be found in the skeleton, with a few exceptions. Elliot Smith and Dawson (1924), for example, suggested carcinoma of the ethmoid and of the rectum as being causal in the production of erosion of the skull base and sacrum in two Byzantine bodies; this is slender evidence, but may be correct. Other cases of primary carcinoma are evidenced by destructive changes in bone. These include Derry's case (1909) of probable nasal carcinoma in a pre-Christian Nubian. Elliot Smith and Derry's case (1910) of sacral erosion in a Nubian male may have been attributable to rectal cancer or chordoma. In all these, a prima facie case can certainly be made.

Evidence of primary malignant tumors in ancient bones is rare. This is not entirely surprising, because even at the present time primary malignant tumors of bone are not common; deaths from them constitute less than 1 percent of all deaths caused by malignant disease. On the other hand, many carcinomas

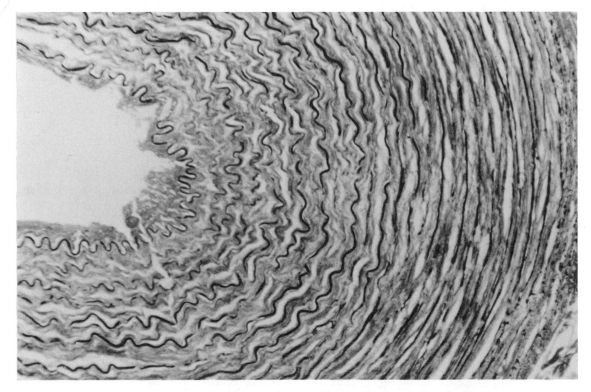

Figure 2.4.   Carotid artery of a male mummy showing fibrosis. Verhoeff elastic–van Gieson.

metastasize to bone in the terminal phase of the disease, and secondary cancer in bone is common in the necropsy room. Osteocartilaginous exostosis or osteochondroma is not uncommon in clinical practice, and cases are known in ancient material. Perhaps the best known is the celebrated specimen of the Fifth-Dynasty ancient Egyptian femur illustrated by Elliot Smith and Dawson (1924) and wrongly diagnosed as osteosarcoma. The contour, absence of periosteal reaction at the base, and lack of spiculation suggest that it is osteochondromatous and quite simple. It is not possible to comment on the two further alleged examples from Fifth-Dynasty graves (Elliot Smith and Dawson 1924), as these are neither described nor illustrated. A Roman pelvic tumor from Alexandria, described by Ruffer and Willmore (1914), was of large size and thought by them to be an osteosarcoma. This must remain in doubt. Rowling (1960) thought that osteosarcoma was possible but not certain. Brothwell (1967) considers the

tumor may have been chondromatous. Certainly cartilage-forming tumors do occur in the pelvis, but are often malignant.

Intracranial meningioma may induce hyperostotic change in the cranium. This has long been known to radiologists. Such a reaction was postulated by Lambert Rogers (1949) in two Egyptian skulls of the First and Twentieth dynasties.

Possibly the most convincing evidence of neoplasm of soft tissues came from Granville, who diagnosed (without histological confirmation) cystadenoma of the ovary, possibly malignant, in a mummy now known to be Ptolemaic. The writer has noticed a small squamous papilloma of the skin in a mummy (Figure 2.7). There are no examples of breast cancer.

KIDNEY DISEASE

Kidney lesions noted by Ruffer (1910a) included unilateral hypoplasia of the kidney; in

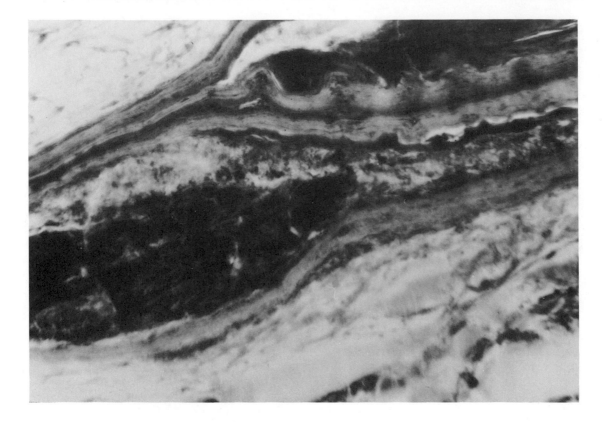

Figure 2.5.   Frozen section of tibial artery of an elderly female mummy showing lipid in an atheromatous plaque. Sudan black.

another Eighteenth- to Twentieth-Dynasty mummy, the kidney showed multiple abscesses with gram-negative bacilli resembling coliforms. Long (1931) described arteriosclerosis in the kidneys of Lady Teye of the Twenty-first Dynasty. Shattock (1909) described and analyzed renal calculi from a Second-Dynasty tomb; oxalates and conidia were noted. A vesical calculus found in the nostril of a Twenty-first-Dynasty priest of Amun contained uric acid covered by phosphates. Ruffer (1910a) described three mixed phosphate–uric acid calculi from a predynastic skeleton.

DISORDERS OF THE ALIMENTARY TRACT

Elliot Smith and Dawson (1924) also refer to the finding of multiple stones in the thin-walled gallbladder of a Twenty-first-Dynasty priestess. Shaw (1938) noted in the canopic preserved gallbladder of an Eighteenth-Dynasty singer that spaces resembling Aschoff-Rokitansky sinuses were present: This suggests chronic cholecystitis. Ruffer (1910a) mentions fibrosis of the liver in a mummy and equates this with cirrhosis, but insufficient evidence is given to evaluate this diagnosis. Little has been written about alimentary disease in mummies. Elliot Smith and Wood Jones (1908) report appendicular adhesions in a Byzantine period Nubian body: These are almost certainly the result of appendicitis. Ruffer (1910a) describes what may well be megacolon in a child of the Roman period and prolapse of the rectum in Coptic bodies. Elliot Smith (1912) mentions two probable cases of scrotal hernia – Rameses V shows a bulky scrotum now empty

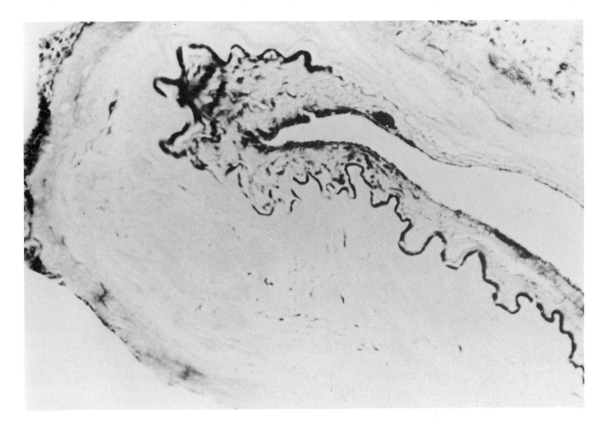

Figure 2.6.   Duplication of internal elastic lamina of tibial artery of an elderly female mummy. Heidenhain's iron hematoxylin.

after evisceration, and the scrotum of Merneptah was excised after death by the embalmers, possibly because of the bulk of a hernia. Anorectal problems seem to have been common, as some of the royal physicians were regarded as shepherds of the royal anus.

RESPIRATORY DISORDERS

Some interesting studies of the lung have been published. Anthracosis in Egyptian mummy lungs was described by Ruffer and by Long (1931). Shaw (1938) reported anthracosis in the lungs of Har-mose of the Eighteenth Dynasty, but Har-mose had also suffered from emphysema and lower lobe bronchopneumonia. Ruffer (1910a) reported pleural adhesions and diagnosed pneumonia in two mummies, one Twentieth Dynasty and

the other Ptolemaic; the latter may have been pneumonic plague, although the evidence is far from complete. Long (1931) reported caseous areas in the lung of a Twenty-first-Dynasty lady. As already indicated, these diagnoses must be accepted with some reserve in view of the possible confusion of molds as leukocytes. With regard to anthracosis, this seems to enhance lung preservation. More recently, silicotic lesions, possibly attributable to inhaled sand, have been noted in mummy lungs.

ACROMEGALY

A fairly rare disease that should be readily recognized is acromegaly. This produces characteristic bone changes. Brothwell (1963) illustrates an ancient Egyptian skull

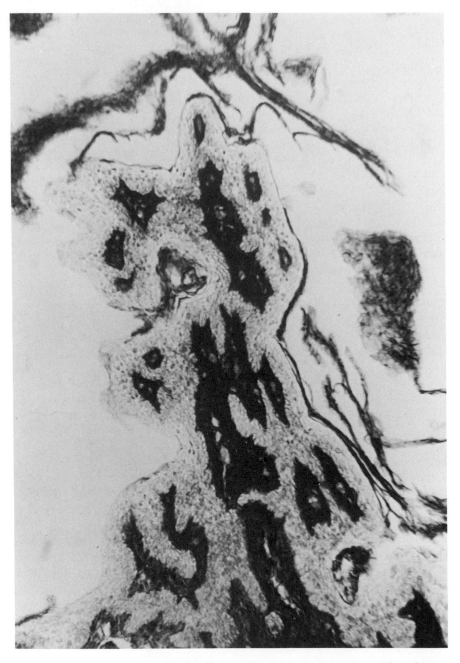

Figure 2.7.   Small squamous papilloma of the skin in a mummy. Phosphotungstic acid–hematoxylin.

that may represent this disease. Aldred and Sandison (1962) gave reasons for their belief that Akhenaten suffered from endocrine disorders. Statues and reliefs show ac- romegaloid facies and eunuchoid obesity. Cameron (1910) describes the bones of two brothers from the Middle Kingdom of ancient Egypt and concludes that the skeleton of

Nekht-Ankh shows eunuchoid changes but also a curious penile appearance suggesting a subincisional operation. It is not certain, however, if this is a genuine lesion. Male Egyptian mummies show circumcision throughout the dynasties until the practice was abandoned in the Christian period.

GYNECOLOGICAL CONDITIONS

Elliot Smith (1912) described lactating breasts in the recently delivered Queen Makere. Williams (1929) reported an observation by Derry that Princess Hehenhit of the Eleventh Dynasty had a narrow pelvis and died not long after delivery with vesicovaginal fistula. Elliot Smith and Dawson (1924) described violent death in an unembalmed 16-year-old pregnant ancient Egyptian girl and postulated illegitimate conception. The *Archaeological Survey of Nubia* revealed a deformed Coptic negress, who died in childbirth as a result of absent sacroiliac joint contracting the pelvis. Vaginal prolapse was also noted in a Nubian specimen by Wood Jones (1908).

CONCLUSION

It will be seen from the reports mentioned in this chapter that many diseases of the present day occurred also in ancient Egypt. There is one notable exception: leprosy, which has not been found before Roman times. Apart from this, the Egyptians before Christ suffered many of the ailments of modern civilized persons.

REFERENCES

Aldred, C., and Sandison, A. T. 1962. The Pharaoh Akhenaten: a problem in Egyptology and pathology. *Bulletin of the History of Medicine* 36:293–316.

Batrawi, A. 1947. Anatomical reports. *Annales du service des antiquités de l'Egypte* 47:97–109.

– 1948. Report on the anatomical remains recovered from the tombs of Akhet- Hetep and Ptah-Irou-ka and a comment on the statues of Akhet-Hetep. *Annales du service des antiquités de l'Egypte* 48:487–97.

– 1951. The skeletal remains from the northern pyramid of Sneferu. *Annales du service des antiquités de l'Egypte* 51:435–40.

Bourke, J. B. 1967. A review of the palaeopathology of the arthritic diseases. In *Diseases in antiquity*, D. Brothwell and A. T. Sandison (eds.). Springfield, Ill.: Thomas.

Brothwell, D. 1963. *Digging up bones.* London: British Museum (Natural History).

– 1967. The evidence of neoplasms. In *Diseases in antiquity*, D. R. Brothwell and A. T. Sandison (eds.). Springfield, Ill.: Thomas.

Brothwell, D., and Powers, R. 1968. Congenital malformations of the skeleton in earlier man. In *The skeletal biology of earlier human populations*, D. Brothwell (ed.). Oxford: Pergamon Press.

Cameron, J. 1910. Report on the anatomy of the mummies. In *The tomb of two brothers*, M. A. Murray, (ed.). Manchester: Sherrat & Hughes.

Czermack, J. 1852. Beschreibung und mikroskopische Untersuchung zweier ägyptischer Mumien. *Akademie der Wissenschaften Wien.* 9:427–69.

Dawson, W. R. 1953. The Egyptian medical papyri in *Science, medicine and history*, E. A. Underwood (ed.). London: Oxford University Press.

Derry, D. E. 1909. Anatomical report. *Archaeological Survey of Nubia*, Bulletin 3.

– 1938. Pott's disease in ancient Egypt. *Medical Press* 197:1.

– 1940–1. An examination of the bones of King Psusennes I. *Annales du service des antiquités de l'Egypte* 40:969–70.

– 1942. Report on the skeleton of King Amenenopet and Har-Nakht. *Annales du service des antiquités de l'Egypte* 41:149–50.

– 1947. The bones of Prince Ptah-Shepses. *Annales du service des antiquités de l'Egypte* 47:139–40.

Elliot Smith, G. 1912. *The royal mummies.* Cairo: Cairo Museum.

Elliot Smith, G. and Dawson, W. R. 1924. *Egyptian mummies.* London: Allen & Unwin.

Elliot Smith, G. and Derry, D. E. 1910. Anatomical report. *Archaeological Survey of Nubia*, Bulletin 6.

Elliot Smith, G., and Ruffer, M. A. 1910. Pott'sche Krankheit an einer Aegyptschen Mumie aus der

Zeit der 21 Dynastie (Um 1000 V. Chr.). In K. Sudhoff (ed). *Zur historischen Biologie der Krankheitserreger*, Heft 3. Leipzig: Giessen.

Elliot Smith, G. and Wood Jones, F. 1908. Anatomical report. *Archaeological Survey of Nubia*, Bulletin 1.

–   1910. Report on the human remains. *Archaeological Survey of Nubia*, Bulletin 2.

Fairbank, T. 1951. *An atlas of general affections of the skeleton*. Edinburgh: Livingstone.

Ferguson, A. R. 1910. Bilharziasis. *Cairo Society Science Journal* 4.

Golding, F. C. 1960. Rare diseases of the bone. In *Modern trends in diagnostic radiology*, J. W. McLaren (ed.). London: Butterworth.

Granville, A. B. 1825. An essay on Egyptian mummies. *Philosophical Transactions of the Royal Society* O: 269.

Gray, P. H. K. 1967. Calcinosis intervertebralis, with special reference to similar changes found in mummies of ancient Egyptians. In *Diseases in antiquity*, D. Brothwell and A. T. Sandison (eds.). Springfield, Ill: Thomas.

Larrey, D. J. 1812–17. *Mémoires de chirurgie militaire et campagnes.*

Lee, S. L. and Stenn, F. F. 1978. Characterization of mummy bone ochronotic pigment. *Journal of the American Medical Association* 240:136–8.

Long, A. R. 1931. Cardiovascular renal disease: report of a case of 3000 years ago. *Archives of Pathology* 12:92–6.

Mitchell, J. K. 1900. Study of a mummy affected with anterior poliomyelitis. *Transactions of the Association of American Physicians* 15:134–6.

Møller-Christensen, V. 1967. Evidence of leprosy in earlier peoples. In *Diseases in antiquity*, D. Brothwell and A. T. Sandison (eds.). Springfield, Ill.: Thomas.

Moodie, R. L. 1923. *Palaeopathology: an introduction to the study of ancient evidences of disease*. Urbana: University of Illinois Press.

Morse, D. 1967. Tuberculosis. In *Diseases in antiquity*, D. Brothwell and A. T. Sandison (eds.). Springfield, Ill: Thomas.

Morse, D; Brothwell, D; and Ucko, P. J. 1964. Tuberculosis in ancient Egypt. *American Review of Respiratory Diseases* 90:524–30.

Rogers, L. 1949. Meningiomas in pharaoh's people: hyperostosis in ancient Egyptian skulls. *British Journal of Surgery* 36:423–6.

Rowling, J. T. 1960. Disease in ancient Egypt: evidence from pathological lesions found in mummies. M. D. thesis, University of Cambridge.

–   1967. Respiratory disease in Egypt. In *Diseases in Antiquity*, D. Brothwell and A. T. Sandison (eds.). Springfield, Ill: Thomas.

Ruffer, M. A. 1910a. Remarks on the histology and pathological anatomy of Egyptian mummies. *Cairo Scientific Journal* 4:1–5.

–   1910b. Note on the presence of "Bilharzia haematobia" in Egyptian mummies of the Twentieth Dynasty (1250–1000 B.C.) *British Medical Journal* 1:16.

–   1911a. Histological studies on Egyptian mummies. *Mémoires sur l'Egypte: Institut d'Egypte* 6(3).

–   1911b. On arterial lesions found in Egyptian mummies (1580 B.C.–525 A.D.). *Journal of Pathology and Bacteriology* 15:453–62.

–   1921. *Studies in the palaeopathology of Egypt.* Chicago: University of Chicago Press.

Ruffer, M. A., and Rietti, A. 1912. On osseous lesions in ancient Egyptians. *Journal of Pathology and Bacteriology* 16:439.

Ruffer, M. A. and Willmore, J. G. 1914. A tumour of the pelvis dating from Roman times (A.D. 250) and found in Egypt. *Journal of Pathology and Bacteriology* 18:480–4.

Salib, P. 1967. Trauma and disease of the postcranial skeleton in ancient Egypt. In *Diseases in antiquity*, D. Brothwell and A. T. Sandison (eds.). Springfield, Ill.: Thomas.

Sandison, A. T. 1955. The histological examination of mummified material. *Stain Technology* 30:277–83.

–   1967a. Degenerative vascular disease. In *Diseases in antiquity*, D. Brothwell and A. T. Sandison (eds.). Springfield, Ill.: Thomas.

–   1967b. Diseases of the skin. In *Diseases in antiquity*, D. Brothwell and A. T. Sandison (eds.). Springfield, Ill.: Thomas.

–   1967c. Diseases of the eyes. In *Diseases in antiquity*, D. Brothwell and A. T. Sandison (eds.). Springfield, Ill: Thomas.

–   1967d. Sexual behaviour in ancient society. In *Diseases in antiquity*, D. Brothwell and A. T. Sandison (eds.). Springfield, Ill.: Thomas.

–   1968. Pathological changes in the skeletons of earlier populations due to acquired disease and difficulties in their interpretation. In *The skeletal biology of earlier human populations*, D. Brothwell (ed.). Oxford, Pergamon Press.

–   1970. The study of mummified and dried

human tissues. In *Science in archaeology*, 2nd ed., D. Brothwell and E. Higgs (eds.). London: Thames & Hudson.

– 1972. Evidence of infective disease. *Journal of Human Evolution* 1:213–24.

Sandison, A. T. and Macadam, R. F. 1969. The electron microscope in palaeopathology. *Medical History* 13:8.

Satinoff, M. J. 1968. Preliminary report on the palaeopathology of a collection of ancient Egyptian skeletons. *Rivista di Antropologia* 55:41–50.

Schultz, A. H. 1939. Notes on diseases and healed fractures of wild apes. Reprinted in *Diseases in antiquity*, D. Brothwell and A. T. Sandison (eds.). Springfield, Ill.: Thomas, 1967.

Shattock, S. G. 1909. A report upon the pathological condition of the aorta of King Merneptah. *Proceedings of the Royal Society of Medicine* (Pathological Section) 2:122–7.

Shaw, A. F. B. 1938. A histological study of the mummy of Har-Mose, the singer of the Eighteenth Dynasty (c. 1490 B.C.). *Journal of Pathology and Bacteriology* 47:115–23.

Wells, C. 1964. *Bones, bodies and disease*. London: Thames & Hudson.

– 1967. Pseudopathology. In *Diseases in Antiquity*, D. Brothwell and A. T. Sandison (eds.). Springfield, Ill.: Thomas.

Williams, H. U. 1929. Human paleopathology. *Archives of Pathology* 7:839.

Wood Jones, F. 1908. The pathological report. *Archaeological Survey of Nubia*, Bulletin 2.

Zorab, P. A. 1961. The historical and prehistorical background of ankylosing spondylitis. *Proceedings of the Royal Society of Medicine* 54:415.

3

Dental health in ancient Egypt

JAMES E. HARRIS
Professor and Chairman,
Department of Orthodontics
University of Michigan
Ann Arbor, Michigan, U.S.A.

PAUL V. PONITZ
Clinical Professor,
Department of Orthodontics
University of Michigan
Ann Arbor, Michigan, U.S.A.

Teeth are the most indestructible of human tissue, and intense interest in them has resulted from two facts: (1) they are often the only biologic record of man and (2), just as important, they are extremely complex morphologically and represent a sophisticated genetic model. Because of the arid climate in Egypt and because of mummification, biologists, anthropologists, and dentists are able to examine not only the teeth but the entire craniofacial skeleton. Further, because soft tissues are well preserved, much has been learned about the tissue directly related to the function of the jaws as well as the tissue that supports the teeth, the periodontium. Dental disease and pathology can be readily interpreted through examination of the immense number of Egyptian mummies available, and an excellent understanding can be obtained of what was the usual state of dental health for these people.

The annual Michigan expeditions since 1965 have examined most of the mummy collections in Egypt, including the Old Kingdom nobles at Giza (3000 B.C.), the New Kingdom nobles and priests at Luxor (1200 B.C.), and the pharaohs and queens of the New Kingdom period at the Egyptian Museum (1500–1000 B.C.).

In addition, the investigation of oral health and disease in ancient Egypt has been greatly enhanced by studies of modern Egypt, especially Nubia (Harris et al. 1970). Not only have the ancestral records of Gebel Adda, Old Nubia (representing almost 2000 years) been studied, but modern Nubian people living in New Nubia at Kom Ombo have been examined over the past 10 years (Holden et al 1970).

From the dental viewpoint, what did the ancient Egyptian look like compared with the people of today? Basically, there has been little change as far as dental disease is concerned. In ancient Egypt, the greatest single problem was attrition, or wear (Figure 3.1). The teeth were rapidly worn down throughout life by the consumption of a coarse diet: Interestingly enough, the pharaohs of Egypt exhibit this wear just as do the farmers of both modern and ancient Egypt. In time, this wear becomes so extensive that the enamel and dentin are eroded away until the pulp is exposed. The living tissue inside the tooth dies, and the empty root canals become a source of chronic infection and abscess. The teeth of Rameses II are excellent examples of the effects of old age, attrition, and ultimate abscesses (Figure 3.2). Ghalioungui and El Dawakhly (1965) concluded that the dental surgeon had drained these abscesses through a hollow reed.

The second greatest problem from the viewpoint of both ancient and modern Egyptians was periodontal, or gum, disease. This disease results in loss of the bony support of the teeth and is often associated with calculus, or tartar, deposits on the teeth. Calculus was often so extensive in the skulls of the ancient Nubians that this deposit frequently held the teeth in place 2,000 years after death. Whereas calculus deposits of any consequence are rarely seen in Americans below 20 years of age, they have been observed in Nubian children in elementary

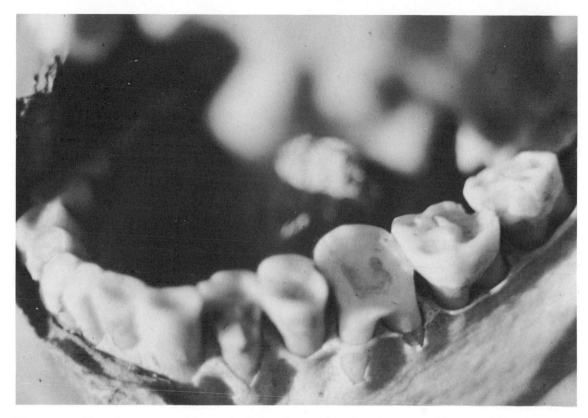

Figure 3.1. *Dental wear or attrition typical of ancient and modern Egyptians. This mandible was found in a Nubian cemetery of A.D. 250.*

school (Holden et al. 1970). The ultimate result is loss of bony support, loose teeth, and deep periodontal pockets or even exposed root bifurcations. This in turn leads to infection, abscesses, and loss of teeth.

Dental caries, or cavities, were far less frequently seen in ancient Egyptians or Nubians, until the latter population was moved to New Nubia near Kom Ombo and the great sugarcane fields. Neither ancient kings, nobles, nor commoners exhibited much dental decay, and where observed, the decay was of the pit-and-fissure variety (top of the tooth) rather than the decay between the teeth so frequently seen in modern civilization. There are two major environmental causes that, one

may speculate, may have resulted in the lack of extensive dental decay. The first is the absence from the diet of refined carbohydrates such as sugar. The second, and equally important, is the extreme wear mentioned earlier, for wear occurs not only on the occlusal surface (top of the teeth), but between the teeth (interproximally) as well. It is between the teeth that dental decay is often initiated in modern Western society. The extensive wear provides a more difficult environment for decay to begin.

A brief comment must be made on dental occlusion or malocclusion and facial types. In general, perhaps owing to extreme wear, the dentition of the ancient Egyptians (Old King-

Figure 3.2. *Lateral cephalogram of the mummy of Rameses II. This pharaoh was very old at death, and the extreme wear of the teeth may be noted with resulting exposure of the pulp chambers and periapical abscesses. Periodontal disease, or loss of bone support of the dentition, is also apparent.*

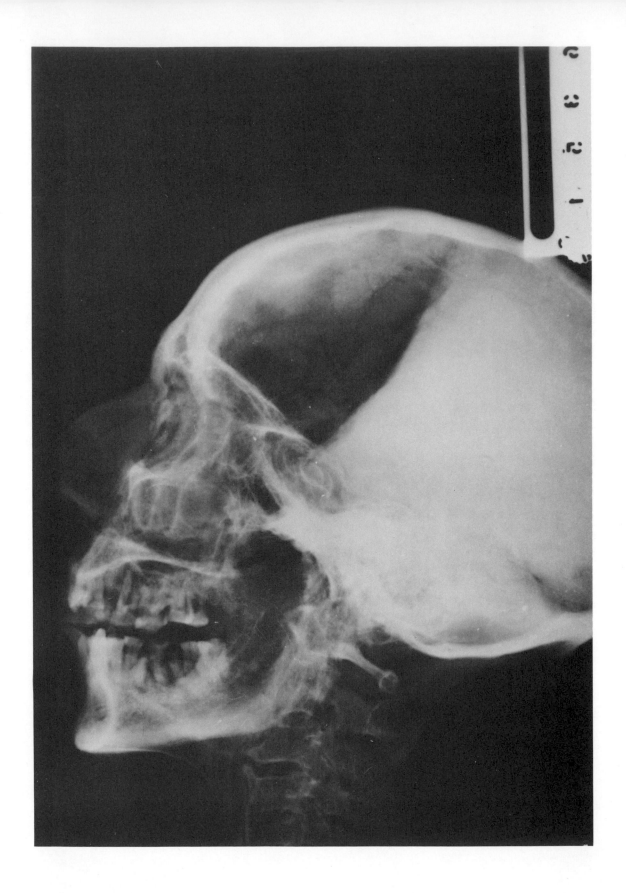

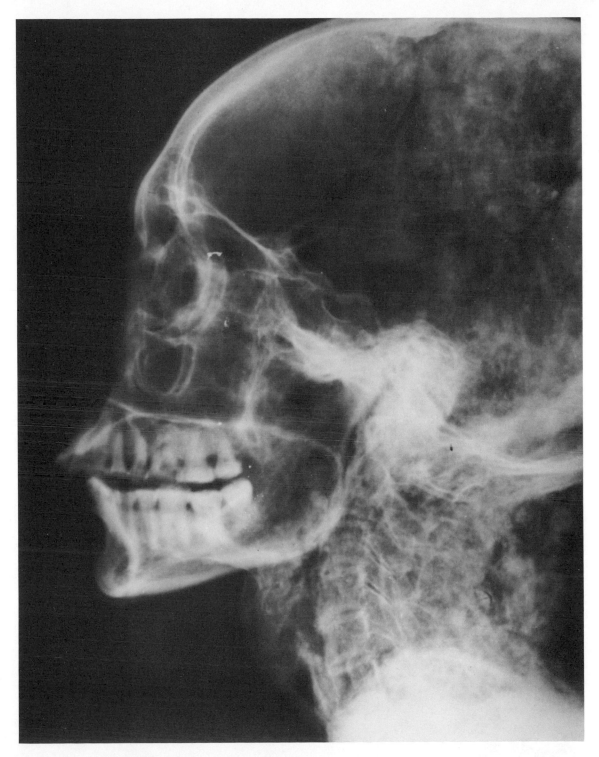

Figure 3.3.   X ray of the mummy of Ahmose Nefertiry illustrating the maxillary prognathism characteristic of the queens of the early Eighteenth Dynasty.

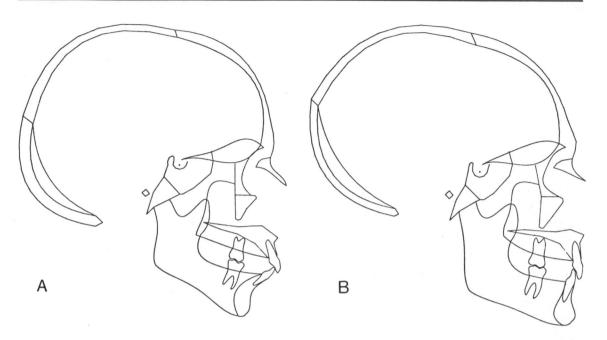

A   B

Figures 3.4.   Computerized tracings of the X rays of the mummies of (a) Tuthmosis I and (b) Tuthmosis IV of the Eighteenth Dynasty. These two facial profiles and anterior dentition demonstrate the remarkable heterogeneity among the pharaohs of the New Kingdom.

dom) and ancient Nubians rarely exhibited the dental crowding or abnormal molar relationships that are frequently observed throughout the world today. Most modern Egyptians have good molar relationships with moderate to severe crowding. However, if the queens of the New Kingdom period (early Eighteenth Dynasty) are examined by X-ray cephalometry, many resemble modern Europeans or Americans, with maxillary, or upper jaw, prognathism. This condition may be either hereditary or environmental – that is, the result of thumb sucking or other oral habits. Queen Ahmose Nefertiry is an excellent example of this type of occlusion (Figure 3.3). In Nubia, this same condition has been observed, but as the result of tongue thrusting or thumb sucking, not as an inherited phenomenon.

The kings of Egypt had little similarity in appearance, contrary to their portrayals by contemporary artists. Computer tracings of X-rays of the mummies of Tuthmosis I and Tuthmosis IV show the considerable

heterogeneity in the mummies of the New Kingdom period as far as craniofacial complex is concerned (Figure 3.4). From the viewpoint of the orthodontist or anthropologist, there is abundant biologic evidence that the pharaohs of the New Kingdom period of Egypt were quite heterogeneous – as variable as any American population. This assumption may be supported by historical records, which indicate that during the New Kingdom period the Egyptian royal family brought princes from many conquered nations into its court, where they were accepted as equals and hence potential consorts (Iskander and Badawy 1965). Except for Tuthmosis I, II, and III, most of the pharaohs of the New Kingdom period had class I (i.e., normal) molar relationships with straight profiles, but neither their craniofacial skeletons nor their soft tissues suggested homogeneity of appearance. In contrast, the nobles of the Old Kingdom studied at Giza and those of the New Kingdom examined at Deir el-Bahri seem quite homogeneous.

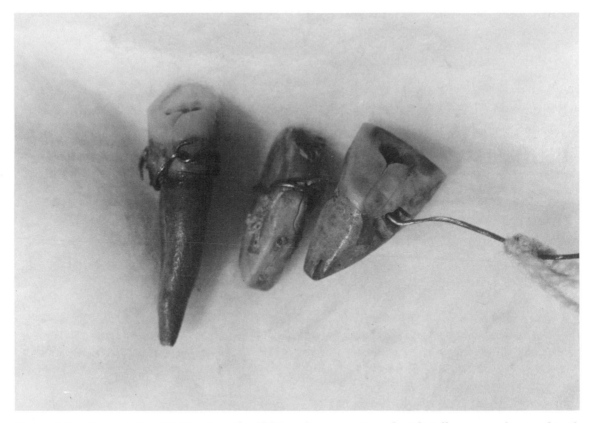

Figure 3.5.   Ancient dental bridge from the Old Kingdom consisting of artificially prepared natural teeth and gold wire.

Finally, a brief mention of the *dental profession* in ancient Egypt. Ghalioungui and El Dawakhly (1965) recall: "Hesy-Re, the oldest person in history who ever carried the title of physician, even in the time of Imhotep, two thousand years before the war of Troy, called himself Chief of Dentists." The time would date back to the Old Kingdom, and there are many medical papyri, such as the famous Ebers papyrus (Ebell 1937), that prescribe medicines to relieve dental pain and even describe how to fix loose teeth. Recently, there has been considerable debate about whether a restorative dental profession existed before the time of the Ptolemaic or Greek period, as no dental prosthesis has been found in a mummy, royal or otherwise (Leek 1967). However, in 1975, Harris, Iskander, and Farid reported the discovery of a dental bridge from

the Fourth Dynasty of the Old Kingdom, about 2500 B.C. (Figure 3.5). This bridge consisted originally of four teeth and replaced an upper left lateral incisor and a central incisor. The bridge consisted of prepared natural teeth, which had very fine holes drilled through them, and gold wire, which was used in a skillful manner to attach the substitute teeth to the abutment teeth. This find, together with the written evidence of the medical papyri, suggests that Ghalioungui and El Dawakhly (1965) were correct when, speaking of the remarkable specialization of medicine, they cited "Men-kaou-Re-ankh who was called a maker of teeth (iry-ibh), to distinguish him from Ni-ankh-Sekhmet who figures on the same stele as a tooth physician."

In short, Egypt has the best recorded

biologic record, documented by history, of any country in the world. Comparisons of the ancient Egyptians with modern Egyptians suggest that attrition and periodontal disease are common to both populations. Only in recent times, with the development of refined sugars, has dental decay become a major problem in urban areas of Egypt. The ancient pharaohs and queens of Egypt show in their dental occlusion and craniofacial skeleton, as revealed by X-rays, the diversity and heterogeneity one might assign to modern Western communities with a history of racial admixture; the Egyptian nobles and the Nubian people present a much more homogeneous facial skeleton with good occlusion, sometimes associated with dental crowding. Except in cities such as Cairo and Alexandria, the types of dental malocclusion seen commonly in the Western world, such as maxillary prognathism, are not inherited, but are associated with oral habits.

REFERENCES

Ebell, E. 1937. *The Papyrus Ebers, the greatest Egyptian medical document* (translation), chapter 89, p. 103. Copenhagen: Levin & Munksgaard.

Ghalioungui, P. and El Dawakhly, Z. 1965. *Health and healing in ancient Egypt*, p. 12. Cairo: Egyptian Organization for Authorship and Translation.

Harris, J. E.; Iskander, Z.; and Farid, S. 1975. Restorative dentistry in ancient Egypt: an archaeologic fact. *Journal of the Michigan Dental Association* 57:401–4.

Harris, J. E.; Ponitz, P. V.; and Loutfy, M. S. 1970. Orthodontic's contribution to save the monuments of Nubia: a 1970 field report. *American Journal of Orthodontics* 58(6):578–96.

Holden, S.; Harris, J. E.; and Ash, M. 1970. Periodontal disease in Nubian children. *International Association of Dental Research Program and Abstract of Papers*, p. 65. Chicago: American Dental Association.

Iskander, Z., and Badawy, A. 1965. *Brief history of ancient Egypt*, 5th ed., p. 207. Cairo: Madkour Press.

Leek, F. 1967. The practice of dentistry in ancient Egypt. *Egyptian Archaeological Journal* 53:51.

4

A classic mummy: PUM II

AIDAN COCKBURN
President, Paleopathology Association
Detroit, Michigan, U.S.A.

WILLIAM H. PECK
Curator of Ancient Art
Detroit Institute of Arts,
Detroit, Michigan, U.S.A.

ROBIN A. BARRACO
Associate Professor of Physiology
Wayne State University School of Medicine
Detroit, Michigan, U.S.A.

THEODORE A. REYMAN
Director of Laboratories
Mount Carmel Mercy Hospital
Detroit, Michigan, U.S.A.

The traditional Egyptian mummy is one on which all the arts of embalming have been employed, the organs have been preserved, the body is wrapped in linen, and everything is contained in a highly decorated sarcophagus. Such was mummy PUM II. It belongs to the Philadelphia Art Museum and was lent to the Paleopathology Association for dissection and study through the courtesy of David O'Connor of the Pennsylvania University Museum – thus the name PUM II, this being the second mummy from that museum. PUM II is now on loan to the National Museum of Natural History, Smithsonian Institution, Washington, D.C., and can be seen there complete with its sarcophagus, and photographs illustrating the autopsy and the finds made during subsequent studies (Figure 4.1).

Little is known of the provenance of this mummy (Figure 4.2). It was probably brought to America about the turn of the century and has been in the possession of the Philadelphia Art Museum since that time, but its origins in Egypt are unknown. The sarcophagus was highly decorated, but lacked the name and any details of the person inside (Figures 4.3 and 4.4).

This chapter gives an overall picture of the autopsy and the findings made during the following years. The work is by no means completed and, indeed, is likely to continue for several more years. This is because new tech-niques are continually being devised and applied to the tissues, wrappings, and resin.

The unwrapping and the autopsy took place on 1 February 1973, as part of a symposium, "Death and Disease in Ancient Egypt," that was held at Wayne State University Medical School, Detroit, Michigan (Cockburn 1973). Radiographic examinations had been made a week earlier at Mount Carmel Mercy Hospital and Hutzel Hospital, Detroit. Specimens of the radiograms and xerograms are presented in Chapter 17. It was seen that the mummy was in good condition. The brain had been removed and replaced with resin, which had formed a pool in the skull before it solidified, appearing in the X ray as a "water level." A diagnosis of fractured skull was also made; this frequently occurs with X rays of mummies that are still wrapped. When the skull was exposed, it was found that these "fractures" were merely scratches in the scalp that appeared as linear defects in the bone on the X ray. Let this be a warning to anyone trying to make a diagnosis on a mummy that is still in its wrappings.

Four packages were seen in the body cavities. No amulets were visible. There was a transitional or sixth lumbar vertebral body that might have made stooping painful and difficult during life. The right fibula and adjoining tibia had a pathological thickening resembling periostitis.

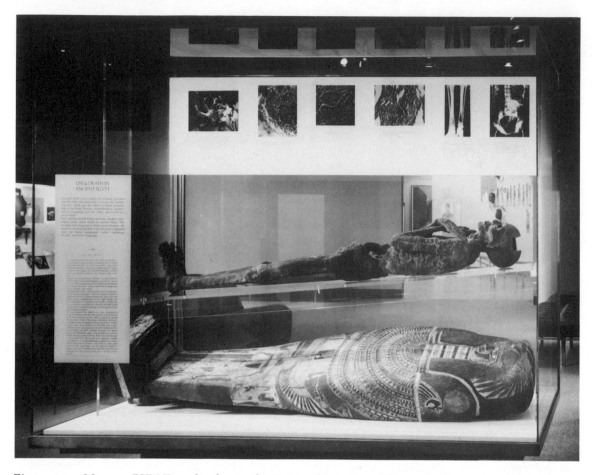

Figure 4.1.   Mummy PUM II on display in the National Museum of Natural History, Washington, D.C. (Photograph by Victor Krantze, Smithsonian Institution)

After the autopsy, more radiographs were taken of the right leg, and polytomographs were made of the skull and temporal bone regions. The polytomographs showed the hole punched through the cribriform plate for the removal of the brain: This had been missed, although it was looked for, in earlier radiological studies.

REMOVAL OF WRAPPINGS AND GENERAL EXAMINATION

There proved to be about 12 layers of linen wrapping of varying qualities of cloth. The outer layers were generally larger sheets or strips of fine weave. Hot liquid resin had been poured liberally over the body at many of the stages, so that most of the wrappings had been converted into a hard, solid mass, which could be removed only with a hammer and chisel or cut through, several layers at a time, with a Stryker saw. After the general broad wrapping had been removed, it was found that limbs and even individual fingers and toes were wrapped separately. As many as nine people worked simultaneously, but it still required almost 7 hours to strip away all the bandages completely.

At a later date, the wrappings were examined by Meryl Johnson (1974). As expected,

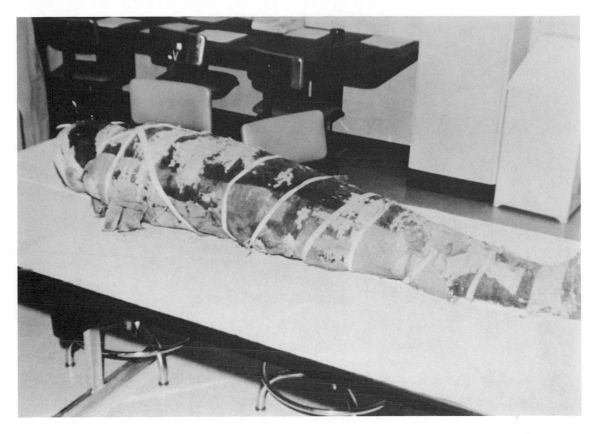

Figure 4.2.   The mummy before unwrapping. (Photograph by Nemo Warr, Detroit Institute of Arts)

the cloth was linen, with a simple weave, but a complete surprise was the discovery of a ball of cotton wrapped between two pieces of linen. It was adherent to the linen and was partially coated with some nondescript material that could have been unguent or tissue juice. This material has not yet been identified (Figure 4.5).

Cotton has not been recorded in the Western World before Christ, the earliest find being from a Roman grave about A.D. 200, though cotton has been found in cultures of the Indus Valley dating back to about 2000 B.C. The interesting question is: How did cotton arrive in Egypt by 200 B.C.? By the time of the Romans, there was considerable trade with India, and indeed the Romans had several trading posts in southern India (Miller 1969). Until the secrets of the monsoon were discovered, sea traffic hugged the coast, with

most traffic taking the route up the Persian Gulf and across the desert to Palestine via Petra. Were the Egyptians using some similar route to import cotton as early as 200 B.C? Perhaps cotton was so rare and valuable that the ball found in PUM II was, as Meryl Johnson has speculated, included as a form of amulet.

THE GENERAL AUTOPSY

The skin and tissues were as hard as plastic and were cut with the Stryker saw. When the saw cut through the resin, the resin burned and gave off a most fragrant odor. The anterior abdominal wall and the incision made by the embalmers in the left side were cut out and removed (Figure 4.6). Inside the abdominal and thoracic cavities there were four packages. Hot resin had been poured in, cov-

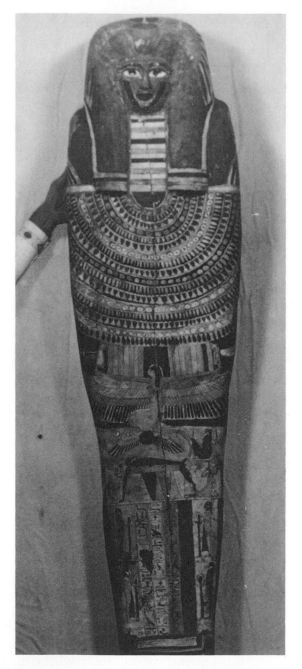

Figure 4.3.   Coffin of PUM II. (Photograph by J. Lawrence Angel, Smithsonian Institution)

ering both the packages and the floors of the thorax, abdomen, and pelvis. The packages were removed with a chisel and were found to be covered with insect pupae (Figures 4.7 and 4.8) preserved by the resin; further study showed that one package contained spleen and some intestine, and the three others contained lung. Some of the aorta and a piece of heart tissue were found in situ, coated with resin. The kidneys and urinary bladder were not seen.

The penis was intact, held in an upright position with support from a small piece of wood (Figure 4.9). There had been no circumcision. The testes were missing, probably having been removed via the pelvis. The right leg was abnormal (Figure 4.10) and is discussed later in this chapter. The feet and hands were in excellent condition. The nails were painted red with henna, and the soles of the feet were white with lime (Figures 4.11 and 4.12).

Radiograms taken before the autopsy had shown a fluid level in the skull, so a window was cut in the cranium above this level. It was found that resin had been poured into the skull through a hole punched through the base of the skull by a tool forced up the left nostril. Presumably, the brain had been removed by a small hook through this same hole and replaced by the resin. Originally, we had supposed that the brain would liquefy and could be drained away. However, this does not happen: Brain tissue, if undisturbed, simply retains its general shape and shrinks to about one-third its volume, as was the case in mummies ROM I, PUM I, PUM III, and PUM IV.

The eyes were intact and on removal appeared well preserved (Figure 4.13). At a later date the temporal bones containing the ears were removed with a circular Stryker saw and taken out through the window in the skull (Figure 4.14). Also at a later date, the anterior part of the lumbar vertebral column was removed in order to search for the spinal cord. The cord had vanished and been replaced by resin trickling down from the skull through the foramen magnum.

Tissues were collected from the location of the thyroid and parathyroids, but later studies failed to reveal these glands. Lewin

Figure 4.4.   Detail from coffin. The Apis bull carrying the mummy. (Photograph by J. Lawrence Angel, Smithsonian Institution)

noted that the color of the mummy changed from a light brown to a darker brown within 24 hours. At the present time, the skin is almost a black brown. On the basis of anatomical studies, Angel estimated the age of the individual as between 35 and 40 years, and his height as approximately 162 cm (5 ft 4 in.).

SPECIAL STUDIES

There has been much speculation regarding the length of time organisms such as seeds or bacterial spores can survive in a dormant state and still become active again. There are reports from the Antarctic of bacteria 10,000

years old recovered in a living state from deep in the frozen soil. It was, therefore, decided to look for living bacteria or fungi inside this mummy. There were two objections to this project: (1) any fungi found might be modern ones that had grown through the mummy from the outside; and (2) the act of cutting open the mummy would contaminate the interior.

In spite of these drawbacks, cultures were taken as soon as the abdominal cavity was opened from areas remote from the opening. All the specimens proved sterile. In two previous mummy autopsies, Reyman had found almost all tissues to be riddled with fungi; in

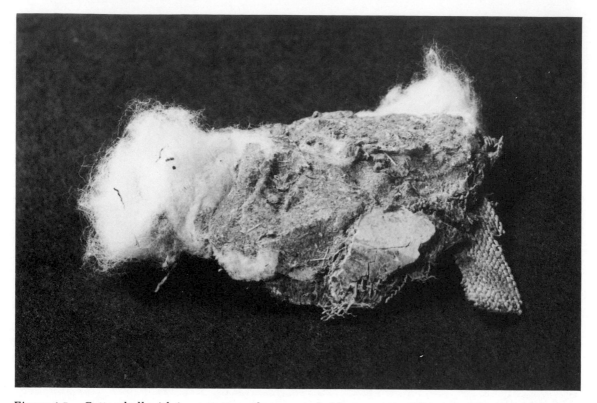

Figure 4.5.   Cotton ball with incrustations that are probably unguents. (Photograph by Nemo Warr, Detroit Institute of Arts)

this case, fungus was absent. The explanation is probably the abundant use of resin. This was obviously very hot and fluid when applied, for it penetrated not only into large cavities, but also into the mastoid air cells via the foramen of the eighth nerve and into the middle ear through a perforation in the eardrum, and even trickled from the skull down the whole length of the spinal canal.

THE RESIN

Specimens of the resin have been distributed to a number of laboratories, and some reports are now available. One scientist was intrigued by the observation that it burned with a most fragrant odor when cut with the Stryker saw and wanted to test this fragrance in his studies of smell. However, when a sample was set on fire in Detroit, the resin melted and burned, giving off a thick black smoke that was most unpleasant!

An analysis of trace elements was undertaken by Nunnelley and his colleagues (1976). Their report, given later in this chapter, indicated that the resin had penetrated all the tissues of the body (Table 4.1).

Another analysis of the resin was made by Coughlin (1977). Using mass spectroscopy, she found that the PUM II resin, or mummy fluid, had completely polymerized into one vast and continuous molecular form. This polymerization is attributable to a combination of aging and the electromagnetic properties of natural botanical products. The result is a kind of organic glass, an intermediate with respect to amber. This condition of being a "glass" has been confirmed by X-ray diffraction. The known use of natron (sodium salts) in later Egyptian glass-making processes – natron is the chemical used for desiccation of the body during the mummification process – suggests that the development of a

Figure 4.6.   The embalmers' incision. The abdomen has been opened, and the layers of wrappings, congealed into one mass, can be seen. (Photograph by Nemo Warr, Detroit Institute of Arts)

Figure 4.7.   Package containing lung covered with fly pupae, which are covered with resin. (Photograph by Nemo Warr, Detroit Institute of Arts)

Figure 4.8.   Embalmed pupa visualized with a scanning electron microscope. × 80. (Courtesy of Peter Lewin, University of Toronto)

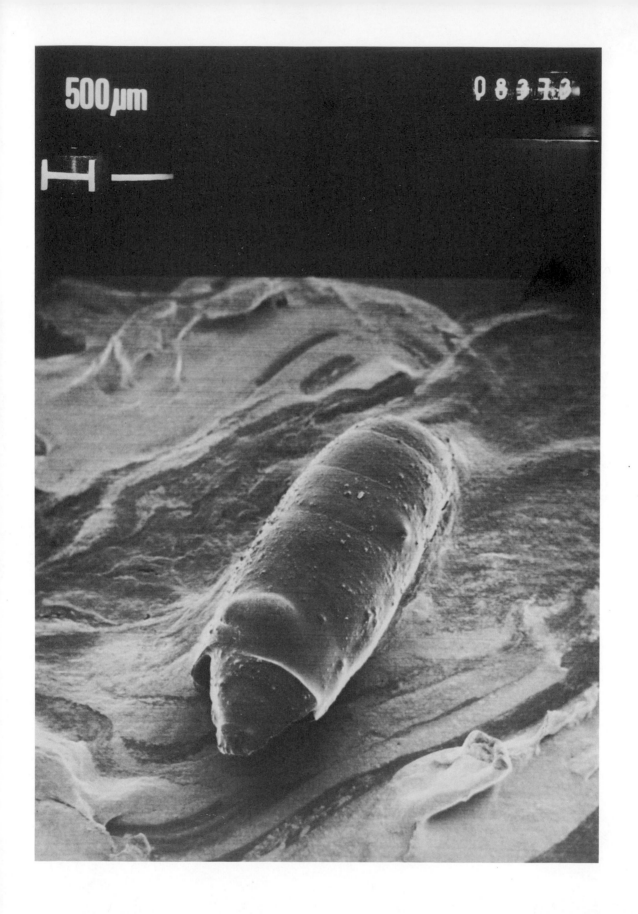

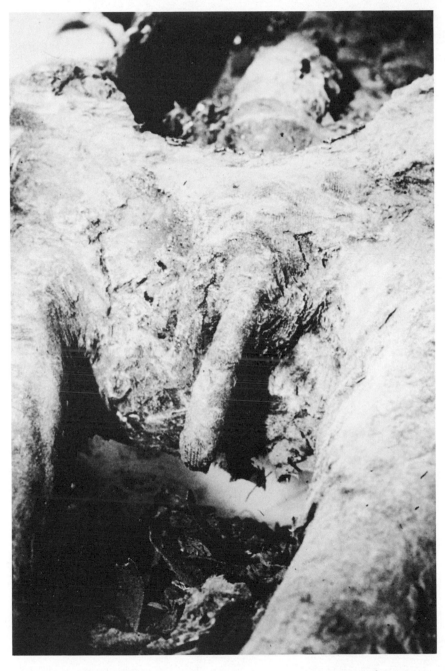

Figure 4.9.   Penis. (Photograph by Nemo Warr, Detroit Institute of Arts)

Figure 4.10.   Lower part of the legs. The right leg is swollen and shows the marks of the bandages. (Photograph by Nemo Warr, Detroit Institute of Arts)

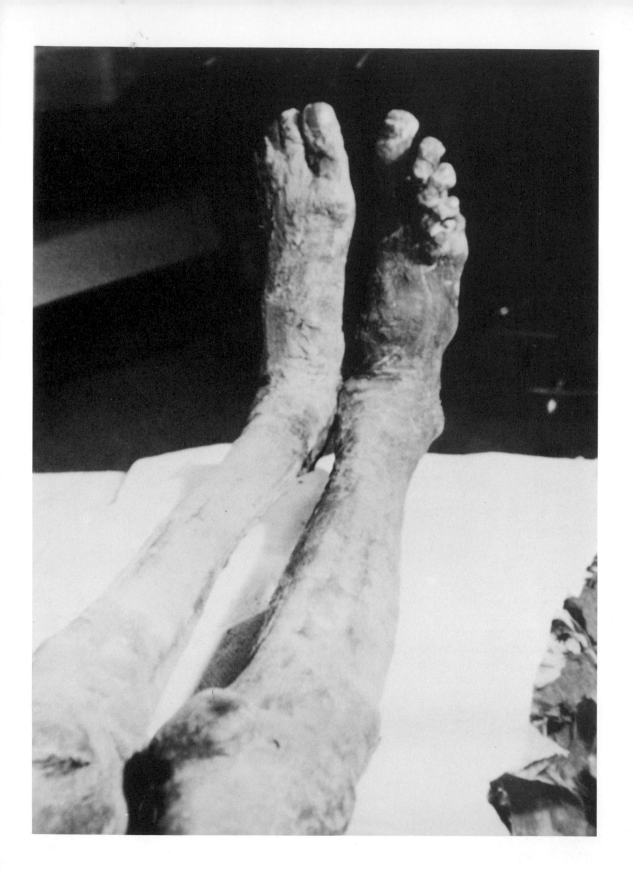

*Figure 4.11.   Hands. The nails are painted red.
(Photograph by Nemo Warr, Detroit Institute of
Arts)*

technology of glass making may have occurred as an offshoot of the mummification process.

The major constituent of PUM II mummy fluid or resin identified by X-ray diffraction techniques is the oil of a coniferous evergreen tree, Coniferae, *Juniperus* (Linnaeus). This genus, *Juniperus*, comprises approximately 35 species of evergreen trees or shrubs whose distribution ranges from the Arctic Circle to Mexico, the West Indies, the Azores, the Canary Islands, North Africa, Abyssinia, and the mountains of tropical East Africa, China and Formosa, and the Himalayas. *Juniperus* is one of the few northern forms that has distributed into the southern hemisphere. These juniper trees have a fragrant wood, and a red to reddish-brown oil is expressed from the

wood, leaves, and shoots. This oil is the major constituent of the PUM II mummy fluid, resin, or glass.

Two additional, although minor, components were separated from the major constituent by means of an organic extraction and filtration technique. One of the additional components identified by thin-layer chromatography 3 is an oil from an aromatic tree, Lauraceae, *Cinnamomum camphora* (Nees & Eberm). This is the camphor tree, a stout, dense-topped tree capable of reaching a height of about 13 m. Its enlarged base, twigs, and bruised leaves have a marked camphor odor. The tree is distributed geographically in Ceylon and Asia. The second additional component is myrrh, the fragrant gum resin exuded from special resin ducts in the bark of the myrrh tree, Burseraceae, *Commiphora myrrha*. This tree ranges in height from 3 to 12 m, and is native to only two parts of the world: southern Arabia and northern Somaliland.

Although minute quantities of other botanical products, such as spices or flowers, may be additional unidentified components, the three components identified as oils of juniper and camphor and the gum resin myrrh are the essential constituents of the PUM II mummy resin.

THE PACKAGES

Peck has described (Chapter 1) how the Egyptians preserved certain organs either in canopic jars or by wrapping them in linen and replacing them in a mummy's abdominal cavity. The process of extracting water with natron shrank the organs considerably, some being only one-tenth their natural weight. Traditionally, organs treated in this way were the lungs, liver, stomach, and intestine.

PUM II had the four customary packages, but Reyman discovered that three of them contained only lung and the fourth spleen plus a tiny piece of intestine. This lent support to the opinion that although the form of mummification had followed traditional lines, it had suffered some debasement. The

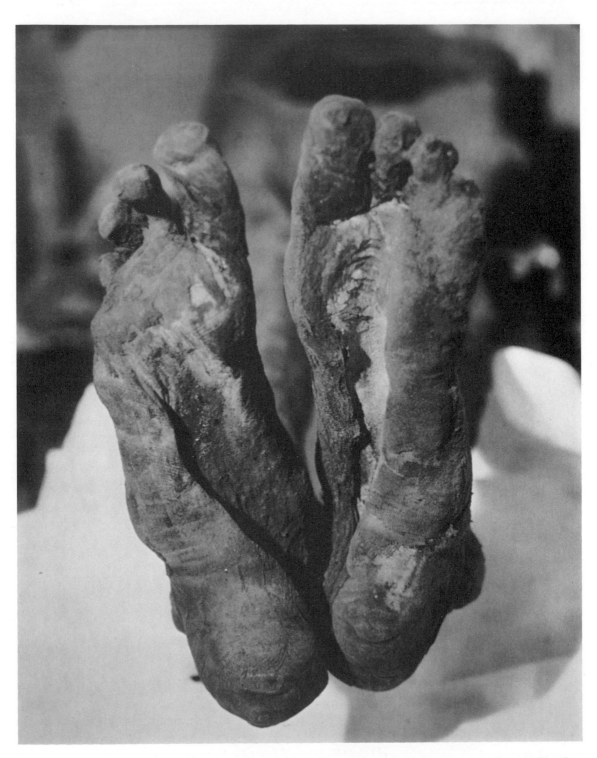

Figure 4.12.　Feet. The feet are painted white with lime. The distortion is postmortem and caused by the wrappings. (Photograph by Nemo Warr, Detroit Institute of Arts)

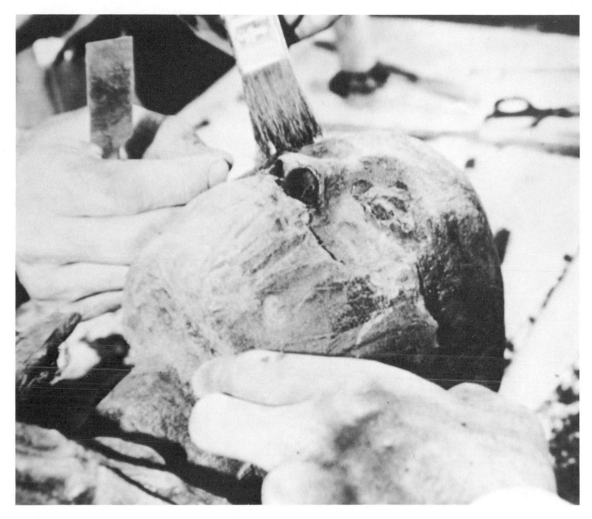

Figure 4.13.   Face emerging. (Photograph by Nemo Warr, Detroit Institute of Arts)

embalming was obviously expensive, yet the embalmers had handled the organs in a slipshod way.

Some of the pupae found embalmed on the packages have been studied by the U.S. Department of Agriculture at its Systematic Entomology Laboratory, Washington, D.C. The entomologists at this laboratory pointed out that certain insects have specialized in breeding on decaying flesh and that Motter, in his studies in 1898, found a wide range of genera in human graves. Indeed, in PUM IV, a child 6 to 8 years of age autopsied in Detroit in 1976 (Chapter 6), the insect larvae within the wrappings had been so numerous that they had penetrated to all parts of the body, including the brain, and had left large holes in the bones and tissues.

In the case of PUM II, the process had not progressed to that point. The insects had certainly laid their eggs on both body and packages, larvae had hatched out and eaten their fill, and some had turned into pupae; but at that point, the embalmers had poured in hot liquid resin, which killed and embalmed them all in an instant (Figures 4.7 and 4.8). Chapter 18 describes the finding of a larva on an eardrum.

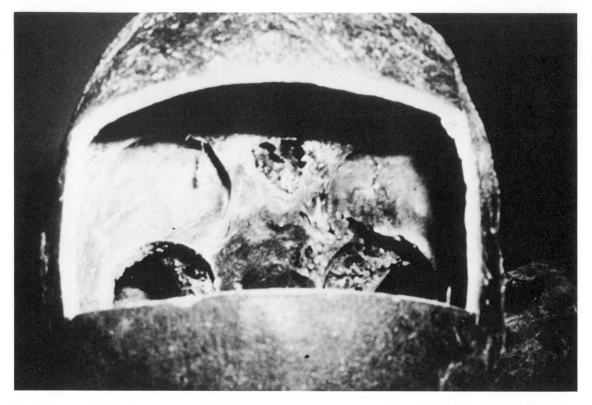

Figure 4.14. *Skull. The openings in the base are two circular ones made by the Stryker saw; the foramen magnum is between; and the opening above this was made by the embalmers. (Photograph by Aidan Cockburn)*

The insects from PUM II so far identified (Cockburn et al. 1975) are: *Dermestes*, probably *frischii*; *Piophila casei*; *Atheta* sp.; and *Chrysomya* sp.

GENERAL HISTOLOGY

The histological report is given in Chapter 15. In brief, it can be said that tissue from PUM II, such as bone, cartilage, and muscle, was found to be in good condition. The skin was preserved and showed intact glandular structure (but without nuclei), hair follicles, and an intact basal layer of epithelium with ghost forms of nuclei and melanin pigment.

The lung tissue from the visceral packages contained intact bronchi and bronchioles with normal cartilage and connective tissue. The pulmonary parenchyma had areas of diffuse and nodular fibrosis. In some sections,

the alveolar septa appeared normal. Within the fibrotic areas, there were anthracotic (carbon) and silicotic (silica) deposits. The silica content of the lung was 0.22 percent; the normal value has an upper limit of 0.20 percent and is usually less than 0.5 percent. These findings indicate that the man had pneumoconiosis, probably from inhaling sand during desert dust storms. Whether he had symptoms of this pulmonary disease is difficult to assess.

One of the visceral packages housed spleen and a small portion of intestine. The spleen, with recognizable capsule and trabeculae, was normal. The intestinal tissue contained a single fragment of partially digested but recognizable meat (muscle) fiber with residual striations. Also present within the tissue was a single parasite egg.

Table 4.1. *Trace element concentrations of PUM II tissues and resin (parts per million dry weight)*

Element	PUM II muscle	PUM II skin	PUM II tendon	PUM II resin	Modern muscle	Mammalian muscle
K	1,500 ± 300	3,500 ± 1,000	1,400 ± 300	580 ± 160	11,600 ± 1,800	10,500
Ca	500 ± 90	83,000 ± 16,000	520 ± 90	760 ± 170	570 ± 80	105
Mn	2.1 ± 0.6	<8	2.9 ± 0.4	7.4 ± 1.3	<1.7	0.21
Fe	145 ± 8	27 ± 4	111 ± 6	600 ± 50	110 ± 30	140
Ni	3.0 ± 0.3	<3	2.6 ± 0.2	7.3 ± 0.7	—	0.008
Cu	4.1 ± 0.3	<2	3.4 ± 0.3	5.8 ± 0.4	1.8 ± .8	3.1
Zn	86 ± 3	37 ± 2	56 ± 2	4.9 ± 0.3	170 ± 30	180
As	0.12 ± 0.006	<0.5	0.06 ± 0.03	0.31 ± 0.08	<0.3	0.16
Se	0.29 ± 0.04	—	0.25 ± 0.03	0.04 ± 0.03	—	2.5
Br	13.9 ± 0.3	6.0 ± 0.3	12.1 ± 0.3	8.9 ± 0.3	—	4
Rb	2.0 ± 0.1	0.6 ± 0.2	1.34 ± 0.08	0.82 ± 0.06	9 ± 5	24
Sr	4.3 ± 0.1	78 ± 2	4.3 ± 0.1	10.1 ± 0.3	0.13 ± 0.006	0.05
Y	0.04 ± 0.03	—	0.08 ± 0.02	0.27 ± 0.4	—	—
Zr	0.36 ± 0.03	<0.9	0.32 ± 0.03	1.82 ± 0.08	—	<0.3
Nb	0.03 ± 0.02	—	0.019 ± 0.016	0.82 ± 0.04	—	—
Mo	1.64 ± 0.05	<0.3	1.32 ± 0.03	4.0 ± 0.1	0.06 ± 0.01	<0.2
Sn	0.57 ± 0.06	<1.6	0.68 ± 0.04	1.37 ± 0.08	—	<0.2
Pb	0.33 ± 0.06	<1.2	0.41 ± 0.04	3.4 ± 0.2	0.8 ± 0.5	<0.2

Dash indicates concentration not measurable.

This has been studied by a number of helminthologists, who are agreed that it probably is *Ascaris*; some state definitely that it is *A. lumbricoides*. This finding was not a complete surprise, for *Ascaris* has already been reported from seven locations in antiquity in Europe. Sometimes, there were millions of ova in the feces, as in the prehistoric salt mines at Hallstatt in Austria.

Both eyes were collected, and the whole section of one revealed the lens to be present, although the cornea had disappeared. The choroid and the ciliary body were intact and contained melanin pigment, but there were no traces of the retina. Large nerves, probably those to the extrinsic eye muscles, were well preserved in the retrobulbar fat and muscle. Portions of the aorta and other vessels were found within the visceral packages; large and small arterioles and arteries also showed areas of intimal fibrous thickening, typical of arteriolar sclerosis. In some of the vessels, partially and completely intact red blood cells could be seen (Chapter 15).

It had been noted by the radiologists that the right leg was abnormal in that there appeared to be periostitis of the fibula and the adjoining part of the tibia. This abnormality was confirmed when the leg was unwrapped, for the right leg was swollen compared with the left. The bandages wrapped around the leg had left distinct marks, suggesting that the leg had been edematous at the time of mummification. The fact that the right toes were curled back led to some dispute, but it was finally agreed that this was a postmortem effect caused by the tightness of the wrappings.

A piece of the affected fibula was removed and sectioned. It was found that a good deal of new but unorganized bone had been laid down in the distal half of the fibula (Chapter 15). The cause of this pathology is unknown.

BIOCHEMICAL STUDIES
It is well known that amino acids are stable for immense periods of time. It is, therefore, not surprising that various workers have re-

ported amino acids and even peptides from Egyptian mummies. However, it had usually been assumed that proteins would break down into component parts over any significant length of time. Cockburn (1963) suggested that this might not necessarily be the case and that gamma globulin and even antibodies might persist under favorable conditions. It was to test this speculation that Barraco set out to isolate proteins from the tissues of PUM II.

Biochemistry is complex and no field for the beginner. The technical details are given step by step in Chapter 19 for those with the learning to follow them. Here it is enough to say that Barraco did indeed succeed in extracting what appeared to be pure protein of a molecular weight of 150,000 (which is the same as for gamma globulin), but that testing indicated it to be biologically inactive. Trace amounts of intact immunoreactive albumen were identified by Reyman.

Another problem to be tackled was the question of whether any given mummy had been treated with natron. Mummies DIA I and PUM II quite obviously had been so treated, but with the remainder in doubt. They just might have been wrapped up in linen without preliminary dehydration. Barraco found that the salt content of the tissues of PUM II was very high, perhaps 10 times that of normal tissues, but the salt content of the uncertain mummies was the same as in normal living tissues. This provides a future test for natron. It also incidentally demonstrated that the natron, like the resin, had penetrated into the tissues all through (Chapter 19).

DATING

Several techniques are available for dating a mummy. First, carbon-14 dating was done by Stukenrath (1974), using linen from the wrappings. The date he obtained was 170 B.C. ± 70 years.

The coffin should give valuable information for dating, but Strouhal (personal communication, 1973) reported that about 10 percent of the 180 mummies he examined in Czechoslovakia appeared to be in coffins not originally intended for them. Angel measured PUM II and concluded that his size was consistent with the assumption that the coffin was made for him. Photographs of the coffin taken by Angel were studied by Fischer (1974), who described it as Greco-Roman. The Apis bull carrying the dead man (Figure 4.4) is a late motif. Fischer also noted that the slight garbling of the hieroglyphs and the absence of any name for the dead man suggest a "stock" coffin rather than one that was custom-made.

An estimate based on cultural features was given by Strouhal (personal communication, 1973). At first sight it appeared that the methods of mummification (packaging the organs, removing the brain, painting the nails with henna and the feet with lime, crossing the arms) indicated a mummy of the Third Intermediate period, perhaps about 700 B.C. However, later evidence showed that the organs were packed carelessly, with three packages containing lung and one spleen and some intestine, instead of the lungs, liver, intestines, and stomach being placed in separate packages. Also circumcision had not been performed. These facts suggested that the methods were a debased form of those used in an earlier period.

The linen wrappings were examined by Johnson (1974), who believed them to be Ptolemaic. In short, all the evidence points to the Ptolemaic period, about 170 B.C.

TRACE ELEMENTS

Tests were made to see whether the concentrations of metals in bone from a vertebra could be estimated by using neutron activation and atomic absorption techniques. It seemed possible that the overwhelming amount of calcium might interfere with the measurements of other metals present in only trace amounts, and this proved to be the case with the neutron activation test; of the 20 metals sought, only the calcium could be measured.

Smith (1974) tested for lead and mercury by

using atomic absorption and found a lead concentration of 0.6 part per million (ppm) and a mercury concentration of 0.43 ppm (dry weight). According to Kehoe (1961), the lead content of modern flat bone averages 6.55 ppm and that of long bone 18.0 ppm, so PUM II had only a fraction of the lead load of modern man. The mercury level in bone from PUM II, however, is about the same as that in modern bone, which ranges from 0.03 to 1.04 ppm with a mean of 0.45 ppm (Goldwater 1972).

Using atomic absorption, Reyman obtained similar results from mummy PUM I. His heavy metal values for soft tissue were (in parts per million): lead, 1.3; copper, 1.9; arsenic, 6.2; and mercury, 0.3. The values for long bone were: lead, 2.5; copper, 2.3; mercury, 0.1; and arsenic, none detected.

Nunnelley and colleagues (1976) reported on the trace elements in PUM II. Specimens of muscle, tendon, and skin were identified visually and separated from a gross sample of mummy tissue. Trace elements were concentrated by ashing the tissue and resin samples at 410°C for 48 hours. Some elements such as arsenic, mercury, and selenium tend to form volatile compounds, and therefore the concentrations of these elements should be regarded as lower limits. Modern tissue samples were dried before ashing and otherwise treated in the same way. The samples were analyzed by X-ray fluorescence. The samples were placed in the collimated X-ray beam from a 3-kW, 60-kV X-ray tube. The characteristic X rays emitted from the samples were detected with a lithium drifted silicon detector. The X-ray intensities were corrected for absorption within the samples and compared with standards to calculate concentrations. X-ray fluorescence is well suited for a trace element scan, as many elements are detected simultaneously.

The results are shown in Table 4.1. Many elements are more concentrated in the resin than in the tissues. This complicates the interpretation of the tissue concentrations. The more exotic elements in muscle and tendon,

such as yttrium, zirconium, and niobium, are possibly attributable entirely to contamination from the resin. Because the resin is so rich in trace elements, it may be possible to develop trace element profiles of other resins to help distinguish the origin or treatment of different resin samples. The skin of PUM II is rich in calcium and strontium (a chemical homolog of calcium). This may be the result of postmortem calcium soap formation in the subcutaneous fat. Skin has a different trace element profile from either muscle or tendon, whereas muscle and tendon have very similar trace element concentrations. This may be because they had similar environments in relation to the resin and that of the skin was different.

Also shown in Table 4.1 are the average results from three modern human muscle samples. These samples were analyzed in the same way as the PUM II samples. Included for comparison in Table 4.1 are typical values for mammalian muscle as reported by Bowen (1966). Compared with modern values, PUM II tissues appear to be low in potassium. Because potassium is an essential component of tissue, the depression in PUM II tissue suggests some removal process – possibly something that occurred during the preparation of the mummy. The modern value for zinc concentration in muscle is about 170 ppm. The zinc concentration in PUM II muscle is about half the modern value. This suggests that PUM II could have had a deficiency in this essential element. However, much caution should be used in describing endemic zinc deficiencies in ancient populations until a larger number of mummies has been studied and the effect of long-term storage on trace element concentrations has been investigated.

Many disease states are correlated with trace element abnormalities. Examples include anemia caused by iron deficiency and vitamin B_{12} deficiency caused by the lack of cobalt porphyrin in the diet. In modern Egypt, retarded growth and the failure of adolescent males to reach sexual maturation have

been associated with endemic zinc deficiency.

Trace element analysis of mummy tissues may offer an opportunity to study long-term trace element body burdens. Studies may also help delineate possible disease states of a given individual before mummification.

DISCUSSION OF THE FINDINGS

Discoveries made during the autopsy of PUM II cover a broad variety of disciplines. Some merely confirm what had been reported earlier, but others are new.

The presence of silica in the lungs was surprising only because it had not been reported before. At certain times of the year, in many areas of Egypt, it is almost impossible to avoid breathing in sand; the air is full of it and so, inevitably, are the mouths and nostrils of the people living there. Shortly after the first report on PUM II appeared, Israeli workers recorded finding sand in the lungs of Bedouins in the Negev, a condition now known as Negev desert lung (1976). A similar condition was reported from a mummy in Manchester, England (1976).

Anthracosis, or carbon particles, in the lung has been reported from most mummies whose lungs have been examined. Air pollution is not a modern development. It was just as severe in antiquity and must have followed quickly after the discovery of fire making, for man would take fire with him into his home, whether cave, tent, hut, or igloo. As none of these had proper chimneys, they would fill with smoke almost to the point of asphyxiation.

Atheromatous disease of the arteries is also a common finding in mummies. Nowadays, a great deal of emphasis is placed on the stress of modern life or on modern diet as factors in the high incidence of this disorder in our present-day industrialized civilization, but the etiological influences were certainly there in the ancient world, and this fact should be taken into account in any theorizing regarding causation.

At the time the autopsy was performed, the perforated eardrum and the disease of the temporal bone in PUM II were the earliest known records of these conditions. Since then, additional finds have been reported by Benitez and Lynn (Chapter 18) and by Horne et al. (1976) in Egyptian mummies, and the same condition was discovered in the Chinese princess described in Chapter 13. In certain areas of the world, temporal bone disease was apparently not uncommon.

Barraco's recovery of protein with a molecular weight of 150,000 is a major step forward in biochemical studies of ancient tissues. It raises the possibility and practicality of demonstrating many large molecules such as gamma globulins and hemoglobins in well-preserved bodies from the past, particularly frozen ones. Once this is done, a whole new frontier of science will open up.

The periostitis of the right fibula and tibia poses a problem that has not yet been solved, though the radiologists suggest that it may be attributable to a chronic condition like varicose veins.

Trace elements were studied by Nunnelley et al. (1976) and by Smith (1974). No conclusions can be derived from examination of a single body, but the pointers for future work are valuable. First, the trace element analysis showed that the resin had penetrated not only the body cavities of PUM II but also the tissues themselves. Smith found that the lead level was only one-tenth that of today, whereas the mercury was at modern concentrations. This suggests that our present-day environment may be polluted with lead, compared with Egypt 2,000 years ago.

The low zinc level in PUM II's muscle may be significant, for it is not uncommon in modern Egypt. Obviously one should not use a single mummy as a basis for conclusions regarding a whole population. Still, the data are interesting.

The cotton ball was a startling discovery. Cotton was grown and used by 2000 B.C. both in South America and in the Indus Valley of India, but it was unknown in the Mediterra-

nean world and Persia until after the time of Christ. Was this cotton ball imported as a valuable object from India or was it grown in Egypt? The find poses a whole series of new questions regarding trade routes and agriculture for which there are as yet no answers.

The analysis of the resin by Coughlin (1977) is another striking piece of work, for she has not only named the components of the fluid by tree, but has even located their sources with some accuracy. In addition, her identification of the resin with glass is a fascinating observation. The idea that the Egyptians' development of glass may have resulted from their experiences in heating together natron and other substances for the purpose of mummification is a new concept and is worth a follow-up study by someone interested in the history of glass making.

Finally, the superb condition of PUM II's tissues is a tribute to the skills and techniques of the Egyptian embalmers. There is little sign of decay. Embodded in "glass," which has perfused every tissue, there is no reason why this body, as long as it is kept in either a dry warm or a cold place, should not continue to survive to the end of time. For a parallel, we can look at insects embalmed in amber from the Baltic Sea: They have been preserved for 30 million years already and could easily survive another 30 million. PUM II, if left undisturbed, could do the same.

REFERENCES

Bowen, H. J. M. 1966. *Trace elements in biochemistry*. New York: Academic Press.

Cockburn, T. A. 1963. *The evolution and eradication of infectious diseases*. Baltimore: Johns Hopkins University Press.

Cockburn, A. 1973. Death and disease in ancient Egypt. *Science* 181:470.

Cockburn, A.; Barraco, R. A.; Reyman, T. A.; and Peck, W. H. 1975. Autopsy of an Egyptian mummy. *Science* 187:1155–60.

Coughlin, E. A. 1977. Analysis of PUM II mummy fluid. *Paleopathology Newsletter* 17:7–8.

Fischer, H. 1974. Quoted in A. Cockburn, R. A. Barraco, T. A. Reyman, and W H. Peck. Autopsy of an Egyptian mummy. *Science* 187:1155–60, 1975.

Goldwater, L. J. 1972. *Mercury: a history of quicksilver*. Baltimore: Fork.

Horne, P. D.; Mackay, A.; John, A. F.; and Hawke, M. 1976. Histologic processing and examination of a 4,000 year old human temporal bone. *Archives of Otolaryngology*. 102:713–15.

Johnson, M. 1974. Quoted in A. Cockburn, R. A. Barraco, T. A. Reyman, and W. H. Peck. Autopsy of an Egyptian mummy. *Science* 187:1155–60, 1975.

Kehoe, R. A. 1961. The metabolism of lead in man in health and disease. *Journal of the Royal Institute of Public Health and Hygiene* 24:1–40.

Miller, J. I. 1969. *The spice trade of the Roman Empire*. London: Oxford University Press.

Nunnelley, L. L.; Smythe, W. R.; Trish, J. H. V.; and Alfrey, A. C. 1976. Trace element analysis of tissue and resin from Egyptian mummy PUM II. *Paleopathology Newsletter* 12:12–14.

Smith, R. G. 1974. Quoted in A. Cockburn, R. A. Barraco, T. A. Reyman, and W. H. Peck. Autopsy of an Egyptian mummy. *Science* 187:1155–60, 1975.

Stukenrath, R. 1974. Quoted in A. Cockburn, R. A. Barraco, T. A. Reyman, and W. H. Peck. Autopsy of an Egyptian mummy. *Science* 187:1155–60, 1975.

ROM I: mummification for the common people

NICHOLAS B. MILLET
Curator, Egyptian Department
Royal Ontario Museum,
Toronto, Ontario, Canada

GERALD D. HART
Physician-in-Chief
Toronto East General Hospital,
Toronto, Ontario, Canada

THEODORE A. REYMAN
Director of Laboratories
Mount Carmel Mercy Hospital,
Detroit, Michigan, U.S.A.

MICHAEL R. ZIMMERMAN
Associate Professor,
Departments of Anthropology and Pathology
University of Michigan,
Ann Arbor, Michigan, U.S.A.

PETER K. LEWIN
Hospital for Sick Children
Toronto, Ontario, Canada

Herodotus mentions that the least expensive form of mummification was that where no treatment was given and the body was simply wrapped in linen. ROM I, autopsied in Toronto in August 1974, was just such a mummy, and it proved to be one of the most interesting studied by our group. The name is shorthand for the Royal Ontario Museum, to which the mummy belongs. The autopsy was an international operation, a demonstration in cooperation among the Toronto Academy of Medicine, the Royal Ontario Museum, and the Detroit group of the Paleopathology Association, with literally dozens of workers involved. The participants are named in reports published elsewhere (Hart et al. 1977; Millet et al. 1978), and those playing major roles are listed at the end of this chapter. The study arose as the result of a lecture on PUM II given by Aidan Cockburn and Theodore A. Reyman in February 1974 at the Academy of Medicine in Toronto. The following day, Eve Cockburn suggested a joint Canadian – United States project, and the day after that Nicholas B. Millet offered to lend a mummy from the collection of the Royal Ontario Museum.

The examination was carried out in a laboratory at the anatomy department in the Medical Sciences Building, University of Toronto. The proceedings were recorded by means of still photography, 16-mm color film, and videotape. The body was unwrapped by Millet and the museum staff in August 1974, and the autopsy was directed by Reyman, Zimmerman, and Lewin. Radiology studies by D. F. Rideout had been completed some days before the unwrapping.

After the autopsy, the tissue specimens were divided among those participants who were interested. This had the advantage of ensuring greater coverage in the search for abnormalities. For example, Lewin reported *Schistosoma haematobium* ova in the liver; Zimmerman found them in the kidney as well as red cells in the bladder; and Reyman discovered ova in the intestine and cirrhosis in the liver. All these related findings gave a fairly clear picture of a single disease that would not have been obtained by a solitary worker.

THE MUMMY

In the winter of 1904–1905 Dr. C. T. Currelly, founder of the Royal Ontario Museum, was attached to the Egypt Exploration Fund's expedition at Deir el-Bahri, across the river from modern Luxor, where he assisted at the excavation of the funerary temple of Menthuhotep II, a king of the Eleventh Dynasty (ca. 2010 B.C.). All the tomb chambers in the temple area had been robbed in antiquity, and

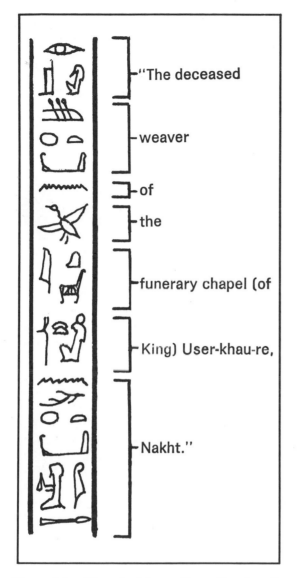

"The deceased

weaver

of

the

funerary chapel (of

King) User-khau-re,

Nakht."

Figure 5.1.   Hieroglyphics on the coffin, describing Nakht. (From Hart et al. 1977)

ner, with planks and small pieces of wood, probably sycamore. The lid, which is carved in anthropoid shape, shows the deceased wearing a long wig striped in blue; his forearms and clenched hands are executed in relief. The whole exterior of the coffin, except the bottom, is covered with a thin coating of gypsum plaster and gaily painted with the usual scenes and inscriptions. Time and the excreta of bats have eroded much of the lid's surface, but the scenes on the box itself are well preserved. The texts inscribed on it (Figure 5.1) tell us that the owner was a male named Nakht, invariably described as "weaver of the *kny*-temple of User-khau-re." User-khau-re was the throne name of the king Setnakht, the first ruler of the Twentieth Dynasty, who died about 1198 B.C. after a brief reign, leaving the throne to his son, the great Pharaoh Rameses III. The latter established a funerary cult for his dead father on the west bank at Thebes, which was carried on in a mortuary chapel of the type called *kny*, a word meaning both "armchair" and "carrying chair" or "sedan chair." The name seems to refer to either a statue of the deceased god-king in a palanquin or an empty thronelike chair over which the spirit of the dead ruler was thought to hover. The chapel's staff of priests and laymen and the regular food offerings for the dead pharaoh's spirit depended on the income from the land with which it was endowed. The fact that the establishment of King Setnakht's chapel included weavers suggests that some of the land was used to grow flax, linen being the common cloth of Egypt in those days. Setnakht's funerary chapel was still being maintained as late as the second regnal year of Rameses IV, about 1164 B.C., and may have continued in being for a century more. The style of Nakht's coffin suggests that it was made in the first half of the twelfth century B.C.

From the historian's point of view, Nakht's mummy was a natural choice for autopsy; there is a good account of its discovery and

many had been used in later times by humbler people. It was in one of these that the coffin of Nakht was found. Although the date the mummy was acquired by the museum is not known, we must assume that its acquisition was directly attributable to Dr. Currelly's presence at the time of the excavation.

The coffin was built in the traditional man-

excavation, and the texts and the style of the coffin provide unusually clear evidence of its date. The simple title "weaver" shows that Nakht was a person of the laboring class and was far more representative of the great mass of the ancient population than most mummies, which tend to be of persons from the middle classes or the aristocracy, because these persons were better able to afford the considerable expense of a traditional Egyptian burial. Finally, we are probably better informed about Nakht's times – the last century of the New Kingdom, whose imperial glories had by this day faded and given way to dreary years of political confusion, moral uncertainty, and spiraling inflation – than we are about any other period of Egyptian history.

Nakht's lower-class family apparently had been able to afford a relatively fine coffin because he had not undergone the expensive process of mummification that included removal of the viscera, but had simply been washed and wrapped in linen (the good preservation of his body was owing entirely to the peculiarly dry and stable Egyptian climate). By omitting mummification, his family had been able to spend more of their presumably slender resources on the coffin itself. The records of this time note that workmen were sometimes given time off to construct the coffin of a deceased member of their guild or group, so cooperative effort may have reduced some of the expense. The cost of decorating such a coffin in Nakht's day was approximately 31 g of silver, representing as much as 10 percent of a working family's yearly income, so his grieving parents must have been willing to sacrifice a great deal to see their young son suitably interred.

Although the body had been carefully wrapped in linen in the traditional manner, the amount of linen used was noticeably less than in many mummies of the period. The bandages were in very good condition and came away easily, chiefly because none of the usual sacral oils had been poured over the body. The manner in which the bandages and filling pads had been arranged suggests familiarity with ritual custom and hints at a professional hand, but the poor people of western Thebes in the twelfth century B.C. may have attained a certain expertise in wrapping their own dead simply from necessity. The wrappings were well preserved in all parts of the mummy except the head, where they had broken through entirely, probably more as a result of the collapse of the cranial vault beneath them than from interference by tomb robbers (Figure 5.2). Several of the pads contained in the wrappings proved to be more or less complete garments; two large, sleeveless tuniclike robes of a type familiar to us from wall paintings and sculpture of the period represent the characteristic male costume (Figure 5.3). Their size is about right for Nakht himself, and his family probably contributed some of his clothing to provide wrapping material. Each piece of cloth was carefully scrutinized for laundry or owner's marks, but none was found.

Surviving records of the community, labor rolls and fragments of official day books, tell us something of the physical conditions under which Nakht lived during his short life. Nakht's house was probably not at Deir el-Medina but nearer his place of burial, in the Asasif valley below the Deir el-Bahri temples, and nearer the cultivation on the east. Since he was in his middle teens when he died, he was probably what the official records of his day describe as a *mnh* or "stripling," a youth of employable age but still unmarried and living in his parents' home and drawing a smaller ration allowance (wages being paid in kind) than the head of a family. The house in which he lived was probably similar to excavated samples, consisting of two or three rooms of mud brick with a flat roof of earth on rafters of palm logs, plus a small courtyard. His diet and general standard of living would have been better than those of most of the peasant cultivators who composed the bulk of the population, and by any reckoning he must have been in the upper

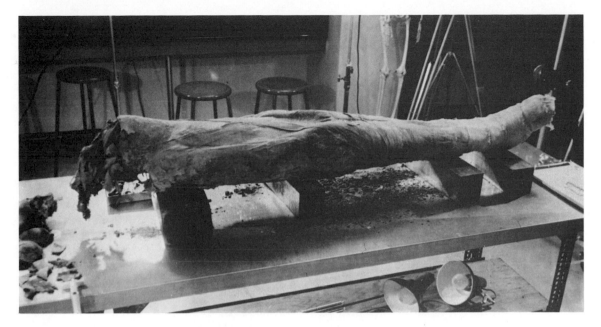

Figure 5.2.   Mummy before unwrapping. Note the broken skull.

ranks of the working class. In addition, he enjoyed the security derived from being attached to a permanent and prestigious institution, which may have exempted him from conscription for military service or forced labor and may have protected him from the economic troubles that afflicted the kingdom in his time.

RADIOLOGIC FINDINGS

The subject was a male teenager, but more exact determination of age was difficult because desiccation had widened some epiphyseal lines and narrowed others, so that the hip joints suggested age 18 and the knees age 14. The left knee showed unfused epiphyses and Harris's lines suggesting episodes of infection or malnutrition during life. The hands were not well shown because they overlapped the pelvis. As far as could be seen, they corresponded with a modern standard of age 14 years, so it was reasonably certain that Nakht was between 14 and 18 years of age at the time of his death. The wisdom teeth were

well developed but unerupted. All the other teeth were accounted for and in good condition, although the upper incisors had fallen into the wrappings postmortem.

The pelvis was of male configuration, but appeared to show protrusio acetabuli. As there was no other evidence of rickets or osteomalacia, this change was probably caused by postmortem bending of an unfused Y-cartilage in the acetabulum. This conclusion was confirmed by a xeroradiograph, which showed the cartilage space where the conventional film did not. It was also later confirmed at autopsy.

X rays of the chest showed a mass in the center and to the right that was too large for a desiccated heart and was assumed to represent the heart and liver adhering together. Patchy opacity in both hemithoraxes suggested the remains of lung tissue adhering to the posterior chest wall. Tomography confirmed the position of these masses within the body cavity and also showed linear opacities – an intact diaphragm – between abdomen and thorax.

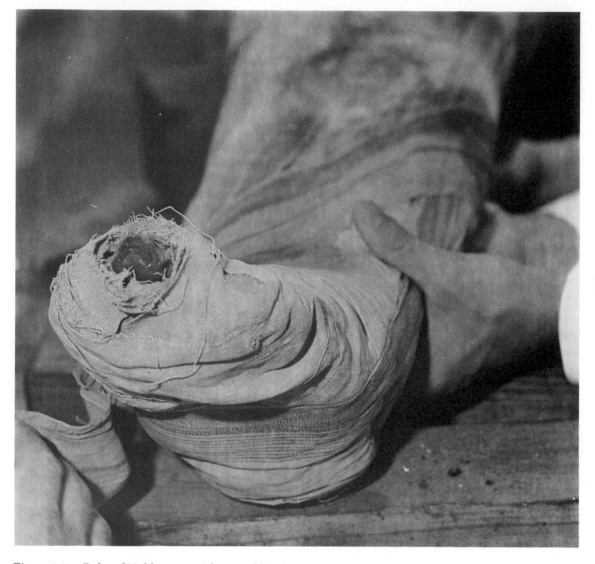

Figure 5.3.   Robe of Nakht wrapped around his feet.

Between the shrunken tissues and the wrappings there was an air-containing gap up to 4 cm across. This indicated that desiccation occurred after wrapping. Harris's lines in the lower 1.5 cm of the distal femoral metaphyses suggested recurrent severe illness in the last 2 years of his life.

Radiological examination gave no evidence that Nakht had been eviscerated or even embalmed. He had desiccated naturally in the dry, hot air of Thebes. Apart from the rather doubtful evidence of Harris's or growth arrest lines, there was no radiological clue to the cause of death.

GROSS ANATOMY

A general study showed that the wrappings of the head had been removed earlier and that the skull had been damaged (Figure 5.2). In

addition, the cranial sutures had not united, and the skull bones had collapsed inwardly and the facial features disintegrated. This gave a first impression of decapitation. However, the pieces of broken bone were recovered from the wrappings.

The body showed no evidence of artificial preservation by means of natron, oils, or bitumen. The body, which was firm, demonstrated an atypical rigor mortis (it could be lifted like a board). The intact body weighed only 5.13 kg, and the overall length was 143.1 cm.

Circumferential measurements of upper and lower extremities before and after unwrapping revealed a shrinkage of 17 cm at the midarm level and of 16.8 cm at the knees. This shrinkage, which could also be seen on the X-ray plates, indicated that the dehydration that had preserved the body had occurred after it had been wrapped.

The skin was a light café-au-lait color and had a tough, leathery consistency. At areas of fissuring, the muscles were visible and had a crumbly, friable consistency. The whorls and ridges on the fingers and toes were preserved. The toeprints were very well preserved and easy to see. The fingernails and toenails were also well preserved.

The autopsy began with a special study of the remnants of the head and neck area. The disturbed wrappings acted as a receptacle to contain the various structures. The skull bones were carefully retrieved and the skull was subsequently reconstructed. On some areas of the frontal and parietal bones the scalp, with hairs 2 mm long, was still adherent. The imprint of the ears was present in the head wrappings, where many loose hairs were also found. The left eye socket was covered with a thin, translucent film on which eyelashes were identified. The teeth showed some wear, but were well preserved and did not show evidence of caries. Three teeth had become dislodged and were later retrieved from under the wrappings of the back and lower extremities.

The cerebral hemispheres were found lying in the base of the skull (Figure 5.4). The left hemisphere weighed 61.6 g and measured 9 by 7 by 2.5 cm. The right hemisphere weighed 66.15 g and measured 8 by 6 by 2.5 cm. The hemispheres were dark brown, firm in consistency, and had a soapy feel. The convolutions were preserved and best seen on the inferior surface. The sigmoid sinus was identified, and dark rusty brown material suggesting dried blood was adherent to the internal lining. The mandible remained in position supported by the cervical structures. There was a postmortem transverse fracture of the neck of the mandible just below the left condyle.

The thorax was opened with the assistance of a Stryker saw. This cut through the leathery skin and the anterior lateral rib margins so that the anterior chest wall could be removed in toto. This gave a unique anatomical demonstration of the pericardium and its attachments (Figure 5.5). The ligaments had been preserved, but the fat had disappeared. The pericardium appeared as a tent tethered between the sternum and the thoracic spine. The lungs had collapsed posteriorly, and it was not possible to distinguish the upper and lower lobes. They were grey black and of a powdery consistency. The aortic arch was transected, and the heart and mediastinum were removed. The heart with the great veins and arteries weighed 17.7 g.

The trachea was identified and removed with the thyroid cartilage. The intact diaphragm was dark brown and its consistency was firm and leathery. It was cut around its periphery and removed. The anterior abdominal wall was then removed in toto. Internally, the wall was connected to the liver by the round ligament. The liver was shrunken, but had retained its shape and had a sharp lower border. It measured 12.5 by 7.5 by 3 cm and weighed 106.3 g. Portions of the gallbladder were removed.

The spleen was present and, taking into account the amount of shrinkage of other organs, was enlarged. It was crumbly and so could not be weighed in toto. The posterior

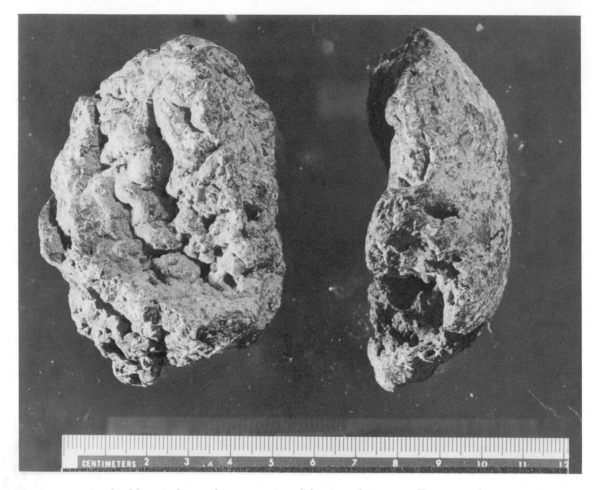

Figure 5.4.   *Cerebral hemispheres showing some of the convolutions well preserved.*

and lateral walls of the splenic bed showed a dark discoloration, possibly representing hemorrhage from a splenic rupture. The intestines were paper-thin and collapsed. The bladder was a saclike hollow structure measuring 6 cm at its widest point. The prostate was not identified. The rectum passed through the pelvic cavity; it was desiccated and very fragile. Removal of the posterior pelvic peritoneum displayed the anatomical details of the sacral plexus excellently preserved. The penis was noncircumcised. The scrotum could be identified, but testicular material was not found.

HISTOLOGIC FINDINGS

The heart and brain were not processed in detail, because they had to be retained for permanent display. The other tissue specimens were rehydrated in modified Ruffer's solution and processed in the same way as fresh tissue. Sections were stained with hematoxylin and eosin, Masson's trichrome, periodic acid–Schiff, acid-fast, Grocott's methenamine silver, elastic tissue, and alizarin red staining techniques, and examined with polarized light (Chapter 15). Histological detail of the tissues was preserved to a

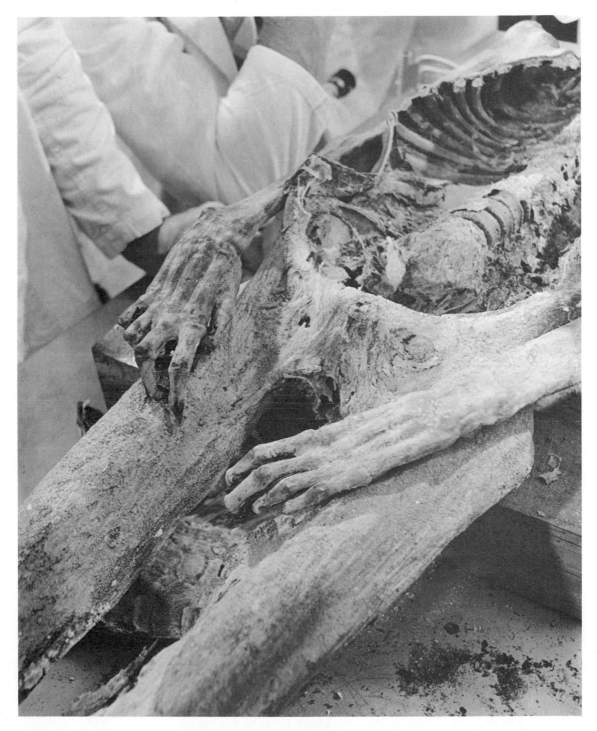

Figure 5.5   Trunk and hands after removal of the organs.

variable degree, although distinct cellular outlines were generally lost. The tissues were infiltrated with numerous saprophytic invaders, both bacterial and fungal, but these organisms could not be cultured.

In the lungs the alveolar architecture was partially preserved, and there was marked deposition of anthracotic pigment within connective tissue. Throughout the tissue, bright birefringent particles were noted; however, the results of silica analyses on lung tissue were normal. Electron microscope microprobe diffraction analysis at University College, Cardiff, Wales, by Dr. E. Pooley, suggests that these particles are granite.

The muscle coats of the intestinal tract at various levels were well preserved. The lumens of both large and small intestines contained numerous ova of both *Schistosoma* spp. and *Taenia* spp. (Chapter 15). Some of the *Schistosoma* ova had large terminal spines. No adult worms of either type were seen, but the *Taenia* ova were clumped together, suggesting the site of degenerated proglottids.

The liver showed preservation of cords of indistinct hepatic parenchyma and a fibrous pattern of early cirrhosis. Thick and thin fibrous septa were common, and the organ contained occasional small nodules of parenchyma. Portal areas contained calcified *Schistosoma* ova with terminal spines similar to those noted in the intestinal lumen. The spleen was poorly preserved and showed heavy postmortem microbial contamination. No malarial pigment was seen. Sections of the gallbladder were poorly preserved. Histologically, the brain was poorly preserved as well. Sections from the myocardium, a coronary artery, and an aortic valve were normal. The various components of the kidney were not well preserved, but several *Schistosoma* ova were seen. The bladder revealed remnants of epithelium. We found no ova or changes suggesting inflammation, but a few well-preserved red blood cells on the mucosal surface suggested hematuria during life.

Postmortem examination of this 3,200-year-old mummy of a teenage Egyptian boy revealed a variety of disease processes. He had at least two types of parasitic infestation, both of which may have produced severe complications.

Numerous ova of *Taenia* spp. were found within the intestinal tract. As the ova of *T. solium* and *T. saginata* cannot be differentiated and no scolex was found, no final decision can be made concerning the type of infestation.

Many schistosomal ova were also found, several with the large terminal spines diagnostic of *S. haematobium* and others without obvious spines. These ova may be degenerated forms or may be *S. mansoni*. Both forms of schistosomiasis are endemic in modern Egypt, and *S. haematobium* has been reported in other Egyptian remains. There was evidence of early cirrhosis of the liver and congestive splenomegaly, possibly with terminal rupture. Schistosomal ova were found in the liver, suggesting that the cirrhosis was secondary to hepatic schistosomiasis. Though *S. mansoni* is the commonest offending parasite, *S. haematobium* may give the same picture in the liver. The calcified ova in the kidney and the presence of blood within the urinary bladder also indicate involvement of this system with *S. haematobium*.

Finding tapeworm ova is also of interest, implying as it does that Nakht ate meat, and probably meat that was not well cooked. Meat fibers have been found in the intestine of another mummy, so ancient Egyptians were probably not strict vegetarians. In any case, the associated malnutrition probably contributed to the youth's demise. The presence of Harris's lines in the leg bones supports this thesis.

An incidental finding, present in almost all ancient remains, is pulmonary anthracosis, which is attributable to environmental pollution from cooking and heating fires and from oil lamps in small rooms. It is also possible that this boy had pulmonary silicosis resulting from inhalation of sand during a sandstorm.

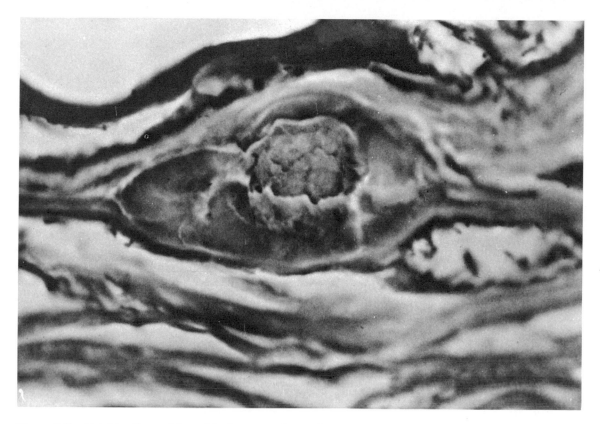

Figure 5.6.   Trichinella *cyst found in intercostal muscle.*

Near the subcutaneous border of an intercostal muscle a small cyst was found, just visible to the naked eye (Figure 5.6). This cyst has the appearance of a parasite. At first, *Cysticercus cellulosae* (the larva of *Taenia solium*) was considered, but these cysts are about 5 mm in diameter and much larger than the muscle fibers. The cyst is small to be that of *Trichinella spiralis*, but it is appropriate when compared with the muscle fibers. The small size could be the result of shrinkage during drying. *Trichinella* infestation, like cysticercosis, is caused by eating inadequately cooked pork, and the finding of this cyst suggests that Nakht ate pork.

EXAMINATION OF TEMPORAL BONES

The temporal bones were examined (Lynn and Benitez 1977) by means of polytomog-raphy and studied under the operating microscope to determine whether ear disease that might have occurred during his lifetime could still be identified (Chapter 18).

The foramen of Huschke was an obvious defect in the temporal bone. As noted by others, its pathological significance is unknown. The ossicles in both temporal bones were dislodged from their normal position in the middle ear cavity. However, they were normally formed and showed no signs of the destructive effects of chronic middle ear disease or mass lesions. It would therefore appear that such ossicular dislocations represented a postmortem artifact. As X rays with polytomography revealed normal air cell system throughout the mastoid regions and no increased density attributable to sclerosis in the area of the oval windows, it seems certain that Nakht did not suffer from chronic middle ear disease or otosclerosis in either ear during

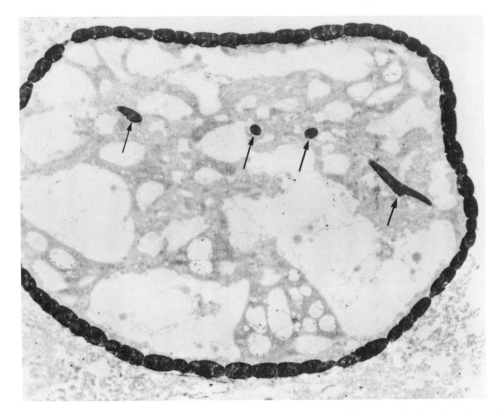

Figure 5.7.   *Electron micrograph showing* Taenia sp. *ova. Three of the hooklets have been cut in cross section; the fourth is seen in full view.* × 5,800.

his lifetime. Furthermore, no bony changes were evident in the inner ears or internal auditory canals, thus indicating the absence of such destructive lesions as neoplasm or congenital malformations.

Examination of each temporal bone with the Zeiss operating microscope confirmed the abnormal position of the ossicles that had been observed radiographically. Although the three ossicles in the right ear were grossly displaced in position, they were normal in form and showed no signs of erosion or destruction, as would have been the case if Nakht had suffered from chronic otitis media, cholesteatoma, or other destructive lesions of the middle ear.

ELECTRON MICROSCOPIC FINDINGS

Tissue samples were taken from the sole of Nakht's right foot and rehydrated in buffered 10 percent formalin. Dry, dustlike intestinal contents were spun in normal saline to obtain a button of material and were rehydrated in Sandison's modification of Ruffer's solution consisting of two parts 5 percent sodium carbonate (aqueous), three parts 96 percent ethyl alcohol, and five parts 1 percent formalin. The epidermis of the skin was well preserved, and cellular components were easily recognized. The intestinal contents contained several ova of *Taenia* spp., which at the ultrastructural level demonstrated in great detail the striated embryophore of an egg with its hooklets (Figure 5.7). Figure 5.8 shows a red cell found in the intestinal contents as visualized with a scanning electron microscope. These observations confirm earlier reports that Egyptian mummified material contains preserved cells with recognizable cytoplasmic organelles.

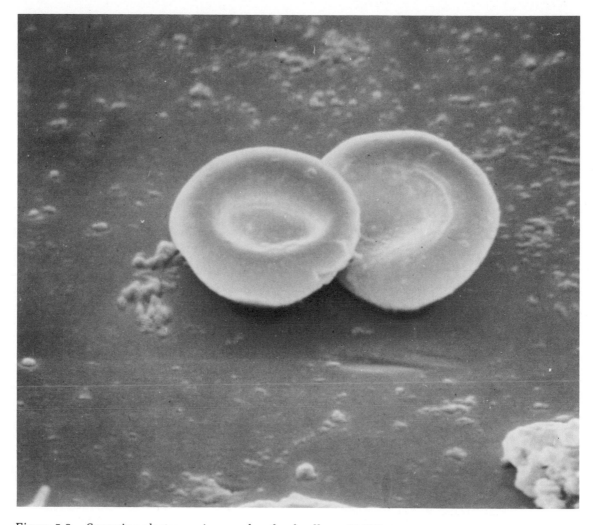

Figure 5.8.   Scanning electron micrographs of red cells. × 23,700.

BLOOD GROUP TESTING

Splenic material and a dark brown substance from the inside of the sigmoid sinus of the boy Nakht were tested with both the serological micromethod (SMM) and the inhibition agglutination tests (IAT). Repeated testing of the splenic material using SMM produced no agglutination, and the procedure was complicated by hemolysis of the absorbed group O cells. However, when splenic material was used in the IAT, a positive result for blood group B was obtained. The sigmoid sinus material showed a positive reaction for blood group B with both SMM and IAT.

There was no explanation for these results until histological sections of the spleen and the sigmoid sinus were studied. The splenic material was heavily contaminated by bacterial and fungal spores, which were hemolytic to the group O test cells. Hence, the presence of B antigen in the splenic material could not be demonstrated by absorption into group O cells, but was detectable by the IAT. Histological sections of the sigmoid sinus showed excellent preservation of red cells and absence of bacterial spores. When tested, this material gave a good source of blood group antigen and the tests were not negated by contaminating hemolytic spores.

The preservation of intact red cells for many centuries is not unique. Zimmerman (1971) has reported preservation of intact red cells in the pulmonary vein of a naturally preserved 2,000-year-old American Indian mummy. At this time, the red cells isolated from Nakht are the oldest known preserved human blood cells. The histological findings make us confident that our testing techniques are bona fide and that Nakht's blood group was B.

ANALYSIS OF PROTEIN EXTRACT

Attempts were made to extract, fractionate, and identify proteins and other macromolecular components in the tissues of Nakht. The prime objective was to detect gamma globulins and the antibodies against infectious agents they may contain. In addition, an attempt was made to identify certain lipids, particularly neutral lipids, to determine the dietary milieu of the young weaver. The following paragraphs describe the degree of preservation of proteins in Nakht's tissues and the nature of the process by which he was mummified.

Because it is known that natron and other salts were used for chemical mummification by the ancient Egyptian embalmers (Lucas and Harris 1962), we studied the relationship of Na and other cation levels in Nakht's tissues to the molecular weight distribution (a measure of degradation) of extracted protein from the same tissues. Desiccated aliquots were weighed, ashed, and solubilized in a $SrCl_2$–HCl solution for analysis of Na, K, Ca, and Mg by atomic absorption spectrophotometry (see Table 19.2).

Figure 19.3B shows the chromatographic pattern for protein (extracted by the same procedures) from freshly autopsied human skeletal muscle and for the extracted protein from Nakht. Peak I represents the high-molecular-weight protein (ca. 150,000 daltons); peak II represents proteins of intermediate molecular weight (ca. 60,000); and peak III represents low-molecular-weight protein (ca. 20,000 or less). In Table 19.2 the various tissue cation levels for Nakht and for freshly autopsied human skeletal muscle are represented as micromoles of cation per gram of dry weight of tissue (gdw). Table 19.2 makes it apparent that Nakht was not natronized or chemically treated with other salts, as tissue cation levels approximate closely those of the freshly autopsied specimen. Further, from Figure 19.3 it is apparent that much of the high-molecular-weight protein from Nakht's tissues has undergone degradation, as can be seen by the shift in chromatographic profile toward the lower-molecular-weight region. Of interest here is a previous report (Barraco 1975) that the degree of preservation of extracted protein from mummified tissue is enhanced with increasing levels of Na in the tissues (natron). The percent amino acid composition of extracted protein from Nakht closely approximates that of myoglobin, the major small-molecular-weight protein (ca. 17,000) in skeletal muscle that may be resistant to degradation.

OTHER STUDIES

When nuclear medical techniques were applied to Nakht, whole-body examination of the intact mummy before unwrapping indicated a high concentration of manganese present in the soil and other contaminants adherent to the wrappings. Various samples were cultured aerobically and anaerobically, without significant results. Electron microscope studies of liver tissue did not demonstrate virus particles. Radioimmunoassay of liver tissue was negative for hepatitis B antigen. Studies of craniofacial characteristics and dentition were also carried out at the faculty of dentistry, University of Toronto.

Computerized transaxial tomography (CTT scan) done on the intact brain at the Hospital for Sick Children showed that the gray and white matter had been preserved and that the lateral ventricles were intact. This procedure made destructive examination of the brain unnecessary. Calcified cysts suggesting cerebral cysticercosis were not demonstrated, and therefore a needle biopsy of affected areas

was not carried out. This is the first time CTT has been used in paleopathology.

The findings reported here are now part of a teaching exhibit on disease in ancient times, sponsored by the Royal Ontario Museum and the Academy of Medicine. The mummy itself is on display at the academy.

DISCUSSION

The importance of ROM I is that the mummy can be located accurately in time and place. To have actual facts like these is a rare luxury when dealing with antiquity, so the data are priceless. We can now state that in the twelfth century B.C., on the bank of the Nile opposite present-day Luxor, there existed a focus of infection of the parasite *Schistosoma haematobium*. This has been proved: Speculations about the epidemiology of the infection in Egypt can begin with this firm base.

The identity of the tapeworm is less certain, for it could be either the pork or the beef parasite. However, the cyst of *Trichinella spiralis* is definitely pork-related, so the tapeworm is probably the same. This raises the interesting subject of eating pork. In the Middle East today, eating the flesh of the pig is taboo, but many people think that in these cases, as with the cow in India, the animal in question was once regarded as sacred. According to Fraser, in his *Golden Bough*, the pig in Egypt was the totem animal of Osiris. Osiris was killed, rose from the dead, reigned over life after death, and was represented as a mummy. The flesh of the pig was forbidden year-round except on the holy day of Osiris. On that day, everyone ate pork, and even the poor who could not afford meat made cakes resembling pigs and ate them instead. Certainly, the findings from ROM I indicate that 3,000 years ago at least one Egyptian ate pork on at least one occasion.

The most likely explanation for the pork taboo is a religious prohibition, probably going back long before Moses to the dawn of Egyptian history. A modern explanation linking it with hygienic reasons and trichinosis can be dismissed. The parasite *Trichinella spiralis* was not discovered until 1834, and its linking with pigs came two decades later. The human disease was not recognized before that time, nor was its association with pigs. The hygienic theory is probably a modern rationalization for an otherwise inexplicable taboo.

ACKNOWLEDGMENTS

Special sections were contributed by the following persons: radiology, D. F. Rideout; trichinosis, U. de Boni, M. M. Lenczner, and John W. Scott; temporal bones, George E. Lynn and Jaime T. Benitez; electron microscopy, Peter K. Lewin and Patrick Horne; blood group testing, Gerald D. Hart, Inge Kvas, and Marja Soots; analysis of protein extract, Robin A. Barraco; nuclear medical techniques, B. N. Ege and K. G. McNeil; electron microscopy, L. Spence; dentition, Arthur Storey and D. W. Stoneman; computerized transaxial tomography, Derek Harwood-Nash.

REFERENCES

Barraco, R. A. 1975. Preservation of proteins in mummified tissue. *Paleopathology Newsletter* 11:8.

Hart, G. D.; Cockburn, A.; Millet, N. B.; and Scott, J. W. 1977. Autopsy of an Egyptian mummy
– ROM I. *Canadian Medical Association Journal* 117:461–73.

Lucas, A., and Harris, J. 1962. *Ancient Egyptian materials and industries*. London: Edward Arnold.

Lynn, G. E., and Benitez, J. T. 1977. Examination of ears: Autopsy of an Egyptian mummy. *Canadian Medical Association Journal* 117:461–73.

Millet, N. B., and Reyman, T. A. *Nakht: the weaver of Thebes. An archeologic and autopsy study*. Toronto: Royal Ontario Museum. To be published.

Zimmerman, M. R. 1971. Blood cells preserved in a mummy 2,000 years old. *Science* 180:303.

6

Egyptian mummification with evisceration per ano

THEODORE A. REYMAN
Director of Laboratories
Mount Carmel Mercy Hospital
Detroit, Michigan, U.S.A.

WILLIAM H. PECK
Curator of Ancient Art
Detroit Institute of Arts
Detroit, Michigan, U.S.A.

The Egyptian embalmers were masters of their art. The arduous trial-and-error method they employed for 2,000 years resulted in a scientific discipline for preserving bodies (Mokhtar 1973). This embalming technique required long hours of toil, expensive medicants, and fine linen wrappings. The costs were necessarily high. Because the majority of people who were mummified in the early dynasties were royalty or were the rich and influential of the time, this had little importance. However, when the poorer segments of the population were finally allowed the privilege of mummification, these costs were prohibitive. For this reason and perhaps others, a significant number of modifications were made in the classical mummification process, most of which appeared to be designed to reduce the cost and the time involved. Although the external form and appearance of the mummy remained the same, the fact is that mummification for the poor became less a preservative and more a symbolic exercise. The rich were given the specialized care demanded by their wealth; the poor were given only what they could afford. With PUM II, we have seen mummification similar to the classical method. In this chapter, two examples will be presented of a common alternative method of embalming: mummification with evisceration per ano. Both mummies were provided by Dr. David O'Connor, Department of Egyptology, Pennsylvania University Museum, and neither had known provenance or coffin. The first mummy, an adult, was designated PUM

III; the second, a child, PUM IV. They will be described separately.

PUM III

RADIOGRAPHY

Before the unwrapping, the mummy was examined radiographically. The body arrived with the head unwrapped and separated from the body at the level of the fifth cervical vertebral body. This separation appeared to have occurred postmortem.

There was a healed fracture of the second left rib. Harris's, or growth arrest, lines were present in the distal femora. A slight dorsal scoliosis was noted.

The chest cavity showed air contrast, and there were central densities that had the general positions of the heart and mediastinal tissue and collapsed lungs. The hemidiaphragms clearly defined the chest cavities inferiorly. In the lower portion of the right hemithorax, lying on or near the diaphragm, was an irregular density with speckled opaque areas.

The contents of the abdominal portion of the trunk were less well defined. There was a density in the right upper quadrant suggesting the liver, conforming to the smooth outline of the diaphragm. Overlying this and present within the remainder of the abdominal cavity were numerous speckled opacities and irregular densities without pattern or identifiable characteristics. In the pelvis, a double shadow contour centrally suggested the urinary bladder with a central air contrast

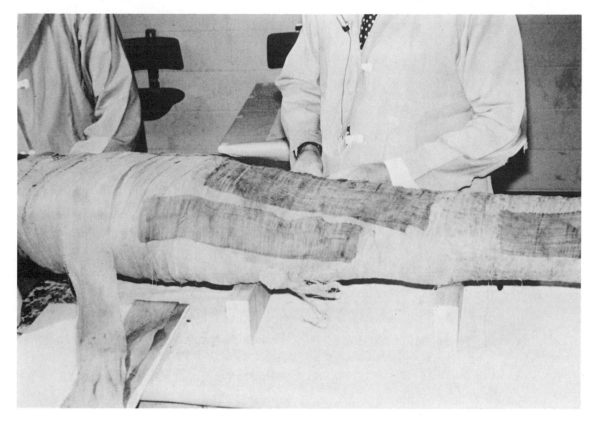

Figure 6.1.   *Dark linen noted in the deeper layers of PUM III's wrappings, used to fill out the body contour. (Photograph by Nemo Warr, Detroit Institute of Arts)*

lumen; uterine shadow was visible directly behind it. Pelvimetry was indeterminate, but the pelvic soft tissue shadows strongly suggested that the mummy was a female. There were numerous irregular wavy air contrast lines within the muscle groups of the thighs, suggesting fissuring of the tissue.

THE WRAPPINGS

The next step was the unwrapping of the mummy. Because the head was already separated from the body and the greater part of the head bandaging was gone, there was no way of knowing how the head had been wrapped. The outer layers of the body wrapping were without decoration and consisted of a medium quality linen apparently torn from larger sheets. Most of these outer bandages

were complete strips with a straight woven end and a fringed end that had been torn from a sheet originally 6 or 7 m long. Whenever possible, these individual linen strips were wrapped around rectangular cardboard as they were removed and labeled according to anatomical site and the layer of the wrapping. With care, it was possible to reconstruct whole sheets or parts of sheets from the strips. Intermixed with the first few layers of outer wrappings were small folded pads of a darker and slightly finer linen (Figure 6.1). These were torn from larger pieces, but to no apparent standard. When unfolded, they were generally rectangular, and some could be matched with others to re-form the original larger pieces. The function of these pads, which were generally on or near the front of

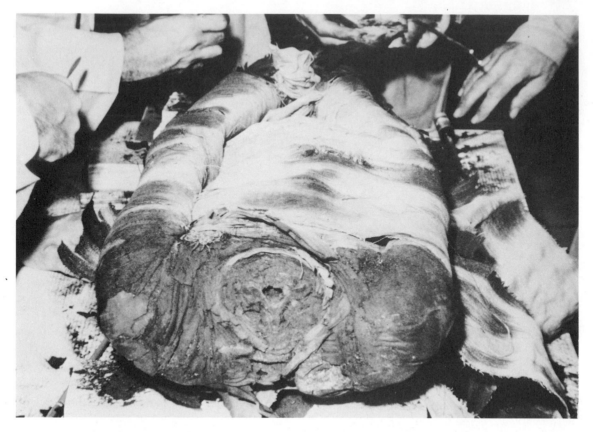

Figure 6.2.   Mummy PUM III viewed from above. The circular bandages around the body enclose the right arm, but not the left. Note the trachea (hole) in the area of the neck from which the head had become separated. The body tissues were so severely degenerated that the arms also became detached inadvertently during the unwrapping. The dark-stained areas on the cloth are the result of chemicals used during the embalming, not of the body fluids. The irregular pieces of cloth between the circular wraps were used to fill out the body contour. (Photograph by Nemo Warr, Detroit Institute of Arts)

the lower body and legs, was to increase the finished size of the body as well as to regularize its outline.

By the fifth layer, the linen material began to show signs of deterioration, possibly from the body fluids, but more likely from liquids used in the wrapping process. After about 10 layers of transverse wrappings, the size of the bandages increased to large sheetlike strips that appeared to have been laid on lengthwise to cover the entire body. At various stages, wood chips, reeds, and fibers were found, all of which were probably accidental inclusions during the wrapping process. Under the large

lengthwise coverings, the wrapping became very irregular, as did the nature of the packing. The spaces between the arms and the body were filled with linen wadding. The legs were bound together with a figure-of-eight continuous bandage, probably to keep the legs together and also to reduce the amount of wadding material necessary. The right arm was bound to the body by a continuous layer of circular bandages; the left arm, at the same stage of wrapping, was not (Figure 6.2). The limbs were individually wrapped, but the fingers and toes were not. The layer of wrapping immediately next to

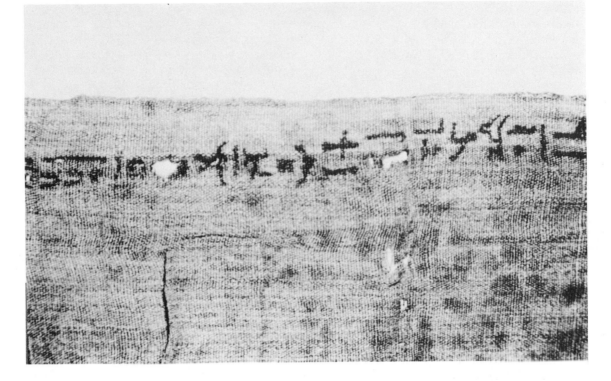

Figure 6.3.   The larger of the two hieroglyphic inscriptions on PUM III's wrappings. See text. (Photograph by Nemo Warr, Detroit Institute of Arts)

the body was crossed over the shoulders and through the pubic area, having been wound around the body from top to bottom. The layers nearest the body were in good condition, suggesting that the intermediate layers, which had partially decomposed, were not affected by body fluids.

The most remarkable part of the wrapping consisted of two pieces of linen that had ink inscriptions (Figure 6.3). The translation of the writing indicates that the owner of the linen was the priest Imyhap, son of Wah-ib-Re. These inscriptions appear not to pertain to the subject mummy, and the wrappings were probably reused fabric. Linen from other areas of the wrapping revealed evidence of wear that seemed to be the result of repeated use.

A disturbing feature was that the ink-inscribed fabric disintegrated within a few days of exposure to air, leaving only holes in

the linen where the inscriptions had been. It was fortunate that several photographs were made of these hieroglyphics.

The time of the mummification has been estimated by Robert Stukenrath of the Carbon Dating Laboratory of the Smithsonian Institution, Washington, D.C., as 835 B.C. (2785 B.P. ± 70 years), following analysis of the cloth wrappings.

GENERAL AUTOPSY FINDINGS

The autopsy was started with some misgiving because of the poor preservation of the body. There was extensive degeneration of the skin of the head, with most of the external parts of the ears missing. Although the external auditory canals appeared normal, there was a perforation of one eardrum, discovered by Lynn and Benitez and described in Chapter 18. The eyes had been either removed or were so severely degenerated that they could

not be identified within the orbits, which contained granular packing. The external nasal bones were deformed, with the tip of the nose pushed posteriorly. The teeth were intact. There were no other apparent fractures. The calvarium was removed, and an irregular 2-cm hole was noted in the cribriform bone communicating with the left ethmoid sinus and left nostril. A small plaquelike mass of brown granular tissue was adherent to the inner aspect of the occipital bone and was thought to represent residual brain. When processed by Jeanne Riddle, intact red and white blood cells were discovered within the degenerated tissue. Her findings are discussed in Chapter 16.

The skin of the body was generally poorly preserved. Numerous large and small fissures were noted that extended into the underlying muscles, particularly over the thighs, buttocks, and lower abdomen. A 7- by 4-cm irregular defect was present in the anterior chest wall that communicated with the right chest cavity and appeared to be postmortem in nature. Flattened, discoid breasts were present on the chest wall, but no nipples could be identified. The external genitalia were female, with poorly preserved labia. Large fragments of the labial, perineal, gluteal, and upper thigh tissue broke away from the body with slight manipulation. The arms were inadvertently separated from the body at the shoulder joints during the unwrapping, owing to very poor tissue preservation. The legs were loosely held at the hip joints. The overall configuration of the body was normal. No flank or abdominal incision was noted. The fingernails were slightly discolored and reddish brown.

The chest plate was removed; the heart, mediastinal connective tissue, and thoracic aorta were intact and were removed en bloc. The lungs were collapsed and adherent to the posterior thoracic wall and to the anterior and lateral aspects of the middle thoracic vertebral bones. They were black, brittle, and fragmented easily during removal. The tissue had a honeycomb appearance, and discrete holes

in the tissue suggested bronchi. Each lung measured approximately 10 by 6 by 1 cm.

The superior surface of the diaphragm was covered by a thin layer of black material, and the right hemidiaphragm contained a central defect through which protruded a piece of irregularly folded cloth that was impregnated with black resinous material. When the abdominal skin was removed, the entire abdominal cavity was found to be packed full of large wads of resin-soaked linen. The liver was displaced and flattened upward and laterally (Figure 6.4). The surface was irregular and the tissue itself was hard and black and without detail. The liver measured approximately 20 by 10 by 4 cm.

The remainder of the abdominal tissue was flattened posteriorly, and no organs could be identified. Tissue was removed from the areas normally occupied by spleen and kidneys. In the pelvis, the urinary bladder and the uterus were removed. The ovoid bladder contained a central lumen, the entire structure measuring 6 by 4 cm. The uterus measured 5 by 3 cm and had tissue on the lateral aspects that ran to the lateral pelvic walls in the position of the broad ligaments. Deep in the pelvis, there was a large defect in the area of the rectum and anus. No intestinal tissue could be identified. Apparently the defect in the area of the anus was the passage used to stuff the linen wads into the abdominal and pelvic cavities. The vertebral column and the spinal canal were normal. No spinal cord was present.

EXAMINATION OF THE HEAD

Following the completion of the autopsy, the head of PUM III was examined in detail by Drs. Richard K. Wesley and Edwin Secord, School of Dentistry, University of Detroit. Their report reads:

Cephalometric analysis of the skull indicated that PUM III exhibits a slight protrusion of the maxilla, compared with modern Caucasian standards, and that the mandible is in a normal relationship to the cranial base. This results in protrusion of the maxilla relative to the mandible – a class II skeletal

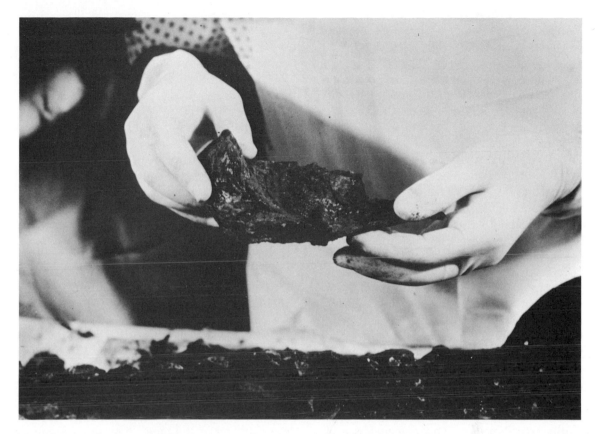

Figure 6.4.   The liver of PUM III as it was being removed. It has the appearance of being carbonized. (Photograph by Nemo Wurr, Detroit Institute of Arts)

relationship. Morphology of the body and ramus of the mandible is normal. The vertical relationship of the body of the mandible to the cranial base is also normal. The upper and lower incisors are in an upright position relative to the mandibular and maxillary arches as well as to each other. The molars are in an end-to-end, or class II, relationship. A 6-mm overbite of the incisors is a result of this class II dental relationship. The facial profile, estimated on the basis of the underlying skeletal and dental pattern (Figure 6.5), includes the well-formed bony chin and prominent nasal bones typical of the eastern Mediterranean people.

Clinical and radiographic evaluation of the dentition indicates severe occlusal attrition as well

as significant interproximal wear. Normal cusp morphology is lacking on all teeth, especially the molars. Severe wear is especially noticeable on the first mandibular and maxillary molars, which usually are larger than the second molar, but in this case are smaller in mesiodistal size. The severe occlusal attrition of the dentition is attributable primarily to a coarse and gritty diet consumed over a prolonged period of time, not to bruxism from contraction of the muscles of mastication. This is evidenced by the lack of antegonial notching of the mandible. A moderate degree of alveolar bone loss is present around all teeth, and no calculus was found. We theorize that this loss of alveolar bone must be secondary to occlusal trauma. There is

Figure 6.5.   a. Soft tissue detail of the facial profile of PUM III, estimated from the bony structure. b. Profile of the head. (a, courtesy of Dr. R. Wesley, University of Detroit Dental School; b, photograph by John Levis, Mount Carmel Mercy Hospital)

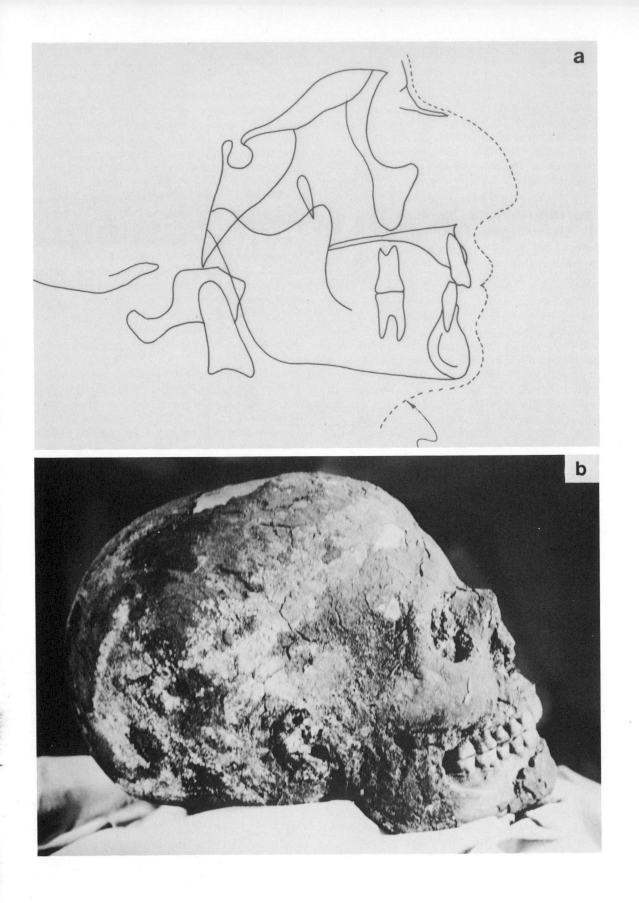

congenital absence of all third molars, and no caries or dental restorations are present. The morphology of the cervical vertebrae indicated that the individual had completed skeletal maturation, and the attrition of the dentition suggested that PUM III was in the third or fourth decade of life at the time of death.

EXAMINATION OF THE BONES

Dr. Michael Finnegan, Osteology Laboratory, Kansas State University, Manhattan, Kansas, submitted this report:

Sex. Recent experiences with mummies have proved interesting in terms of assessing the sex of the individual. Even after the unwrapping is complete, the sex may not be obvious in terms of the soft parts remaining. For example, do we have a penis and a scrotum flattened up against the pelvis or do we have the two labia minora elongated and flattened up against the pelvis? Under such conditions, it is best to look at the bony skeleton to determine the sex of the individual.

The width of the greater sciatic notch was utilized to ascertain the sex of this individual on the basis of X-ray examination of the skeleton before the unwrapping. The greater sciatic notch of the innominate bone was extremely broad, suggesting female. Care must be used in taking many of the angular measurements for skeletal determination of sex from radiographs because the exact position of the skeleton during X-ray examination is not always known. When the greater sciatic notch is seen as extremely broad, it is a very good indicator of femaleness, even in an X ray. The converse of this, however, cannot be used alone to determine maleness. On the basis of width of the greater sciatic notch, the mummy was judged to be female. The width of the subpubic arch is also a reliable indicator of sex, but the angle of the pelvis hinders such an approach in a radiograph. A well-developed preauricular sulcus was noted in the X rays on both sides of the sacrum, and in each case the sulcus was eroded, suggesting femaleness.

The skeleton in general suggests a female individual. Once the skeleton was unwrapped and autopsied, the subpubic angle was found definitely to be female. Based on Phenice's criteria (1969), the fine morphology of the pelvis was also female. The left femur head diameter was 40.5 mm, which would suggest a female individual.

Age. At the time of the autopsy, I had suggested that the appearance of suture closures suggested an age of 25 to 30 years, based on the limited amount of sutures that could be seen on the cranium owing to covering of soft tissue. However, as Brooks (1955) has pointed out, suture closure is relatively unreliable as an age indicator. At the end of the autopsy, the pubic symphyses were cut away and returned to our laboratory so that we could assess age based on the pubic symphysis. Also, a section of the right femur was taken for use in determining age by thin-section techniques. Examination of the pubic symphyses placed them in Todd's (1921) phase 7, suggesting an age of 35 to 39 years. Thin sections and microradiographs produced and read by Dr. D. J. Ortner of the Smithsonian Institution suggested an age at death of 42 years, but Dr. Ortner feels that this estimate may be a little high.

Stature. The stature of the individual was calculated by measuring from the heel up to the farthest extent of the vertebrae and combining that length with the head and neck height measured from the detached head of the mummy. This totaled 153 cm. However, because of the dehydration and subsequent shrinking of the mummy, primarily at the loci of the intervertebral discs, we measured the left femur physiological length (412 mm) in order to apply stature formulations to reconstruct possible live stature. Using the Trotter and Gleser (1958) formula for white females, a stature of 155.86 cm is obtained.

Those measurements considered reliable are listed in Table 6.1.

WEIGHT ESTIMATION

Applying Dr. Finnegan's data to what is known regarding the water content of the human body, the living weight of the mummy was calculated. The estimated length of the body was 156 cm. The normal range of weight for a nonobese female of this height is between 43.2 and 52.3 kg (95 and 115 lb). The dry weight and the mummified weight would be very nearly the same. The dry weight of the human body has been calculated to be 25 to 30 percent (Sunderman and Boerner 1949). The weight during life then would be approximately 37.8 to 45.4 kg (83.2 to 99.8 lb). Again, assuming a degree of caloric malnutri-

Table 6.1. *Measurements and indices of PUM II (in millimeters)*[a]

Head length	182
Head breadth	142
Minimum frontal breadth	93
Bizygomatic breadth	130
Total face height	110
Nose height	55
Cephalic index	78.02
Frontoparietal index	65.49
Total facial index	84.62
Left femur physiological length	412
Left femur maximum head diameter	40.4
Right femur midshaft diameters	
Mediolateral	23.6
Anteroposterior	23.3

[a] These are standard measurements used in physical anthropology.

tion and considering the weight lost with the removed organs, this calculation is certainly reasonable.

HISTOLOGICAL EXAMINATION

An insect larva was found within the body cavity during the autopsy, and electron micrographs were taken by Dr. Jeanne Riddle (Chapter 16). These photographs were sent to G. C. Steyskal and J. M. Kingsolver of the Systematic Entomology Laboratory at the U.S. National Museum, Washington, D.C., who identified the larva as *Thelodrias contractus* Motschulsky, family Dermestidae, a cosmopolitan species.

The tissues obtained during the autopsy were processed using the methods outlined in Chapter 15. Unfortunately, the poor preservation noted grossly was even more evident microscopically. Even those organs that were identifiable at the time of the autopsy were virtually without histological detail. A single exception was a small sample taken from the lateral aspect of the left breast. The tissue was very friable and crumbled during the processing. However, within this disintegrated specimen, there was a rounded, 1-cm nodule that remained intact (Figure 6.6). The nodule was composed of connective tissue – not fat, which would be more typical of normal breast tissue. Within the connective tissue, there were irregular cystlike spaces, some of which contained large, partially preserved cuboidal cells with recognizable nuclei. These had the appearance of epithelial cells. The overall configuration and residual microanatomy strongly suggested that this was a fibroadenoma of the breast. The woman's age, the size of the tumor, and its lateral position in the breast supported this thesis.

SUMMARY: PUM III

PUM III was a female with an estimated age of 35 years at the time of death in approximately 835 B.C. She was 156 cm tall and weighed approximately 41 ($\pm$ 4) kg. The wrappings were modest and parts were reused linen. Two small hieroglyphic inscriptions were present within the wrappings. She had sustained a fracture of the left second rib that had healed. All third molars were absent congenitally. Partial evisceration had been performed per ano, and the brain had been extracted through the left nostril. The body tissues were poorly preserved with the exception of a small fibroadenoma, a benign tumor of the left breast.

PUM IV

RADIOGRAPHY

From its size, the second mummy was obviously a child. Before the unwrapping, the mummy was studied radiographically. As with PUM III, the study was performed by Dr. Karl Kristen and is considered in more detail in Chapter 17. His report follows:

The skeleton appears to be normally developed. Within the wrappings, posteriorly, there is a body-length opaque, slatlike inclusion that has the density of wood. There are bilateral basal and inferior occipital fractures without evidence of healing. The nature of the skull fractures is not apparent and may represent postmortem artifact or damage. There are no visible healed fractures and no developmental anomalies. No Harris's lines are visible. The epiphyses are open and there are sev-

eral unerupted teeth, including some of the central incisors. The age of the child is estimated to be 8 years.

THE WRAPPINGS

The exterior condition of the mummy made it obvious that great care would be needed in the removal of the bandaging. A loose, ropelike cord was wrapped around the body from neck to feet, concentrated around the ankles. When removed, this proved to be a net, probably a fishnet, made of string in a square-knot pattern (Figure 6.7). The purpose or significance of this is not known. The outside wrapping consisted of a painted covering that would normally be termed a shroud, except that it did not completely cover the body, but extended from the top of the head to the ankles and did not meet in the back. It was made of coarse linen, assembled from several pieces in a neat patchwork, an unusual feature for the last decorated piece placed on a mummified body. The area of the textile that had covered the face was completely destroyed, so no painted portrait remained. The rest of the covering was in a badly damaged and fragmentary condition, but traces of paint remained. These were interpreted as geometric patterning and, in the lower left-hand corner, a figure of a seated Anubis jackal. This covering was stitched to the lower sides of the torso and across the top of the head with a thin strip of twisted linen cloth.

Under the net and the painted covering, the first layer of wrapping consisted of narrow linen strips that crossed the body at right angles to the trunk (horizontal). This was thinly applied and did not conceal the next layer, which was composed of thin strips running from the feet to the neck (vertical). Then the pieces of linen became larger, almost random in their application, of no standard size, and of varied quality. Some had been torn roughly from large shapes, and some showed obvious signs of wear, as if they had been used for cleaning or other rough work. Little attempt had been made to pad out the mummy to a lifelike shape, with the exception of a small concentration of material below the rib cage. Some additional folded material had been applied to the sides of the head to round it out. Interspersed with the partial layers of this wrapping were loose cord ropes. These had been applied in a diagonal spiral to the body and seem to have been used to hold the wrappings in place.

After the "random" wrapping had been removed, the body was found to have been tied to a full-length board. The board was broken at its midportion about the level of the pelvis. Linen strips had been used at the neck and at the sides, probably to straighten the body out. It was now seen to have been dressed in a short tunic, a simple flat garment with a neckhole cut in the middle, placed over the head of the mummy and extending down the front and back of the body to about waist level. The neckline of this garment had been rolled and stitched. It was decorated with two vertical bands of near black color, which extended completely down the front and the back, passing over the shoulders on either side. These bands were not part of the original weaving of the linen material, but had been added to the fabric in a tapestry weave. Near the two edges of this simple garment were borders of the same color as the solid bands, but the borders were composed of four threads with space left between them.

Beneath the decorated tunic, the body was enveloped in what appeared to be a shroud but proved to be three complete full-length garments with neckholes. These may have been made expressly for the burial, as the edges of the front and back were not hemmed, the neckholes were crudely stitched, and the

Figure 6.6.   Round, 1-cm piece of tissue from breast of PUM III, consisting of connective tissue without fat (a). Elastic tissue. Within small spaces are cells that resembled epithelial cells (b). The tissue is thought to be a fibroadenoma. Elastic tissue. (Photographs by John Levis, Mount Carmel Mercy Hospital)

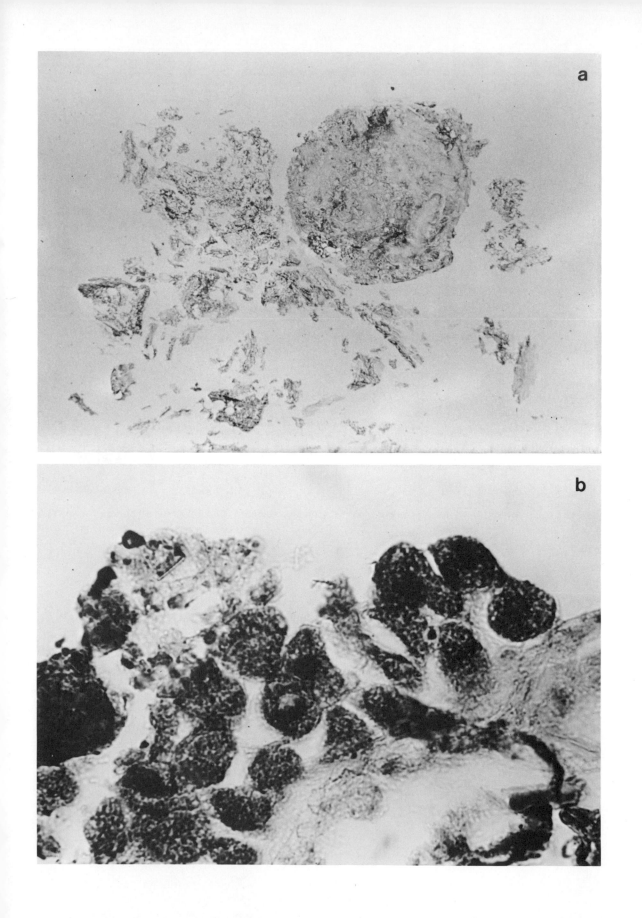

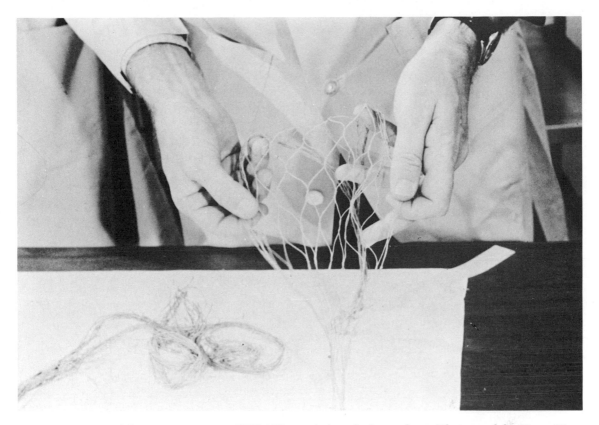

Figure 6.7. Part of the outer wrapping of PUM IV, consisting of a knotted net. (Photograph by Nemo Warr, Detroit Institute of Arts)

fabric rather narrow. In contrast, the decorated tunic could well have been used in life. Around the neck and wrists, next to the body, knotted cords of one strand were found. A small fragment of plaster cartonnage was discovered in the "random" wrapping, located deep enough inside to suggest that it was not part of the original decoration. The paint preserved on it suggested it may have been a sliver from the eye decoration of a face mask. The general impression was that the mummy had been prepared from the contents of the embalmer's scrap bag, and the cartonnage fragment swept up with the other material.

On the basis of this examination, the mummy was dated as Roman period, probably in the first or second century A.D.

GENERAL AUTOPSY FINDINGS

At the time of autopsy, the body weighed approximately 3.5 kg and measured 106 cm from crown to heel and 65 cm from crown to rump; the head circumference was 50.7 cm (Figure 6.8). The body length of 106 cm, even considering some shrinkage owing to dehydration, is small by modern standards for an 8-year-old. Calculation of the weight was not possible, as total evisceration had been performed and the body tissue was so severely degenerated. The skin, which was very dark brown, was poorly preserved and full of holes varying in size from 1 mm to 1 cm. There were numerous small beetles and insect larvae present on almost all areas of the skin. These were scarab beetles, similar to those found

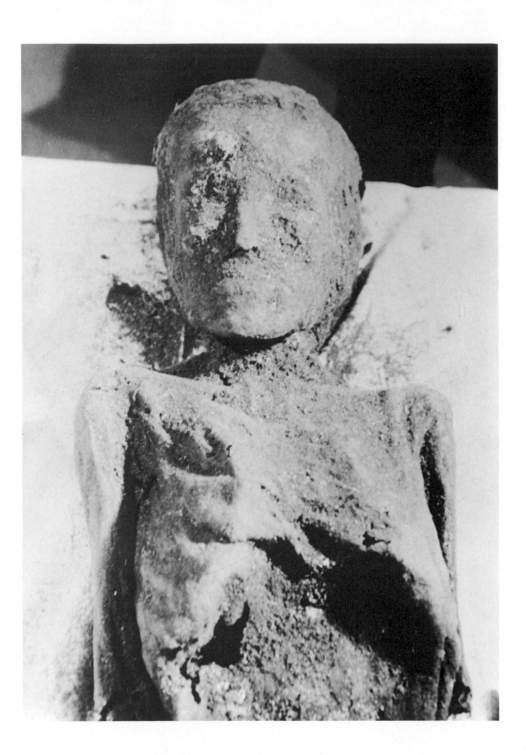

Figure 6.8. Upper part of the body of PUM IV, showing the granular packing material and the extensive degeneration of the tissue. (Photograph by Nemo Warr, Detroit Institute of Arts)

Figure 6.9.   Abdominal and chest cavities of PUM IV, packed with granular material that had the appearance of sawdust. (Photograph by Nemo Warr, Detroit Institute of Arts)

during previous mummy autopsies. The mummy's eyes were depressed into the orbits by cloth packing and a black tarry substance; they had assumed a cup-shaped configuration, but appeared to be otherwise intact. The external ears were poorly preserved, but the external canals appeared normal. The visible teeth were well preserved, with no recognizable caries, and the anterior incisors appeared to be milk, or baby, teeth. The head was partially covered by dark hair, thickly matted with a dark brown to black material. There were no specific skin lesions or other external abnormalities. The external genitalia were those of an immature male: The penis measured 5 by 1 cm and was uncircumcised. The scrotum was present, but it was not certain whether the testes were also there. There was

a large defect in the rectal area that contained a moderate amount of granular packing material and a wad of cloth, 5 cm in diameter, which partially sealed the defect, implying mummification per ano.

The anterior trunk wall and then the anterior chest wall were removed; no flank or abdominal incision was present. There were no recognizable organs in the body cavities. Instead, the cavities were filled with large masses of granular packing material, which may be sawdust (Figure 6.9). This was similar to that noted on the skin, under the wrapping and between the legs, and in the anal defect. Large numbers of dried insect larvae were also found. Sections through the neck revealed residual structures resembling the esophagus and/or trachea. The spinal canal

Figure 6.10.   *Two cerebral hemispheres found loose within the cranial cavity of PUM IV when the calvarium was removed. (Photograph by Nemo Warr, Detroit Institute of Arts)*

was opened, but no trace of the cord or other structures was seen. When the calvarium was removed, two large brown granular masses were found within the posterior fossae (Figure 6.10). These appeared to be the cerebral hemispheres, measuring approximately 10 by 4 cm each, and markedly desiccated. The posterior and inferior portions of both occipital bones were fractured, probably after death. The temporal, parietal, and frontal bones appeared normal, although there was slight separation of the sagittal suture posteriorly. Small fragments of spinal ligaments were found on the odontoid process. During the dissection, the tissue had degenerated so severely that virtually every bone of the thoracic cage, spinal column, and pelvis became loose and separated. There was no evidence of trauma or other pathologic process, and determination of cause of death was not possible.

The tissue removed during the autopsy was so severely degenerated that processing for histological examination was not successful. When placed in the rehydrating solution, it crumbled into amorphous sediment in the bottom of the container. No detail was observed in the cell block preparations of this material (see Chapter 15).

SUMMARY: PUM IV
The data on the mummified body of a male child indicated that he was 8 to 10 years of age at the time of death, probably in the first century A.D. His height was approximately 106 cm, short by modern standards for a child of

this age. The wrappings and filler were crude. The body had been eviscerated per ano, although the brain had not been removed. The body was so severely degenerated that little gross and no microscopic detail persisted.

SUMMARY: PUM III AND PUM IV

The mummification of these two bodies was less than classical – indeed, was haphazardly done. Both bodies had been eviscerated per ano, but to varying degrees. PUM III had partial abdominal evisceration and removal of the brain, but no organs had been removed from the chest. PUM IV had total abdominal and thoracic evisceration, but the brain had not been removed. In both instances, the wrapping had been done in much the same manner. The outer wrappings were circular bandages around the body, and the deeper layers consisted of larger sheets or pieces of clothing. Various artifacts and assorted fragments of unrelated wrapping and decorative material had been included in the deeper wrappings. PUM III had been treated sparingly with resin and the cavities packed with resin-soaked linen wads. PUM IV had been packed with what appeared to be sawdust mixed with a colorless oily substance. Both bodies had apparently been poorly dehydrated, with subsequent severe tissue degeneration. PUM III had been mummified in approximately 835 B.C. and it appeared that more care had been taken with the preparation of the mummy. PUM IV perhaps demonstrated the ultimate debasement of the mummification process. Both bodies were small for their age and may reflect poor nutrition or the effects of disease. This may have been a result of their low socioeconomic status, which is suggested by the method of mummification employed.

REFERENCES

Brooks, S. T. 1955. Skeletal age at death: the reliability of cranial and pubic age indicators. *American Journal of Physical Anthropology* 13:567–90.

Mokhtar, G.; Riad, H.; and Iskander, Z. 1973. *Mummification in ancient Egypt.* Cairo: Cairo Museum.

Phenice, T. W. 1969. A newly developed visual method of sexing the os pubis. *American Journal of Physical Anthropology* 30:297–301.

Sunderman, E. W., and Boerner, F. 1949. *Normal values in clinical medicine,* p. 649. Philadelphia: Saunders.

Todd, T. W. 1921. Age changes in the pubic bone: II, The pubis of male Negro-white hybrid; III, The pubis of the white female; IV, The pubis of the female Negro-white hybrid. *American Journal of Physical Anthropology* 4:1–70.

Trotter, M., and Gleser, G. 1958. A re-evaluation of estimation of stature based on measurements taken during life and of long bones after death. *American Journal of Physical Anthropology* 16:79–123.

PART II
Mummies of the Americas

Mummies and mummification practices in the southwestern and southern United States

MAHMOUD Y. EL-NAJJAR
New Mexico State University
Las Cruces, New Mexico, U.S.A.

THOMAS M. J. MULINSKI
University of Idaho
Moscow, Idaho, U.S.A.

The origin of mummification practices is not precisely known. It appears, however, at least among American Indian tribes, that they may have resulted from the belief in a future life. Grave offerings and the various cultural and personal artifacts recovered indicate that such beliefs did exist. Mummification practices, both in Egypt (Chapter 1) and in the Americas, were of two types: artificial and natural. Artificial mummification flourished and was more prevalent in ancient Egypt. Natural mummification, on the other hand, is the predominant kind in the dry areas of the New World. In North America, almost all desiccated bodies recovered thus far are from rock shelters, caves, and overhangs. Mummies found in these localities are usually in a sitting position, tightly flexed, with the arms and knees drawn to the chest and the head bent forward.

In the New World, mummification is known to occur in three main regions: the southern and southwestern United States, the Aleutian Islands, and Peru. This chapter is limited to a discussion of mummification practices and a tabulation of the whereabouts of mummies in the southern and southwestern United States (Figure 7.1).

SOUTHWESTERN UNITED STATES

Most of our information regarding mummification practices comes from the study of burial techniques and of the extensive mortuary offerings made by the prehistoric natives of the New World. In the American Southwest, these natives are known to anthropologists as *Anasazi*, which is the Navajo word for "ancient people" and is applied to the prehistoric inhabitants of the plateau area of the American Southwest; this includes the drainage of the Rio Grande, and the San Juan, Little Colorado, Upper Gila, and Salt rivers, much of Utah, and some of eastern Nevada. Southwestern American mummies come from three main localities: northeastern, east-central, and southern Arizona. From northeastern Arizona, mummies have been recovered from Canyon de Chelly and Canyon del Muerto, Vandal Cave, and Painted Cave.

Canyon de Chelly and its major tributary, Canyon del Muerto, have yielded some of the best preserved desiccated bodies in the New World. The majority of these (n = 10) is housed at the American Museum of Natural History, New York City. No studies, either histological or anthropological, have been done on them. Four of these mummies are being studied at Case Western Reserve University, Cleveland, Ohio. A partial mummy is at the Human Variation Laboratory, Arizona State University, Tempe, and the naturally desiccated body of a Pueblo child is now being examined further at the Department of Anthropology in the same university.

The earliest of the Anasazi are known as the

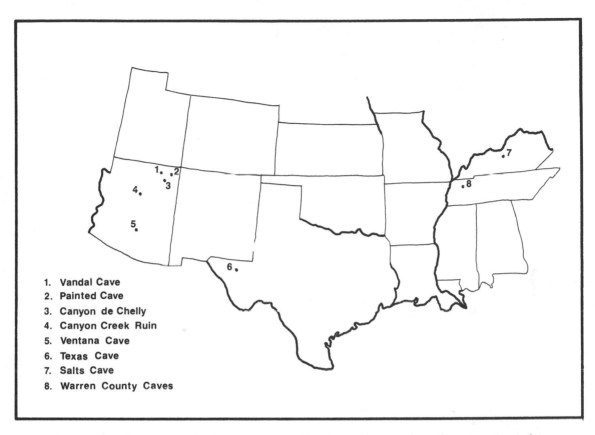

1. Vandal Cave
2. Painted Cave
3. Canyon de Chelly
4. Canyon Creek Ruin
5. Ventana Cave
6. Texas Cave
7. Salts Cave
8. Warren County Caves

Figure 7.1.   Sites where mummies have been found in the southern and southwestern United States. (Map by Timothy Motz, Detroit Institute of Arts)

Basket Makers, a seminomadic group of hunters and gatherers who lived in the caves and rock shelters of the area between A.D. 100 and 700. The Basket Makers survived in an unusually harsh environment. They had no cotton, no wool, no pottery, and no draft animals. Their diet consisted mainly of corn and squash, though hunting small game animals and gathering nuts and seeds provided enough in the way of food supplements to ensure their biological survival. The Basket Makers were short and had coarse, black hair with a tendency to be wavy, little body hair, and brown skin (Wormington 1973).

In general, the desiccated bodies of these inhabitants were placed in pits or stone-lined cists that had originally been constructed for storage. Occasionally, however, a body was left in a corner on the floor of a cave or was placed in a crevice. This was probably done immediately after death occurred, before the body had stiffened. It is possible that the small size of the cist led to the custom of flexing, in which the knees are drawn up to the chest and the arms are extended at the side. The desiccated bodies were usually wrapped in fur blankets, but occasionally tanned deerskins were used. Bodies of infants and small children were wrapped in a padlike mass of soft fiber made from the leaves of yucca plants and shrouded either in fur, skin, or feather-cloth blankets (Figures 7.2 through 7.4). Mortuary offerings included baskets, sandals, beads and ornaments, weapons, digging sticks, cone-shaped pipes, and a variety of personal possessions.

In her discussion of mummification among the Basket Makers, Wormington concludes:

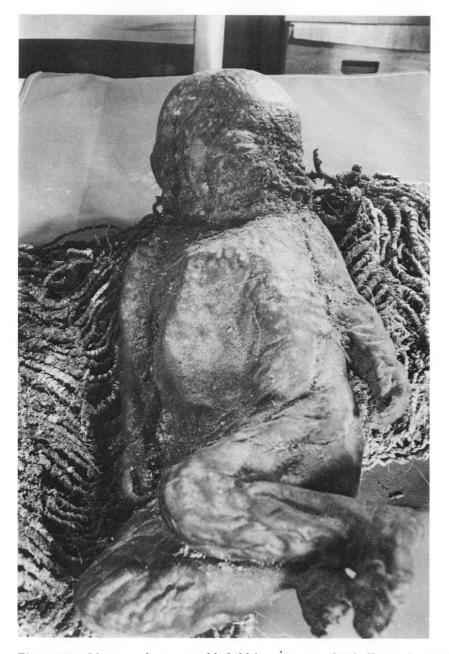

Figure 7.2.   *Mummy of a 1-year-old child from Canyon de Chelly, Basket-Maker period No. 2, A.D. 300–500.*

Almost every body is found wrapped in a blanket of fur and it is probable that these served as wraps and blankets for living as well as shrouds for the dead. The manner in which these coverings were constructed is most ingenious. Strings were made of yucca fibers, then tied together in close parallel rows, producing a light warm climate. Sometimes they were ornamented with borders made of cords which had been wrapped with strips of bird skin. Some mantles of tanned deer skin were also made

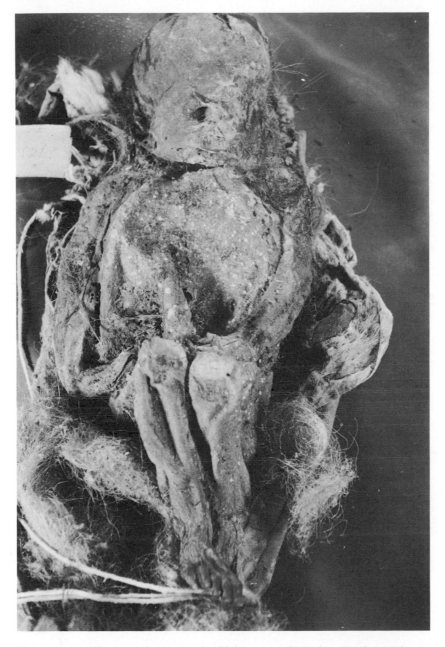

Figure 7.3.   Mummy of a 7-month-old fetus in rabbit skin, Basket-Maker period No. 2, A.D. 300–500.

and it may be that there were some woven robes, for a few fragments of woven cloth have been found. [Wormington 1973]

Reporting on mummies from Ventana Cave, southern Arizona, Haury (1950) found two types of burial practices: flexed and ex-tended. In flexed bodies, the arms are folded across the chest; in extended bodies they lie at the sides or rest on the abdomen. Haury fur-ther states that the degree of flexure varies, from doubling up the legs without drawing them up to the chest to tight flexing, the latter

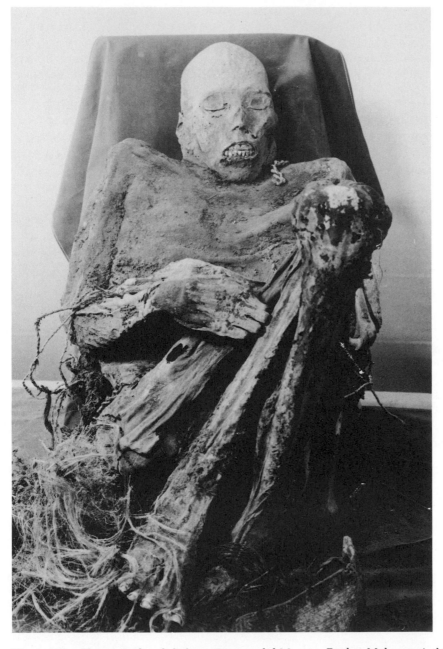

Figure 7.4.   Young male adult from Canyon del Muerto, Basket-Maker period No. 2, A.D. 300–500.

being more common. In the former cases, there is a binding about the legs. Here, as elsewhere in the Southwest, the mummies were entirely the result of natural desiccation. The author concludes: "There is no evidence whatever that evisceration or other artificial means of preserving were known."

Among the unique finds associated with Basket-Maker mummies is a pair of unworn sandals. Apparently, these ancient nomads

believed that the dead person required a new pair of sandals for the time when he would rise again and walk. These sandals were woven of cord made from the fibers of yucca and apocyum, a plant related to the milkweed. They were double-soled, somewhat cupped at the heel, had a square toe, and were usually ornamented with a fringe of buckskin or shredded juniper bark (Wormington 1973).

One of the most unusual mummies recovered from the American Southwest is a young Pueblo child approximately 3 years of age. The Pueblos were the agriculturist descendants of the Basket-Makers and lived in large communal houses in the same area between A.D. 700 and 1300. The child died during the eleventh century (El-Najjar et al. 1975). Its desiccated body was laid flat on an elaborate cradleboard with a cottonwood-bark sunshade in place around the head. There was also a worn textile fragment round the neck and a bracelet round the right wrist. This burial contrasts markedly with other burials recovered from the same site. None of the others had a grave cover or had as many or as elaborate grave goods. The child's death has been attributed to severe anemia (El-Najjar and Robertson 1976). The diagnosis was based on macroscopic, radiographic, and histochemical analysis. In the earlier paper, El-Najjar et al. conclude:

Of particular interest is the fact that this apparent three-year-old child was still on a cradleboard. If we interpret the association correctly, the child was unable to walk, perhaps even unable to participate in normal infant behavior, and may well have been mentally retarded. We noticed no evidence of fractures or broken bones, nor did we see any other pathological features. [El-Najjar et al. 1975]

Four mummies from Canyon del Muerto are now being examined at the Department of Anthropology and Institute of Pathology, Case Western Reserve University, Cleveland. These are an adult male, an adult female, a young child, and a fetus. They are dated around A.D. 100–300. Autopsies have been performed on two of these mummies, and bone and soft tissues are being processed.

The mummies from Vandal Cave and Painted Cave, northeastern Arizona, as well as those from east-central and southern Arizona are all stored at the Arizona State Museum, Tucson. Pertinent information on each mummy is provided in Tables 7.2 through 7.8. Before each site and its mummies are discussed below, it is necessary to point out two things. First, when the burial number of a mummy is not known, the Arizona State Museum (ASM) catalog number of the specimen is used in its place in order to distinguish individuals. The catalog number always appears in parentheses. Second, the age and sex of each mummy were estimated by T. M. J. Mulinski and Dr. Walter H. Birkby, physical anthropologist at the museum.

Of the 43 mummies listed in Table 7.1 only 18 (41.9 percent) are complete. The rest are partial mummies of varying degrees of completeness. There are 29 (69.4 percent) subadults and 14 (32.6 percent) adults. Of the latter, 4 are definitely or probably males, 6 are definitely or probably females, and 4 cannot be sexed. With regard to dating, it can be safely assumed that all mummies are from prehistoric times, although none is older than 2,000 years. In fact, the oldest ones that can be dated with some degree of assurance are only 1,500 years old.

All the dehydrated remains at the ASM collection have become mummified through natural desiccation. These individuals were buried in either caves or rock shelters, where the extremely dry conditions allowed the soft tissues to dry out before putrefaction could destroy them. In connection with this, it is interesting to note that the integument of the head is usually the first to disintegrate. For example, more than one mummy in the collections has generally intact skin on the torso and extremities, but the covering of the head, especially of the face, has not been preserved or has a significantly lesser percentage of skin still intact. This phenomenon is undoubtedly

Table 7.1. *Sites with mummies in the collections of the ASM*

Site name	Site number[a]	Total number of burials	Number of mummies in ASM collections	Time period[b]	Cultural affiliation[b]	Archeological reference
Vandal Cave	Ariz. E:7:1	11	8	A.D. 500–700 A.D. 1150–1250	Anasazi	Haury (1936)
Painted Cave	Ariz. E:7:2	3	2	A.D. 1150–1250	Anasazi	Haury (1945)
Canyon Creek Ruin	Ariz. V:2:1	40	5	A.D. 1300–1350	Anasazi	Haury (1934)
McCuen Cave	Ariz. W:13:6	21(?)	10	Undoubtedly prehistoric	?	None
Ventana Cave	Ariz. Z:12:5	39	11	A.D. 1000–1400[c]	Hohokam	Haury (1950)
Texas Cave	Texas 0:7:3[d]	2(?)	2	?	?	None
Miscellaneous	Various places in Arizona and Colorado	?	5	Probably prehistoric	?	None

[a] Except when noted, sites are designated according to the system employed by the Archeological Survey of the ASM (Wasley 1964).
[b] Information on the time period and cultural affiliation of Vandal Cave, Painted Cave, Canyon Creek Ruin, and Ventana Cave comes from Haury (personal communication).
[c] Ventana Cave was occupied off and on for thousands of years, beginning more than 10,000 years ago. However, all the burials except three (nos. 20, 35, and 36), and all the mummies, date from the period A.D. 1000–1400.
[d] This designation is according to the system employed by the no-longer-existing Gila Pueblo Archeological Foundation.

related to differences in the thickness of the soft tissues surrounding the skull and postcranium.

Vandal Cave (Table 7.2), 8 mummies. The site had two main occupations, a Basket-Maker III (A.D. 500–700) and a Pueblo III (A.D. 1150–1250). Burials 7, 9, and 10 are from the earlier occupation; burial 2 is from a later level. The exact provenance of the other 4 mummies is not known. They are, however, from the same general area of the site. The Vandal Cave mummies are among the best preserved at the ASM collection.

Painted Cave (Table 7.3), 2 mummies. Both Painted and Vandal caves are located some 30 km north of Canyon de Chelly. The major occupation of the site was during Pueblo III times (A.D. 1150–1250), which is

when the burials took place. Another mummy recovered from Painted Cave is on permanent loan to the Amerind Foundation, Dragoon, Arizona.

Canyon Creek Ruin (Table 7.4), 5 mummies. Canyon Creek Ruin is located in the western half of the Fort Apache Indian Reservation in east-central Arizona. The site is from the period A.D. 1300–1350.

McCuen Cave (Table 7.5), 10 mummies. McCuen Cave is located in the southern half of eastern Arizona. The site is prehistoric, although little information is available on its exact date of occupation.

Ventana Cave (Table 7.6), 11 mummies. Although some are incomplete, preservation of these mummies is very good. Ventana Cave is located in the Castle Mountains of southern Arizona in the Papago Indian

Table 7.2. *Mummies from Vandal Cave*

Burial number	Burial position	Condition of mummy	Sex	Age (yr)	Comments
2	Flexed	Complete, but no nails or hair present	Female(?)	Old adult, (>40)	Part of shroud still covering legs
7	Flexed	Complete, but no nails present; small amount of hair present	?	Infant (1–2)	Bracelet still in place on right wrist
9	Flexed	Complete, but no nails present; very small amount of hair present	Male(?)	Adult	Part of burial blanket still covering lower torso
10	Semiflexed	Apparently complete; nails and fair amount of hair present	?	Adult	Body still almost entirely covered with burial blanket and encrusted with soil
?(0–485)	Flexed/semiflexed	Complete; nails, but only very small amount of hair present	?	Infant (0.5–1.0)	Buried on flexible cradleboard
?(0–487A)	Flexed	Complete, although head detached from rest of body; some nails and very small amount of hair present	?	Infant (1.5–2.5)	Cranium with occipital deformation
?(0–487B)	?	Partial; only right forearm and hand and right leg and foot present	?	Infant (0.5–1.5)	
?(0–487C)	?	Partial; only lower part of left leg and foot present	?	Infant (birth–0.5)	

Reservation. This cave has a long history of interrupted use from over 10,000 years ago to the present century. All the mummies recovered, however, date from A.D. 1000–1400.

Texas Cave (Table 7.7), 2 mummies. Texas Cave is located roughly 80 km northwest of Toyah, Texas, at the western end of the state. Little information is available from this site. Of the two mummies, one is complete and one partial.

Other localities (Table 7.8). In addition to the above, 5 other mummies at the ASM come from various places in Arizona and Colorado. No exact provenance is known. The only noteworthy feature is a rather extreme example of lambdoid deformation exhibited by the cranium from Slab House Ruin.

Very few biological studies on the mummies in the ASM collections have been carried out. The only known published report on any of this material is by Gabel (1950), who examined the human remains from Ventana Cave but made no specific studies of the mummies. On the other hand, two unpublished studies on the Ventana Cave mummies are mentioned by Haury (1950). One of these was concerned with paleopathology:

The recovery of the mummies in Ventana Cave aroused considerable interest among some of the personnel of the Indian Service as to the possibility of tracking down diseases. Dr. Joseph D. Aronson, then special investigator in tuberculosis for the service, and engaged in research on valley fever (coccidioidomycosis), arranged to have the mummies x-rayed. The incidence of valley fever among Papago is very high and it was hoped that some

Table 7.3. *Mummies from Painted Cave*

Burial number	Burial position	Condition of mummy	Sex	Age (yr)
?(0–514)	Semiflexed(?)	Partial: left leg and foot missing, part of right foot missing; anterior walls of thoracic and abdominal cavities almost completely disappeared; nails absent; some hair present	?	Infant (0.5–1.5)
?(0–515)	Flexed	Complete, but skin rather worm-eaten in appearance; nails and some hair present	Male	Child (3.5–4.5)

Table 7.4. *Mummies from Canyon Creek Ruin*

Burial number	Burial position	Condition of mummy	Sex	Age (yr)	Comments
13	Extended(?)	Indeterminate, but no skin left on exposed part of head	?	Infant (1–2)	Body still wrapped in shroud(s); buried on cradle-board
20	?	Partial: only bones of right leg and both feet present, with some tissue also present	?	Adult	Osteitis observable on tibia and fibula; both feet in sandals
22	Extended(?)	Indeterminate, but does not appear to be much tissue left	?	Probably infant	Body still wrapped in shroud(s)
32	?	Partial: skeleton disarticulated; very little tissue remaining	?	Infant (birth–0.5)	Body still wrapped in shroud(s); buried on cradle-board
33	?	Partial: skeleton disarticulated and incomplete; with no skull and very little postcranium remaining; very little tissue left	?	Fetal–newborn	

evidence might be found to show whether or not the earlier residents in the area were afflicted, too. The findings were negative, at least to the extent that no trace of the disease was revealed if it was present and nothing significant appeared from a pathological standpoint either. [Haury 1950]

A second study involved a search for ABO antigens in tissue samples:

In 1947, Mr. Edward L. Breazeale, then with the Division of Laboratories, Arizona State Department of Health (now Assistant Agricultural Chemist, University of Arizona), undertook blood group tests of Ventana Cave mummy tissue. Samples from ten mummies were examined by two different methods: (1) by absorption and cross agglutination studies, and (2) by extracting the tissue with normal saline and using the extracted fluid as the antigen against known type A and B cells. The results produced by these two methods are in perfect agreement. Of the ten samples analyzed, nine were type "O" (burials 2, 3, 5, 6, 8, 9, 11, 15, 31) and one was type "AB" (burial 16). [Haury 1950]

One other research project should be mentioned. Birkby has undertaken an examination of the head hair from several mummies for ectoparasites. His results, which are unpublished, are presented in Table 7.9. Out of 18 individuals with a sufficient amount of hair to be analyzed, 8 (44.4 percent) had head lice (*Pediculus humanus capitis*). In all instances, only the nits were found.

Table 7.5. *Mummies from McCuen Cave*

Burial number	Burial position	Condition of mummy	Sex	Age (yr)	Comments
Mc:1–4	?	Partial: body jumbled mass, but some tissue present	?	Fetal–newborn	
?(0–483)	Extended	Complete, although head detached from rest of body; a few strands of hair present	?	Infant (1–2)	Body still somewhat covered by shroud(s)
?(0–493)	?	Partial: only some lumbar vertebrae, pelvis, and lower extremities present; only small amount of tissue present	Male	Adult	
?(0–494)	?	Partial: head, upper extremities (except for left humerus), and feet missing; fair amount of tissue still remaining	Female	Adult	
?(0–500)	Extended	Complete; abundant amount of hair present	?	Infant (birth–0.5)	Body still wrapped in shroud(s); buried on cradleboard
?(0–502)	Flexed	Complete, although head detached from rest of body and skin not well preserved; no hair present	?	Infant (1–2)	
?(0–503)	Extended	Complete; some hair present	?	Fetal–newborn	Body still wrapped in shroud(s); buried on cradleboard
?(0–512)	?	Partial: only head, some vertebrae, and rib fragments present; some tissue preserved, but no hair present	Female	Adult	Cranium exhibits lambdoid deformation
?(0–750A)	?	Partial: only both legs and feet present; some tissue preserved	?	Adult	Feet in sandals
?(0–750B)	?	Partial: only left leg and foot present; some tissue preserved	?	Adult	Foot in sandal

SOUTHERN UNITED STATES

Another collection of naturally mummified prehistoric American natives comes from Kentucky. The best known of these is Little Alice. Little Alice's desiccated body was recovered by two local men in 1875 near what is now known as Mummy Valley. On the limestone slab where the body was found, the following inscription appears:

Sir I have found one of the Grat wonder of the World in this cave Whitch is a muma "Can All Seed hear after" found March the 8 1875

T.E. lee J l lee "an Wd Cutliff" discuvers [*sic*, Robbins 1971]

Of the original discovery, Watson cites the following published account, which appeared in a newspaper clipping:

It was lying on the ledge up against the wall of the cave with a pile of ashes and half-burned sticks in front. A bowl, pipe, several pairs of moccasins made of grass and bark, some pieces of an exceedingly light wood, flints and arrow points, etc., were all about. [Watson 1969]

Table 7.6. *Mummies from Ventana Cave*

Burial number	Burial position	Condition of mummy	Sex	Age (yr)	Comments
3	Flexed	Complete, although right upper extremity disarticulated at elbow; some hair present	Female	Old adult (>50)	Most of head hair present is gray; calculus present on anterior mandibular teeth; periodontal disease evident
5	?	Partial; skeleton fairly complete, but disarticulated; some tissue remaining, including very small amount of hair	?	Young child (5–6)	
6	Extended	Complete, although left upper extremity disarticulated at elbow; some hair present	Female	Young child (4–5)	Wooden block wrapped with textiles placed under head of this individual when buried
9	Extended	Complete; small amount of hair present, but no nails	Male	Adult	Nose plug and earrings still present; most of torso and upper extremities covered with shroud; grave goods included skin quiver containing cord with attached shell, nose plug, projectile points, four bone awls, human-hair wig, cactus-spine needle, and several fragments of preserved sandals, and miscellaneous cotton cloth fragments
11	Extended	Complete; nails and hair absent	Female	Adult	Body covered with cotton robe when buried
15A	?	Partial: skeleton fairly complete, but disarticulated; very little tissue remaining; extremely small amount of hair present	?	Infant (1.5–2.5)	
15B	?	Partial: only left forearm and hand present; some tissue remaining	?	Young child (5–6)	
16	?	Partial: only skull, some cervical vertebrae, and right talus present; some tissue remaining, including very small amount of hair	Male(?)	Old adult (>40)	
24	?	Partial: skeleton fairly complete, but somewhat disarticulated; some tissue remaining	?	Infant (0.5–1.5)	Body placed in twined bag and buried in grass nest
25	Semiflexed	Complete; fair amount of hair present	?	Probably infant/young child	Body still almost entirely covered with shroud
29	Semiflexed	Complete, although head detached from rest of body; fair amount of tissue remaining; very small amount of hair remaining	?	Infant (1.0–1.5)	Body buried in fur robe shroud

Table 7.7. *Mummies from Texas Cave*

Burial number	Burial position	Condition of mummy	Sex	Age (yr)	Comments
?(0–740)	?	Partial: skeleton fairly complete, although somewhat disarticulated; some tissue remaining	?	Infant (birth–0.5)	
?(0–759)	Extended	Complete; large amount of hair present	?	Infant (birth–0.5)	Body still covered with burial blanket; buried on mat

Table 7.8. *Mummies from several localities in the Southwest*

Location	Burial position	Condition of mummy	Sex	Age (yr)	Comments
Slab House Ruin, Duggagei Canyon, Arizona (0–200)	?	Partial: legs and feet missing; no skin left anteriorly and only small amount remaining posteriorly	?	Infant (1–2)	Cranium exhibits rather extreme lambdoid deformation
Yellow Jacket Canyon, Colorado (0–245)	?	Partial: skeleton fairly complete, but almost completely disarticulated; some tissue remaining	?	Infant (1.5–2.5)	
Cliff House, Tonto Basin, Arizona (0–498)	?	Partial: skeleton fairly complete, but for most part disarticulated; some tissue remaining	?	Infant (birth–0.5)	
Cottonwood area, Arizona (0–501)	?	Partial: most of face and lower extremities missing	?	Infant (birth–1)	
Duggagei Canyon, Arizona (0–511)	?	Partial: only skull, some cervical vertebrae, and innominates present; small amount of tissue present	Female	Adult (15–20)	

Little Alice was bought by a man named Morrison, who described her as follows:

The little girl turned to stone, the most interesting and wonderful of all cave phenomena; a little girl, petrified or mummified by the action of the cave air; a mummy that was found in Salts Cave in 1875; that during the 47 years since the discovery, it was exhibited in the Smithsonian Institution and at various other places. [Watson 1969]

Little Alice was displayed in commercial caves for many years after the original discovery. In 1958 she was brought to the University of Kentucky, where detailed studies were made. The presence of external genitalia showed the desiccated body to be that of a young male about 9 to 10 years of age! Little Alice is now known as Little Al. According to Robbins (1971), the body is in an excellent

Table 7.9. *Findings of examination of head hair from certain North American Indian mummies in ASM collections for head louse*, Pediculus humanus capitis

Site	Number of mummies with lice	Number of mummies without lice
Vandal Cave	0	3 (burials 7, 9, 10)
Painted Cave	2 (0–514, 0–515)	0
McCuen Cave	0	3 (Mc:2, 0–500, 0–503)
Ventana Cave	6 (burials 3, 5, 9, 15A, 16, 25)	2 (burials 6, 29)
Texas Cave	0	2 (0–740, 0–759)
Total	8	10

Source: W. H. Birkby (unpublished data).

state of preservation, except for slight fungus as a result of its exposure to the outside atmosphere. Radiocarbon dating using abdominal and lower thoracic tissue produced an age of 1,960 ± 160 years. On the basis of cultural and physical anthropological data, Robbins concluded that Little Al may have belonged to a group of Woodland Indians who were the recent human occupants of the cave.

Several other desiccated bodies have been found in the Mammoth Cave area. Most of these were discovered by saltpeter miners early in the nineteenth century in Short Cave. Between 1811 and 1815, at least four mummies were found (Meloy and Watson 1969). The Mammoth Cave mummy known as Fawn Hoof was found in Short Cave in 1813. She was sitting in a stone box grave, of the kind commonly found in Tennessee and neighboring counties in Kentucky to the south of Short Cave. Physically, the body was well preserved; the flesh dry, hard, and dark in color. According to Robbins (1974), Fawn Hoof was dressed in several finely fashioned skin burial garments and was accompanied by a variety of grave goods. Fawn Hoof is the only one of the mummies that definitely seems to have been accompanied by grave goods (Watson 1969). According to Watson (1969), Fawn Hoof and several items found with her were given to the American Antiquarian Society in

Worcester, Massachusetts, about 1817. She was then exhibited at the U.S. National Museum in 1876. Her body has been dissected, and the clean bones are stored at the Division of Physical Anthropology, Smithsonian Institution.

Another mummy, known as Scudder mummy, was also recovered from Short Cave. Deerskin wrappings on the body and deerskin items found with it indicated that it was from the same population as Fawn Hoof (Robbins 1974). According to Robbins, the Scudder mummy, thought to be an adolescent boy, showed evidence of a fracture of the occipital bone which may have contributed to his death.

The remains of a mummy known as Lost John were recovered from Mammoth Cave in 1935. The desiccated body was found lying partially crushed under a boulder. Apparently, Lost John was the victim of a prehistoric mining accident. Neumann (1938) believes that the miner was kneeling when the boulder fell, its impact forcing him to fall on his right side. The cultural items found near the body indicate that this individual was involved in mining activities at the time of death (Robbins 1974).

According to Robbins, the body is well preserved, with flesh and internal organs present except for areas where rodent activities are

evident. Lost John was a male in his forties. Textile material, evidently some sort of a blanket or robe of open twined weave, was tied with a braided cord around the body, and a mussel-shell pendant was suspended from the neck by a piece of two-strand twisted cord. A crude limestone hammer, bundles of reeds tied with grass, sticks, parts of gourds, a fragment of bagging, a stout pole which probably was used as a ladder, some hickory nuts, and human excrement are the only other materials in the cave (Neumann 1938). On the basis of cultural and geological evidence, Neumann suggests a date of about 500 years since John's death. Lost John has not been studied since the work of Neumann, but detailed analysis, including histologic and radiographic examination, are now being made by Dr. Louise Robbins of the University of North Carolina, Greensboro.

Mummies have also been found in caves in Tennessee. Holmes (1891–1892) reported on two mummies found in "a copperas cave" in Warren County, West Tennessee. The bodies, a male and a female, were discovered in 1810. Both had been placed in large cane baskets and buried in the cave floor. The female, like the bodies in Short Cave, was wrapped in a succession of materials, including hides, a feather cloak, and a piece of plain textile. According to Holmes, a scoop net, a moccasin, and a mat–all made of bark thread–were also found. In addition, Holmes states: "She had in her hand a fan formed of the tail feather of a turkey."

According to Robbins (1974) the interment of the Short Cave mummies exhibits a pattern similar in some ways to mummies found in Tennessee, though it is different in others. In both areas mummies were wrapped in deerskin and accompanied with grave goods. Robbins further suggests that the Tennessee mummies differ in that some were disarticulated at the hips before being wrapped or dressed. She states:

Following the wrapping, they were placed upright in woven baskets. Wrapping the body before burial, and the kind of wrapping used, implies that the Short Cave and Tennessee mummies have come from one population occupying a broad geographic area.... If the Short Cave mummies were part of the Tennessee population, it is curious that the former were not disarticulated at the hips before burial. Why particular members of the population were selected for burial in caves is another interesting question that leads to speculation concerning the social structure of the people, a question that the existing evidence is quite inadequate to answer. [Robbins 1974]

CONCLUSION

As already stated, the origin of mummification practices among New World natives remains unknown. Why particular individuals were placed in caves and others buried in an open shelter is an unanswered question. It is possible that only persons of high social and/or economic status or warriors were left to dry. Wrapping the bodies may have been done to protect them or because the dead would need the wrappings in a future life. For example, the contemporary American natives (Navajos) who inhabit Canyon de Chelly and Canyon del Muerto never venture close to these caves, which are known to them as *chindi*, or "haunted houses." It is common practice among these people to abandon the house when an individual dies. In the New World, grave offerings may have had more than one purpose. In addition to the belief in a future life, the dead man's possessions may no longer have been used by other members of the community because of magicoreligious practices. Regardless of these questions, however, New World mummies offer a rich resource for future research to scholars in many disciplines. Only a small number of studies have been done, most of them long before the present revival of interest in human paleopathology and the utilization of advanced analytical techniques. The mummies at the Arizona State Museum and the American Museum of Natural History will be excellent subjects for future autopsies.

REFERENCES

El-Najjar, M. Y.; Morris, D. P.; Turner, C. G.; and Ryan, D. 1975. An unusual pathology with high incidence among the ancient cliff-dwellers of Canyon de Chelly. *Plateau* 48:13–21.

El-Najjar, M. Y., and Robertson, A. 1976. Spongy bones in prehistoric America. *Science* 193:141–3.

Gabel, N. E. 1950. The skeletal remains of Ventana Cave. In *The stratigraphy and archeology of Ventana Cave*, E. W. Haury (ed.), pp. 473–520. Tucson: University of Arizona Press.

Haury, E. W. 1934. *The Canyon Creek ruin and the cliff dwellings of Sierra Ancha*. Medallion Papers, No. 14. Globe, Ariz.: Gila Pueblo.

–   1936. Vandal Cave. *Kiva* 1(6):1–4.

–   1945. *Painted Cave, northeastern Arizona*. Dragoon, Ariz.: Amerind Foundation.

–   1950. *The stratigraphy and archeology of Ventana Cave*. Tucson: University of Arizona Press.

Holmes, W. H. 1891–2. Prehistoric textile art of the eastern United States. In *Thirteenth annual report of the Bureau of American Ethnology*, pp. 3–55. Washington, D.C.: Smithsonian Institution.

Meloy, H., and Watson, P. J. 1969. Human remains: "Little Alice" of Salts Cave and other mummies. In *The prehistory of Salts Cave, Kentucky*, P. J. Watson (ed.), pp. 65–9. New York: Academic Press.

Neumann, G. K. 1938. The human remains from Mammoth Cave, Kentucky. *American Antiquity* 3:339–53.

Robbins, L. M. 1971. A woodland "mummy" from Salts Cave, Kentucky. *American Antiquity* 36(2):201–6.

–   1974. Prehistoric people of the Mammoth Cave area. In *Archeology of the Mammoth Cave area*, P. J. Watson (ed.), pp. 137–62. New York: Academic Press.

Wasley, W. W. 1964. *The archeological survey of the Arizona State Museum*. Tucson: University of Arizona Press.

Watson, P. J. 1969. *The prehistory of Salts Cave, Kentucky*. Reports of investigation no. 16. Springfield: Illinois State Museum.

Wormington, M. 1973. *Prehistoric Indians of the Southwest*. Colorado Museum of Natural History, Series 7. Denver: The Museum.

8

Aleutian and Alaskan mummies

MICHAEL R. ZIMMERMAN
Associate Professor of Pathology and Anthropology,
The University of Michigan, and
Pathologist, Wayne County General Hospital,
Detroit, Michigan, U.S.A.

Mummies from Alaska are unique in several respects, owing to the climatic extremes of this area. The frigid climate of Alaska proper has resulted in the production of frozen mummies with remarkable preservation of histologic detail. In contrast, the cool damp climate of the Aleutian Islands would seem to be poorly suited to natural mummification. Cultural practices have supervened in this area to produce mummies.

A 1,600-YEAR-OLD FROZEN ESKIMO MUMMY

In October 1972 the frozen body of a woman washed out of a low beach cliff at Kialegak Point on Saint Lawrence Island in the Bering Sea. The Kialegak site is on the Southeast Cape of Saint Lawrence Island, which is about 60 km from Russia and 200 km from mainland Alaska. Occupation of the island can be traced back more than 2,000 years.

The body was found by three Eskimo hunters, the Gologergen brothers of the village of Savoonga. They felt that the body would be of interest to scientists and reburied it in the tundra, below the permafrost level, which in that area is 5 to 10 cm below the surface (Collins 1933). In the summer of 1973, visiting National Park Service anthropologist Zorro Bradley was notified of the find and taken to the burial site. With the permission of the Eskimos of the island, Bradley and the Gologergens exhumed the body, placed it in a bag, and transported it to Northeast Cape. From there it was flown to Nome and on to Fairbanks, where it was stored in the freezer facilities of the federal Arctic Health Research Center (no longer in operation).

Using the facilities of the research center (as arranged by Dr. Robert Rausch), the author and George S. Smith, of the National Park Service and the Anthropology Department of the University of Alaska, performed a complete autopsy (Zimmerman and Smith 1975). The body was thawed at room temperature, the process taking 24 hours. Tattooing noted on the arms indicated some degree of antiquity, as this practice had been discontinued on Saint Lawrence Island by the 1930s (Geist 1928). Tissue from psoas and quadriceps femoris muscle was radiocarbon dated at two separate institutions. The Smithsonian Institution Laboratory date was A.D. 405 ± 70 years (SI-1656) and the University of Pennsylvania Laboratory date was A.D. 370–390 ± 90 years (P-2090, I-7584). This dating placed the body in the Old Bering Sea Phase on Saint Lawrence Island (A.D. 200–500) (Birket-Smith 1959).

Examination of the tattoos (Smith and Zimmerman 1975), which were confined to the arms, was undertaken in an effort to provide an archeologic date. The tattooing on the right forearm, much clearer than on the left, was visible on the dorsal aspect of the forearm, hand, and fingers, starting 90 mm below the elbow. The tattooing on the right forearm consisted of rows of dots with alternating lines (Figure 8.1). The dots measured approximately 1 mm across and the solid lines were approximately 3 mm in width. The total length covered by the forearm tattooing was about 100 mm.

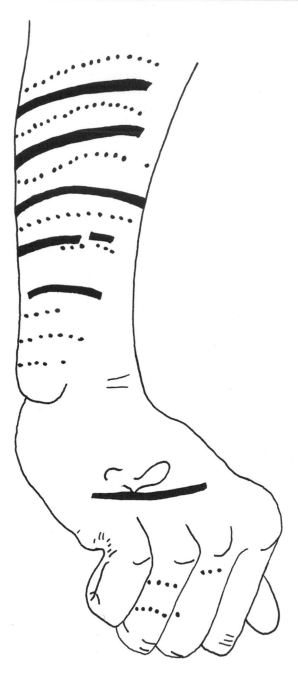

Tattooing on the dorsal aspect of the right hand was at first too faint to be clearly seen. Upon examination with infrared film, it was seen to be a "flanged heart" shape attached to a horizontal line. There were also two rows of dots on the second and third fingers. The coloration of the tattooing was dark blue to black.

The tattooing on the left arm was more elaborate than that found on the right (Figure 8.2). Its coloration was the same. The left arm was more desiccated than the right, making observation more difficult. The most proximal solid line had attached to it four designs that resembled the "flanged heart" seen on the right arm, but smaller. Each of these designs was attached to the main line by a vertical line. The tattooing on the dorsal aspect of the left hand was very difficult to distinguish, even with the use of infrared photography. Piecing together the many photos taken, the design on the back of the left hand seemed to consist of an oval, which may or may not have been complete, with a line extending laterally from its proximal border. Within the oval there may have been another oval, but this could not be verified (Figure 8.2). There was also tattooing on the second and fourth fingers consisting of two rows of dots. The tattooing on the forearm started 80 mm below the elbow and was 146 mm in length.

The process of tattooing on Saint Lawrence Island is described by Otto Geist in a letter to Dr. Charles Bunnell dated 1928, a portion of which follows:

Some of the St. Lawrence Island Eskimo women and girls have beautifully executed tattoo marks. These are made free hand although sometimes an outline is traced before the tattooing takes place. The pigment is made from the soot of seal oil lamps which is taken from the bottom of tea kettles or similar containers used to boil meat and other food over the open flame. The soot is mixed with urine, often that of an older woman, and is applied with steel needles. Two methods of tattooing are practiced. One method is to draw a string of sinew or other thread through the eye of the needle. The

Figure 8.1. *Pattern of tattooing on the right forearm of a 1,600-year-old frozen body from Saint Lawrence Island, Alaska. (Drawing courtesy of George S. Smith, University of Alaska)*

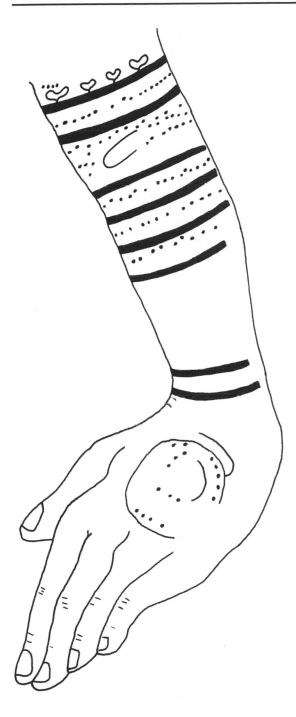

Figure 8.2. *Pattern of tattooing on the left fore-arm of a 1,600-year-old frozen body from Saint Lawrence Island, Alaska. (Drawing courtesy of George S. Smith, University of Alaska)*

thread is then soaked thoroughly in the liquid pigment and drawn through the skin as the needle is inserted and pushed just under the skin for a distance of about a thirty-second of an inch when the point is again pierced through the skin. A small space is left without tattooing before the process is again repeated. The other method is to prick the skin with the needle which is dipped in the pigment each time. [Geist 1928]

Collins (1937), in illustrating decorative motifs of Old Bering Sea Style 2, has a drawing of a gorget-like ornament with a motif very similar to the tattooing design found here (Figure 8.3). Collins also shows four designs that could be along the same lines as the "flanged hearts" (Figure 8.4). These are also from the Old Bering Sea Phase. Collins also states that a similar design occurs on a dart socket piece (Okvik) which he bought on Little Diomede Island and which is illustrated on Plate 14-5 of "Archaeology of St. Lawrence Island, Alaska."

Thus the artistic motifs of the tattooing correlate with the radiocarbon dates in placing this individual within the Old Bering Sea Phase of Alaskan prehistory.

The body (Figure 8.5) appeared to be quite well preserved. As is usual with bodies long dead, the skin was dark brown. The subject appeared to have been an elderly woman; no external male genitalia were visible, and the breasts were atrophic. The unclothed body weighed about 25 kg and showed mild scoliosis. No incisions, scars, or decubitus ulcers were seen. Some brown hair was found on the vertex of the scalp. The right side of the face was partly crushed. Several teeth were missing, as was the left lower leg. No pathological changes were seen in the protruding distal left femur.

Standard Y-shaped and intermastoid incisions were made. The internal organs were somewhat desiccated, but were generally comparable in appearance to those of cadavers used for anatomical dissection. The body was that of a postmenopausal woman; atrophic female internal genitalia were identified. Gross pathological changes were

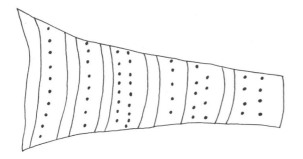

Figure 8.3. Decorative motif of the Old Bering Sea Style 2 type, as found on a gorget-like ornament. (From Collins 1937:82)

found in several viscera. There was a moderate degree of coronary atherosclerosis, but no evidence of myocardial infarction, acute or healed. The lower lobes of both lungs showed fibrous adhesions to the chest wall and diaphragm, and the lungs contained heavy deposits of anthracotic pigment. The smaller bronchi of both lungs were packed with moss (later identified as *Meesia triquetra*), forming casts of the bronchi. A calcified carinal lymph node was found. Moderate scoliosis and aortic atherosclerosis were present. The brain was a crumbling brown mass.

The tissues were somewhat desiccated, a process which continues even in the frozen state. The tissues were rehydrated with Ruffer's solution, embedded in paraffin, and sectioned as any fresh tissue would be.

Sections of the coronary arteries clearly showed the atheromatous deposits that had been seen grossly (Figure 8.6). The myocardium was less well preserved; striations and, as is usual in mummified tissue, nuclei were not seen. The lungs showed the patchy deposition of anthracotic pigment observed in modern patients with centrilobular emphysema (Figure 8.7). The alveolar architecture was generally preserved; many of the alveoli appeared to be coalescent, although some of this change may be postmortem artifact. Some moss fibers were seen in the bronchi and were associated with hemorrhage (Figure 8.8). The liver showed clearly the distinction between the parenchymal

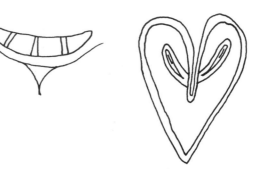

Figure 8.4. Decorative designs of the Old Bering Sea Style 2 type that are similar to the "flanged heart." (From Collins 1937:82)

cells and the portal triads, particularly with the trichrome stain. The cells contained a brown pigment that failed to stain for iron and bile. This almost certainly represents lipofuscin. The thyroid contained well-preserved follicles, and the colloid took the specific iron stain.

The calcified carinal lymph node contained numerous concentric areas of fibrosis with central calcification (Figure 8.9). These were interpreted as healed granulomas; identical lesions were seen in the spleen and possibly in the meninges, where the calcified lesions were much smaller and may represent phleboliths. Examination with polarized light revealed only minute and insignificant amounts of silica, and results of staining for acid-fast bacilli were negative. Stains for fungi revealed many weakly staining budding yeast cells and hyphal filaments. The morphology was that of a *Candida* species. A screening conjugate reagent for *Candida* sp.

Figure 8.5.   *The frozen body of an Eskimo woman. The scoliosis is clearly visible.*

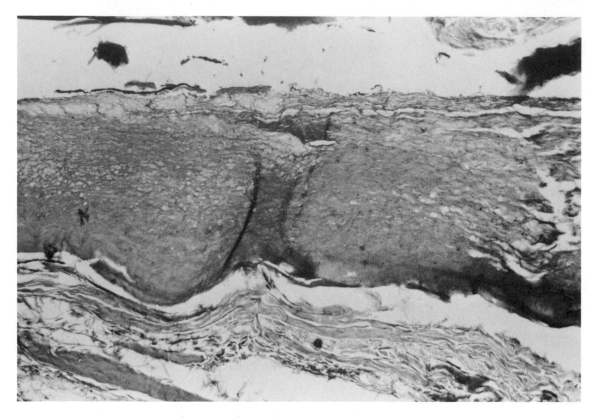

Figure 8.6.   *Coronary artery showing atherosclerotic plaque. Hematoxylin and eosin.* × 95.

stained the hyphae, but not the yeast cells. The same fungi were found in other tissues, such as the diaphragm. Fluorescein-labeled *Histoplasma capsulatum* antiglobulins that had been absorbed with cells of *Candida albicans* did not demonstrate *H. capsulatum.*

Examination of the feces revealed the ova of a fish trematode, *Cryptocotyle lingua* (Figure 8.10). The ova of this parasite have been reported in modern Eskimos by Rausch et al. (1967), but the adult helminth has not been identified in man.

The conclusion from the gross findings was that this elderly woman had been trapped in her semisubterranean house by a landslide or earthquake, and had been buried alive and asphyxiated. This conclusion was based on several facts. The body was unclothed, and Eskimos are unclothed only in their houses; when burial is deliberate, they are clothed. In view of the preservation of the body, one would have expected any clothing to have been preserved also.

Aspiration of foreign material into the bronchi is known to occur in accidental inhumation and has been demonstrated in persons buried in heaps of coal (Gonzalez et al. 1954). The microscopic finding of hemorrhage associated with the moss fibers in the bronchi is consistent with asphyxiation. It is not unusual for red blood cells to be preserved for extended periods. Preserved erythrocytes have been reported in the tissues of Peruvian (Allison 1975) and North American Indian (Zimmerman 1973) mummies. Microscopic fracture of the right temporal bone was also seen, with associated hemorrhage indicating that this was a true antemortem fracture, thus confirming the role of trauma in this woman's death (J. Benitez, personal communication).

Paleopathology includes an interest in the

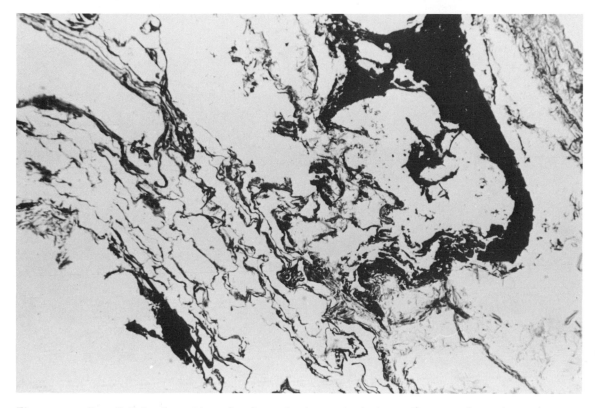

Figure 8.7. Centrilobular deposition of anthracotic pigment in the lungs. Hematoxylin and eosin. × 37.5.

history of disease processes. This Eskimo woman, far removed from the stresses of modern technological society, suffered from coronary artery disease – a process that has been well documented as far back as dynastic Egypt, by both historical (Breutsch 1959) and anatomical (Sandison 1970) evidence. The present case not only confirms the antiquity of the process of coronary atherosclerosis, but also exhibits its occurrence in a preliterate society.

The finding of severe anthracosis can be attributed to a lifetime spent around open cooking and heating fires. Similar findings have been reported in several mummies (Brothwell et al. 1959; Sandison 1970; Zimmerman et al. 1971). Air pollution, at least on a local level, is not a recent phenomenon.

Several of the organs also showed a healed granulomatous process. Tuberculosis is considered to have been nonexistent in Alaska prior to its introduction by the Russians in the early eighteenth century. Of the fungi pathogenic for man that produce a granulomatous reaction, only histoplasmosis is thought to occur in Alaska, but information on its distribution is far from complete (Comstock 1959). Less than 1 percent of modern Eskimos have a positive cutaneous reaction to histoplasmin (Sexton et al. 1949; Comstock 1959). *Histoplasma capsulatum* was not demonstrated in the tissues of this woman. However, the distribution of the granulomas is most consistent with the diagnosis of healed histoplasmosis. The *Candida* species that was found, both in the granulomas and elsewhere in the body, undoubtedly is a postmortem invader. The weak staining of the fungi indicates that the contamination occurred some considerable time in the past, probably shortly after death.

In summary, this elderly woman is thought

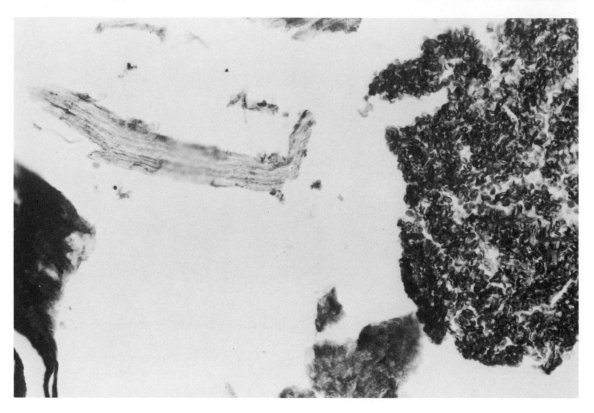

Figure 8.8.   Aspirated moss fiber associated with hemorrhage in bronchial lumen. Masson's trichrome.
× 95.

to have suffered a traumatic death some 1,600 years ago. There was gross and microscopic evidence of skull fractures, and the finding of aspirated moss in the bronchi associated with hemorrhage suggests that accidental burial and suffocation played a significant role in her death. Other pathological changes documented included coronary atherosclerosis, scoliosis, anthracosis, and emphysema, and probable healed histoplasmosis. Radiocarbon dating and archeologic evaluation of tattoos on the body correlated in giving an approximate date of A.D. 400.

AN ALEUTIAN MUMMY

The mummy reported on in this section was collected in 1938 from Kagamil Island, in the central part of the Aleutian chain, by Dr. Aleš Hrdlička of the Division of Physical Anthropology, U.S. National Museum (1945). The individual was an Aleut, probably of the immediate pre-Russian era (prior to 1740). There were insufficient archeologic data for more precise dating, and the material was too recent to be dated by the radiocarbon method (Willis 1970).

The cold and damp climate of the Aleutian Islands would appear ill suited to the practice of mummification, which is generally based on desiccation. Hrdlička (1945) attributes the development of mummification by the Aleuts to a reluctance to part with the deceased; Laughlin (n.d.) points to the anatomic interests of the Aleuts in conjunction with their desire to preserve and use the spiritual power residing in the human body. The Aleuts studied comparative anatomy, using the sea otter as the animal most like man, conducted autopsies on their dead, and had an extensive

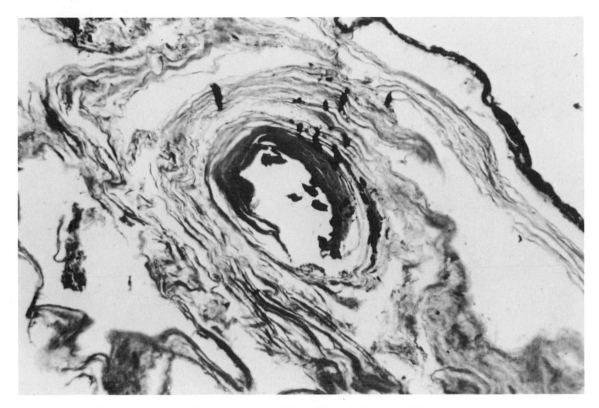

Figure 8.9. Fibrocalcific granuloma in a carinal lymph node. Hematoxylin and eosin. × 95.

anatomic vocabulary. Mummification as an Aleut funerary practice was an extension of their pragmatically oriented culture.

The technique of mummification varied with the social status of the deceased (Jochelson 1925; Veniaminov 1945; Dall 1945). The bodies of hunters and tribal leaders were eviscerated through an incision in the pelvis or over the stomach. No chemicals were used, but fatty tissues were removed from the abdominal cavity, which was stuffed with dry grass. The body was then put in running water, which completed the removal of fat, leaving only skin and muscle. The body was then bound with the hips, knees, and elbows flexed. This position has been variously explained as an imitation of the fetal position, an attempt to economize on space, or an effort to prevent the dead from returning and harming the living. Jochelson (1925) rejects these interpretations in pointing out that the flexed position was the habitual leisure posture of the Aleuts. The binding of the mummy bundle is properly considered an effort to maintain the deceased in a comfortable position.

The flexed body was then air-dried by carefully and repeatedly wiping off exuded moisture. When drying was complete, the cords were removed and the mummy was wrapped in its best clothes, usually a coat of aquatic bird skins. It was encased in a waterproof coat of sea lion intestines and then various layers of seal, sea lion, or otter skins and perhaps some matting. The entire bundle was then tied with a braided sinew cord and removed to a burial cave. There the mummies were placed on platforms or suspended from the ceiling to avoid contact with the damp ground. The cave in which the mummy described in this chapter was found was heated by a volcanic vent, creating a preservative warm, dry atmosphere.

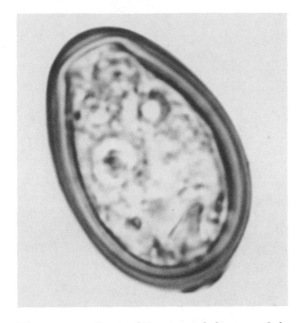

Figure 8.10.   Ovum of Cryptocotyle lingua, *a fish trematode, found in the feces.* × 600.

These caves were probably used only for a few hundred years before the Russian contact of the early eighteenth century. An Aleut tale (Jochelson 1925), explaining the use of the warm cave, tells of a rich headman, Little Wren, who lived near the cave on Kagamil Island. His young son was accidentally killed by his brother-in-law. In the subsequent funeral procession, the boy's pregnant sister slipped on a rock and suffered a fatal miscarriage. As the season was snowy and cold, the chief decided to place the bodies in the nearby cave, which had been used previously for storage. The chief declared that the cave would become a mausoleum for his entire family, and when he died of grief shortly afterward, he was interred there with all his possessions.

Dr. Hrdlička removed some 50 mummies from the warm cave in 1938 (it is thought that the body of Little Wren had been removed in 1874 by Captain E. Hennig of the Alaska Commercial Company). Except for blood group determinations (Candela 1939), the mummies remained undisturbed at the Smithsonian until a group directed by the author examined one in 1969. It was fortunate that the mummy selected for study was apparently that of a common man, as it had not been eviscerated.

Radiologic examination of the 112-cm-long, coffin-shaped, fur-wrapped bundle revealed the outlines of the heart and lungs. The brain appeared as an occipital opacity. Pathologic changes were limited to minimal arthritic changes in the vertebral column and evidence of dental attrition and periodontal disease. A number of radiopaque masses were seen in the left side of the abdomen.

The wrappings were removed sequentially. The outer five were animal skins, probably sea otter. The innermost layer was an eider-down parka, composed of numerous bird-skins sewn together with the feathers on the inside, with a spotted fur collar. No incision was seen in the body.

The individual was an adult male of indeterminate age. The weight was approximately 10 kg, and the overall length of the body was 165 cm. The skin was dark brown, dry, and leatherlike. The body was flexed as shown in Figure 8.11.

The face was partially covered by a birdskin, probably a cap that had slipped down. There was some balding, and the hair appeared singed, suggesting the body had been suspended over a fire for desiccation. A mustache and full beard were present.

There was a full complement of teeth, in normal occlusion. Except for the shrinkage of mummification, the neck, chest, abdomen, and genitalia were unremarkable. No skin incisions were seen. Lodged between the left forearm and left side of the abdomen was an empty birdskin pouch.

A relatively standard postmortem examination was possible, using a Stryker electric autopsy saw to remove the rigid tissues of the anterior chest and abdominal wall. The thoracic viscera were found to be intact and were removed. Gross pathologic change was limited to a few pleural adhesions and consolidation of the lower lobe for the right lung.

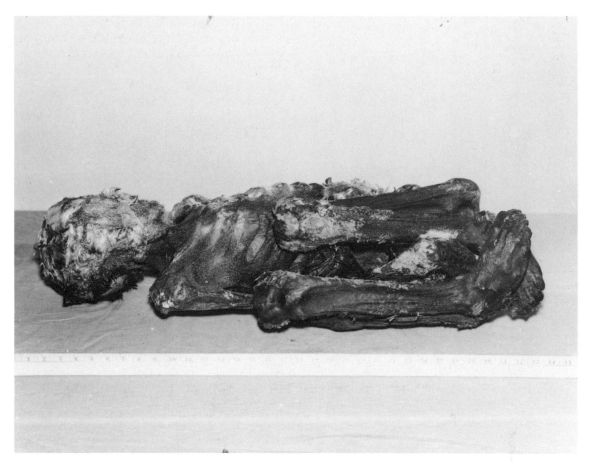

Figure 8.11. Aleutian mummy, unwrapped, with birdskin cap covering right side of face and birdskin pouch between left forearm and abdomen.

The abdominal viscera were poorly preserved, only the distal large intestine remaining intact, being filled with coprolites.

The rectum was slightly dilated; it measured 7 cm in diameter and was filled with fecal material. The abdominal aorta and iliac vessels were well preserved and easily identified. A firm yellow plaque measuring 2 by 1 cm was noted in the right iliac artery, but there was no other gross evidence of atherosclerosis.

The skull was examined after removal of the calvarium. The bone and the dura were found to be intact. The latter was thin and transparent in the frontal area and thicker and opaque elsewhere. Upon removal of the dura, the brain tissue was found to be shrunken into the posterior fossa of the cranial cavity, the major portion of which was empty (Figure 8.12). The brain was roughly rectangular; it measured 14 by 10 by 5 cm, and was covered with a fine crystalline material.

The upper left maxillary dentoalveolar process was removed en bloc. Extreme dental attrition was present to a point slightly beyond the interdental contacting tooth surfaces. Heavy dental calculus and periodontal bone loss were evident.

The pulmonary architecture was generally well preserved, although cellular detail was lost. A moderate amount of interstitial black anthracotic pigment was noted throughout.

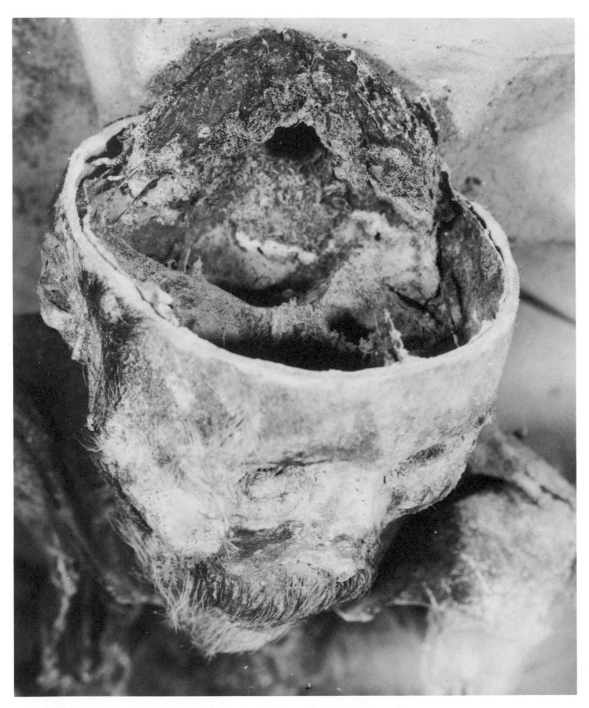

Figure 8.12. *Position of brain in posterior midline, as seen after removal of the calvarium.*

Microscopic examination of the aorta revealed preservation of the three layers of the wall and of the elastic tissue. There was no calcification or atheromatosis, but cellular detail was absent. There was excellent preservation of the general architecture of the iliac artery and vein, including the elastic tissue and a venous valve. The single atherosclerotic plaque noted on visual inspection was composed of cholesterol crystals and contained minute calcific foci.

Sections of skin from the abdominal wall, eyelid, ear, and lower lip showed only connective tissue, a few structures suggestive of blood vessels, hair follicles and shafts, and in the ear, well-preserved cartilage. The inferior labial artery was partially collapsed and exhibited a poorly stained plaque that was considered to be consistent with atherosclerotic intimal changes. A section from the right thigh showed connective tissue, skeletal muscle, and a few areas of pigmented epidermis. Multiple hair shafts were noted in each of several follicles.

The Schneiderian membrane of the antral floor consisted of connective tissue covered by a thin (12μ) amorphous basophilic layer of epithelium. Polarized light elucidated the perivascular connective tissues and the perpendicular fibers (Sharpey's) of the periosteum.

The dental pulp chambers were thoroughly desiccated and contained only scattered strands of unrecognizable filamentous elements. No odontoblasts were seen. Nonetheless, the hard tooth structures were in an excellent state of preservation. Ground tooth sections revealed complete histologic properties of enamel and dentin. The Hunter-Schreger bands and the incremental lines of Retzius were clearly visualized in the enamel. The tubular nature of the dentin was perfectly preserved, and the incremental lines of Von Ebner and Owen were seen. Secondary dentin was present, measured about 350μ in thickness, and was separated from the reparative (tertiary) dentin by a prominent basophilic line. The reparative dentin was 1 to 2 mm in thickness under the worn cuspal areas where the attrition had abraded through the enamel into the dentin of the teeth. In spite of the severe attrition, structurally sound enamel remained in the intercuspal areas. Because of the excellent reparative response to attrition and the absence of periapical dental disease, it was assumed that the teeth were vital.

Several special studies were performed.

The blood group, determined on cancellous bone of the femoral head by a modification of the agglutination-inhibition test described by Springer et al. (1957) was O.

Cultures of the lung tissue failed to reveal viable organisms.

Analysis of skin, heart, kidney, brain, and muscle revealed almost total preservation of protein content. Enzyme analyses showed absence of activity of lactic dehydrogenase, creatine phosphokinase, glutamic pyruvic transaminase, glutamic oxalacetic transaminase, and alkaline dehydrogenase. The analyses were done on homogenates of dry tissue (20 mg per milliliter of normal saline).

Neutron-activation analysis of lung, hair, fingernails, skin, and retroperitoneal tissue revealed no unexpected nuclides. The sodium activity in all the samples was greater than anticipated from previous studies of dried human tissues (H. B. Gardner 1969, personal communication), and the right lower lobe showed some increase in ^{82}Br as compared with the remainder of the lung.

The crystalline areas were subjected to X-ray crystallographic analysis, which revealed them to be composed of acid-ammonium-sodium-phosphate-hydrate and apatite, a calcium–phosphate compound.

Examination of the coprolites was negative for parasites and the ova of parasites. Chemical analysis revealed the coprolites to be composed of ammonia and phosphates; as there was no calcium, their radiopacity was a function of density. The coprolites were soluble in a wide range of organic and inorganic solvents, including chloroform, acetone, ethanol, water and dilute acids, and alkalis.

The reduction of the lower lobe of the right lung to a consolidated mass was the most remarkable gross pathologic change in the mummy. Histologic examination confirmed the destruction of the parenchyma in this area and revealed the presence of free gram-negative bacilli and clumps of material that took a red color with the Brown-Hopps stain. Multiple small aggregates of crystalline material, often containing gram-negative bacilli, were found, not only in the right lower lobe, but in the other lobes of both lungs, the heart, the trachea, and the retroperitoneum.

There are several possible explanations for the appearance of the right lower lobe and the crystalline areas. A component of postmortem change is indisputable; the question is one of degree. Do these areas represent antemortem disease or are they attributable entirely to postmortem change?

The posterior midline position of the remains of the brain indicates that the body was in the supine position during the postmortem period of liquefaction and subsequent desiccation. Although the condition of the lower lobe of the right lung could be a manifestation of postmortem autolysis, one would expect, given the supine position of the body, that a change of this nature would involve the posterior portions of both lungs. The mummification of the other lobes is evidence against autolysis. Conversely, the failure of one lobe to mummify implies a predisposing factor, such as antemortem disease.

Analysis of crystalline material present in the lungs and viscera revealed it to be inorganic, and not to be confused with adipocere. No adipocere was seen grossly, although the foggy and hazy conditions common in the Aleutians are said to favor the formation of adipocere (Evans 1963). Adipocere results from the postmortem autolysis of body fats and consists primarily of palmitic, stearic, and hydroxystearic acids (Mant 1957). Evans (1962) notes that the crystals of adipocere are found only in tissues containing fat and not in such structures as the lung and trachea. It is apparent that the crystalline material under discussion is not adipocere, but probably represents postmortem mineralization of areas containing discrete aggregates of gram-negative bacilli.

There are two possible explanations for the presence of the bacteria in the crystalline foci. One is that the bacilli were present throughout the tissues and were preserved only in the areas of mineralization. Many gram-negative and gram-positive cocci were scattered throughout the tissues, and it is difficult to imagine a process that would preserve bacilli selectively in one area and cocci in another. The more plausible explanation is that the crystalline foci represent antemortem bacterial abscesses. The preservation of bacteria for 300 years is not unusual; bacteria have been stained in the intestinal contents of a 4,000-year-old Egyptian mummy (Ruffer 1921). The bacteria may have had a role in the process of mineralization by invoking a mechanism similar to that which results in adipocere (Evans 1963). Proteolytic bacterial enzymes may produce a localized acidic environment conducive to the deposition of calcium salts, especially if supersaturation resulted from desiccation. However, in the absence of experimental studies, a discussion of the process of mineralization remains speculative.

The distribution of the crystalline areas also suggests antemortem abscesses. As the bacteria in these foci appear to be the same as the gram-negative bacilli in the right lower lobe, one can infer that the terminal illness was lobar pneumonia (possibly caused by *Klebsiella pneumoniae*), with septicemia and multiple visceral abscesses.

Postmortem changes have altered the picture considerably. Although they might have caused all the changes described, this appears to be improbable for the reasons already given. The involvement of the tracheal cartilage can be explained by postmortem invasion by the bacilli.

Anthracotic pigment found in the lungs can be attributed to the culinary habits of the Aleuts. Until recently they prepared their food over an open seal-oil fire, which filled

their homes with smoke. Indeed, ocular changes in the Aleuts, noted by early visitors, were attributed to the smoke (Petroff, quoted in Hrdlička 1945), and modern visitors have found it impossible to live in Aleut houses for the same reason (T. D. Stewart, personal communication). The lungs showed changes consistent with moderate emphysema and bronchiectasis, probably of the same origin. Tobacco may be ruled out as a cause, as the use of tobacco was unknown before the advent of the Russians.

Severe masticatory dental stresses must have existed, because of the marked dental attrition, the increased thickness of the lamina dura, the prominent hypercementosis of the tooth roots, and the significant deposits of tertiary dentin.

The presence of periodontal disease was manifested by deposits of heavy dental calculus, periodontoclastic bone changes, and migratory protrusion of the anterior teeth.

There was no evidence of impacted teeth, supernumerary teeth, missing teeth, dental caries, or malocclusion disorders. Similarly, the maxillary bone and antrum exhibited no pathologic changes. A possible atheromatous plaque was noted in the inferior labial artery.

Sections of the major blood vessels showed only mild focal atherosclerosis of the iliac vessels. This finding, combined with the roentgen evidence of mild arthritic changes, enabled us to estimate the age of the subject at the fourth or fifth decade.

There were two interesting incidental findings. Chondrocytes appeared to be well preserved, even at the ultrastructural level (Yeatman 1971). An inexplicable finding was the presence of several hair shafts in single follicles in a section of skin (a nonhuman characteristic). As the cadaver was separated from the fur wrappings by an eiderdown parka, this section could not be one of adherent animal skin.

The negative findings are also noteworthy. There was no evidence of trauma, and no foreign material or organisms were seen in the pulmonary alveoli; this evidence ruled out accidental death by drowning. No poisons were found in the tissues analyzed by neutron activation. Results of staining fungi and tubercle bacilli were negative.

The blood group, determined on bone from the femoral head, was type O. The science of paleoserology is still in a state of evolution (Lengyl 1975), but blood groups have been successfully determined in varied mummified material (Boyd and Boyd 1939; Candela 1939; Thieme et al. 1956). Candela (1939) typed 30 of the Aleutian mummies when they arrived in Washington, D.C., by use of vertebral bone corings. The blood group distribution was: 11 O, 11 A, 6 B, and 2 AB. These results are in contrast to the prevalence of type O in Eskimos and American Indians (although some type A is found among Northwest Coast Indians). Candela noted that the Aleuts have an almost identical blood type distribution to that of Eastern Siberian tribes, but he felt that the number of individuals typed was too small to draw any valid conclusions regarding the origin of the Aleuts.

In summary, examination of a 200- to 300-year-old Aleutian cadaver mummified by desiccation suggested that the cause of death was lobar pneumonia caused by a gram-negative bacillus, possibly complicated by septicemia and diffuse metastatic abscesses. The abdominal viscera were not preserved, and the role of intraabdominal disease in the death of the subject could not be assessed. Other findings included pulmonary anthracosis and mild atherosclerosis. Severe masticatory dental stresses were attested to by marked dental attrition, increased thickness of the lamina dura, prominent hypercementosis of the dental roots, and significant deposits of tertiary dentin. Periodontal disease was manifested by deposits of heavy dental calculus, periodontoclastic bone changes, and migratory protrusion of the anterior teeth. No adipocere was seen, and there was no evidence of death from trauma, drowning, or poisoning.

On 7 October 1978 another Aleutian mummy dated about 1700 was examined at the Pea-

body Museum at Harvard. It had been in the museum for about 100 years, having been in the group removed by Captain Hennig from Kagamil Island in 1874. The body was covered with the original wrappings of sea lion skin lined with sea otter fur. The mummy was that of a woman in the sixth decade of life. Preliminary studies have suggested healed pleuritis and renal tubular necrosis. She also suffered from middle ear disease and head lice. Other studies are in progress (Zimmerman 1979).

REFERENCES

Allison, M. A.; Klurfeld, D.; and Gerszten, E. 1975. Demonstration of erythrocytes and hemoglobin products in mummified tissue. *Paleopathology Newsletter* 11:7–10.

Birket-Smith, K. 1959. *The Eskimos*, 2nd ed. London: Methuen.

Boyd, L. G., and Boyd, W. C. 1939. Blood group reactions of preserved bone and muscle. *American Journal of Physical Anthropology* 25:421–34.

Breutsch, W. L. 1959. The earliest record of sudden death possibly due to atherosclerotic coronary occlusion. *Circulation* 20:438–41.

Brothwell, D. R.; Sandison, A. T.; and Gray, P. H. K. 1959. Human biological observations on a Guanche mummy with anthracosis. *American Journal of Physical Anthropology* 30:333–47.

Candela, P. B. 1939. Blood group determinations upon the bones of thirty Aleutian mummies. *American Journal of Physical Anthropology* 24:361–83.

Collins, B. 1937. Archeology of St. Lawrence Island, Alaska. *Smithsonian Institution Miscellaneous Collections* 96(1).

Collins, H. B. 1933. Prehistoric Eskimo culture of Alaska. In *Explorations and field work of the Smithsonian Institute in 1932*. Washington, D.C.: Smithsonian Institution.

Comstock, G. W. 1959. Histoplasmin sensitivity in Alaskan natives. *American Review of Tuberculosis and Pulmonary Disease* 79:542.

Dall, W. H. 1945. Quoted in *The Aleutian and Commander islands and their inhabitants*, A. Hrdlička (ed.), pp. 184–91. Philadelphia: Wistar Institute.

Evans, W. E. 1962. Some histological findings in spontaneously preserved bodies. *Medicine, Science and the Law* 2:153–64.

–   1963. Adipocere formation in a relatively dry environment. *Medicine, Science and the Law* 3:145–53.

Geist, W. W. 1928. Diary. University of Alaska archives.

Gonzalez, T.; Vance, M.; Helpern, M.; and Umberger, C. J. 1954. *Legal medicine: pathology and toxicology*. New York: Appleton.

Hrdlička, A. 1945. *The Aleutian and Commander islands and their inhabitants*. Philadelphia: Wistar Institute.

Jochelson, W. 1925. *Archeological investigations in the Aleutian Islands*. Washington, D.C.: Carnegie Institute.

Laughlin, W. S. n.d. The use and abuse of mummies. Unpublished manuscript. University of Connecticut, Storrs.

Lengyl, I. A. 1975. *Paleoserology*. Budapest: Akademiai Kiado.

Mant, A. K. 1957. Adipocere: a review. *Journal of Forensic Medicine* 4:18–35.

Petroff, I. 1945. Quoted in *The Aleutian and Commander islands and their inhabitants*, A. Hrdlička (ed.), p. 174. Philadelphia: Wistar Institute.

Rausch, R. L.; Scott, E. M.; and Rausch, V. R. 1967. Helminths in the Eskimos in western Alaska, with particular reference to Diphyllobothrium infection and anemia. *Transactions of the Royal Society of Tropical Medicine and Hygiene* 61:351–7.

Ruffer, M. A. 1921. *Studies in the paleopathology of Egypt*. Chicago: University of Chicago Press.

Sandison, A. T. 1970. The study of mummified and dried human tissues. In *Science in archeology*, 2nd ed., D. Brothwell and E. Higgs (eds.), pp. 490–502. New York: Praeger.

Sexton, R. L.; Ewan, J. R.; and Payne, R. C. 1949. Determination of the specificity of histoplasmin and coccidioidin as tested on 365 Aleuts of the Pribilof Islands. *Journal of Allergy* 20:133–5.

Smith, G. S., and Zimmerman, M. R. 1975. Tattooing found on a 1600 year old frozen, mummified body from St. Lawrence Island, Alaska. *American Antiquity* 40:434–7.

Springer, G. F.; Rose, C. S.; and Gyorgy, P. 1957. Blood group mucoids: their distribution and growth-promoting properties for *Lactobacillus bifidus* var. Pen. *Journal of Laboratory and Clinical Medicine* 43:532–42.

Thieme, F. P.; Otten, C. M.; and Sutton, H. E. 1956. A blood typing of human skull fragments from

the pleistocene. *American Journal of Physical Anthropology* 14:437–44.

Veniaminov, I. 1945. Quoted in *The Aleutian and Commander islands and their inhabitants*, A. Hrdlička (ed.), pp. 182–4. Philadelphia: Wistar Institute.

Willis, H. E. 1970. Radiocarbon dating. In *Science in archeology*, 2nd ed., D. Brothwell and E. Higgs (eds.), pp. 46–57. New York: Praeger.

Yeatman, G. 1971. Preservation of chondrocyte ultrastructure in an Aleutian mummy. *Bulletin of the New York Academy of Medicine* 47:104–8.

Zimmerman, M. R. 1973. Blood cells preserved in a mummy 2000 years old. *Science* 180:303–4.

Zimmerman, M. R. 1979. Harvard mummies: a preliminary report. *Paleopathology Newsletter*, 25:5–8.

Zimmerman, M. R., and Smith, G. S. 1975. A probable case of accidental inhumation of 1600 years ago. *Bulletin of the New York Academy of Medicine* 51:828–37.

Zimmerman, M. R.; Yeatman, G. W.; Sprinz, H.; and Titterington, W. P. 1971. Examination of an Aleutian mummy. *Bulletin of the New York Academy of Medicine* 47:80–103.

9

Mummies of Peru

JAMES M. VREELAND, JR.
Department of Anthropology
University of Texas, Austin, Texas, U.S.A.

AIDAN COCKBURN
President, Paleopathology Association
Detroit, Michigan, U.S.A.

ANTHROPOLOGICAL AND HISTORICAL PERSPECTIVES

JAMES M. VREELAND, JR.

Peruvian mummies have been the object of anthropological and historical interest for more than four centuries. In 1560, long before Egyptian pharaohs were put on public display in Cairo's Museum of Archaeology, curious Europeans had already been queuing up in Lima's San Andrés Hospital to view several of the marvelously preserved mummies of Peru's legendary Inca kings.[1] Struck by what seemed to them an idolatrous, but fascinating, custom, the early Spanish chroniclers of Andean culture noted that the practice of mummifying principal lineage heads and local chiefs was widespread in western South America. Today, studies of pre-Hispanic mortuary practices draw heavily on these richly detailed ethnohistorical accounts, as well as on the wealth of cultural and biologic materials preserved in the desertic coastal zone of Peru. Here, despite the absence of written history until the arrival of Pizarro in 1532, the archeologic record of mummification is now 6,000 years old.

Although the origins of this practice still remain unclear, naturally mummified bodies[2] occur in Peruvian graves before Andean societies became sedentary and stratified; they may well have provided models for subsequent experimentation with methods of artificially preserving human flesh. The importance of specialized techniques to retard decay of the remains of local secular and theocratic elite individuals probably increased with the emergence of complex societies and clearly represents an intensification of the ancient Andean practice of ancestor worship (Trimborn 1969:116). Venerated as "living corpses," the mummified bodies of clan ancestors or chiefs often served as community or tribal fetishes, and in the case of the Inca rulers, as historical gods. By the end of the pre-Hispanic epoch, grave goods accompanying the mummy of a high-status figure often included the mummified bodies of his wives, retainers, and slaves (Steward 1948:10).

Although the term *mummy* is repeatedly used to describe the often extraordinarily well-preserved human remains recovered from Peruvian cemeteries, there is in fact little agreement on what constitutes a Peruvian mummy and how it was actually produced. The term *mummification* will be used here to refer to all natural and artificial processes that bring about the preservation of the body or its parts. Such methods include drying by air, sun, or fire (with or without evisceration); covering with plastic materials (such as clay); filling body cavities with plant or other materials; and embalming with chemical or other substances (Dérobert and Reichlen, n.d.:8).

Three principal types of mummification can be identified in pre-Columbian America (Dawson 1928b; Comas 1974):

Type I. Natural mummification, caused by a number of factors (either singly or in combination) such as dryness, heat, cold, or absence of air in the burial unit or grave

Type II. Intentional natural mummification, brought about through the intentional exploitation or deliberate enhancement of

natural processes, such as those listed above

Type III. Artificial mummification, produced by a variety of techniques including evisceration, fire-and-smoke curing, and the application of such embalming substances as resins, oils, herbs, and other organic materials

In Peru, the combined archeologic and ethnohistorical evidence indicates that the large majority of mummies known are of types I and II.[3] The skin, when preserved, has generally been modified to a tough, almost leathery consistency. Although connective tissue frequently remains, most or all of the internal organs have disintegrated to a fine powder, often filling much of the abdominal cavity. However, mummified bodies with nearly intact intestinal tracts have been found in several coastal regions where natural preservation was particularly favorable (Stewart 1973:44; Allison et al. 1974).

PHYSICAL AND CULTURAL
ENVIRONMENT

Geographically, the Peruvian, or central Andean, cultural area consists of the coast and highlands of Peru and the adjacent highlands of Bolivia, southern Ecuador, and parts of the north coast and highland regions of Chile (Figure 9.1). The Peruvian littoral, a narrow desert zone crosscut at nearly regular intervals by fertile river valley oases, supported a number of densely populated regions, some of which grew to the status of large chiefdoms and states. Despite the periodic saturation of the surface air layer and occasional winter drizzles (*garúa*), precipitation is negligible and rarely penetrates more than a few centimeters into the ground. Not all areas are, however, equally conducive to the preservation of organic materials; some sites may have been specifically chosen as burial precincts because of their optimal conditions for preservation. The preservative effect of the uppermost soil horizon, when enriched with certain salts, has long been cited as an important additional factor in the mummification process (Rivero and von Tschudi 1854; Mead 1907).[4]

In contrast to the coast, the highland zone is far less homogeneous, characterized by a series of complex gradients of climate and vegetation. In the treeless high-altitude valleys and plateaus, or altiplano, a marked rainy season delivers as much as 800 mm of precipitation annually. Near- or below-freezing temperatures are recorded through much of the year, especially on the upper slopes (over 4,300 m), where frost is an almost nightly occurrence. Both the rarefied atmosphere and cool temperatures of the altiplano were doubtless key factors in mummification processes.

Any description of Andean mummification practices must move in two dimensions: time and space. Prehistoric Andean cultural chronology is customarily broken down into seven major periods, beginning with a long preceramic sequence dating back over 20 millennia and closing with the Spanish conquest of the Inca empire in 1534 (Table 9.1). The first ceramic period, beginning about 1800 B.C., also marks the emergence of the "Peruvian cultural tradition" (Bennett 1948), characterized by the appearance of widespread maize agriculture, irrigation, terracing, complex religious iconography, and marked ancestor worship (Willey 1971:88).

Following this so-called Initial period, a series of three Horizon styles developed, each typified by a complex of more or less homogeneous traits or features, separated by two Intermediate periods when regional cultures eclipsed and superseded the unifying Horizon styles (Rowe 1967). Most authorities generally agree that the three pan-Peruvian cultures evolved from three highland sites: Chavín de Huantar, Huari or Tiahuanaco, and finally Cuzco. From these centers, certain diagnostic styles are seen to have spread outward through most of the central Andean area. However, increasingly persuasive arguments supporting a tropical forest inception for the first of these highland cultural

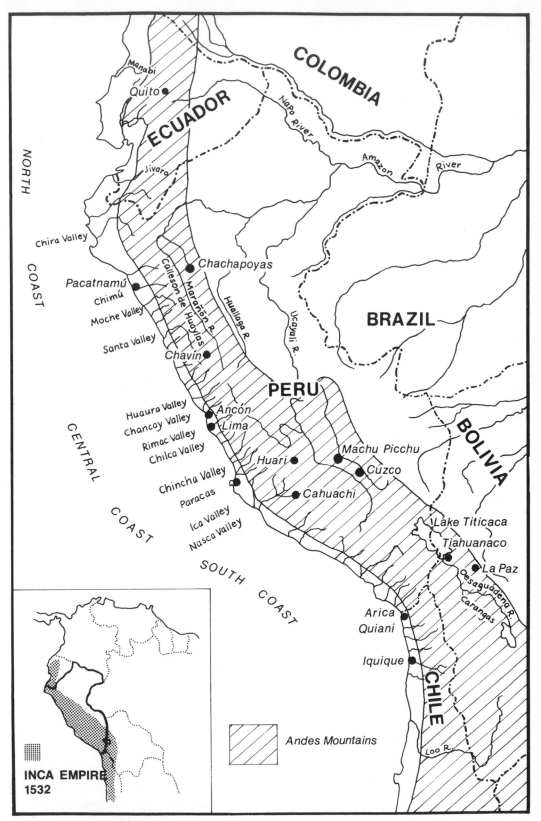

Figure 9.1.   Central Andean cultural area.

Table 9.1. *Major cultural periods and phases for the Central Andes*

Period	Time	Culture
Colonial	A.D. 1534	Spanish Conquest
Late Horizon	A.D. 1476	Inca Empire
Late Intermediate	A.D. 1000	Ica, Chimú, Chancay
Middle Horizon	A.D. 600	Huari, Tiahuanaco
Early Intermediate	200 B.C.	Nasca, Moche, Paracas-Necropolis
Early Horizon	900 B.C.	Paracas-Cavernas, Chavín
Initial	1800 B.C.	Arica, Santo Domingo
Preceramic	10,000 B.C.	Tres Ventanas
	20,000 B.C.	Ayacucho

matrixes have induced some investigators to look east of the Andes for archeological evidence of the origins of Peruvian civilization in this moist lowland region (Lathrap 1974).

PRECERAMIC AND INITIAL PERIODS (20,000–900 B.C)

Although numerous burials containing skeletons wrapped in skins, hides, and vegetable-fiber fabrics have been described for the coastal regions, and to a much lesser extent for dry highland cave sites, little evidence of mortuary practices involving mummification occurs before the fifth millennium B.C. Engel (1970, 1977) recovered four naturally mummified bodies, two adults and two juveniles, from Tres Ventanas Cave in the upper Chilca Valley (4,000 m), dating from about 4000 to 2000 B.C. In contrast to the tightly flexed adult bodies, placed on their sides, the position of the subadults was semiflexed, lying on their backs. The bodies had been wrapped in camelid mantles or cloaks bearing traces of a red pigment and were found with fragments of netting and twined and looped fabrics. The skin and hair appear sufficiently well preserved by the high-altitude tomb matrix that we might justly term these individuals the oldest mummies so far reported from South America.

The earliest mummified remains from northern Chile are quite different and indicate that some coastal fishing societies 5,000 years ago practiced a variant of type III mummification (Figure 9.2; see Uhle 1918; Skottsberg 1924; Nuñez 1969; Munizaga 1974). Special attention had been given to the preparation of mummies of infants, described by Bird at the site of Quiani:

In all cases the viscera and brains appear to have been removed; the legs, arms and body reinforced by sticks inserted under the skin or in the flesh; the faces coated with thin clay and painted; a wig of human hair fastened over the head; and [frequently a] sewn leather casing wrapping the body. [Bird 1943:246]

The presence of several coats of paint on a similar mummy from Punto Pichalo suggests that the body had not been buried immediately following death, but may have been stored or displayed for a considerable time before final interment.[5] In 1917 Skottsberg, working at the preceramic cemetery of Los Gentiles in Arica, found a bundle containing the remains of two infants similarly mummified (Skottsberg 1924:32–37).

A second variety of mummification existed during this same period that appears to represent a form of secondary burial. The extended cadaver was flayed, and the skin was replaced with a thin coating of clay or "cement" and then wrapped in a reed matting. Bodman (in Skottsberg 1924) analyzed such

Figure 9.2.   Detail of elaborately prepared (Type III) 5000-year-old mummy of child from Quiani, Chile. Body is coated with thin layer of clay and some paint, wrapped with birdskin "cloak," with wig of human hair over face. Twined reed matting and cordage is not shown. Specimen #41.1/5780, American Museum of Natural History, New York.

a brownish "cement" and found it to contain sand and a certain agglutinate, mixed with an unidentified material. Although Skottsberg reported that the mummy had been eviscerated, no clear evidence of evisceration or embalming was found by Alvarez Miranda in similar interments at Arica (Alvarez Miranda 1969:189).

EARLY HORIZON TO EARLY INTER-MEDIATE PERIODS (900 B.C.–A.D. 600)

The abundant archeologic information from the Paracas–Ica area of the south coast of Peru indicates that by the end of the Early Horizon period a major transformation in burial practices had taken place in that region. At least as early as 400 B.C., sedentary farming communities had developed a keen interest in enhancing the preservation of the dead, wrapped in an upright position inside "mummy bundles" up to 1.5 m in height. The most spectacular finds of this period include 429 mummy fardels recovered from the Necropolis at Paracas, of which approximately one-half have been opened (Tello 1929). Nearly all the bodies examined were elderly males showing a distinctive type of cranial deformation (Tello 1959, 1980; Vreeland 1978a).[6]

Seated in a coiled basket or gourd, the flexed cadaver was generally covered with a simple cotton shroud and wrapped with plain-weave cloth, often alternated with polychrome, patterned fabrics widely acclaimed for their intricately embroidered designs. Naturally mummified remains of parrots, cavies, foxes, dogs, cats, and deer have also been found in Paracas period mummy bundles. In the larger bundles, four discrete layers of ceremonial garments and wrapping cloths normally occur (Yacovleff and Muelle 1934; Bennett 1938; Tello 1959).

The subject of a heated debate for decades, convincing evidence of artificial mummification (type III) is generally indirect and exceedingly difficult to recognize in archeological contexts. Strong (1957:16) has argued that the presence of large areas of calcined earth and ashes at the temple complex of Cahuachi suggests that the site had been used as a massive mummy-processing area about 2,000 years ago. No mummies were recovered from Cahuachi, but bundles belonging to the same cultural period were found at the Necropolis by Mejía Xesspe in 1927–1928. Following his examination of the largest and best preserved units, Tello concluded:

After extracting the viscera and a great part of the muscles, the body has been subjected to a special mummifying treatment. At times the head has been removed from the body, the brain tissue being

extracted through the foramen magnum. The thorax is opened nearly always across the sternum and the lungs and heart pulled out.... In certain cases they have made incisions in the extremities to pull out the muscles.... The body is [then] subjected to a process of mummification through the use of fire and perhaps various chemical substances, as indicated by the carbonized appearance of certain parts of the body, and by the salty efflorescences of the chemical substances employed. [Tello 1929:131–135]

Unfortunately no large intact bundles are now available against which to check Tello's findings. Unable to determine ABO blood groups from muscle tissues extracted from Paracas mummies opened by Tello, Candela (1943) attributed the negative results to the presence of some gummy, resinous substances, which, he suggested, had served as preservatives. In an independent study, Yacovleff and Muelle argued that mummification of the Paracas cadavers had been caused by natural desiccation (type II) without any artificial treatment:

To explain the preservation of these mummies, it is not necessary to revert to an hypothesis involving the use of fire and certain chemical substances, because the physical conditions of the place suffice to impede the decomposition of organic material.... The nearly complete absence of vegetation for twenty kilometers round about is due to the perpetual aridity; the rich salinity of the dry soil and the constant on shore wind; the relative height of the cemeteries above possible ground water; the constant action of the sun – all this makes special treatment for the preservation of the bodies unnecessary. [Yacovleff and Muelle 1932:48]

This view is also shared by Mahler and Bird (n.d.), who examined a single large Necropolis bundle in 1949. They found no evidence of foreign material inside the body cavity; the brain had not been removed, and no appreciable dehydration of the tissues appeared to have occurred before burial.

Recent examinations of smaller Necropolis cadavers also have failed to produce any convincing evidence of artificial preservation (Allison and Pezzia 1973; Rivero de la Calle

1975; Vreeland 1976). The leathery consistency and dark brown color of the skin, drawn tight over the upper portions of the skeleton, indicate that the bodies probably had been desiccated intentionally, at least in some cases. Nonetheless, sufficient body fluids or putrefaction products resulted from autolytic decomposition to have induced extensive rotting in the lower sections of the bundles. The internal soft tissues in most cases had been reduced nearly to powder, and only bits of fibrous connective tissues remained attached to the walls of the body cavities and extremities. Furthermore, the presence of large numbers of pupa cases of necrophagous insects attests to the lack of any immediate and complete mummification treatment, especially one requiring the use of fire and certain embalming substances such as those described by Tello. However, tissues taken from one small bundle recently opened and presently under microscopic examination show what appear to be artificially preserved, fire-dried, and in places, burned tissues, found in association with several small pieces of charcoal (Vreeland 1978a:41). Nevertheless, in the absence of further studies of large units, the theory that the presumably elite individuals wrapped in such bundles had in fact been accorded a complex, type III mummification must remain moot.

Little is known regarding mummification practices from other regions during the Early Intermediate period, but available archeological information suggests that a pattern distinct from that described for the south coast occurred in other areas. On the north coast, the Mochica buried their high-status dead in cane coffins placed in chamber tombs filled with diverse grave offerings (Larco Hoyle 1946:170). One exceptional (and very late) tomb excavated at the site of Pacatnamú yielded three mummies, all fully extended and placed on their backs (Ubbelohde-Doering 1966).[7] One of these, an 18-year-old woman, was found with skin "like fine old parchment and black hair" and tattooing on the right forearm. In contrast to the type

III treatment described during the previous period on the Chilean north coast, no evidence of artificial mummification has been reported from postceramic periods in that area.

MIDDLE HORIZON PERIOD (A.D. 600–1000)

During this period, two highland kingdoms, emanating from the sites of Huari and Tiahuanaco, spread differentially through much of the central Andes, bringing with them significant changes in mortuary practices that, in some areas, persisted until 1534. Although no mummies clearly dating from this period have been reported in the damp altiplano region, several excellently preserved bundles have been found on the coast (Reiss and Stübel 1880–1887; Allison et al. 1974). Unfortunately, no comprehensive study of these remarkable fardels has yet been made, but recent examinations of several large coastal Huari bundles from the Ica–Nasca area provide a general conception of the mortuary treatment involved.

Tightly flexed and covered in a cloth shroud or poncho, the cadaver was generally seated on a firmly rolled and coiled cotton disc about 50 cm wide. A more or less cylindrical bundle was then built up around the body by alternating layers of lint cotton with plain-weave textiles and with tightly interworked cord superstructures. Other packing materials, such as grass, reeds, and leaves, were also used. Protruding from the top of the "body" was a slightly conical false head constructed of alternating fiber and fabric layers, frequently decorated with metal or shell "eyes," "nose," a human-hair wig, and a woven cap or sling headband. A tapestry tunic was then placed over the completed, bottle-shaped bundle (Vreeland 1977; Figure 9.3).

Examinations of the human remains from the central and southern Peruvian coast areas (Allison and Pezzia 1973; Vreeland 1976) and northern Chile (Le Paige 1964:56) showed no evidence of type III mummification. Desiccation by means of natural agents had probably been utilized in some cases. Preservation was also enhanced by the tightly wrapped and highly absorbent cotton layers, which would have helped draw off body decomposition fluids. This process is demonstrated in radiographs taken of the bundles before they were opened; the radiopaque oval areas below the skeleton correspond to the cotton discs, apparently hardened by the absorption of internal moisture (Vreeland 1976; Figure 9.4).

LATE INTERMEDIATE PERIOD (A.D. 1000–1476)

Despite the decline of the Huari and Tiahuanaco cultural influences in the central Andes, throughout much of Peru numerous characteristics of Middle Horizon mortuary practices continued, with regional variations, until the Spanish Conquest. The most common form of interment continued to be the bundle burial, except in the eastern sierra, where the body was cut in sections or cremated (Tello 1942:120). In the western sierra the cadaver was wrapped in several ways: in a cactus fiber net, in twisted grass cords, or in a deer- or camelid-hide bag. The body was always tightly flexed in adult burials, with knees drawn up against the chest, hands opened flat over the face, and arms and legs bound in place with additional cords (Figure 9.5). The wrapping cords often covered the entire mummy, with the exception of a rectangular opening over the face and a smaller aperture for the toes (Figure 9.6). Although complete mummies are rarely preserved in the moist highland zones, hair, usually faded from its natural black to a reddish color, is often encountered. The hair of male mummies appears to have been kept relatively short, and louse eggs are quite common (Villar Córdoba 1935:227).

The mummy burial also predominated in the coastal areas, where the bundles were normally topped with false heads and masks in Middle Horizon style (Figure 9.7). Some of the larger units weigh over 100 kg (Waisbard

Figure 9.4.   Composite print of radiograph taken
of a Middle Horizon bundle similar to the one
shown in Figure 9.3, showing tightly flexed mummy with metal offering in mouth (a), stone or metal
bracelets on wrists (b), and seated on a cotton disc
(c) heavily impregnated with radiopaque body decomposition fluids.

and Waisbard 1965:82). In the northern
sierra, the mummy bundle was covered with
a layer of mud to form a conical structure and
was crowned with a modeled and painted
clay mask simulating the human head. Some
of the larger mummy casings are from 1 to 1.5
m high, weighing over 125 kg (Savoy
1970:167).

Mummies on the north Peruvian coast were
generally extended and fully clothed (as in
the Early Intermediate period Mochica burials), in contrast to the flexed and shroud-
covered mummies from other coastal regions
(Figures 9.8 and 9.9).[8] Preparation of these
large, plain-weave cotton shrouds required
considerable amounts of energy and raw material. One large central coast bundle contained an estimated 265 km of single-ply
yarn, requiring some 4,000 hours of spinning
and plying time, used to weave the 60 m[2] of
cotton wrapping shrouds alone. Another very
large and well preserved unit from the Lima
area, containing the remains of an 18-to-20-
year-old woman, had been wrapped with
over 150 kg of cotton cloth; one of these textiles, having a complete warp length of 40.40
m, is the longest single-web fabric known
from prehistoric America (Vreeland 1978b).

LATE HORIZON PERIOD
(A.D. 1476–1534)
The final period of the Andean prehistoric
sequence begins with the effective consolidation of the Inca empire and concludes with
the destruction of its highland capital, Cuzco,
by the Spanish in 1534 (Rowe 1946). Although the burial ritual accompanying the
Inca elite appears to depart from preexisting
patterns, mortuary practices evidenced
throughout most of Peru show general regional variations characteristic of the preceding periods (Vreeland 1980). Supplementing
the very limited available archeological data
from the highlands is a rich corpus of

Figure 9.3.   High-status Middle Horizon mummy bundle. Anthropomorphic features include false head
and hair, face mask, cap, headband, necklace, and poncho shirt. Probably from Nasca area. Collection
Museo Nacional de Antropología y Arqueología, Lima.

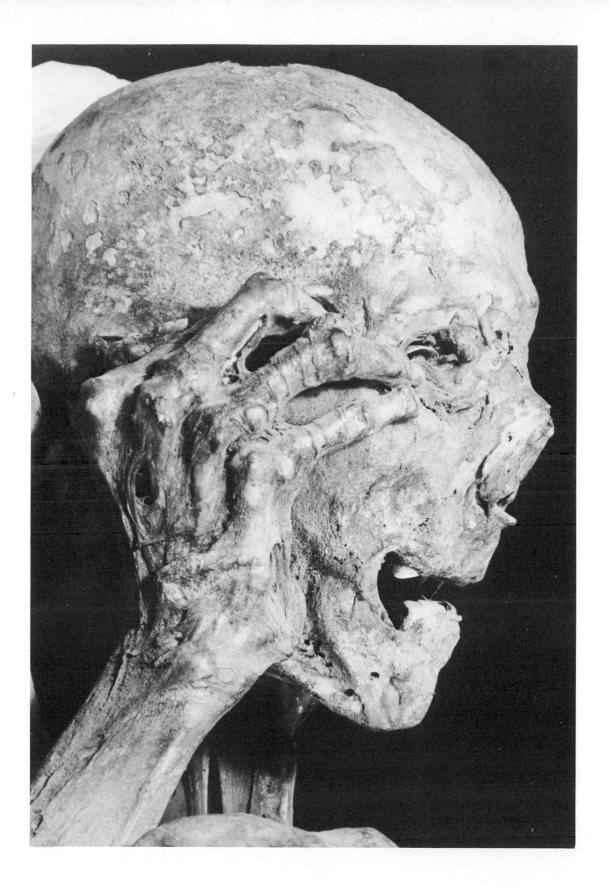

ethnohistorical information written during the early part of the Colonial period. These sources indicate that the practice of mummification was largely conditioned by two factors: local traditions and social rank or status of the deceased. One of the fullest accounts of this regional variation is provided by the Indian chronicler Guamán Poma, who not only described, but also illustrated at the beginning of the seventeenth century the prevailing customs of the four Inca *suyos* ("quarters") constituting the Inca empire (Guamán Poma [1613] 1956:451–456).[9]

In Condesuyo (i.e., the central highlands) the body of the deceased was placed either in a burial tower or in a sepulcher located on a high mountain ridge or peak. Guamán Poma noted that the cadaver was eviscerated and that certain balsamic substances were used to preserve it. In the Yungas (lowlands or coastal) regions, on the other hand, the body was covered with a simple cotton funeral shroud, then wrapped in cloth or cord ropes, forming a netlike superstructure. The upper portion of the mummy bundle was painted or decorated to suggest the human head and face. Apparently the viscera and sometimes even the flesh were removed from the bones and placed in freshly made ceramic vessels buried next to the mummy bundle.

An early account of type III mummification is also given in a document written about 1580, describing the mortuary ritual of the Pacajes, an ethnic group inhabiting the Bolivian altiplano region to the southwest of Lake Titicaca: "The manner in which these Pacajes bury their dead is to remove the viscera, and to throw them in a pot which they bury next to the cadaver, bound with ropes of straw.... The deceased was buried with the best clothing and plenty of food" (Jiménez de la Espada 1965:339). This practice is in part corroborated by Ponce Sanginés and Linares Iturralde (1966), who examined 10 mummies from the Bolivian province of Carangas.

Three had been eviscerated through an incision made in the abdominal wall. The cadavers were subsequently mummified by the naturally dry, cool atmosphere of the Bolivian altiplano, where it appears that this type of mummification may have been extensively practiced in the Late Horizon period and possibly earlier.

Social status also conditioned mortuary practices to a significant degree. According to the Jesuit chronicler Blas Valera, the common Indian was generally buried with the few possessions he owned in a simple grave in outlying community fields. In contrast, a member of the *kurakas*, or regional nobility, was often interred in a multiroom sepulcher with certain of his wives, servants, and others selected to serve him in the afterlife. These victims were sacrificed (some not unwillingly) and "embalmed" in the same fashion as the *kuraka* (Valera [1609] 1945:14–15). Guamán Poma ([1613] 1956:428) added that whereas gold or silver offerings were placed in the mouths of the *kurakas*, clay offerings were customarily used in the interments of common Indians.[10] In Chachapoyas, burial in conical clay casings apparently was reserved for the principal members of local descent groups; individuals of lower status were simply buried in the ground.[11]

Funeral ceremonies following the death of an Inca sovereign were probably the most elaborate rituals of their kind performed in prehistoric Peru (Figures 9.10 and 9.11). The combined ethnohistorical descriptions, though not uniform in detail, indicate that the cadaver of the king was placed on a special seat or throne in a flexed position, arms crossed over the chest, and head positioned over the tightly drawn-up knees. Bits of silver or gold were placed in the mouth, fists, and on the chest. The body was then dressed in the finest vicuña cloth and "wrapped in great quantities of cotton, and the face covered."[12]

One month after death, the body of the Inca

Figure 9.5. *Elderly male mummy, not eviscerated. Note cotton fiber in orbits and remains of string laced between fingers. No provenance; probably coastal Late Intermediate to Late Horizon period. Collection Museo Nacional de Antropología y Arqueología, Lima.*

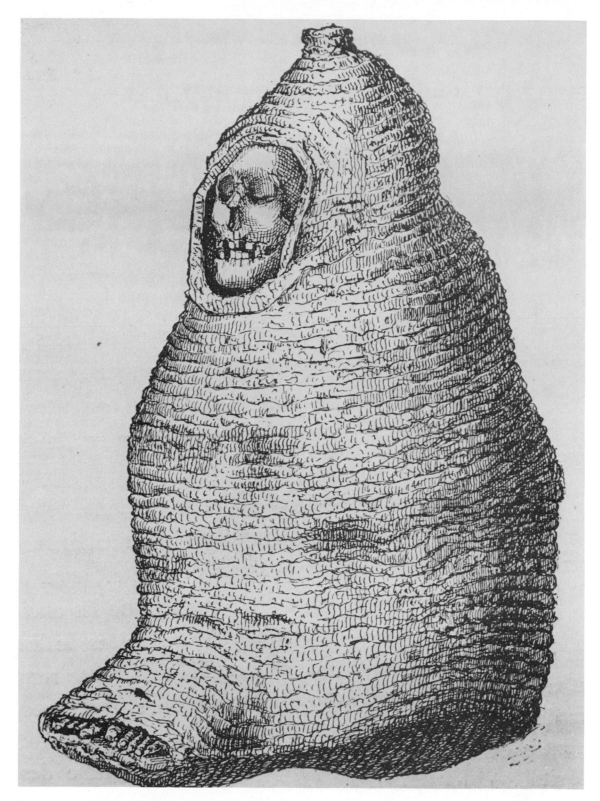

Figure 9.6. Sketch of highland mummy wrapped in cords with apertures for hands and feet. Cuzco area, probably Late Intermediate or Late Horizon period. (After Wiener 1880)

Figure 9.7. Large mummy bundle from Lima area, Late Intermediate period. False head and face mask follow typical Middle Horizon pattern. Collection Museo Nacional de Antropología y Arqueología, Lima.

was placed in the royal funeral sepulcher, or *pucullo*, along with great quantities of fine cloth, woven by special craft personnel expressly for the royal funeral. Tributed or bestowed from the four *suyos* of the vast empire, this exquisite cloth was either folded and placed next to the body or under the funeral shrouds, or burned. Food and drink were also

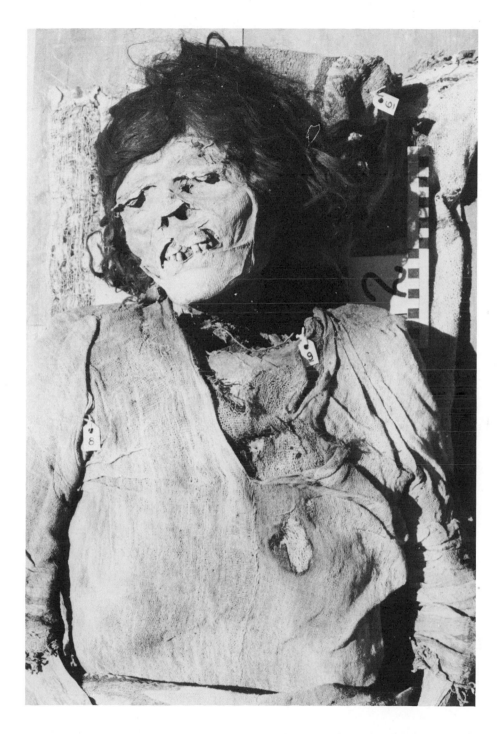

Figure 9.8.   Classic Lambayeque period (ca. A.D. 800–1200) mummy from Collus, Lambayeque Valley, north coast of Peru. Dressed with fabrics decorated in Middle Horizon style, the mummy was interred in extended position typical of mortuary practices on the coast prior to Middle Horizon influence there. Brüning Museum, Lambayeque.

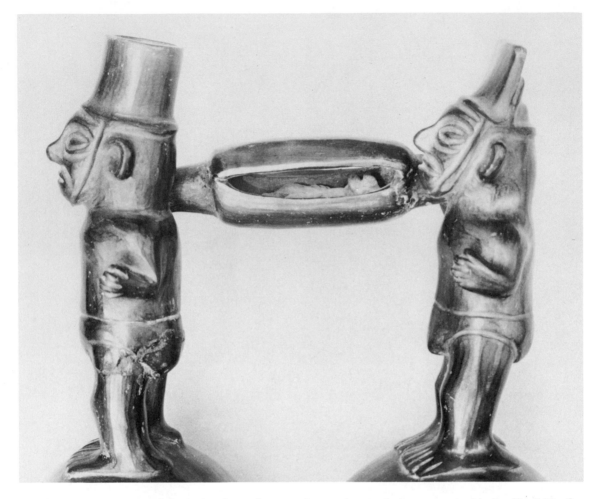

Figure 9.9.   *Pottery vessel depicting funeral procession and extended mummy inside "coffin." North Peruvian coast, Late Intermediate period. Collection Museo Nacional de Antropología y Arqueología, Lima.*

included with the dead ruler's armaments, symbols of power and office, and bags containing all his used clothing, nail parings, hair, and even the bones and corn cobs upon which he had once feasted. Llamas were sacrificed, as were some of the Inca's principal wives, concubines, and retainers.[13] Upon conclusion of the ceremonies in Cuzco, the royal mummies were reclaimed by the lineage groups of the dead king and were cared for by male and female attendants. These specially appointed custodians knew not only when to give food and drink to the king's mummy, but also acted as spokesmen

for the dead ruler's personal desire. They carried out routine chores such as whisking the flies from the mummy's brow, changing and washing its clothing, calling in visitors with whom the Inca wished to "speak," and lifting the bundle when its occupant needed to "urinate" (Polo de Ondegardo [1554] 1916a:124; Pizarro [1571] 1939:294–295; Imbelloni 1946:190). Ordinarily none but these professional cult personnel was permitted to look on the royal mummies, except when these relics were removed from their sepulchers and exhibited in Cuzco during certain religious and state ceremonies, such as the two solstice

NOBIENBRE
AIA MARCAI
quilla

la fiesta delos defuntos

festivals and the coronation of Inca rulers (Molina [1575] 1916; Estete [1535] 1938:54–56).

The earliest known account of an Inca royal mummy is that recorded by Pizarro's secretary, Pedro Sancho del la Hoz, who less than 10 years after the Conquest described the mummy of Huayna Capac (d. 1525) as being nearly intact, wrapped in sumptuous cloth, and "missing only the tip of the nose" (Sancho [1543] 1938:183). Garcilaso de la Vega, who saw several royal mummies collected by the Spanish licentiate Polo de Ondegardo in Cuzco in 1559, provides a more detailed description:

The bodies were so intact that they lacked neither hair, eyebrows nor eyelashes. They were in clothes just as they had worn when alive, with *llautus* ["bands"] on their heads but no other sign of royalty. They were seated in the way Indian men and women usually sit, and their eyes were cast down. . . . I remember touching the finger of Huayna Capac. It was hard and rigid, like that of a wooden statue. The bodies weighed so little that any Indian could carry them from house to house in his arms or on his shoulders. They carried them wrapped in white shrouds through the streets and plazas, the Indians dropping to their knees, making reverences with groans and tears, and many Spaniards removing their caps. [Garcilaso de la Vega [1609, pt. I, bk. V, chap. XX] 1963)

The bodies of at least three Inca kings and the ashes of another, as well as the bodies of two *coyas* ("queens"),[14] were sent to Lima in 1560 by Polo. Some 20 years later, the Spanish priest Acosta noted that they were still "wonderfully preserved, causing great admiration" among the people of that city (Acosta [1590, bk. V, chap. VI] 1954:146).

Despite these provocative accounts, little information on the actual mummification process employed by the Incas can be gleaned from the available ethnohistorical reports. Both Garcilaso de la Vega and Guamán Poma stated that the bodies of Inca rulers (and their principal wives) were embalmed, but except for the special properties of the natural highland environment, no preservation processes are mentioned. Blas Valera, on the other hand, specifically described a type III treatment, including both evisceration and embalming. This was effected with a variety of balsam brought from the province of Tolú.[15] When applied with some unspecified substances, Valera remarked, "the body thus embalmed lasted four to five hundred years" (Valera [1609] 1945:14). He further noted that when Tolú balsam was unavailable, the embalmers resorted to a kind of bitumen prepared with an unidentified material that preserved the flesh, apparently with some success. Acosta also reported that the bodies of the royal mummies had been "dressed with a certain bitumen" (see Imbelloni 1946:193). Father Cobo described a mummy that was "so well cured and prepared that it seemed to be alive, its face was so well formed and complexion full of color," and then went on to tell us how this was done: "The preservation of the face . . . was effected by means of a piece of calabash placed under each cheek, over which the skin had become very taught and lustrous, with false open eyes" (Cobo [1653, bk. XII, chap. X] 1964:65).

Unfortunately, the accounts of these three Jesuit fathers, Cobo, Acosta, and Valera, are all somewhat late to be completely reliable and may not in fact represent entirely independent sources. It is unlikely that Cobo ever saw a royal mummy. Garcilaso de la Vega's narrative was written in Spain some 40 years after he witnessed the mummies in Cuzco as a young man. Valera, like Acosta, may well have seen the Inca mummies brought back to Lima, where he began his novitiate in 1568. But his papers, lost in 1596, are known to us only through the writings of Garcilaso de la Vega, remarkable for their impressionistic detail rather than their veracity. Furthermore, the descriptions of Acosta and

Figure 9.10. Inca period mummy being borne on a stretcher during November "festival of the dead." (From Guamán Poma [1613] 1956)

CONQVISTA
DEFVNTOGVAINACAPAC
INGA- ILLAPA

collcuan aenterrallo al cuzco -

Fuaen eldefunto dequito
aenterralle aʃubobedaireal
del cuzco

al

Valera presumably refer to mummification procedures applied only to Inca nobility and are hardly confirmed by examinations of highland mummies recovered from the Cuzco area.

The use of certain herbs and plant materials in the embalming process, mentioned by the Augustinian friar Ramos Gavilán ([1621] 1976:73) in reference to Colla burial practices, has also been suggested by Cornejo Bouroncle (1939:108), who claims to have identified the remains of the fragrant plant muña[16] stuffed inside a number of "lower-status" Inca period cadavers in the Cuzco museum. Lastres (1953:73), after examining mummies in both the Cuzco and Lima archeology museums, found no evidence of evisceration and was of the opinion that mummification had been caused predominantly by natural agencies, with the occasional use of herbs and balsam applied to the skin. On the other hand, McCreery (1935) examined about 20 mummies in Cuzco and stated that they were all rather well preserved, of all ages and sexes, and that no evidence of artificial mummification could be found. From these conflicting reports it is clear that additional research is urgently needed to document adequately the nature and range of mummification practices in Late Horizon burials.

HUMAN SACRIFICE, TROPHY HEADS, AND HUMAN TAXIDERMY

Although ritual sacrifice and mummification appear to be closely related, especially in the later pre-Hispanic periods, space precludes all but the briefest review of human sacrifice and the practice of preserving the body or its parts. Types I and II mummification have been cited in reference to burials (usually of juveniles) in the 6,000-m and higher zones of the central Andes, in what clearly appear to be cases of human sacrifice. The most celebrated of such interments is that found at Cerro El Plomo in 1954, when the frozen body of an 8-

to 10-year-old boy, dressed in a camelid wool poncho, was recovered with several metal ornaments and figurines from an Inca period sepulcher (Medina Rojas 1958).[17] Mummies of several other subadults have been reported in similarly inaccessible high-altitude funeral cairns (Schobinger 1966), where the youths had probably been made intoxicated with the alcoholic corn beer *chicha* or with the narcotic coca leaf (*Erythroxylon coca*), and either sacrificed by strangulation or simply left there to die (Ramos Gavilán [1621] 1976:26).

On the Peruvian coast, mummy bundles containing the remains of dismembered bodies also suggest human sacrifice, possibly to obtain entrails for purposes of divination (Allison et al. 1974; Walter 1976). A different motive is indicated by the partial immolation found in a recently opened unit from the Lima area, dated to the Late Intermediate period. The bundle contained a pair of well-calloused, naturally mummified feet and lower legs, torn off at the knees (Vreeland 1978b). In a bundle from the same area Jiménez Borga found a mummified left foot (Waisbard and Waisbard 1965:82, pl. ix). Both bundles were relatively large and constructed in a manner identical to those containing entire mummified bodies.

Type III mummification of sacrificed individuals is amply documented for the late pre-Hispanic periods by several early ethnohistorical references. On the eve of the Spanish entry into northern Peru in 1531, Estete reported that in the coastal Ecuadorian village of Pasao (Manabí province), the Indians flayed the bodies of the dead and burned off the remaining muscle tissue. The skin was then "dressed like a sheep's hide" and stuffed with straw. Thus "crucified," the body, with arms crossed, was hung from the temple roof (Estete [1535] 1938:207–208). The Chanca Indians of the central Peruvian highlands, as well as the Incas, practiced a similar kind of human taxidermy, displaying the stuffed

Figure 9.11. *"Mummified" bodies of Inca king Huayna Capac, his queen, and a retainer carried on a litter from Quito to Cuzco for burial. (From Guaman Poma [1613] 1956)*

skins of their war captives in temples, making their stomachs into drums (Métraux 1949: 408). According to Estete ([1535]1938:208–209), the heads of the dead in Pasao were also mummified with certain balsamic substances and shrunken in a process similar to that described for the Jíbaro of the eastern Ecuadorian mountain slopes (Tello 1918; Figures 9.12 through 9.16).

SUMMARY AND CONCLUSIONS
Why Peruvian dead were artificially mummified is a question that cannot be comprehensively answered until more archeological and ethnohistorical information is made available. Based on the present evidence, one can tentatively state that religious, cult, and magical motives, some clearly derived from political and economic contexts, were all contributing factors in Andean mummification practices. As noted earlier, artificial enhancement of soft tissue preservation, which occurs naturally under proper conditions, and the desire to maintain the body in a more or less lifelike state, presumably represent an intensification of the mortuary cult and an extension of the widespread Andean custom of ancestor worship. The aperture over the face in the wrappings of many highland mummies, tubes leading from the tomb to the ground surface, washing and changing the ritual mummy clothing, and the false heads and face masks of the coastal bundles from the Middle Horizon period on, signify further elaboration of the concept of the "living dead," enabling, in effect, the deceased to participate in its own mortuary cult.[18] The mummified corpse thus represented a physical entity with "human" characteristics, and also served as an intermediary for symbolic communication between the worlds of the living and the dead. In some areas such active and periodic participation in ritual activity also presented a threat to the conservation of the mummy bundle and its contents. Mummification therefore may also have provided a solution to the problem of body disarticulation during transport and display.[19]

During the Late Horizon period, mummies of the Inca kings served an important state function. Preserved by royal privilege in sanctuaries on family estates, the mummies not only provided unequivocal testimony to the previous existence of those rulers, but also ensured their continued worship as semideified clan ancestors, whose cult was maintained by their descendants and specially appointed retainers. In a similar manner, possession of the mummified body of an important clan ancestor or of a political or religious figure constituted a powerful religious talisman that could often be exploited by the lineage or community owning it.

If the origins of Andean artificial mummification still remain obscure, it is nonetheless likely that natural mummification (Figure 9.17) may have directly inspired the prehistoric Peruvians to experiment with intentional mummification for a variety of cultural reasons, with motives growing more varied as complex societies emerged. The possibility that some forms of artificial preparation may have been introduced from the tropical lowlands to the east of the Andes should not be eliminated. A variant of type II treatment, appearing relatively early in the pre-Hispanic sequence on the north coast of Chile, frequently is mentioned in later periods, especially on the periphery of the central Andean area (northern Chile, the Bolivian highlands, parts of Ecuador and southern Colombia, and the Peruvian montaña), where contacts with tropical forest cultures apparently were more continuous and extensive. Whatever their origins, if complex embalming techniques had been utilized by the Cuzco elite for religious and political reasons, these practices soon fell into disuse and were forgotten as a result of the Spanish campaign to baptize Inca nobility quickly and forcibly.

In summary, the range of possible agencies or processes noted for each of the three types of mummification considered here are:

Type I. Natural mummification
  1. Perpetually dry or frozen tomb matrix
  2. Hot (coastal) or cold (highland) temperatures throughout much of the year

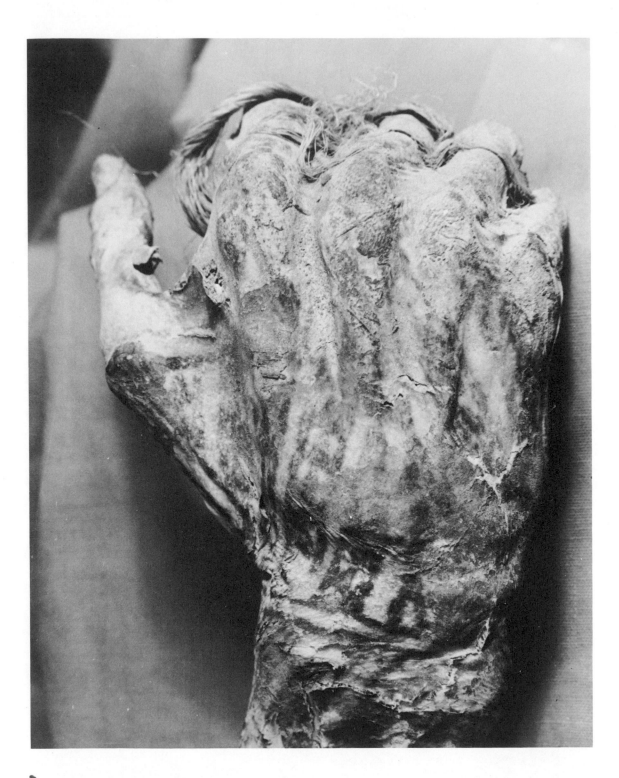

Figure 9.12. Mummified hand showing tattooing. Probably central coast, Late Intermediate period. Collection Museo Peruano de Ciencias de la Salud, Lima.

3. Anaerobic tomb environment (e.g., direct burial in sand)
4. Local soil characteristics (salinity, alkalinity, etc.)
5. Highly absorbent substances (e.g., sand) in direct contact with the cadaver

Type II.   Intentional natural mummification
1. Body intentionally dried by rarefied and/or cold sierra atmosphere
2. Body intentionally desiccated by warm coastal climate
3. Body intentionally wrapped with materials of highly absorbent nature (e.g., cloth, cotton fiber, leaves, grass)
4. Intentional location of cemeteries in areas having favorable natural conditions for preservation of organic materials

Type III.   Artificial mummification
1. Evisceration of internal organs and/or other soft tissues
2. Replacement of soft tissues with plastic materials (e.g., clay)
3. Removal of all or parts of skeletal material and replacement with sticks, grasses, or other reinforcing materials
4. Fire desiccation
5. Smoke curing
6. Use of bitumen, balsam, or other resinous substances
7. Filling of body cavity with herbs or other materials having antiseptic chemical properties

DISEASES
AIDAN COCKBURN

In proportion to the number of its mummies, Peru has not been investigated in depth for evidence of disease in ancient populations. Most research on mummy bundles has been concentrated on the textiles, not on the human bodies inside. This is particularly true of the grave robbers, so it is not uncommon to find an ancient cemetery plundered, with bones scattered unwanted on the desert surface. In Egypt there is a long history of careful dissection of bodies and diagnosis of pathologic conditions, but in Peru, investigation has been limited to the efforts of a handful of devoted scientists, who often worked under difficult conditions. Interest has sharpened within the past decade or two. In Lima, a group of physicians, disturbed at the lack of attention being given to the wealth of material available, founded the world's first museum of paleopathology. The museum, now renamed Museum of Health Sciences (Museo Peruano de Ciencias de la Salud) has an excellent building and a good collection. There are enormous possibilities for future research, although funds are badly needed for a sustained program. Under its new director, Dr. Fernando Cabieses, the future looks promising. In Ica, Marvin Allison and Alejandro Pezzia have been doing first-class work for a decade and have made some brilliant discoveries, with far-ranging implications.

Much of the data that follow are drawn from these sources.

TUBERCULOSIS
The antiquity of tuberculosis in the Old World has been well documented (Chapter 2). In the New World, the subject is more debatable, but there appears to be clear evidence that bone tuberculosis at least was present in pre-Columbian times (Morse 1961). A clear-cut case from Peru was described by Garcia Frias (1940), but the body, unfortunately, was not clearly dated: It was an Inca mummy, and these were prepared right into Colonial times.

Before discussing the antiquity of tuberculosis in the New World, it is necessary to analyze the meaning of the term. Basically, it refers to a condition in a person or an animal

Figure 9.13.   *Adult male mummy with cranial deformation. Note tattooing on wrist, cords used to bind cadaver in flexed position and textile impressions from mortuary shroud on right knee. Probably from coast, Late Intermediate to Late Horizon period. Collection Museo Nacional de Antropología y Arqueología, Lima.*

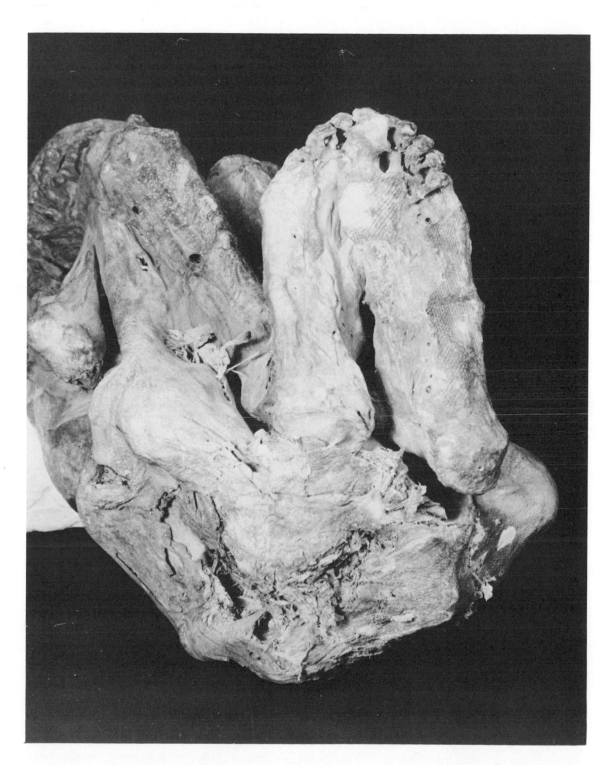

Figure 9.14. Posterior view of adult male mummy showing no signs of evisceration. Probably coastal, Late Intermediate to Late Horizon period. Collection Museo Nacional de Antropología y Arqueología, Lima.

caused by an acid-fast-staining mycobacterium, which produces a granulomatous tissue reaction. There are several mycobacteria that can cause lesions of this type, so the discovery of some granulomatous abnormality does not necessarily prove the presence of what today we call "tuberculosis." In man, granulomatous lesions in bone can be the result of infections by mycobacteria of man, cattle, birds, or even what appear to be free-living organisms from water and mud. Some forms of tuberculosis can also be caused by fish or snake pathogens. In addition to the variety of organisms that can cause "tuberculosis," the status of the human host can also alter the clinical picture. Populations exposed to human tuberculosis for the first time differ sharply in their reactions from those that have had centuries of saturation. After centuries of exposure, the urban dweller is apt to develop a chronic pulmonary form; the rural African or Eskimo, infected for the first time, suffers a rapidly progressive, often fatal, glandular type.

These points must be kept in mind when considering the case of tuberculosis reported by Allison and Pezzia (1973). The body was disinterred with its burial goods from the Nasca area in the department of Ica, and is dated to between A.D. 200 and 800; carbon-14 dating gave a time of around A.D. 700. The body, that of an 8-year-old child, was sitting in a hunched position on an adobe pad made to fit its body contours in life (Figure 9.18). Radiography showed Pott's disease (tuberculosis of the spine) involving the first, second, and third lumbar vertebrae. As so often happens in this condition, pus from the lesion had tracked from the spine down the psoas muscle, producing a psoas abscess and leaving a hollow sack about 5 cm in diameter. The abscess cavity contained a layer of dried caseous material. The organs of the throat and abdomen were removed, and it proved possible to identify, then rehydrate in Ruffer's solution, the heart, lungs, right kidney, liver, and spleen. On inspection with a hand lens, small white nodules resembling tubercles were found on the lung, pleura, and pelvis of the kidney. Histologically, little structure could be recognized at the cellular level. Staining by the Ziehl-Neelsen method revealed many clumps of acid-fast bacilli in those lesions.

We have here a clear-cut case of bone and miliary tuberculosis with a psoas abscess, which belongs to an era and location far removed from any Old World contact. Unfortunately, the source of the infection – whether human, mammal, bird, or other – cannot be determined, so the case cannot definitely be claimed as one of the "human" types found in both Old and New Worlds today.

HOOKWORM INFESTATION

A favorite game among medical historians is arguing about how infectious diseases spread. The battle over Columbus and the first outbreak of syphilis at Naples in 1493 has been progressing merrily almost ever since the epidemic. A similar debate, although less heated, has involved hookworm: Was it brought over from Africa by infested slaves, or was it already present in the Americas before the voyage of Columbus?

The matter has now been settled, for Allison et al. (1974b) found adult hookworms (*Ancylostoma duodenale*) in the small intestine of a Tiahuanaco mummy from a gallery burial in Peru dated between A.D. 890 and 950 (Figure 9.19). The mummy was opened in southern Peru in 1960, and pieces of intestine were studied. Examination with a 20-power dissecting microscope showed the worms attached to the intestinal lumen. They were subsequently photographed at 100 enlargement by scanning electron microscope. Details of the parasites' heads and buccal cavities were clearly visible, leaving no doubt about the identification. One worm had two large teeth and a rudimentary one, indicating that the species was *A. duodenale*.

It is highly probable that hookworms were also imported from Africa during the period of the slave trade, but the native Indians must have been infested long before this occurred. The parasites were probably carried to Peru

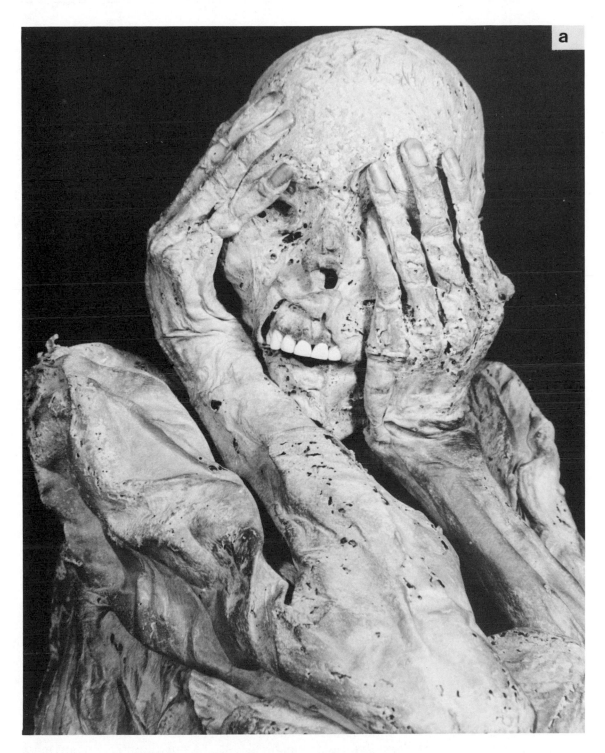

Figure 9.15. a. Adult male mummy, not eviscerated. Note extensive insect damage and remains of cord bindings between fingers. b. Lateral view, showing trephined skull and cephalic material inside. Possibly

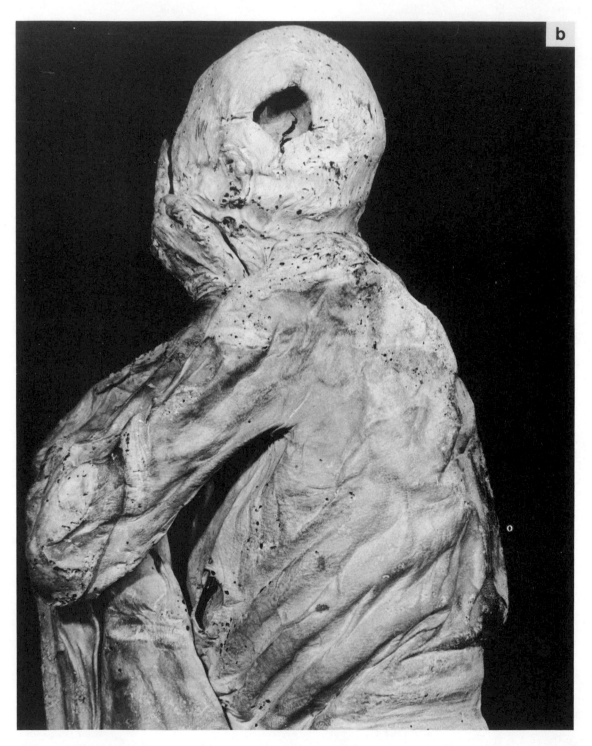

b

highland, Late Intermediate to Late Horizon period. Collection Museo Nacional de Antropología y Arqueología, Lima.

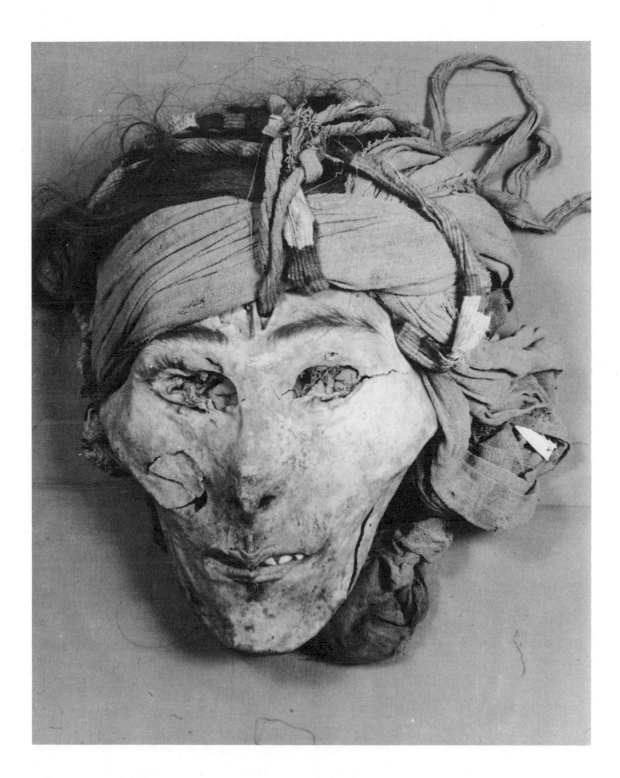

Figure 9.16. Mummified trophy head showing perforation in forehead and cactus spine pinned through lips. Nasca area, Early Intermediate period. Collection Museo Peruano de Ciencias de la Salud, Lima.

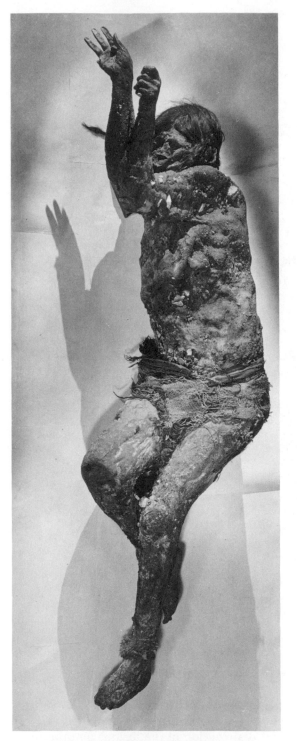

Figure 9.17.   *Four-century-old copper miner mummified naturally by copper salts. Chuquicamata, Chile. (Courtesy of American Museum of Natural History, New York)*

by the original migrants from Asia, who brought them over the Bering land bridge during their wanderings in the Ice Age.

VERRUGA, OR CARRIÓN'S DISEASE

Carrión's disease is confined to Peru and neighboring countries. It is caused by the microorganism *Bartonella bacilliformis*, is transmitted by sandflies, and is confined to certain valleys in the high mountains. It appears in two forms: The first is Oroya fever, an acute disease with destruction of the blood and a high mortality; the second is an unpleasant but not fatal skin infection, verruga, characterized by multiple fungating skin lesions. The disease is named after a medical student called Carrión, who showed that the two diseases were different forms of one infection by inoculating himself with crusts of verruga and afterward dying of Oroya fever.

A case of verruga was identified by Allison et al. (1974a) in a mummy found in the Nasca area. This is not in the area in which *Bartonella* infection is endemic, but the burial was of the Tiahuanaco culture and it is known that Nasca was invaded by Tiahuanaco people who would have had to pass through areas where the disease is endemic during the course of the invasion. The body had been sacrificed by being cut in half at the lower lumbar vertebrae: The bottom half was missing. The skin was found to have a rash on the back, arms, and legs, and was covered with small nodules varying from pinhead to pea size. On autopsy, the left part of the thoracic cavity proved to be empty, with no trace of the heart or lung. The left hand was severed at the wrist and restored to normal size by immersion in Ruffer's solution for a week. At the end of this time, a series of different types of skin lesion was readily seen. There were vesicles as well as pendulous tumorlike lesions, areas of healing as well as of excoriation. Evidently this was a process of long duration showing numerous stages of the disease. On section, some of the tumors proved granulomatous. Giemsa stain (but not Gram's) revealed clumps of organisms in the lesions and blood vessels. Pictures of 8,000 magnification by

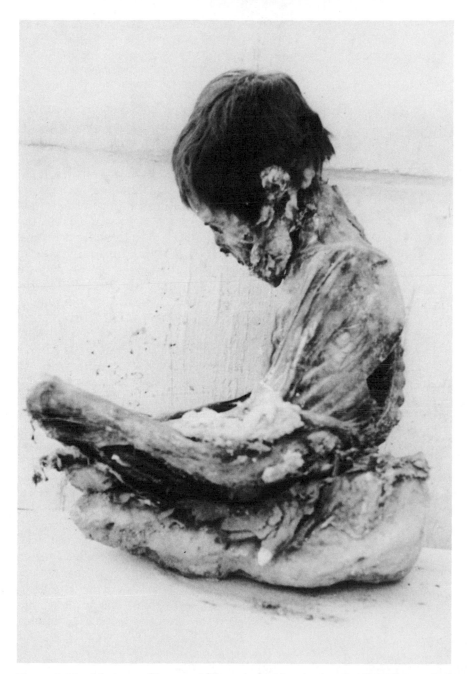

Figure 9.18.   Mummy of 8-year-old Peruvian with spinal and miliary forms of tuberculosis. Ica Museum.

the scanning electron microscope showed bacilli with a single polar flagellum. *Bartonella* stains badly with Gram's stain and well with Giemsa's and has one or more flagella at the tuft of one pole of the body. A diagnosis of Carrión's disease in the verruga stage was therefore made, based on the nature and distribution of the lesions and the identification of organisms similar morphologically to *B. bacilliformis*.

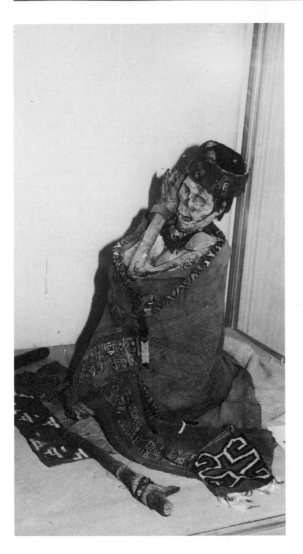

Figure 9.19   Mummy with hookworm infestation. Ica Museum.

TREPANATION

Trepanation is the removal of a piece of the skull while the individual is still living and without penetration of the underlying tissues. It was practiced until very recent times, and all the evidence shows that a large proportion of the patients survived, often undergoing repeated openings of the skull. The literature is reviewed by Lisowski (1967) and Margetts (1967). The practice was apparently worldwide, and many examples have been found in Peru. The modern operation, which uses an improved tool, is called trephination.

This is a surprising operation to be undertaken, and many guesses have been made about why it would have been done. These discussions and theories are beyond the scope of this section, but I can add from personal experience that the operation is perfectly practical. In 1934, as a final-year medical student, I worked as assistant to a brain surgeon in the north of England. One of my tasks was to remove the skull bone so the surgeon could tackle the tumor in the brain below. Providing the patient had a local anesthetic for incision of the skin, he felt little discomfort while I drilled four holes in his skull with a hand drill and burr, then cut the bone between them with a Gigli (fretwork) saw. Those were the days before antibiotics and modern technologies; all 15 patients died later – but from advanced brain tumors, not from my crude surgeries. My attempts were probably not too far removed from those of the trepanist of antiquity.

According to Allison et al. (1976), more than half of all the trepanned skulls in the world have been found in Peru. The list of finds is a long one, and an incomplete record of publications includes McGee (1894), Muñiz and McGee (1894), Hrdlička (1914), Moodie (1927), Rogers (1938), Quevedo and Sergio (1943, 1945a,b), Weiss (1949, 1958), Rocca (1953), Grana et al. (1954), Stewart (1956), Lastres and Cabieses (1960), MacCurdy (1970), and Allison et al. (1976).

Four main techniques were used in trepanation: scraping away the bone (Figure 9.20), boring a hole with a drill, making a circular hole with some cutting method, and making straight sawing cuts to outline an oblong hole. Allison et al. (1976) examined 288 skulls from the Ica area and found that 24 had been trepanned or treated surgically; 13 showed evidence of the fracture that had led to the trepanation. Of the 24, 11 died, but 13 survived the operation for at least 6 to 8 weeks. Obviously, this was a fairly common

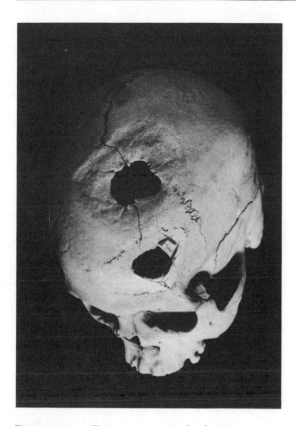

Figure 9.20.   Trepanation. Peabody Museum, Harvard CL-1. One operation has been done by the scraping technique and has healed; the second was done by a groove-cutting method. (Courtesy of Peabody Museum)

operation in the Ica region and met with a fairly high rate of success.

THYROID DISEASE
Thyroid disease was discovered in a 30-year-old female of the Nasca culture in the Ica valley. Carbon-14 dating placed her death about 94 B.C. The possibility of thyroid disease was suggested before autopsy, when radiologic examination showed a 1-cm area of pathologic calcification in the thyroid region. The calvarium of the skull was twice the normal thickness and there was calcification of the aorta.

Dissection of the neck revealed the thyroid gland, which usually cannot be found in mummies. The gland contained two large calcified areas and several smaller focuses, surrounded by thick tissues. Histological examination disclosed occasional scattered follicles filled with thick colloid material.

The diagnosis of thyroid disease rests on the physical findings of the gland, skull, and aorta, which are common in hypothyroidism (Gerszten et al. 1976).

BLOOD GROUPS OF PERUVIAN MUMMIES
The blood groups of Peruvian mummies have long attracted considerable attention, mainly in the hope that light might be shed on the evolution and movements of people in antiquity. The work was first pioneered on a large scale by Boyd and Boyd (1933, 1937, 1959) using the agglutination-inhibition technique. Coombs introduced the mixed-cell agglutination technique, and this was used by Otten and Florey on Chilean mummies, as described in Chapter 20. Allison et al. (1976) tested 111 mummies with both methods plus induction of antibody. They found that all three techniques detected the presence of H(O) antigen. All blood groups (A, B, AB, and O) have now been identified in Peruvian mummies, but the distributions differ in various ages. The B and AB groups were present in early pre-Columbian Indians, but the B group became almost extinct in Colonial mummies and is very rare in more recent Indian populations. Boyd (1959) has suggested that the ancestors of modern Indians carried the B group with them in their wanderings over the Bering land bridge, but that this B group was gradually eliminated by natural selection. Allison et al. (1976) think there is merit in this suggestion. They also point out that the western cultures of South America probably originated with small family groups that were valley-oriented, so that examination of material from different valleys might help in tracing migration patterns.

POROTIC HYPEROSTOSIS AND CRIBRA ORBITALIS
These pathologic conditions of the cranium are commonly found in ancient bodies from

many parts of the world, although they are rarely seen today. Porotic hyperostosis is primarily a hypertrophy of the bone marrow of the cranium that results from severe anemia in childhood. The vault is greatly thickened, and the true bone thinned, so that the marrow is visible. Cribra orbitalis is a particularly baffling manifestation of the upper plate of the orbital cavity, as bone marrow does not normally occur there. Because radiography reveals a "hair-on-end" picture that is typically seen in modern-day thalassemia, some physical anthropologists and physicians working on specimens from the eastern Mediterranean claim that the condition indicates severe falciparum malaria. According to current theory, the abnormality of the blood that produces thalassemia also protects against falciparum malaria and the disease is therefore common in highly malarious areas. However, as the presence of malaria in ancient Peru is highly unlikely (though the matter is debatable), the protective effect would not seem to have been a factor in that country.

Both porotic hyperostosis and cribra orbitalis were prevalent in pre-Columbian Peru. The average mummy autopsy does not reveal their presence unless a radiogram is made. Normally the skull is not scalped, nor are the eyeballs removed for examination of the socket. However, there are large numbers of skulls in collections as well as strewn over the desert around cemeteries, and these show without question that the ancient peoples of Peru suffered from severe anemias that resulted in the growth of their bone marrows and the pathologies known to us today as porotic hyperostosis and cribra orbitalis.

In October 1976, the Paleopathology Association held a symposium in Detroit to review the causes of these conditions (Cockburn 1977). Although the results were not altogether conclusive, it became clear that iron deficiency was the chief contributing factor. Other factors, such as inadequate protein in the diet, may have played a part, but the essential cause was a lack of iron. This brings us to diet: It is always people living on a vegetarian diet who are vulnerable. The Peruvians had few domestic animals, except for members of the camel family at high altitudes, and these could not survive successfully at lower levels. The plains people had an exceedingly good source of fish in the ocean, brought there by the Humboldt Current, and in addition hunted the sea lions ("sea wolves") along the coast. In general, though, animal protein provided only a small part of the ancient Peruvian diet. These nutritional deficiencies are recorded for us today as cases of porotic hyperostosis and cribra orbitalis.

Other conditions leading to lack of iron in the body include loss of blood. Such parasitic infestations as schistosomiasis, with its hematuria and/or melena, and hookworm, with its bleeding into the intestine, can cause severe iron deficiency and anemia, although in modern times these have not been related to any bone pathology resembling porotic hyperostosis. Vitamin deficiencies (e.g., scurvy and rickets) have also been suggested as causative factors, but this theory is now discredited. In Mediterranean countries thalassemia is a very possible cause, for it does produce a picture identical to porotic hyperostosis and is found even today. The matter is not yet settled. As far as Peru in antiquity is concerned, however, thalassemia was probably not a factor.

An example of a skull from Peru showing porotic hyperostosis is shown in Figure 9.21. The literature on porotic hyperostosis up to 1966 has been carefully reviewed by Angel (1967); the views he expresses tilt strongly in support of the thalassemia theory. Most probably the condition in Peru is the result of nutritional deficiencies and not malaria.

ACKNOWLEDGMENTS

Part of the research upon which the section "Anthropological and Historical Perspectives" is based was carried out with an Organization of American States research grant (No. 73-38/910) between 1974 and 1975, in Lima's Museo Nacional

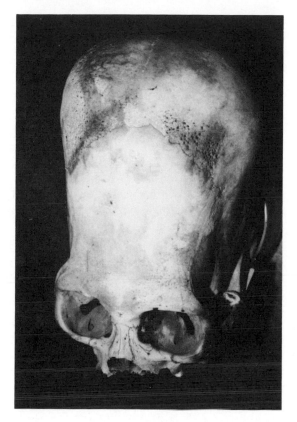

Figure 9.21.   Mummy with porotic hyperostosis. The skull was deformed deliberately in childhood. Museo Peruano de Ciencias de la Salud, Lima.

de Antropología y Arqueología. I would like to thank Dr. Luis G. Lumbreras, director of that institution, and the personnel of the departments of textile and physical anthropology, for the continued support that made this project possible. Valuable technical assistance during the first 2 years of study was provided by Prof. T. Mejía Xesspe, Dr. Germán Sánchez Checa, and Dr. Tamotsu Ogata. Many students, too numerous to name individually, helped in analyzing the textile materials recovered. Additional financial support was provided through the University Research Institute and the University of Texas Latin American Archaeology Program (Austin); scientific instruments and laboratory supplies were generously provided by Bausch & Lomb Corporation, Rochester, New York, and Hummel Chemical Company, South Plainfield, New Jersey.

Valuable substantive and editorial suggestions on previous drafts have been made by R. P. Schaedel, J. B. Bird, and N. Kaufman, whose critical observations are gratefully acknowledged here. I would also like to thank Dr. Fernando Cabieses, director of the Museo Peruano de Ciencias de la Salud, and Dr. Frédéric Engel, director of the Centro de Investigación de Zonas Aridas (Lima), for permission to examine and photograph mummies in those collections.

NOTES

1.   Acosta ([1590, bk. V, chap. VI] 1954:146); Riva-Agüero (1966:397).

2.   Naturally mummified individuals in non-burial contexts have also been reported from several areas in the central Andes. Sénechal de la Grange found the naturally mummified body of a woman in a pre-Columbian copper mine in Chuquicamata, Chile, where the body was accidentally preserved in part by the action of the copper salts (in Boman 1908:757). Imbelloni (1956:282) mentioned that a number of mummies of individuals who died in cave accidents are presently found in various Argentine museums. In 1903 a child was found buried in, and subsequently mummified by, a natural salt formation in the Argentine altiplano (Casanova 1936:264).

3.   Outside the central Andean area, evidence of artificial mummification is extensive (see in particular Steward 1949:721 and Linné 1929). Among the northern Andean and circum-Caribbean peoples, the body of the chief was desiccated, placed in a hammock and later cremated (Caquetío); disemboweled, desiccated, and kept as an idol or buried with several wives (Antillean Arawak); desiccated, temporarily buried, then roasted and reburied (Pititú). In southern Panama it was customary to bury the common people in the ground, but to desiccate and preserve the remains of the chief, which were slung in a hammock and mummified over a slow-burning fire (Lothrop 1948:147). The Ecuadorian Quijo eviscerated, smeared with tar, and smoked the bodies of their dead chiefs, and then filled the abdominal cavity with jewels (Steward and Métraux 1948:655). Dawson (1928a:73–74) examined two mummies from Colombia, both of which appeared to him to have been eviscerated – one via a perineal incision, the other through a cut in the abdominal wall – and then smoke-cured.

4.   In such arid zones almost anything buried in the sand will dry out to a naturally mummified

state. For example, during the restoration of a series of Republican period forts constructed around Lima during the war with Chile, a number of Chilean soldiers were recovered from mass graves scooped in the sand. With buttons still shiny, documents intact, and soft external tissues remarkably preserved after nearly 100 years, the cadavers were sent back to Santiago, where they purportedly were returned to their families. Naturally mummified animal carcasses have even served as landmarks to travelers on old Peruvian roads (Rivero and von Tschudi 1854:209).

5.   Bird (1943:246). This form of mummification (type III) appears to extend as far south as Iquique and may have continued briefly into the succeeding cultural phases (Schaedel 1957:21–25 and Appendix IV; cf. Nuñez 1969:130 ff.).

6.   Tello (1929:120–125) posited that the slightly earlier Cavernas phase mummies at Paracas contained predominantly the remains of adult females and that all the mummies recovered had deformed crania (40 percent of them trephined) and probably had belonged to a single, stratified ethnic group.

7.   Hollow cane tubes have been found leading from the tomb to the ground surface to permit, it is speculated, the passage of food and drink to the "living corpse" buried in the shaft tomb below (see Larco Hoyle 1946; Trimborn 1969:119–120). Similar tubes (*ushnu*) have been reported in ethnohistorical literature bearing on Late Horizon Inca burial practices (Zuidema, n.d.).

8.   Extended burials are also reported from the far northern Chira Valley, as well as from the Huaura, Chancay, and Rimac Valleys on the central coast, and Chincha on the south coast of Peru (see Mead 1907:8; Williams 1927:1; Tello 1942:121; Reichlen 1950; Cieza de Leon [1553, bk. I, chap. LXIII] 1959:312; and Buse 1962:261).

9.   Original dates for the ethnohistorical sources are given in brackets.

10.   Gold palatal offerings in Peruvian mummies are found at least as early as the Salinar phase of the Early Intermediate period on the north Peruvian coast (see Larco Hoyle 1946).

11.   This status distinction is perhaps no more graphically illustrated than in the case of the burial of a Chachapoyas chief named Chuquimis, who was convicted of regicide after allegedly having poisoned the Inca king Huayna Capac in 1525. Chuquimis died before the death sentence could be carried out, but desiring to desecrate the noble Chachapoyas lineage completely (and thereby set an example for other would-be assassins), the Incas ordered that the mummified cadaver of Chuquimis be exhumed from its clay casing and buried in the ground "like any other common Indian" (Espinoza Soriano 1967:246).

12.   Acosta ([1590, bk. V, chap. VII] 1954:147); Garcilaso de la Vega [1609, pt. I, bk. V, chap. XXIX] 1963:190); Cobo ([1653, bk. XIV, chap. XIX] 1964:274–275.

13.   Garcilaso de la Vega [1609, pt. I, bk. VI, chap. V] 1963:199). The burial of the Inca king Huayna Capac was supposed to have been accompanied by the sacrifice of no less than 1,000 individuals throughout the 80 Inca provinces (Polo [1567] 1916b:9). Polo also informs us that part of Atahualpa's ransom delivered to his Spanish captors at Cajamarca was obtained through the looting of jewels and precious metals from Huayna Capac's mummy in Cuzco (Polo [1561] 1940:154).

14.   There is considerable confusion among the ethnohistorical sources regarding how many, and precisely which, of the royal mummies were actually discovered by Polo. Imbelloni (1946), after examining the various testimonies, tends to favor the version given by Garcilaso de la Vega, holding that the mummified remains of three kings (Huiracocha, Tupac Yupanqui, and Huayna Capac) and two queens (Mama Runtu and Mama Ocllo) were recovered from the villages around Cuzco, where they were still the object of great veneration in 1559. The identification and ultimate fate of the mummies will probably never be known, but informative and fascinating studies have been published by some of Peru's eminent historians, including Juan Toribio Polo (1877) and José de la Riva-Agüero (1966). It is certain, however, that the royal mummies brought to Lima were not provided Christian reburials (Von Hagen 1959:189, note 2). Rather, after some two decades of exhibition, they were probably unceremoniously buried in a patio or courtyard of the Hospital San Andrés, Lima, sometime after 1580 (see Riva-Agüero 1966:397).

15.   "Tolú" probably refers to Santiago de Tolú, located in the archbishopric of Cartagena, Colombia (Bernal 1970), "donde se coge muy oloroso bálsamo, sangre de drago, y otras resinas, y licores medicinales" (Vázquez de Espinoza [1628] 1948:294). Peruvian balsam probably originated in Colombia, where it is typed as *Myroxylon balsamum* or *M. toluifera*, known to have magnificent

antiseptic properties. It grows in the montaña and moist lowland regions of tropical South America and is also found in Peru at Tarapoto, Pozuzo, Húanuco, Loreto, Huallaga, Iquitos, and other regions (MacBride 1943:241–242). As early as 1887, Peruvian balsam was found interred with a mummy at the Necropolis at Ancón (Safford 1917:22), a sample of which is now in the National Museum of Natural History (catalog no. 132613), where spectrographic analysis has recently been conducted on it. However, it should be pointed out that the hardened, blackish resin had been stored inside a *crescentia* gourd container and was not recovered directly from the mummified bodies.

16. *Mithostachis mollis?* In the sierra of Huancayo, Lima, and Ancash, among other regions not high enough to permit the customary freeze-drying of potatoes, muña leaves are used as a preservative for these tubers. The plants are found between 2,400 and 3,700 m along permanent waterways where they grow in an uncultivated state (E. Cerrate, personal communication, Museo de Historia Natural, Lima, 1977).

17. The nematode *Trichuris trichiura* was found in the boy's intestine (Stewart 1973:46).

10. This concept is partially illustrated by an incident recorded by Bennett (1946:618) several decades ago. In the center of an underground family mausoleum used since pre-Hispanic times in northern Chile, a table had been set up, and around it the mummies of the family ancestors were arranged. A member of the family who was seriously ill was carried into the chamber, seated at the table, and surrounded by offerings of food and gifts. A ceremonial dance was then given by his relatives in order to help him "die well."

19. Perhaps such a situation is suggested in a pre-Hispanic north Peruvian coast myth of an important local *kuraka* who died en route to his homeland after an extended stay in Cuzco as an Inca prisoner. His corpse was mummified over a funeral pyre at Pacatnamú and borne home on a litter by his retainers (Kosok 1965:175–176).

REFERENCES

Anthropological and historical perspectives

Acosta, J. de. 1954. *Historia natural y moral de las Indias* [1590]. Madrid: Biblioteca de Autores Españoles.

Allison, M. J., and Pezzia, A. 1973, 1974. Preparation of the dead in pre-Columbian coastal Peru.

Paleopathology Newsletter, 4:10–12; 5:7–9.

Allison, M. J.; Pezzia, A.; Gerszten, E.; and Mendoza, D. 1974. A case of Carrión's disease associated with human sacrifice from the Huari culture of southern Peru. *American Journal of Physical Anthropology* 41(2):295–300.

Alvarez Miranda, L. 1969. Un cementerio precerámico con momias de preparación complicada. *Rehue* 2:181–90.

Bennett, W. C. 1938. If you died in old Peru. *Natural History* 41(2):119–25.

– 1946. The Atacameño. In *Handbook of South American Indians*, vol. 2, pp. 599–618. Bureau of American Ethnology, Bulletin 143. Washington, D.C.: Smithsonian, Institution.

– 1948. The Peruvian co-tradition. In *A reappraisal of Peruvian archaeology*, W. C. Bennett (ed.), pp. 1–7. Menasha: Society for American Archaeology.

Bernal, S. 1970. *Guía bibliográfica de Colombia*. Bogotá: Universidad de los Andes.

Bird, J. B. 1943. Excavations in northern Chile. *Anthropological Papers* 38(4):173–318.

Boman, E. 1908. *Antiquités de la région andine de la République Argentine et du désert d'Atacama*. Paris: Imprimerie Nationale.

Buse, H. 1962. *Perú 10,000 años*. Lima: Colección "Nueva Crónica."

Candela, P. B. 1943. Blood group tests on tissues of Paracas mummies. *American Journal of Physical Anthropology* 1:65–8.

Casanova, E. 1936. El altiplano andino. *Historia de la nación Argentina* 1:251–75.

Cieza de Leon, P. de. 1959. *The Incas of Pedro de Cieza de Leon* [1553]. Edited by V. Von Hagen; translated by H. de Onis. Norman: University of Oklahoma Press.

Cobo, B. 1964. *Historia del Nuevo Mundo* [1653]. Madrid: Biblioteca de Autores Españoles.

Comas, J. 1974. Orígenes de la momificación prehispánica en America. *Anales de Antropología* 11:357–82.

Cornejo Bouroncle, J. 1939. Las momias Incas: Trepanaciones cráneas en el antiguo Perú. *Boletín del Museo de Historia Natural Javier Prado* 3(2):106–15.

Dawson, W. R. 1928a. Two mummies from Colombia. *Man*, no. 53 (May), pp. 73–4.

– 1928b. Mummification in Australia and in America. *Journal of the Royal Anthropological Institute* 58:115–38.

Dérobert, L., and Reichlen, H. n.d. *Les Momies*. Paris: Editions Prisma.

Engel, F. 1970. La Grotte du mégatherium à Chilca et les écologies de Haut-Holocène Péruvien. In *Echanges et communications*, J. Pouillon and P. Miranda (eds.). The Hague: Mouton.

– 1977. Early Holocene funeral bundles from the central Andes. *Paleopathology Newsletter* 19:7–8.

Espinoza Soriano, W. 1967. Los señores étnicos de Chachapoyas. *Revista Histórica* 3:224–332.

Estete, M. de. 1938. Noticia del Perú [ca. 1535]. In *Los cronistas de la Conquista*, H. Urteaga (ed.), pp. 195–251. Paris: Biblioteca de Cultura Peruana.

Garcilaso de la Vega, I. 1963. *Commentarios reales de los Incas*, part 1, vol. 133. [1609]. Madrid: Biblioteca de Autores Españoles.

Guamán Poma de Ayala, F. 1956. *Nueva corónica y buen gobierno* [ca. 1613]. L. F. Bustios Galvez (ed.). Lima.

Imbelloni, J. 1946. Las momias de los reyes cuzqueños. *Pachakuti IX, Humanior (sección D)* 2:183–96.

– 1956. *La segunda esfinge indiana*. Buenos Aires: Librería Hachette.

Jiménez de la Espada, M. (ed.). 1965. *Relaciones geográficas de Indias*, vol. I, pp. 183–5, Madrid: Biblioteca de Autores Españoles.

Kosok, P. 1965. *Life, land and water in ancient Peru*. New York: Long Island University Press.

Larco Hoyle, R. 1946. A cultural sequence for the North Coast of Peru. In *Handbook of South American Indians*, vol. 2, pp. 149–75. Bureau of American Ethnology, Bulletin 143. Washington, D.C.: Smithsonian Institution.

Lastres, J. B. 1953. El culto de los muertos entre los aborígenes Peruanos. *Perú Indígena* 4(10–11):63–74.

Lathrap, D. W. 1974. The moist tropics, the arid lands, and the appearance of great art styles in the New World. *Special Publications* 7:115–158. Museum of Texas Tech University, Lubbock.

Le Paige, G. 1964. El precerámico en la cordillera atacameña y los cementerios del período agro-alfarero de San Pedro de Atacama. *Anales de la Universidad del Norte*, no. 3.

Linné. S. 1929. *Darien in the past*. Göteborgs Kung. Vetenskaps-och Vitterherts, Samhälles Handlingar, Fjärde Följden, Series A, Band 1(3). Göteborg.

Lothrop, S. K. 1948. The archeology of Panama. In *Handbook of South American Indians*, vol. 4, pp. 143–67. Bureau of American Ethnology, Bulletin 143. Washington, D.C.: Smithsonian Institution.

MacBride, J. F. 1943. The flora of Peru. In *Botany*, vol. 12, part 3, no. 1. Chicago: Field Museum of Natural History.

Mahler, J., and Bird, J. B. n.d. Mummy 49: A Paracas Necropolis mummy. Manuscript in possession of the authors, American Museum of Natural History, New York.

McCreery, J. H. 1935. The mummy collection of the University of Cuzco. *El Palacio* 39(22–4):118–20.

Mead, C. W. 1907. *Peruvian mummies and what they teach*. Guide leaflet 24. New York: American Museum of Natural History.

Medina Rojas, A. 1958. Hallazgos arqueológicos en el Cerro El Plomo. In *Arqueología Chilena*. Santiago: Centro de Estudios Antropológicos, Universidad de Chile.

Métraux, A. 1949. Warfare, cannibalism, and human trophies. In *Handbook of South American Indians*, vol. 5, pp. 383–409. Bureau of American Ethnology, Bulletin 143. Washington, D.C.: Smithsonian Institution.

Molina, C. de (of Cuzco). 1916. Relación de las fábulas y ritos de los Incas [1575]. In *Colección de libros y documentos referentes a la historia Peruana*, ser. I, no. 1, H. Urteaga and C. A. Romero (eds.), pp. 1–103. Lima.

Munizaga, J. R. 1974. Deformación craneal y momificación en Chile. *Anales de Antropología* 11:329–36.

Nuñez, A. 1969. Sobre los complejos culturales Chinchorro y Faldas del Morro del norte de Chile. *Rehue* 2:111–42.

Pizarro, P. 1939. Relación del descubrimiento y conquista de los reinos del Perú [1571]. In *Los cronistas de la Conquista*, ser. 1, no. 2, H. Urteaga (ed.), pp. 265–305. Paris: Biblioteca de Cultura Peruana.

Polo, J. T. 1877. Momias de los Incas. In *Documentos Literarios del Peru*, 10:371–8.

Polo de Ondegardo, J. 1916a. Errores y supersticiones de los indios [1554]. *Colección de libros y documentos referentes a la historia del Perú*, ser. 1, no. 3, H. Urteaga and C. A. Romero (eds.). Lima.

– 1916b. Instrucción contra las ceremonias y ritos que usan los indios conforme al tiempo de

su gentilidad [1567]. *Colección de libros y documentos referentes a la historia del Perú*, ser. 1, no. 3, H. Urteaga and Romero (eds.). Lima.

– 1940. Informe . . . al licenciado Briviesca de Muñatones . . . [1561]. *Revista Histórica* 13:125–96.

Ponce Sanginés, C., and Linares Iturralde, E. 1966. *Comentario antropológico acerca de la determinación paleoserológica de grupos sanguíneos en momias prehispánicas del altiplano boliviano*. Publication no. 15. La Paz: Academia Nacional de Ciencias de Bolivia.

Ramos Gavilán, A. 1976. *Historia de nuestra señora de Copacabaña* [1621]. La Paz: Academia Boliviana de la Historia.

Reichlen, H. 1950. Etude de deux fardeux funéraires de la côte centrale du Pérou. *Travaux de l'Institut Français d'Etudes Andines* 1:39–50.

Reiss, J. W., and Stübel, M. A. 1880–7. *The necropolis of Ancón in Peru: A contribution to our knowledge of the cultures and industries of the empire of the Incas, being the results of excavations made on the spot*. Translated by A. H. Keene. Berlin: Ascher.

Riva-Agüero, J. de la. 1966. *Obras completas de José de la Riva-Agüero: vol. V, Sobre las momias de los Incas*. Lima: Pontífica Universidad Católica del Perú.

Rivero, E. M., and von Tschudi, J. J. 1854. *Peruvian antiquities*. Translated by F. Hawks. New York: Barnes.

Rivero de la Calle, M. 1975. Estudio antropológico de dos momias de la cultura paracas. In *Ciencias*, ser. 9, Antropología y Prehistoria no. 3. Havana.

Röseler, P. 1975. Huac'a und mallqui-Röntgenanatomische und pathologische Studien zum präspanischen Peru. *Röntgenstrahlen* 32.

Rowe, J. H. 1946. Inca culture at the time of the Spanish Conquest. *Handbook of South American Indians*, vol. 2, pp. 183–330. Bureau of American Ethnology, Bulletin 143. Washington, D.C.: Smithsonian Institution.

– 1967. Introduction. In *Peruvian archaeology: selected readings*, J. H. Rowe and D. Menzel (eds). Palo Alto: Peek.

Safford, W. E. 1917. Food-plants and textiles of ancient America. In *Proceedings of the 19th International Congress of Americanists* (1915), pp. 12–30. Washington, D.C.: The Congress.

Sancho, P. 1938. Relación para SM de lo sucedido en la conquista y pacificación de estas provincias de la Nueva Castilla y de la calidad de la tierra [1543]. In *Los cronistas de la Conquista*, ser. 1, no. 2, H. Urteaga (ed.), pp. 117–193. Paris: Biblioteca de Cultura Peruana.

Savoy, G. 1970. *Antisuyo*. New York: Simon & Schuster.

Schaedel, R. P. 1957. Informe general sobre la expedición a la zona comprendida entre Arica y La Serena. In *Arqueología Chilena*, R. P. Schaedel (ed.). Santiago: Universidad de Chile.

Schobinger, J. (ed.). 1966. La "momia" del Cerro El Toro. *Anales de Arqueología y Etnología* (Suppl.) 21.

Skottsberg, C. 1924. Notes on the old Indian necropolis of Arica. *Meddelanden fran Geografiska Forenningen i Göteborg* 3:27–78.

Steward, J. H. 1948. The circum-Caribbean tribes. In *Handbook of South American Indians*, vol. 4, pp. 1–41. Bureau of American Ethnology, Bulletin 143. Washington, D.C.: Smithsonian Institution.

– 1949. South American cultures, an interpretive summary. In *Handbook of South American Indians*, vol. 5, pp. 669–772. Bureau of American Ethnology, Bulletin 143. Washington, D.C.: Smithsonian Institution.

Steward, J. H., and Métraux, A. 1948. Tribes of the Peruvian and Ecuadorian montaña. In *Handbook of South American Indians*, vol. 3, pp. 535–656. Bureau of American Ethnology, Bulletin 143. Washington, D.C.: Smithsonian Institution.

Stewart, T. D. 1973. *The people of America*. New York: Scribner.

Strong, W. D. 1957. Paracas, Nazca and Tiahuanaco cultural elements. *Memoirs of the Society for American Archaeology* 13.

Tello, J. C. 1918. Es uso de las cabesas humanas artificialmente momificadas y su representación en el antiguo arte peruano. *Revista Universitaria* 1:477–533.

– 1929. *Antiguo Perú*, part 1. Lima: Excelsior.

– 1942. *Orígen y desarrollo de las civilizaciones prehistóricas andinas*. Lima: Librería Gil.

– 1959. *Paracas*, part 1. Lima: Empresa Gráfica Scheuch.

– 1980. *Paracas*, part 2, *Cavernas y necrópolis*. Lima: Universidad Nacional Major de San Marcos.

Trimborn, H. 1969. South Central America and the Andean civilizations. In *Pre-Columbian*

American religions, E. O. James (ed.). History of Religion Series. New York: Holt.

Ubbelohde-Doering, H. 1966. *On the royal highways of the Inca*. New York: Praeger.

Uhle, M. 1918. Los aborígenes de Arica. *Revista Histórica* 6:5–26.

– 1975. La momia peruana. *Indiana 3*. [1898]. Ibero-Amerikanisches Institut, Berlin: Preussischer Kulturbesitz, 189–97.

Valera, B. 1945. Las costumbres antiguas del Perú y la historia de los Incas [ca. 1590]. In *Los pequeños grandes libros de historia americana*, ser. 1, vol. 8, F. Loayza (ed.). Lima: Miranda.

Vázquez de Espinosa, A. 1948. Compendio y descripción de las Indias Occidentales [1628]. *Smithsonian Miscellaneous Collections* 108 (whole volume).

Villar Córdoba, P. E. 1935. *Arqueología del departamento de Lima*. Lima.

Von Hagen, V. (ed.). 1959. *The Incas of Pedro de Cieza de Leon*. Norman: University of Oklahoma Press.

Vreeland, J. M. 1976. Second Annual Report: Proyecto de investigación textil "Julio C. Tello." Research report presented to the Secretariat for Technical Cooperation, Organization of American States, Washington, D.C.

– 1977. Ancient Andean textiles: clothes for the dead. *Archaeology* 30(3):166–78.

– 1978a. Paracas. *Américas* 30(10):36–44.

– 1978b. Prehistoric Andean mortuary practices: a preliminary report from central Peru. *Current Anthropology* 19(1):212–14.

– 1980. Prácticas mortuorias andinas: perspectivos teóricas para interpretar el material textil prehispánico, vol. 3. Lima: Actas del Tecer Congreso del Hombre y de la Cultura Andina.

Waisbard, S., and Waisbard, R. 1965. *Masks, mummies and magicians*. Edinburgh: Oliver & Boyd.

Walter, N. P. 1976. A child sacrifice from pre-Inca Peru. Paper presented at the annual meeting of the Southwestern Anthropological Association, San Francisco, April 1976.

Weiner, C. 1880. *Pérou et Bolivie: récit de voyage*, Paris: Librairie Hachette.

Willey, G. R. 1971. *An introduction to American archaeology*, vol. 2, *South America*. Englewood Cliffs, N.J.: Prentice-Hall.

Williams, H. U. 1927. Gross and microscopic anatomy of two Peruvian mummies. *Archives of Pathology and Laboratory Medicine* 4:26–33.

Yacovleff, E., and Muelle, J. C. 1932. Una exploración de Cerro Colorado. *Revista del Museo Nacional* 1(2):31–102.

– 1934. Un fardo funerario de Paracas. *Revista del Museo Nacional* 3(1–2):63–153.

Zuidema, R. T. n.d. Shaft tombs and the Inca empire. Manuscript in possession of the author. University of Illinois, Urbana.

Diseases

Allison, M. J., and Pezzia, A. 1973. Documentation of a case of tuberculosis in pre-Columbian America. *American Review of Respiratory Diseases* 107:985–91.

Allison, M. J.; Pezzia, A.; Gerszten, E.; and Mendoza, D. 1974a. A case of Carrión's disease associated with human sacrifice from the Huari culture of southern Peru. *American Journal of Physical Anthropology* 41:295–300.

Allison, M. J.; Pezzia, A.; Hasega, L.; et al. 1974b. A case of hookworm infestation in a pre-Columbian American. *American Journal of Physical Anthropology* 41:103–5.

Allison, M. J.; Houssaini, A. A.; Castro, N.; Munizaga, J.; and Pezzia, A. 1976. ABO blood groups in Peruvian mummies. *American Journal of Physical Anthropology* 44:55–62.

Angel, L. J. 1967. Porotic hyperostosis or osteoporosis symmetrica. In *Diseases in antiquity*, D. Brothwell and A. T. Sandison (eds.). Springfield, Ill.: Thomas.

Boyd, W. C. 1959. A possible example of action of selection in human blood groups. *Journal of Medical Education* 34:398–9.

Boyd, W. C., and Boyd, L. G. 1933. Blood grouping by means of preserved tissues. *Science* 78:578.

– 1937. Blood grouping tests on 300 mummies. *Journal of Immunology* 32:307–9.

Cockburn, E. (ed.). 1977. *Porotic hyperostosis: an inquiry*. Monograph no. 2. Detroit: Paleopathology Association.

Garcia-Frias, J. E. 1940. La tuberculosis en los antiquos Peruanos. *Actualidad medical peruana* 5:274.

Gerszten, E.; Allison, M. J.; Pezzia, A.; and Klurfeld, D. 1976. Thyroid disease in a Peruvian mummy. *Medical College of Virginia Quarterly* 12:52–3.

Grana, F.; Rocca, E. D.; and Grana, L. R. 1954. *Las trepanaciónes craneanas en el Perú en la época prehispánica*. Lima: Imprenta Santa Maria.

Hossaini, A. A., and Allison, M. J. 1976. Paleoserology studies: ABO and histocompatibility antigens in mummified American Indi-

ans. *Medical College of Virginia Quarterly* 12:67–73.

Hrdlička, A. 1914. Anthropological work in Peru in 1913 with notes on the pathology of the ancient Peruvians. *Smithsonian Miscellaneous Collection* 61(12).

Lastres, J. B., and Cabieses, F. 1960. *La trepanación del cráneo en el antigua Perú.* Lima: Imprenta de la Universidad Nacional Major de San Marcos.

Lisowski, F. P. 1967. Prehistoric and early historic trepanation. In *Diseases in antiquity,* D. Brothwell and A. T. Sandison (eds.). Springfield, Ill.: Thomas.

MacCurdy, C. G. 1970. Surgery among the ancient Peruvians. *Art Archeology* 7:381–94.

Margetts, E. L. 1967. Trepanation of the skull by the medicine-men of primitive cultures. In *Diseases in antiquity,* D. Brothwell and A. T. Sandison (eds.). Springfield, Ill.: Thomas.

McGee, W. J. 1894. Primitive trephining illustrated by Muñiz Peruvian collection. *Bulletin of the Johns Hopkins Hospital* 5:1–23.

Moodie, R. L. 1927. Injuries to the head among pre-Columbian Peruvians. Studies in paleopathology, 21. *Annals of Medical History* 9:91–102.

Morse, D. 1961. Prehistoric tuberculosis in America. *American Review of Respiratory Diseases* 85:489.

Muñiz, M. A., and McGee, W. J. 1894–5. *Primitive trephining in Peru.* Bureau of American Ethnology, Annual Report 16, pp. 7–72. Washington, D.C.: Smithsonian Institution.

Quevedo, A., and Sergio, A. 1943. La trepanación incan en la región del Cuzco. *Revista del Museo Nacional* 12:1–18.

– 1945a. La trepanación incan en la región del Cuzco. *Revista del Museo Nacional* 13:1–7.

– 1945b. La trepanación incan en la región del Cuzco. *Revista del Museo Nacional* 14:1–10.

Rocca, E. 1953. *Traumatismos encefalocraneanos.* Lima: Imprenta Santa Maria.

Rogers, S. L. 1938. The healing of wounds in skulls from pre-Columbian Peru. *American Journal of Physical Anthropology* 23:321–40.

Stewart, T. D. 1956. Significance of osteitis in ancient Peruvian trephining. *Bulletin of the History of Medicine* 30:293–320.

Weiss, P. 1949. *La cirugía del cráneo entre los antiguos Peruanos.* Lima: Imprenta Santa Maria.

– 1958. *Osteológica cultural prácticas cefálicas,* part 1. Lima: Imprenta Universidad Nacional Major de San Marcos.

PART III
Mummies of the world

10

Bog bodies of Denmark

CHRISTIAN FISCHER
Director, Silkeborg Museum
Silkeborg, Denmark

Translated by KIRSTINE THOMSEN

In general, bog bodies are considered a northwestern European phenomenon reflecting special forms of sacrifice or punishment common among Germanic peoples around the time of the birth of Christ. The German researcher Dieck (1965) has shown that from Norway in the north to Crete in the Mediterranean and from Ireland in the west to Russia in the east bog bodies number more than 1,400 (that is the 1968 figure; today the number is closer to 2,000) and cover a period from 9000 B.C. to World War II.

If we consider this large distribution area as one, the bodies can be seen as a natural phenomenon and not as the practice of a special, culturally influenced group. Of more than 1,400 known bog bodies or parts thereof (20 percent of the finds consist of isolated legs, arms, and heads), less than one-third can be dated. About 100 may be dated to the Stone Age and the older Bronze Age (5000–1000 B.C.) and a similar number to the period up to A.D. 400. From the period A.D. 400–1500 about 100 are known. From A.D. 1500 until today, about 50 have been recorded (Lund 1976).

THE PRESERVATIVE QUALITIES OF THE BOG

The reason for the preservation of the bog bodies (and of other organisms also) lies in the special physical and biochemical makeup of the bog, above all the absence of oxygen and the high antibiotic concentration. The manner in which the body was deposited is also of great importance – for example, placed in the bog in such a way that air was rapidly excluded. It is important not only that the bog water contained a high concentration of antibiotics but also that the weather was cold enough (less than 4°C) to prevent rapid decomposition of the body. If the body had been deposited in warm weather, one can assume that the presence of anaerobic bacteria in the intestinal system would have had a destructive effect on the interior of the corpse before the liquid of the bog could penetrate the body.

The bogs in which bodies are found can be roughly divided into three types: raised bogs (acid), fens (containing lime), and transitional types. The raised bog holds the greatest interest for us because the bodies found there are so well preserved that they are hardly different from when they were deposited; in the fen and the transition areas all the soft parts have generally disappeared, and the body is found as skeleton or adipocere.

The flora of the raised bog is sparse and dominated by peat moss (sphagnum), the leaves of which are constructed in a special manner. Only a few of the cells contain chlorophyll and are able to carry out the process of photosynthesis. The other cells, which lie among those containing chlorophyll, are dead and empty. They are constructed of cellulose and connected with one another and their surroundings through pores. These "empty" cells have an extraordinary ability to absorb water; for this reason, the raised bog feels wet to walk on even in dry summers. It is a widely held misconception that the raised bog acts as a sponge that sucks up groundwater. This is not true. The deepest lying peat layers are so compressed that they totally pre-

vent the passage of any water. In addition, it has been shown by laboratory experiments that the height to which water can be raised by capillarity is here only 40 to 50 cm.

The water of the raised bog is therefore not mineral-containing groundwater, but rainwater. The raised bog holds water because of its ability to absorb surface water and the ability of the underlying peat to retain it. In addition, the surface of the bog is dotted with mounds so that rainwater does not run off. The nutrient bases that the peat moss requires – Ca, Mg, Na, and K – come from atomized seawater, which is carried into the atmosphere; the amount of nutrient salts therefore increases the closer one gets to the sea.

Measurements have shown that the true annual growth of the raised bog amounts to approximately 15 mm, but pressure from the layers above results in the bog's profile showing only 6.4 mm growth per year. This compression is the causative factor preventing oxygen from reaching the underlying layers, and the oxygen lack is the reason the peat and the organisms in it cannot be destroyed by oxygen-dependent bacteria. In addition, raised peat is one of the most acid soil types one can find: The sphagnum produces sulfuric acid in very small quantities, and in the peat humic acid is produced by the action of anaerobic bacteria on lignin. Because the acids are mixed only with "buffer-poor" neutral precipitation, it takes a very small amount of acid to lower the pH value to the 3.6 to 4.0 of the Danish raised bog. These acids further act as antibiotics and cause an acid tanning of the skin, as can be seen in the Grauballe man.

Today, very few untouched raised bogs are found in Western Europe, for the peat has been utilized as fuel since before the birth of Christ. Indeed, many bog bodies are placed in peat diggings of the Iron Age: For example, a wooden peat spade from the Iron Age was found just beside the Tollund man.

DANISH BOG BODIES

The distribution of bog bodies is extensive geographically as well as in time. If one fo-cuses on Denmark, however, most of the bog bodies are undated, owing to the fact that they were found before such scientific methods as pollen analysis and carbon-14 dating were in use. It is quite characteristic also that the bog bodies are naked or accompanied only by simple, uncharacteristic articles of clothing difficult to date. Of the finds that can be dated, either scientifically or culturally, almost all are from the period around the birth of Christ, approximately 400 B.C. to A.D. 400, which is known in Danish archeological terminology as the pre-Roman or Roman Iron Age.

Fortunately, a number of the best preserved Danish bog bodies are among those that have been dated and carefully examined, all of them excavated within the first 10 years following World War II. This was a period during which peat was still being dug for fuel, but, unfortunately for bog body research, this practice has now ceased. Among the Danish bog bodies that were scientifically excavated and examined, we must make special mention of the three bodies from Borremose found in 1946, 1947, and 1948; the Tollund man of 1950; and the last one on the scene, the Grauballe man of 1952.

BODIES FROM BORREMOSE

The three bodies from Borremose were excavated in a stretch of boggy land that surrounds a unique fortified Iron Age village dating back approximately to the birth of Christ (Figure 10.1).

Borremose body 1946 (Figure 10.2) was found 2 m beneath the surface of the bog in an upright, strongly contracted position, which can best be explained by assuming that the body had been placed in the bog in a sitting position; the upper body had then, by its own weight, sunk down toward the thighs. A birch stick had been placed above the body: This is often seen with bog bodies and is undoubtedly intended to prevent the corpse from "going again" (i.e., returning to life). The find was excavated by S. Vestergaard Nielsen, Director of the Aars Museum, and Knud Thorvildsen of the National Museum (1947).

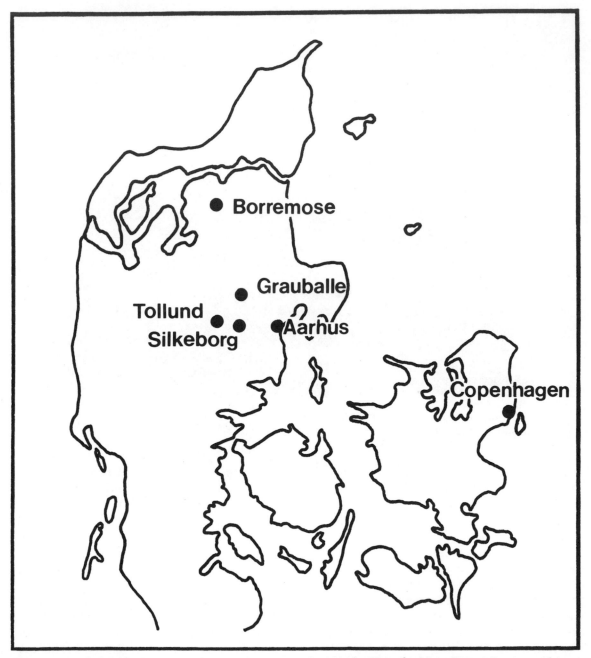

Figure 10.1.   Locations of bog bodies found in Denmark. (Map by Timothy Motz, Detroit Institute of Arts)

The body was that of an adult man who had been quite small, scarcely more than 155 cm tall. Because of the acidity of the bog, the bones were decalcified and soft, but covered with skin, sinews, and muscle tissue. The face, lower body, sex organs, hands, and feet were especially well preserved. The left eye was intact; at the beginning of the excavation it was closed, but by the end it had opened: the eyeball protruded, in color yellow white,

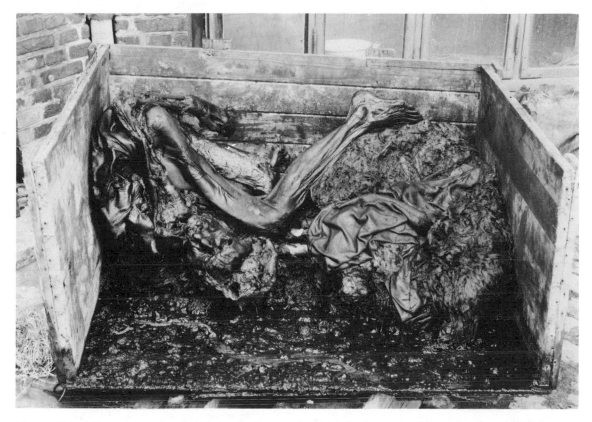

Figure 10.2.   Borremose man 1946 lying in the crate in which he was moved from the place where he was found. National Museum, Copenhagen.

with a dark iris. The nose and lower part of the face were undamaged. The tongue was well preserved. The teeth, of which only a few were left, were loose. Scattered beard stubble 6 mm in length was found on the upper lip, cheeks, and chin, corresponding to about 2 days' growth. The back of the head had been crushed, so that the cranium was open and the brain matter visible. The edge of the break was not new; that is to say, the head had been crushed before the body was deposited in the bog (Figure 10.3).

Of the inner organs, the intestine and its contents were partly intact. The large intestine contained his last meal (about 40 ml.), which consisted of about 65 percent spurry and 25 percent pale willowweed. Some animal tissue was found, which in all probability came from the intestine itself, and also a few

short animal hairs, most probably hair from mice, which are always found in grains and seed (Brandt 1951).

The arm bones of the Borremose man were whole, but a break was found in the right thigh just above the knee. The shaft of the bone protruded through the skin on the underside of the thigh. The sharp edges of the fracture indicated that the break had occurred before the bones were decalcified by the acid water of the bog. The hands were well preserved, narrow, and finely formed, giving the impression that they had not been used for rough work. The fingernails had dissolved. The feet had high insteps, and the toes were in normal position. The toenails were loose; they were attractively formed, curved regularly, and had been cut or trimmed.

A rope 94 cm long and 1 cm thick was tied

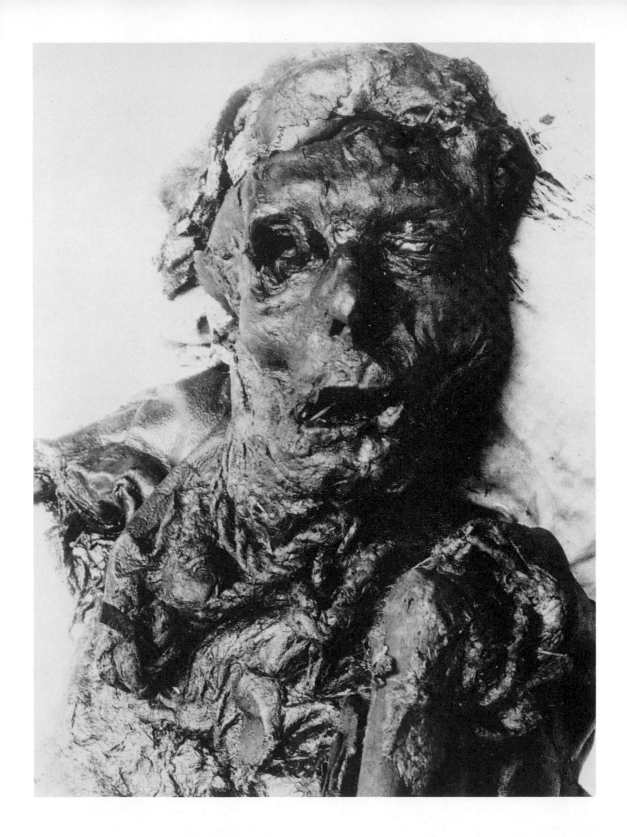

Figure 10.3.   Borremose man 1946, with the noose still around his neck. National Museum, Copenhagen.

around the neck and pulled so tightly that the neck measured only 37 cm in circumference. This rope was very likely the cut-off end of the cord with which the man was hanged. Whether the mutilations already mentioned – the broken skull and thigh – were inflicted before or after the hanging, or whether they had been a contributory cause of death, could not be ascertained. Apart from the rope, the body was naked, but two fur cloaks, almost alike, lay by the feet. These were, presumably, his only articles of clothing, corresponding to the Germanic style of dress mentioned by Caesar: "They wore only short fur cloaks but were otherwise naked."

The pollen test conducted at the National Museum indicated that the man was buried in the bog within the first 200 years after the birth of Christ. The Borremose body 1946 is dated by the carbon-14 method to 650 B.C. ± 80 years; that is, to the final period of the Bronze Age. None of the carbon-14 datings mentioned in this chapter have been calibrated.

Borremose body 1947, which was found about 1 km from the 1946 body, was too decomposed for the sex to be established. The body had been placed on its stomach in an Iron Age peat digging. Above the body in the peat digging lay sticks and heather twigs. The upper part of the body was naked; the hips and legs were covered with a woolen blanket, a shawl, and a ragged piece of cloth (Thorvildsen 1952). Some parts of the body were greatly decomposed. The stomach and part of the lower portions of the body were completely gone; part of the upper portion of the body had also disappeared; the head, the esophagus, and the left shoulder, complete with arm and hand, were preserved. The back of the head was pressed flat, and the skull was crushed so that the brain matter was visible. The hair, which had been darkened by the bog, was short – only 3 to 7 cm long. Around the neck was an ornament consisting of a leather string with an amber pearl and a pierced conical bronze disk. The right leg was

broken 10 cm below the knee joint; the break was not new, but must have occurred when the body was deposited. It cannot be determined with certainty if it was inflicted before or after death. The flattening of the skull was probably caused by pressure from the peat.

Near the upper part of the body, there were some small bones belonging to a baby, which may indicate that a mother and her new born infant had been deposited in the bog together. The cause of death could not be established. The date was established from the remains of a clay vessel of pre-Roman type that lay in immediate contact with the body. Pollen analysis of the body showed that it dates back to about the birth of Christ. The carbon-14 method placed the date at 430 B.C. ± 100 years.

In 1948, continued digging revealed still another bog body, 2 m under the surface in an Iron Age peat digging like the others. The dead person was a plump woman, lying on her stomach. Her right arm was bent at the elbow, and her right hand was resting on her chin. The left hand was propped against the left shin; the left leg was bent at the knee and pulled vigorously up toward the abdomen. The back of her head was intact, showing distinct signs of scalping. The face was crushed, and this must have been done before the body was deposited in the bog, for pieces of the skull were mixed with small pieces of brain matter. This could not have been caused by the pressure of the peat, as the back of the head would also have been crushed and this was not the case. Most of the scalp with hair on it had been loosened from the skull by the scalping and lay above it in the bog. The hair appeared to have been of medium length. The excavation was carried out by museum conservator B. Brorson Christensen.

The dead woman had been laid in a rectangular woolen blanket, 175 by 115 cm, but was otherwise naked. It is possible that the blanket had acted as a kind of skirt. The carbon-14 method dated the body to 530 B.C. ± 100 years, and a carbon-14 test used on the mate-

Figure 10.4.   Tollund man at the National Museum, Copenhagen. Lars Bay, Silkeborg Museum.

rial, where the humic acid had not been extracted, indicated 610 B.C. ± 100 years. The woman had therefore lived during the transition period between the Bronze Age and the Iron Age. Pollen analysis indicated that the body had been deposited about the time of the birth of Christ.

THE TOLLUND MAN

The Tollund man (Figure 10.4) is among the best preserved ancient bodies that are still extant, and his unbelievably well preserved head is without equal (Thorvildsen 1951).

He was found in central Jutland in May 1950 when two farmers, digging for peat, reached 2.5 m below the surface. A displacement in the peat layer showed that he lay in an Iron Age peat digging, which also contained a short wooden spade of the Iron Age type.

After calling the police, staff from the Silkeborg Museum, acting on the advice of Professor P. V. Glob, decided to send the body to the National Museum. It was still lying in the peat where it had been discovered, so the whole block was crated and shipped to Copenhagen. There the final excavation and examination took place under the direction of museum conservator Knud Thorvildsen.

The body was that of an adult male approximately 1.60 m tall; it lay in a natural sleeping position resting on its right side. The body was slightly contracted, the knees completely pulled up, and the arms bent so that the left hand lay under the chin and the right near the left knee – a position often encountered in bodies from ordinary Iron Age graves. One must assume that he had been placed in that position on purpose, which makes him different from other bog bodies, whose place-

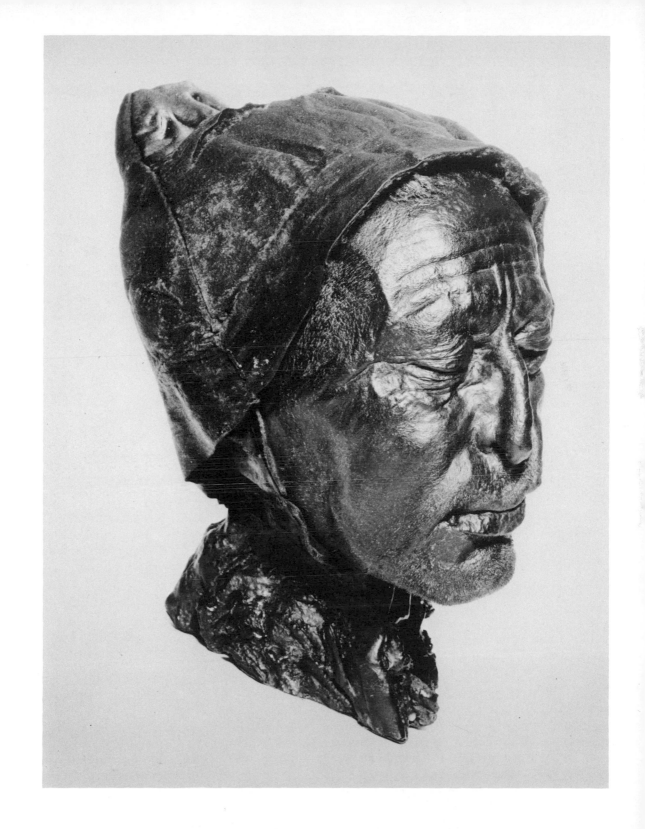

Figure 10.5.   Tollund man's head after conservation. Lars Bay, Silkeborg Museum.

ment is more haphazard. He must have been placed in a sleeping position either within the first 8 to 12 hours after death (before rigor mortis had developed) or 1 to 3 days later (the time when rigor mortis would have worn off).

The skin and most of the soft parts were intact: Only the skin of the hands and their soft parts had disappeared; the arms and legs were partially dissolved. The skin of the up-turned side had begun to disappear, but the rest of the body was intact; the sex organs were well preserved. The best preserved part was the head (Figure 10.5), with its com-posed, sleeping expression. The wrinkled forehead and closed eyelids and lips were in such good condition that the man looked like a living person who had fallen asleep. The hair was completely preserved; it was cut short, but in no particular style. There was 1 to 2 mm of beard stubble on the upper lip, the chin, and the cheeks, which represents about 48 hours' growth, unless it is caused by post-mortem contractions of the skin. The eye-brows were also intact.

Dissection showed that the inner organs (heart, lungs, and liver) were very well pre-served, but unfortunately they were not ex-amined further. The alimentary canal, with stomach and small and large intestines, was intact. This was examined by Dr. Bjovulf Vimtrup and Dr. Kay Schaurup, anatomists, and Dr. Hans Helbaek, botanist, for evidence of the dead man's last meal. The contents of the stomach and small intestine were modest, only 0.5 and 10 ml; the large intestine con-tained 260 ml. This indicates that the meal was consumed 12 to 24 hours before death.

The body was naked, except for a narrow leather belt and a pointed hood made from calfskin, with the hair on the inside. The hood was kept on the head by a thin thong under the chin. Even while the corpse was lying in situ, it could be seen that a rope made from two braided thongs was pulled tightly around its neck. One end of the rope was tied to form an open loop through which the other end was pulled, so that it circled the neck. The free end of the rope measured 1.25 m and had

been sharply severed, so it had probably been longer. The rope was pulled so tightly around the neck that it had left a visible furrow in front and on the side of the throat, though there was no mark in the place where the loop was fastened. In order to find out whether the cause of death was ordinary hanging with dislocation of the axis or slower choking (for the rope was undoubtedly the cause of death), the neck and head of the corpse were exam-ined and X-rayed by Dr. Chr. I. Bastrup. His examination led him to believe that the cervi-cal vertebrae had not been dislocated and the axial process had not broken, as one would have expected if the man had been hanged by falling with the noose around his neck. Un-fortunately, most of the bones and teeth were so decalcified that they were difficult to evaluate. As far as could be observed, the wisdom teeth had erupted and the teeth were in good shape. The eruption of wisdom teeth indicates that the man was at least 22 years old. The X-ray pictures showed also that the brain had shrunk in a peculiar fashion (Figure 10.6). As a result of the decalcification, sev-eral bones were also bent and modeled in a peculiar way.

Analysis of the stomach contents showed that they consisted of a pure vegetarian por-ridge or gruel without any trace of animal content. According to the examining physi-cians, animal remains ought to have been present if they had been ingested just prior to death, considering the excellent state of pres-ervation of the body. The porridge was a combination of grain and weed seeds, to which had been added fat from linseed (*Linum usitatissimum* L.) and gold of plea-sure (*Camelina linicola* Sch. & Sp.). The grain varieties were naked and covered barley (*Hordeum tetrastichum* Kcke., var. *nudum*, and *H. tetrastichum* Kcke.), and the most im-portant of the weeds was knotweed (*Polygonum lapathifolium* agg. and *P. per-sicaria* L.). The examination showed that the water drunk with the meal must have been bog water as small leaves of sphagnum were found among the stomach contents. The Tol-

lund man had also suffered from the intestinal parasite *Trichuris* in rather large numbers (over 100 in a 24- by 24-mm slide preparation). The composition of the meal helps to confirm the dating of the Tollund man as the relationship between naked and hulled barley corresponds to the time of the birth of Christ, and linseed and spelt, which were also found in the stomach, are first reported in the Danish flora about 400 B.C. (Helbaek 1951). By the carbon-14 method, the Tollund man is dated to 210 B.C. ± 40 years.

Unfortunately, after the detailed examination was concluded, preservation was attempted for the head only. Apart from the bog body from Rendswühren in North Germany, which was preserved in 1871 by smoking it at the local butcher's shop, no one at this time had any experience in preserving bog bodies. The method used consisted of first replacing the bog water contained in the cells with distilled water containing formalin and acetic acid. The liquid was then changed to 30 percent alcohol, and later to 99 percent with the addition of toluol. Finally, the head was placed in pure toluol, which was generally saturated with paraffin. The paraffin was afterward replaced by carnauba wax heated to different temperatures. This preservation process made it possible to retain all the correct proportions and the facial expression, but the head as a whole shrank about 12 percent; nevertheless, it is today the best preserved head of any person from ancient times. The rest of the body was dried up and thus partly destroyed.

Two feet were kept, however, one in water and one in formalin. In 1976, the foot that had been kept in water was dried and treated with wax: This caused it to shrink about 25 percent. The foot that had been kept in formalin was preserved by freeze-drying: This produced a fine result with practically no shrinking.

THE GRAUBALLE MAN

The last well preserved bog body is that of the Grauballe man, who was found in 1952 about 20 km east of the place where the Tollund

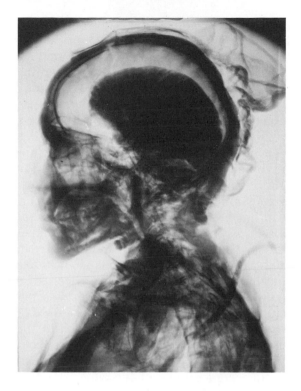

Figure 10.6. X ray of the Tollund man's head. The brain has shrunk, but is still preserved.

man was found and 10 km north of the town of Silkeborg (Glob 1956). While digging for peat in a small bog, the cutters accidentally struck a body that was clearly placed in an Iron Age peat digging. The size of the digging could not be ascertained, as the greater part of the surrounding peat was gone when the people from the museum arrived. The corpse was placed in the digging slightly on a slant. It lay on its chest with the left leg almost stretched out and the right leg and arm bent. The pressure of the peat had deformed the head slightly (Figure 10.7).

Professor P. V. Glob requested that a crate be constructed around the corpse and the peat in which it lay, as had been done with the Tollund man. The crate with its contents was then taken to the Museum of Prehistory in Aarhus, where the examination could be continued and preservation procedures carried out.

A plaster cast impression was taken to re-

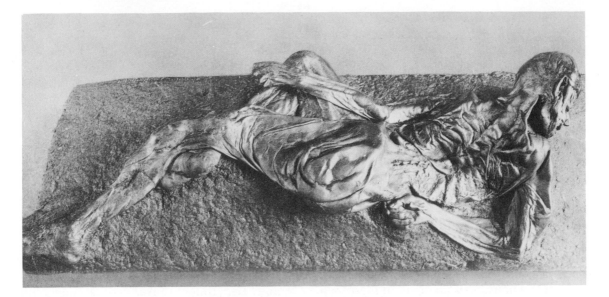

Figure 10.7.   Grauballe man at the Museum of Prehistory, Aarhus. Forhistorisk Museum, Aarhus Kommunehospital.

cord the body's position and size in case shrinkage occurred during the preservation process. A pathological, anatomical, and forensic examination of the body was performed by Professor Willy Munck (1956). Professor Carl Krebs and Dr. Erling Ratjen X-rayed the body (1956). External examination showed that the head was bent slightly backward and turned a little to the right. On the left side of the forehead there was a slightly curved depression 10 cm in diameter. The left earlobe was well preserved and, close to the opening to the ear, some irregular defects were seen, which probably occurred after death. The eyes were tightly shut and the eyeballs completely flattened and dried up. The color of the iris could not be determined, but the eyes probably had been quite dark. There were no eyebrows present. The mouth, which was slightly open, appeared to have been quite large. Teeth found in the upper jaw were 4+ and +3, 4, 5; in the lower jaw −1, 2, 3, 4, 5. The teeth were very black and worn on the masticating surface.

On the front of the neck was a large wound whose upper edge began 5 cm below and 3 cm behind the right ear. It ran upward and forward, a little above the lower jaw, and the edges were smooth, except for a notch in the middle of the lower side. The tongue was shriveled, but so well preserved that its shape could be recognized, including the tip. The epiglottis could be seen, but the gullet and the esophagus had been severed by the throat wound. The lungs seemed not to be fastened to the wall of the chest. A large, soft, red mass, approximately 15 by 10 cm and covered by a capsule was found at the site of the liver. The intestine with its contents was removed. In the scrotum a flat solid body was recognizable, presumably a testis. The stomach, spleen, pancreas, suprarenal gland, kidney, ureter, and bladder were not recognizable.

There was some reddish hair on the top of the head and a few strands of beard on the upper lip and chin, approximately 2 or 3 mm to 1 cm in length. A microscopic examination of the hair from the head indicated that it was of medium thickness and probably had been dark: The reddish color was attributable to the action of the bog water. The hair of the beard was somewhat thicker than that of the head.

The purpose of the X-ray examination was to look for signs of violence, to determine the age of the man, if possible, and to decide

whether the bone system showed traces of disease. There were no signs of any serious illness marking the bone system. There was a skull fracture in the right temple region, and the left shin bone was fractured. It could be seen that the skull was partly flattened and depressed toward the middle, and at the back of the crown there was a 2- by 1-cm break in conjunction with a linear fracture. The other skull deformations were attributed to the pressure of the surrounding peat mass. The break on the tibia was oblique: The fracture line began 10.5 cm under the knee on the outside of the under part of the leg, traveling downward and inward to the inner side 14 cm from the knee. The fracture had been open. There was no sign of callus formation on the ends of the bones and no break in the fibula. Both the skull fracture and the leg fracture must have been inflicted immediately before or after death. The break in the tibia without an accompanying break in the fibula suggests a direct blow to the shin. A fall or similar accident would, as a rule, have caused a break in the fibula.

A second group of deformations and fractures must be attributed to postmortem displacement resulting from the pressure of the peat.

X-ray pictures of the thoracic portion of the spinal column indicated the beginning of rheumatoid arthritis, a disease that seldom occurs before the age of 30.

The cause of death was undoubtedly the long cut from ear to ear, which was so deep that it had severed the gullet. The wound must have been inflicted by another person: Its direction and appearance rule out the possibility of suicide or of an injury inflicted after death. Whether the man was knocked unconscious before his throat was cut could not be determined. The skull fracture seems to have been caused by a blunt instrument. It could not be determined whether the oblique fracture of the tibia had occurred before or after death.

The teeth were examined by two dentists, Friis and Warrer (Glob 1965). In the upper jaw there were 7 teeth in their proper positions, in the lower jaw only 5; but 14 sockets for other teeth were clearly visible, and 9 teeth were found in several different parts of the mouth. The teeth were very small. One tooth had been lost long before death, as the socket had healed. Some other teeth showed periodontitis and caries, which at times must have caused toothache. One bad tooth had caused malocclusion. The wisdom teeth had not erupted.

The preservation of the right hand and foot was so good (Figure 10.8) that the papillae lines could be clearly seen. To find out whether these lines corresponded with what is found in people living today, Inspector C. H. Vogelius Andersen of the Police Criminology Department in Aarhus examined both the fingerprints and the footprints (1956). The right thumbprint was the clearest and could immediately be classified as a whorl pattern, a so-called double-curve pattern; the right middle finger showed an ulnar loop pattern. The fingerprint expert came to the conclusion that if one had had a fingerprint card index from the time of the Grauballe man, it would have been easy to identify him! The two fingerprint patterns occur with a frequency of 11.2 and 68.3 percent, respectively, in the present-day Danish population. In other words, there was nothing irregular about this fingerprint, so far one of the oldest found anywhere in the world.

The alimentary canal contained food remains from the man's last meal. The content of the intestine was approximately 610 ml, more than double the amount of food found in the Tollund man (270.5 ml). How far these volumes correspond to the amount of food eaten cannot be estimated. Food remains of the same type were evenly distributed throughout the entire digestive canal, from stomach to intestinal outlet, indicating that the Grauballe man must have eaten just before he was killed. The examination was conducted by Dr. Hans Helbaek (1959), who also examined the Tollund man. In contrast to the Tollund man, the Grauballe man had eaten

Figure 10.8.   Grauballe man's right hand. Forhistorisk Museum, Aarhus Kommunehospital.

animal matter as well as vegetable matter, for 15 tiny fragments of greatly dissolved bones were found. A bone specialist, Conservator Ulrik Møhl, was of the opinion that the bones might have come from the ribs of a small pig. Thus it can be shown that the Grauballe man had eaten meat along with the vegetable soup or porridge. The latter consisted of cultivated seeds as well as weed seeds, just like the meals the Tollund man and the Borremose man had consumed before death. The soup of the Grauballe man had consisted of barley (*Hordeum tetrastichum* Kcke., var. *nudum* and *H. tetrastichum* Kcke.), seeds of knotweed (*Polygonum lapathifolium* agg. and *P. persicaria* L.) and soft bromegrass (*Bromus mollis* L.), and small quantities of wheat (*Triticum dicoccum* Schuble) and oats (*Avena sativa* L.). In addition, there were seeds of over 50 varieties of weeds, some of them certainly gathered on purpose, but the rest probably harvested by chance along with the cultivated varieties. Two of the plants are, furthermore, found near the coast (from Grauballe to the coast is approximately 50 km); the presence of seeds of these two plants must indicate that either a trade in foodstuffs operated or the Grauballe man had been near the coast immediately before his death. In addition, two small stones and a small piece of charcoal were found in the intestines, indicating that he did not chew his food very thoroughly. Like the Tollund man, he suffered from intestinal worms (*Trichuris*).

As with the Tollund man, there was no trace of fresh fruit, vegetables, herbs, or berries, which one would expect at a time of year when they are available. Death, therefore,

Table 10.1. *Dates of eight Danish bog bodies determined by carbon-14 method*

Body name	Specimen number[a]	Date	Source of information
Grauballe man			Tauber (1956)
Sample with humic acid	K-503A	A.D. 310 ± 100 yr	
Sample with humic acid extracted	K-503B	55 B.C.	
Borremose 1946	K-2813	650 B.C. ± 80 yr	Thorvildsen (pers. comm. 1977)
Borremose 1947	K-1395	430 B.C. ± 100 yr	Thorvildsen (pers. comm. 1977)
Borremose 1948			
Sample with humic acid	K-2108A	610 B.C. ± 100 yr	Thorvildsen (pers. comm. 1977)
Sample with humic acid extracted	K-2108B	530 B.C. ± 100 yr	Thorvildsen (pers. comm. 1977)
Huldremose	K-1395	A.D. 30 ± 100 yr	Munksgård (1973)
Krogens Møllemose	K-2132	A.D. 80 ± 100 yr	Munksgård (1973)
Haraldskjaermose	K-2818	450 B.C. ± 80 yr	author's data
Tollund man			
Sample A	K-2814A	250 B.C. ± 55 yr	author's data
Sample B	K-2814B	180 B.C. ± 50 yr	author's data
Average		210 B.C. ± 40 yr	author's data

[a] Identification number assigned by the National Museum of Denmark, Carbon-14 Dating Laboratory.

probably occurred during either the winter or early spring in the case of both the Grauballe and Tollund men.

After this thorough examination it was, fortunately, decided to attempt to preserve the whole body. Dissection had shown that the acid bog water had acted as a tanning agent on the skin, and it was decided to complete the tanning nature had begun. The process used is technically known as pit tanning; it lasted 18 months. The body was placed in an oak tub, and the tanning solution was continually concentrated through the gradual addition of oak bark (875 kg was used). Following this treatment, the body was soaked in a solution of Turkey red oil and then dried. This process prevented any change taking place in either the appearance or the size of the Grauballe man.

CARBON-14 DATING OF DANISH BOG BODIES

As mentioned before only about 10 percent of the Danish bog bodies have been dated, almost all to the period from 500 B.C. to A.D. 400. Eight were dated by the carbon-14 method at the National Museum. Data on these bodies are given in Table 10.1.

CULTURAL CONDITIONS IN DENMARK AND NORTHERN GERMANY DURING THE EARLY IRON AGE, 500 B.C.–A.D. 400

The period of these dated bog bodies, in Danish archeological terminology, is pre-Roman Iron Age (earlier called Celtic Iron Age) and Roman Iron Age. Both periods are prehistoric.

The pre-Roman Iron Age (500 B.C.) is the first Iron Age period in Scandinavia and northwestern Germany and is simultaneous with the end of the Hallstatt culture in central Europe and the La Tène culture (also called Celtic culture). Although central Europe is dominated by these widespread geographic cultures, the influence of the Germanic peoples who lived in northern Europe does not seem to have been very marked. During the pre-Roman period, Denmark was a peasant community, with people and animals living under the same roof, each at its own end, in rectangular houses measuring about 15 by 5 m. The houses were sometimes isolated, but are also found in small villages. Some houses were larger, indicating a social division within the community. Cremation was the

usual funerary custom; the cleansed bones were placed in a clay urn, often in a graveyard. In bog burials, sacrificial objects were placed above the body – jewelry and clothing accessories, clay vessels with food, sometimes even, as at Dejbjerg and Rappendam, carts, and in one instance at Hjortespring, a war canoe, 13.5 m in length, filled with war gear.

Around the time of the birth of Christ, during the reign of Emperor Augustus, the Romans pushed the boundaries of the empire outward to the Danube and the Rhine, but the emperor's plan to extend the boundary farther north by subduing the Germanic tribes ended abruptly when the Roman commander Varus, in A.D. 9, was totally defeated by the Germanic chieftain, Arminius, only 300 km south of the Danish border. From then on the border remained along the Rhine and the Danube until A.D. 260, when the empire collapsed.

Extensive trade developed between the Roman Empire and the Germanic community, giving rise to the name Roman Iron Age (A.D. 0–400). This followed on the end of the pre-Roman Iron Age without any obvious cultural change. These trade connections were the source of many Roman accounts of the Germanic peoples, of which Tacitus' *Germania* (A.D. 98) was certainly the most significant, although he himself probably never visited Germany. These Roman sources are the subject of considerable dispute, for they usually seek not only to give a description of the Germans, but to picture the Romans and their deeds in a favorable light in contrast to the more primitive Teutons. Archeological material is much more reliable, consisting of grave finds, settlements, and sacrificial finds.

The reason people from the Iron Age were executed and then placed in a bog has been disputed at great length, and it is far from decided whether this was regular, routine punishment, perhaps followed by sacrifice to the gods in the bog, or a straightforward sacrifice. There is also the possibility that some of the bodies represent victims of murder or robbery.

The punishment theory has its origin in the work of Tacitus, for he says: "The coward, the unwarlike, the man stained with abominable vices, is plunged into the mire or the morass, with a hurdle put over him." The sacrifice theory also originates with Tacitus. In his description of rites connected with the worship of Ertha, or Mother Earth, he concludes with the words: "Afterwards the car, the vestments and the divinity herself are purified in a secret lake. Slaves perform the rite, who are instantly swallowed up by its waters."

These quotations do not explain why corpses were placed in bogs and lakes without the grave goods and food that other people had in normal graves, but they do raise the possibility that the reason may have been punishment or as a sacrifice to the gods. As the Danish and the northwestern German groups of bog bodies include men, women, and children, both whole bodies and parts, usually showing signs of violence (crushed limbs or skulls) over and above what would have been necessary for execution or sacrifice, it is difficult to imagine that these people were selected merely to be given to the gods. Perhaps they had, in one way or another, come into conflict with society and been executed.

More detailed examination of all the bog bodies found in Denmark and northern Germany should enable us to arrive at an explanation. The most important thing is exact dating. The carbon-14 technique can be used on even the most poorly preserved bodies, as long as it is chemically possible to remove humic acid and whatever has been used for preservation. The time of year the executions took place is also significant. The Tollund man, the Grauballe man, and possibly, the Borremose man 1946 were killed either during winter or in early spring; this implies a seasonal condition for the sacrifice. The papillae patterns on the feet of both the Tollund man and the Grauballe man and those on the hands of the Grauballe man indicate that these persons were not manual workers, but had a superior position in the community, perhaps chieftains or priests, possibly both.

However, lacking standards of comparison, it is impossible to reach a firm conclusion.

There is an interesting find from Rappendam on Sjaelland (Kunwald 1970). Carts or parts of them were offered as part of the sacrifice, along with a woman about 35 years old, though only her skeleton remains. The connection between carts and human sacrifice is mentioned in Tacitus' description of ceremonies for the fertility goddess Ertha, and Professor P. V. Glob believes that most of the Danish and northern German bog bodies are sacrifices to this goddess.

As well as giving a possible explanation of the legal and religious structure of society, the bog bodies offer an unusual chance of getting to know how people in the Iron Age lived. And the most fascinating aspect may be the discovery that an Iron Age person, if dressed in modern clothes, would look just like a person living today.

BOG BODIES ON EXHIBITION IN DENMARK AND NORTHWESTERN GERMANY

A list has been compiled by Dieck (1965). The largest collection is in the Landsmuseum, Schleswig, Germany, where the man from Rendswühren, who was found in 1871, is displayed. He was naked except for his left leg, which was covered by a piece of leather with the hair side in. His head was covered by a rectangular woolen blanket and a fur cap. The cause of death was a triangular hole in the forehead. He was preserved by smoking. At this museum also there are two bodies, a man and a woman, which were found in 1952 only 5 m apart. Both were dated to the period just after the birth of Christ, but it cannot be proved whether they were deposited at the same time. The woman, who died at about the age of 14, had been drowned. Her head and limbs are well preserved, but her chest is considerably dissolved. The hair had been light blonde; on one side of the head it was only 2 mm in length, but on the other side it was 4 to 5 cm. She was naked apart from a fur collar

and a cover over her eyes; this latter may possibly have been a cover for the mouth.

The man's skin and the hair on his head were preserved, but the bones were totally decalcified. The hair was presumed to be dark, going gray; its length was 2 to 2.5 cm. He is estimated to have been middle-aged. He was most probably choked by a hazel stick which lay around his neck. Both bodies were covered by thick branches and the girl with a rock as well.

The Gottorp Museum has the Damendorf man, but this body was washed out and only the skin and the hair remain. He was naked, covered by a cape and other apparel, including shoes. A 2-cm-wide split in the chest area tells us the cause of death. Also at this museum there is the Dätgen man and a single decapitated head; both have a characteristic hairstyle, with the long hair gathered in a knot. Tacitus mentions that the Germanic tribe, the Suevi, wore their hair this way.

The Tollund man is on exhibition at the Silkeborg Museum and the Grauballe man at the Museum of Prehistory in Aarhus; both museums are in Jylland. The National Museum in Copenhagen is planning an exhibition of the bodies from Borremose, but it is not yet known when the exhibition will be ready. There are other bodies in storage at the National Museum, but these were not preserved and as a result have shriveled up. They may, however, repay further study.

Since this chapter was written, another specimen has been recognized, after being in the National Museum since 1938. This is the Elling woman, found in the peat only 90 m from the Tollund man. She was about 30 years old and was dressed in a sheepskin cloak with the furry side next to the body. Without doubt, she had been hanged like the Tollund man. Carbon-14 dating placed her at 210 B.C. ± 70 years, which is almost the same as the Tollund man.

At an examination performed by Silkeborg Museum staff in 1976 it was found that the body had dried up into a mummy since its original discovery in 1938. No signs of dis-

ease were found, so the woman had apparently been quite healthy. Her hair was in good condition and was elaborately braided. Presumably she, like the Tollund man, had been sacrificed to the gods.

REFERENCES

Brandt, I. 1951. Planterester i et moselig fra Borremose. *Aarbøger for Nordisk Oldkyndighed og Historie* 1950:342–51. English summary.

Dieck, A. 1965. *Die europäischen Moorleichenfunde.* Neumünster: Karl Wachholtz Verlag.

Glob, P. V. 1956. Jernaldermanden fra Grauballe. *KUML* 1956:99–113. English summary.

– 1965. *Mosefolket.* Copenhagen: Gyldendal. Also published in English.

Helbaek, H. 1951. Tollundmandens sidste måltid. *Aarbøger for Nordisk Oldkyndighed og Historie* 1950:311–41. English summary.

– 1959. Grauballemandens sidste Måltid. *KUML* 1958:83–116. English summary.

Krebs, C., and Ratjen, E. 1956. Det radiologiske fund hos moseliget fra Grauballe. *KUML* 1956:138–50.

Kunwald, G. 1970. Der Moorfund in Rappendam auf Seeland. *Prähistorische Zeitschrift* 45: 42–88.

Lange-Kornbak, G. 1956. Konservering af en Oldtidsmand. *KUML* 1956:155–9. English summary.

Lund, A. A. 1976. Moselig. *Wormianium* 1976.

Munck, W. 1956. Patologisk-anatomisk og retsmedicinsk undersøgelse af Moseliget fra Grauballe. *KUML* 1956:131–7.

Munskgård, E. 1973. *Oldtidsdragter.* Copenhagen: Nationalmuseet. English summary.

Tacitus. *Germania.* In *Complete Works.* New York: Modern Library, 1942.

Tauber, H. 1956. Tidsfaestelse af Grauballemanden ved kulstof-14 maling. *KUML* 1956:160–3.

Thorvildsen, E. 1952. Menneskeofringer i Oldtiden. *KUML* 1952:32–48. English summary.

Thorvildsen, K. 1947. Moseliget fra Borremose i Himmerland. *Nationalmuseets Arbejdsmark* 1947:57–67.

– 1951. Moseliget fra Tollund. *Aarbøger for Nordisk Oldkyndighed og Historie.* 1950:302–9. English summary.

Vogelius Andersen, C. H. 1956. Forhistoriske Fingeraftryk. *KUML* 1956:151–4. English summary.

11

Mummification in Australia and Melanesia

GRAEME L. PRETTY
Senior Curator of Anthropology and Archaeology
South Australian Museum
Adelaide, Australia

ANGELA CALDER
Honorary Associate in Anthropology
South Australian Museum
Adelaide, Australia

Any description of mummification in Australia and Melanesia is made difficult by the irregularity of the evidence about funerary customs. There are two reasons for this. First, the accuracy of ethnographic reports has varied considerably, both spatially and temporally. Second, the variability and complexity of mortuary practices in these regions have continued to defy systematization. Mummification was considered to be only one component in a spectrum of cultural rituals associated with death and was practiced by relatively few tribal groups.

The most adequately documented area for this practice is the Torres Strait Islands. Ethnographic accounts for the entire region, however, have proved exceptionally valuable sources of information about the techniques involved and the diverse motives for the preservation of the dead. In Australia, study of the corpse was instrumental in determining the suspected agent of death; by contrast, Melanesian customs were concerned more with maintaining the physical integrity of the deceased. Similarly, preservation techniques ranged from simple procedures involving natural desiccation by solar processes to more complex methods of smoke drying corpses.

A further problem has been created by vagueness in the definition of *mummification* and the use of the word. Whereas accidental preservation of exposed bodies by desiccation occurred widely and randomly, purposeful preservation of bodies, or true mummification, was seldom recorded for Australia or Melanesia.

Despite these limitations, the evidence for mummification in this part of the world deserves serious attention. First, it refines our knowledge of mummification as a globally distributed custom. Second, this type of research also provides uncontaminated source material for biological assessments of indigenous populations in regions where hybridization has recently occurred.

DISTRIBUTION

In Aboriginal Australia a great range of mortuary rituals was employed in the disposal of the dead. The processes involved were either *simple* (primary disposal) or *compound* (secondary disposal) (Hiatt 1969:104). Simple disposal was characterized by the use of only one procedure, at a specific time, and was generally spread across the south of the continent. Compound disposal, of which mummification or desiccation was occasionally a part, was widespread in the northern part of Australia. Desiccated and partly decayed bodies have been randomly reported from widely scattered localities, as well as preserved parts of bodies such as hands and organs (Howitt 1904:459–460; Mathews 1905; Dawson 1928). Although smoking and drying techniques for preserving bodily parts were widespread, deliberate mummification was restricted to five main areas (Berndt and Berndt 1964:392–394; Figure 11.1). These were the lower Murray River and Adelaide Plains (Angas 1847a; Flower 1879; Taplin and Meyer, in Woods 1879:20–21; 198–200; Tolmer 1882:273; Stirling 1893; Basedow 1925; Elkin 1954:313); Gippsland

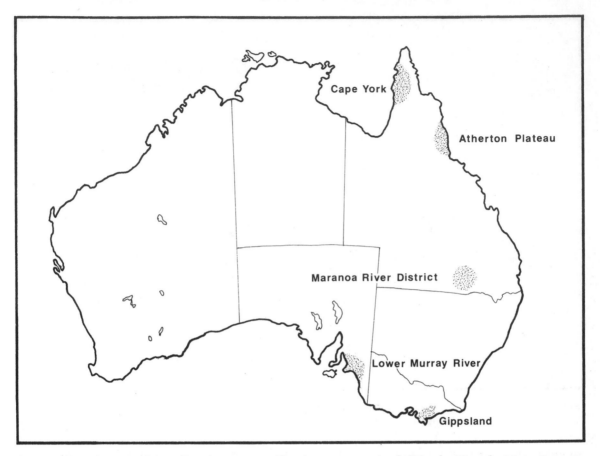

Figure 11.1.   Areas in Australia where mummification was practiced. (Map by Timothy Motz, Detroit Institute of Arts)

(Howitt 1904:459–460); the Maranoa district in southeastern Queensland (Howitt 1904:467–468; Hamlyn-Harris 1912b; Bull 1965); the Atherton Plateau near Cairns (Roth 1907:366–403; Hamlyn-Harris 1912b); and Cape York (McConnel 1937).

In Melanesia the preserving of fleshed and integumental parts and whole bodies appeared to be an extension of a widespread tradition of conserving relics of the dead as memorials. The best known example of this practice was the Melanesian penchant for preserving heads, usually as trophies. Deliberate mummification of whole bodies was, however, performed in several distinct localities (Pretty 1969; Figure 11.2). One area where the practice was common until very recent times was the Central Highlands in a

locality inhabited by the Kukukuku tribes (Simpson 1953:163–166; Bjerre, 1956:85–89; Pretty 1972:20–28). Numerous references exist to mummification in surrounding areas (Rhys 1947:148; Le Roux 1948:747–755) as far south as coastal Port Moresby, west as far as the Wahgi valley, and toward the north coast where the Buang tribes smoke dried bodies (Vial 1936:37; Girard 1957).

Another main center was in West New Guinea, where the custom extended along part of the southern shore of Geelvinck Bay and into the ranges east of Wissel Lakes (Held 1957:177–178; J. V. de Bruijn 1963; personal communication). Closer to Torres Strait, in an area enclosed by the tract of swampy, low-lying country that extends from the lower Fly River across into Irian Jaya, there were fre-

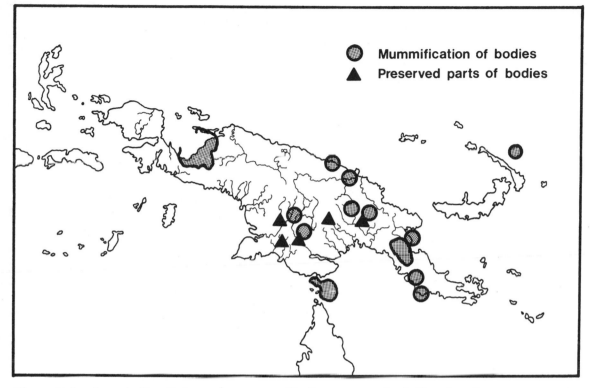

Figure 11.2.   Areas in New Guinea where mummification was practiced. (Based on Pretty 1969; map by Timothy Motz, Detroit Institute of Arts)

quent references to the preservation of sections of bodies and occasionally entire corpses (Haddon 1935:332–333).

A third important center for mummification was the Torres Strait (Pretty 1969; Figure 11.3). There the practice was confined to the eastern islands (Pretty 1969:35), which Haddon distinguished culturally from the western group (Haddon 1935:322). However, most reports have concerned Darnley Island (Flower 1879; d'Albertis 1880; Sengstake 1892) and the mummy collected by Macleay in particular (Hamlyn-Harris 1912a; Elliot Smith 1915; Dawson 1924; Fletcher 1929; Abbie 1959; Pretty 1969).

The distribution of mummification in Melanesia was not restricted exclusively to the aforementioned areas. As mummification was not an obligatory ritual for many tribes, its practice was sporadic. This behavioral flexibility was demonstrated at the 1963 Mount Hagen Agricultural Show, when a community from Laiagam, not known as practitioners of mummification, brought in attendance the preserved body of one of their fight leaders. While alive, this leader had expressed a desire to see the show, but died 2 years prematurely. His kinsmen had preserved his body and carried it to the show in accordance with his wishes (S. G. Moriarty, 1965; personal communication).

ORIGINS

Prehistorians consider that island Southeast Asia was the area from which Australia and

Figure 11.3.   Areas in the Torres Strait where mummification was practiced. (Based on Pretty 1969; map by Timothy Motz, Detroit Institute of Arts)

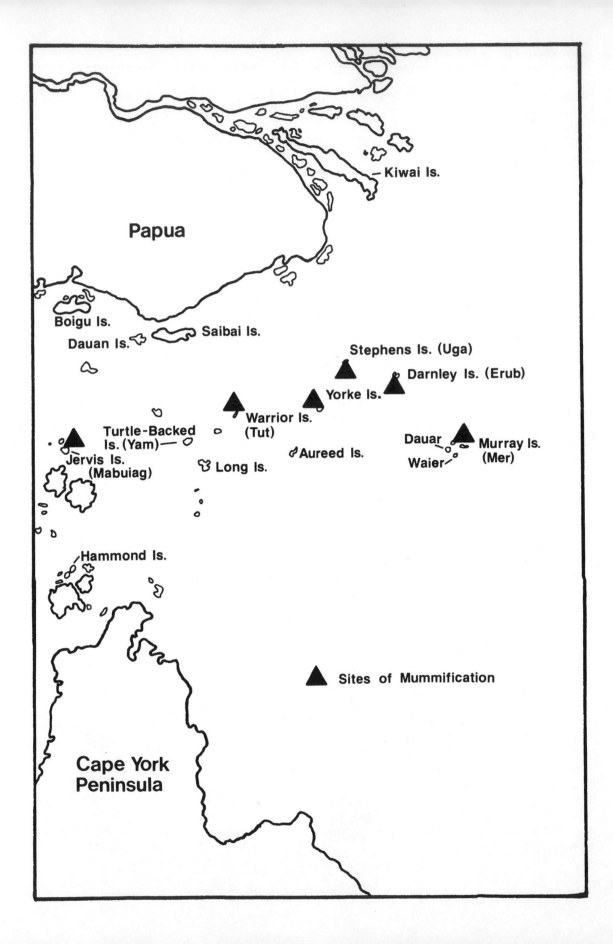

Papua

Kiwai Is.

Boigu Is.

Dauan Is.

Saibai Is.

Stephens Is. (Uga)

Darnley Is. (Erub)

Yorke Is.

Warrior Is. (Tut)

Turtle-Backed
Is. (Yam)

Jervis Is.
(Mabuiag)

Long Is.

Aureed Is.

Dauar

Waier

Murray Is.
(Mer)

Hammond Is.

▲ Sites of Mummification

Cape York
Peninsula

Melanesia were first settled (Mulvaney 1969). This event occurred some 40,000 years ago during the last glacial epoch and at a time when the lowered sea levels resulted in an extension of the main land masses and the linkage of Tasmania and New Guinea with the Australian continent. Archeological research has shown that since first settlement, Aboriginal societies have been adapting to changing local conditions and that much social and cultural diversity has ensued (Mulvaney and Golson 1971). Such variation has occurred within both Australia and New Guinea, the latter being subjected to a greater range of outside influences owing to its proximity to both Southeast Asia and the Pacific.

Unlike other regions, which also had extensive antiquities, Australia and Melanesia have no mummies of any great age. In this region, deliberate mummification was never intended to prolong the corpse's existence for more than a few years at most. Therefore, the identification of ancient mummies from prehistoric skeletal remains has proved extremely difficult.

One promising aspect, which may be linked with the origins of mummification but which has not yet been adequately examined, is the antiquity and distribution of compound funerary modes. All known specimens of historic mummies were from compound contexts. Compound burial practices in this region extended over a long period of time. However, to date there is no information to suggest that simple primary inhumations were as ancient as compound methods.

Prehistorians are presently unable to determine what funerary customs were practiced by the ancient people of Trinil, Modjokerto, and Ngandong in Java because the remains recovered have come from disturbed stratigraphical contexts. However, the finds from Niah (Howells 1973:177) and the interral of cremated ashes at Mungo (Bowler et al. 1970) were indicative of compound disposal modes. Careful excavations have also distinguished secondary from primary burial at Roonka in South Australia.

There, preliminary interpretations suggest that this practice may have been widespread throughout southern Australia during the period from 7000 to 4000 B.P. (Pretty 1977). The presence of compound disposal is indicative of a milieu in which mummification may have occurred, but extremely rigorous archeological observation is necessary before such conclusions can be definitively reached.

Information about burial customs became more detailed from the late sixteenth to midtwentieth centuries, during which period European explorers, traders, administrators, missionaries, and scholars made ethnographic comments about the region. None of these reports were sufficiently comprehensive to give insight into the origins of mummification, and few provided explanations for the distinctiveness of the practice in such widely scattered localities.

Taplin was one author who sought explanations for the origins of the practice by studying the Narrinyeri, a confederation of aboriginal tribes who lived at the mouth of the Murray River (Taplin, in Woods 1879). He explained the distinctiveness of mummification to this community by relating it to a unique mythic hero, Ngurundere, from whom the tribes claimed their descent. This custom and a sense of community cohesion were reinforced by tribal sanctions that held all other tribal groups as enemies. The Narrinyeri, then, represented a people who, in a region characterized by simple disposal methods, practiced a highly complex mode of funerary disposal that incorporated mummification (Angas 1847b). Other scholars of the Narrinyeri have used this evidence for mummification as a tool for interpreting the prehistoric settlement patterns of the lower Murray River (Stirling 1911; Pretty 1977).

SOCIAL AND RELIGIOUS BELIEFS

The stimulus for preserving whole bodies, parts of bodies, or human bones in Australia and Melanesia was generally attributable to one or more of the following sentiments:

1. A simple compensatory and restorative reaction to the loss of a relative or friend and the sense of grief this caused
2. The performance of a formally sanctioned duty to a certain class of person
3. The fulfillment of obligations to avert mischief by the spirits of the dead
4. The bearing of symbols of status; for example, the jawbone of her dead husband borne by a wife as a sign of widowhood

In the majority of instances where the motives for purposeful mummification were recorded, the custom accorded largely with the first of these sentiments – the playing out of a compensatory reaction to grief. Apparently, mummification in this region was neither a formal observance nor a mode of treatment conferred upon all members even of a certain class. The process and motives were limited by temporal considerations, for once the mourning period with its associated rites was over, the corpse was finally disposed of by procedures such as burial, cremation, or exposure to the elements.

This range of reasons for preserving the physical remains of the dead bore parallels to similar motives that governed several other aspects of mortuary practices, many of which were incorporated into mummification rites. Mourning observances often involved the demonstration of grief by self-inflicted injury or token cannibalism of the corpse. In some areas, inquests were held to divine the inflictor of death by a study of the deceased's entrails, resulting frequently in long-term vendettas.

This plethora of funerary behaviors was inevitable in small-scale societies whose religious systems provided no transcendental dimension to death. Death was a disaster that ranged in severity according to the social status of the deceased. Small children, women, and senile persons were generally of less importance than warriors or leaders. In such an emotional situation the feelings of grief and revenge felt by relatives often determined the funerary procedures. Mummification was an exceptional treatment conducted under special circumstances. It was an expensive process and placed considerable demands upon kinfolk.

The desperate and vindictive attitude to death of these societies and the consequent motives for mummification stood in marked contrast to other regions, such as ancient Egypt or Peru, where mummification was performed to celebrate beliefs about the afterlife. On the other hand, the actual mode of disposal corresponded to practices elsewhere, in that selection for mummification was based on the status of the deceased.

Although basic social and religious beliefs concerning the disposal of the dead were similar in Australia and Melanesia, there were distinguishing features in both regions. These differences were based more on a change of emphasis than on any single observable factor. In Australia the aim of mummification was to obtain an accommodation between the dead and the living; in Melanesia the aim was generally to prolong the physical presence of a dead relative in defiance of decomposition.

TECHNIQUES

In Australia the techniques used generally involved the exposure of the corpse in a tree or on a specially constructed platform (Figure 11.4). Initially the body was tied into a sitting position, often with limbs flexed against the chest or abdomen. The corpse was left in the open until the tissues were desiccated by the sun, and then deposited in the branches of a tree or on a wooden platform. In some localities, preservation was accelerated by sewing all the orifices closed and smoking the body while it was on the platform. Sometimes the body cavity was opened and the intestines removed in order to refine the drying process. Putrefying fluids were often collected and used in allied rites. When dry, the epidermis was peeled off and the pale corpse

Figure 11.4.   *Mode of preparing the bodies of the warriors slain in battle among the tribes of Lake Alexandrina. After a fight is over, the corpses of the young men who have been killed are set up cross-legged on a platform, with the faces toward the rising sun. The arms are extended by means of sticks; the head is fastened back; and all the apertures of the body are sewed up. The hair is plucked off, and the fat of the body, which had previously been taken out, is mixed with red ocher and rubbed all over the corpse. Fires are then kindled underneath the platform, and the friends and mourners take up their position around it, where they remain about 10 days, during the whole of which time the mourners are not allowed to speak; a guard is placed on each side of the corpse, whose duty it is to keep off the flies with bunches of emu feathers or small branches of trees. The weapons of the deceased are laid across his lap, and his limbs are painted in stripes of red and white and yellow. After the body has remained several weeks on the platform, it is taken down and buried; the skull becomes the drinking cup of the nearest relation. Bodies thus preserved have the appearance of*

mummies; there is no sign of decay; and the wild dogs will not meddle with them, though they devour all manner of carrion.

was smeared with red ocher and grease or decorated by painting totemic designs on its face and chest. Body hair was either cut off or pulled out and used for making waistbands and other personal ornaments (Berndt and Berndt 1964:392ff).

In Melanesia the methods employed varied slightly between the inland regions and the coastal and island areas. The most detailed accounts have come from Torres Strait (Haddon 1912; 1935; Hamlyn-Harris 1912a; MacFarlane, in Haddon 1935:325–326). In Torres Strait, the dead body was set apart for a few days. It was then taken out from shore in a canoe, where the swollen outermost epidermal layer was stripped away. Viscera were removed through an incision in the side between the ribs and the hips and then thrown into the sea. The abdominal cavity was filled with pieces of palm pith and the incision sewed up. The brain was removed after screwing an arrow through the back of the neck and into the foramen magnum. The body was then returned to the shore, lashed to a rectangular wooden framework, and hung up to dry behind a grass screen. Punctures were made at the knees and elbows and in the digital clefts on the hands and feet to drain off bodily fluids. The tongue, palmar tissue, and soles of the feet were stripped off and presented to the spouse.

Months later, when the drying process was complete, the mummy was decorated. Artificial eyes of shell with pupils of black beeswax were cemented into the orbits, the earlobes were decorated with tufts of grass and seeds, and the wrists and ankles were sheathed in bands of palm fronds. The whole body was then given a coating of red ocher. The loins were covered, in the case of a woman with a petticoat, and in the case of a man occasionally with a shell pubic ornament. The decorated mummy, on its frame, was then tied to the center post of the bereaved spouse's house and when, in the course of time, it fell to

pieces, the head only was retained.

On mainland New Guinea, the techniques of mummification differed in several details. For example, preservation of the body was obtained largely by smoking. In Torres Strait, drying in the open was the major agent. Although a fire was kept constantly burning near the mummifying corpse, this was considered to be more for the deceased's comfort than for preservation. Additional differences in technique were found in the Tauri–Lakekamu watershed area of central New Guinea. Evisceration was not practiced here, and the body and limbs were arranged in a squatting position, except during smoking, when the hands were tied to a house beam. The origins of the techniques of smoke drying are not definitely known, but the preservation of meat and game employing similar smoking processes was known by the Wahgi and Tauri (Kukukuku) tribes.

EXAMPLE OF A MELANESIAN MUMMY: TORRES STRAIT SPECIMEN IN THE MACLEAY MUSEUM, SYDNEY

The Macleay Museum in the University of Sydney contains the preserved body of a male Torres Strait islander (Figure 11.5). It was collected from Darnley Island in 1875 by the Australian zoologist Sir William Macleay. In 1914, during a session of the British Association for the Advancement of Science, it became a focus of attention when put forward by Professor Grafton Elliot Smith as a conclusive demonstration of ancient Egyptian influence on Oceanic culture. For various reasons, no comprehensive description of the mummy appeared until 1969, when Pretty published a detailed examination of the specimen and reviewed the controversy surrounding Elliot Smith's diffusionist hypothesis (Pretty 1969). The Macleay specimen is an excellent example of the Melanesian technique of mummification, and Pretty's report is the basis of the description presented below.

Figure 11.5. *Torres Strait mummy in the Macleay Museum, University of Sydney. (From Pretty 1969)*

General description.   The mummy is that of an adult male, 177 cm in length with a breadth at the shoulders of 35 cm. It was lashed to a wooden frame of two verticals and eight crosspieces at the head, under the armpits, and below the knees. The body must have been dried in a vertical position, as its weight, suspended from under the shoulders, caused sagging at the breast and the head to sink into a deep cleft formed by the clavicles, which raised the shoulders to a position level with the ears. Facial features were restored and the whole body coated with red ocher. The skin, where exposed, was parchment yellow except on the back of the left hand, where some of the original melanin, dark gray in color, was retained. The mummy's state of preservation was good, though there had been some postmortem insect attack around the jaw, mastoids, upper part of the breast, abdomen, and posterior surfaces of the legs. Loss of some of the distal phalanges of the left foot probably occurred during the mummification process, as the remaining bones were exposed and splashed with red ocher.

The head.   The head (Figure 11.6) was shaved and the crown painted an even black from the forehead around and behind the ears and across the lambdoid region, where the cranium lay against the topmost crosspiece of the frame. No hair was found, the only potential sample from the left cheek being identified as a vegetable fiber. Like the rest of the body, the face was thickly coated with red ocher and the features considerably restored. The shrunken orbits were filled with a black mastic that had discolored the overlying red ocher to gray and had cracked the painted surface along its junction with the facial tissue. The artificial lentoid eyes were shell, and measured 4 by 1.5 cm. Emphasis was added to each eye by setting a spot of black resinous material in the center to imitate the pupil and by ringing the margins with flattened strips of the same substance to represent eyelids. The nasal septum had been pierced in life, but the hole plugged postmortem with a rod of coral-line material broken off flush with the nostrils. The mouth had opened and the lips had parted, exposing the teeth. Both upper central incisors were displaced after death but were located radiographically, one being lodged in the posterior pharynx and the other in the chest.

An interesting feature of the head was the double strand of rolled two-ply twine that ran across the center of the oral cavity. This was inserted to fix the jaw firmly to the head and made demands on the skill of the mummifiers as the twine was passed between lower lip and teeth, over the teeth and across the oral cavity, and joined with the other end of the string at the base of the jaw. This technique was well known and is beautifully illustrated by a decorated skull (Figure 11.7) from Hammond Island, Torres Strait, which is housed in the South Australian Museum (registration number A.17907). In this specimen and in some other mummies, notably the one described by Flower (1879:391), there was a supplementary lashing binding the ramus to the zygoma on each side. This was unnecessary in the Macleay specimen because the tissues surrounding the jaw were well preserved.

Only part of the ears were present, but sufficient remained to show that the lobes were intact at death. From them hung threaded strings of seeds broken in half and threaded through the ends. They were identified as Job's tears (*Coix lacrymi*). The brain had been removed, and examination from behind showed a vertical incision at the nape of the neck. Probing demonstrated that the cranial cavity was empty, and radiographs showed that the articulation of the foramen magnum and the first cervical vertebra had been disrupted.

The techniques for keeping the head upright were complex. A vertical stance during drying would incline the head to fall forward on the chest, but this was prevented by two devices. First, a double strand of flat three-ply plaited twine was passed across the forehead again and across the other side, being finally

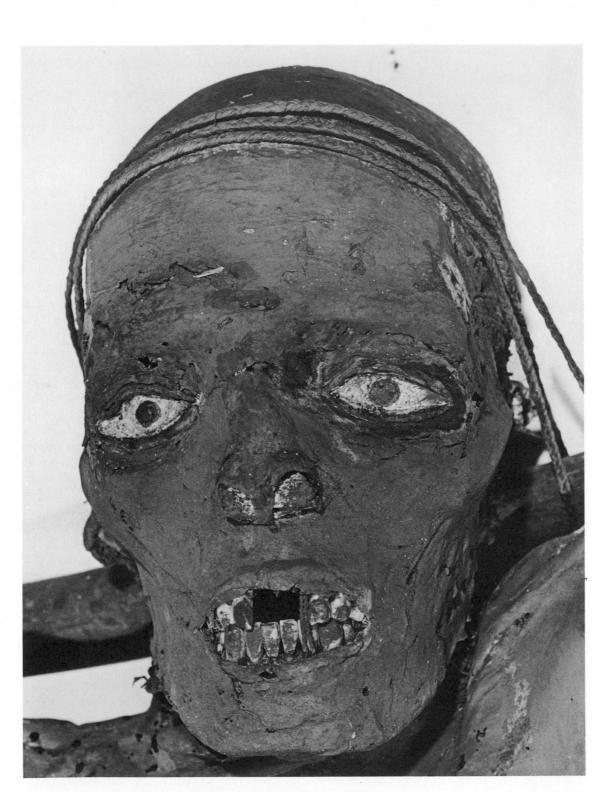

Figure 11.6.   *Head of the Macleay mummy. (From Pretty 1969)*

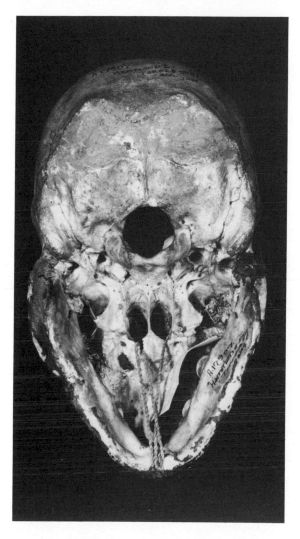

Figure 11.7.   Hammond Island cranium (A.17907)
showing attachment of mandible to skull by lashing. (From Pretty 1969)

colored gray about the region of the eyes. The
significance of the punctures remains prob-
lematical.

Thorax and abdomen.   Both thorax and abdomen were painted with red ocher more
thickly on the front than on the back. A curious feature of this region was the absence of
lateral shrinkage of the skin, leaving no protruding outlines of ribs either ventrally or dor-
sally. X rays confirmed that all ribs were present. The navel was discernible and penis and
scrotum complete. There was no pubic hair.

The abdomen was clothed with a dress
consisting of a fibrous material passed between the legs and kept in place by a
waistband. The waistband had a zigzag pattern woven into it and was wound twice
around the body and knotted at the front. The
tie-ends were of blue calico, a reminder of the
islanders' frequent contact with Europeans
prior to Macleay's visit. Tucked into the
waistband front and back were two lengths of
teased-out or beaten fibers identified as coming from the bark of a species of fig. Each
length had been doubled over, creased and
drawn together to form a knob, then stuffed
into a waistband behind, while the free ends
had been carried out across the abdomen and
tucked under the waistband in front. Behind
and within the bark fiber was a length of
twine, decorated at intervals of 6 to 8 cm with
fronded insertions of plant material from a
species of ginger. Darnley Island men normally went naked, but pubic coverings were
known for other male mummies (Flower
1879; Hamlyn-Harris 1912a).

On the left flank was an incision 8.5 cm
long, which had been sewed together with
rolled two-ply twine by a running suture and
finished by knotting at each end. Ethnographic accounts were explicit about an incision at this site for removal of the viscera and,
for the Macleay specimen, this was confirmed
by X rays that were consistent with complete
removal.

The limbs.   Close examination of the limbs
showed that the body had been punctured at
the joints to allow putrefying fluids to escape,

knotted under the crosspiece. Second, the
chin was probably supported by a small
wooden prop, as the skin was pinched and
depressed both beneath the chin and on the
breast just above the sternum. This latter
method was recorded by both Haddon (1912)
and Hamlyn-Harris (1912a), but lashing the
head to the top crosspiece was unusual.

Curious features were two punctures in the
skin of the forehead at each end of the temples. The surrounding tissues had been dis-

thus confirming the observations of Haddon and Hamlyn-Harris. Also as a result of drainage and drying, shrinkage had occurred, which necessitated retying the lashings at fewer sites. Primary lashing sites, distinguished from the furrows and wrinkling they had left on the skin, were situated in positions suited to the binding of a body in a vertical position. The hands had originally been lashed to the frame across the wrists, so that they lay with the palms flat against the frame rather than facing the sides of the body. The feet were originally lashed to the double crosspiece at the base of the frame, which also served to support the feet. Shrinkage through drying had pulled both feet a few centimeters above the basal crosspiece, but their former contact with the base of the frame was shown by the presence of strips of skin which had remained across the soles of the feet. Other primary lashing sites were located underneath the knees and slightly above present lashing sites.

Drainage was effected by making holes in the skin at the joints, and mourners further helped to expel putrefying fluid by stroking and kneading the limbs. In the Macleay specimen, drainage punctures were identified at the knees and on the back of the left hand where the skin was drawn back. Elsewhere they proved difficult to identify because of fractures and deterioration of the tissues. There were breaks in the skin of the hand between the first and second fingers and on the right foot between the hallux and the second toe. Another hole was located on the inside of the right arm at the elbow. Longitudinal wrinkling and folding of the skin at the joints, in particular at the knees, was suggestive of manual drainage.

Both fingernails and toenails were absent, consistent with ethnographic records of the removal of nails and palmar skin, which were peeled off and presented to the spouse.

The limbs were decorated at the ankles and wrists with bands of palm leaf, 4 cm wide. Each band consisted of two partly overlapping strips of leaf secured by wrapping and a small flat knot. These had been put on before the mummy was given its coating of red ocher, as the skin beneath was free of paint. *Radiographic and gastroscopic examinations.* An X-ray examination of the mummy unexpectedly showed a close scatter of granular opacities in the region of the lower abdomen. From a radiological viewpoint there was doubt about whether this represented the stuffing of the abdominal cavity substances or arose from radiopaque material in the coating of the red ocher itself. Weighed against this latter interpretation was the complete covering of the entire body with red ocher and the fact that the X-ray opacities were confined to the abdominal region. To resolve the problem, a series of tests was undertaken on samples of red ocher.

The first problem was to determine the composition of the paint that was smeared on the mummy. According to Hamlyn-Harris it was a mixture of coconut oil and ocher, but Haddon claimed the mixture to be of ocher and human grease. The first tests on the ocher were made from a mainland Australian sample and mixed with coconut oil and lard. This mixture was smeared onto tracing paper and X-rayed on a 10-cm thickness of Masonite board using an 80-milliamps-per-second ray with an intensity of 58 kilovolts from a distance of 80 cm. The plates were blank, showing the red ocher to be transradiant.

On specialist radiological advice the experiment was repeated, as the X-raying of a layer of red ocher on 10-cm Masonite slabs was held to be an unsatisfactory comparison with the ocher on the mummy. The contrasts achieved between a layer of ocher on 10-cm Masonite and 10 cm of Masonite alone were held to be so slight as to escape detection. A similar mass of balsa, being transradiant, was considered to be a truer comparison. In the repeat test, a sample of red ocher confirmably from Saibai Island in Torres Strait was used. Three mixtures were prepared, ranging from fine powdered to medium grained and coarse granules. The radiographs were at 63 kV, exposed at 2, 4, and 6 seconds from a distance of

127 cm. The plates showed the red ocher to be transradiant except where it was in compact grains or masses, thus supporting the hypothesis that radiopacity would depend on the size of particles and their specific gravity, but affirming the essential X-ray transparency of red ocher.

Reexamination of the coat of ocher covering the belly did not reveal any granular surface texture, but the paint had been thickly applied to this area, so the possibility that the ocher was affecting the radiographs still remained. To decide the issue and find out what the abdominal cavity contained, a gastroscope was inserted deep into the body from an insect hole near the right armpit. With it, a series of 5-mm color transparencies was made of the body cavity. Their interpretation was limited by uncertainties, first about the precise locus of the photos within the cavity, and second by the absence of a scale. However, the photos revealed that the abdominal cavity was empty of viscera and contained several lengths of a vinelike plant buried in a matrix of dusty grass and granular material. The lengths of vine stem appeared to be circular in section and occasionally had thorns or stems branching out from them. The matrix was mostly organic, but was assumed to be crystalline in places, as it reflected the light from the camera flash. Among the matrix were some curious smooth-textured ovoid balls of unknown material, some of which adhered to the body wall and some of which had broken away from it. Pock-marked depressions appeared in the interior body wall where these ovoid balls had broken away. The balls were not punctured and, as they were unlikely to have been insect cocoons, their composition remained doubtful. They may have resulted from the drying of fatty materials inside the body cavity in combination with the granular material with which the cavity had been stuffed. The gastroscopic examination demonstrated, without destroying any tissues, that the body had been emptied of viscera through an incision in the side and the cavity packed with lengths of a light vegetable stem and some earthy debris.

Paleopathology. The most striking radiographic finding was the irregular thickening of many of the long bones. Changes were observed in the left radius, left humerus, right clavicle, right ulna, both femora, both tibiae, and the right fibula. In general the changes consisted of an irregular thickening of the cortex involving the shafts of the long bones and containing irregular areas of erosion that were of both a sclerotic and a destructive type, as seen by the formation of cloacae. The ends of the long bones were normal in appearance, as were the joints. The changes in the right tibia and fibula were especially interesting in that the lesions, which extended over several centimeters, occurred at the same level in both bones. The left radius and tibia were severely affected, whereas the left ulna and fibula were spared. These osteological changes are consistent with the diagnosis of yaws, but no complementary skin lesions were detected on the mummy's external surface.

No evidence of any bony defects or fractures within the skull or facial bones was detected, nor was there any indication of degenerative arthritis in the skeleton as a whole.

EXAMPLES OF AUSTRALIAN MUMMIES
Although several Aboriginal mummies have been recorded, none has been described as thoroughly as the Macleay Museum specimen. Preserved bodies and organs of Australian Aborigines were collected from early Colonial times onward. That recovered from Adelaide by Sir George Grey and presented to the Royal College of Surgeons (London) in 1845 was one of the earliest known. Flower briefly described this specimen as an adult male bound in a sitting position (Flower 1879:393–394). The legs were flexed in line with the sides of the thorax and abdomen and the forearms were crossed, with each hand resting on the opposite foot. All orifices were

sewed with a running suture, but the viscera had not been removed. The body had been decorated with ocher and the mouth stuffed with emu feathers. Ironically, Flower directed the body to be unfleshed and the skeleton preserved so that the specimen would be more instructive! Unfortunately, even this has since been destroyed when the college was damaged during World War II.

A more detailed account of another South Australian mummy found at Rapid Bay near Cape Jervis was given by Tindale and Mountford (1936:487–502). The specimen was excavated from Kongarati Cave and was found in association with a wide range of molluscs, some wooden fire-making implements, a bone point, fragments of netting, and a kangaroo-skin cloak. The mummy was that of an elderly female, buried in a slate-lined cist grave in a flexed position. The right side had been oriented in a northerly direction. In view of the charred nature of the spine and hips and the presence of fatty matter on the right side, the authors concluded that the body had been smoke-dried. A proportionate amount of decomposition had ensued to parts of the right side, and the chest was distorted as if the lateral walls of the thorax had been crushed. This was explained by the local custom of forcing out the dying breaths of an individual by jumping on the thorax.

A third notable South Australian mummy was located in cliffs above the Murray River at Fromm's Landing (Sheard et al. 1927:173–176). The preserved corpse of an infant about 2 years of age was found lying on a net bag filled with long grasses. The body had been placed on its left side in a crouched position and was covered with a further layer of loose grasses and a wallaby hide. Hackett's examination of the body indicated that a depression on the right parietal was probably caused by a fracture, but this had not been sufficiently serious to cause death as there were osteological signs of healing. Both this specimen and that found in Kongarati Cave are housed in the South Australian Museum.

Figure 11.8.   Mummy of Ngatja, from the upper Russell River, North Queensland. (From Klaatsch 1907)

Other equally interesting mummies have been found elsewhere in Australia. Klaatsch (1907) has described his ruthless acquisition of the body of Ngatja (Figure 11.8); its present whereabouts are uncertain. An Aboriginal mummy from Morphett Vale is stated to be in the ethnographic collection of the Berlin Museum (Tindale 1974:55), and several specimens are housed at the Queensland Museum in Brisbane. The latter were all found in Queensland and have never been adequately examined, although Hamlyn-Harris gave a cursory account of them in his 1912 paper (1912b).

CONCLUSION

The nature of mummification techniques in New Guinea, Torres Strait, and Australia was such that the duration of a preserved body was only temporary. Concern expressed for a dead relative would not continue much beyond one or two generations, which eliminated the necessity of seeking more durable preserving methods. Consequently, as the practice of mummification lapsed with intensified European contacts, and with the introduction of Christianity in particular, the number of mummified specimens available to science declined. For this reason, scholars have had to rely more upon ethnographic and historical sources for details about funerary modes in Australia and Melanesia. One important advantage in such reports is their status as first-hand accounts of the rationale for mummification. The motives recorded and techniques described were as varied as the cultures in which mummification was practiced, a fact that runs contrary to the original diffusionist hypothesis proposed by Elliot Smith (Pretty 1969).

Ethnographic and historical records, however, are rarely sufficiently complete to enable prehistorians to reconstruct ancient mummification practices in detail, and further investigations of the few surviving specimens using modern scientific and medical acumen are required. More research about the disposal of the dead is especially necessary for clarifying traditional Australian Aboriginal thought.

It is clear that a great deal of uncertainty still remains about mummification in this region. There exists, moreover, important undescribed material whose study is long overdue and whose description will help dispel some of these obscurities. Similarly, the contribution of information from such specimens is of immense value to human biological and paleopathological research. The contribution of mummified material to research about indigenous peoples prior to recent hybridization will be extremely important. The authors have attempted to present a comprehensive survey of Australian and Melanesian mummification with a view to indicating appropriate directions for future research.

REFERENCES

Abbie, A. A. 1959. Sir Grafton Elliot Smith. *Bulletin of the Postgraduate Committee on Medicine, University of Sydney* 15(3):101–50.

Albertis, L. M. 1880. *New Guinea: what I did and what I saw there.* London: Sampson Low, Marston, Searle and Rivington.

Angas, G. F. 1847a. *Savage life and scenes in Australia and New Zealand.* London: Smith, Elder.

– 1847b. *South Australia illustrated.* London: Thomas McLean.

Basedow, H. 1925. *The Australian aboriginal.* Adelaide: F. W. Preece.

Berndt, R. M., and Berndt, C. H. 1964. *The world of the first Australians.* Sydney: Ure Smith.

Bjerre, J. 1956. *The last cannibals.* London: Michael Joseph.

Bowler, J. M.; Jones, R.; Allen, M.; and Thorne, A. G. 1970. Pleistocene human remains from Australia, a living site and human cremation from Lake Mungo, Western New South Wales. *World Archaeology* 2(1):39–60.

Bull, J. 1965. The sleepers in the ranges. *People* 16(11):12–13.

Dawson, W. R. 1924. A mummy from Torres Straits. *Annals of Archaeology and Anthropology* 11:87–96.

– 1928. Mummification in Australia and America. *Journal of the Royal Anthropological Institute* 58:115–38.

Elkin, A. P. 1954. *The Australian aborigines: how to understand them.* Sydney: Angus & Robertson.

Elliot Smith, G. 1915. On the significance of the geographical distribution of the practice of mummification: a study of the migrations of peoples and the spread of certain customs and beliefs. *Memoirs and Proceedings of the Manchester Literary and Philosophical Society* 59:1–143.

Fletcher, J.J. 1929. The society's heritage from the

Macleays: part 2. *Proceedings of the Linnaean Society of New South Wales* 54(3):185–272.

Flower, W. H. 1879. Illustrations of the mode of preserving the dead in Darnley Island and in South Australia. *Journal of the Anthropological Institute* 8:389–95.

Girard, F. 1957. Les peintures rupestres Buang, district de Morobé, Nouvelle Guinée. *Journal de la société des océanistes* 13:4–49.

Haddon, A. C. 1908. Sociology, magic and religion of the eastern islanders. In *Reports of the Cambridge Anthropological Expedition to Torres Straits*, vol. 6. Cambridge: Cambridge University Press.

– 1912. Arts and crafts. In *Reports of the Cambridge Anthropological Expedition to Torres Straits*, vol. 4. Cambridge: Cambridge University Press.

– 1935. General ethnography. In *Reports of the Cambridge Anthropological Expedition to Torres Straits*, vol. 1. Cambridge: Cambridge University Press.

Hamlyn-Harris, R. 1912a. Papuan mummification, as practised in the Torres Straits islands, and exemplified by specimens in the Queensland Museum collections. *Memoirs of the Queensland Museum* 1:1–6.

– 1912b. Mummification. *Memoirs of the Queensland Museum* 1:7–22.

Held, G. J. 1957. *The Papuas of Waropen*. The Hague: Nijhoff.

Hiatt, B. 1969. Cremation in Aboriginal Australia. *Mankind* 7(2):104–14.

Howells, W. 1973. *The Pacific Islanders*. Wellington: Reed.

Howitt, A. W. 1904. *The native tribes of South-East Australia*. London: Macmillan.

Klaatsch, H. 1907. Some notes on scientific travel amongst the black population of tropical Australia in 1904, 1905, 1906. *Report of the Eleventh Meeting of the Australasian Association for the Advancement of Science, 1907*, pp. 577–92. Adelaide: The Association.

Le Roux, C. C. F. M. 1948. *De bergpapoea's van Nieuw-Guinea en hun woongebied*. Leiden: Brill.

Mathews, R. H. 1905. *Ethnological notes on the Aboriginal tribes of New South Wales and Victoria*. Sydney: White.

McConnel, U. H. 1937. Mourning ritual among tribes of Cape York Peninsula. *Oceania* 7(3):346–71.

Mulvaney, D. J. 1969. *The prehistory of Australia*. London: Thames & Hudson.

Mulvaney, D. J., and Golson, J. (eds.). 1971. *Aboriginal man and environment in Australia*. Canberra: Australian National University Press.

Pretty, G. L. 1969. The Macleay Museum mummy from Torres Straits: a postscript to Elliot Smith and the diffusion controversy. *Man* 4(1):24–43.

– 1972. Report of an inspection of certain archaeological sites and field monuments in the territory of Papua and New Guinea. Typescript. South Australian Museum.

– 1977. The chronology of the Roonka Flat: a preliminary consideration. In *Stone tools as cultural markers: change, evolution, and complexity*, R. V. S. Wright (ed.), pp. 288–331. Canberra: Australian Institute of Aboriginal Studies.

Rhys, L. 1947. *Jungle pimpernel: the story of a district officer in central Netherlands New Guinea*. London: Hodder & Stoughton.

Roth, W. E. 1907. Burial ceremonies and disposal of the dead, North Queensland. *Records of the Australian Museum* 6(5):365–403.

Sengstake, F. 1892. Die Leichenbestattung aus Darnley Island. *Globus* 61(16):248–9.

Sheard, H. L.; Mountford, C. P.; and Hackett, C. J. 1927. An unusual disposal of an aboriginal child's remains from the Lower Murray, South Australia. *Transactions and Proceedings of the Royal Society of South Australia* 5:173–6.

Simpson, C. 1953. *Adam with arrows: inside New Guinea*. Sydney: Angus & Robertson.

Stirling, E. C. 1893. Report on inspection of Aboriginal mummy chambers at two localities on the Coorong, South Australia, 1893. Field notes in Department of Anthropology research file, South Australian Museum.

– 1911. Preliminary report on the discovery of native remains at Swanport, River Murray, with an inquiry into the alleged occurrence of a pandemic among the Australian aboriginals. *Transactions of the Royal Society of South Australia* 35:4–46.

Tindale, N. B. 1974. *Aboriginal tribes of Australia, their terrain, environmental controls, distribution, limits and proper names*. Berkeley: University of California Press.

Tindale, N. B., and Mountford, C. P. 1936. Results

of the excavations of Kongarati Cave, near Second Valley, South Australia. *Records of the South Australian Museum* 5(4): 487–502.

Tolmer, A. 1882. *Reminiscences of an adventurous and chequered career at home and at the Antipodes.* London: Sampson Low, Marston, Searle & Rivington.

Vial, L. G. 1936. Disposal of the dead among the Buang. *Oceania* 7(1):63–8.

Woods, J. D. 1879. *The native tribes of South Australia.* Adelaide: Wigg.

12

Japanese mummies

KIYOHIKO SAKURAI
Department of Archaeology
Waseda University
Tokyo, Japan

TAMOTSU OGATA
Department of Anatomy
Niigata University School of Medicine
Niigata, Japan

Translated by R FREEMAN
Waseda University, Tokyo, Japan

A RESEARCH AND CULTURAL HISTORY

KIYOHIKO SAKURAI

In a country of high humidity, such as Japan, the belief that mummification could not, and did not, exist would not be altogether unfounded, but rather more a matter of common sense. However, through our investigations we have been able to establish that mummification was in fact a very old custom in our country and one that was practiced right up until the early part of the twentieth century.

It is recorded that the great priest Kūkai (Kōbō-daishi, A.D. 774–835), who is famous not only for the studies of esoteric Buddhism he made while in China, but also for founding the Shingon sect of Japanese Buddhism, became mummified upon his death at Mount Kōya (a sacred mountain of this sect). In addition, during the eleventh and twelfth centuries there were many priests who voluntarily attempted self-mummification.

In existence today are the twelfth-century mummies of the Fujiwara family, a powerful clan of northeastern Japan, and, dating from the seventeenth to the twentieth century, the mummies of numerous priests. Apart from those of the Fujiwara family, all Japanese mummies are those of priests who voluntarily sought self-mummification.

The Japanese idea of mummification, practiced in accordance with Buddhist principles, was subject to a strong Chinese influence. In China, Buddhists had long been practicing mummification of the dead; these mummies were known as *nikustin* ("of the body or flesh"). The act of self-immolation in order to become a mummy was termed *nyūjō* ("entering into Nirvāna"). Priests who became mummies were given the title *nikushin-butsu* ("a Buddha of the body") or *nyūjō-butsu* ("a Buddha of Nirvāna") and were worshipped and respected in the same way as the Buddhist statuary.

In this regard, there was a great difference between the principles of mummification in China and Japan and those practiced in Egypt and South America.

This section describes mummies unique to Japan under the following headings: (1) mummies as recorded in literature, (2) existing mummies, and (3) the principles of mummification.

MUMMIES AS RECORDED IN LITERATURE
The number of mummies mentioned in Japanese literature is quite large. I should like here to introduce only a few examples.

In the *Genkō-shakusho* (completed in 1322) – a history of Japanese Buddhism and collection of priests' biographies covering a period of more than 700 years up to 1273 – it is recorded that in A.D. 1003 the priest Zōga attained *nyūjō* at the age of 87. In compliance with his will his body was placed in a large barrel, buried for 3 years, and then exhumed. At that time he was found to be in a state of perfect preservation.

This procedure of placing the body in a large barrel or earthenware urn and burying it for 3 years followed by exhumation was a

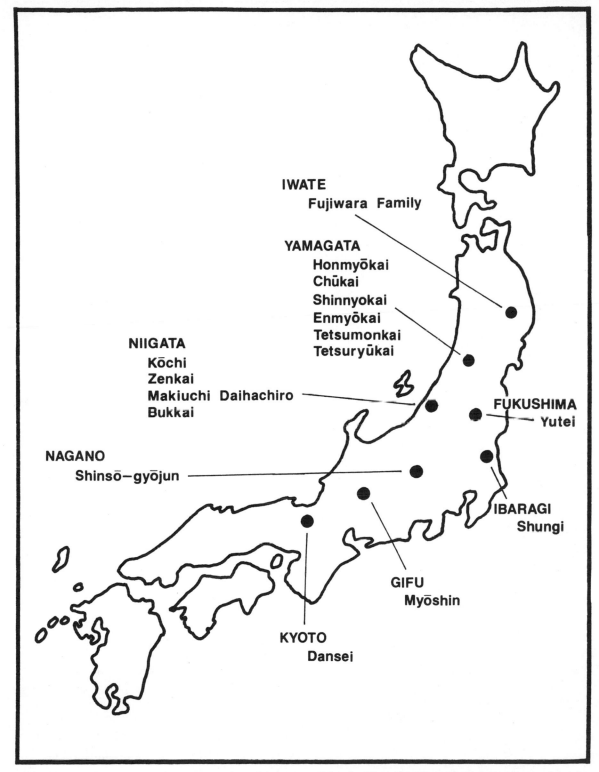

IWATE
Fujiwara Family

YAMAGATA
Honmyōkai
Chūkai
Shinnyokai
Enmyōkai
Tetsumonkai
Tetsuryūkai

NIIGATA
Kōchi
Zenkai
Makiuchi Daihachiro
Bukkai

FUKUSHIMA
Yutei

NAGANO
Shinsō–gyōjun

IBARAGI
Shungi

GIFU
Myōshin

KYOTO
Dansei

Figure 12.1. *Locations of mummies found in Japan. (Map by Timothy Motz, Detroit Institute of Arts)*

Table 12.1. *Mummies of the Fujiwara family*

Relationship	Name	Date of mummification	Age at death (yr)
Grandfather	Fujiwara Kiyohira	1128	73
Father	Fujiwara Motohira	1157	Unknown
Son	Fujiwara Hidehira	1187	66
Grandson	Fujiwara Yasuhira	1189	23

method of mummification that had been practiced in China from the fifth to sixth centuries up until modern times. The *Genkō-shakusho* indicates that the practice was brought to Japan during the eleventh century.

However, whether or not the mummy of priest Zōga ever became the object of worship cannot be ascertained.

The *Genkō-shakusho* also contains an account of the priest Rinken who attained *nyūjō* in 1150. It is recorded that upon mummification he was duly enshrined at Mount Kōya. Thus it can be seen that by the middle of the twelfth century the practice of worshipping mummies of those who attained *nyūjō* had become established in Japan.

EXISTING MUMMIES

In Japan there are 19 mummies in existence. Figure 12.1 shows their geographical distribution. The following paragraphs describe several of the mummies in detail.

Mummies of the Fujiwara family (preserved at the Chūson-ji Temple, Hiraizumi City, Iwate prefecture). The 4 mummies enshrined in the Konjiki-dō (the Golden Hall) of Chūson-ji Temple are listed in Table 12.1. Whether these 4 examples became mummified through natural processes or by embalming is difficult to judge, for their internal organs have been devoured by rats. Different opinions are held by different scholars. However, there is general agreement that some form of embalming must have been carried out.

The Fujiwara family, a powerful clan of northeast Japan, created in this remote region a culture comparable to that of the capital of those days. It was undoubtedly to ensure the permanent preservation of the remains of the leaders of this powerful family that mummification was carried out.

Mummy of priest Kōchi (preserved at the Saishō-ji Temple, Teradomari City, Niigata prefecture). Born in Shimo-osa (the modern Chiba prefecture), Kōchi became a priest at Renge-ji Temple in the village of Ōura, which is in his native district. Later he departed on a pilgrimage to various provinces. His journeys took him to the northernmost parts of Honshū before he traveled to, and settled at, the temple of Saishō-ji in Echigo (the modern Niigata prefecture). There he attained *nyūjō* in A.D. 1363. His mummy soon became an object of faith and is displayed annually on 2 October, when it is worshipped by numerous believers.

The head of this mummy is yellowish brown in color. Its soft tissues have dried and adhered, although those of the face have almost completely fallen away, exposing the bone. The skin and tissues covering the area from the head down to the small of the back have been well preserved, but there are large cavities made by rats in the stomach and chest. In literature of the nineteenth century, this mummy is recorded as being in a state of perfect preservation. It is believed that deterioration advanced rapidly following the Meiji restoration and the subsequent neglect of their traditional practices by the Japanese.

Mummy of priest Tetsumonkai (preserved at the Chūren-ji Temple, Asahi Village, Yamagata prefecture). This is a compara-

Figure 12.2.   Mummy of priest Tetsumonkai.

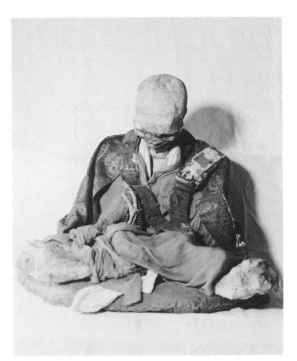

Figure 12.3.   Mummy of priest Tetsuryūkai.

tively well preserved example (Figure 12.2). Tetsumonkai was an ordinary day laborer who, in his youth, killed a samurai in a fight over a woman and subsequently fled to the sanctuary of Chūren-ji Temple on the slopes of Mount Yudono, a sacred mountain of the Shingon sect. Upon becoming a priest, he built roads and constructed bridges for the benefit of the local inhabitants and visited various provinces in order actively to engage in the propagation of Buddhism. In his final years he settled at Chūren-ji Temple, where, for a period of 3 years he abstained from the five cereals (rice, barley, corn, millet, and beans). Upon his attainment of *nyūjō* in 1829, priests surrounded his body with numerous large candles, drying it out by means of the candles' heat.

Mummy of priest Tetsuryūkai (preserved at the Nangaku-ji Temple, Tsuruoka City, Yamagata prefecture).   This person, a disciple of Tetsumonkai, attained *nyūjō* in 1868 (Figure 12.3). Although it was his wish to attain a state of mummification, he became sick and died during the course of his ascetics. Other priests buried him in an underground stone chamber (cist) beneath the temple, but

later exhumed and embalmed him so that, unlike other mummies, his internal organs have been removed. To do this an incision about 18 cm in length was made across his lower abdomen and then sewed up with linen thread. X-ray examination shows that the abdominal cavity had been packed with lime, indicating that this mummy owes its preservation entirely to embalming rather than to natural processes.

Mummy of priest Bukkai (preserved at the Kannon-ji Temple, Murakami City, Niigata prefecture).   This, the most recent of Japanese mummies, is that of a priest who practiced asceticism at Mount Yudono before attaining *nyūjō* in 1903. In accordance with his will, priests constructed an underground chamber and enshrined his body. After 3 years they were supposed to exhume and mummify it. But in Japan at that time exhumation became forbidden by law; so that Bukkai remained buried just as he was.

In 1961 we excavated his tomb. At about 1 m beneath the stone slab covering the grave,

Figure 12.4. Burial chamber of priest Bukkai. The ceiling stone and front wall have been removed.

there was a skillfully constructed chamber of hewn stone, measuring 1.25 m along the sides by 2 m in depth (Figure 12.4). Near the floor of this chamber was a shelf of iron bars upon which the strong wooden coffin had been placed. As the body had been in the earth for a long time, although parts of it had mummified, the remainder had become a skeleton.

PRINCIPLES OF MUMMIFICATION

With the exception of those of the Fujiwara family, all Japanese mummies are of priests who achieved mummification through their own volition, even though other priests had to assist the process of transformation. What was it that inspired them to undertake this? It was a principle that developed from the Maitreya faith – faith in Maitreya-bodhisattva, the Buddha of the future.

They believed that 5,670,000,000 years following Śākyamuni's (Buddha's) attainment of Nirvāna, Maitreya will appear in this world for the salvation of all sentient beings. As priests, they wanted to assist Maitreya when the time arrived. They believed that in order to do this, they should await his coming in their earthly form – that is, as mummies. This was the principle, and although there arose many legends (starting with Kūkai) of famous priests who attained mummification, quite a large number did in fact achieve this condition. For both the mummies described in literature and those that are in existence today, the principle behind their mummification was tied to the Maitreya faith.

Following the seventeenth century, as the Maitreya faith became combined with the Shūgendō or mountain asceticism of Sangaku-sūhai (a primitive form of mountain worship), many priests of the lower orders, living in the strict feudal society of the Tokugawa era, turned to mummification as a form of self-assertion.

CONCLUSION

In this section I have endeavored to explain something about Japanese mummies. The method by which mummification was achieved is not yet clearly understood. However, from records and from tradition, the following process may be deduced:

1. By gradually reducing the body's intake of nutrition over a long period, the body's constitution was altered to one that was strongly resistant to decomposition. Abstaining from the five cereals was for this purpose.
2. After death the body was interred for 3 years in an underground stone chamber (cist); it was then exhumed and dried.

As explained previously, the mummies of the Fujiwara family appear to have undergone some form of embalming. Where did knowledge of this art of mummification come from?

In the book Kitaezo-zusetsu (An Illustrated Book of North Ezo), compiled by the early

nineteenth-century explorer, Māmiya Rinzō, there is an account of a custom among the Ainu of Sakhalin of mummifying their chieftains. Also, Aleut mummies have been discovered on Kagamil Island in the Aleutian chain. Although the practice of mummification cannot be found among the Ainu of Hokkaidō, the fact that many of the mummies that exist today come from the northern part of Japan suggests that mummification as it developed in these northern regions is an influence that must be taken into consideration.

RESEARCH BY NATURAL SCIENCE
TAMOTSU OGATA

Books and articles on Japanese mummies have been published in great numbers (Andō 1961). We, the members of the Group for Research on Japanese Mummies, have been doing extensive research, primarily on the mummies of people who devoted themselves to Buddhism. The following discussion is restricted to mummies whose general examination was completed between 1959 and 1969.

To the author, the most interesting question is whether these remains were artificially mummified. The investigating commission has carried out a great deal of research on the famous mummified remains of four generations of the Fujiwara clan. Their efforts to decide whether the mummification was artificial or natural produced two opposing points of view (Furuhata 1950; Hasebe 1950; Suzuki 1950). The mummies we examined are quite different from those of the Fujiwaras, many of which show obvious traces of treatment. An investigation of these mummies therefore will suggest the solutions regarding the question of artificial mummification in Japan.

RESEARCH MATERIALS AND METHODS OF INVESTIGATION

The nine mummies discussed here are all Japanese males who devoted themselves to Buddhism. The dates of their deaths range from the period of the Northern and Southern dynasties to the Meiji era – that is, from 1363 to 1903. Their deaths were caused by starvation as a result of asceticism and by illness after asceticism. The remains show various stages of mummification, and one is almost a skeleton. Osteology, craniology, and somatology were applied according to the mummies' conditions. X-ray examination was also used on occasion. The reconstructed statures are calculated by Pearson's formulas. Fingerprint studies were done sometimes by using alginate impression material to make a plaster model and sometimes by the naked eye at autopsy. Blood groups were determined by an absorption test against anti-A, anti-B, and anti-O agglutinin serum, using skin and/or muscle tissue.

A lightweight, portable X-ray machine was used.

RESULTS

All mummies but one were considerably damaged by rats and insects.

Kōchi Hōin. Most of the remaining skin is well preserved, but almost no viscera remain, and there is no evidence that the brain and viscera had been removed. It is therefore impossible to decide whether artificial methods of mummification had been used. The mummy is in a sitting position. It weighs 4.7 kg and is 159.9 cm in height. Many of the teeth had fallen in his lifetime. It is reported that he died at the age of 82. X-ray examination shows the closure of the principal sutures of the skull is not complete. The bones of the lower limbs are more developed than those of the upper limbs. The vertebral column is highly kyphotic. Several sparrow-egg-sized shadows, the origins of which are unknown, are observed in the internal surface of the right parietal bone. Cervical spondylosis and osteophytes are seen on the anterior walls of the bodies of the fifth and sixth cervical vertebrae. Lumbar spondylosis with osteophytes is seen on the anterior walls of the bodies of the second and fifth lumbar vertebrae; the intervertebral disc spaces are narrow. The blood group is AB.

Shungi Shōnin. The skin is relatively well preserved and there are remains of hair. Although there are no viscera, there is no evi-

dence that the brain or viscera were removed. The mummy weighs 4.1 kg and is 157.2 cm in height. It looks rather like a turtle, because it bent forward when put into the cavity of a seated stone image of *Amida-Nyorai* (Amitabha Buddha). Its hands are clasped in prayer. The cervical vertebrae show luxation. The alveolar part of the mandible and the alveolar process of the maxillary bone are highly atrophied. All the teeth were lost in his lifetime. It is impossible to measure the orbit because both upper and lower eyelids are mummified in a closed state. X-ray examination shows that the closure of the principal sutures of the skull is moderately complete. The vertebral column is highly kyphotic. It is reported that Shungi Shōnin died at the age of 78. The bones of the upper and lower limbs are particularly well developed. The right thumb shows a radial loop fingerprint; the blood group is O. The bodies of the sixth and seventh cervical vertebrae show spondylosis with osteophytes.

Zenkai Shōnin (Figure 12.5). Most of the skin is well preserved, and the hair, penis, and scrotum remain. Most of the surface is brittle, and the color of the body is yellowish brown. Although there is no evidence that the brain or viscera were extracted, only the penis and scrotum can be found. Zenkai wears the same clothes and a pair of tabi as at the time of his death. His hands are clasped, and the wrists are tied together with string. A straight, deep furrow is seen from the right to the left lumbar region through the umbilical region at a right angle to the axis of the body. It seems to be a ligature mark that was made when the body was tied with string to a prop on its back to make it assume the desired posture for mummification, thus maintaining an exceedingly good sitting position. The soft parts may have decomposed at one time, but are well preserved. The lower legs are relatively short. The body weighs 7.0 kg and is 160.3 cm in height. The blood group is O. The upper and lower eyelids of both eyes are closed. In spite of its good posture, the vertebral column is highly kyphotic. Most of the molars of the maxillary bone fell off in his

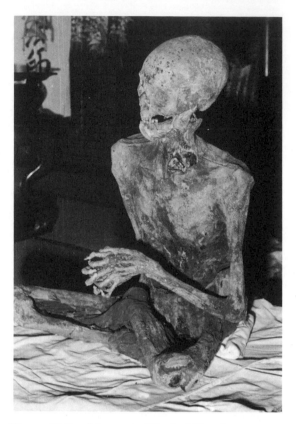

Figure 12.5.   Mummy of Zenkai Shōnin.

lifetime, and the bone has become highly atrophied. X-ray examination shows that the closure of the principal sutures of the skull is almost complete. It is reported that he died at the age of 85. X rays reveal many fairly large, oval shadows around the right internal surface of the occipital bone, the right orbit, both maxillary sinuses, and the vertebral column, but their origin is uncertain.

Chūkai Shōnin (Figure 12.6). The skin is well preserved, but the soft parts of the lower body have largely disintegrated. There is no evidence that the brain and viscera were extracted. The mummy is 159.1 cm in height, weighs 6.0 kg, and has blood group A. It is in a sitting position. Several ligature marks can be seen in the cervical region, on the right frontal chest wall, and from the right lumbar region to the left through the umbilical region at a right angle to the axis of the body. Old pieces of cloth are found attached to the im-

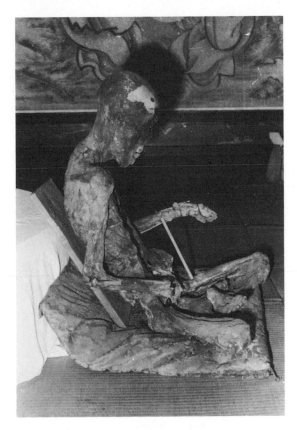

Figure 12.6.   Mummy of Chūkai Shōnin.

pressions caused by the rope. This suggests that Chūkai's clothed body had been tied with string to prevent it from falling forward in the mummification process. The upper and lower lids of both eyes are closed, and most of the soft parts under the headgear are decomposed. The facial part below the headgear is painted with something black (presumably Chinese ink), and most of the upper lip and some of the lower lip are shaped and mended with black substance. A board is tied to the back of the mummy to keep it in a good sitting position, and the left forearm is supported by a piece of wood to keep it forward and holding a bamboo stick. The mending of the face and the attempts to maintain the mummy's posture were performed after the second stage of treatment and will not, therefore, be discussed here. X rays show the closure of the principal sutures of the skull to be moderately

complete. Dental attrition is light, and the alveolar process of the maxillary bone and the alveolar part of the mandible are not much atrophied. According to the temple records, Chūkai died at the age of 58. The lower parts of the legs are relatively short, and the whole body is of slender frame. On X rays, spondylosis can be seen in the bodies of the fifth and sixth cervical vertebrae and the thoracic vertebrae. There are obvious osteophytes and osteosclerosis caused by osteoarthritis on the margin of the acetabulum. There is slight kyphosis of the vertebral column. The temple records say that Chūkai's corpse was smoke-dried, which probably explains its black brown color.

Shinnyokai Shōnin.   The skin is relatively well preserved but looks black brown, possibly because the body was smoke-dried during mummification. The skin surface is covered with innumerable white spots. The body weighs 6.0 kg and is 156.9 cm tall. The blood group is AB. The inferior aperture of the pelvis is wide open, and part of the diaphragm and urinary bladder remain, but there is no evidence that the brain and viscera were extracted. The mummy is in a crooked sitting posture, almost falling backward. The vertebral column is remarkably kyphotic, and a deep furrow is seen from the right lumbar region to the left through the umbilical region at a right angle to the axis of the body. A shallow furrow is also seen in the fifth intercostal space of the right side of the chest wall parallel to the ribs. These furrows may have been formed when the body was tied to a prop on its back at the time it was mummified. X rays show the closure of the principal sutures of the skull to be moderately complete. All the teeth had dropped out in his lifetime. The alveolar process of the maxillary bone and the alveolar part of the mandible are atrophied. Shinnyokai is reported to have died at the age of 96. The skeleton is delicate. The lower parts of the legs are relatively short. X rays show five oval shadows on the left scapula, but their origin is uncertain. Spondylosis can be seen in the lower thoracic vertebrae, and

Figure 12.7.   Mummy of Enmyōkai Shōnin.

the body of the twelfth thoracic vertebra has been deformed into a wedge shape. In front of it, a new bone with a sharp margin is observed. It is not known whether the new bone was formed as the result of a compression fracture during his lifetime or because of spondylosis. There is a postmortem fracture in the second lumbar vertebra. The bodies of the third, fourth, and especially the fifth lumbar vertebrae have been deformed into a wedge shape. The acetabulum protrudes slightly into the pelvis.

Enmyōkai Shōnin (Figures 12.7 and 12.8). The soft parts are brittle but thicker than those of the other mummies. This body has also kept its original shape better than the others, though the lower part is not well preserved. The skin is atrophied, and there is no sign that the brain and viscera were extracted. The area from the right cheek to the upper and lower lips has been mended, and the surface is painted black brown, as is the rest of the body. The upper and lower limbs were painted black brown (presumably by persimmon tannin) after being tied with twine in several places at a right angle to the axes. The parts beneath the twine are not painted and look light black brown. As the body had possibly been smoke-dried first, it was undoubtedly painted during the second treatment. X rays show the closure of the principal sutures of the skull to be almost complete. There is slight dental attrition. Enmyōkai is said to

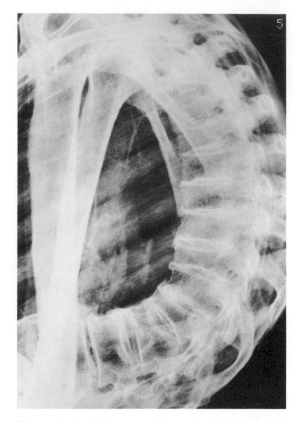

Figure 12.8.   Mummy of Enmyōkai Shōnin.

have died at the age of 55. The vertebral column is highly kyphotic and has the shape of a bow. The body is 164.7 cm tall and weighs 6.8 kg. X-ray examination shows spondylosis in the thoracic and lumbar vertebrae. Both the twelfth thoracic vertebra and the first lumbar vetebra, constituting the apex of kyphosis, have deformed into a wedge shape. In the hip joints, osteoarthritis is seen, but there is no sign of osteoporosis. Because of its corpulence it is believed that the body became mummified after it had decomposed.

Tetsumonkai Shōnin. This mummy has suffered serious damage. The remaining skin is black brown, and hair also remains. There is no evidence that the brain and viscera were extracted. Part of the diaphragm and the left lung remain. Temple records report that the body was smoke-dried using big candles immediately after death. This suggests that the

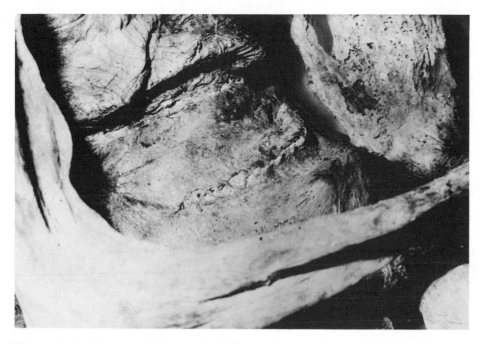

Figure 12.9.   Mummy of Tetsuryūkai Shōnin, showing the incision, extending from the right to the left iliac regions through the hypogastric region, with a continuous suture applied.

color of the skin may possibly be attributable to soot. A distinctive feature of this mummy is that the soft part of the face, from the frontal to the mandible part, has disintegrated and seems to have been mended with something black. It was painted over with yellow paint, then probably with black Japanese lacquer. Dental attrition is fairly advanced, and in X rays, the closure of the principal sutures of the skull is nearly complete. It is said that Tetsumonkai died at the age of 62. Long symmetrical ligature marks can be seen on both sides of the chest. They are like those of a cord used to tuck up kimono sleeves. They were probably left when the upper part of his body was tied to a prop on his back to make it assume a sitting position during the first stage of mummification. It is said that Tetsumonkai extracted his left eyeball and cut off his external genital organs himself. A soft part, which may have been his scrotum, was preserved separately. Tetsumonkai's blood group is B, and this was also the blood group of the scrotum. It may be assumed, therefore,

that this scrotum belonged to him. The mummy is 162.1 cm tall and weighs 5.0 kg. The bones of the lower limbs are well developed. The vertebral column, especially the lower lumbar vertebrae, is highly kyphotic. The left thumbprint, and possibly the right also, show an ulnar loop. In X rays, many oval shadows, origin unknown, are observed in the nasal cavity and in the right and left maxillary sinuses. Spondylosis can be seen from the eighth to the eleventh thoracic vertebrae. In the body of the twelfth thoracic vertebra, a compression fracture, possibly posthumous, can be clearly seen.

Tetsuryūkai Shōnin (Figure 12.9). This mummy is very well preserved because it was treated by a special method. The skin, especially of the face and head, is black because the body was dried over fire. A curved incision approximately 18.0 cm long runs from the right to the left iliac region through the hypogastric region. The thoracic, abdominal, and pelvic cavities are filled with lime powder and some has spilt from the inferior aper-

ture of the pelvis. The incision is continuously sutured. The forearms are relatively long. The body, 159.2 cm tall, is in a sitting position. The vertebral column is highly kyphotic. Slight dental attrition can be seen in X rays; the closure of the principal sutures of the skull is almost complete. It is said that Tetsuryūkai died at the age of 62. The mummy weighs 15.0 kg, heavier than the other mummies, but this is because of the weight of the lime powder. The left thumbprint possibly shows an ulnar loop; the index finger a whorl; and the middle finger possibly a whorl. The right thumbprint possibly shows an arch and the middle finger possibly an ulnar loop. The blood group is A. Osteoporosis is observed in the vertebrae. Osteophytes are found on the upper margins of each acetabulum, and osteoarthritis is also seen. Periostitis can be seen on the medial side of the upper part of the left femur. There is slight osteoarthritis in the left patellofemoral joint. The cortexes of the medial sides of both tibial shafts have become rough. It is uncertain whether this is attributable to the posttraumatic periosteal reaction.

Bukkai Shōnin. Immediately after his death in 1903, Bukkai was put in a wooden coffin and was placed in a stone room with special devices; we excavated his body in 1961. Many bones were separated at the joints. The soft parts had decomposed and were attached to the bones like dirt, but some skin of his back was mummified. There is no evidence that the brain and viscera were extracted. The body was probably in a sitting position when it was placed in the coffin, but it was not so at the time of our excavation. Four teeth were left on the mandible. The alveolar part of the mandible was atrophied, and the roots of the teeth were exposed. X rays show the closure of the principal sutures of the skull to be moderately complete. Bukkai's age at death is said to be 76. The body is 158.2 cm in height and weighs 7.2 kg. The lower leg bones are sturdy. Linea aspera of the femur is well developed; the tibia is nearly platycnemic; and the fibula is thick. These facts suggest that

Bukkai made ample use of his lower limbs. In X rays all the vertebrae, from the second cervical vertebra down to the fourth lumbar vertebrae, show various deformations attributable to ankylosing spondylitis, which results in a bamboo spine. An abnormal finding is that the right fifth rib and left fourth rib are bifurcated where they touch the cartilage. In addition, bulging or spinous cortical thickening is seen in the interosseous crest of the right radius and ulna and the head of the left fibula.

LIVING ORGANISMS FOUND INSIDE AND OUTSIDE THE BODIES

All mummies except Bukkai, who was buried in the ground, had been damaged by rats, although the extent of the damage varies. In particular, the viscera were eaten, and the destruction was accelerated by rat excreta. Flies generally got into the bodies during the period of decomposition with moisture. They laid eggs in the remains either at the time of death or during mummification. The season of death and the condition of the mummies when they were placed on the ground can be deduced from the pupal sloughs of the flies that remain in the bodies. It should be noted that the lunar calendar was used before the fifth year of the Meiji era (1872) and the solar calendar after that. Among those who were mummified on the ground and in the cold season, flies were observed in Zenkai and Kōchi, although the records say that the priests' deaths occurred in January and October, when such flies as *Lucilia, Sarcophaga,* and *Calliphora* were not likely to be active. It seems, therefore, that either the viscera were kept damp until spring or the dates of death were recorded incorrectly. It is recorded that Shungi died on 15 February, was placed in a hermetic place after 17 days, and then was taken out. However, the finding of pupal sloughs of *Sarcophaga* suggests that this body had been placed in a spot where there were many flies or that it was taken out of the hermetic place with wet viscera. It is recorded that Chūkai died on 21 May and was

taken from the ground after having been buried. Pupal sloughs of *Lucilia* found in the remains indicate that the viscera were still moist either when he died or when he was taken out from the ground. Pupal sloughs of *Fannia canicularis* adhered to Shinnyokai (the date of his death is uncertain) and those of *Lucilia* to Enmyōkai, who died on 8 May. This implies that these priests' viscera were still damp when their bodies were placed on the ground. If the record of the temple of Enmyōkai is correct, the priest's viscera must have been in a condition that allowed flies to adhere to them immediately after the priest died. As these examples show, pupal sloughs of flies can be a valid and important clue in determining whether the dates of death shown in records correspond to dates suggested by the condition of the body, as well as in understanding the whole process of mummification.

DISCUSSION

The mummies discussed in this section were not natural mummies, nor were they found by accident. Their mummified condition can be seen with the naked eye. They were made and enshrined in temples by people who wanted to preserve them and who believed in the religious ideas of Buddhism. It can be said, therefore, that all were mummified intentionally. Unlike Egyptian mummies, the viscera do not seem to have been extracted, except for the mummy of Tetsuryūkai in 1879 or 1880 in the Meiji era. This was treated by making an incision, filling the cavity of the body with lime powder, and applying a continuous suture. Most of the viscera of the seven mummies other than those of Tetsuryūkai and Bukkai were eaten by rats, but some viscera obviously remain in two of them. There is no evidence that brains were extracted.

The most important stage of mummification is the first one, because it is then that the dead body is made into a mummy. In many cases, mummies were treated several times, but the later forms of treatment are not dis-

cussed here. According to the records, some bodies were smoke-dried in the first treatment, and indeed, black soot could be observed on their skins. The most notable feature of these mummies is their posture, which is related to the tenets of Buddhism. Many were mummified with the lower and upper eyelids closed. The position of the lower limb is different in each mummy. In some, the joints were broken and the bones separated, so that the original position remains uncertain, though it was probably a sitting position. In many mummies, the heads were bent forward and in some their palms were pressed together in prayer, as in the case of Zenkai, whose wrist joints were tied with string. However, the original forms were destroyed as time passed, and the upper limbs are now in various attitudes.

Unlike the Egyptian method, neither oil nor resin was applied during the first stage of mummification, although, in Tetsuryūkai's case, filling the body cavity with lime powder helped good preservation. Modern medical techniques had presumably been introduced by the time this work was performed. People capable of applying a continuous suture must have been exceedingly skillful.

Fingerprints could be investigated in only three cases. This had nothing to do with the period of the mummies, but was the result of damage by rats and insects and of corrosion. Blood groups were determined for all except Enmyōkai and Bukkai. Blood grouping was not possible in Enmyōkai's case because of decomposition and in Bukkai's case because of corrosion of almost all the soft parts.

The mummies are enshrined as objects of religious worship at their respective temples, which were far removed from cities. It was therefore necessary to take X rays with a lightweight, portable machine. In addition, in the case of the mummies with soft parts remaining on the bones, special devices were used to emphasize the bones. Senile spondylosis was found in almost all. Bukkai exhibited ankylosing spondylitis. Some remains had osteoarthritis in the hips and

knees. Some had periostitis or periosteal thickness of the cortex. In some cases, fracture of the vertebral bodies was reported, but this may have been a posthumous modification. Many oval shadows were found in the skulls and on the scapulas and the vertebral column, but the origin of these is uncertain.

As is obvious from this discussion, these nine mummies were in the process of rapid deterioration. From the very beginning of the examination, the author, together with Sōjirō Maruyama, has aimed at achieving a better state of preservation for the mummies. Preventive measures against microorganisms, insects, and rats were taken, and decomposing parts were tied together with twine and fixed with glue.

CONCLUSION

This investigation of nine Japanese mummies dating from medieval times (1363) to the present (1903) has shown that many were mummified by artificial means.

ACKNOWLEDGMENTS

The cultural history presented in the second part of this chapter is derived from the following sources: Mummies in Japan (1961) by the late Professor Dr. Kōsei Andō, the previous head of our group, and "Research by Cultural History" by the late Professor Andō and Professor Kiyohiko Sakurai in Research of Japanese Mummies.

As reported in the second section of this chapter, its author took part in anatomical and anthropological studies of mummies. The following staff of Niigata University School of Medicine contributed information in their professional fields: the late Professor Emeritus Dr. Shungo Yamanouchi and Professor Dr. Rokurō Shigeno for medicolegal studies, Professor Emeritus Dr. Shūei Nozaki for review of X rays of mummies, and Professor Emeritus Dr. Sachū Kōno for observations on alterations of bones seen in X rays. Together with Dr. Ryūhei Homma, the author dealt with living organisms found in and around mummies.

Thanks are due every professor and doctor for his help. The author wishes to thank Mr. Akira Matsumoto, lecturer of Nihon University, for his help during this investigation. The author would like to express his gratitude also to the following people for their help: the staff of Shōnai Hospital, Tsuruoka City, Yamagata prefecture, for radiographic studies; Professor Dr. Susumu Saitō, Mr. Yoshinobu Ikegami, and the late Dr. Hiromasa Muraki for identifying species of living organisms; and Mr. Sōjirō Maruyama for helping mend the mummies.

Gratitude must also be expressed for the encouragement and assistance of the priests of the respective temples.

REFERENCES

Andō, K. 1961. Mummies in Japan. Tokyo: Mainichi Newspapers. In Japanese.

– 1969. Preparation of Nyūjō mummies. In Research of Japanese mummies, Group for Research of Japanese Mummies (ed.), pp. 83–95. Tokyo: Heibonsha. In Japanese.

Andō, K., and Sakurai, K. 1969. Research by cultural history: Extant Japanese mummies. In Research of Japanese mummies, Group for Research of Japanese Mummies (ed.), pp. 21–82. Tokyo: Heibonsha. In Japanese.

Furuhata, T. 1950. Blood groups, fingerprints and teeth of four generations of the Fujiwara clan. In The Chūsonji Temple and four generations of the Fujiwara clan, pp. 45–66. Tokyo: Asahi Shimbun. In Japanese.

Hasebe, K. 1950. Various questions on the remains. In The Chūsonji Temple and four generations of the Fujiwara clan, pp. 7–22. Tokyo: Asahi Shimbun. In Japanese.

Suzuki, H. 1950. Anthropological observations on the remains. In The Chūsonji Temple and four generations of the Fujiwara clan, pp. 23–44. Tokyo: Asahi Shimbun. In Japanese.

13

Miscellaneous mummies

ANTONIO ASCENZI
Professor of Morbid Anatomy
University of Rome
Rome, Italy

AIDAN COCKBURN
President, Paleopathology Association
Detroit, Michigan, U.S.A.

EKKEHARD KLEISS
Professor Emeritus
Department of Embryology
Universidad de los Andes
Mérida, Venezuela

THE INFANT MUMMY OF UAN MUHUGGIAG

ANTONIO ASCENZI

The mountain range called Tadrart Acacus, situated in southern Libya (Fezzan), extends to the east of the Ghat oasis and covers a surface of about 6,000 km². A large area of it was explored by Mori during the winter of 1958–1959 on the occasion of his fourth paleoethnological expedition (Mori 1960). Almost at the end of the mission, in the Tagzelt Valley a deposit was discovered under a natural shelter called Uan Muhuggiag. Like the other shelters in the zone, the walls and ceiling are covered by more than 100 rock paintings, indicating that the site was actively used during the prehistoric era. The shelter faces south and is situated at the base of a tall rock wall that borders the southeastern end of the Teshuinat's wadi. At its entrance, the shelter is about 40 m long by 3.5 m high. Its depth does not exceed 4 m.

The height is not uniform, but progressively decreases to 90 cm at the level of the inner wall. In contrast with the other decorated shelters in the same zone, the floor here is a mixed deposit of sand and randomly distributed ashes. The surface of the floor is situated 1 m above the bed of the wadi and is separated from it by bulky rocks that have facilitated the settlement of the deposit (Figure 13.1).

In order to examine the structure of the deposit, an excavation was made measuring 160 by 80 cm and running east and west. In this way it was established that the deposit reached a depth of about 1 m and rested on sandstone flags. On the western side of the excavation were found fragments of long bones from animals and of bone tools. The eastern side of the excavation revealed a clear stratification caused by alternating layers of coals, ashes, and fibrous matter. Under the lowest layer of coals, the sandstone floor showed an intentional circular excavation 25 cm in diameter but only 3 cm deep. Here a spherical object was found, completely masked by a layer of randomly distributed vegetable fibers, measuring approximately 25 cm in diameter. When the vegetable fibers were removed, the mummy of a child appeared, almost completely wrapped in an envelope of animal skin and bearing a necklace of little rings made from the shells of ostrich eggs (Arkell et al. 1961).

Dating by the carbon-14 method was carried out by Professor E. Tongiorgi at the University of Pisa using two different types of samples: the lowest coal layer of the deposit and the envelope of animal skin. The first sample was 7,438 ± 220 years old; the second 5,405 ± 180 years old.

As far as the soft tissues are concerned, the child was in a good state of preservation, very much like that usually found in mummies obtained by drying. The body was in an unusual position, with an extreme flexion of the trunk and a forced rotation of the head to the right side. The upper right arm was extended and adduced somewhat posteriorly, that is, behind the trunk. The right forearm was flexed. The legs were in a squatting position near the head, but it was not possible to deduce their exact position because a partial dislocation had occurred (Figure 13.2).

Figure 13.1.   The shelter of Uan Muhuggiag. (Courtesy of F. Mori)

A careful examination of the abdominal walls revealed a long incision in the anterior wall, apparently for the purpose of removing the thoracic and abdominal viscera, of which no vestige could be seen. This agrees with the conclusion that the corpse's extremely flexed position could not have been assumed unless the trunk had previously been eviscerated. Between the edges of the abdominal incision lay a very irregular and bent cone 10 cm long. Its maximum circumference reached 6 cm. It was arranged in such a way that only its apex protruded from the incision. The structure consisted of a light and porous conglomerate of black mold mixed with granules, probably vegetable seeds.

The region around the external genital organs was badly preserved, so that the sex remained undetermined.

Anthropological research consisted of descriptive morphological, anthropometric, radiological, histological, and chemical examination. By these means it was possible to deduce that the child had negroid characteristics and that at the time of death it was about 30 months old. The skull showed the following main features: pentagonoid form with sharp occipital heel, protuberant forehead of infantile type, dolichocephaly near to mesocephaly, camecephaly, and tapeinocephaly. The face revealed an obvious prognathism, with a Camper's angle of 70°.

A careful examination of soft tissues and a radiological investigation of the skeleton failed to reveal any change that could be interpreted as responsible for death.

Samples of soft tissues prepared from the scalp and the trunk were hydrated, fixed in formol, embedded in paraffin, and sectioned for examination under the optical microscope. Hydration in saline solution showed that imbibition was similar to that occurring

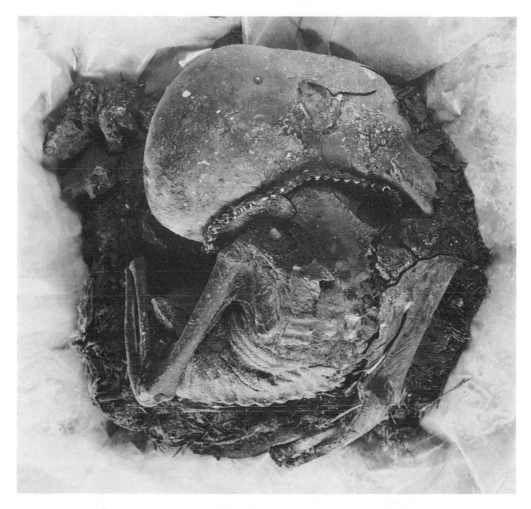

Figure 13.2.   The infant mummy of Uan Muhuggiag. (Courtesy of F. Mori)

when air-dried tissues receive this treatment. Microscopic examination revealed that tissue structure was not well preserved. The tissue was in no way different from that of a Peruvian mummy chosen as a control, in which mummification was carried out by drying process. There is evidence, therefore, to support the view that in the child of Uan Muhuggiag mummification was obtained by drying after removing thoracic and abdominal viscera.

This conclusion emphasizes that the spontaneous preservation of bodies in warm and dry environments probably played a major role in suggesting to ancient populations the

ways in which artificial mummification could be obtained. This was also the view of Pettigrew (1834), De Morgan (1896), and Elliot Smith and Dawson (1924) regarding the origin of mummification practices among the ancient Egyptians.

THE FROZEN SCYTHIANS OF SIBERIA
AIDAN COCKBURN

The Scythians were a nomadic pastoral people living in southern Russia north of the Black Sea. They formed part of a group of similar peoples inhabiting the vast, treeless

Figure 13.3.   Burial sites of the Scythians. (Map by Timothy Motz, Detroit Institute of Arts)

territories that extend from Europe to China north of Greece, Persia, and India, and include Siberia (Figure 13.3). The term *Scythian* is used to cover the cultures of all these peoples. The key feature of all these cultures was the horse.

Herodotus visited Olbia in the course of his travels and wrote an account of the Scythians. His work is the basis of half the information in this section. The Scythian form of mummification is described by Herodotus, but few of these bodies have survived. The account is fairly explicit, assuming that the process used for preserving the king's body was also followed for the common people in preparation for the 40 days of exhibition.

The tombs of their kings are in the land of the Gerrhi, who dwell at the point where the Borysthenes is first navigable. Here, when the king dies, they dig a grave, which is square in shape, and of great size. When it is ready, they take the king's corpse, and having opened the belly, and cleaned out the inside, fill the cavity with a preparation of chopped cypress, frankincense, parsley-seed, and anise-seed, after which they sew up the opening, enclose the body in wax, and, placing it on a waggon, carry it about through all the different tribes. On this procession each tribe, when it receives the corpse, imitates the example which is first set by the Royal Scythians; every man chops off a piece of his ear, crops his hair close, makes a cut all round his arm, lacerates his forehead and his nose, and thrusts an arrow through his left hand. Then they who have the care of the corpse carry it with them to another of the tribes which are under the Scythian rule, followed by those whom they first visited. On completing the circuit of all the tribes under their sway, they find themselves in the country of the Gerrhi, who are the most remote of all, and so they come to the tombs of the kings. There the body of the dead king is laid in the grave prepared for it, stretched upon a mattress; spears are fixed in the ground on either side of the corpse, and beams stretched across above it to form a roof, which is covered with a thatching of twigs. In the open space around the body of the king they bury

one of his concubines, first killing her by strangling, and also his cup-bearer, his cook, his groom, his lackey, his messenger, some of his horses, firstlings of all his other possessions, and some golden cups; for they use neither silver nor brass. After this they set to work, and raise a vast mound above the grave, all of them vying with each other and seeking to make it as tall as possible.

When a year is gone by, further ceremonies take place. Fifty of the best of the late king's attendants are taken, all native Scythians – for, as bought slaves are unknown in the country, the Scythian kings choose any of their subjects that they like, to wait on them – fifty of these are taken and strangled, with fifty of the most beautiful horses. When they are dead, their bowels are taken out, and the cavity cleaned, filled full of chaff, and straightway sewn up again. This done, a number of posts are driven into the ground, in sets of two pairs each, and on every pair half the felly of a wheel is placed archwise; then strong stakes are run lengthways through the bodies of the horses from tail to neck, and they are mounted upon the fellies, so that the felly in front supports the shoulders of the horse, while that behind sustains the belly and quarters, the legs dangling in midair; each horse is furnished with a bit and bridle, which latter is stretched out in front of the horse, and fastened to a peg. The fifty strangled youths are then mounted severally on the fifty horses. To effect this, a second stake is passed through their bodies along the course of the spine to the neck; the lower end of which projects from the body, and is fixed into a socket, made in the stake that runs lengthwise down the horse. The fifty riders are thus ranged in a circle round the tomb, and so left.

Such, then, is the mode in which the kings are buried: as for the people, when any one dies, his nearest of kin lay him upon a waggon and take him round to all his friends in succession: each receives them in turn and entertains them with a banquet, whereat the dead man is served with a portion of all that is set before the others; this is done for forty days, at the end of which time the burial takes place. [Herodotus, *Persian Wars*, Book IV:71–73]

In the 1860s, the archeologist V. V. Radlov excavated two large burial tombs in the Altai Mountains and found, among other specimens, some fur garments in excellent condition. In 1927 a Soviet archeologist started to investigate a stone mound near Shiba in the Altai Mountains. The finding of metal, wood, and bone stimulated further work on similar mounds in the region.

In 1929, S. I. Rudenko opened a mound in the Pazyryk Valley, and this was so rewarding that he continued with other mounds, for another four decades, except during World War II. The reason these graves were so important lies in the fact that they had partially filled with ice, which preserved soft materials like flesh, skins, and clothes.

A typical tomb has been described by Artamonov (1965). The builders first dug a rectangular pit about 5 m deep and 8 m wide. Inside this, they built a double-walled framework of wood, filling in the space between the walls with soil or rock. At one end, a space outside this framework was left open to receive the horses that were sacrificed in the funeral rites. The body of the king and his wife or concubine were placed in the inner chamber in a coffin made from a hollow tree trunk, together with a variety of household goods and treasures. The whole was then roofed in with beams and bark and covered with soil. The tomb was finished by laying a stone platform about 40 to 50 m in diameter. The completed tomb rose 4 to 5 m above ground level.

Water soaked through into the tomb, or possibly moisture condensed from the air, and in those cold conditions froze to form layers of ice over the contents. The tombs had been robbed, probably by the Turks who invaded the Altai Mountains after the third century B.C. The bodies inside had been mutilated in a search for ornaments; arms and legs had sometimes been amputated, and in one instance, the heads of both a man and a woman had been cut off. In spite of this, the bodies were in a good state of preservation, with skin and hair still intact (Figures 13.4 through 13.6). During the excavations, the ice was removed by pouring in hot water and then removing it with the thawed-out ice.

The bodies had been embalmed by a process not unlike that of the Egyptians, except

Figure 13.4.   *Mummy of a leader of the fifth Pazarokovsky cavalry, fifth to fourth century B.C. Excavated by S. I. Rudenko, 1949. (Courtesy of Imperial Ermitage of Lenin, Ministry of Culture, USSR)*

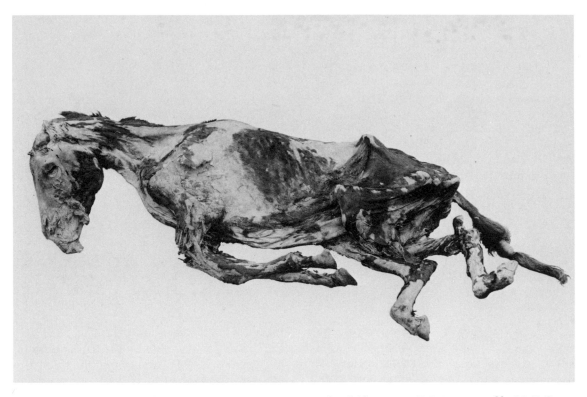

Figure 13.5.   *Mummified horse of the first Posiranov cavalry, fifth century B.C. Excavated by M. P. Groznov. (Courtesy of Imperial Ermitage of Lenin, Ministry of Culture, USSR)*

that natron was not used. The brains, internal organs, and sections of muscle had been removed, and the cavities stuffed with grass or hair to maintain the shape of the body. The skin was then sewed up with thread made of hair or tendon. The heads were wholly or partially shaved, but before burial hair had been attached artificially to the women and beards to the men.

The horses buried with the king were not touched by the robbers and were recovered in excellent condition.

Most of our information comes from the work of Rudenko in five barrows of the Scyth-

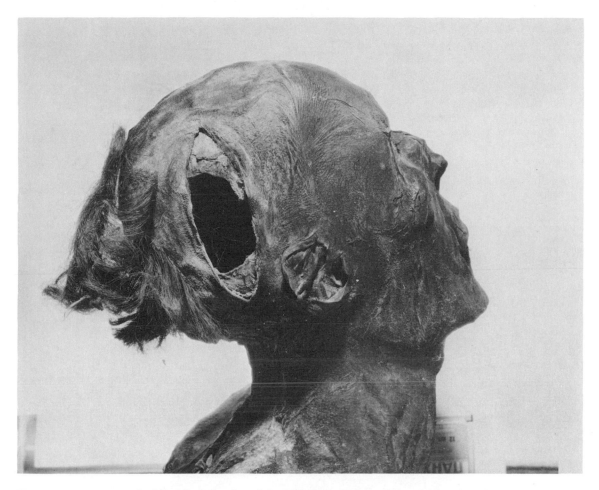

Figure 13.6.   *Mummified head of a leader of the fifth Posiransky cavalry, fifth to fourth century B.C. Excavated by S. I. Rudenko, 1949. (Courtesy of Imperial Ermitage of Lenin, Ministry of Culture, USSR)*

ian period in the Altai Mountains. The Pazyryk barrows were looted in antiquity, but enough specimens remained to provide an adequate picture of the life and death of the chieftains buried there. The work has been published in English (1970) and is the source for the following account.

Burials took place in spring, in early summer, or later in autumn. This required the bodies to be preserved until burial could take place, but apparently only the chiefs and nobles were embalmed. Whether the common people were kept for burial at special times of the year is not known.

The three key features of full mummifica-

tion were (1) removal of the entrails, (2) slitting of the limbs, sometimes with excision of the muscles, and (3) trephination of the skull and removal of the brain. To remove the entrails, an incision was made in the abdominal wall from the ribs down to the iliac bone. After the organs had been removed, the incision was sewed up with sinews, except that the woman in one barrow had been sewed with black horsehair cord. The treatment of the muscles varied from one body to another, but in general the slits in the skin of the limbs followed a common pattern. The cuts were made on the inner aspects of the arms and legs and ran deeply down the full extent of

the limbs to the hands or feet. Presumably some preservative had been introduced, possibly salt, but this has not been identified. In other parts, such as the buttocks or shoulders, punctures with the point of a knife had been made, also probably for the introduction of a preservative. In some of the bodies, the muscles had been removed and replaced by padding, such as horsehair. This leads Rudenko to ask if these could have been examples of ritual eating of the body after death. Herodotus reports this custom as being quite common and says that it was habitual among certain tribes to kill off their old people and eat part of the flesh. These events were happy occasions with rejoicing; unfortunate persons who died of illness were buried without being eaten and were lamented by relatives and friends because they did not live long enough to be killed.

Removal of the brain was a common practice. In barrow 2 of the Pazyryk barrows, the skin over the parietal bone in the man had been cut away and pulled back, then crude tools such as chisel and mallet had been used to knock out an irregular disc of bone. The brain had then been removed and the cavity filled with soil, pine needles, and larch cones. The plate of bone that had been struck out had been replaced, and the skin sewed back with a cord of twisted horsehair.

The custom of burying a horse with its rider was very ancient and persisted into recent times in one form or another. The Turks maintained it into the first millennium A.D., and in the east Altai it persisted among the Telesi up to the end of the nineteenth century. The horses found in tombs in the Altai were presumably the favorite steeds of the buried princes and kings, although some kings had numerous horses buried with them.

GUANCHE MUMMIES FROM THE CANARY ISLANDS
AIDAN COCKBURN

The Canary Islands were invaded in 1402 and subsequently colonized by the Spaniards. They were occupied by an indigenous people called the Guanche, of unknown origin, who practiced mummification for their aristocracy. At the time of the occupation, there were probably thousands of mummies in ancestral caves, but nearly all have been lost and today only a few are left, scattered in museums around the world.

The Canary Islands are volcanic and rise from the ocean floor in very deep water. There is no possibility that they were at any time linked to West Africa, from which they are separated at the closest point by 107 km of ocean. This, of necessity, implies that the original immigrants had boats big enough to sail the open sea. The voyages of vessels like *Kon Tiki* have shown that such journeys can indeed be made. Irish monks with frail craft made of skin discovered Iceland and may even have reached America, so it does not seem unlikely that groups of peoples passed from Africa to these islands during an early period of man's history. At the time of discovery, the islanders did have boats, but these were rare and communications between the islands must have been infrequent. Tenerife itself, being 12,000 ft high, can be seen from a long distance.

Whatever method was used, the fact is that the crossing was made not only once but probably a number of times. It seems obvious that several races colonized the island, for the dominant groups were white with fair hair; others lower on the social scale were much darker, with black or brown hair. Their language was a mixture of Berber, Arabic, and other tongues. Who they were is a mystery. There have been numerous speculations, starting with Cro-Magnon man, but most are only guesses. While in West Africa, I heard legends of a king who heard of land to the west and sailed away with his tribe to look for it. This was told to me in connection with America, but the Canaries seem much more likely.

Hooton (1925) gave as his opinion that the first settlers probably came from Africa south of Morocco during the Neolithic period,

bringing with them sheep and goats, but no knowledge of cultivated cereals. They survived in a relatively unmixed form on the island of Hierro. The invaders probably came from the Anti-Atlas and Atlas regions and penetrated only the southern islands. They were probably brunet whites, whose sole cultivated cereal was barley. About the same time another race of tall blond whites arrived. They were very warlike and probably came from the Atlas ranges of Morocco and Algeria. A fourth invasion affecting the eastern islands has been named "Mediterranean" and came when bronze was already in use in the eastern Mediterranean. These people probably spoke Berber.

Almost certainly, the islands were visited by the Phoenicians, for they established colonies along the West African coast; Hanno sailed that way in his famous voyage from Carthage, and others seem to have made the complete trip around Africa about 600 B.C. Other casual visitors in recent times would have been Arabs and Berbers.

BURIAL PRACTICES

The common people of the Canaries were buried either singly or in groups in simple trenches or holes, covered with dry stones. These tombs are indistinguishable from those of the Sahara.

The leaders and aristocrats were mummified and placed in mortuary caves. In 1526 Thomas Nichols was shown a cave in Guimar containing 300 to 400 mummies (quoted by Hooton 1925), and the Guanches said at that time that there were more than 20 such caves on Tenerife, but only 4 or 5 of these have been found. A famous cave between Arico and Guimar was explored in 1770 and contained 1,000 mummies.

These caves were plundered, partly to supply an important ingredient in medical preparations popular in Europe at that time, and partly by the modern natives, who destroyed them. Very few of these mummies survive today. There are no known complete ones in the islands themselves.

The embalming processes in Tenerife were recorded by two early visitors to the island, Espinosa and Abreu Galindo (see Hooton 1925). There was a professional class of morticians who were social outcasts, just as in Egypt. The bowels of the body were removed and the body washed twice a day. Espinosa says that the embalmers forced down the mouth a concoction made of mutton grease and powders of various sources, and Galindo records that the body was anointed with sheep butter, then sprinkled with a powder of dust of decayed pine trees and pumice stone. The incision in the abdomen was semilunar below the ribs, and sometimes all the viscera were taken out. The brain was removed from the head by means that are not clear, and the cavity was filled with a mixture of sand, ground pine bark, and juice of mocan.

There is a good deal of doubt on many points, especially the means of removing the brain, but it seems generally agreed that drying the body in the sun took 15 days. One observer says the body was smoked at night. After this time, the corpse was sewed up and wrapped in leather from sheep specially chosen for the purpose. Many other skins were piled on top, some being wrapped round the body.

Some of the mummies were enclosed in coffins made to fit them and constructed from tree trunks hollowed out for the purpose; others were arranged in rows simply in the skins in which they had been wrapped. The bodies were then placed in inaccessible mortuary caves. A few of the remaining Guanche mummies are in the Museum of Archaeology and Ethnology, Cambridge, England, and one was made available for examination to Brothwell et al. (1969). Their report is a detailed one and should be consulted for specific items. Briefly, it can be said that there was a remarkable similarity in the process of embalming to that practiced in Egypt. This has been known for a long time, but the present study confirms it. The position of the body, arms, and hands, and the presence of

subcutaneous packing for the tissues are characteristic of mummies of the Twenty-first Dynasty of Egypt. However, the skull had been crushed, so that details regarding either the condition of the brain or its absence were not available, nor were any packages of organs discovered, either inside or outside the body.

The mummy had been radiographed by Scales in 1927, and he had reported evidence of rickets. In the later study doubt was cast on this diagnosis, and an alternative one of osteoarthritic disease of a degenerative type was substituted. The more recent films failed to demonstrate any opaque material in the skull apart from fragments of bone.

The tissues were not as well preserved as in Egyptian mummies, but were good enough to reveal certain features. Among these was anthracosis of the lungs, an indicator of exposure to smoky fires. Air pollution existed on the Canary Islands long before modern civilization arrived there.

THE MARQUISE OF TAI
AIDAN COCKBURN

Of all the mummies known to the authors, this seems to be by far the best preserved. We have not seen it in the flesh, but an hour-long 16-mm movie of the autopsy in color was lent to us by the Chinese government, and this left no doubt about the incredibly good condition of the tissues. The elasticity of the skin was such that when a finger was pressed against it, the skin rebounded to erase the depression formed, and the joints could be flexed without apparent damage. No other mummy described in the literature has similar capacities (China Reconstructs 1973; Group for Research on the Han Cadaver of Mawangtui 1976).

The lady is said to be the wife of Litsang, the marquis of Tai, who was chancellor of the principality of Changsha, Hunan, in the early Western Han dynasty, about 2,100 years ago. When she died, it is said that she was buried according to the Chinese *Book of Rites*.

There are many tombs in the Changsha area of Hunan, most of which were looted by robbers in the 1930s and 1940s. However, two huge mounds remained untouched until the present government began the task of excavation. It was a formidable undertaking, but after 4 months of digging the workers finally reached the tomb at the bottom. It was worth the effort, for there was a series of coffins, one inside the other, the whole covered with 5 tons of charcoal to soak up any water that might penetrate the thick layer of clay surrounding the coffins. The gorgeously decorated coffins lay inside each other, until finally, in the center, was the dead woman.

A description of all the art treasures found does not belong here. Some can be seen in the *National Geographic Magazine* of May 1974. It is the body that interests us. This had been immersed in a solution containing some mercury salts and then sealed hermetically, so that the fluid was still present 2,100 years later. The Chinese scientists who examined the body came to the conclusion that the exclusion of air had been primarily responsible for the preservation of the body. They claim that aerobic organisms would quickly exhaust all the oxygen present and produce an anaerobic environment in which decay could not take place. The mercury solution would play a minor disinfecting role. This may be correct, but no one has yet shown that the simple exclusion of oxygen will preserve a dead body as well as this. Perhaps the mercury compounds were more important than was thought by the investigators.

Apart from this, the autopsy revealed a great deal. The Lady Ch'eng was about 50 years old, somewhat obese, 154 cm tall, and weighed 34.3 kg. As mentioned earlier, the skin was elastic and in good condition, as was the brownish black hair. She had 16 teeth, with some crowns badly worn. One important point was a perforation of the right eardrum, this being the earliest recorded instance of such pathology, except for mummy PUM II, who was probably living a little earlier. The brain had shrunk to about one-third

normal size, which is roughly the same as in Egyptian mummies where the brain is present. All the internal organs were intact and in situ, even the pulmonary plexus of the vagus nerve, the thoracic duct, and the artery to the appendix. In the stomach and intestines there were 138 muskmelon seeds which the lady must have eaten shortly before dying.

The causes of death were twofold. First, a gallstone about the size of a bean completely obstructed the lower end of the common bile duct, and this must have caused excruciating pain. Second, her arteries had many atheromatous plaques, as well as showing arteriosclerotic changes, and her coronary arteries were particularly affected. The walls of the left coronary artery were severely involved, about three-quarters of the lumen being blocked by disease. The diagnosis was that she suffered from severe biliary colic that resulted in coronary thrombosis and death. It is very rare for the cause of death to be diagnosed in an ancient body, apart from accident or violence.

Other pathologic conditions were identified. There was evidence of tuberculosis infection, indicated by calcified tuberculosis foci in the upper lobe of the left lung; blood fluke (*Schistosoma japonicum*) ova were found in the connective tissue of the liver and the walls of the rectum; ova of whipworms (*Trichuris trichiura*) and pinworms (*Enterobius vermicularis*) were present in the intestines; the fourth intervertebral space was narrowed and had a bony outgrowth that could have caused severe back and leg pains. Paintings of the lady show her walking with a stick, and this is the probable explanation. Her right forearm was deformed as the result of a fracture that had not been properly treated. Altogether, the autopsy shows a picture of an elderly lady suffering from the traumas and hazards of her age and time.

The tissues of the body, although shrunken, were in good condition. The brain had disintegrated into a crumbling mass, but the different layers of the abdominal wall were clearly discernible and the contours of the abdominal organs visible. Under the electron microscope, details of some structures could be distinguished. Her blood group was found to be A.

Some elaborate studies on the body tissue were conducted by anonymous Chinese scientists; the paper was published in 1976 and authorship listed as the Group for Research on the Han Cadaver of Mawangtui. Nothing of note was uncovered, apart from striations on muscles, collagen fibers, and amino acids. It is unfortunate that the body was preserved with formaldehyde after discovery. In view of the fact that it had survived for 2,000 years, was it really necessary to do this immediately? Could the body not have been kept as it was or at least placed in a refrigerator? We paleopathologists must have closer contacts with the archeologists who are the first to discover these bodies; unless we do, biochemical and immunologic studies will be hampered at the source.

Still, the Chinese did remarkably well with their marquise of Tai; perhaps someday scientists from the Western World will be able to see this most remarkable body from antiquity.

HEAD HUNTERS OF THE AMAZON
EKKEHARD KLEISS

Head hunting is undoubtedly one of the oldest forms of primitive warfare. To behead the defeated enemy was a custom common in ancient Assyria and could be observed still in the twentieth century among savage tribes in many parts of the world. Paleolithic heads discovered in Bavaria, carefully decapitated and buried separately from the bodies, suggest that head hunting already existed in prehistoric times.

In Europe, before World War I, Montenegrins used to cut off the head of a victim, which was then carried by a lock of hair as a trophy of victory. Later on, during the Balkan war of 1912–1913, this form of

head hunting was replaced by cutting off the nose and the upper lip with the mustache. Both cases indicate the great importance attributed to hair, which plays a predominant role in the magic concepts behind these cruel customs.

Of the forms of head hunting that are so widespread all over the world, we can only quote some examples. Herodotus tells about head hunters in Asia who did their grisly work during or after the battles. In the first half of the twentieth century, many hill tribes in northeastern India, especially in Assam, as well as several tribal groups in Burma, in the Malay Archipelago (e.g., the Ibans and formerly also the Kadazans of Borneo) and in Indonesia in general, still devoted themselves to all kinds of human sacrifices, combined nearly always with decapitation and frequently with cannibalism. The Igorots and Tagalogs of Luzon, Philippines, abandoned such practices in the middle of the twentieth century. The Asmat and other southwestern Papuans reverted during World War II to tribal fighting and head hunting, which they had formerly given up, and the practice still persists locally in New Guinea as well as in other parts of Oceania, sporadically associated with cannibalism.

Several tribes in Nigeria and other African tribal groups frequently carry out human sacrifices similar to the head-hunting customs of Indonesia, along with cannibalistic rites. North American Indians, when fighting against the white settlers, took the scalp rather than the whole head; they considered the hairy part more important than the rest because the soul was located in the hair.

Finally, of the South American head hunters, who most likely came from the Caribbean area and extended their domains to pre-Andean and trans-Andean regions, Jibaros of the basin of the High Amazon River deserve our attention on account of their skill in preparing shrunken heads or *tsantsas* (Figure 13.7). These great warriors not only exerted a strong influence during the last cen-

turies on other tribes in Ecuador and Peru, but also resisted successfully all attempts to civilize them or to suppress head hunting. Although officially the preparation of tsantsas is strictly forbidden, nobody can really enforce the law in the jungle, and ancient customs persist even in modern times.

The Indonesians and other head hunters dried, smoked, or mummified the whole head of their foe, sometimes preserving tattoo marks and even the actual features of the victim. They also skinned the head and painted the skull with ash, chalk, and ocher, in this way preparing trophies of a macabre beauty. All these are more or less in the original size of the human head. The Jibaros, however, shrink the tsantsa to the size of a fist or of the head of a small monkey, maintaining during all this reduction the original features, like a portrait or a caricature.

The preparation of a tsantsa is a laborious process. First, the head has to be cut as near to the trunk as possible, preserving all the skin of the neck. If the victim was slain near the village of the head hunters, the preparation starts immediately, and this, according to the opinion of some experts, gives the best results. On the other hand, transportation of the head through the jungle during several days undoubtedly produces a certain degree of putrefaction, which facilitates the next step, the separation of the skin from the skull. After receiving authorization from the chief of the tribe to make the tsantsa (this is given in a solemn ceremony), the head hunter, sometimes assisted by more experienced fellow tribesmen, makes an incision in the midline of the scalp, from the crown to the occipital bone or even down the whole dorsal portion of the neck. Then he separates the scalp from the roof of the skull. This is quite simple, as the tissues can be separated easily with the bare hands. The halves of the scalp hang outward like two inverted sacks. The more complicated part of the dissection is to separate the skin from the bones of the face, a process for which the Indians use sharp bamboo

Figure 13.7.   Tsantsas. (Courtesy of Dr. Etta Becker-Donner, Völkerkundemuseum, Vienna)

knives, shells, or flint stones. Sometimes they break the bones to facilitate their extraction. Great care must be taken to preserve the eyelids, lips, nose, and ears. The preparation of the neck is similar.

The process of shrinking the head is another toilsome task. If the head was fresh, the natural retraction of the skin at once reduces the mask to half its original size or less. In any case, to avoid further decay, the head is put into a bowl for several days with a decoction of plants, probably rich in tannin and other coagulating agents that will preserve and shrink the tissues at the same time. Sometimes the prepared skin is boiled in these extracts of plants and barks. Then the openings of the mouth and eyes are sewed with plant fibers, often fixed to small painted wooden sticks; with religious rites this closure avoids the evil spirits, execrations, or other calamities that might come out of the orifices. The cut in the scalp and neck is also sewed,

practically disappearing under the abundant hair. The head has now been transformed into a sort of sack, open only where the neck was severed from the trunk.

Into this sack, the Jibaro carefully pours hot sand that has been heated in the shard of a used pottery vessel (both details are important for the success of the preparation); some tribal groups also use three hot pebbles taken from the next river and rolled around in the inside of the sack. This ritual is to avoid any evil reaction from the spirit of the victim. At the same time, it burns away any excess of connective and other tissues, thus helping to shrink the skin in a proportioned way. On the outside, the skin of the face is frequently anointed with vegetable oils or fat, and the features are constantly modeled as their size is gradually reduced. Because of the smoke from special plants or woods, the plant extracts, and the powdered charcoal, this whole procedure produces the dark color typical of tsantsas, even if the victim had been a white man (this can be proved, at least in some cases, by the hair distribution – e.g., mustache – and by other details of the features).

During the whole time the head is being prepared, the head hunter must observe strict rules of fasting and other rituals, otherwise the spiritual success of the procedure would be in danger. When the work is finished, there is a big fiesta for the tribe, with special purification rites for the successful head hunter and introduction ceremonies for the tsantsa, which nearly always belongs to the whole clan as a sort of talisman.

This ideological framework demonstrates clearly that head hunting in general and the making of tsantsas in particular are not at all the result of bloodthirst or cruelty, but of spiritual concepts, as is typical of primitive religions.

Although we cannot discuss here all the details of the ideas or beliefs that induced primitive men to practice head hunting, basically there is a concept of the existence of a material soul. This soul matter has its seat in the head and can be stored and added to the existing stock of an individual or of the whole tribe. Its possession transfers to the owner of the head trophy certain qualities of the victim: strength, courage, sagacity, and so on. Similar ideas are associated with cannibalism, because the eaten parts of the foe's body transfer analogous qualities to the eater, an idea still believed by many primitive populations. Human sacrifices of any kind, including head hunting, are intimately related to fertility rites (the cycle of life), to the initiation of boys to manhood, to better status in the other world, where the head of the victim will assure his services to the owner, and to immanent or real power in the widest sense of the word – for example, related to the building of a new long house or the launching of a war canoe. The special importance of hair has been mentioned already and may be compared to the Biblical story of Samson and the use of amulets, arm rings, and necklets made from the hair of slain foes and therefore of magic virtue. In the British Museum in London, there is the tsantsa of a sloth, considered by the Jibaros and other South American tribes as the forefather of mankind, probably because of its hairy aspect. Because of the reduced size of the tsantsa itself, the hair of the head, which was practically never cut during the life span of the victim, seems extremely long and is frequently braided and/or adorned with feathers or beads.

Because of the commercial interest in tsantsas all over the world, fake ones made from the heads of monkeys or with the hairy skin of other animals abound on the market. In some cases, especially when they were made by the Indians themselves, their identification as fakes is rather difficult. However, even in such specimens, we must admire the masterly skill of primitive men.

REFERENCES

The infant mummy of Uan Muhuggiag
Arkell, A. J.; Cornwall, J. W.; and Mori, F. 1961.
    Analisi degli anelli componenti la collana della

mummia infantile di Uan Muhuggiag. *Revista di Antropologia* 48:161.

De Morgan, J. 1896. *Recherches sur l'origine de l'Egypte.* Paris: Leroux.

Elliot Smith, G., and Dawson, W. 1924. *Egyptian mummies.* London: Allen & Unwin.

Mori, F. 1960. Quarta missione paletnologica nell'Acacus (Sahara Fezzanese). *Ricerca Scientifica* 30:61.

Mori, F., and Ascenzi, A. 1959. La mummia infantile di Uan Muhuggiag: osservazioni antropologiche. *Rivista di Antropologia* 46:125.

Pettigrew, T. 1834. *History of Egyptian mummies.* London: Longmans.

The frozen Scythians of Siberia

Artamonov, M. I. 1965. Frozen tombs of the Scythians. *Scientific American* 212:101–10.

Herodotus. *Persian Wars,* Book IV, pp. 71–3. Quoted in S. I. Rudenko. *Frozen tombs of Siberia: the Pazyryk burials of Iron Age horsemen.* Berkeley: University of California Press, 1970.

Rudenko, S. I. 1970. *Frozen tombs of Siberia: the Pazyryk burials of Iron Age horsemen.* Translated by M. W. Thompson. Berkeley: University of California Press.

Guanche mummies from the Canary Islands

Brothwell, D. R.; Sandison, A. T.; and Gray, P. H. K. 1969. Human biological observations on a Guanche mummy with anthracosis. *American Journal of Physical Anthropology* 30:333–48.

Hooton, E. A. 1925. The ancient inhabitants of the Canary Islands. *Harvard African Studies* 7:40–5.

The Marquise of Tai

Group for research on the Han Cadaver of Mawangtui, Shanghai Institute of Biochemistry, Academic Sinica, and Hunan Medical College. 1976. The state of preservation of the cadaver of the Marquise of Tai found in the Han tomb no. 1 in Mawangtui near Changsha as revealed by the fine structure of the muscle and other tissues. *Scientia Sinica* 19:557–72.

1973. Study of a body 2,000 years old. *China Reconstructs* 22(10):32–4.

Head hunters of the Amazon

Cranstone, B. A. L. 1961. *Melanesia: a short ethnography.* London: British Museum.

De Graff, F. W. K. 1923. The head-hunters of the Amazon. In *Encyclopaedia Britannica.*

Durham, E. 1923. Head-hunting in the Balkans. *Man* 18.

Heine-Geldern, R. v. 1924. *Kopfjagd und Menschenopfer in Assam und Birma.* Vienna: Anthropologischen Gesellschaft.

Hodson, T. C. 1912. Head-hunting among the hill tribes in Assam. *Folklore* 20.

Karsten, R. 1923. *Blood revenge, war and victory feasts among the Jibaro Indians of Eastern Ecuador.* Smithsonian Institution, bulletin 79. Washington, D.C.: Smithsonian Institution.

– 1935. The head-hunters of Western Amazonas: the life and culture of the Jibaro Indians of eastern Ecuador and Peru. Helsingfors: Akademische Buchhandlung.

Kleiss, E. 1966. Tsantsas: ein Mythus wird Maskottchen. *Wiener Tierärztliche Monatsschrift* 53:482.

– 1967. Zum Problem der natürlichen Mumifikation und Konservierung. *Zeitschrift für Morphologie und Anthropologie* 59:204.

Kleiss, E., and Simonsberger, P. 1964. *La parafinización como método morfológico.* Mérida: Universidad de los Andes.

Paredes Borja, V. 1963. *Historia de la medicina en el Ecuador.* Quito: Casa de la Cultura Ecuatoriana.

Sowada, A. (O.S.C.) 1968. New Guinea's fierce Asmat: a heritage of headhunting. In *Vanishing people of the earth.* Washington, D.C.: National Geographic Society.

Thomson, C. J. S. 1924. Shrunken human heads. *Discovery.*

PART IV
The study of mummies

14

Physical anthropology: determining sex, age, and individual features

J. LAWRENCE ANGEL
Curator of Physical Anthropology
National Museum of Natural History
Smithsonian Institution
Washington, D.C., U.S.A.

Wrappings and iron-hard skin protect mummies. They speak less clearly than do skeletons to those of us trained to read bones. In both cases we have to learn the proper language: careful comparative observation, measurements, and photography for bones; for mummies, anatomical observation and comparative radiography, for which we now need new standards. X rays of teeth, epiphyses, and pelvis in a mummy tell the age at death just as clearly as the skeleton does, and more clearly than do the fragmentary bones we often have to read. Growth arrests and many disease processes show clearly in X rays; bone robusticity and size are clear; but, especially in a wrapped mummy, body build and even sex can be confusing as seen radiographically. Dating is based on cultural artifacts or on carbon-14 analysis of wrappings or of collagen from skin or an extracted leg bone.

Usually what we need to know about a mummy are sex, age, race, stature, build, occupation, parity (number of births), state of health, and any culturally relevant congenital or genetic variations.

SEX

The bigger cross section of the adult female's true pelvis (birth canal) depends on extra cartilage-formed growth of the pubic bone, the lower ilium, and usually also the sacrum. This apparently stems from the increased production of estrogen at puberty, just as the steroid hormones in general are starting to cause closure of the epiphyses, of the cartilage joining the three parts of each hip bone at the hip socket, and later of all the other pelvic epiphyses. These events occur in females 1 to 2 years before they do in males. Such growth timing and hormone differences produce in females a high ischiopubic index (with consequent wide subpubic angle), a wide sciatic notch, and a long and wide preauricular sulcus (Washburn 1948). These indicators (Table 14.1) successfully differentiate the sexes in 85 to 95 percent of cases. A high pelvic brim–posterior iliac index and a low first sacral vertebral–total sacral breadth index are less reliable indicators of femaleness (Flander 1978). None of these are well visualized in the ordinary anteroposterior X-ray film of an intact mummy. Using anteroposterior and lateral X rays in a classic study of 600 Londoners, Young and Ince (1940) published sex differences for most of these and for other less reliable traits. The sample consisted of pregnant women and hospital staff at a time just before the ban on unnecessary exposure to radiation.

Phenice (1969) describes for the female ischiopubic ramus a thin line for the crus

Table 14.1. *Radiographic indicators of sex, based on appearance of skeleton*

Indicator	Characteristic in female	Characteristic in male	Overlap
Shape of true pelvis	Big, curved, cylindrical	Small, funnel-shaped	ca. 15%
Ischiopubic index (pubic length × 100/ischial length)			
Whites	100 or less	84	87–90
Blacks	95	80	83–86
Subpubic angle (degree)	94	76	ca. 82–88
Appearance of ventral pubis	Muscle-free triangle	Adductor brevis reaches edge	
Appearance of dorsal pubis	Birth scars sometimes	No birth scars	
Appearance of phallic crus attachment	Thin line	Flat surface, lipped	Variable
Index: iliac brim × 100/ dorsal ilium			Very variable
Adult	89	72 or less	75
Newborn	99	85–100	89
Appearance of sciatic notch	Wide	Narrow	ca. 10–20%
Appearance of preauricular sulcus	Slight to marked	Absent or trace	Slight
Index: S1 breadth × 100/ total sacral breadth			
Whites	45 or more	40 or less	38–46
Blacks	49	43 or less	43–49
Size of femur head (mm)	43.8	49.7	41.5–45.5
Size of humerus head (mm)	42.7	48.8	40.5–44.5
Size of femur shaft, AP view (mm)	26.9	31.1	24–32
Size of fibula shaft, transverse view (mm)	8.8	12.0	8.5–12.1
Appearance of skull	Small, toothy	Bigger, muscular	
Appearance of brow ridges	Absent to slight	Slight to marked	ca. 30%
Appearance of occipital crest	Absent to moderate	Slight to marked	ca. 40%
Appearance of mastoids	Small	Big	ca. 50%
Appearance of chin	Small, median	Big, broad	ca. 50%
Degree of prognathism	More	Less	ca. 50%
Size of first molar (mm)	9.7	10.7	ca. 50%

Source: Data from Borovansky (1936), Washburn (1948), Young and Ince (1940), Krogman (1973), Beyer (1976), Phenice (1969), Straus (1927), Stewart (1968), and our own records.

clitoridis and above this a triangular muscle-free area formed by pubic growth medial to the adductor-muscle boundary and blending with the wider female ventral bevel of the bony symphysis (Todd 1920; Gilbert and McKern 1973). This extra pubic growth also produces a more curved ramus bounding the wider subpubic space and a wider separation of pubic tubercles. This last detail is visible in the usual X ray. To visualize the others, special X-ray equipment is needed, with the beam aimed from below and in front and the film either behind or inside the true pelvis; such a film also shows the presence or absence of scars of childbirth. The usual X ray deceives. The radiologists who examined PUM II identified the mummy as female (Figure 14.1), apparently because the subpubic angle in the oblique view looks wide and the hunched shoulders look narrow compared

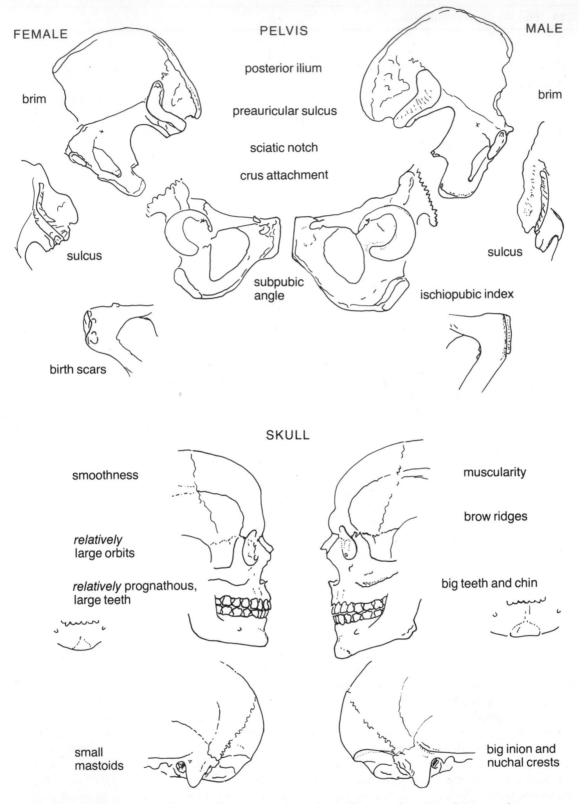

Figure 14.1. *Pelvic and skull sex differences (and overlap), illustrated in United States whites.*

Table 14.2. *Age (years) at which various levels of tooth development occur in males and females*

Tooth	Follicle	Platform	Skeletal eruption	Occlusal level	Root closure	Range (platform stage)
M_1						
Female	0	4.0	5.7	6.9	10.6	3.1–4.9
Male	0+	4.3	5.8	6.9	10.3	
$P_1(3)$						
Female	1.8	7.2	9.7	10.3	12.5	6.0–8.1
Male	2.2	7.4	10.1	10.9	13.0	
M_2						
Female	3.2	8.7	10.7	11.8	14.6	7.3–10.0
Male	3.5	8.9	11.2	12.7	15.0	

Source: Garn et al. (1958).

with the hip breadth (although PUM II lacks the wide trochanteric spread characteristic of females). The three physical anthropologists (Angel, Lasker, and Strouhal) who saw the X rays and the mummy before unwrapping predicted maleness because the femoral heads were big (50 mm, allowing 12 percent for X-ray magnification), the pelvic brim was absolutely narrow, there was no lucency to suggest a preauricular sulcus, the hands and feet were relatively large, and the skull had male traits.

Skull indicators of sex show more overlap than pelvic ones (Figure 14.1). Over half of white adult males and about 3 percent of females have pronounced to extreme glabella and brow ridges; 11 percent of males and 77 percent of females have absent to slight glabella and brow ridges; medium brow ridges, characteristic of 36 percent of males and 21 percent of females, are poor indicators of sex (Borovansky 1936). Mastoid size and general strength of crests for neck muscles show rather more overlap. Because growth ends almost 2 years earlier in females than in males, the female face is relatively small. Female chin height is 12 percent below the male mean, although female vault size is only 7 percent less, tooth size only 3 to 6 percent less, and orbit size the same as in males; this makes the female look a bit toothy and large-eyed! However, there is a considerable overlap in these traits, of the order of 30 percent.

Because most skeletons discovered are fragmentary and mummies are complete, sex determination on the basis of X-ray examination ought to be easier in a mummy, even before unwrapping. Nevertheless, sex determination is still difficult because of the variation produced by individual gene-hormone balance.

AGE AT DEATH

The patterns of ossification and of tooth formation (Tables 14.2 through 14.5) are clear by X ray and to some extent by visual inspection – for example, through a half-open mouth or through eroded skin at or over a joint. Newborn or premature infants are obvious from size and proportions. Occasionally the desiccated body of an anencephalic "monster" of Native American origin found in a cave has helped to feed the UFO cult. Size alone is a poor guide to age. Sick children – for example, those with severe anemia – are retarded in size growth and often in bone age as well; in such cases, tooth calcification may be a better guide, although it too is altered to some extent by disease and malnutrition. On the other hand, adolescents may be adult in size.

Table 14.3.   *Age (months) at which milk teeth erupt in males and females*

	I_1	I_2	C	M_1	M_2	Standard deviation
Female						
Upper jaw	9.6	11.9	20.1	15.7	28.4	2.0–4.3
Lower jaw	7.8	13.8	20.2	15.6	27.1	2.1–4.2
Male						
Upper jaw	9.1	10.4	18.9	16.0	27.6	1.5–4.4
Lower jaw	7.3	13.0	19.3	16.2	25.9	1.6–3.8

Source: Robinow (1942).

Table 14.4. *Age (years) at which permanent teeth erupt in males and females*

	I_1	I_2	C	$P_1(3)$	$P_2(4)$	M_1	M_2	(M_3)
Female								
Upper jaw	7.2	8.2	11.0	10.0	10.9	6.2	12.3	20.4
Lower jaw	6.3	7.3	9.9	10.2	10.9	5.9	11.7	20.8
Male								
Upper jaw	7.5	8.7	11.7	10.4	11.2	6.4	12.7	20.4
Lower jaw	6.5	7.7	10.8	10.8	11.5	6.2	12.1	20.5
Standard deviation	0.8	0.9+	1.3	1.5−	1.6+	0.8	1.4−	2.4

Source: Hurme (1957).

For them and for subadults about 16 to 20 years old, bone age is the vital criterion, as it is in the mummy of King Tutankhamun (Harris and Weeks 1973).

During the early twenties closure of the epiphyses of the ilium, ischium, vertebral bodies, transverse processes, and ribs, and of the medial clavicle, is a good guide. Up to age 55 or 60 the changing pubic symphysis is the best indicator (Todd 1920; McKern and Stewart 1957; Gilbert and McKern 1973), as shown in Figures 14.2 through 14.5. If a section of femur shaft is available, one can estimate age on the basis of the state of cortical bone remodeling, as in Kerley's (1965) method of counting "fragments" versus complete osteons, now further refined by Don Ortner and D. Ubelaker. Also helpful is careful inspection of individual osteon structure (Jowsey 1960; Ortner and von Endt 1971): With increasing age, osteon formation is less and less regular, with a denser surface at the periphery and next to the now more open central canal.

In old age (55 and over), osteoclasts remove more bone, especially endosteally, than osteoblasts can lay down as outer lamellae or in osteons, and the result is osteoporosis (Frost 1966; Garn 1970). Lack of exercise in old age also contributes to the retraction of spongy bone from the ends of long bones and in vertebrae (visible by X ray), and this atrophy too makes fractures more likely. Underlying this structural aging of bone, and also of artery walls and of skin (Montagna 1965), is the gradual density increase of collagen fibrils

Table 14.5.   Bone growth changes with age

Age (yr)	Upper extremity	Female	Male	Lower extremity	Female	Male	Axial skeleton	Female	Male
Prenatal				Calcaneus, talus, knee	4.5–9.5 f mo	4.5–9.5 f mo			
Birth	Humerus head	0+ mo	0+ mo	Cuboid	3 mo	3+ mo	Mandible fuses at chin		
	Capitate	6 mo	9 mo	Femur head	6 mo	9 mo	Vertebral arches fuse at midline from below up	0.5–1.5 yr	0.5–1.5 yr
	Humerus greater tuberosity			Tibia ankle	9 mo	9 mo	Fontanel at bregma disappears		
1	Hamate	9 mo	1 yr	Fibula ankle					
	Radius wrist								
2	Metacarpals			Metatarsals			Axis midline fuses	2+ yr	2+ yr
	Lunate								
3	Medial epicondyle	3.25 yr	6+ yr	Patella	2.5 yr	3.75 yr	Vertebral arches fuse to bodies	3 yr	3 yr
				Greater trochanter	2.5 yr	3.75 yr	Cervical		?
4	Radius proximal	3.75 yr	5.5 yr	Fibula proximal	2.5 yr	3.75 yr			
5	Scaphoid	5.5 yr	6.5 yr	Calcaneus tuberosity	5.5 yr	7.5 yr	Occiput pars lateralis to lumbar	4–6 yr	4–6 yr
	Ulna wrist						Axis and atlas	6 yr	6 yr
6	Ulna proximal	7.5 yr	10 yr				Occiput condyle area	5–7 yr	5–7 yr
7									
8	Pisiform	8 yr	11 yr	Talus posterior tuberosity	8 yr	9.5 yr			
9	Trochlea elbow	8.5 yr	9.5 yr	Tibia tuberosity	9.5 yr	11.5 yr			
				Lesser trochanter	9.5 yr	11.5 yr			
10	Lateral epicondyle	9+ yr	11.5 yr	Talus tuberosity fuses	10 yr	12 yr			
11				Ischiopubic ramus occ joins	11.5 yr	13.5 yr			
12	Glenoid rim, subcoracoid, medial clavicle, scapular borders			Iliac crest	12.5 yr	15 yr	Epiphyses of vertebrae and rib tubercles		
				Ischial tuberosity	12.5 yr	15 yr			
Puberty									
13	Elbow	13 yr	15 yr	Acetabulum	13 yr	14.75 yr	Spheno-occipital synchondrosis	15 yr(?)	17+ yr
14	Medial epicondyle	14 yr	15.75 yr	Calcaneus tuberosity	13.5 yr	15 yr			
15	Metacarpals I	14 yr	16 yr	Metatarsals	14.5 yr	15.5 yr			
	Phalanges II–V	14 yr	16.5 yr	Ankle	15 yr	16 yr			
16	Wrist	16 yr	18 yr	Hip	17.5 yr	17.5 yr	Sacral vertebrae 2–5	17 yr(?)	18.5 yr
17									
18	Scapular epiphyses	15–17 yr	17–19+ yr	Knee	18 yr	18.5 yr	Sternum corpus	18 yr (?)	19 yr
19				Iliac crest	19 yr(?)	19.5 yr			
20	Shoulder	19.5 yr	20 yr	Ischial tuberosity	20 yr(?)	21 yr	Epiphyses of vertebrae ribs, sacrum (except for mid thoracic lag)	19.5 yr(?)	20.5 yr
21							Epiphyses of midthoracic		21+ yr
22									
23							Sacral vertebrae 1–2	22 yr(?)	23.5 yr
24	Medial clavicle (may remain open even after 30 yr)	24 yr (?)	24.5 yr						

Shown above the horizontal line representing puberty are the times of appearance of centers of ossification of epiphyses and of hand and foot bones as seen in radiographs, and the first fusions in the axial skeleton. Range of individual variation is about 2 years.

Shown below the horizontal line (except for the posterior tuberosity of the talus) are the times of fusion of epiphyses to shafts as seen either in X-rays or directly. Peripheral gaps remain around the central area of fusion for up to 6 months, occasionally longer. Range of individual variation is from 1 to 5 years, increasing with age and with skewing toward older ages. Diet and environmental stimulation may slow or speed the rate of maturation.

Source: Fiftieth percentile ages from data of Western Reserve subjects (Mainland 1945, quoting Todd, Francis, Greulich, Pyle) and Korean War dead (McKern and Stewart 1957).

1　　　　　　　**18–19**
Adolescent *ridge and furrow,* rugged rounded ridges and sharp grooves like waves of molten metal. No margins or bevel.

2　　　　　　　**20–21**
Foreshadows ventral bevel.
New finely textured bone posteriorly fills grooves, hints dorsal margin. Bony nodules can occur superiorly.

3　　　　　　　**22–24**
Start of ventral bevel (rarefied rounded strip) and of dorsal margin. Flux: progressive obliteration of ridge and furrow system.

4　　　　　　　**25–26**
Dorsal margin clear.
Lower margin rounded.
Ventral bevel advances.
Ridges blur with porosity.

5　　　　　　　**27–30**
Upper end starts with bony nodules that may foreshadow ventral rampart.
Lower end clearer.

6　　　　　　　**30–35**
Ventral rampart developed, increasing definition of extremities; pubic face and ventral aspect still granular. No lipping.

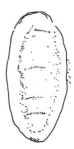

7　　　　　　　**35–39**
Bony outgrowths, especially at sacrotuberous, inguinal, gracilis attachments. Face flatter with less activity.

8　　　　　　　**39–44**
Smooth face inactive. Oval outline and defined rim, unlipped, less granular resting face. Ventral aspect inactive.

9　　　　　　　**45–50**
Rim develops around face, lipped at dorsal and ventral edges and *angular* in section.

10　　　　　　　**50–ca. 65**
Irregularity and *erosion.* Erosion and erratic growths on face with breaking down of ventral margin and rarefaction upper ventral.

10+　Over 65
Irregular breakdown, *pitted,* spongy.

Figure 14.2. Schematic representation of Todd's age phases of the pubic symphysis face. Each of the half dozen regions of the symphysis can change independently. There is a 5- to 10-year leeway in any of these processes, and the variation increases with age.

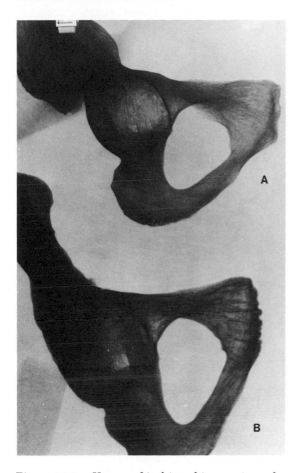

Figure 14.3. *X rays of ischiopubic area in male and female Egyptians. a. Female about 35 years old. Note at symphysis muscle-free triangle and irregular ventral rampart. b. Male about 19 years old. Note lack of triangle. Note also ridge and furrow formation with trabeculae supporting each ridge.*

from middle age onward (35 and over) (Cobb 1952). This thickening of fibrous tissue in bone, especially next to strong tendons and ligaments like those encasing the intervertebral disc substance, leads to exostoses. The bone-forming periosteum is raised by dammed-up tissue fluid under it, a process increased by stress. Such arthrosis and arthritis increase in middle and old age, especially in vertebral bodies and facets (Stewart 1968). The least accurate indicator of age is

skull suture closure (Todd and Lyon 1924; Figure 14.6), which is very hard to see at all in radiographs. Relative tooth wear is no better.

Figures 14.3 through 14.5 show new standards for the estimation of age on the basis of the roentgenographic appearance of the pubic symphysis. Note that in the twenties each wave on the face surface has its own radiating spongy trabecula. These trabeculae vanish as the face develops the ventral bevel and the various elements of the border and rim, which are hard to see by X ray. Old-age skin wrinkling and sagging (neck, breasts, buttocks) may show radiographically, and skin should be ageable histologically (Montagna 1965). We all know people in their seventies who act as if they were 50, as well as vice versa, and the skin and fibrous tissue parallel such contrasts between biological and calendar age. There are also age-rate differences among systems. Weathering of the skin should be visible in mummies, perhaps with class as well as sex differences.

PARITY

The stresses of later pregnancy and childbirth make characteristic posterior bone scars at the pubic symphysis visible as lucent areas next to the denser pubic face rim and later deformation of pubic tubercles (Putschar 1931, 1976; Stewart 1957, 1970; Angel 1972). The relaxation of ligaments also affects sacroiliac joints; the preauricular sulcus deepens, develops crossbars, and broadens, apparently from hormonally mediated bone growth. Here we have no standards, and the change is hard to see radiographically.

RACE

Race traits are as clearcut as sexual traits if we are comparing averages of archeological populations, but for mummies, as for forensic finds, we usually must deal with individuals, who may be either hybrid or intermediate according to the history of their ethnic group. A

typical Nubian mummy should have more black traits than one from Lower Egypt, but may not; a typical Peruvian mummy should be slightly less Mongoloid than an Aleut, but may not be. Well-developed shovel incisors are a classic marker for both Mongoloids and native Americans, but they occur in 5 percent of whites and not in 100 percent of Asiatic-origin peoples. Therefore, unless one is dealing with an unidentified and stripped mummy from a circus sideshow, cultural traits of wrappings, evisceration, position, and any artifacts are as good guides to ethnic belonging as are biological traits.

STATURE AND BUILD

Direct measurement of total length, with either anthropometer or steel tape with accurate blocks, in an extended mummy is probably accurate if one allows 1 cm for scalp thickness and about 1.5 cm for the heelpad, as desiccation at the joints should counterbalance the increase in cadaver length over standing height. With the anthropometer one can measure all long bones except the femur, subtracting 1 to 2 cm for cartilage, and then use the appropriate Trotter-Gleser formula (1952, 1958) to estimate stature. Bone lengths determined by X ray are equally accurate if one knows the magnification factor. The general tendency is to exaggerate. For analysis of body build, an extended mummy is misleading, because usually the shoulders are hunched and therefore narrowed and because desiccation shrinks both muscle and fat. Hip breadth, trunk length, hands, feet, and head are measurable, so any striking deviation from average build will be clear.

Analysis of postural details such as lumbar curve, femoral neck angulation and reaction area, tibia plateau tilt at the knees, flexion facets at the ankles, and strength of muscles needs special X-ray views, usually lateral. Stress from the occupation may also be harder to analyze in a mummy than in a skeleton that is complete.

CONGENITAL AND GENETIC VARIATIONS

These range from such obvious and rare traits, often dominant, as lobster-claw appendages, polydactyly, cleidocranial dysostosis, hip dislocation, and chondrodystrophy (Sorsby 1953) to carpal fusions, shifts in vertebral counts, cervical rib, and atlas arch, as well as many small tooth and skull traits not always possible to see by X ray. Sometimes postnatal stress facilitates the expression of a potentiality, as in a separated arch (spondylolysis) of the fourth, or more often the fifth, lumbar vertebra, which is really a fatigue fracture related to inherited build (Nathan 1959). This may be hard to see by X ray unless lipping and perhaps slipping of the fifth lumbar vertebra signal its occurrence.

HEALTH AND DISEASE

The best hard tissue indicators of a healthy childhood are a true pelvic index above 90, as even mild suboptimal nutrition lets the pelvic brim deform fractionally (Nicholson 1945); lack of growth arrest lines on tooth enamel (linear hypoplasia) or in spongy ends of long bones (Steinbock 1976); tall stature in relation to the potential expected; and teeth resistant to caries and loss. One cannot say that a youth dying prematurely of disease was as healthy as if he had survived to age 80, so for both individual and group, longevity is the really critical index of adult health (Angel 1975).

Bone responds to disease organisms just as it does to any stress (Putschar 1960): by atrophy through action of osteoclasts if the stress is unremitting or even necrosis if the blood supply is cut off, or by hypertrophy if stress is intermittent with time for recovery, or by a mixture of both processes. An example can be seen in the spots or rings of necrosis surrounded by crater rings or nodules of extra-hard bone, giving the snailtrack or ser-

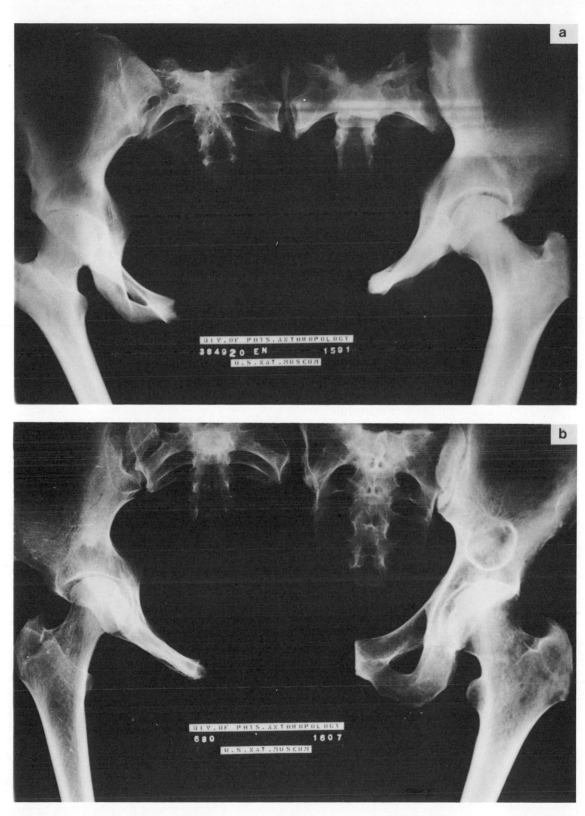

Figure 14.4. X rays of male and female pelves of difference ages. Anteroposterior views. a. No. 384920, white female, 17 years; No. 19, white male, 19 years. b. No. 680, white female, 30 years; No. 1607, white

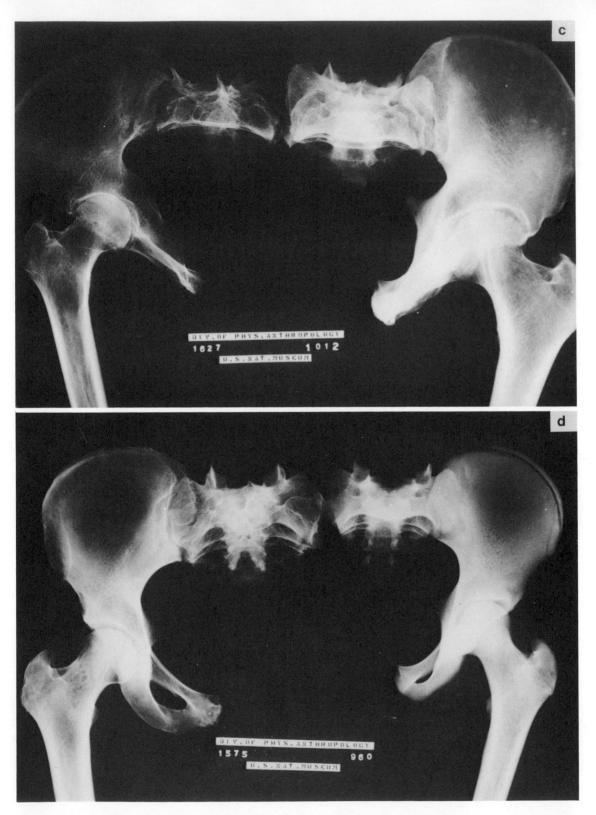

male, 32 years. c. No. 1627, black female, 90 years; No. 1012, black male, 74 years. d. No. 1575, white female showing birth scars, 44 years; No. 1591, black male with unfused epiphyses, 19 years.

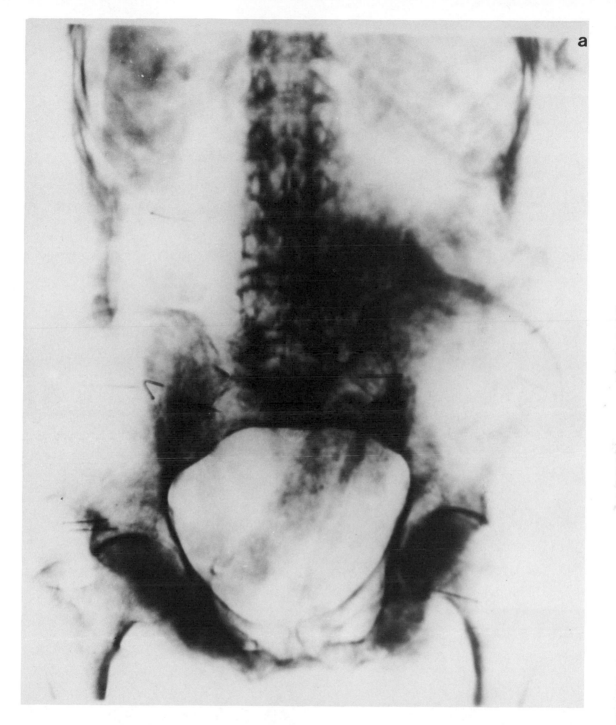

Figure 14.5. X rays of pelvic areas of two Egyptian mummies: an adult male, PUM II (a) and an adult female (b). (Courtesy of Dr. Stuart L. Wheeler, University of Richmond)

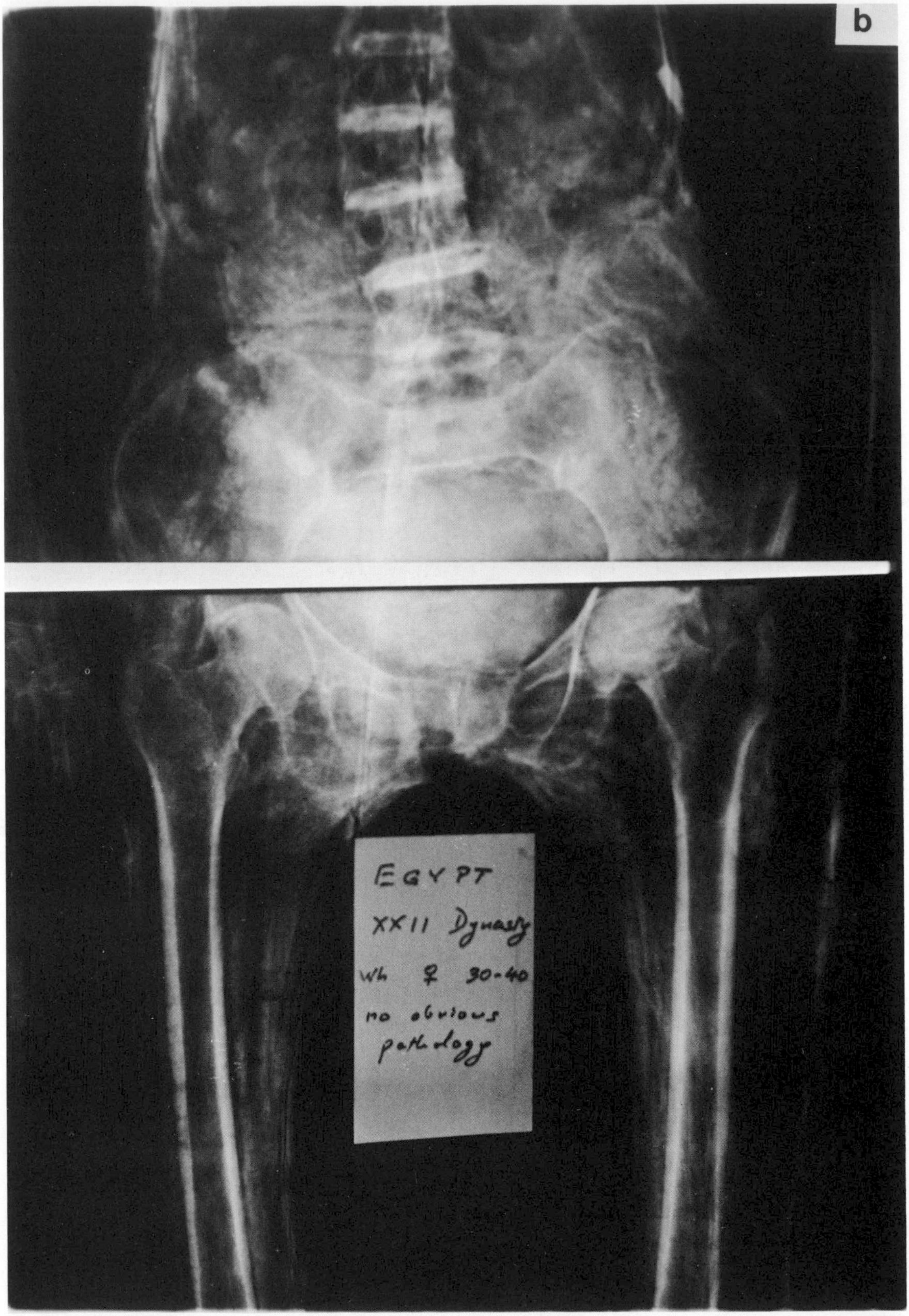

EGYPT
XXII Dynasty
Wh ♀ 30-40
no obvious
pathology

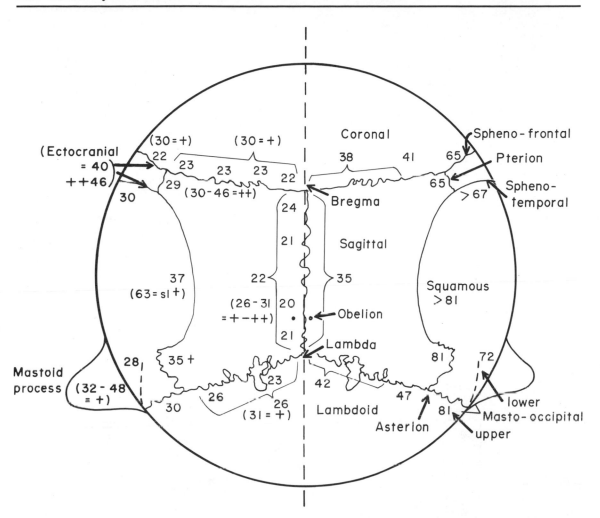

Figure 14.6.   *Endocranial skull suture closure. The numbers indicate approximate age at time of closure.*

piginous formations and caries sicca of tre-
poneme diseases in the skull vault (Hackett
1976) and the spindle-shape but pitted enlargement of long bone shafts. Paget's disease, because of the huge increase in blood
flow through one or more bones, destroys cortical and hugely increases spongy bone.

According to Møller-Christensen (1967)
leprosy produces a spindle-tip wasting of
metapodials without the formation of new
bone and enlargement of phalangeal vascular
foramina   until   necrosis   takes   over.
Rheumatoid collagen disease also ends in
necrosis – irregular, but at joints rather than

in shafts of bones (Resnick 1976). Active
tuberculosis destroys bone around the nests
of bacteria in tubercles (Hackett 1976; Steinbock 1976), and in Pott's disease vertebrae
often collapse (Morse 1967). Recovery involves rebuilding of solid bone, fusing the
collapsed vertebral bodies into the kyphotic
mass of the hunchback.

Finally, the space-occupying cell nests of
carcinoma metastases spreading from pelvis
to skull through the trunk skeleton along the
vertebral venous plexus, as shown by Batson
(1942), cause little rough spherical holes near
the diploic veins. The wild bone marrow

nests in multiple myeloma often look punched-out and make the skull X ray resemble a negative plum pudding. A benign giant cell tumor, mainly lytic, can produce a balloon of empty thin bone. On the other hand, true bone cancer causes mostly hypertrophy, often in a radial pattern in sarcoma (Brothwell 1967).

All the above are rare diseases, though dramatic. The commoner reactions of bone to disease are hypertrophic (Steinbock 1976): an irregular subperiosteal layer often of woven bone, laid down in inflammation from many infections, either direct or blood-borne, including periostitis from staphylococcal stimulus (streptococcal infections may produce lysis); the denser bone of hypertrophic arthritis, which is less a disease than a raising of periosteum by blocked tissue fluid caused by stress plus age or a dense remodeling of fatigued joint surface; porotic hyperostosis or the enlargement of marrow space in any of the anemias, well seen by X ray (Angel 1967). Bone reaction to osteomyelitis is also mainly hypertrophic once the stage of drainage abscess (cloaca) and involucrum is passed. Bone reaction to particular organisms is not constant over time, but can change as the parasite evolves and the host-parasite balance changes (Cockburn 1963).

TRAUMA

Healed fractures seem too obvious to mention, but fractures of vertebrae, of skull vault, and childhood (often greenstick) fractures with minimal displacement may be harder to see by X ray than they are grossly in dry bone. There is enough prehistoric–historic–modern change in fracture locations and types (Angel 1974) to make them interesting. Osteomyelitis from a penetrating wound of bone or from compound fracture, and periostitis from a wound of cortical bone, are usually clear by X ray (Steinbock 1976), but small wounds and bruises, for example those of the outer table of the skull, are much easier to see in dry bone.

CONCLUSIONS

In the usual anteroposterior radiographs taken of extended mummies, criteria for determining sex and age are rather restricted. Special views enhance the estimates: subpubic view of pelvis and profile of skull. Most other skeletal details are clear by X ray, except for those small changes in facets, crests, and angles at joints that relate to posture, gait, and occupation.

REFERENCES

Angel, J. L. 1967. Porotic hyperostosis or osteoporosis symmetrica. In *Diseases in antiquity*, D. Brothwell and A. T. Sandison (eds.), pp. 378–89. Springfield, Ill.: Thomas.
–   1972. Ecology and population in the eastern Mediterranean. *World Archaeology* 4:88–105.
–   1974. Patterns of fractures from Neolithic to modern times. *Anthropologiai Közlémenyek* 18:9–18.
–   1975. Paleoecology, paleodemography and health. In *Population, ecology, and social evolution*, S. Polgar (ed.), pp. 167–90. Chicago: Aldine.
Batson, O. V. 1942. The rôle of the vertebral veins in metastatic processes. *Annals of Internal Medicine* 16:38–45.
Borovansky, L. 1936. Pohlavní rozdíly na lebce človĕka. Prague: Czech Academy of Arts and Sciences, II class. English summary in *Anthropologie* 16:129–33, 1938.
Brothwell, D. 1967. The evidence for neoplasms. In *Diseases in antiquity*, D. Brothwell and A. T. Sandison (eds.), pp. 320–45. Springfield, Ill.: Thomas.
Brothwell, D., and Sandison, A. T. 1967. *Diseases in antiquity: a survey of the diseases, injuries and surgery of early populations.* Springfield, Ill.: Thomas.
Cobb, W. M. 1952. Skeleton, In *Cowdry's problems in aging*, 3rd ed., A. I. Lansing (ed.), pp. 791–856. Baltimore: Williams & Wilkins.
Cockburn, T. A. 1963. *The evolution and eradication of infectious diseases.* Baltimore: Johns Hopkins.
Flander, L. B. 1978. Univariate and multivariate methods for sexing the sacrum. *American Journal of Physical Anthropology* 49:103–10.

Frost, H. M. 1966. *The bone dynamics in osteoporosis and osteomalacia.* Springfield, Ill.: Thomas.

Garn, S. M. 1970. *The earlier gain and later loss of cortical bone.* Springfield, Ill.: Thomas.

Garn, S. M., Lewis, A. B., Koski, K., and Polacheck, D. L. 1958. Sex differences in tooth development. *Journal of Dental Research* 38:561–7.

Gilbert, B. M., and McKern, T. W. 1973. A method for aging the female os pubis. *American Journal of Physical Anthropology* 38:31–8.

Hackett, C. J. 1976. *Diagnostic criteria of syphilis, yaws and treponarid (treponematoses) and of some other diseases in dry bones.* Berlin: Springer.

Harris, J. E., and Weeks, K. R. 1973. *X-raying the pharaohs.* New York: Scribner.

Hurme, V. O. 1957. Time and sequence of tooth eruption. *Journal of Forensic Sciences* 2:377–88.

Jowsey, J. 1960. Age changes in human bone. *Clinical Orthopaedics* 17:210–18.

Kerley, E. R. 1965. The microscopic determination of age in human bone. *American Journal of Physical Anthropology* 23:149–63.

Krogman, W. M. 1973. *The human skeleton in forensic medicine,* 2nd ed. Springfield, Ill.: Thomas.

Mainland, D. 1945. *Anatomy as a basis for medical and dental practice.* New York: Hoeber.

McKern, T. W., and Stewart, T. D. 1957. *Skeletal age changes in young American males.* Technical report EP–45. Natick, Mass.: Quartermaster Research and Development Command.

Møller-Christensen, V. 1967. Evidence of leprosy in earlier peoples. In *Diseases in antiquity,* D. Brothwell and A. T. Sandison (eds.), pp. 295–306. Springfield, Ill.: Thomas.

Montagna, W. (ed.). 1965. *Aging: advances in biology of skin,* vol. 6. Elmsford, N. Y.: Pergamon Press.

Morse, D. 1967. In *Diseases in antiquity,* D. Brothwell and A. T. Sandison (eds.), pp. 249–71. Springfield, Ill.: Thomas.

Nathan, H. 1959. Spondylolysis, its anatomy and mechanism of development. *Journal of Bone and Joint Surgery* 41A:303–20.

Nicholson, G. 1945. The two main diameters at the brim of the female pelvis. *Journal of Anatomy* 79:131–5.

Ortner, D. J., and von Endt, D. W. 1971. Microscopic and electron microprobe characteriza-

tion of the sclerotic lamellae in human osteons. *Israel Journal of Medical Sciences* 7:480–2.

Phenice, P. W. 1969. A newly developed visual method of sexing the os pubis. *American Journal of Physical Anthropology* 30(2):297–302.

Putschar, W. G. J. 1931. *Entwicklung, Wachstum, und Pathologie der Beckenverbindungen des Menschen mit besonderer Berücksichtigung von Schwangerschaft, Geburt, und ihrer Folgen.* Jena: Gustav Fischer.

– 1960. General pathology of the musculoskeletal system. In *Handbuch der allgemeinen Pathologie,* vol. 3, pt. 2, F. Büchner, E. Letterer, and F. Roulet (eds.), pp. 363–488. Berlin: Springer.

– 1976. The structure of the human symphysis pubis with special consideration of parturition and its sequelae. *American Journal of Physical Anthropology* 45:589–94.

Resnick, D. 1976. Rheumatoid arthritis of the wrist: the compartmental approach. *Medical Radiography and Photography* 52:50–88.

Sorsby, A. (ed.). 1953. *Clinical genetics.* London: Butterworth.

Steinbock, R. T. 1976. *Paleopathological diagnosis and interpretation: bone diseases in ancient human populations.* Springfield, Ill.: Thomas.

Stewart, T. D. 1957. Distortion of the pubic symphyseal face in females and its effect on age determination. *American Journal of Physical Anthropology* 15:9–18.

– 1958. The rate of development of vertebral osteoarthritis in American whites and its significance in skeletal age identification. *Leech* 28:144–51.

– 1968. Identification of the skeletal structures. In *Gradwohl's legal medicine,* 2nd ed., F. E. Camps (ed.), pp. 123–54. Bristol: Wright.

– 1970. Identification of the scars of parturition in the skeletal remains of females. In *Personal identification in mass disasters,* T. D. Stewart (ed.), pp. 127–35. Washington, D.C.: Smithsonian Institution (and Department of the Army).

Straus, W. L. 1927. The human ilium: sex and stock. *American Journal of Physical Anthropology* 11:1–28.

Todd, T. W. 1920. Age changes in the pubic bone. I. The male white pubis. *American Journal of Physical Anthropology* 3:285–334 (old series).

Todd, T. W., and Lyon, D. W., Jr. 1924. Endocranial suture closure, its progress and age relation-

ship. Adult males of white stock. *American Journal of Physical Anthropology* 7:325–84 (old series).

Trotter, M., and Gleser, G. C. 1952. Estimation of stature from long bones of American whites and Negroes. *American Journal of Physical Anthropology* 10:463–514.

– 1958. A re-evaluation of estimation of stature based on measurements of stature taken during life and of long bones after death. *American Journal of Physical Anthropology* 16:79–124.

Young, M., and Ince, J. C. H. 1940. A radiographic comparison of the male and female pelvis. *Journal of Anatomy* 74:374–85.

Washburn, S. L. 1948. Sex differences in the pubic bone. *American Journal of Physical Anthropology* 6:199–208 (new series).

15

Processing of mummified tissue for histological examination

THEODORE A. REYMAN
Director of Laboratories
Mount Carmel Mercy Hospital
Detroit, Michigan, U.S.A.

ANN M. DOWD
Technical Supervisor, Histology Laboratory
Mount Carmel Mercy Hospital
Detroit, Michigan, U.S.A.

The ingredient basic to all mummification is the drying or dehydration of the tissue at a rate greater than the decomposition of that tissue. Once dehydration occurs and most of the water has been removed from the tissue, the action of bacteria, fungi, and autolytic enzymes is stopped. The drying then "fixes" the tissue so that further degradation of its chemical components is arrested. In ancient Egypt and other areas of similar climate, the drying was expedited by the hot environment. Equally good, however, was the very cold climate of Alaska or Siberia, where tissue decay was arrested for the most part by freezing, and dehydration occurred at a more leisurely pace. The end result was the same.

Continued desiccation through the centuries has a variable effect on the tissue, as has the method of mummification. The tissue from Egyptian mummies is generally very dry, quite hard, and brittle. The Peruvian mummies and the frozen Alaskan mummies, when thawed, have tissue that is softer and sometimes has a greasy texture. Depending on how fast the tissue was dehydrated and frozen, varying states of preservation are apparent even on gross examination of the tissues from these mummies. However, the gross appearance of tissue may be quite misleading. Tissue that appears well preserved grossly may be very poorly preserved microscopically and vice versa. At least a few sections should be processed even if the material seems unrewarding on gross inspection. We have routinely processed at least one section of every piece of tissue we have obtained from

autopsy or from dissection of isolated body parts. For the neophyte paleopathologist and paleohistotechnologist, experience in handling and interpreting the tissue changes is important training even if the tissue is poorly preserved. A great number of artifacts may be present within this type of material, and identifying them constitutes a significant learning process.

The variety of tissue specimens obtained from an autopsy depends on both the method of mummification and the degree of preservation of the body. Many mummies have been eviscerated, partially or totally. The organs that were removed may be present with the body as visceral packages, in separate containers, or not at all. Some mummies have all the organs intact. When preservation has been good, the organs are identifiable grossly. In poorly preserved bodies, organ identification may be impossible, and specimens should be taken from anatomic locations where the organs should have been. In our series, we have grossly and microscopically identified the following tissues and organs: skin with adnexal structures, bone, fibrous and hyaline cartilage, tendon, subcutaneous tissue, fat, arteries, veins, nerves, skeletal muscle, heart, lung, trachea, spleen, intestine, brain, liver, urinary bladder, uterus, breast, and an intact eye. Although the nature of many of these organs was obvious, some were not identifiable when viewed grossly. Because of this, each piece of tissue should be separately and accurately labeled at the time it is removed from the body or body part. We

have done this by assigning a number to each specimen, then making a master list with a description to correspond to each item. For example, if the right lung is the fifth specimen removed from the mummy, it is labeled "specimen 5" and so indicated on the master sheet. If a portion of the tissue was taken from the right upper lobe, it is more specifically labeled for example, "5a" and the appropriate entry made on the master sheet. The desirability of knowing the exact anatomic location of a given section may not become obvious until after the slide has been viewed with the microscope. If proper identification has not been maintained, it may then be too late to verify the section's source.

After the preliminary identification and selection of tissue, the next step is to cut the tissue into suitable sections for processing. Because it is so hard and brittle, care must be taken in attempting to cut it, as it may easily break into fragments. If trimming is not necessary, it should not be attempted. However, there is a size limitation, and the sections should be no larger than standard sections of fresh surgical or autopsy tissue. We have limited the maximum size to 2.5 cm². This size rehydrates easily and also fits on the standard glass microscopic slide. If the tissue is quite brittle, it may be more easily trimmed to size after rehydration has taken place. Bone and cartilage can be cut to appropriate thickness with a band saw. In our experience, hand sawing is too traumatic to the tissue. Visceral packages from Egyptian mummies are often heavily laden with resin and are exceedingly difficult to cut. The band saw has been used effectively to cut these packages as well. The remaining tissue, such as preserved parenchymal organs, skeletal muscle, and other soft tissue, may be cut with sharp scalpels. Invariably, some fragmentation occurs but this cannot be avoided. Larger pieces of tissue from organs or body parts can be rehydrated intact, but this is considerably more time-consuming and not usually necessary. The sectioned tissue should be placed into clean, separate, and properly labeled containers. If

three sections have been taken from an organ, they should be put into three separate containers. They may be stored that way or taken directly to the next step – rehydration.

The tissue is practically devoid of water, and before impregnation can begin, water must be returned to the tissue at the cellular level. Because of the residual lipids present within the tissue, a lipid solvent, or wetting agent, enhances the diffusion of water into the tissue. Alcohol is quite suitable and freely diffuses into the tissue at the same time as the water. Our rehydrating solution is that recommended by Sandison (1955) and consists of the following:

95 percent ethyl alcohol	30 volumes
1 percent aqueous formalin	50 volumes
5 percent aqueous sodium carbonate	20 volumes

Equally good results have been obtained with a 0.25 percent solution of trisodium phosphate (Van Cleve and Ross 1947). We have used the latter solution primarily for tissue that is to be frozen and stained for fats, as it has interfered less with the freezing and sectioning.

The use of formalin and alcohol appears to be the best general method for rehydrating the tissue. The alcohols may act by creating new chemical bonds (Zugibe 1970) in this harshly treated tissue and may enhance the staining reactions.

Containers should be large enough to hold at least 20 times as much rehydrating fluid as tissue. Occasional gentle agitation facilitates rehydration, which normally takes about 24 hours for the smaller sections. Small, less compact specimens may become soft in 4 to 6 hours, but very dense tissue may require as long as 72 hours. Most of the tissue is dark brown to black and may contain white crystalline deposits that appear to be adipocere, a mixture of waxes (Evans 1962). This should not be confused with crystals of natron, the salt compound used during mummification: the latter are found on the skin surface and in the body cavities, not deep inside the tissue.

Within minutes of the tissue's being placed in the rehydrating fluid, a dark pigment begins to leach out into the solution. If this pigmentation is severe by the end of 12 hours, the fluid should be decanted and replaced with fresh. The nature of this pigment is not known. It is not indican, a putrefactive pigment, but is probably some form of bilirubinoid or a similar pigment. In the case of Egyptian mummies, there may be black resinous material on the surface of the tissue that often settles to the bottom of the container and may cause the tissue to become glued there. If resinous material is seen, the container should be replaced by a clean one and the fluid also changed.

During rehydrating, the tissue may swell to over twice its dry thickness. The sections should be tested frequently and gently with a pin until they are softened throughout. Once the tissue has softened, the rehydrating fluid should be decanted and replaced with equal parts of fresh rehydrating fluid and 95 percent ethyl alcohol. Small internal zones of the tissue may not be properly rehydrated if care is not taken at this time. If the fluid has not completely permeated the whole specimen, impregnation will be incomplete and subsequent sectioning and staining will be inadequate. If there are any doubts, the rehydration should be allowed to continue for several hours. Occasionally, the more poorly preserved tissue specimens disintegrate even with nominal periods of rehydration: this is unavoidable. Even with well preserved specimens, small fragments often break off the main tissue section. The material in the bottom of the container should not be discarded, for cell block preparations (De-Girolami 1968) can be made from it and can be processed and viewed with the microscope.

When the tissue is adequately softened, final trimming can easily be done, if necessary. However, excessive handling of the rehydrated specimen should be discouraged, as it is still friable and may break apart readily.

Bone and other calcified tissue should be rehydrated before decalcification. Mild decalcifying agents should be used. We have used Cal-Ex (Fisher Scientific Company) and formic acid–sodium citrate (Luna 1968), but EDTA solutions may be equally satisfactory (Zugibe 1970). With small fragments of very soft bone, frequent checks should be made for softness. Excessive decalcification may disintegrate the specimen (Zugibe 1970). Decalcification time varies considerably – from only a few hours to as long as 5 days if the bone is dense. The use of a single station on the tissue processor for oscillation during decalcification may decrease the time required (Russell 1963).

We have modified the techniques of Sandison (1957) and Russell (1956) to fit the daily routine of this laboratory. The double-impregnation technique used employed 2 percent celloidin in methyl benzoate and paraffin with a 56.5 °C melting point. The procedure can be performed manually or with automated equipment such as the Autotechnican Duo (Technicon Corporation, Tarrytown, New York). Other tissue processors should be readily adaptable to these schedules. The Technicon Corporation recommends against using amyl acetate in the Autotechnicon Ultra because of its low flash point. Various other volatile solvents are also normally used, so we have located the Autotechnicon Duo under a large exhaust hood and have had no trouble.

If the tissue processor is not in use overnight, the 48-hour short method is preferred (Table 15.1). If the processor is used for routine tissue specimens every night, then the longer 5-day semiautomated method can be used (Table 15.2) or the whole infiltration performed completely manually. The disc used on the tissue processor for the 48-hour method is illustrated in Figure 15.1. None are needed with the 5-day method as the baskets are advanced manually. Once the operator is familiar with the techniques, modifications in the timing sequences can be made to conform better to individual laboratories.

We have cut all sections on a standard rotary microtome (American Optical Corporation, Buffalo, New York), but a sliding mic-

Table 15.1. *Method for double impregnation of Egyptian mummy tissue with paraffin and celloidin[a] in 48 hours using Autotechnicon Duo tissue processor*

Time	Station number	Solution	Time in each (hr)
Day 1, disc I			
8 A.M.	1	80% Ethyl alcohol	2
10 A.M.	2	8% Phenol in 95% ethyl alcohol	4
2 P.M.	3	Absolute ethyl alcohol I	2
4 P.M.	4	Absolute ethyl alcohol II	2
6 P.M.	5	Absolute ethyl alcohol and amyl acetate (equal parts)	1
7 P.M.	6	Amyl acetate I	4
11 P.M.	7	Amyl acetate II	4
3 A.M.	8	2% Celloidin in methyl benzoate I[a]	5
Day 2, disc II			
8 A.M.		Change solution in beakers	
8 A.M.	8	2% Celloidin in methyl benzoate II	4
12 Noon	9	2% Celloidin in methyl benzoate III	9
9 P.M.	10	Benzene	1
10 P.M.	11	Paraffin I (MP 56.5°C)	5
3 A.M.	12	Paraffin II (MP 56.5°C)	5
			24
Day 3			
8 A.M.		Embed in paraffin (56.5°C)	

[a] Put 75 g celloidin into 500 ml absolute alcohol overnight. Add 500 ml diethyl ether and stir or shake solution occasionally until celloidin is completely dissolved (2–4 days). Add this to 3 kg (2,740 ml) of methyl benzoate.

rotome could be used instead. The sections were cut at 6 μ, mounted on standard glass slides (25 by 75mm) with a thin layer of egg albumen, and heat-dried. Care should be taken when floating the sections on the water bath as they tend to come apart. We have prevented much of this by lowering the water bath temperature to 45 °C with the 56.5 °C-melting-point paraffin. Although celloidin has helped prevent pulling apart and wrinkling of the sections, these problems have not been completely eliminated.

Because little or no nuclear detail persists in this tissue, hematoxylin and eosin staining generally has not been used. The connective tissue components are, however, often well preserved, and we have routinely used the following staining procedures:

1. Periodic acid–Schiff (PAS) (Luna 1968)
2. Masson's trichrome (Luna 1968)
3. Mallory's phosphotungstic acid–hematoxylin (Luna 1968)

4. Alcian blue (Luna 1968)
5. Verhoeff's elastic tissue–van Gieson (Culling 1974)

Oil red O and a variety of other lipid stains (Zugibe 1970) are used in certain instances, as dictated by the microscopic appearance of the routine stains. If lipid stains were anticipated, we have used trisodium phosphate (0.25 percent) as the rehydrating solution and then cut cryostat sections. This gave better results with the lipid stains (Plate 7).

Many artifacts have been present within the tissues. Some of these were exogenous – that is, material introduced by the embalming process. Others were attributable to the method of preserving the tissue. These were reflected in the staining reactions and probably resulted from the extreme drying, inadequate rehydration, or chemical decomposition of the tissue itself during mummification. Other changes in the tissue may be produced by the method of processing. The use

Table 15.2. *Semiautomated[a] 5-day method for double impregnation of Egyptian mummy tissue with paraffin and celloidin using Autotechnicon Duo tissue processor*

Time	Station number	Solution	Time in each (hr)
Day 1			
8 A.M.	1	8% Phenol in 95% ethyl alcohol	2
10 A.M.	2	Absolute ethyl alcohol I	2
12 P.M.	3	Absolute ethyl alcohol II	2
2 P.M.	4	Absolute ethyl alcohol and amyl acetate (equal parts)	2
4 P.M.		Remove tissue and store in alcohol–amyl acetate in glass jar overnight	
Day 2			
8 A.M.	1	Amyl acetate I	6
2 P.M.	2	Amyl acetate II	2
4 P.M.		Remove tissue and store in amyl acetate in a glass jar overnight	
Day 3			
8 A.M.	1	2% Celloidin in methyl benzoate I	8
4 P.M.		Remove tissue and store in 2% celloidin in methyl benzoate in a glass jar overnight	
Day 4			
8 A.M.	1	2% Celloidin in methyl benzoate II	8
4 P.M.		Remove tissue and store in 2% celloidin in methyl benzoate in a glass jar overnight	
Day 5			
8 A.M.	10	Benzene	1
9 A.M.	11	Paraffin I (MP 56.5°C)	3
12 Noon	12	Paraffin II (MP 56.5°C)	4
4 P.M.		Embed in paraffin (MP 56.5°C)	

[a] Basket advanced manually rather than with disc.

of buffered formalin and alcohol may introduce some artificial changes in the tissue (Culling 1974). Differences in the application of the staining procedures may induce some variations, but these have usually been identified by the use of controls – that is, fresh surgical or autopsy tissue of known type and staining reaction. In those instances where identification of specific substances would be valuable, highly selected rehydration, fixation, infiltration, and staining procedures should be employed to demonstrate those substances best. A variety of reference texts is available for this type of problem.

Because of the disappearance of nuclear and cytoplasmic detail, the histological identification of tissue, organ, and ultimately disease processes has been made on the basis of connective tissue patterns. A familiarity with these patterns in fresh surgical or autopsy tissue of similar type is required. In the better preserved specimens, the organ, tissues, and pathological processes may be obvious at first glance. In those specimens that have been poorly preserved, even knowing the organ from its anatomical location may not allow one to make the correct identification unequivocally.

The most hardy connective tissue elements, in terms of being well preserved and maintaining normal tinctorial properties, are elastic fibers. Several techniques are available for histochemical identification of these fibers; we have had the most success with Verhoeff's elastic tissue–van Gieson stain. Phosphotungstic acid–hematoxylin and al-

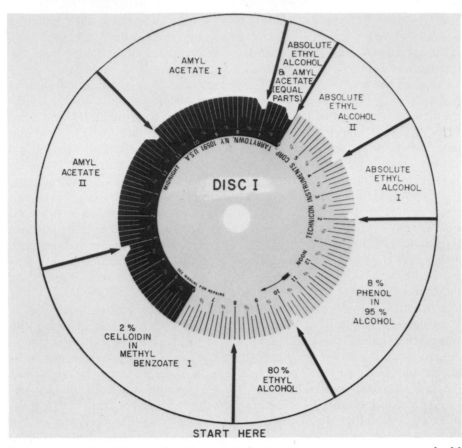

Figure 15.1.   *Metal disc used on the Autotechnicon Duo tissue processor, punched for the time intervals for the first 24 hours of the 48-hour method. See Table 15.1.*

dehyde fuchsin methods demonstrate these fibers as well, but the Verhoeff method has given the best differentiation of the fibers from the surrounding tissue (Sandison 1963). According to Thompson (1966), the hematoxylin in the Verhoeff elastic tissue–van Gieson stain is probably attached to the elastic fibers as a hematein–metal ion complex that reacts as a cationic dye with sites on the elastic fibers, staining them blueblack (Plate 5). The van Gieson picric acid–acid fuchsin counterstain acts as an anionic dye and stains the collagen red. In well-preserved mummified material, this has generally been the case. Dense collagenous tissue in some instances has stained as elastic tissue, probably as a result of the degradation of the chemical composition of the tissues into similar

reactive units. Normally the phosphotungstic acid–hematoxylin stain (Thompson 1966) imparts a red color to collagen and elastic fibers. In our material, collagen and elastic fibers have stained bluish red or purple. This may be a result of incomplete rehydration, but more likely has been caused by chemical change in the tissue sites responsible for the binding of the dye.

With the Masson trichrome stain (Thompson 1966), the collagen stains blue and muscle and cytoplasm red (Plate 6). Elastic fibers stain weakly red or not at all. In some mummified material, the collagen has taken on a blue red or purple color. The variability of the staining reaction with the trichrome and phosphotungstic acid–hematoxylin techniques in mummified tissue

probably resulted from intrinsic chemical rearrangement in the fibers and other tissue components. The inappropriate staining of collagen with the elastic tissue stains suggests a similar change in the reactive or binding sites on the fibers and implies a degradation of the fibers themselves. The result of this change, whatever its nature, is seen not only as inappropriate or paradoxical staining, but as less vivid staining as well. In poorly preserved tissue, all selective staining properties may be lost, the tissue staining uniformly poorly or not at all. This drastic change has been most common in muscle, less common in collagen and reticulum, and uncommon in elastic fibers (Plate 8).

This variability can also be seen with PAS and alcian blue techniques. The PAS technique stains a variety of carbohydrate- and protein-containing substances (Thompson 1966). Tissue components that normally react with the staining technique are numerous and include basement membranes, epithelial mucins, reticulum of lung, cartilage matrix, nucleus pulposus, collagen, and elastic fibers. The degree of staining within this group is variable and is thought to depend on the amount of hexose in the chemical makeup of the tissue. Although mummified tissue has been generally PAS-positive, the degree of staining was reduced from that of fresh surgical or autopsy tissue of the same type. This was particularly true in cartilage, both hyaline and fibrous types. Alcian blue (Thompson 1966), the exact chemical nature of which is a secret, stains acid mucopolysaccharides in the method we use. Structures that would normally stain with alcian blue are connective tissue matrix, chondrocytes and lacunar rims, perichondrium, and matrix in the media of elastic arteries. Hyalouronic acid, chondroitin sulfate, and similar substances in cartilage and vessels are probably responsible for much of the staining in these areas. In the mummified tissue, alcian blue staining activity has been markedly reduced, even more so than PAS. Colloidal iron staining (Luna 1968) can also be used for detecting acid mucopolysaccharides and has been similarly reduced in intensity in these tissues.

The other group of substances that may be demonstrated histochemically are the lipids. In human tissue, depending on its type, a variety of lipid compounds may be present and can be identified by the use of appropriate staining techniques (Zugibe 1970). These include glycolipids, fatty acids, neutral fats, phospholipids, and cholesterol. As we can see in Table 15.3, oil red O stains all the lipid compounds. Differentiation of most of these substances can often be accomplished with three other stains: PAS, Nile blue sulfate, and luxol fast blue. For cholesterol, the Schultz stain may be used.

Although mummified tissue may be as much as 25 percent lipid by weight, the amount of residual reactive lipid that can be identified by the appropriate staining techniques may be meager. Once the tissue has been passed through the lipid solvent solutions during processing, most of the residual lipid has been dissolved out of the tissue. Small amounts of bound lipid may persist, however. If the demonstration of lipids was anticipated, we have rehydrated the tissue in 0.25 percent trisodium phosphate and cut frozen sections on the cryostat. Storage in alcohol or prolonged storage of the rehydrated tissue in 10 percent buffered formalin may leach lipid from the tissue. Some of the lipid stains use alcohol or acetone as the solvent and vehicle for the stains, and these chemicals may also leach some of the lipid from the tissue during the staining procedure (Zugibe 1970). We have followed the suggestion of Zugibe (1970) and used propylene glycol to help prevent this.

Although the preservation and staining reactions of the various tissue components in mummified tissues are somewhat predictable, there may be considerable variation. As a rule, the more metabolically active a tissue was during life, the more rapidly and com-

Table 15.3. *Staining reaction of various lipids in human tissue*

Stain	Neutral fats	Fatty acids	Glycolipids	Phospholipids	Cholesterol
Oil red O	+	+	+	+	+
Periodic acid–Schiff	–	–	+	+	–
Luxol fast blue	–	–	–	+	–
Nile blue sulfate	+ (red)	+ (blue)	–	–	–
Schultz	–	–	–	–	+

Source: Zugibe (1970).

pletely it degenerates after death. In most mummified specimens, it would be expected that bone, cartilage, and connective tissue fibers would be preserved, but that very little epithelial or parenchymal tissue would survive with intact histological detail. Less than optimal preservation may be seen as areas of irregular or inappropriate staining. However, even in poorly preserved specimens some parenchymal structures may stain well enough to allow identification despite a lack of cellular detail. In other instances, this type of tissue is so poorly preserved that it will not take the stain and appears colorless or yellow in a given specific preparation and sometimes by all staining methods. On the other hand, connective tissue generally is well enough preserved so that it stains adequately, if not normally, even in poorly preserved tissue. Unstained degenerated parenchyma or epithelium may be identified only because it has a certain spatial relationship to the connective tissue in the microscopic sections. When the material has been so poorly preserved that the connective tissue components do not stain, the microarchitecture also has been too poorly preserved to be of diagnostic value. We cannot emphasize too strongly the need to have a sound idea of normal connective tissue and parenchymal patterns and, if necessary, to process sections of similar fresh tissue for comparison. As it exists, interpretaion of histological patterns in mummified tis-

sue is a challenge even to the experienced histopathologist and should not be attempted by the amateur microscopist alone.

The following discussion relates some of our specific experience with staining reactions in mummified tissue and some of the microscopic variations we have seen in the individual tissue components. Although most of our observations are on Egyptian mummy tissue, they are generally applicable to other (Alaskan and Peruvian) mummies.

Skeletal, cardiac, and smooth muscle has been seen in varying stages of preservation. In some instances, it has stained normally and residual cross-striations have been noted in skeletal muscle (Plate 1). When it has been less well preserved, the muscle stained orange to orange red with Masson's trichrome technique, and striations have not been found. Poorly preserved muscle has not stained at all, has appeared yellow, has been amorphous or vacuolated, or has consisted of masses of needlelike fibrillar structures (Plate 2). Occasionally tissue that could not be muscle by virtue of its anatomical location has had a similar appearance, suggesting that this represents the ultimate in chemical degradation of these tissues.

Skin that has not been eroded by harsh embalming techniques may have recognizable epithelium present. This has been seen as layered keratin, and occasionally there have been ghostlike basal cells present with mela-

nin pigment and faded, slightly staining nuclei. Sweat and sebaceous glands have been identified, but the cytoplasm of the cells was granular, degenerated, and without nuclei. Hair shafts have persisted intact.

Epithelial cells and parenchymal cells in the viscera usually have been poorly preserved. They appeared as amorphous zones within the connective tissue components and were without cytoplasmic or nuclear detail. The spatial relationships were often maintained, and the degenerated cellular components have been stained red with Masson's trichrome (Plate 3). In less well preserved specimens, this type of tissue has been yellow and unstained. However, in some sections the vascular structures were identified, and other connective tissue patterns allowed adequate interpretation and identification of the parenchymal zones.

Bone and cartilage have usually been well preserved, even when soft tissue was not. In bone, haversian systems, lacunae, and cement lines were easily recognized, and occasional osteocyte remnants have been seen to persist. The areas of cortical and trabecular bone were readily identified. Cellular marrow has not been preserved, though occasional fat cells were found in the better specimens. The fibrils and matrix of cartilage were also usually well preserved. Lacunae and occasional partially preserved chondrocytes have been seen (Plate 4). Cartilage from various sites such as joints, intervertebral discs, bronchi, trachea, and ear were all identifiable. In bone and cartilage, there has been decreased staining with PAS and alcian blue when compared with fresh tissue; this probably resulted from degradation of the glycoprotein and polysaccharide matrix of these tissues. This may account for the brittleness and lack of cohesion in these specimens.

Brain, although identifiable grossly, has always been severely degenerated microscopically. Occasional PAS-positive granular areas have been noted and may represent partially preserved phospholipid or, less likely, lipochrome or similar pigment. No other histological detail has been seen, even in grossly very well preserved specimens. Nerves have been identified in tissues, but this has been unusual in our experience. The one eye we have processed had preserved choroid, uveal tract, lens, and melanin pigment, but no retinal cells. The eye appeared normal microscopically.

Sections of the heart have shown good preservation of cardiac muscle as well as connective tissue, but we have not been able to find intact striations. In one specimen of heart with a coronary artery, the vessel was normal. Sections of the aorta have shown good preservation, and we have identified atherosclerotic plaques similar to those seen in modern sections (Figure 15.2). The plaque stained with oil red O and there were clear zones within the connective tissue of the plaque that appeared to be the site of degenerated foam cells.

Lung has been identified in sections by its connective tissue pattern and bronchial cartilage, alveolar septa, and vascular structures. There has commonly been deposition of black anthracotic pigment within the connective tissue. Using polarized light, particles typical of silica have occasionally been seen (Figure 15.3). In other instances, there has been a large amount of birefringent material in the degenerated lung sections that did not conform to connective tissue deposits. Care must be taken before deciding that this is silica. A variety of endogenous substances may be birefringent: for example, uric acid, some phosphates, and a number of lipid substances in crystalline form. Analysis of ashed lung tissue for free silica by atomic absorption spectroscopy or other more sophisticated techniques (Tapp et al. 1975) may be necessary to establish the material as silica.

Sections of liver, spleen, intestine, kidney, and urinary bladder have often contained connective tissue that was sufficiently preserved to allow proper identification. Even organs removed during the embalming process and found within packages returned to the visceral cavities or in separate containers

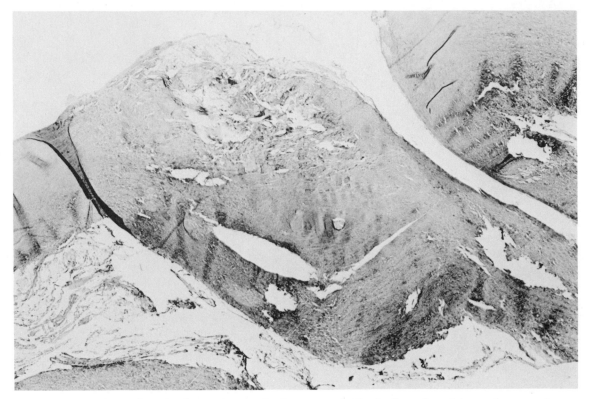

Figure 15.2.   *Atherosclerotic plaque present in the aorta, similar to those found in modern specimens. The plaque contains clear spaces that appear to be the sites of degenerated foam cells. Trichrome.*

have been well enough preserved for proper interpretation. In situ organs, on the other hand, have been so severely degenerated that there was no microscopic detail persisting, and the only method of identification has been by their anatomical size, shape, and location within the body.

Blood vessels, both arteries and veins, have often been identified. Usually there has been no difficulty distinguishing between the two. Generally the elastic arteries and those smaller arteries with formed connective tissue or muscular walls have been easily recognized. Red blood cells have not uncommonly been found within the lumens of these vessels, either as partially preserved masses or occasionally as well-formed, typical biconcave discs (Figure 15.4). In two cases, recognizable white blood cells have been found in mummified tissue more than 2,000

years old. Riddle et al. (1976) have convincingly demonstrated virtually intact white blood cells and red blood cells by scanning electron microscopy in tissue from an Egyptian mummy. Lewin (personal communication, 1975) has seen white blood cells in sections from a 3,200-year-old Egyptian mummy. We have seen both cases and there can be little doubt that the structures are white blood cells. Many investigators have seen red blood cells in mummified tissue (Zimmerman 1973). A word of caution, however, must be given about interpreting structures as being blood cells – or any intact cell, for that matter. In processing tissue, there is always a possibility that extraneous fresh tissue fragments may be introduced into the block or onto the slide prior to staining. A few red or white blood cells, small fragments of epithelium, or other tissue components may

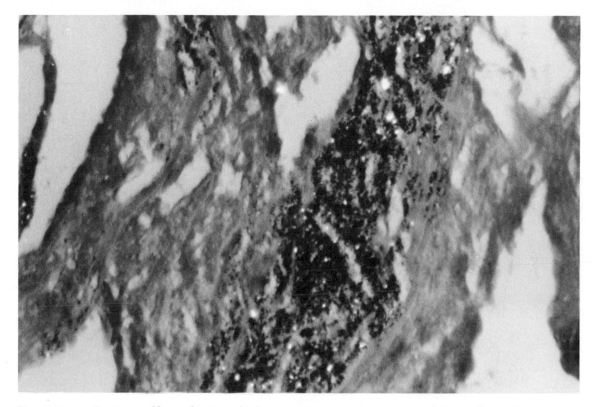

Figure 15.3. *Sections of lung showing thick connective tissue zones containing not only black carbon pigment but bright particles typical of silica. Trichrome.*

be inadvertently carried over onto the slide. Usually, the proper identification of these cells is prompt, as they are so well preserved, are free floating, and are not connected to any of the obvious mummified tissue. They may be seen overlying the ancient tissue or may be present in only one of several serial sections. In these situations, they should be considered contaminants until proved otherwise. In most mummified tissue, there have been large numbers of fungal spores and mycelia. The spores in particular may be mistaken for red blood cells. They have often been the proper size, have stained as do red blood cells, and in one case we have seen, had the concave or biconcave shape of red blood cells, but, in truth, were fungal spores. In this last instance, the nature of the structures was not determined until electron microscopic exam-

ination showed that they were formed spherules with central lumens.

Parasitic ova generally are very hardy structures and tend to be preserved. We have identified three different parasitic ova: *Schistosoma haematobium*, *Taenia* sp., and *Ascaris* sp. (Figure 15.5). All have been found within the lumen of intestines. When they have become degenerated or calcified, their identification has been less certain. In all instances, the ova should be measured with an ocular micrometer. If there is any question about their exact nature, they should be submitted to an expert parasitologist for verification.

Various other artifacts have been found within the tissue. Fragments of vegetable and plant material used in the embalming process may be seen in the prepared tissue sections.

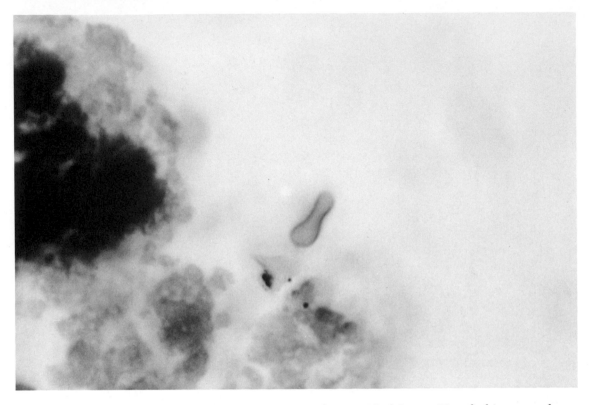

Figure 15.4.   *Well-preserved red blood cells in sections of mummified tissues. Note the biconcave shape of the erythrocyte. Trichrome.*

These structures have been easily recognized by their geometric configuration (Figure 15.6). If there is any doubt about the appearance of this type of material, a small piece of cork from the surgical cutting board should be processed, as it has the same appearance. These cellulose plant walls should be birefringent with polarized light. During mummification, the lipid material present within the tissue has sometimes been converted to long-chain waxes and has appeared as masses of somewhat radially arranged tissue deposits that have been strongly PAS-positive (Figure 15.7). Typically, it has also been birefringent with polarized light. These deposits have been found in subcutaneous fat or within the substance of parenchymal organs and probably represent adipocere (Evans 1962). In the subcutaneous fat, soaps of calcium may have been formed as a result of degeneration of this tissue. In some instances, the fat cell outlines have been intact, but the cell contents have often undergone granular degeneration and have been recognized only because of their positions within the sections. The fat in female breast tissue has had a similar appearance. Other extraneous materials noted in microscopic sections have been insect parts and granular debris of unknown origin.

A major advance in the histopathological interpretation of mummified tissue has been a method used by Riddle whereby paraffin sections may be removed from a slide after selected areas have been identified. These sections have been appropriately processed and transmission electron microscopy performed on them (Chapter 16). This technique

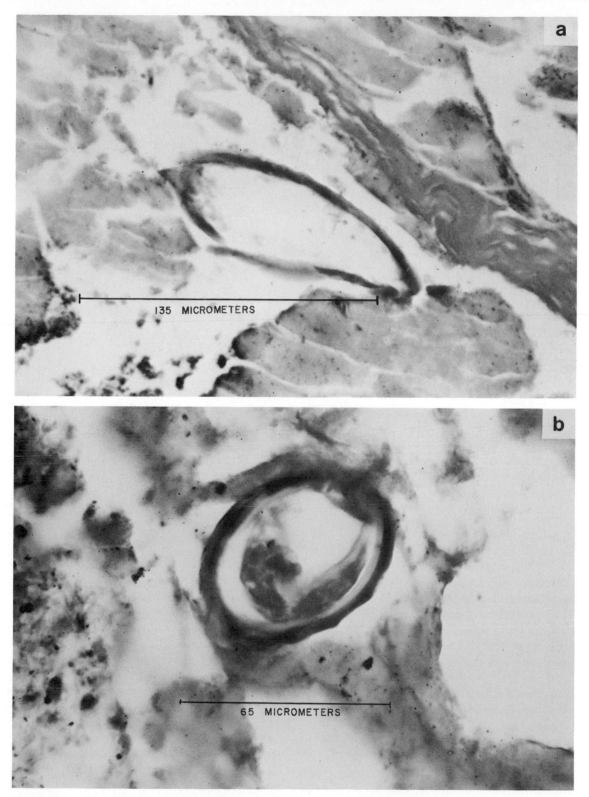

135 MICROMETERS

65 MICROMETERS

Figure 15.5.   Parasites in mummified tissue. Ova of Schistosoma haematobium (a), Ascaris (probably)
lumbricoides (b), and Taenia sp. (c) have been identified. Note the hooklets in the Taenia ova. The small,

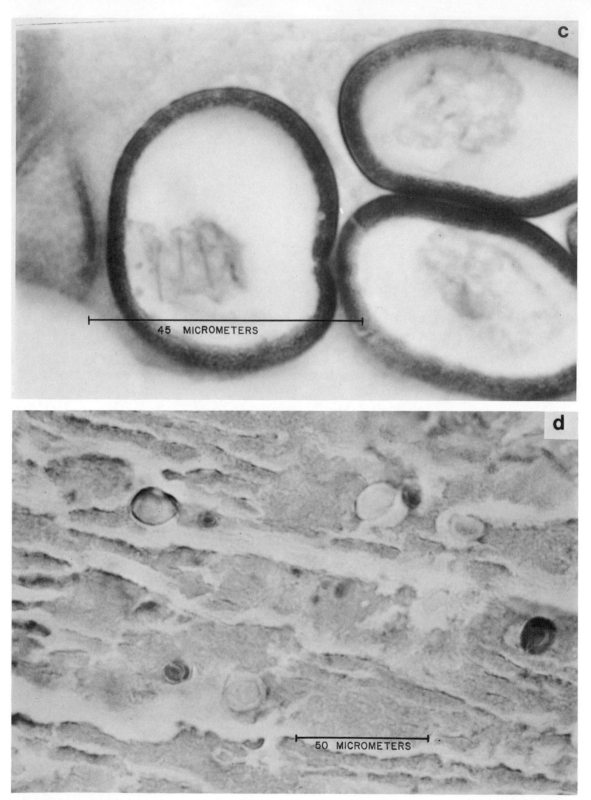

round laminated structures (d) are unfertilized Taenia ova. All suspected ova should be measured directly with an ocular micrometer.

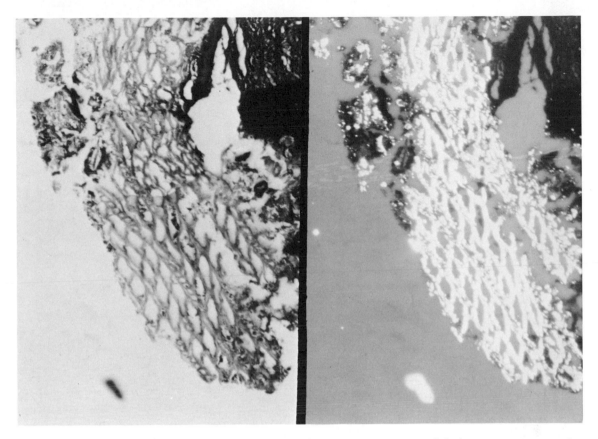

Figure 15.6.   *Plant material, identifiable by its usual geometric structures (left) having a microscopic appearance similar to cork. With polarized light, the plant cell walls are birefringent (right). Trichrome.*

should enable us to identify more correctly the various structures we have seen by light microscopy in sections of mummified tissue.

However, this is the final step. Careful handling, processing, and gross and microscopic scrutiny of this ancient tissue are prerequisites. With some forethought and ingenuity, most modern technological methods can be used in the study of disease and disease processes in ancient man.

ACKNOWLEDGMENTS

This report was supported in part by the Mount Carmel Research and Education Corporation, Detroit, Michigan.

REFERENCES

Culling, C. F. A. 1974. *Handbook of histopathological techniques*, 3rd ed., pp. 421–2. London: Butterworth.

DeGirolami, E. 1968. Plasma-thrombin technic in diagnostic cytology. *Bulletin of Pathology* 9(1):27.

Evans, W. E. 1962. Histological findings in spontaneously preserved bodies. *Medicine, Science and the Law* 2:153–64.

Luna, L. G 1968. *Manual of histologic staining: methods of the Armed Forces Institute of Pathology*, 3rd ed., p. 8. New York: McGraw-Hill.

Riddle, J. M.; Ho, K.; Chason, J. L.; and Schwyn, R. C. 1976. Peripheral blood elements in an Egyp-

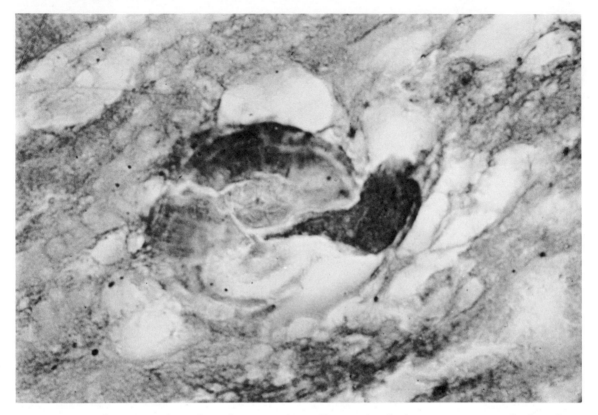

Figure 15.7.   Adipocere, a waxy degradation product of human lipids that occurs in many bodies, demonstrating a somewhat radially striate appearance in tissue. This material is also birefringent with polarized light. Trichrome.

tian mummy: a three dimensional view. *Science* 192:375.

Russell, N. L. 1956. A rapid double-embedding method for tissues. *Medical Laboratory Technology* 13:484–5.

– 1963. A rapid method for decalcification of bone for histological examination using the "histette." *Medical Laboratory Technology* 20:299–301.

Sandison, A. T. 1955. The histological examination of mummified materials. *Stain Technology* 30:277–83.

– 1957. The preparation of large histological sections of mummified tissues. *Nature* (London) 198:597.

Tapp, E.; Curry, A.; and Anfield, C. 1975. Sand pneumonoconiosis in an Egyptian mummy. *British Medical Journal* 2:276.

Thompson, S. E. 1966. *Selected histochemical and histopathological methods.* Springfield, Ill.: Thomas.

Van Cleve, H. J. and Ross, J. A. 1947. A method for reclaiming dried zoological specimens. *Science* 105:318.

Zimmerman, M. R. 1973. Blood cells preserved in a mummy 2000 years old. *Science* 180: 303–4.

Zugibe, F. T. 1970. *Diagnostic histochemistry.* St. Louis: Mosby.

16

A survey of ancient specimens by electron microscopy

JEANNE M. RIDDLE
Director of Research
Division of Rheumatology
Henry Ford Hospital
Detroit, Michigan, U.S.A.

The literature contains a limited number of reports that illustrate the application of electron microscopy to the investigation of ancient specimens. The transmission electron microscope has been used to study the fine structural details of samples of bone (Allison et al. 1974; Ascenzi 1963; Isaacs et al. 1963; Little 1960; Lynn and Benitez 1974; Race et al. 1968; Shackleford and Wyckoff 1964; Wyckoff et al. 1963), cartilage (Yeatman 1971), skin (Leeson 1959; Lewin 1967, 1968; Lewin and Cutz 1976), muscle (Macadam and Sandison 1969), sclera (Macadam and Sandison 1969), peripheral blood elements (Riddle et al. 1976), and a bacterium (Allison et al. 1974). Even fewer reports demonstrate the usefulness of scanning electron microscopy. Surface features of samples of ancient teeth (Lamendin 1974), hair (Chiarelli et al. 1970–1971), peripheral blood elements (Riddle et al. 1976), and an insect (Lynn and Benitez 1974) have been published to date.

This survey extends the existing information. Scanning electron microscopy is used to identify the composition of ancient textiles and to classify insects. Also a new technical approach that allows a single ancient specimen to be viewed with a continuum of microscopy has been developed.

SCANNING ELECTRON MICROSCOPY

PRINCIPLES OF OPERATION
The source of illumination is electrons, produced by energizing a tungsten wire filament with a specific amount of current. The resultant beam of electrons is focused, demagnified into a probe, and scanned across the surface of the specimen in an ordered pattern. Specimens studied in the scanning electron microscope are usually coated with a thin layer of gold. Interactions between the probe of electrons and the coated surface of the specimen result in the emission of secondary electrons. These are collected, amplified, and displayed on a cathode ray tube. This final image is photographed. The major advantage of the scanning electron microscope is its increased depth of focus. The surface topography of specimens appears to be displayed three dimensionally.

Specimens studied in this survey include peripheral blood elements, textiles, human hair, and insects.

PERIPHERAL BLOOD ELEMENTS
A group of peripheral blood elements was found when a brown mass located in the skull of PUM III (a female Egyptian mummy dated to 835 B.C. ± 70 years) was examined (Riddle et al. 1976).

Preparative procedure. This specimen was prepared as follows for examination with the light microscope:

1. The specimen was placed in 10 percent formalin for 4 days.
2. Dehydration with ethanol was accomplished as follows:
   a. 70 percent, 72 hours

b. 80 percent, 4 hours
c. 95 percent, 4 hours
d. 95 percent, overnight
e. 100 percent, 1 hour
f. 100 percent, 3 hours
3. The infiltration and embedding scheme
    followed the schedule detailed.
    a.   Zylene, 2 hours
    b.   Melted paraffin (56° to 58° C),
         3 hours
    c.   Fresh melted paraffin (56° to 58° C),
         overnight
    d.   Embedment in paraffin
4. Sections 6 μ thick were cut on an ordi-
    nary microtome, stained with hema-
    toxylin and eosin, and mounted.
5. Erythrocytes and leukocytes of various
    types were observed. Each leukocyte was
    located, classified, and photographed.

After the light microscopic study was
completed, sections 6 μ thick were prepared for scanning electron microscopy.

1. The coverslip was loosened and removed
    by immersing the slide overnight in
    xylene.
2. Any remaining mounting media (Per-
    mount) were dissolved by exposing the
    section for 10 minutes in three separate
    changes of fresh xylene.
3. The section was allowed to air dry.
4. The glass microscope slide was scored
    and broken. Only that portion with the
    section attached remained intact.
5. This piece of glass slide was glued to an
    aluminum stub with the section exposed.
6. The section was coated with a thin layer
    of gold (approximately 200 Å thick) using
    the sputtering procedure.
7. The surface topography of the peripheral
    blood elements was displayed using an
    ETEC Autoscan scanning electron micro-
    scope operated at 20 kV.

Results: At the level of the light microscope,
erythrocytes (Figure 16.1) as well as leukocytes including the neutrophil (Figure 16.2),

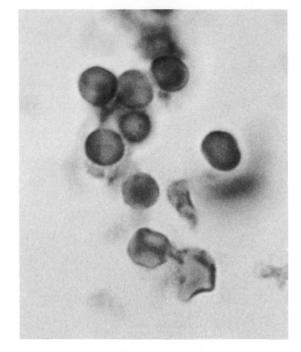

Figure 16.1.   *Erythrocytes in the cranial mass
found in PUM III, viewed by light microscopy.
Hematoxylin and eosin.* × 3,700.

eosinophil (Figure 16.3), and lymphocyte
(Figure 16.4) were easily identified.

In the scanning electron microscope, fibrin
strands with which the various peripheral
blood elements were enmeshed were obvious
(Figure 16.5). The erythrocyte population
consisted of intact cells. They ranged in
diameter from 3.4 to 4.3 μ. Some erythrocytes
were spherical; others nearly retained their
typical biconcave configuration (Figure
16.6), and still other red blood cells were
markedly misshapen. The exterior surfaces of
the erythrocytes were smooth, wrinkled, or
had a pebbled appearance.

Neutrophils were identified most frequently in the white blood cell population.
They were round to oval in shape and had an
average diameter of 5.7 by 6.1 μ. Their surface
topography exhibited a relatively smooth exterior with only a few scattered, short projections (Figure 16.7). Occasionally, the in-

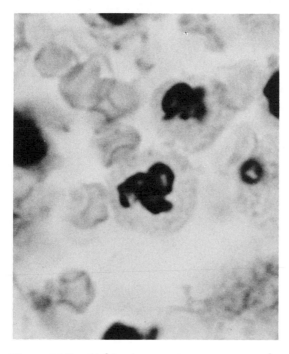

Figure 16.2.   Light microscopic appearance of an ancient neutrophil. Hematoxylin and eosin. × 4,820.

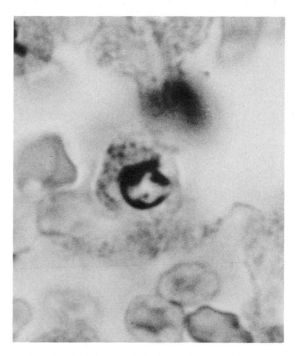

Figure 16.3.   Eosinophils from PUM III. × 4,820.

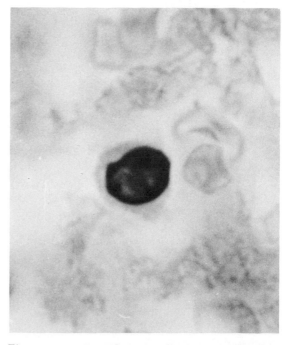

Figure 16.4.   Lymphocytes displaying their typical characteristics of a nucleus and scant cytoplasm. × 4,820.

terior of a neutrophil was visible. We then observed the general outline of the nucleus and other cytoplasmic organelles. In the cytoplasm there were many small, rounded structures with a mean diameter of 0.2 μ. The size, shape, and location of these organelles were compatible with known characteristics of the specific granulation of the neutrophil, a major identifying feature of this type of leukocyte (Figure 16.8).

Another type of granulocyte, the eosinophil, was identified by its cytoplasmic granules, which were round to oval in shape and on the average measured 0.4 by 0.6 μ in diameter (Figure 16.9).

Lymphocytes were smaller, measuring from 4.1 to 4.7 μ across. Their exterior surface was either smooth or roughened by an occasional microvillous projection or broad fold (Figure 16.10).

A few small rounded structures that measured approximately 1.2 μ in diameter were

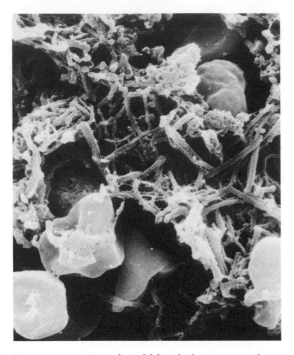

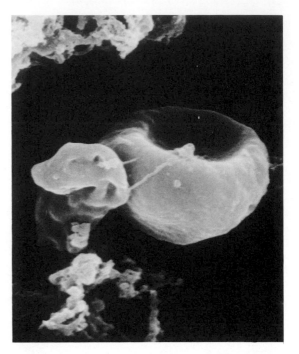

Figure 16.5.   *Peripheral blood elements in the cranial mass from PUM III, enmeshed in a network of fibrin.* × 5,070.

Figure 16.6.   *Erythrocytes that have partially retained their biconcave shape.* × 10,150.

either individually attached to the fibrin network or adhered as a cohesive aggregate. Their exterior surface demonstrated a few short, blunt pseudopods (Figure 16.11). These morphologic features were consistent with the interpretation that these entities were blood platelets.

The various types of leukocytes in this ancient specimen were classified using the same nuclear and cytoplasmic criteria we would utilize for the identification of their modern counterparts. This suggests that no major structural alterations have occurred in peripheral blood elements for at least 2,700 years. Erythrocytes and leukocytes were reduced in size by about 50 percent; however, the blood platelets remained within their normal size range of from 1 to 3 μ. The specific granules of both the neutrophil and the eosinophil were not significantly decreased in size.

Whether this focal collection of erythrocytes, leukocytes, platelets, and fibrin represents an antemortem subdural hematoma or a postmortem blood clot is debatable.

TEXTILES

The scanning electron microscope is well suited for studying the pattern of weaving and the single fibers from which individual threads of a fabric are composed. The ancient textiles included in this survey are wrappings from PUM II (a male Egyptian mummy dated to 170 B.C. ± 70 years), a fibrous ball found among the wrappings of PUM II, wrappings from PUM III, and several Peruvian burial blankets from the Huari culture (pre-Inca period). Appropriate reference fibers were also investigated.

Preparative procedure: A representative area of the textile or a few individual fibers were processed as follows:

1.   The top of an aluminum stub was covered with double-coated tape (Scotch brand, No. 410, 12.7 mm wide, 3M Company).

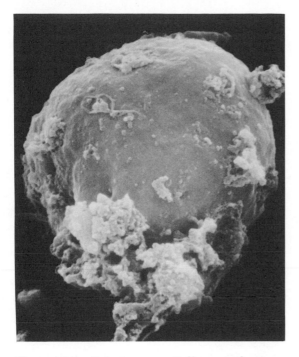

Figure 16.7. Exterior, essentially smooth, topography of neutrophils. × 10,150.

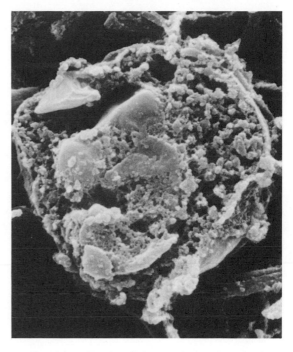

Figure 16.8. Nuclear lobes and cytoplasmic granules observed inside the neutrophil by scanning electron microscopy. × 10,150.

2. The specimen was mounted by pressing it against the tape.
3. The exposed surface of the specimen was coated with a thin layer of gold.
4. The surface topography of both the textiles and individual fibers was examined using an ETEC Autoscan scanning electron microscope operated at 20 kV.

Results: Wrappings from PUM II consisted of a fabric that was constructed from relatively fine threads. Single threads of this fabric were composed of slender, relatively straight fibers that exhibited prominent jointlike markings (Figure 16.12). These fibers were identical to reference flax fibers, thereby identifying the ancient textile as a linen fabric.

The fibrous ball was made up of randomly oriented individual fibers. These fibers were straight and displayed a repeating, inherent twist (Figure 16.13), which corresponded exactly to the three-dimensional appearance of reference cotton fibers. This small cotton ball was apparently used as some sort of pad, perhaps during the mummification process, as it was coated with a thin layer of gray claylike material and separately wrapped in a piece of unusually fine linen. This small amount of cleaned but unspun cotton is reportedly the earliest found in use in Egypt or in the Western World as a whole (Johnson 1974).

Certain of the wrappings from PUM III demonstrated crystalline formations associated with the surface of individual threads. These concretions were present in two forms: flat, oval masses and spicular crystals.

The Peruvian burial blankets were of varying textures. One blanket was composed of relatively thin threads, whereas two others were made up of heavier yarns that showed

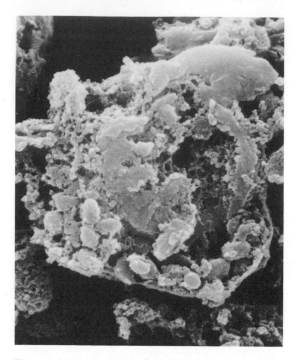

Figure 16.9.   Interior of eosinophils, containing large, oval specific granules. × 10,150.

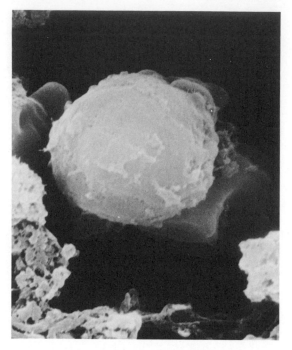

Figure 16.10.   Surface of lymphocytes, revealing a few folds. × 10,150.

color variations from light tan to dark brown. Examination of the individual fibers from which all the blankets were composed revealed a straight fiber that displayed a repeating, inherent twist, identifying these Peruvian burial blankets as cotton textiles.

HAIR

The microarchitecture of a few hairs from the head of PUM IV, a mummified male child from Egypt, was investigated by scanning electron microscopy.

Preparative procedure: Several single hairs were processed.

1. Hairs were fixed overnight in 6.5 percent buffered glutaraldehyde (Sabatini et al. 1964).
2. They were transferred to buffered sucrose and stored for 24 hours at 4 °C.
3. The hair was then dehydrated in ethanol:
    a. 70 percent, 15 minutes
    b. 80 percent, 15 minutes
    c. 95 percent, 15 minutes
    d. 95 percent, 15 minutes
    e. 100 percent, 30 minutes
    f. 100 percent, 1 hour
4. Each hair was air-dried.
5. An aluminum stub was covered with double-coated tape.
6. The hairs were mounted by gently pressing them against the tape.
7. The exposed hair was coated with a thin layer of gold.
8. The surface topography of the hairs was examined.

Results: Human hairs are keratinous filaments covered by scalelike plates (epicuticles). These represent an extremely flattened modification of keratinized epidermal cells. The lines of epicuticle borders normally run transversely and are spaced at regular intervals. The borders of the epicuticles are generally smooth with only a few serrations.

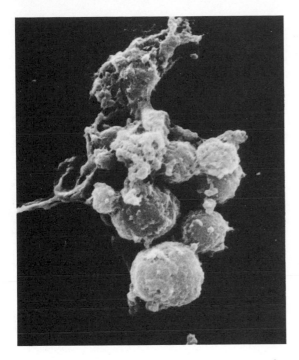

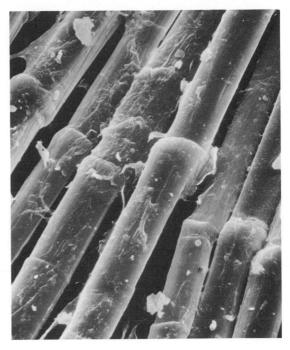

Figure 16.11. *Blood platelets seen in conjunction with the fibrin network.* × 10,150.

Figure 16.12. *Fibers (flax) from the wrappings of PUM II. Note the straight appearance and prominent jointlike markings.* × 600.

Hairs from the head of PUM IV showed prominent scalelike plates. Epicuticular borders were oriented transversely and spaced at fairly regular intervals. Individual epicuticles demonstrated raised edges and pitted surfaces.

INSECTS

Scanning electron microscopy provides the capability of viewing adult insects and their larval forms three dimensionally. Different areas of the specimen can be photographed so that the mouthparts, number and type of appendages, and microarchitecture of the outer covering can be displayed in detail. Compilation of this information can provide necessary clues for identification and classification of ancient insects. The majority of specimens included in this survey were obtained from PUM III. One specimen was found associated with PUM IV.

Preparative procedure: Each insect was either examined in the original state of preservation or, if necessary, its outer surface was cleaned.

1. For cleaning, the insect was placed in chloroform overnight. The chloroform was replaced with 60 percent ethanol for 30 minutes. Loosened resin and debris were gently teased from the surface of the specimen under a dissection microscope.

2. An aluminum stub was covered with double-coated tape.

3. The insect was mounted by gently pressing either its dorsal or ventral surface against the tape. Careful attachment of the specimen allowed the insect to be detached after one surface had been studied, turned over, and the opposite surface viewed, in case structural features on both the top and the bottom of the specimen were required for its positive identification.

Figure 16.13.   Cotton fibers from PUM II. Note the repeating inherent twist. × 1,010.

Figure 16.14.   Surface details of a dermestid larva. × 65.

4. The exposed surface of the insect was coated with a thin layer of gold.
5. Representative micrographs were produced using an ETEC Autoscan scanning electron microscope operated at 20 kV.

Results: Several scanning electron micrographs depicting the general topography of each insect were obtained and supplied to Dr. George Skeyskal, Research Entomologist, United States Department of Agriculture.

The specimen found among the linen wrappings of PUM III was identified by Dr. John M. Kingsolver as a dermestid larva (Thelodrias contractus Motschulsky, family Dermestidae). Survey of this larval form (Figure 16.14) included the detailed morphology of its mouthparts, lateral appendages (Figure 16.15), and hairlike surface projections (Figure 16.16).

Another larval form found in the ear canal of PUM III was classified by Dr. Raymond Gagne as a fly larva (Chrysomya albiceps Wiedemann, family Calliphoridae), a blowfly that is still common in the Mediterranean region.

A third specimen found embedded in the flesh of the pelvic region of PUM III had wings. It was identified as a beetle Necrobia sp., probably the species Necrobia refipes De Geer, commonly known as the red-legged ham beetle in the family Cleridae, subfamily Corynetinae. Because the head portion of the insect was missing, the species designation is questionable.

Numerous insects were observed grossly within the fleshy portions of the body of PUM IV. The dorsal view of one intact example of these insects demonstrated definite wings and the ventral view (Figure 16.17) showed distinct mouthparts and two compound eyes. Dr. R. E. White identified this insect as a member of the Anobiidae family known as the drug-store beetle, Stegobium paniceum (Linnaeus).

Figure 16.15.   Intact lateral appendages of the dermestid larva. × 250.

Figure 16.16.   Hairlike projections, an outstanding feature of the exterior of the dermestid larva. × 980.

TRANSMISSION ELECTRON MICROSCOPY

PRINCIPLES OF OPERATION

The source of illumination is electrons, produced by energizing a tungsten wire filament with a specific amount of current, frequently 50 kV. The resultant beam of electrons passes through a series of electromagnetic lenses and interacts with the specimen. The final image is composed of many different densities. This image is viewed on a phosphercoated surface, as electrons evoke the emission of a visible type of illumination when they strike this material. The final image is directly recorded on negative film, which can subsequently be used to produce permanent black-and-white prints. The major advantage of the transmission electron microscope is its resolution capability of from 2 to 5 A.

Examination of ancient tissues by transmission electron microscopy allows us to evaluate the degree of cellular stabilization achieved by the various methods of preservation. Disease processes can also be investigated at a subcellular level of organization.

PERIPHERAL BLOOD ELEMENTS

The specimens used for this investigation were the same 6 μ sections from PUM III. *Preparative procedure:*

1.   Each 6 μ section, previously stained by hematoxylin and eosin as well as coated with gold, was placed under a dissecting microscope.
2.   All the sectioned material and gold coating around the focal collection of peripheral blood elements was scraped away.
3.   The piece of glass slide with the selected sectioned material attached was placed in a 6.25–cm watch glass.

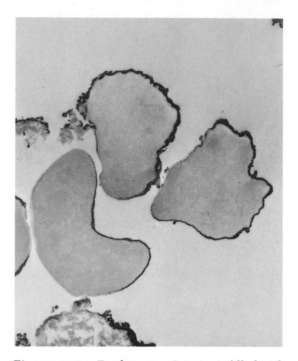

Figure 16.17.   Ventral view of a beetle collected from the scalp of PUM IV. × 36.

Figure 16.18.   Erythrocytes. Interior is filled with a homogeneous material. × 10,660.

4. The sectioned material was covered with different solvents and embedding media as follows:
    A. 100% percent propylene oxide (15 minutes, 2 changes)
    b. 90 percent propylene oxide + 10 percent Maraglas (30 minutes, 1 change)
    c. 80 percent propylene oxide + 20 percent Maraglas (30 minutes, 1 change)
    d. 60 percent propylene oxide + 40 percent Maraglas (30 minutes, 1 change)
    e. 50 percent propylene oxide + 50 percent Maraglas (30 minutes, 1 change)
    f. 100 percent Maraglas (60 minutes, 3 changes)
5. The sectioned material covered by 100 percent Maraglas was placed in a 60°C oven overnight.

6. The glass slide was separated by exposing the bottom of the watch glass to liquid nitrogen for 30 seconds.
7. The polymerized Maraglas and sectioned material were peeled off the microscope slide.
8. Excess embedding medium was removed by using a coping saw and then a single-edged razor blade.
9. The embedded sample was attached to a previously polymerized block in Maraglas.
10. Ultrathin sections were cut using an LKB Ultratome and a diamond knife.
11. Ultrathin sections were double-stained with lead citrate (Reynolds 1963) and uranyl acetate (Watson 1958). They were viewed in an RCA EMU – 4 transmission electron microscope.

Results: Individual strands of fibrin exhibited a distinct periodicity, which measured from 170 to 190 Å. The periodicity of fibrin

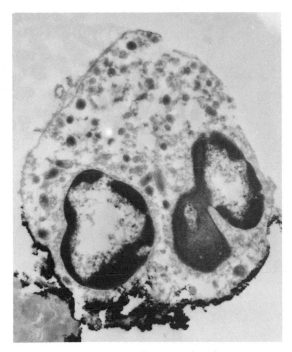

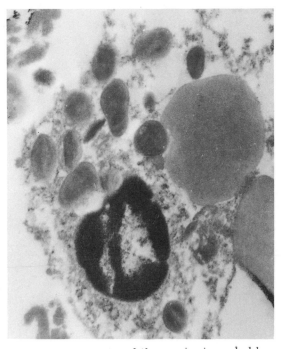

Figure 16.19. Neutrophils in ultrathin sections, displaying nuclear lobes and membrane-bound granules. × 11,380.

Figure 16.20. Eosinophil granules bounded by a limiting membrane. Some contained a central dense crystalloid. × 17,650.

processed using modern techniques measures from 200 to 250 Å.

The plasma membrane of erythrocytes was visible only occasionally, because their surfaces were covered with gold. Erythrocytes demonstrated both concave and wavy outlines. The interiors of these red blood cells were filled with a homogeneous material of moderate electron density, presumably hemoglobin. No intracellular inclusions were seen (Figure 16.18).

The major features that characterize neutrophils – lobed nucleus and prominent cytoplasmic granules – were present. Nuclear lobes showed peripheral chromatin clumping, a limiting membrane, and nuclear pores. Individual neutrophilic granules ranged from round to oval in shape, were limited by a membrane, and contained a homogeneous material that varied in denseness (Figure 16.19).

The outstanding feature of eosinophils was their cytoplasmic granules. Eosinophilic granules were limited by a membrane, and some displayed a typical central, dense crystalloid, surrounded by less dense homogeneous material (Figure 16.20).

Lymphocytes demonstrated typical nuclear ultrastructure such as marked chromatin clumping, nuclear pores, and a double nuclear membrane. The outer nuclear membrane was studded with ribosomes. Cytoplasmic features included free ribosomes, mitochondria, and profiles of rough endoplasmic reticulum (Figure 16.21).

COMBINED MICROSCOPIC APPROACH

A series of techniques was developed to study cells and tissue components within ancient specimens by a continuum of microscopy. The following example illustrates the method's advantages. A small group of structures that

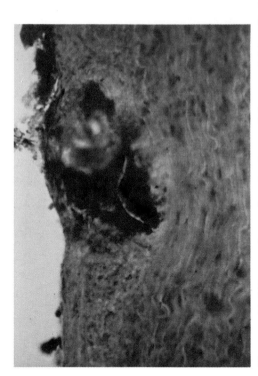

Plate 5.   Differentiation of collagen and elastic fibers by the elastic tissue–van Gieson stain. Elastic fibers appear as black, wavy lines. The defect near the top is a small lipid plaque.

Plate 6.   Mummified skeletal muscle stained red with Masson's trichrome, indicating good preservation. The pale horizontal (originally blue) band is perimysial connective tissue; the clear defects are fracture artifacts produced during sectioning of the tissue.

Plate 7.   Lipid in an atherosclerotic plaque of the aorta is identified using Oil Red O stain and is seen as the intensely red area in this section. The pale wavy lines near the bottom are unstained elastic fibers. This area corresponds to the intimal defect seen in Plates 5 and 8.

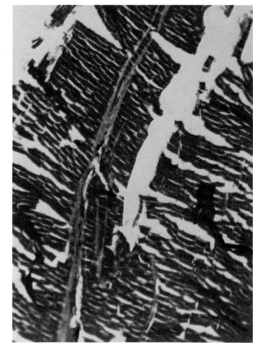

Plate 8.   This is the same area of the aorta as seen in Plates 5 and 7. Collagen is blue and smooth muscle is red. Some of the elastic fibers have stained irregularly red as well. Trichrome

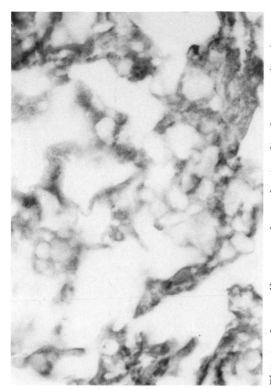

Plate 1. Residual cross-striations in skeletal muscle, indicating good preservation. Phosphotungstic acid–hematoxylin.

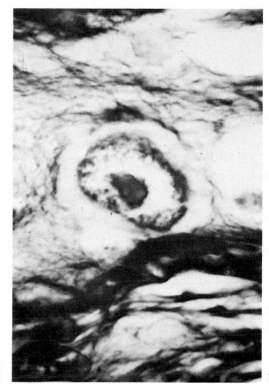

Plate 2. Less well preserved muscle visualized as vacuolated masses of poorly staining tissue. Phosphotungstic acid–hematoxylin.

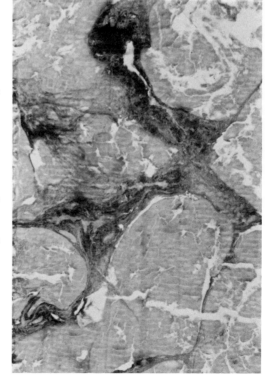

Plate 3. Section of liver, demonstrating the relationship of the connective tissue (seen as the dark connecting bands) to the parenchyma (the pale nodular zones). The parenchymal cells are degenerated and appear amorphous. Elastic tissue–van Gieson.

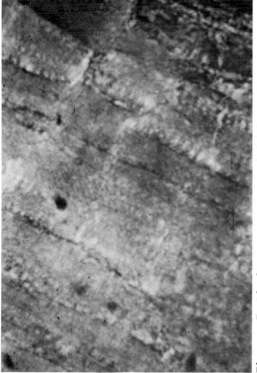

Plate 4. Well-preserved cartilage, indicated by the persistence of a virtually intact chondrocyte from an intervertebral disc. Trichrome.

REFERENCES

Allison, M. J.; Pezzia, A.; Gerszten, E.; and Mendoza, D. 1974. Case of Carrión's disease associated with human sacrifice from the Huari culture of southern Peru. *American Journal of Physical Anthropology* 41:295–300.

Ascenzi, A. 1963. Microscopy and prehistoric bone. In *Science in archeology*, D. Brothwell and Higgs (eds.), pp. 526–38. New York: Praeger.

Chiarelli, B.; Fuhrman, A. C.; and Massa, E. R. 1970–71. Nota preliminare sulla ultrastruttura dei capelli di mummia egiziana al microscopio elettronico a scansione. *Rivista di Antropologia* 57:275–8.

Isaacs, W. A.; Little K.; Currey, J. D.; and Tarlo, L. B. H. 1963. Collagen and cellulose-like substance in fossil denture and bone. *Nature* 197:192.

Johnson, M. 1974. *Paleopathology Newsletter* 5:3–4.

Lamendin, H. 1974. Observations with SEM of rehydrated mummy teeth. *Journal of Human Evolution* 3:271–4.

Leeson, J. D. 1959. Electron microscopy of mummified material. *Stain Technology* 34:317–20.

Lewin, P. K. 1967. Palaeo-electron microscopy of mummified tissue. *Nature* 213:416–17.

– 1968. The ultrastructure of mummified skin cells. *Canadian Medical Association Journal* 98:1011–12.

Lewin, P. K., and Cutz, E. 1976. Electron microscopy of ancient Egyptian skin. *British Journal of Dermatology* 94:573–6.

Little, K. 1960. The matrix of old and osteoporotic bones. *Proceedings of the European Regional Conference on Election Microscopy* 2:791.

Lynn, G. E., and Benitez, J. T. 1974. Temporal bone preservation in a 2600-year-old Egyptian mummy. *Science* 183:200–2.

Macadam, R. F., and Sandison, A. T. 1969. The electron microscope in palaeopathology. *Medical History* 13:81–5.

Race, G. J.; Fry, E. L.; Matthews, J. L.; Wagner, M. J.; Martin, J. H.; and Lynn, J. A. 1968. Ancient Nubian human bone: a chemical and ultrastructural characterization including collagen. *American Journal of Physical Anthropology* 28:157–62.

Reynolds, E. S. 1963. The use of lead citrate at high pH as an electron opaque stain in electron microscopy. *Journal of Cell Biology* 17:208.

Riddle, J. M.; Ho, K.; Chason, J. L.; and Schwyn, R. C. 1976. Peripheral blood elements found in an Egyptian mummy: a three-dimensional view. *Science* 192:374–5.

Sabatini, D. D.; Miller, F.; and Barrnett, R. J. 1964. Aldehyde fixation for morphological and enzyme histochemical studies with the electron microscope. *Journal of Histochemistry and Cytochemistry* 12:57–71.

Shackleford, J. M., and Wyckoff, R. W. G. 1964. Collagen in fossil teeth and bone. *Journal of Ultrastructure Research* 11:173–80.

Watson, M. L. 1958. Staining of tissue sections for electron microscopy with heavy metals. II. Application of solutions containing lead and barium. *Journal of Biophysical and Biochemical Cytology* 4:475–9.

Wyckoff, F. W. G.; Wagner, E.; Matter, P.; and Doberenz, A. R. 1963. Collagen in fossil bone. *Proceedings of the National Academy of Sciences of the United States* 50:215–21.

Yeatman, C. W. 1971. Preservation of chondrocyte ultrastructure in an Aleutian mummy. *Bulletin of the New York Academy of Medicine* 47:104–8.

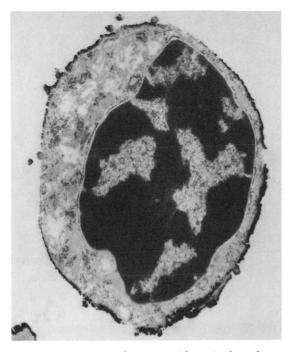

Figure 16.21.   Lymphocytes with typical nucleus. Their cytoplasm showed mitochondria and free ribosomes. × 19,570.

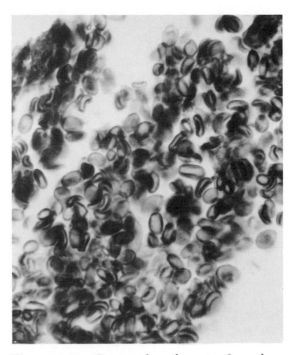

Figure 16.22.   Presumed erythrocytes from the stomach of PUM II. × 725.

appeared to be erythrocytes by light microscopy was found in a section of the stomach from PUM II (Figure 16.22). Examination of these structures with the scanning electron microscope revealed intact structures that were round to oval in shape, exhibited a pebbled surface and a concave central area, and measured 7.4 by 4.0 μ. Their interior displayed a hollow central area limited by a moderately thick wall (Figure 16.23). By transmission electron microscopy, these structures demonstrated a central area that was either "empty" or contained membranous components and was surrounded by a wall that had outer and inner limiting membranes. This entity clearly did not have the characteristic features of an erythrocyte when viewed by electron microscopy. By applying this series of microscopic studies, a misconception was corrected and other possibilities for the identity of these structures, such as yeasts or fungal spores, were suggested.

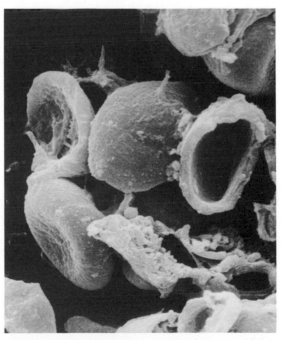

Figure 16.23.   Scanning electron micrograph of the presumed erythrocytes, revealing a hollow central area. They are actually fungal spores. × 4,860.

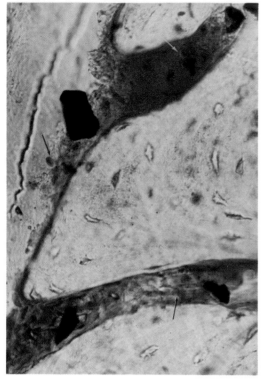

Plate 11. Undecalcified horizontal section from the right temporal bone of a 1,600-year-old frozen Eskimo (Saint Lawrence Lady). The haversian systems contain red blood cells (black arrows) and white cells (white arrow). Villanueva stain.

Plate 12. Undecalcified horizontal section from the left temporal bone of PUM II. There is excellent preservation of the vascular channels. Villanueva stain.

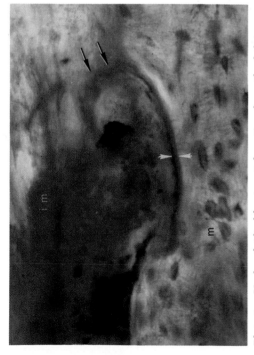

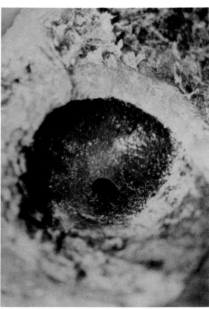

Plate 9. Undecalcified horizontal section from the left temporal bone of PUM II, showing an active osteoid seam (white arrows). Resorption foci are identified by Howship's lacunae (black arrows). There is an area of incompletely mineralized bone (im). The cells below the osteoid seam are osteocytes with nuclei; the clear area around them is mineralized bone matrix (m). Villanueva stain.

Plate 10. Right eardrum of a 2,000-year-old Egyptian mummy, PUM II. There is a thin coating of resin covering the entire drum surface. An oval perforation of about 10 percent of the pars tensa is seen in the posteroinferior quadrant. Its smooth and well-circumscribed margins suggest it resulted from middle ear infection.

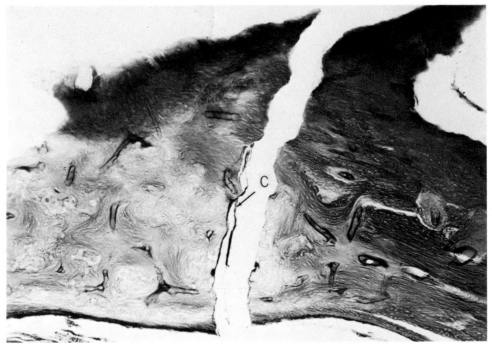

Plate 13.   Histological detail of fracture line shown in Figure 18.7. There are small strands of connective tissue (C) in between the margins, indicating that this is not an artifact. Hematoxylin and eosin.

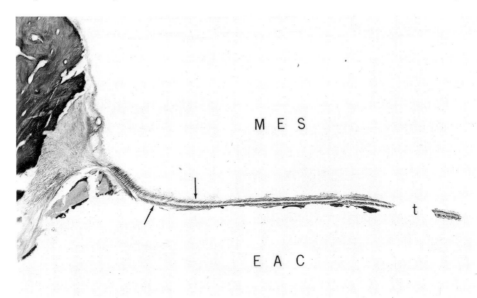

Plate 14.   Photomicrograph of horizontal section through remnants of the right eardrum of a 4,000-year-old Egyptian mummy, ROM II (temporal bone processed without rehydration). Note histological detail of fibrous layer (arrows) including postmortem tear (t). MES, middle ear space; EAC, external auditory canal. Hematoxylin and eosin.

topsy, we discovered the material to be hardened resin that had been introduced in a liquid state into the cranial cavity through the defect in the left ethmoid sinus and cribriform plate (see Chapter 4). Laminographic examination of the ethmoid sinuses showed the defect in the bony trabeculae. Because the embalming had apparently been performed with the body in the supine position, when the resin solidified it assumed this characteristic pattern. The best example of identification of intact organs was seen in ROM I. No evisceration had been performed at the time of mummification, and all the organs were in situ. Dr. David Rideout was able to identify most of the major organs radiographically. At autopsy, his interpretation was confirmed; his findings are considered in detail in Chapter 5.

A somewhat similar situation was noted following the radiographic examination of PUM III. The pleural cavities were filled with air and the lungs appeared as small, slightly dense, symmetrical paravertebral zones. No intrinsic calcification could be identified. The heart was seen as an ill-defined retrosternal density. The aorta was not seen, and no vascular calcification was noted. There was clear delineation of the thoracic and abdominal cavities by the diaphragm.

Examination of the abdomen revealed a crescent-shaped structure in the right upper quadrant, thought to be shrunken liver. In the pelvis, two central rounded densities could be seen and were identified as urinary bladder and uterus. Within the abdominal cavity there were large irregular densities that contained numerous opacities of varying sizes that approached the density of bone. A sharply outlined but otherwise similar appearing structure protruded past the diaphragm into the thoracic cavity. Although we were tempted to call this intestine, we could not explain all the granular densities. At autopsy, the heart and lungs were present as were remnants of the diaphragm. The right upper quadrant mass was liver, and the pelvic structures were the urinary bladder and uterus. However, the irregular densities in the ab-

dominal cavity were linen wads stuffed into the body during the mummification (Chapter 6). The structure protruding into the chest was of similar material.

Another form of soft tissue found with mummies are the organs that have been removed during mummification and then returned to the body as visceral packages. Two packages were found in DIA I, but examined radiographically later (DIA II). This examination showed that one was very likely soft tissue, and histological examination revealed that it was composed of degenerated but recognizable muscle tissue, perhaps heart. The second package had a peculiar wavy internal configuration with numerous granular densities, similar in type to that seen in the abdomen of PUM III. When sectioned, the packages were found to consist of wads of resin-soaked linen, presumably used to seal the embalmer's incision.

In PUM II, radiographic examination of the chest and abdomen revealed the presence of four elongated, cylindrical masses, one in the right hemithorax and three crisscrossed in the pelvis (Figure 17.9). We thought these were visceral packages, but could not determine radiographically if indeed there was tissue in them. Subsequently, histological examination showed that they contained lungs, spleen, and intestine. Much of the density of the packages was attributed to the several layers of linen impregnated with resin that had been used to wrap the dehydrated organs.

CONCLUSION

As one tool in the investigation of mummies, radiographic examination has proved its worth. In several instances, abnormalities probably would have gone unnoticed without prior radiographic examination. Significant data, both medical and anthropological, have been derived from its use.

REFERENCES

Bourke, J. B. 1967. A review of the paleopathology of the arthritic diseases. In *Diseases in an-*

tiquity, D. Brothwell and A. T. Sandison (eds.), pp. 352–70. Springfield, Ill.: Thomas.

Brothwell, D.; Molleson, T.; Gray, P. H. K.; and Harcourt, R. 1970. The application of x-rays to the study of archeological materials. In *Science in archeology*. D. Brothwell and E. Higgs (eds.), pp. 513–25. New York: Praeger.

Chapman, F. H. 1972. Vertebral osteophytosis in prehistoric populations of central and southern Mexico. *American Journal of Physical Anthropology* 36:31–8.

Gray, P. H. K. 1967. Calcinosis intervertebralis, with specimen reference to similar changes found in mummies of ancient Egypt. In *Diseases in antiquity*, D. Brothwell and A. T. Sandison (eds.), pp. 20–30. Springfield, Ill.: Thomas.

Greulich, W. W., and Pyle, J. I. 1959. *Radiographic atlas of skeletal developmental of the hand and wrist*, 2nd ed. Stanford: Stanford University Press.

Harris, J. E., and Weeks, K. R. 1973. *X-raying the pharaohs*. New York: Scribner's.

Hoerr, N. L.; Pyle, J. I.; and Francis, C. C. 1962. *Radiographic atlas of skeletal development of foot and ankle*. Springfield, Ill.: Thomas.

Pyle, J. I., and Hoerr, N. L. 1955. *Radiographic atlas of the skeletal development of the knee*. Springfield, Ill.: Thomas.

Schine, H. R.; Baensch, W. E.; Friedl, E.; and Uehlinger, E. 1951. *Roentgen diagnostics*. New York: Grune & Stratton.

Wells, C. 1967. A new approach to paleopathology: Harris's lines. In *Diseases in antiquity*, D. Brothwell and A. T. Sandison (eds.), pp. 390–404. Springfield, Ill.: Thomas.

Radiographic examination of mummies with autopsy correlation

KARL T. KRISTEN
Department of Radiology
Mount Carmel Mercy Hospital
Detroit, Michigan, U.S.A.

THEODORE A. REYMAN
Director of Laboratories
Mount Carmel Mercy Hospital
Detroit, Michigan, U.S.A.

The use of radiography in the study of mummies is well established (Brothwell et al. 1970; Harris and Weeks 1973). These techniques enable us to learn a great deal about the mummy without using invasive methods. This chapter reports the results of the radiographic examination of six Egyptian mummies and a few separate parts (Table 17.1) and the correlation of the radiographic findings with the pathologic conditions noted at the time of autopsy.

Absence of motion artifact and the freedom to increase the exposure dose allow a more optimal use of high-resolution and contrast techniques than can be achieved with live patients. Long tube–film distance decreases geometric magnification. This in conjunction with long roll films facilitates recording and viewing of the whole skeleton, or at least a larger portion thereof than could be seen using the usual format dimensions (Figure 17.1). Also, with the mummy secured to a movable cradle, we have utilized fluoroscopy to determine whether previously identified masses within the body cavities are fixed or movable.

Artifacts within the wrappings or within the body cavities can hinder evaluation of anatomic structures. Body section radiography (laminography) has been particularly useful here. Ultrathin section laminography (microtomography) has been quite useful in the evaluation of the temporal bones, paranasal sinuses, and facial structures.

The line-edge enhancement phenomenon of xerography may be particularly useful in the study of soft tissues and osseous trabecular patterns.

SKULL

The assessment of the skull and upper spinal segments follows standard interpretation. Cranial vascular imprint patterns may be seen. The inner and outer tables of the calvarium, the sella, the paranasal sinuses, and the mastoid air cells usually can be identified without difficulty. The examination of the maxilla, the mandible, and the dental apparatus may present positional problems because the head and neck cannot be moved in the ordinary manner (see Chapter 3). Laminographic study of the temporal bone for the auditory structures is considered in Chapter 18.

We have seen three cases in which the radiographic changes were interpreted as skull fractures. The first case was a separate adult male head from the Detroit Institute of Arts (DIA IV) in which there were bilateral comminuted temporal fractures involving the left zygoma and maxilla (Figure 17.2). The characteristics were those of blunt force injury. The fractures were not typical of those that might result from postmortem damage caused by mishandling. Additionally, soft tissue was pressed into the defect between the bone fragments and there was amorphous soft tissue clinging to the inner aspect of the fracture site. Normally, this portion of the cranial cavity is devoid of any soft tissue. We attempted to identify histologically blood that

Table 17.1. *Data on Egyptian mummies and separate parts*

Designation[a]	Historical period[b]	Sex[c]	Age (yr)[d]	Specimen[e]
DIA I	1220 B.C.	Male	30–35	Mummy
DIA II	1220 B.C.	Male	30–35	Visceral packages from DIA I
DIA III		Male(?)	Adult	Vertebrae only
DIA IV		Male (?)	Adult	Head only
PUM I	892 B.C.	Male	30–35	Mummy
PUM II	170 B.C.	Male	30	
	1200 B.C.	Male	14–18	Mummy
PUM III	835 B.C.	Female	20–25	Mummy
PUM IV	?100 B.C.	Male	8–10	Mummy
	A.D. 100			

[a] DIA, Detroit Institute of Arts; PUM, Pennsylvania University Museum; ROM, Royal Ontario Museum.
[b] Determined by radiocarbon dating or available historical data; DIA III and DIA IV could not be dated.
[c] As determned grossly at autopsy.
[d] Estimated on basis of combined anthropological, radiographic, and historical data.
[e] Radiographic examination of DIA I, DIA II, DIA III, and PUM II by John Wolfe, M.D., Hutzel Hospital, Detroit; PUM I by Wallace Miller, M.D., Hospital of University of Pennsylvania, Philadelphia; ROM I by David Rideout, M.D., Princess Margaret Hospital, Toronto; PUM II, PUM III, PUM IV, and DIA IV by Karl Kristen, M.D., and Kenneth McGinnis, M.D., Mount Carmel Mercy Hospital, Detroit; PUM IV by Joseph Reed, M.D., Childrens Hospital of Michigan, Detroit.

would have been present as residual stainable iron had this been an antemortem fracture. Unfortunately, the material was so degenerated that this was not possible, and the diagnosis of antemortem fracture has remained tentative.

The second case was PUM IV, a young boy who had been totally eviscerated, though the brain was still present. Extensive fractures involved the inferior and posterior parts of the occipital bones (Figure 17.3). There was also a reducible separation of the sagittal suture posteriorly. There was a small amount of amorphous soft tissue adherent to the inner aspect of the occipital bones, particularly on the left side, but this showed an almost perfect gravitational pattern and probably resulted from postmortem seepage of fluids and degenerated tissue into the posterior fossa. The mummy had been supported by a body-length wooden strut placed within the wrappings. The posterior aspect of the occipital bones rested on this board. The board had been broken in its midportion about the level

of the pelvic brim, and this suggested that the mummy had been dropped or otherwise mishandled and that the skull fracture may well have resulted from this postmortem trauma.

The third case, PUM II, represents an example of apparent fracture. The films revealed several straight and some sharply angular lines thought to represent linear fractures in the parietal bones (Figure 17.4). At autopsy, no skull fractures were found, but there were several deep scratches in the skin of the scalp that had been made postmortem, giving the radiographic appearance of linear fractures in the skull.

Except for the modest amount of bone resorption thought to be the result of periodontal inflammatory disease, no maxillary or mandibular osseous abnormalities have been found. PUM IV had several unerupted teeth, including the central lower incisors. In PUM III there was congenital absence of all third molars.

Determination of age by radiographic techniques in mummies is similar to its determi-

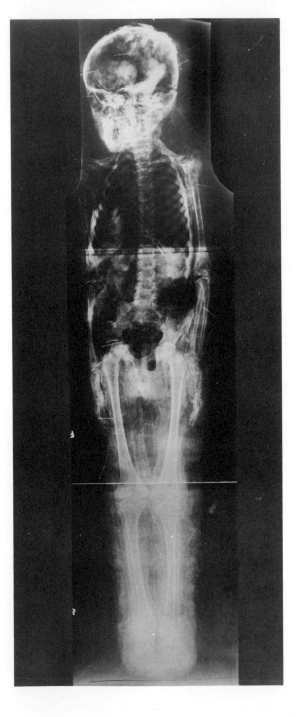

Figure 17.1.   *Large-format radiograph of the mummy PUM IV, produced using a 3-m tube–film distance, a long film changer, and roll film.*

nation in modern populations (Pyle and Hoerr 1955; Greulich and Pyle 1959; Hoerr et al. 1962), though Harris states that dental development may be as much as 2 years behind in the ancient Egyptian population (Harris, personal communication 1976). For example, based on dental pattern, general bone size, and lack of any evidence of epiphyseal closure (Figure 17.5), the age of PUM IV was originally thought to be 6 to 8 years. Harris's data suggest that he may have been about 2 years older.

SPINE, RIBS, AND PELVIS

The bones of the spinal column usually have been readily visualized. Varying degrees of osteophyte production have been noted, from none in the young boy of 8 years to minimal amounts in the male mummies in the age range of 30 to 35 years. These changes have corresponded to those described in many ancient human remains (Bourke 1967; Chapman 1972). They are equivalent in size and amount to similar changes seen today and probably occur for the same reasons. One mummy, PUM II, had a sixth lumbar, or transitional, vertebral body that showed minimal wedging and slight anterior spondylolisthesis. These changes were accompanied by exaggerated lumbar lordosis and degenerative arthritic changes, indicating an unstable configuration that may have prevented him from doing strenuous labor.

In the 8-year-old boy (PUM IV) there was apparent calcification of the intervertebral discs that appears to represent a postmortem artifact, as described by Gray (1967) (Figure 17.6). Whether the increased density of the intervertebral disc in these situations is attributable to calcium is not known. The possibility of absorption of calcium from natron during mummification has been mentioned (Gray 1967), but seems unlikely. The diffusion of calcium ions after death from the bone of the vertebral bodies into the intervertebral disc has more appeal. We have analyzed two intervertebral discs, one from a mummy

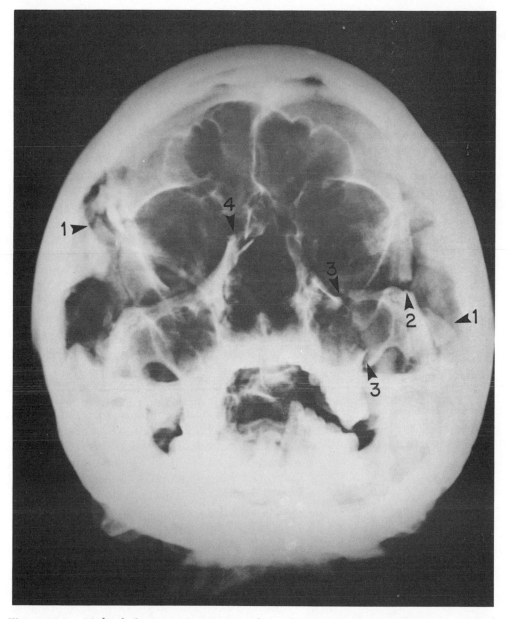

Figure 17.2.   Multiple fractures (arrows) involving the temporal bones (1), the left zygoma (2) and maxilla (3), as well as the right nasal bone (4).

without the change and one from a mummy with the change. The total calcium content for these specimens was 0.3 percent and 0.6 percent by weight. The difference is small and does not appear to support the theory that significant amounts of calcium are present in those intervertebral discs with increased opacity, a density that approaches that of the

Figure 17.3.   Bilateral occipital fractures (1) as well as suture separations (2) visible in the frontal (a) and lateral (b) projections of the skull of PUM IV. Note the shrunken cerebral hemispheres, which are seen as ill-defined densities (*).

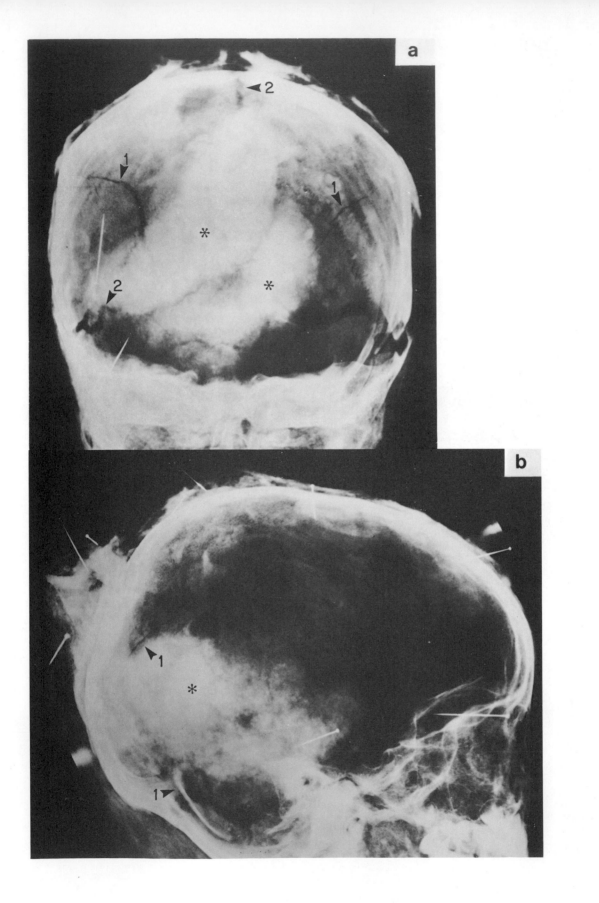

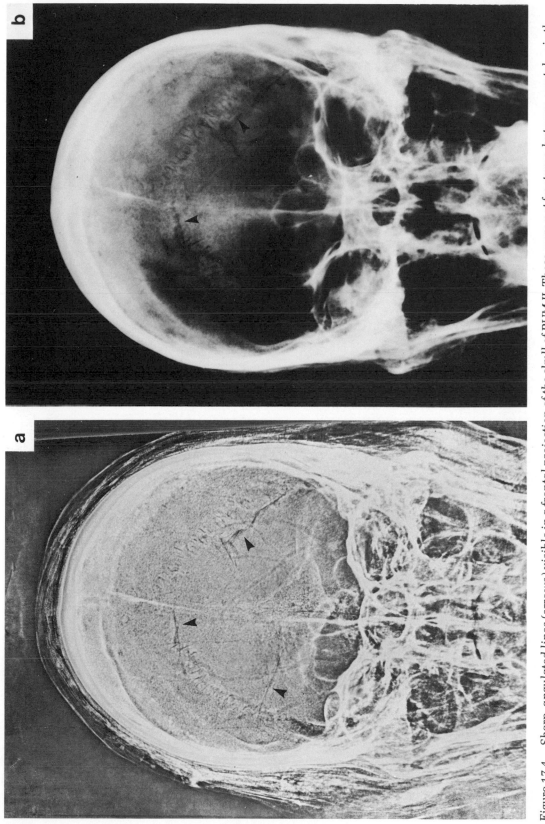

Figure 17.4. Sharp, angulated lines (arrows) visible in a frontal projection of the skull of PUM II. These were not fractures, but were scratches in the skin of the scalp. Note the detail in the wrappings recorded on the xerogram (a), but obscured on the X-ray film (b).

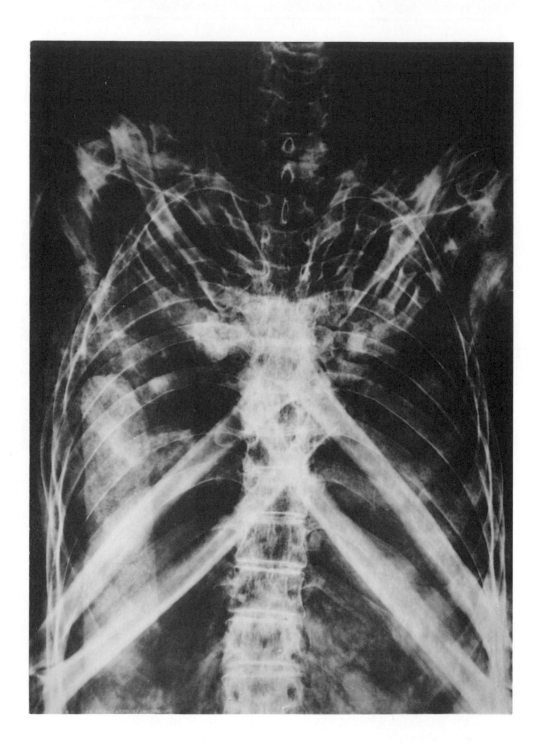

Figure 17.9.   Elongated, opaque cylindrical object in the right hemithorax and right abdominal cavity, which proved to be a visceral package containing the dehydrated organs. Three other similar packages were found in the pelvis.

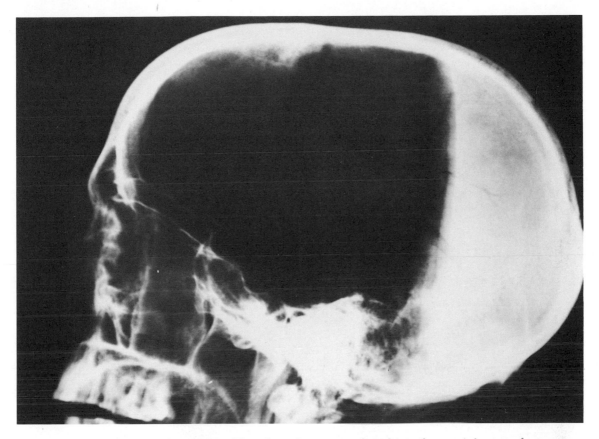

Figure 17.8.   Meniscus effect, attributable to liquid resin introduced into the cranial cavity during embalming, which hardened with the body in the supine position.

sue. Skeletal muscle has been markedly reduced in volume by the loss of water. In the mummies with considerable degenerative changes, air contrast planes have been seen between large muscles; occasionally, severe fissuring was seen as irregular, transversely oriented air contrast defects. If significant dehydration occurred after wrapping, large air spaces may be seen between the wrapping and the shrunken tissues.

Because not all mummies have been eviscerated, some or all of the viscera may be in their normal anatomic locations. Though the size, shape, and density may not conform to living examples, the overall appearance and location have often allowed identification.

Cranial soft tissue has usually been lacking in these mummies, with the exceptions of the skin covering the head, and in some instanc-

es, the presence of the cerebral hemispheres within the cranial cavity. When present, these structures were detected radiographically. The cerebral hemispheres have been present in three (PUM I, PUM IV, ROM I) of the seven skulls we have seen. Oddly enough, in all three the size of the cerebral hemispheres was similar, each measuring about 10 by 5 cm. Because the shrunken brain is loose within the skull, it settles by gravity to the posterior fossa during examination in the supine position. Radiographically, the hemispheres were seen as ill-defined, slightly opaque masses lying over the inner aspect of the occipital bones (Figure 17.3). Another mummy (PUM II) showed a much more dense posterior fossa opacity that had a fluid level effect (Figure 17.8). In our naiveté, we thought that this represented brain. At au-

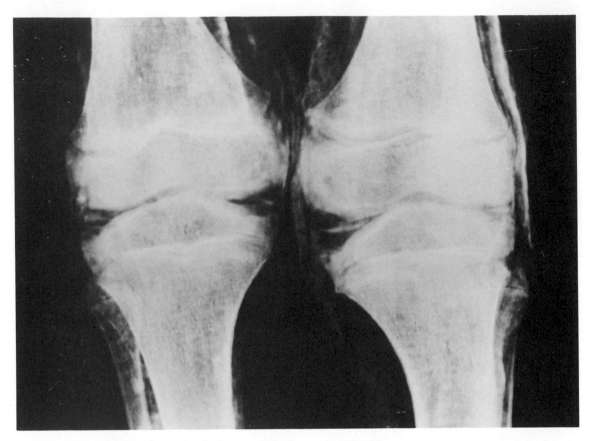

Figure 17.5.   Open epiphyses in the bones of the knees in PUM IV.

bone. However, two cases are insufficient for valid conclusions to be drawn, and more analyses will be necessary.

Several mummies have shown freshly fractured ribs and costochondral and costovertebral separations. These have been assumed to be postmortem in nature. One mummy, PUM III, had a healed fracture of the second left rib laterally. There was abundant callus, and microscopic examination showed the tissue had all the characteristics of a benign reparative process. No extranumerary or cervical ribs and no other intrinsic osseous abnormalities have been noted in our studies.

The pelves in each of the five mummies so examined have been normal. Attempts to determine the sex of the mummy utilizing pelvimetry have not been rewarding. Including

measurements of the femoral neck, condyle angles, and other parameters (Schine et al. 1951) may be more helpful. However, pelvimetry coupled with analysis of soft tissue detail permits the most accurate conclusions. In DIA I, no examination was done, but it was unnecessary because we knew from historical data that the body was male. PUM I was so poorly preserved that virtually every joint in the skeleton was separated and pelvimetry and analysis of bone size were considered unreliable. At autopsy, a shrunken penis established the sex of the mummy. PUM II was thought to be a female, based on equivocal identification of a gynecoid pelvis. The soft tissue detail revealed an elongated opaque area between the shrunken thighs that was originally thought to be extraneous material

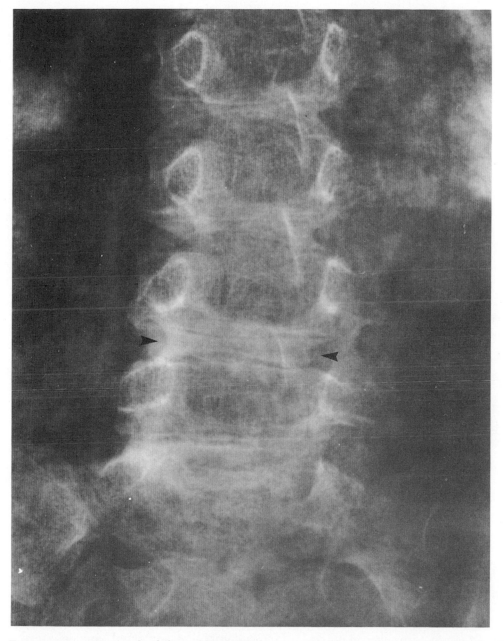

Figure 17.6. *Intervertebral disc of PUM IV, showing a density approaching that of the adjacent verte-bral bodies and suggesting calcification. Chemical analysis of the disc did not reveal an inordinately high calcium content.*

between the wrappings because its radiodensity was much greater than that of the adjacent soft tissue. At autopsy, we found that this density was in fact a 10-cm penis, apparently uncircumcised, and supported partially erect by a small wooden slat. The confusion arose because of the wood and its added density. ROM I was known from historical data to be a

teenage boy, and radiographic interpretation concurred (see Chapter 5). PUM III was a young female. Pelvimetry and other bone measurements were indeterminate, but pelvic soft tissue detail showed the presence of a uterus and urinary bladder and the sex was established on this basis. Autopsy confirmed these interpretations. PUM IV, a young boy, was so diagnosed not on the basis of bone configuration but by the fact that the penis and scrotum could be seen on the films. In this instance, the density of these structures was similar to that of the other surrounding soft tissue.

EXTREMITIES

The evaluation of the upper extremities before unwrapping has posed some problems because of the fixed positioning of the arms and hands. We have not found any abnormalities in the upper extremity bones. However, any suspicious findings should be reevaluated after autopsy when the arms can be moved or a film cassette can be interposed between the extremity and the body. Often the clavicles and the scapulae have been displaced inward and upward owing to the binding of the arms close to the body by the wrappings. Occasionally, one or several joints have been disarticulated owing to postmortem degeneration.

The lower extremities posed less of a technical problem, though superimposition of the bones on lateral views can be troublesome. PUM II had a distorted foot with the toes severely flexed and turned medially, but this appeared to be an embalming artifact attributable to tight wrapping in this area. The bones and joints of the foot were entirely normal radiographically. The only abnormality we have seen was in the right fibula and tibia of the same mummy. At autopsy, the right leg was noticeably larger than the left and appeared swollen. Radiographically, the right fibula was thickened in the medial aspect of the middle and lower thirds with blurring of the normally sharp cortical contour

(Figure 17.7). There was no apparent medullary component to this lesion. Opposite this area, a similar but very localized zone was noted on the lateral aspect of the tibia at the junction of the middle and lower thirds. The osseous reaction appeared to be periosteal in location in both bones. Microscopic examination showed the lesion to be a nonspecific osteoblastic periostitis. The differential diagnosis includes vascular stasis changes and localized inflammation, but the cause remains uncertain.

Growth arrest lines, or Harris's lines, may be seen as transverse bars radiographically in the metaphysis near the epiphysis in the long bones and have been particularly well seen in the distal femur or tibia (Wells 1967). The lines result from the temporary arrest of growth of the epiphyseal cartilage in children during episodes of acute stress such as malnutrition and acute illnesses. These lines have been present in three of the five mummies examined for this change (Figure 17.7). In some instances, the growth arrest lines have been quite faint, and in the cases where they were not noted, such lines may have been obscured by the cloth covering the mummy. Although the lines are nonspecific indicators, their presence may correlate with subsequent findings from the autopsy. The long bones should be reexamined after unwrapping so that better radiographic study can be achieved.

SOFT TISSUE

Residual soft tissue in the mummies has taken two forms. The first and most common has been the tissue such as skin, subcutaneous fat, and muscle that remained on the body following mummification. These structures have had varying degrees of opacity and could be identified readily. The formation of calcium soaps in subcutaneous tissue owing to postmortem degeneration of the tissue may give irregular density to that tissue, but it has never been great enough to interfere with visualization of underlying bone or soft tis-

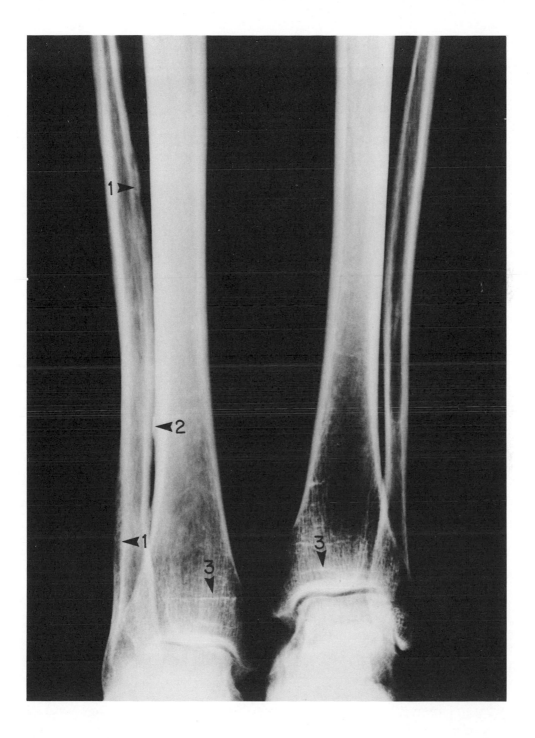

Figure 17.7.   Right fibula of PUM II, showing a wavy irregular contour with loss of sharp cortical margin (1). A small similar area is seen on the lateral tibia (2). Note the horizontal growth arrest lines present in both distal tibiae (3).

Temporal bone studies

JAIME T. BENITEZ
Director, Division of Otoneurology and
the Temporal Bone Research Laboratory
William Beaumont Hospital
Royal Oak, Michigan, U.S.A.

GEORGE E. LYNN
Professor of Audiology
Wayne State University School of Medicine
Detroit, Michigan, U.S.A.

Temporal bones from ancient populations, particularly Egyptian mummies, which were often meticulously embalmed, are usually in a good state of preservation and therefore offer an excellent opportunity for the histopathological investigation of inflammatory, neoplastic, and traumatic ear conditions as well as metabolic disorders involving the temporal bone. Furthermore, with the application of modern techniques, research can be carried out on the haversian systems, associated vascular channels, and osteon activity.

X-RAY STUDIES

X-ray study is the first procedure carried out in the investigation of ancient temporal bones prior to removal of the bones from the skull. Radiographic studies have been enhanced in the past decade with the development of a special technique, polytomography, which allows the survey of tissues in a millimeter-by-millimeter fashion. X-ray polytomography of the temporal bones makes it possible to detect middle ear malformations, otic bony capsule abnormalities, or changes in the internal auditory canal. However, it should be emphasized that certain radiographic findings may be misleading, and final interpretation depends on macroscopic examination and histological studies. For example, with the temporal bones of PUM II, there was a definite increase in density throughout the polytomograms of the left temporal bone as compared with the right, which suggested the possibility of extensive bone disease involving the left specimen. However, macro-

scopic examination revealed that the increased density was caused by the presence of large amounts of resin in the air cells of the mastoid and cavities of the middle and inner ears and that there were no anatomical abnormalities in the middle ear and mastoid of these specimens. Furthermore, histological study of undecalcified sections showed no evidence of metabolic bone disease, which could also have caused the unusual X-ray findings. Clearly, the increased radiological density, which might have been attributed to premortem temporal bone disease, was, in fact, the result of a large amount of resin infiltration into the left ear of PUM II during the embalming procedure.

MACROSCOPIC STUDIES WITH ZEISS OPERATING MICROSCOPE

After removal of the temporal bones by means of the block or Schuknecht method (1968), specimens such as that shown in Figure 18.1 are studied with the Zeiss operating microscope, which allows magnification up to 40 times. Observations are recorded in a systematic fashion regarding the preservation or absence of soft tissues in the external ear canal, the surface of the petrous pyramid, and the internal auditory meatus. A detailed description is made of the eardrum or its remnants. The posterior portion of the tegmen tympani is removed with fine rongeur forceps so that the contents of the middle ear cavity may be inspected. The arcuate eminence of the superior surface of the petrous bone is removed with an electric saw to reveal the superior and posterior semicircular canals.

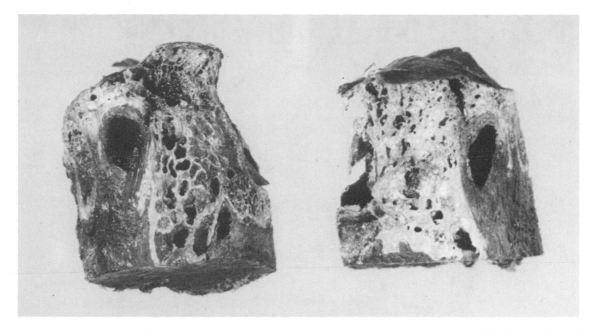

Figure 18.1.　Temporal bones of a 2,000-year-old Egyptian mummy, PUM II, removed by the method of Schuknecht. Specimens include the external auditory canal, middle ear, mastoid process, and the petrous pyramid, which houses the inner ear.

(This portion of the petrous bone is then processed by the undecalcified method.) Color photographs are taken of significant findings of the external canal, eardrum, mastoid process, middle ear cavity, and petrous pyramid. We use a Zeiss Ikon camera (ICAREX 35S) attached to a side viewer that operates with a synchronized flash.

Macroscopic studies by the method described above have been carried out on 12 temporal bones from 6 Egyptian mummies, 19 temporal bones from skeletal remains from 19 Egyptian mummies, a pair of temporal bones from a Nubian skull, 55 temporal bones from the skulls of 41 mummies from Peru, and one set of temporal bones from the skull of an Eskimo from Saint Lawrence Island frozen since her death around A.D. 400. Studies of these specimens have revealed evidence of antemortem disease as well as postmortem abnormalities.

Exostosis of the external auditory canal has been a common finding in material studied by Gregg and Bass (1970) and Hrdlička

(1934), but we have not found a single case in any of our specimens.

Figure 18.2 shows an example of a normal mastoid process from a Peruvian mummy from the Ica culture, A.D. 1200–1400. Pneumatization and preservation of the air cell system is excellent, and there is no evidence that mastoid disease ever occurred. An excellent example of an abnormal mastoid process from ROM II is shown in Figure 18.3. The normally pneumatized bone has been replaced by very dense sclerotic tissue as a sequela of chronic mastoid infection occurring sometime during the individual's lifetime. The opposite mastoid revealed similar findings.

Eardrums are rarely observed in ancient temporal bones because of rapid disintegration after death; however, preservation of eardrums or remnants was found in three Egyptian mummies (PUM II, PUM III, ROM II) and in the Eskimo from Saint Lawrence Island. The right eardrum of PUM II (Lynn and Benitez 1974) showed a small oval perfora-

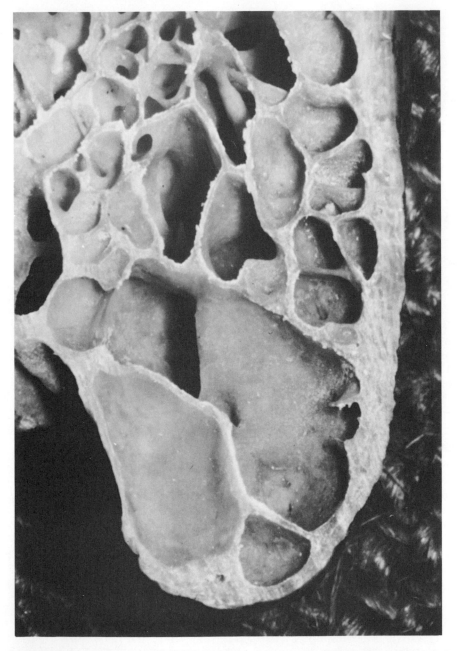

Figure 18.2.   Left mastoid process of a Peruvian mummy from the Ica culture, A.D. 1200–1400. Pneumatization and preservation of the air cells is excellent, with no evidence of disease.

tion in the posteroinferior quadrant occupying approximately 10 percent of the pars tensa (Plate 10). The left ear of PUM III revealed a large oval perforation occupying about 40 percent of the posterior portion of the membrane.

These perforations have smooth and well-circumscribed margins suggesting they were caused by middle ear infection occurring during life. The irregular margins of the remnants of the right eardrum of ROM II appear to

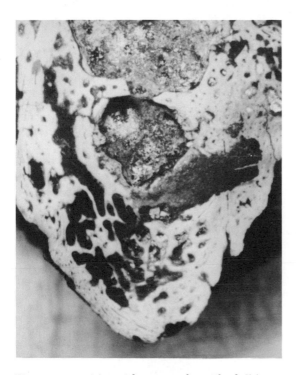

Figure 18.3. *Mastoid process from the left temporal bone of a 4,000-year-old Egyptian mummy, ROM II, showing severe sclerosis as a result of chronic mastoid infection.*

be postmortem artifacts. However, the well-circumscribed margins of the remnants of the right eardrum of PUM III appear to be the sequela of an inflammatory process.

Intact ossicles found in their normal anatomical position with preservation of the ligaments were seen in the Egyptian mummies PUM II and PUM III. Most of the temporal bone specimens we have studied so far have shown displaced but intact ossicles in the middle ear cavity with no evidence of disease process. Today, fixation of the stapes in the oval window by otosclerosis is a common condition affecting the mobility of the stapes; however, we have not seen this disorder in the 90 temporal bones studied so far.

The internal auditory meatus was occluded with resin in the temporal bones of PUM II. In the left temporal bone of the Eskimo from Saint Lawrence Island, there were remnants of the eighth nerve (Figure 18.4).

In the left external auditory canal of PUM II, a beetle larva measuring about 1.5 by 0.2 mm was found embedded in resin at the medial end close to the rim of the eardrum. It was prepared for examination by scanning electron microscopy (Figure 18.5) and later identified as Staphylinidae, genus probably *Atheta*, by D. M. Anderson, Systematic Entomology Laboratory, U.S. Department of Agriculture.

HISTOLOGICAL STUDIES WITH UNDECALCIFIED SECTIONS

A great advantage of studying undecalcified bone sections is that it provides a superior method for evaluation of bone remodeling, which is a tissue function and can be expected to change with aging and disease. Therefore, in our paleopathological investigations, we routinely remove the upper portion of the petrous bone, which includes the arcuate eminence with the arc of the superior semicircular canal, for such studies. Other portions of the temporal bone can also be removed for these studies without damage to the rest of the specimen that is to be processed by routine histological techniques after decalcification. For histological control on undecalcified sections, a similar specimen is removed from a fresh temporal bone. The specimens are fixed in 80 percent alcohol and processed by the method for undecalcified bone (Frost 1958; Villanueva and Frost 1961; Villanueva et al. 1964).

Light microscope study of sections revealed excellent preservation of the haversian systems in the specimens from the temporal bones of PUM II (Benitez and Lynn 1975). The vascular channels were clearly seen, and, amazingly, they were similar in appearance to those of the specimen from fresh temporal bone that was used as a control. Sections from the temporal bones of PUM II appeared to have a yellowish color that could be detected even after staining (Plate 12). This was attributed to infiltration of resin.

A striking finding in the sections from

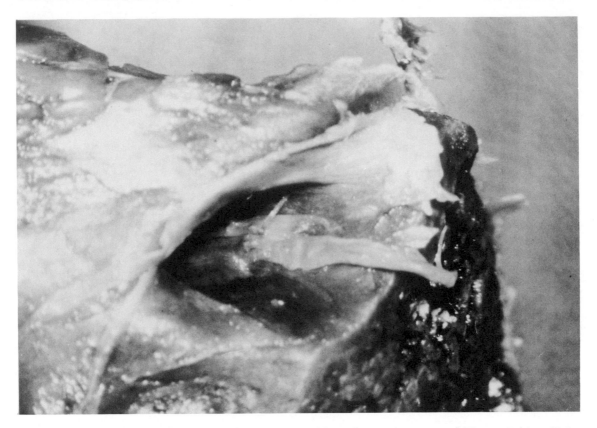

Figure 18.4.   Internal auditory canal of the left temporal bone from a 1,600-year-old frozen Eskimo (Saint Lawrence Lady), with remnants of the eighth nerve.

PUM II was the presence of osteocytes in the lacunae with good preservation of the nucleus, which is remarkable for a tissue over 2,000 years old. When specific stains are used, for example, the polychrome bone stains of Villanueva (1964, 1974), osteoid seams can be unequivocally demonstrated and the activity of the osteon may be determined. In the sections from PUM II, we were able to identify osteoid seams. Plate 9 shows an active osteoid seam from the left temporal bone. In the osteon, there were areas of incomplete mineralized bone and areas of resorption (Howship's lacunae); the thickness of the seam is about 5 μ. This is a low-turnover type of bone, but without evidence of metabolic bone disease.

Undecalcified sections from the specimens of the 1,600-year-old frozen Eskimo showed a highly mineralized type of bone, but no evidence of metabolic bone disease. There was good preservation of the vascular channels, and blood cells were identified (Plate 11).

HISTOLOGICAL STUDIES AFTER DECALCIFICATION

The histological preparation of the human temporal bone includes several procedures, which we shall describe with modifications that we believe are convenient for the processing of ancient specimens.

Fixation.   Rehydration of mummified tissues has been recommended for histological studies, in particular for soft tissues (Ruffer 1921). For the temporal bones, we prefer fixation in 10 percent neutral buffered formaldehyde as the first step, which at the same

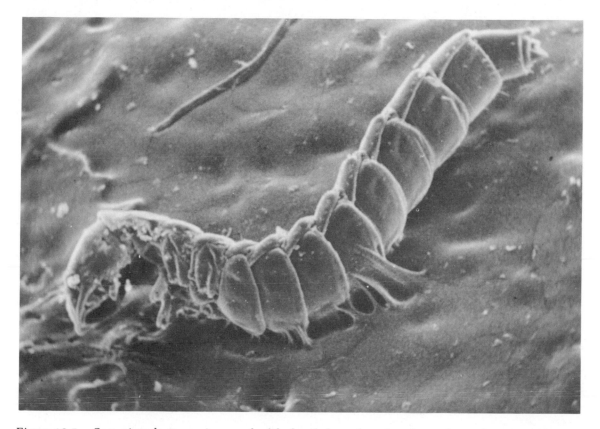

Figure 18.5.   Scanning electron micrograph of the beetle larva found on the eardrum of mummy PUM II. It was identified as Staphylinidae, genus probably Atheta. × 55.

time produces hydration of the specimen. The fixative solution should be prepared at the time of use by mixing 100 ml of 37 to 40 percent formaldehyde solution, 900 ml distilled water, 4 g sodium phosphate (monobasic), and 6.5 g sodium phosphate (anhydrous dibasic). The temporal bones are placed for 2 weeks separately in 500-ml glass jars with screw-on lids containing about 450 ml of the solution. A change of solution is performed on the second day. The jars may be shaken once or twice a day to improve penetration of the fixative. Any changes noted on the anatomical structures of the specimens should be recorded.

Decalcification.   In ancient specimens, the use of strong acid solutions for decalcification should be avoided. We use a solution of ethylenediaminotetraacetic acid (EDTA) at a

pH of 7.2 to 7.8. It is obtained as Versene 100, which is an aqueous solution of the tetrasodium salt of EDTA and can be purchased from Fisher Scientific Company. The 1.0 M Versene stock solution is prepared with 10,500 ml of Versene 100, 7,500 ml of distilled water, and 900 ml of glacial acetic acid. With some modifications, we follow the technique described by Gussen and Donahue (1965). Decalcification is carried out in a 0.66 M Versene solution prepared with 666 ml of Versene stock solution and 333 ml of distilled water. Because distilled water has a pH of approximately 6, the resulting solution will have a pH between 7.2 and 7.8. If the pH is below 7.2, a little Versene stock solution may be added to bring the pH back to between 7.2 and 7.8.

After fixation, each specimen (without

washing in water) is placed in a separate 500 ml jar containing about 450 ml of the decalcifying solution, which is changed twice a week. The period of time required for decalcification is 16 weeks. The end point of decalcification may be more precisely determined by X rays. When decalcification is complete, each temporal bone is placed in a very dilute solution (0.1 to 0.2 M) of Versene, and the solution is changed once every 3 days, three times in all. This dissolves any calcium versenate crystals that may have precipitated within the tissues.

Dehydration.   We modified the dehydration time periods as follows: 40 percent alcohol, 10 minutes; 60 percent, 10 minutes; 80 percent, 10 minutes; 95 percent, 25 minutes; absolute alcohol and ether (1:1), 20 minutes.

Embedding.   Parlodion dissolved in ether–alcohol is used for embedding the specimens. It is similar to celloidin, but dissolves in ether–alcohol more readily, and the parlodion-embedded specimens do not dry as much as the celloidin-embedded specimens, so that a better consistency can be achieved for sectioning. The specimens are placed in a solution of 2 percent parlodion for 1 week. This is followed by a 4 percent solution for 2 weeks, then a 6 percent solution for 3 weeks. Specimens should be oriented so that the opened areas of the bones (epitympanum and superior semicircular canal) are turned upward in an effort to allow air bubbles that form within the bones to rise.

Hardening.   Specimens are placed in 12 percent parlodion solution for 4 weeks, with the superior surface of the petrous bone facing the bottom of the jar. To maintain the specimen in the middle of the jar, it can be suspended with surgical silk (3–0); this helps the orientation of the bone in the desired angle of section. After the 4-week period, the jar top is unscrewed and slipped 0.3 cm back from one side of the jar where it is allowed to rest for about 7 to 8 hours per day, from the beginning to the end of the workday. When the specimen has hardened in this manner for 2 to 3 weeks, the parlodion reaches a concen-

tration that is no longer sticky but is still quite soft and pliable; when this occurs the upper layer of the parlodion around the edge and down the side of the jar is trimmed away with a scalpel blade, and 100 ml of chloroform is poured on top of the block. The top is placed on the jar securely and the specimen left in this manner for 1 week.

Mounting.   When the block has been covered in chloroform for 1 week, the peripheral part of the parlodion is trimmed away with a scalpel blade and the block containing the specimen is removed from the jar. Further trimming is necessary to obtain the approximate shape and orientation desired for cutting. For sectioning on the horizontal plane, the inferior part of the block (which was facing the top of the jar) is cut flat with a scalpel blade in a plane that is parallel to the flat upper portion of the block. Trimming should be performed so as to leave about 1 cm of parlodion around the specimen and about 1.5 cm at the bottom; thus a rectangular block is obtained. While trimming, one should bear in mind that the remainder of the superior semicircular canal is to be positioned in the vertical plane. Other anatomical landmarks used for orientation of the specimen are the internal and external auditory canals. To obtain horizontal sections in the plane of the modiolus of the cochlea, a counterclockwise rotation is given to the specimen so that the posterior surface of the petrous bone is in the vertical plane. If the malleus is present and in its anatomical position, this rotation places the long process also in a vertical plane (Figure 18.6). When the specimen has been trimmed so that this orientation is achieved, the inferior portion of the block is softened in an ether–alcohol solution for about 1 minute. A mounting block is placed in the solution alongside the specimen. The wet portions of the specimen and the mounting block are next dipped in a small amount of 12 percent parlodion, then pressed firmly together and any gaps filled with 12 percent parlodion. The mounted specimen is placed in 150 ml of chloroform for 2 days. This solution is then

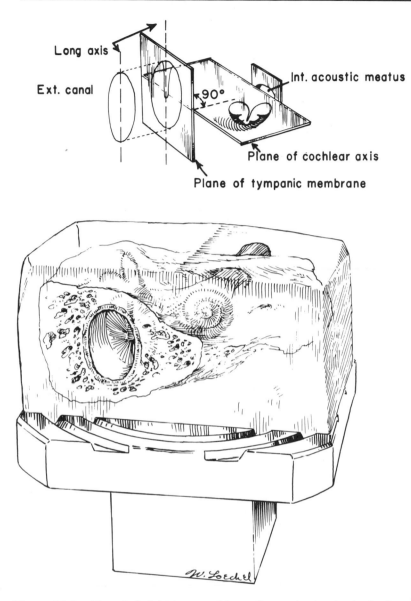

Figure 18.6. *Mounted right temporal bone for sectioning in the horizontal plane. Slight counterclockwise rotation has been given to the specimen to bring the cochlea into a horizontal plane. The relationship of the external and internal ear canals is shown above.*

replaced with 80 percent alcohol. The specimen may be sectioned on a sliding microtome within 1 hour's time or stored in this solution indefinitely.

Sectioning and Staining.   The microtome is set to cut at 20 μ, obtaining about 400 sections from each specimen. All sections are placed in 80 percent alcohol solution on pieces of onionskin paper that are consecutively numbered. Every tenth section is placed in a separate dish for staining. Conventional staining of the temporal bone is usually performed with hematoxylin and eosin. As the cutting proceeds, color photographs are taken of the specimen with the operating microscope at midmodiolar level and at other levels show-

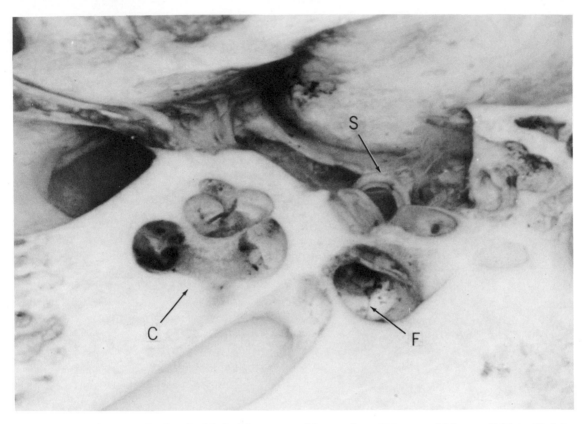

Figure 18.7.   Photograph of embedded right temporal bone of a 1,600-year-old frozen Eskimo (Saint Lawrence Lady). Sectioning is at midmodiolus level. There is a fracture line in the vestibule (F) extending through the niche of the oval window. The stapes bone (S) is evulsed. The cochlea (C) and otic capsule around it are normal in appearance.

ing interesting findings. In the right temporal bone of the 1,600-year-old frozen Eskimo, a fracture line was found in the vestibule of the inner ear extending through the niche of the oval window with luxation of the stapes (Figure 18.7). There was histological evidence that this fracture is not an artifact (Plate 13), and it is consistent with fractures of the skull and maxilla on the right side found by Zimmerman and Smith (1975) at autopsy.

Mounting of sections. Sections are mounted on slides in a configuration that centers the important anatomical parts of the temporal bone consistently. The outline of a slide is made on a piece of paper and the desired location of the cochlea marked; the slides are placed over this pattern and the sections

aligned with the pattern. All temporal bone sections are mounted in the slide with the left anatomical side of the section placed on the right side of the slide. The posterior anatomical side of the temporal bone section is placed in the anterior portion of the slide.

Comparison of rehydrated and unrehydrated bone. The histological processing of ancient temporal bones offers a challenge to otopathologists and temporal bone technicians. We contemplated the possibility that rehydration could disintegrate tissues essential for the interpretation of certain otopathological conditions, and so we were anxious to carry out a comparative study. The temporal bones from a 4,000-year-old Egyptian mummy skull provided us with this unique

opportunity. As a result of the kind invitation of Patrick Horne, Pathology Department, Banting Institute, Toronto, Canada, a pair of temporal bones was removed by one of us (GEL) from the skull of ROM II following the method of Schuknecht. After removal of the arcuate eminence for study in an undecalcified state and for macroscopic evaluation with photography of both specimens in our Detroit laboratory, the left temporal bone was returned to the Banting Institute for histological study by Horne et al. (1976). They rehydrated the specimen for 1 week in Sandison's modification of Ruffer's solution (1955). Fixation was carried out in 10 percent neutral buffered formaldehyde solution for 2 weeks and decalcification by the method of Gussen and Donahue (1965) with EDTA. The right temporal bone was processed in our laboratory without preliminary rehydration of the specimen. Our first step was fixation in 10 percent neutral buffered formaldehyde solution for 2 weeks, followed by decalcification by the method of Gussen and Donahue (1965) with EDTA. The dehydration time was shortened as described above.

We had an opportunity to study serial sections of both specimens. The general state of structural integrity was similar. Artifactual microfractures of bony tissues produced during decalcification were noted on both embedded specimens, and they were in fact microscopically similar. However, a better preservation of soft tissues was found in the right temporal bone processed without rehydration; a remnant of the basilar membrane was present throughout all turns of the cochlea of the inner ear, whereas in the left specimen, processed with rehydration, a remnant of the basilar membrane was found only in the basal turn. Moreover, in the rehydrated left ear, a good portion of the tympanic membrane disintegrated and histological detail at the level of the perforation was lost, whereas in the unrehydrated right ear, there was excellent preservation of the remnants of the tympanic membrane including the small artifactual postmortem tears (Plate 14).

SUMMARY AND CONCLUSIONS

Techniques have been described for the study of ear disease that may have occurred during the lifetime of ancient people and that may still be preserved in their temporal bones.

The recommended procedures for obtaining maximum information include X rays with polytomography, dissection and macroscopic observation with thorough photographic documentation, and histopathologic studies with undecalcified and decalcified material. The traditional procedure of rehydration for histological studies may be omitted to lessen the amount of tissue disintegration that may occur during preparation of the specimen for sectioning. However, fixation in 10 percent neutral formaldehyde is essential prior to decalcification. Conventional decalcification with trichloroacetic acid cannot be used because it produces disintegration of the tissue. A solution of EDTA is preferred.

With modern laboratory techniques, pre- and postmortem abnormalities involving the mastoid, eardrum, and various structures of the middle and inner ears can be identified in ancient temporal bones, and findings contribute new information about the nature of diseases in ancient cultures.

ACKNOWLEDGMENTS

This research is supported in part by General Research Support Grant RR–05384, Wayne State University School of Medicine (Gel), and a grant from William Beaumont Hospital, Royal Oak (JTB).

REFERENCES

Benitez, J. T., and Lynn, G. E. 1975. Temporal bone studies: findings with undecalcified sections in a 2,600-year-old Egyptian mummy. *Journal of Otology and Laryngology* 89:593–9.

Frost, H. M. 1958. Preparation of thin undecalcified bone sections by rapid method. *Stain Technology* 33:273–6.

Gregg, J. B., and Bass, W. M. 1970. Exostosis in the external auditory canals. *Annals of Otology, Rhinology and Laryngology* 79:834–9.

19

Paleobiochemistry

ROBIN A. BARRACO
Associate Professor, Department of Physiology
Wayne State University School of Medicine
Detroit, Michigan, U.S.A.

The most abundant ancient specimens have always been skeletons; consequently, the major investigative effort for paleopathologists has focused on gross observation, morbid anatomy, histology, and to some extent, blood-typing studies. Most discoveries have consisted of lesions of tuberculosis, leprosy, and yaws. However, diseases, particularly infectious diseases, rarely leave detectable traces in bones. The study of surviving soft tissues is required for the comprehensive analysis of ancient diseases. In this regard, the biochemical analysis of ancient tissues may prove a useful tool for the investigation of ancient diseases. Because the chemical composition of an organism provides a record of the interaction of that organism with its environment, biochemical studies, along with archeological and other paleobiological approaches, constitute a coordinated effort to clarify the relationship of some diseases – in particular, infectious diseases – to geography, heredity, and diet and to learn about the subsequent adaptation of people and culture to a disease environment.

This approach should, however, be exercised with caution, as a detectable chemical abnormality will neither explain a disease process nor provide specificity concerning its cause. The abnormal is only a specialized aspect of the normal and, thus, a reactive state of the individual without a precise relationship between any cause and a disease process (Johnson 1966). The identification of particular biochemical constituents, in excessive or deficient amounts, does not necessarily establish constant relationships to single diseases. For example, the particular type of hemoglobin associated with sickle cell anemia is found in some individuals without anemia or evidence of disease. The same reasoning applies to genetics: The realization of genetic potential is not equivalent to having the potential, and abnormalities of genetic origin can be duplicated by nongenetic processes. These considerations emphasize the importance of using collaborative multidisciplinary and contemporary statistical approaches in paleopathology.

This chapter surveys the types of paleobiochemical analysis used to discern a reactive pattern or pathologic process in ancient tissues. The analytic principle in paleobiochemistry is the same as in every other science: first, qualitative, and second, quantitative, analysis.

MACROMOLECULAR CONSTITUENTS IN ANCIENT TISSUES

PROTEINS

At present, attempts are being made to extract, fractionate, and identify protein material in ancient tissues. The prime motivation for analyzing proteins lies in the possibility of detecting gamma globulins and the antibodies against infectious agents they may contain. Also, certain proteins, such as hemoglobin, may provide genetic markers for certain pathological conditions in heterogeneous populations and may suggest something of their geographical and ethnic or racial progression since ancient times.

Gussen, R., and Donahue, D. 1965. Decalcification of temporal bones with tetrasodium edetate. *Archives of Otolaryngology* 82:110–14.

Horne, P. D.; MacKay, A.; Jahn, A. F.; and Hawke, M. 1976. Histologic processing and examination of a 4,000-year-old human temporal bone. *Archives of Otolaryngology* 102:713–15.

Hrdlička, A. 1934. Ear exostosis. *Smithsonian Miscellaneous Collection* 93:1–100.

Lynn, G. E., and Benitez, J. T. 1974. Temporal bone preservation in a 2,600-year-old Egyptian mummy. *Science* 183:200–2.

Ruffer, M. A. 1921. *Studies in the Paleopathology of Egypt.* Chicago: University of Chicago Press.

Sandison, A. T. 1955. The histological examination of mummified material. *Stain Technology* 30:277–83.

Schuknecht, H. F. 1968. Temporal bone removal at autopsy. *Archives of Otolaryngolory* 87:33–41.

Villanueva, A. R.; Hattner, R. S.; and Frost, H. M. 1964. A tetrachrome stain for fresh mineralized bone sections useful in the diagnosis of bone disease. *Stain Technology* 39:87–94.

Villanueva, A. R., and Frost, H. M. 1961. Rapid method for obtaining hematoxylin and eosin biopsy sections of bone for table diagnosis. *American Journal of Clinical Pathology* 36:54–9.

Villanueva, A. R. 1974. A bone stain for ostcoid seams in fresh unembedded mineralized bone. *Stain Technology* 49:1–8.

Zimmerman, M. R., and Smith, G. S. 1975. A probable case of accidental inhumation of 1,600 years ago. *Bulletin of the New York Academy of Medicine* 51:828–37.

Although very little information is available about the chemical and physicochemical state of preservation of protein and other macromolecules in mummified tissue, amino acid analysis has been reported for structural proteins of calcified fossils, bone, and hair (Abelson 1957; Wyckoff et al. 1963; Ho 1966; Jope 1967; Miller and Wyckoff 1968; Akiyama and Wyckoff 1970; Gillespie 1970) and of muscle tissue from Egyptian mummies (Lausarot et al. 1972). Further, there is evidence that apparently native proteins persist many years under a variety of preservation conditions. Biologically active proteins have been reported to survive in dried bloodstains exposed to the atmosphere for 10 years (Ducos et al. 1962), in whole blood stored at room temperature in the dried state for up to 40 years (Keilin and Wang 1947; Sensabaugh et al. 1971a,b), in seeds and spores germinating after dormant periods of as long as 2,000 years (Keilin 1959), and in 10,000-year-old samples of Alaska permafrost (Skujins and McLaren 1968). In addition, the long-term stability and antigenicity of material in preserved biological specimens have been demonstrated by blood-typing studies on mummified tissue from Egypt, the Aleutian Islands, the United States, and South America (Boyd and Boyd 1937; Candela 1939; Otten and Flory 1964; Lippold 1971; Stastny 1974).

SALTS

Because natron and other salts were used for chemical mummification by the ancient Egyptian embalmers (Lucas and Harris 1962), the study of the relationship between the degree of preservation of proteins in mummified tissue and tissue cation levels may provide useful archeological information concerning the nature of the mummification process. The cataloging of this kind of information would be useful for characterizing the class or status of the individual and the period and flux of traditions for the ancient society investigated. Also, the characterization of the mummification process for individual paleobiological

specimens may help to standardize paleopathological findings, as the ability to detect diseases varies with the degree of preservation of the ancient specimen.

LIPIDS

The lipid content of ancient tissues is a potentially great source of information in several respects. First, the tendency for lipids to accumulate and be stored within the body in relatively stable form in discrete cellular and subcellular components makes them excellent markers for the integrity of tissues long after death. Second, the qualitative composition of lipid populations in biological material correlates strongly with diet and, to a lesser extent, with more general features of life style, such as exposure to sunlight and extent of physical exertion. Third, lipids provide excellent macromolecular markers for certain pathological conditions, many of which are inherited metabolic defects. Fourth, lipids undergo gradual chemical changes, such as autoxidation of unsaturated fatty acids, which may be time-dependent and therefore of possible value as a biological dating tool.

Lipids and tissue integrity. Lipids have a discrete subcellular distribution. Neutral lipids and the lipid-soluble vitamins are primarily constituents of the cell cytoplasm; phospholipids and sphingolipids are structural components of membranes. The extent to which each type of lipid is recoverable from ancient tissues may reflect (within the limits of obvious chemical modifications) the state of preservation of different subcellular components and organelles.

Lipids, diet, and life style. Because some lipids and lipidic vitamins (A, E, D, and K, principally) are not synthesized in human tissues, the precise nature of these components reflects the dietary availability of them or their precursors. The obvious and best known correlation is that between the amount of polyunsaturated fats ingested and the extent of saturation in the stored fatty acids of triglycerides. A similar correlation

probably exists for the fatty acid components of certain phospholipids and sphingolipids. The presence of tocopherol and many of the isoprenoid derivatives also reflects their availability in the diet. Lipids may also be indicative of the life status and life style of the person from whose tissue they are extracted. For example, the relative distribution of lipoprotein densities in the blood is a function of both age and sex. The amount of various vitamin D precursors might vary in proportion to the amount and intensity of sunlight exposure. (A comparison of tissues from an Egyptian mummy and a frozen Eskimo might be particularly instructive in this regard.) By integrating life status, age, sex, and life style information derived by archeological and other disciplinary approaches with the biochemical information on lipid composition, it should be possible to make some distinctions between the strictly environmental, as opposed to cultural, determinants of the physicochemical characteristics of the ancient tissues under study.

Paleopathology of lipids.   Lipids can provide remarkably reactive signs of disease. Some of these reflect dietary deficiencies. The absence of certain carotene derivatives from some ancient tissues, but not from others, might suggest visual defects in the person from which the former was obtained. Certainly the correlation of lipid composition patterns with other pathological evidence might lead to enlightening conclusions about causative factors in ancient society. In addition, the lipids are peculiarly characterized by susceptibility to certain inborn errors of metabolism that invariably have nonrandom geographical and ethnic distributions (the various gangliosidoses are prominent examples). The possibility exists that evaluation of lipid patterns from ancient tissues will reveal the presence of these defects and suggest something of their geographical and ethnic progression since ancient times.

Lipids as a supplementary dating tool.   The temporal parameters of autooxidation of unsaturated fatty acids have not been studied over the range of time that would be possible with specimens of ancient tissue. Another even less investigated area concerns the degradation of conjugated moieties attached to lipids, such as the carbohydrate portion of sphingolipids, over long periods of time under conditions of mummification. These and similar chemical properties might serve as a supplementary tool for biological dating. Only a detailed analysis of the lipid composition of ancient tissues can verify or reject the feasibility of such a technique.

Overall value of lipid studies.   Lipids can be extracted from ancient tissue samples by a comprehensive protocol to maximize the yield of neutral lipids, phospholipids, sphingolipids, lipidic vitamins, and lipoproteins. They can be fractionated into each of the above categories, and each group can be analyzed for quantitative composition. The end result is a detailed profile of the lipid composition suitable for analysis within the context of information generated by other paleobiologic areas of investigation. As a result, a detailed description of the lipid composition of ancient tissues could provide:

1. Information on the diet of the ancient people and, in conjunction with other types of evidence, the extent to which that diet was based strictly on availability or on cultural and perhaps ritualistic determinants as well
2. Information on the life style and life status of the person in a given ancient society
3. An indication of the state of preservation of different components of ancient tissue – information that will advance understanding of the science of tissue preservation and macromolecular stability in nonliving tissue over long periods
4. Information on the presence and cause of diseases involving lipids
5. Possible information on the geographical and ethnic occurrence of certain heritable maladies in ancient times
6. A basis for evaluating the feasibility of

using lipid chemistry as a biological dating tool

In a more general sense, the overall significance of the study of lipids reflects that of chemical studies on ancient tissues as a whole: namely, the use of standard chemical methodologies to provide information and informed speculation on the biological, cultural, and medical characteristics of the populations and environment of ancient times.

AMINO ACID DATING OF PRESERVED BIOLOGIC SPECIMENS

The standard method for determining the age of most fossils and other organic material is by the carbon-14 technique, which is based on the ratio of carbon 12 to carbon 14 in the specimens. Because carbon 14, a radioactive isotope, decays at a known rate and is not replenished following death of an organism, the proportion of ordinary carbon to carbon 14 slowly increases. However, there are some difficulties with the standard error of the carbon 14 clock (Baxter et al. 1974). Further, the reliability of carbon 14 dating diminishes substantially with samples 50,000 years old or more.

Recent evidence suggests that the amino acid racemization reaction can be used to estimate the age of organic material (Bada and Protsch 1973). Only L-amino acids are usually found in the proteins of living organisms, but over time they undergo slow racemization, producing the nonprotein D-amino acids. Thus, the proportion of D- to L-amino acids in paleobiological material steadily increases with time. However, the racemization reaction is dependent on temperature, and some estimate of the temperature history of a paleobiological specimen must be available for accurate dating. Amino acid dating requires much smaller quantities of material than radiocarbon dating, and because amino acids racemize at different rates, a chosen amino acid can be used for dating such mate-

rial in a specific age range; this extends the applicability of the technique beyond that of radiocarbon dating.

However, some difficulties persist with the amino acid dating technique. For example, there remains some uncertainty about the effect of pH variation on the racemization rates of amino acids in preserved biologic materials. Further, it is often difficult to obtain a reliable temperature record of the ancient specimen. For more recent specimens (ca. A.D. 500 – B.C. 2000), as is usually the case with mummies, this is even more critical. The temperatures at which most paleobiological specimens are found result in small D/L ratios even for amino acids such as aspartic acid and alanine, which racemize substantially faster than most amino acids. Consequently, it is necessary to detect small amounts of D-amino acid. For this reason, gas chromatography of amino acid mixtures has proved far more sensitive than the traditional ion-exchange chromatographic methods (Barraco 1975a).

Finally, if the relationship between rate of racemization and temperature can be established from a series of rate constants for each amino acid at different temperatures, this would permit calibration of amino acid dating without the use of radiocarbon standard curves, avoiding the incorporation of radiocarbon standard error into the estimation of date by amino acid racemization. This would result in an absolute rate constant for the racemization of individual amino acids at a given temperature. Further, the estimation of age of a given paleobiological specimen could be calculated from approximately 20 data points, representing the estimations of D/L ratios from separate amino acid racemization reactions (not including glycine but including the epimerization of L-isoleucine to the nonprotein amino acid D-alloisoleucine).

TECHNIQUES FOR BIOCHEMICAL ANALYSIS OF ANCIENT TISSUES

In the biochemical studies of six mummies autopsied by our Detroit group (DIA I, PUM I,

PUM II, PUM III, ROM I, Saint Lawrence Eskimo), it became apparent that not only did the nature of the mummification process vary greatly (from the extremely deliberate and meticulous mummification of PUM II to the apparently unintended and natural mummification of the Saint Lawrence Eskimo), but also the degree of preservation of macromolecular constituents extracted from these tissues varied greatly. Consequently, a series of studies was initiated to evaluate not only the persistence and heterogeneity of macromolecular constituents in mummified tissue, but also the relationship between the degree of preservation of these materials in ancient tissues and the nature of the mummification process. Other paleobiological and paleohistological findings from some of these mummies have already been reported (Lewin et al. 1974; Lynn and Benitez 1974; Cockburn et al. 1975; Reyman et al. 1976; Zimmerman 1978; Barraco et al. 1977).

Various types of tissues were used for the biochemical studies. Some samples were taken from the internal aspects of the organ packages (usually located in the abdominal and thoracic cavities of the intentionally preserved bodies) or from the organs in situ (in the case of naturally mummified specimens). However, the following methods and results are intended to apply as general guidelines for the biochemical analysis of mummified tissues. Because ancient relics are difficult to obtain, certain tissues (e.g., spleen, brain) are often not available; also, the preservation condition of some samples is unsatisfactory.

PROTEIN REHYDRATION AND
EXTRACTION PROCEDURES
The tissue samples are cut into small pieces (2 to 3 mm) and placed in various rehydrating mediums. Portions of the rehydrated tissue are used for histology, and the remaining samples are used for biochemical analyses. The methods for rehydration of the mummified tissue were evolved after considerable trial and error. Each tissue requires specific rehydration and extraction conditions depending on its type, its degree of preservation, and the desired macromolecular material, so that standard techniques in protein and lipid chemistry can be used for subsequent analysis. The most satisfactory results for most tissues are obtained when the tissue is rehydrated in 2.5M NaCl (25 ml per gram dry tissue) for 30 hours at 25 °C in a closed vessel. Using this technique, 30 hours later the solution typically appears dark brown, with a sludge at the bottom representing the rehydrated tissue. For unnatronized tissues the rehydrating solution has much less color. The dark color from natronized tissue (i.e., PUM II) is probably attributable to the resins and oils used by the ancient Egyptian embalmers (Lucas and Harris 1962). The total contents of the vessel are then homogenized. The resulting suspension is filtered through cheesecloth and recentrifuged. The supernatant (S_1) is decanted and lyophilized; the pellet (P_1) is stored. Following lyophilization, S_1 is resuspended in 5 mM sodium phosphate buffer, pH7 (10 ml per gram dry tissue).

At this point, two separate procedures are usually employed for the extraction and purification of the protein material (Figure 19.1). Details of these procedures have been reported previously (Barraco et al. 1977). In the first procedure, S_1 is treated with sequential series of solvents to isolate and purify the protein material. In the second procedure, S_1 is ultracentrifuged to form a second supernatant (S_2) and a second pellet (P_2). Further, S_2 is desalted on a Sephadex G–25 column, forming DS_2. DS_2 is subsequently fractionated on Sephadex G–100, forming three major fractions: DS_2–I, DS_2–II, and DS_2–III (Figure 19.3 C). Then polyacrylamide gel electrophoresis (PAGE) is conducted on each fraction (Figure 19.4).

Figure 19.1.   *Procedures for extraction and purification of protein material from mummified tissue.*

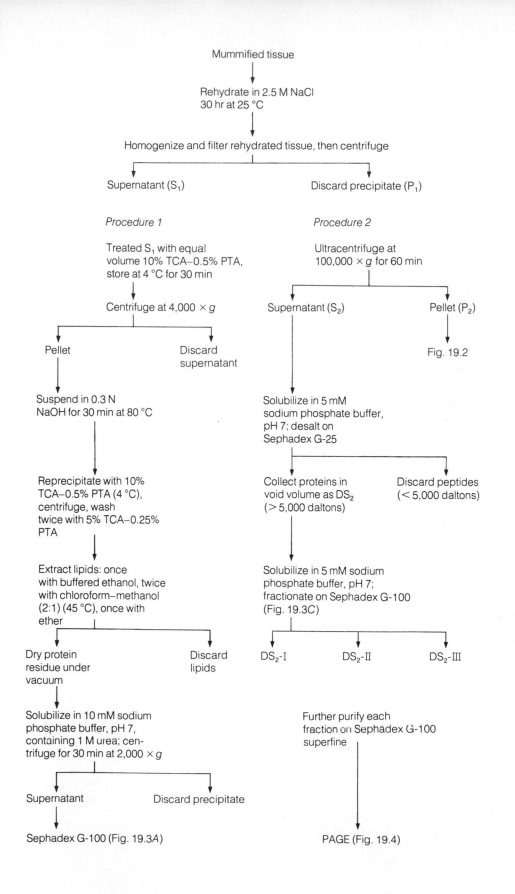

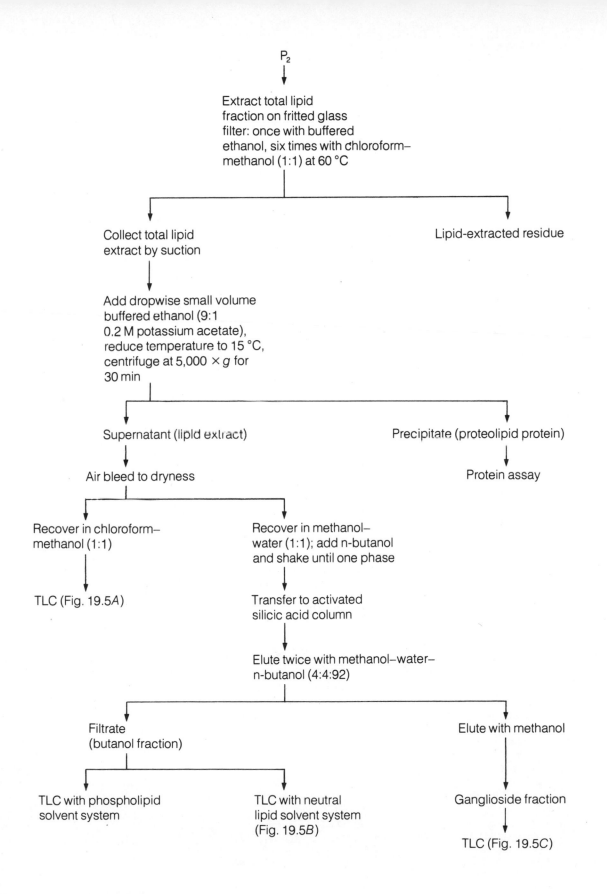

P$_2$

Extract total lipid
fraction on fritted glass
filter: once with buffered
ethanol, six times with chloroform–
methanol (1:1) at 60 °C

Collect total lipid
extract by suction

Lipid-extracted residue

Add dropwise small volume
buffered ethanol (9:1
0.2 M potassium acetate),
reduce temperature to 15 °C,
centrifuge at 5,000 × g for
30 min

Supernatant (lipid extract)

Precipitate (proteolipid protein)

Air bleed to dryness

Protein assay

Recover in chloroform–
methanol (1:1)

Recover in methanol–
water (1:1); add n-butanol
and shake until one phase

TLC (Fig. 19.5A)

Transfer to activated
silicic acid column

Elute twice with methanol–water–
n-butanol (4:4:92)

Filtrate
(butanol fraction)

Elute with methanol

TLC with phospholipid
solvent system

TLC with neutral
lipid solvent system
(Fig. 19.5B)

Ganglioside fraction

TLC (Fig. 19.5C)

EXTRACTION PROCEDURES AND ANALYTICAL METHODS FOR LIPID MATERIAL

Portions of P_2 are weighed and placed on a medium porosity fritted glass filter previously rinsed with acetone–water. P_2 is treated with a sequence of solvents, and the total lipid extract is fractionated by thin-layer chromatography (TLC) or further partitioned into gangliosidic, neutral lipid, phospholipid, or proteolipid fractions (McCluer et al. 1962), which can be separately analyzed by TLC (Figure 19.5). The plates are sprayed with reagents to visualize neutral lipids, phospholipids, cholesterol and its esters, and gangliosidic sialic acid. The chromatograms are analyzed by densitometry (Canalco, Model G) to standardize mobilities and preliminarily screened by the iodine vapor test. Also, portions of the total lipid extract and the lipid-extracted residue are assayed for protein, neutral hexose, and sialic acid. Details of these procedures have been reported previously (Barraco et al. 1977).

ANALYTICAL TECHNIQUES FOR PROTEIN MATERIAL

Quantitative estimation of protein is done by the method of Lowry et al. (1951) and the biuret assay (Itzhaki and Gill 1964). PAGE is conducted under basic conditions (Davis 1964) and acidic conditions (Brummel and Montgomery 1970). Portions of DS_2 are solubilized in 0.05 M NH_4HCO_3 and rechromatographed on Sephadex G–25 fine (column dimensions 30 by 0.9 cm; eluant 0.5 M NH_4-HCO_3, pH 7.8; flow rate 3.6 ml cm^{-2} hr^{-1}) where the excluded proteins in the void volume (>5,000 daltons) are used for amino acid analysis (Beckman M–121 Micro column with AA Computing Integrator, AΛ–20 resin) before and after hydrolysis (6 N HCl, 22 hours, 110 °C) by stepwise elution according to a modified procedure of Moore and Stein (1963) (Table 19.1).

DETERMINATION OF TISSUE CATION CONTENT

For the cation studies, tissue samples (2 to 3 mm) are rinsed briefly with methanol to remove surface salt and are lyophilized. Desiccated aliquots are weighed, ashed, and solubilized in a $SrCl_2$–HCl solution for analysis of Na, K, Ca, and Mg by atomic absorption spectrophotometry (Putney and Bianchi 1974) (Table 19.2).

EXPERIMENTAL RESULTS

Most of the following results, particularly the lipid studies, are derived from the analysis of PUM II, the best preserved of the six mummies autopsied. However, the salient features of these studies apply to the entire range of mummies analyzed.

PROTEINS

Depending on the type of tissue used and the degree of preservation of the mummy, the yield of protein in the extracted sample has varied from 5 mg per gram of desiccated mummified tissue for rectus abdominis from ROM I to 30 mg per gram for spleen from PUM II. In fact, all tissue samples analyzed from the series of six mummies have yielded proteinogenous material, except for what appeared to be dried blood located in bone marrow samples from PUM II. The majority of the extracted protein is found in P_2, which consists primarily of membrane material.

Proteins precipitated from S_1 are fractionated on Sephadex G–100 in a urea buffer system (Figure 19.3A and B) and DS_2 is chromatographed on Sephadex G–100 in a nonurea buffer system. The molecular weight distribution in both gel filtration systems is approximately 5,000 to 130,000 daltons. This molecular weight range for the extracted protein is confirmed by PAGE in the presence of sodium dodecyl sulfate (SDS) (Barraco et al.

Figure 19.2. Procedure for desalting S_2 fraction on Sephadex G-25 column.

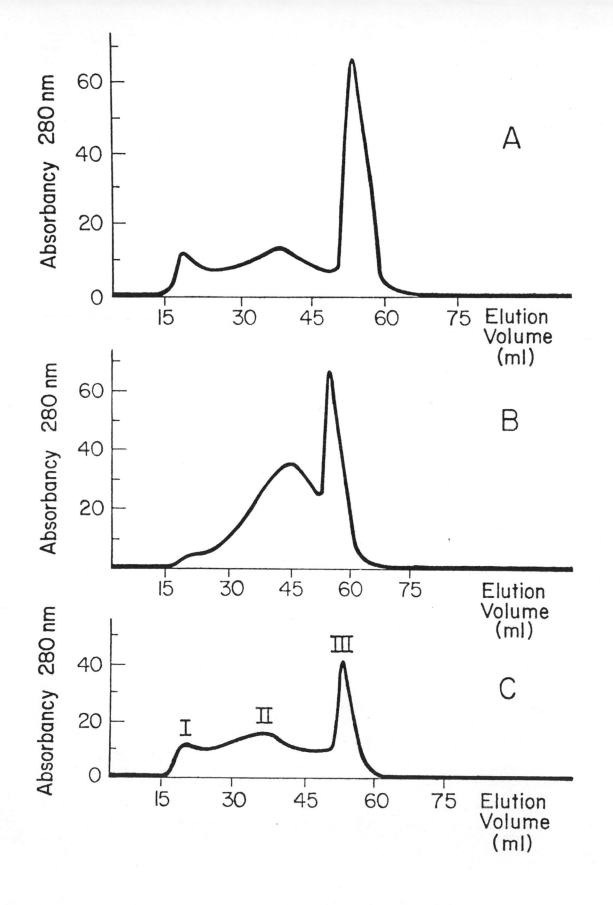

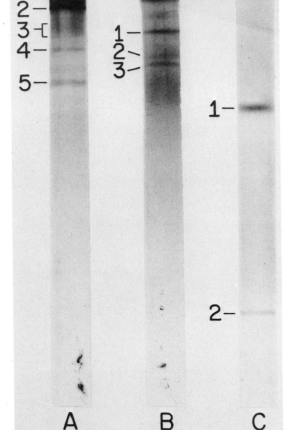

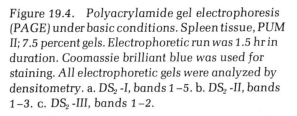

nents from the tissue samples of spleen from PUM II using PAGE with basic conditions. Electrophoresis of rectus abdominis tissue, also from PUM II, under identical conditions results in the separation of fewer components. Using PAGE under acidic conditions, much of the material does not enter the gel, and it has not been possible to produce any major band separation for a variety of tissues from several of the mummies.

LIPIDS

Almost half the dry weight of P_2 is removed by the lipid extraction. A significant amount of protein soluble–chloroform–methanol is detected in the total lipid extract, indicating the presence of intact proteolipid material. The sugar assays indicate the presence of significant quantities of bound neutral hexose in the gangliosidic fraction of the total lipid extract and the lipid-extracted residue. However, significant quantities of bound sialic acid are detected only in the lipid-extracted residue, indicating the persistence of sialic acid in glycoproteins but not gangliosides. Although separation of some gangliosidic material on TLC can be achieved (Figure 19.5), the resolved components do not stain with resorcinol reagent and do not separate into any recognizable pattern, further indicating the degradation of gangliosides. These results held true for P_2 derived from spleen tissue even though splenic tissue is rich in gangliosides and other glycolipids (Weigandt and Bucking 1970).

Figure 19.4.   *Polyacrylamide gel electrophoresis (PAGE) under basic conditions. Spleen tissue, PUM II; 7.5 percent gels. Electrophoretic run was 1.5 hr in duration. Coomassie brilliant blue was used for staining. All electrophoretic gels were analyzed by densitometry. a. DS_2-I, bands 1–5. b. DS_2-II, bands 1–3. c. DS_2-III, bands 1–2.*

1977). PAGE conducted under basic conditions reveals a significant degree of heterogeneity (Figure 19.4). It is possible to separate and consistently identify 10 compo-

AMINO ACIDS

Table 19.1 shows the amino acid distribution, as percent of total, for hydrolyzed proteins from rectus abdominis from PUM II (natronized) and ROM I (unnatronized). To rule

Figure 19.3.   a. *Gel chromatography of S_1 from PUM II (rectus abdominis) on Sephadex G-100 (eluant: 10mM sodium phosphate buffer, pH 7, containing 1 M urea). b. Gel chromatography of DS_2 from ROM I (rectus abdominis) on Sephadex G-100 under same conditions as a. c. Gel chromatography of DS_2 from PUM II (splenic tissue) on Sephadex G-100 (eluant: 5 mM sodium phosphate, pH 7). The resulting chromatography produced three fractions DS_2-I, DS_2-II, DS_2III. Each fraction was rechromatographed separately on Sephadex G-100 superfine. Subsequently, each purified fraction was subjected to PAGE.*

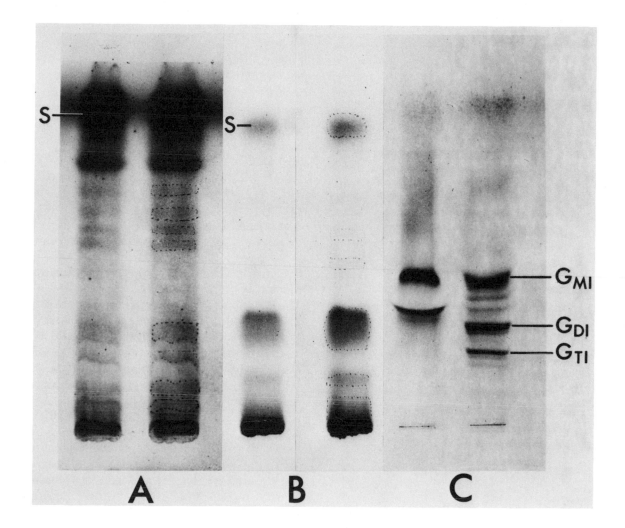

Figure 19.5.   Thin-layer chromatography (TLC) of lipid fractions from PUM II. a. TLC of total lipid fraction (not including proteolipids); 20- by 20-cm plate. First solvent system: heptane–ether–methanol–acetic acid (90:20:2:3). Second solvent system: chloroform–methanol–H_2O (70:30:4). General detection reagent: sulfuric acid–dichromate. Sterol and sterol ester reagent: sulfuric acid–acetic acid. With the sulfuric acid–dichromate reagent, 11 components were detected, 1 of which stained for sterols (designated S). b. TLC of neutral lipid fraction (butanol filtrate; McCluer et al. 1962), 20- by 20-cm plate. Solvent system: heptane–ether–methanol–acetic acid (90:20:2:3). With the surfuric acid–dichromate reagent, 6 major components were detected, 1 of which stained for sterols (designated S). c. TLC of ganglioside fraction (methanol filtrate; McCluer et al. 1962), 20- by 20-cm plate. Solvent system: n-propanol–H_2O–conc. NH_4OH (6:2:1). Resorcinol reagent was used for detection of gangliosidic sialic acid: left, spleen gangliosides; right, pigeon brain gangliosides as standards extracted with the same procedure. Major gangliosides designated: G_{MI}, monosialoganglioside; G_{DI}, disialoganglioside; G_{TI}, trisialoganglioside). With the sulfuric acid–dichromate reagent, 2 major spleen gangliosides were detected. However, both components stained negligibly with resorcinol reagent, indicating they are asialogangliosides or other unidentified lipids.

Table 19.1. *Amino acid distribution of hydrolyzed proteins extracted from rectus abdominis of two mummies from the series of six, expressed as percent of total*

Amino acid	PUM II	ROM I
Lysine	3.9	4.1
Histidine	1.2	2.0
Arginine	6.1	3.7
Aspartic acid	11.5	13.1
Threonine	3.6	5.0
Serine	4.7	5.0
Glutamic acid	13.2	18.8
Proline	14.9	4.7
Glycine	12.4	12.0
Alanine	17.1	11.6
Half cystine	—	—
Valine	4.0	4.7
Methionine	—	1.7
Isoleucine	1.2	3.2
Leucine	4.4	4.5
Tyrosine	0.6	3.5
Phenylalanine	1.2	2.4

Dash indicates undetectable at this level of sensitivity.

out contamination by free amino acids, amino acid analysis is also conducted on unhydrolyzed portions of DS_2; however, no free amino acids are detected in the DS_2 fraction. The amino acid analysis for PUM II approximates the amino acid composition of normal mixed tissue proteins except for the absence of half cystine and methionine. However, there appears to be a smaller amount, as percent of total, of basic amino acids, indicating a possible point of preferential degradation of residues in the proteins. This is also evident, particularly with arginine, for ROM I. Further, the high glycine-proline content for PUM II indicates the possible persistence of collagen in the tissues. In fact, electron microscopic analysis of rectus abdominis tissue from PUM II indicates the ubiquitous presence of masses of striated fibrous material resembling collagen (Cockburn et al. 1975). On the other hand, there appears to be little collagen present in the tissue samples from ROM I.

Table 19.2. *Cation content (micromoles per gram dry weight) of tissue from six mummies and of one fresh autopsy specimen*

Specimen	Na	K	Ca	Mg
Human autopsy	216.9	206.2	9.6	27.5
ROM I	187.0	291.2	7.4	1.1
Eskimo	16.3	3.8	371.2	68.7
PUM I	73.1	122.5	23.4	10.9
PUM II	747.0	57.5	11.5	6.9
DIA I	2,488.0	253.8	74.1	33.8
PUM III	345.0	143.0	5.6	3.4

These results can be compared with those of a previous study involving extracted material from an unnatronized Twelfth-Dynasty mummy in which the collagen content was apparently high (Lausarot et al. 1972). In fact, except for the low amount of lysine, which is reported to undergo preferential degradation with time (Lausarot et al. 1972), the percent amino acid composition of extracted protein from ROM I closely approximates that of myoglobin, which may be the major small-molecular-weight protein (ca. 17,000) in skeletal muscle resistant to degradation following desiccation under natural conditions.

CATION STUDIES

In Table 19.2 the various cation levels in tissue samples from the series of six mummies are represented as micromoles of cation per gram of dry weight of tissue. To test the procedures, specimens of freshly autopsied human rectus abdominis are also analyzed by atomic absorption. Table 19.2 shows that ROM I, PUM I, and Saint Lawrence Eskimo were not natronized, as the tissue cation levels approximate closely those of the freshly autopsied specimen. On the other hand, it appears that PUM II and DIA I were natronized (primarily Na salts), and PUM III was partially natronized. Also, the high content of Ca in the muscle of the Saint Lawrence Eskimo indicates the possibility of high lime content in the soil at the burial site.

PRESERVATION OF PROTEINS

One of the major considerations for the paleopathologist is evaluating the degree of preservation of the tissues and the nature of the mummification process. Because it was known that natron and other salts were used for chemical mummification by the ancient Egyptian embalmers (Lucas and Harris 1962), the relationship of Na and other cation levels in the mummified tissue with the molecular weight distribution (a measure of degradation) of extracted protein from the same tissues was investigated. These studies indicate that the degree of preservation (relative distribution of high-molecular-weight protein) of extracted protein varies in relation to the levels of Na in the tissues (natron). For example, compare the molecular-weight-distribution profiles of PUM II and ROM I (Figure 19.3 A and B). In fact, the molecular-weight-distribution profiles for extracted protein from five mummies indicate that natronized tissue undergoes less degradation and the degree of preservation of extracted protein correlates closely with the levels of Na in the tissues (Barraco 1975b, 1978).

CONCLUSIONS

These studies indicate that protein and lipid material can be extracted from mummified tissues and processed using various preparative and analytical techniques. There is evidence that conjugated proteins with carbohydrate and lipid prosthetic groups remain stable and extractable as glycoprotein and proteolipid. Further, it appears that some protein of relatively high molecular weight (ca. 130,000) remains intact, although much of the protein has been degraded with a marked attenuation of heterogeneity. Similar results were reported for protein material extracted from neck tissue of a 4,000 year old Egyptian mummy (Lausarot et al. 1972). In that study, the extracted protein migrated as a single component in a phosphate–SDS PAGE system with a relative mobility corresponding to an apparent molecular weight of 60,000.

In the studies using PAGE under acidic conditions, the extracted protein demonstrates negligible mobility and heterogeneity. Along with the amino acid analyses, these results indicate the possibility that basic amino acid residues in preserved biological specimens undergo preferential degradation. In fact, a net increase in the negative charge of surviving proteins from 8-year-old dried human blood has been reported (Sensabaugh et al. 1971a,b). Also, from the amino acid composition of Egyptian mummy structural protein, it was determined that almost exclusively the basic amino acids undergo chemical modification (Lausarot et al. 1972). This has important implications for recently developed techniques that date organic material using amino acid racemization reactions (Bada and Protsch 1973; Bada et al. 1974).

Greater heterogeneity appears to persist in the lipid than in the protein material. The amount of lipid extractable from S_2 appears negligible. However, from P_2 the total lipids can be fractionated into gangliosidic, neutral lipid, phospholipid, and proteolipid components. The neutral lipid fraction and the total lipid extract demonstrate a persistence of heterogeneity. Using TLC it is possible to identify 6 major components in the neutral lipid fraction and 11 components in the total lipid fraction from the spleen of PUM II. Further, the neutral lipid and total lipid fractions stain positive for cholesterol and sterol esters (Figure 19.5). This indicates the persistence of neutral lipids such as sterols, fatty acids, tocopherols, and triglycerides in mummified tissue. The precise qualitative characteristics of the lipid material can be further evaluated by gas chromatography and other analytical techniques.

These studies also indicate that tissue samples from organ packages appear to be better preserved than tissue from the carcass. This observation is supported not only by histological analysis, but also by the fact that the recovery of extracted proteins (> 5,000 daltons) is greater for these tissues. This may be attributable to a difference in natronization

procedures (Lucas and Harris 1962). The organs, once removed by embalmers, may have been soaked in vessels containing liquid natron, while the remaining carcass, too large for easy submersion in a vessel containing liquid natron, may only have been packed with crystals of natron. Finally, the studies evaluating the relationship between tissue cation levels and preservation of macromolecular tissue components may provide useful archeological information concerning the characterization of the mummification process for different periods and the standardization of paleopathological findings, as the ability to detect disease may vary with the degree of preservation of the ancient specimen.

REFERENCES

Abelson, P. 1957. Some aspects of paleobiochemistry. *Annals of the New York Academy of Sciences* 69:276–85.

Akiyama, M., and Wyckoff, R. W. G. 1970. The total amino acid content of fossil Pecten Shells. *Proceedings of the National Academy of Sciences of the United States* 67:1097–1100.

Bada, J. L., and Protsch, R. 1973. Racemization reaction of aspartic acid and its uses in dating fossil bones. *Proceedings of the National Academy of Sciences of the United States* 70:1331–4.

Bada, J. L.; Schroeder, R. A.; and Carter, G. F. 1974. New evidence for the antiquity of man in North America deduced from aspartic acid racemization. *Science* 184:791–3.

Barraco, R. 1975a. Amino acid dating. *Paleopathology Newsletter* 9:8–9.

– 1975b. Preservation of proteins in mummified tissue. *Paleopathology Newsletter* 11:8–9.

– 1978. Preservation of proteins in mummified tissue. American Journal of Physical Anthropology 48:487–91.

Barraco, R. A.; Reyman, T. A.; and Cockburn, T. A. 1977. Paleobiochemical analysis of an Egyptian mummy. *Journal of Human Evolution* 6:533–46.

Baxter, M. S.; Aitken, M. J.; Clark, R. M.; and Renfrew, C. 1974. Calibration of the radiocarbon time scale in "Matters Arising." *Nature* 249:

Boyd, W. C., and Boyd, L. G. 1937. Blood grouping tests on 300 mummies. *Journal of Immunology* 32:307–19.

Brummel, M. C., and Montgomery, R. 1970. Acrylamide gel electrophoresis of the S-sulfo derivatives of fibrinogen. *Analytical Biochemistry* 33:28–35.

Candela, P. B. 1939. Blood group determinations upon the bones of thirty Aleutian mummies. *American Journal of Physical Anthropology* 24:361–83.

Cockburn, T. A.; Barraco, R. A.; Reyman, T. A.; and Peck, W. H. 1975. Autopsy of an Egyptian mummy. *Science* 187:1155–60.

Davis, B. J. 1964. Polyacrylamide disc gel electrophoresis. *Annals of the New York Academy of Sciences* 121:321–404.

Ducos, J.; Ruffie, J.; and Varsi, M. 1962. Mise en évidence des antigènes Gm^a Gm^b Gm^x dans les taches de sang sec. *Vox Sang* 7:722–31.

Gillespie, M. M. 1970. Mammoth hair: stability of α-keratin structure and constituent proteins. *Science* 170:1100–1.

Ho, T. Y. 1966. The isolation and amino acid composition of the bone collagen in Pleistocene mammals. *Comparative Biochemistry and Physiology* 18:353–8.

Itzhaki, R., and Gill, D. 1964. A microbuiret method. *Analytical Biochemistry* 9:401–5.

Johnson, L. C. 1966. In *Human paleopathology*, S. Jarcho (ed.), pp. 68–81. New Haven: Yale University Press.

Jope, M. 1967. The protein of brachiopod shell. II. Shell protein from fossil articulates: amino acid composition. *Comparative Biochemistry and Physiology* 20:601–5.

Keilin, D. 1959. Problem of anabiosis for latent life: history and current concepts. *Proceedings of the Royal Society*, series B 150:149.

Keilin, D., and Wang, Y. L. 1947. Stability of haemoglobin and of certain endoerythrocytic enzymes in vitro. *Biochemistry Journal* 41:491–500.

Lausarot, P. M.; Ambrosino, C.; Favro, F.; Conti, A.; and Massa, E. R. 1972. Preservation and amino acid composition of Egyptian mummy structure proteins. *Journal of Human Evolution* 1:489–99.

Lewin, P.; Mills, A. J.; Savage, H.; and Vollmer, J. 1974. Nakht, a weaver of Thebes. *Rotunda* 7:15–19.

Lippold, L. K. 1971. The mixed cell agglutination method for typing mummified human tissue. *American Journal of Physical Anthropology* 34:377–83.

Lowry, O. H.; Rosebrough, N. J.; and Farr, A. L. 1951. Protein measurement with the Folin phenol reagent. *Journal of Biological Chemistry* 193:265–75.

Lucas, A., and Harris, J. . 1962. *Ancient Egyptian materials and industries*, pp 270–326. London: Edward Arnold.

Lynn, G. E.; and Benitez, J. T. 1974. Temporal bone preservation in a 2,600-year-old Egyptian mummy. *Science* 183:200–2.

McCluer, R. H.; Coram, E. H.; and Lee, H. S. 1962. A silicic acid absorption method for the determination of ganglioside sialic acid. *Journal of Lipid Research* 3:269–74.

Miller, M. F., and Wyckoff, R. W. G. 1968. Proteins in dinosaur bones. *Proceedings of the National Academy of Sciences of the United States* 60:176–8.

Moore, S., and Stein, W. H. 1963. Chromatographic determination of amino acids by the use of automatic recording equipment. In *Methods in enzymology*, vol. 6, S. Colowick and N. O. Kaplan (eds.), pp. 819–31. New York: Academic Press.

Otten, C. M., and Flory, L. L. 1964. Blood typing of Chilean mummy tissue: a new approach. *American Journal of Physical Anthropology* 21:283–5.

Putney, . W., and Bianchi, C. P. 1974. Site of action of dantrolene in frog sartorius muscle. *Journal of Pharmacology and Experimental Therapeutics* 189:202–12.

Reyman, T. A.; Barraco, R. A.; and Cockburn, T. A. 1976. Histopathological examination of an Egyptian mummy. *Bulletin of the New York Academy of Medicine* 52:506–16.

Sensabaugh, G. F., Jr.; Wilson, A. C.; and Kirk, P. L. 1971a. Protein stability in preserved biological remains. I. Survival of biologically active proteins in an 8-year-old sample of dried blood. *International Journal of Biochemistry* 2:545–57.

1971b. Protein stability in preserved biological remains. II. Modification and aggregation of proteins in an 8-year-old sample of dried blood. *International Journal of Biochemistry* 2:558–68.

Skujins, J. J., and McLaren, A. D. 1968. Persistence of enzymatic activity in stored and geologically preserved soils. *Enzymologia* 34:213–25.

Stastny, P. 1974. HL-A antigens in mummified pre-Columbian tissues. *Science* 183:864–6.

Wiegandt, H., and Bucking, H. W. 1970. Structure and function of ganglioside. *European Journal of Biochemistry* 15 287–91.

Wyckoff, R. W. G.; Wagner, E.; Matter, P.; and Doberenz, A. R. 1963. Collagen in fossil bone. *Proceedings of the National Academy of Sciences of the United States* 50:215–18.

Zimmerman, M. R. 1978. Age determination of an Alaskan mummy: morphologic and biochemical correlation. *Science* 201:11–12.

Paleoserology

RAYMOND L. HENRY
Professor of Physiology
Wayne State University School of Medicine
Detroit, Michigan U.S.A.

Genetic determinants are important ancillary factors in the classification of man and his migration history. Antigens of the ABO Landsteiner designation are most often considered because they are distributed throughout body tissue (Landsteiner and Witt 1926). Positive results from typing identify specific antigens present and enable more precise physical characterization of human populations in antiquity. Such characterization can then be compared with present-day populations in the same geographic area. Determination of blood types provides further data for or against hypotheses of spatial and temporal migrations of man. Information about blood group and other antigens from ancient tissues should serve to supplement morphological characterization; it should not be used independently. These data in conjunction with archeological accounts, historical documentation, and ethnographic information can greatly benefit the anthropologist.

Since the initial study of Boyd and Boyd (1933), several attempts have been made to identify the antigen-antibody reactions of ancient tissue. Early investigations were aimed at the tissue antigens, as antibody induction was not satisfactory. However, more recent technology for antibody production seems to be rendering more reliable results. Precipitin tests, agglutination inhibition, mixed-cell agglutination, and antibody induction with precipitin tests have been employed for the serologic study of ancient tissues. Certain pitfalls and inconsistencies in these methods should be recognized to allow yields of useful data from investigative procedures. Anyone attempting serological work should avail

himself of the extensive experience of the United States military (Camp et al. 1973). Particular attention should be given to the section on pitfalls encountered in performing blood-grouping tests, in Annex S of the monograph.

Designation of major ABO system blood groups depends on the antigen present on red blood cells or other tissue cells. Ordinarily those specific antigens are identified with specific known antibodies in standard sera developed against known ABO antigens. This procedure is the direct typing system, whether precipitation or agglutination techniques are employed. In some instances, it may be possible to identify antibodies against ABO antigens in the specimen to be tested. These procedures are known as reverse, or indirect, testing. Anti-A agglutinin should be present in specimens of the B group, and anti-B in specimens of the A group. Both anti-A and anti-B agglutinins should be present in group O individuals, and neither is present in group AB persons. Positive results in these tests strongly suggest presence of that particular antibody; however, absence of antibody in ancient serum or tissue still leaves open the question of group identity because the specific antibodies may have deteriorated or been otherwise lost. The ABO system of blood grouping is unique in having naturally occurring isoantibodies always reciprocal to the antigens present on cells of that individual. Isoantibodies have been reported for other blood group systems, such as Rh-Hr, Kell, I, Ss, and Wr[a]; however, such systems are so rare that testing ancient tissues for anything except the ABO system is not war-

ranted. This chapter deals with application of data so far obtained as well as critical evaluation of the techniques employed. Within the discussion, detailed descriptions of methods are presented.

Attention to ABO antigens has gained favor because these antigens are generally distributed throughout body tissues so that bone, hair, skin, muscle, or whatever happens to be available may yield information. Reports concerning negative data should be viewed with caution. Positive data, on the other hand, appear to be more reliable.

Boyd and Boyd (1937) tested 300 mummy specimens and reported a surprising excess of AB types. That the absorptive power of mummified material was present for A and B types was not questioned. However, group A substance was suggested as being perhaps less stable than B type. Nevertheless, this alone could not account for the large number of AB types. Failure of tissue to remove either anti-A or anti-B did not necessarily identify the blood group as O because of the possible disappearance of one or both agglutinogens. The finding of A and B groups in predynastic Egyptian material suggested to the authors that blood groups were not as recent as supposed in 1937, but rather supported postulations of anthropoid ancestors of man.

Brues (1954) discussed selection and genetic drift of the ABO allelomorphs. Early assumptions were made that only the O gene originally existed and that A and B arose and spread at different times, producing the current distribution among living populations. Frequencies of the A gene showed large ranges among populations having reasonably homogeneous morphological features. Investigation of Australian aborigines revealed evidence for vigor of genetic drift under primitive conditions exceeding that in population admixtures. Sufficient data from ancient tissues may substantiate some of these claims. A basic question seems to be whether any selective value can be attached to the ABO blood group genes or to the genotypes formed by them. Distribution of the types in ancient tissues, such as Egyptian mummies, coupled with identification of diseases in these ancient populations, would be of interest.

PRECIPITIN TEST

Precipitin tests depend upon the presence of free antigens of the ABO groups. The precipitate, or interfacial ring, develops if antigens are present to react with antibodies formed in rabbits against laked human blood of known ABO groups. Boyd and Boyd (1937) found that antiserum against human muscle did not prove successful for precipitin tests. These researchers tested 94 Egyptian mummies, with only 4 positive and 1 doubtful result. They found only 2 positive and 7 doubtful cases in 159 American ancient specimens. Some of the positive-reacting specimens also gave nonspecific precipitations with normal serum.

PROCEDURES

These procedures are based on Boyd and Boyd (1937).

1. Extract 0.05 g of material with 0.4 to 0.5 ml saline at 3 °C for 48 hours.
2. Remove supernatant and dilute 1:4 to reduce specific gravity below that of serum.
3. Test against immune rabbit serum prepared by injecting laked human blood of known blood group.
4. An interfacial ring of precipitin marks a positive reaction.

PRECAUTIONS NOTED

Failure of the supernatant from prepared mummified specimens to form a precipitin ring with known antisera to types A or B does not necessarily mean a positive reaction for type O unless precipitation occurs with anti-O serum. Either A or B antigen or both may have deteriorated or be too weak to react. In like

manner, a negative reaction to type O antiserum does not rule out type O for the specimen. Boyd and Boyd (1937) suggested that A may be less stable than B over a long period of time.

MIXED-CELL AGGLUTINATION TEST

Because insufficient blood samples are preserved in ancient mummified tissues, the precipitin tests are not readily adaptable. Coombs et al. (1956) described a method to identify blood group antigens on tissues devoid of blood. This has been termed the mixed-cell agglutination test. The procedure is based on the idea that tissue cell surfaces contain blood group antigens capable of capturing and retaining specific antibodies. Suspensions of isolated tissue cells incubated with erythrocytes of specific A, B, or O will show aggregates induced by interactions of the specific antigen with the specific antibody. The method was successfully used by Otten and Florey (1963) in typing Chilean mummy tissue. Later Lippold (1971) utilized the technique of mixed-cell agglutination to demonstrate the continuity of ABO types between living populations and aboriginal specimens from prehistoric populations of the Aleutian Islands, southwestern United States, Peru, and Chile. In addition to dried tissues, skeletal remains, especially from vertebral bodies, appear to give adequate results with this agglutination test.

Stastny (1974) reported on HL-A antigen profiles in pre-Columbian mummies from 500 to 2,000 years old. Patterns of reactions to the mixed-cell agglutination test were consistent with known relations between HL-A antigens. Distribution was similar in present-day descendants of the ancient populations studied. Allison et al. (1976) found mixed-cell agglutination of fresh or mummified tissue to be equally reliable as agglutination inhibition and induction of antibody production. However, agglutination inhibition would not establish the presence of H(O) antigen.

PROCEDURES
These procedures are based on Otten and Florey (1963).

1. Mummified tissues are shredded and washed gently in saline to remove dust and dirt. Shreds are cut into small pieces, microblended, and incubated at 37 °C for 45 minutes in 2.5 percent trypsin solution to isolate the cells by digestion.
2. Clumps of material are removed, and the free cell suspension is washed three times in physiologic saline by centrifugation and decantation.
3. Washed cells are incubated in the refrigerator at 4 to 5 °C for 1 hour in 5 percent bovine albumin and then washed once again in saline.
4. Cell suspensions are divided into three aliquots. Each is incubated for 2 hours at 4 to 5 °C, one with 2 drops of anti-A antiserum, one with 2 drops of anti-B antiserum, and one with 2 drops of saline extract of *Ulex europaeus* seeds. (The last is prepared by washing 20 to 30 ml fresh seeds in tap water followed by physiologic saline. Seeds are ground in a blender by adding saline until the mixture has the consistency of thick pea soup. The mixture is incubated for 4 to 6 hours at 37 °C and stored in the refrigerator for 24 hours. The mixture is spun at high speed to separate out the clear greenish yellow fluid. A polypore filter may then be used. Sodium azide to 0.01 percent is added to prevent bacterial growth. Specificity will be improved if the mixture is left in the refrigerator for 2 days. The white precipitate that forms can be removed by centrifugation. Titer should be at least 1:32 using pooled group O erythrocytes. Retiter if the reagent is stored frozen.)

5. Red cells of A, B, and O are trypsinized with 1 percent crude trypsin in saline. One part trypsin to nine parts of 2 percent cell suspensions are prepared and incubated 10 minutes at 37 °C, followed by washing the cells until the supernatant shows no agglutination of detector erythrocytes (usually three washings). Each tissue cell aliquot prepared in step 4 is subdivided into three aliquots. Trypsinized A_1 erythrocytes (1 drop) are added to one aliquot of each set, B erythrocytes to the second, and O erythrocytes to the third. The mixtures are incubated at 37 °C for 10 minutes.

6. The nine tubes from step 5 are centrifuged lightly for 1 minute to improve the microscopic detection of red cell aggregates near the bottom. A clearly defined positive will have only one tube of the nine showing aggregates. A or B tissue cells may also demonstrate H specificity in theory. (B erythrocytes with low or no H receptors should be selected as detector cells.)

PRECAUTIONS NOTED

Hair fragments or fragments of hair follicles tend to absorb agglutinins nonspecifically and should be avoided. The inner epidermal layer (stratum granulosum) is more endowed with A, B, and H antigens than other layers of skin. Buccal, lingual, esophageal, or gastric mucosa, or any area of mucus-secreting cells of the respiratory tract, increases success in typing (Szulman 1960, 1962). However, the nonsecretor phenotype may have low levels of mucus-bound antigens. High levels of H antigen could lead to false assignment of blood group O. However, the H antigen is more abundantly distributed in O types and therefore minimizes erroneous assignments.

Negative results may appear if the tissue had undergone sufficient bacterial decomposition. The most common bacteria are of the species *Bacillus*, which act only on H(O) substance (Iseki 1962). However, it is assumed that all antigen types would be equally destroyed so that AB types would not be confused with A or B. Fresh cadaver tissue of known blood groups should be used as controls. All materials used in the test (forceps, scissors, etc.) should be sterilized before each test.

AGGLUTINATION-INHIBITION TEST

Appropriate known red cell types agglutinate according to the unknown antisera tested and deductions are made from the test results. These reactions can be inhibited by first mixing the antisera with specific antigens. If the investigator utilizes known red cell types and known antisera then unknown antigens can be detected by inhibition of the agglutination reaction. Specific antigens will inhibit agglutination by like known antigens only. Testing of tissue cells, even if ancient, depends on the recognition that cells of many tissue types contain surface antigens of the ABO systems. Direct precipitin tests would depend on removal of the antigens from unknown tissue, and the titer endpoints would be difficult to determine. Incubation of tissue with known antisera and removal of the absorbed complexes with tissue material, however, still allows good titers of nonabsorbed antibodies with detectable reduction of the specific antibodies that were absorbed.

PROCEDURE

This procedure is based on Allison et al. (1976).

1. Serial double dilutions of anti-A and anti-B sera are made from 1:2 through 1:2,048. These dilutions are titered with 1 drop of 2 percent saline suspensions of group A cells (in anti-A tubes) and group B cells (in anti-B tubes). The last tubes showing strong agglutination (1+) are taken as endpoints. An eightfold concentration of the titers is used to dilute the original antisera.

2. Tissues to be tested are pulverized by grinding with sand, and 500 mg are di-

vided equally into two tubes. The tissue mixture is moistened with 4 to 5 drops of sterile physiologic saline, and 0.9 ml of diluted antiserum (anti-A or anti-B) is added. Tubes are stoppered and left at 5 °C and 10 °C for 48 hours with shaking at 12-hour intervals.

3. Tubes from step 2 are serofuged until a clear supernatant is obtained. Supernatants are diluted to the original 0.9 ml volume with saline.

4. Serial dilutions from 1:2 to 1:64 are made from the A and B supernatants. One drop of 2 percent group A (or group B for that set) is added to each tube. Tubes are shaken five times by hand.

5. Mild centrifugation assists in reading agglutination. Doubtful agglutinations are read microscopically.

6. If group A, but not group B cells agglutinate, the tissue is considered type A. Conversely, type B tissue is identified if group B but not group A agglutinates. No agglutination in either A or B leads to a designation of type AB tissue; agglutination of both A and B points to type O tissue.

PRECAUTIONS NOTED

Failure of tissue supernatants to absorb A and B agglutinins would lead one to assign the tissue to group O, but this result could also be obtained when A and B antigens are absent owing to destruction as well as by genetic design. The positive finding of A, B, or AB antigens, however, can be taken as valid.

Titer reduction may occur with deterioration of antisera during the test period if proper controls from cadavers of known types are not run simultaneously. Improper refrigeration or bacterial growth could alter results. In addition, nonspecific absorption may occur. False positive agglutination may occur if tissue matter is present in the serofuged supernatant from pulverized cell samples. Even inert materials such as charcoal, kaolin, and benzonite can cause significant antibody

titer reduction (Theime and Otten 1957). Variable particle size of different bone specimens can affect the absorption rate and degree. Fine ground bone quickly loses its antigenicity if not properly sealed in a vacuum container. In addition, a multitude of plant and animal contaminants can result in false inhibition reactions (Salazar Mallen 1951).

Boyd and Boyd (1937) early recognized that antibody serum strength was important for proper results. The absorbing power of mummy tissue is usually very weak, and not all agglutinin can be removed from serum with a high antibody titer. They recommend beginning with a dilution of the antibody serum to 1:8 rather than 1:32, as used by Allison et al. (1976). Caution should be noted with fine powdered bone because it will nonspecifically absorb both anti-A and anti-B antibodies of the dilute serum and the specimen may be thought to be AB. If the sera are too strong, insufficient absorption may occur and the specimen may be incorrectly called type O (Candela 1936).

When both anti-A and anti-B are removed, indicating AB tissue antigens, tissue can be tested for inhibition of diluted anti-chicken serum against chicken red cells. No removal of these antibodies supports immunological specificity. When mummified tissue appears to be group O, the question of whether Schiff's nonsecretor agglutinogens are present becomes important. Boyd and Boyd (1937) tested known nonsecretors and found the proper agglutinogens in muscle tissue.

ANTIBODY INDUCTION WITH PRECIPITIN TESTS

Allison et al. (1976) introduced an antibody-induction test for A, B, and H antigens by immunizing adult New Zealand white rabbits by injecting the antigens mixed with Freund's adjuvant. Highest titers were reached at 3 weeks, and blood group determinations agreed with those based on the agglutination-inhibition technique. The antibody-induction method, however, offers the

advantage of H antibody production and measurement that are in agreement with those obtained by the mixed-cell agglutination technique.

PROCEDURE
Based on Allison et al. (1976).

1. Freund's adjuvant is prepared by grinding 50 mg of heat-killed human tubercle bacilli with 1.0 ml sterile mineral oil. This mixture is stirred in 29 ml of light mineral oil.
2. Mummy tissue (ratio of 100 mg to 2.0 ml sterile 0.9 percent saline) is allowed to soak for 4 days or longer. Adjuvant (0.5 ml per 100 mg tissue in 2.0 ml saline) is added to the tissue suspension and stored at refrigerator temperature for 2 days with periodic shaking.
3. Before injection of rabbits, blood is collected from ear veins and the serum tested to ensure that no initial A or B agglutinins are present. Rabbits are then injected intramuscularly with 0.5 ml of antigen-adjuvant once weekly for 3 weeks. Control animals are injected with 0.5 ml saline plus adjuvant.
4. Rabbits are bled prior to each injection to determine antibody titers. Three sets of eight tubes each are prepared from rabbit serum. Each set is subjected to a twofold serial dilution. A 2 percent solution of human red blood cells of known A, B, or O type is prepared and 1 drop of each type is added to a set of tubes, respectively. All tubes are shaken gently and checked for agglutination macroscopically and microscopically. A threefold rise in titer is considered significant for detection of specific agglutinins.

PRECAUTIONS NOTED
As in other techniques, the antigen for A, B, and H(O) must be present for antibody induction in rabbits. Absence does not imply that the antigen was not there earlier. However, this technique has the advantage of giving a positive result for identification of O type in-

dividuals. The relatively long incubation periods of 4 days, 2 days, and so on require careful attention to sterile techniques so that bacterial or other degradation of the desired molecules does not occur. The final step in the procedure is a precipitin-type test. The same precautions for selection of A, B, and O type red cells and reading the degree of precipitation should be recognized.

HL-A ANTIGEN SYSTEM OF TYPING

This technique utilizes a modified cytotoxicity inhibition employing antigens of the HL-A system. The procedure, therefore, is an agglutination-inhibition test. Stastny (1974) found a similarity between prehistoric and present-day American Indian populations in terms of 14 HL-A antigens. The HL-A profile of coastal Peruvian mummies matched well that of American Indians in general. They resembled the Ixils of Guatemala and the Quechuas of the Peruvian highlands with respect to W28 and W27 antigens and the Aymaras of Chile with respect to HL-A5 antigens. This was, admittedly, a small group of mummies tested, but the work does reflect the usefulness of application of agglutination inhibition to typing other than the ABO system.

PROCEDURE
This procedure is based on Stastny (1974).

1. Tissue selected is shredded in a homogenizer and passed through a stainless-steel sieve to obtain a fine powder. Each powder (200 mg) is extracted twice in 2 ml sterile saline with constant stirring at 37 °C for 30 minutes.
2. The two extracts are combined and dialyzed against distilled water at 4 °C. Protein concentration is determined by standard methods.
3. Each sample is freeze-dried and dissolved in fresh rabbit serum to 8 mg per milliliter. Working solutions of 2, 4, and 6 mg per milliliter are made for testing.
4. Antiserums against known HL-A anti-

gens are selected and diluted to the optimal concentration (highest dilution that will still destroy 100 percent of target cells). Target lymphocytes are obtained from typed reference donors. Lymphocytes are prepared by flotation on Ficoll-Hypaque and frozen in small portions with liquid nitrogen.

5. Fluorescein histocompatibility is used for analysis of cytotoxicity.

6. Fluorescein-labeled lymphocytes remain visible if inhibition of cytotoxic antibodies occurs. This inhibition is expressed as a percent of control lymphocytes without antibody. Tissues with at least 50 percent inhibition are considered positive for the presence of that particular antigen.

PRECAUTIONS NOTED

Some variation between duplicate determinations exists so at least two samples should be tested. Controls should consist of antiserum without mummy antigen, antiserum with mummy antigen but without complement, complement alone, and antigen alone. Standard inhibition with lymphocytes known to contain HL-A antigen should also be tested. Skin is probably the best tissue to test, but several different tissues should be run.

Glycoproteins, polysaccharides, lipopolysaccharides, and streptococcal M1 protein have been reported to inhibit antibodies to the HL-A series (Hirata and Terasaki, 1970; Mittal et al. 1973). Contamination may be a factor in some of the tests. Antigens also may have been denatured over time, so negative results should not be considered conclusive.

M-N GROUPS

M-N antigens are rather labile and should be tested for as near to time of blood sample drawing as possible. Matson and Roberts (1949) took careful precautions for refrigeration and transport to preserve the samples for testing. Boyd and Boyd (1934a, 1934b, 1937) were not able to demonstrate M-N antigens in muscle tissue even with fresh or freshly dried cadaver material. It appears that none of the procedures described above, with possible usefulness to identify other groups, can have any hope of detecting M-N groups from mummified tissue.

REFERENCES

Allison, M. J.; Houssaini, A. A.; Castro, N.; Munizaga, J.; and Pezzia, A. 1976. ABO blood groups in Peruvian mummies: an evaluation of techniques. *American Journal of Physical Anthropology* 44:55–62.

Boyd, W. C.; and Boyd, L. G. 1933. Blood grouping by means of preserved muscle. *Science* 78:578–9.

– 1934a. Group specificity of dried muscle and saliva. *Journal of Immunology* 26:489–94.

– 1934b. An attempt to determine the blood groups of mummies. *Proceedings of the Society of Experimental Biology and Medicine* 31:671–2.

– 1937. Blood grouping tests on 300 mummies. *Journal of Immunology* 32:307–19.

Brues, A. M. 1954. Selection and polymorphism in the ABO blood groups. *American Journal of Physical Anthropology* 12:559–606.

Camp, F. R., Jr.; Conte, N. F.; and Brewer, J. R. 1973. *Military blood banking 1941–1973: a monograph*, pp. 2–24. Fort Knox, Ky.: U.S. Army Medical Research Laboratory.

Candela, P. B. 1936. Blood-group reactions in ancient human skeletons. *American Journal of Physical Anthropology* 21:429–32.

Coombs, R. R. A.; Bedford, D.; and Rouillard, L. M. 1956. A and B blood-group antigens on human epidermal cells demonstrated by mixed agglutination. *Lancet* 270:416–63.

Hirata, A. A.; and Terasaki, P. I. 1970. Cross-reactions between streptococcal M proteins and human transplantation antigens. *Science* 168:1095–6.

Iseki, S. 1962. Blood group specific decomposing enzymes from bacteria. *Bibliotheca Haematologica* 13:215–18.

Landsteiner, K.; and Witt, D. H. 1926. Observations on the human blood groups. *Journal of Immunology* 11:221–47.

Lippold, L. K. 1971. The mixed cell agglutination method for typing mummified human tissue. *American Journal of Physical Anthropology* 34:377–83.

Matson, G. A.; and Roberts, H. J. 1949. Distribution of the blood groups, M-N and Rh types among Eskimos of the Kuskokwin Basin in western Alaska. *American Journal of Physical Anthropology* 7:109–22.

Mittal, K. K.; Terasaki, P. I.; Springer, G. F.; Desai, P. R.; McIntire, F. C.; and Hirata, A. A. 1973. Inhibition of anti-HL-A alloantisera by glycoproteins, polysaccharides and lipopolysaccharides from diverse sources. *Transplant Proceedings* 5:499–506.

Otten, C. M.; and Florey, L. L. 1963. Blood typing of Chilean mummy tissue: a new approach. *American Journal of Physical Anthropology* 21:283–5.

Salazar Mallen, M. 1951. Estudio immunologico de restos oseos antiguos. *Gaceta Médica de México* 81:122–7.

Stastny, P. 1974. HL-A antigens in mummified pre-Columbian tissues. *Science* 183:864–6.

Szulman, A. E. 1960. The historical distribution of blood group substances A and B in man. *Journal of Experimental Medicine* 111:785–800.
1962. The historical distribution of blood group substances in man as disclosed by immunofluorescence. *Journal of Experimental Medicine* 115:977–96.

Theime, F. P.; and Otten, C. M. 1957. The unreliability of blood typing aged bone. *American Journal of Physical Anthropology* 15:387–97.

Index

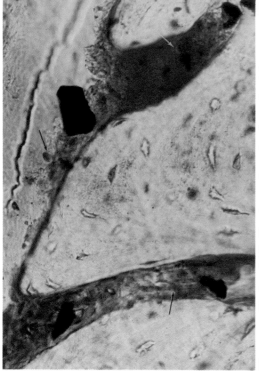

Plate 11. Undecalcified horizontal section from the right temporal bone of a 1,600-year-old frozen Eskimo (Saint Lawrence Lady). The haversian systems contain red blood cells (black arrows) and white cells (white arrow). Villanueva stain.

Plate 12. Undecalcified horizontal section from the left temporal bone of PUM II. There is excellent preservation of the vascular channels. Villanueva stain.

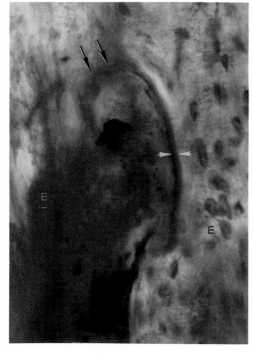

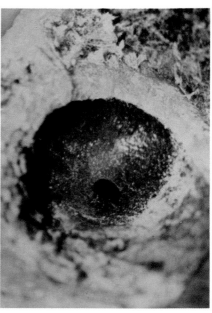

Plate 9. Undecalcified horizontal section from the left temporal bone of PUM II, showing an active osteoid seam (white arrows). Resorption foci are identified by Howship's lacunae (black arrows). There is an area of incompletely mineralized bone (im). The cells below the osteoid seam are osteocytes with nuclei; the clear area around them is mineralized bone matrix (m). Villanueva stain.

Plate 10. Right eardrum of a 2,000-year-old Egyptian mummy, PUM II. There is a thin coating of resin covering the entire drum surface. An oval perforation of about 10 percent of the pars tensa is seen in the posteroinferior quadrant. Its smooth and well-circumscribed margins suggest it resulted from middle ear infection.

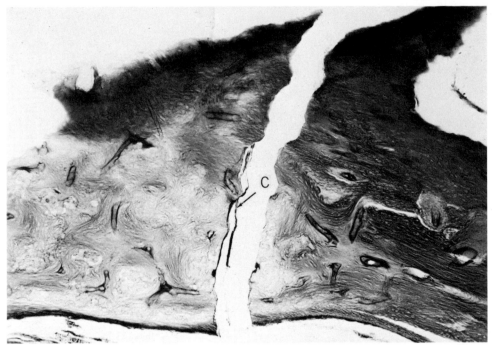

Plate 13. *Histological detail of fracture line shown in Figure 18.7. There are small strands of connective tissue (C) in between the margins, indicating that this is not an artifact. Hematoxylin and eosin.*

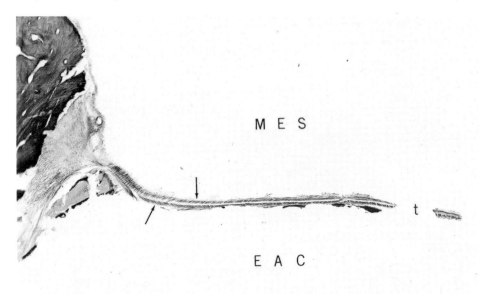

Plate 14. *Photomicrograph of horizontal section through remnants of the right eardrum of a 4,000-year-old Egyptian mummy, ROM II (temporal bone processed without rehydration). Note histological detail of fibrous layer (arrows) including postmortem tear (t). MES, middle ear space; EAC, external auditory canal. Hematoxylin and eosin.*

17

Radiographic examination of mummies with autopsy correlation

KARL T. KRISTEN
Department of Radiology
Mount Carmel Mercy Hospital
Detroit, Michigan, U.S.A.

THEODORE A. REYMAN
Director of Laboratories
Mount Carmel Mercy Hospital
Detroit, Michigan, U.S.A.

The use of radiography in the study of mummies is well established (Brothwell et al. 1970; Harris and Weeks 1973). These techniques enable us to learn a great deal about the mummy without using invasive methods. This chapter reports the results of the radiographic examination of six Egyptian mummies and a few separate parts (Table 17.1) and the correlation of the radiographic findings with the pathologic conditions noted at the time of autopsy.

Absence of motion artifact and the freedom to increase the exposure dose allow a more optimal use of high-resolution and contrast techniques than can be achieved with live patients. Long tube–film distance decreases geometric magnification. This in conjunction with long roll films facilitates recording and viewing of the whole skeleton, or at least a larger portion thereof than could be seen using the usual format dimensions (Figure 17.1). Also, with the mummy secured to a movable cradle, we have utilized fluoroscopy to determine whether previously identified masses within the body cavities are fixed or movable.

Artifacts within the wrappings or within the body cavities can hinder evaluation of anatomic structures. Body section radiography (laminography) has been particularly useful here. Ultrathin section laminography (microtomography) has been quite useful in the evaluation of the temporal bones, paranasal sinuses, and facial structures.

The line-edge enhancement phenomenon of xerography may be particularly useful in the study of soft tissues and osseous trabecular patterns.

SKULL

The assessment of the skull and upper spinal segments follows standard interpretation. Cranial vascular imprint patterns may be seen. The inner and outer tables of the calvarium, the sella, the paranasal sinuses, and the mastoid air cells usually can be identified without difficulty. The examination of the maxilla, the mandible, and the dental apparatus may present positional problems because the head and neck cannot be moved in the ordinary manner (see Chapter 3). Laminographic study of the temporal bone for the auditory structures is considered in Chapter 18.

We have seen three cases in which the radiographic changes were interpreted as skull fractures. The first case was a separate adult male head from the Detroit Institute of Arts (DIA IV) in which there were bilateral comminuted temporal fractures involving the left zygoma and maxilla (Figure 17.2). The characteristics were those of blunt force injury. The fractures were not typical of those that might result from postmortem damage caused by mishandling. Additionally, soft tissue was pressed into the defect between the bone fragments and there was amorphous soft tissue clinging to the inner aspect of the fracture site. Normally, this portion of the cranial cavity is devoid of any soft tissue. We attempted to identify histologically blood that

Table 17.1. *Data on Egyptian mummies and separate parts*

Designation[a]	Historical period[b]	Sex[c]	Age (yr)[d]	Specimen[e]
DIA I	1220 B.C.	Male	30–35	Mummy
DIA II	1220 B.C.	Male	30–35	Visceral packages from DIA I
DIA III		Male(?)	Adult	Vertebrae only
DIA IV		Male (?)	Adult	Head only
PUM I	892 B.C.	Male	30–35	Mummy
PUM II	170 B.C.	Male	30	
	1200 B.C.	Male	14–18	Mummy
PUM III	835 B.C.	Female	20–25	Mummy
PUM IV	?100 B.C.	Male	8–10	Mummy
	A.D. 100			

[a] DIA, Detroit Institute of Arts; PUM, Pennsylvania University Museum; ROM, Royal Ontario Museum.
[b] Determined by radiocarbon dating or available historical data; DIA III and DIA IV could not be dated.
[c] As determned grossly at autopsy.
[d] Estimated on basis of combined anthropological, radiographic, and historical data.
[e] Radiographic examination of DIA I, DIA II, DIA III, and PUM II by John Wolfe, M.D., Hutzel Hospital, Detroit; PUM I by Wallace Miller, M.D., Hospital of University of Pennsylvania, Philadelphia; ROM I by David Rideout, M.D., Princess Margaret Hospital, Toronto; PUM II, PUM III, PUM IV, and DIA IV by Karl Kristen, M.D., and Kenneth McGinnis, M.D., Mount Carmel Mercy Hospital, Detroit; PUM IV by Joseph Reed, M.D., Childrens Hospital of Michigan, Detroit.

would have been present as residual stainable iron had this been an antemortem fracture. Unfortunately, the material was so degenerated that this was not possible, and the diagnosis of antemortem fracture has remained tentative.

The second case was PUM IV, a young boy who had been totally eviscerated, though the brain was still present. Extensive fractures involved the inferior and posterior parts of the occipital bones (Figure 17.3). There was also a reducible separation of the sagittal suture posteriorly. There was a small amount of amorphous soft tissue adherent to the inner aspect of the occipital bones, particularly on the left side, but this showed an almost perfect gravitational pattern and probably resulted from postmortem seepage of fluids and degenerated tissue into the posterior fossa. The mummy had been supported by a body-length wooden strut placed within the wrappings. The posterior aspect of the occipital bones rested on this board. The board had been broken in its midportion about the level

of the pelvic brim, and this suggested that the mummy had been dropped or otherwise mishandled and that the skull fracture may well have resulted from this postmortem trauma.

The third case, PUM II, represents an example of apparent fracture. The films revealed several straight and some sharply angular lines thought to represent linear fractures in the parietal bones (Figure 17.4). At autopsy, no skull fractures were found, but there were several deep scratches in the skin of the scalp that had been made postmortem, giving the radiographic appearance of linear fractures in the skull.

Except for the modest amount of bone resorption thought to be the result of periodontal inflammatory disease, no maxillary or mandibular osseous abnormalities have been found. PUM IV had several unerupted teeth, including the central lower incisors. In PUM III there was congenital absence of all third molars.

Determination of age by radiographic techniques in mummies is similar to its determi-

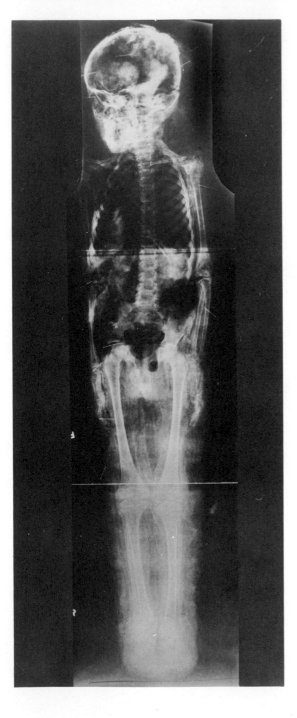

Figure 17.1.   Large-format radiograph of the mummy PUM IV, produced using a 3-m tube–film distance, a long film changer, and roll film.

nation in modern populations (Pyle and Hoerr 1955; Greulich and Pyle 1959; Hoerr et al. 1962), though Harris states that dental development may be as much as 2 years behind in the ancient Egyptian population (Harris, personal communication 1976). For example, based on dental pattern, general bone size, and lack of any evidence of epiphyseal closure (Figure 17.5), the age of PUM IV was originally thought to be 6 to 8 years. Harris's data suggest that he may have been about 2 years older.

SPINE, RIBS, AND PELVIS

The bones of the spinal column usually have been readily visualized. Varying degrees of osteophyte production have been noted, from none in the young boy of 8 years to minimal amounts in the male mummies in the age range of 30 to 35 years. These changes have corresponded to those described in many ancient human remains (Bourke 1967; Chapman 1972). They are equivalent in size and amount to similar changes seen today and probably occur for the same reasons. One mummy, PUM II, had a sixth lumbar, or transitional, vertebral body that showed minimal wedging and slight anterior spondylolisthesis. These changes were accompanied by exaggerated lumbar lordosis and degenerative arthritic changes, indicating an unstable configuration that may have prevented him from doing strenuous labor.

In the 8-year-old boy (PUM IV) there was apparent calcification of the intervertebral discs that appears to represent a postmortem artifact, as described by Gray (1967) (Figure 17.6). Whether the increased density of the intervertebral disc in these situations is attributable to calcium is not known. The possibility of absorption of calcium from natron during mummification has been mentioned (Gray 1967), but seems unlikely. The diffusion of calcium ions after death from the bone of the vertebral bodies into the intervertebral disc has more appeal. We have analyzed two intervertebral discs, one from a mummy

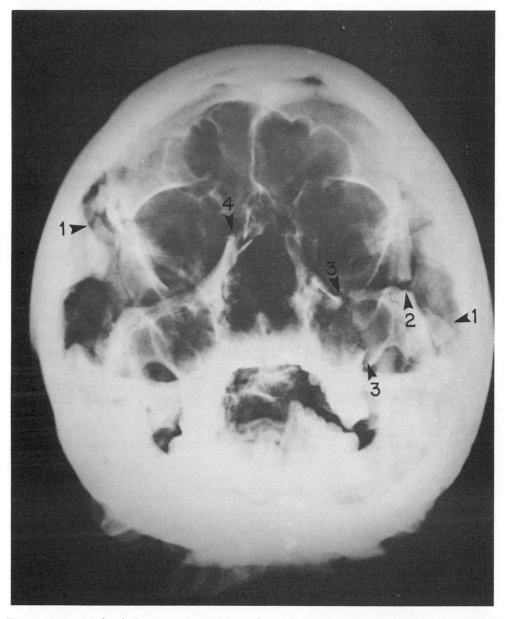

Figure 17.2.   Multiple fractures (arrows) involving the temporal bones (1), the left zygoma (2) and maxilla (3), as well as the right nasal bone (4).

without the change and one from a mummy with the change. The total calcium content for these specimens was 0.3 percent and 0.6 percent by weight. The difference is small and does not appear to support the theory that significant amounts of calcium are present in those intervertebral discs with increased opacity, a density that approaches that of the

Figure 17.3.   Bilateral occipital fractures (1) as well as suture separations (2) visible in the frontal (a) and lateral (b) projections of the skull of PUM IV. Note the shrunken cerebral hemispheres, which are seen as ill-defined densities (*).

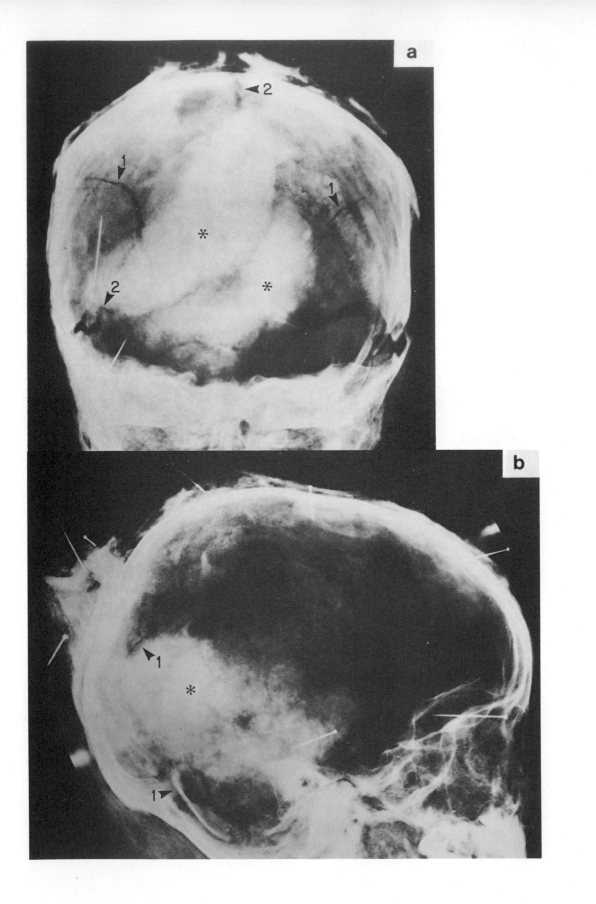

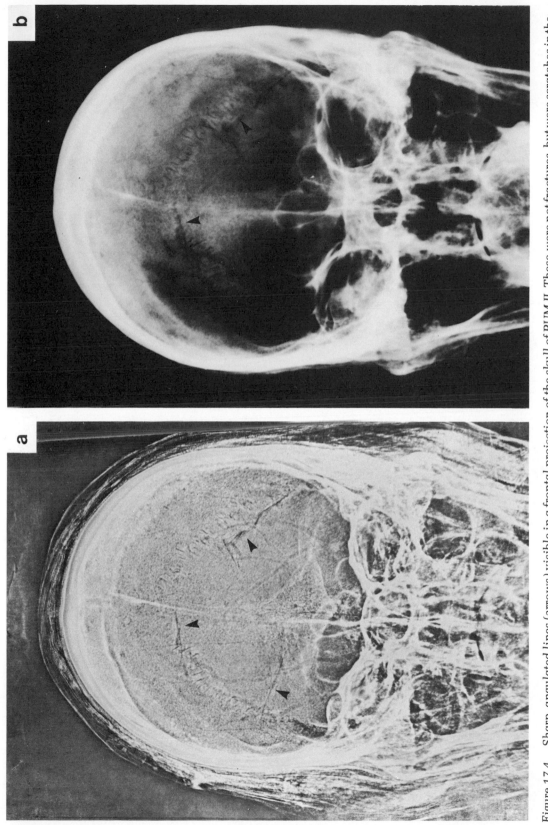

Figure 17.4. Sharp, angulated lines (arrows) visible in a frontal projection of the skull of PUM II. These were not fractures, but were scratches in the skin of the scalp. Note the detail in the wrappings recorded on the xerogram (a), but obscured on the X-ray film (b).

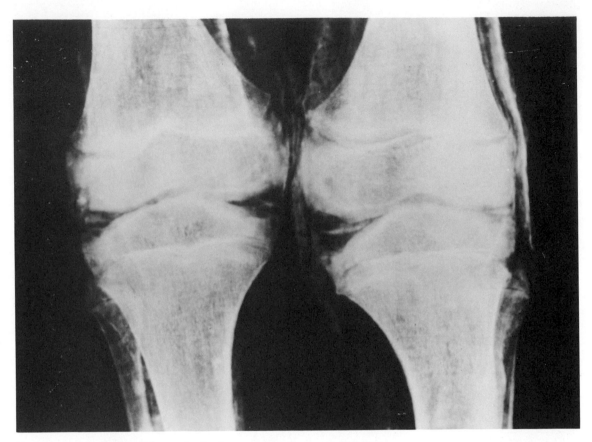

Figure 17.5.   Open epiphyses in the bones of the knees in PUM IV.

bone. However, two cases are insufficient for valid conclusions to be drawn, and more analyses will be necessary.

Several mummies have shown freshly fractured ribs and costochondral and costovertebral separations. These have been assumed to be postmortem in nature. One mummy, PUM III, had a healed fracture of the second left rib laterally. There was abundant callus, and microscopic examination showed the tissue had all the characteristics of a benign reparative process. No extranumerary or cervical ribs and no other intrinsic osseous abnormalities have been noted in our studies.

The pelves in each of the five mummies so examined have been normal. Attempts to determine the sex of the mummy utilizing pelvimetry have not been rewarding. Including

measurements of the femoral neck, condyle angles, and other parameters (Schine et al. 1951) may be more helpful. However, pelvimetry coupled with analysis of soft tissue detail permits the most accurate conclusions. In DIA I, no examination was done, but it was unnecessary because we knew from historical data that the body was male. PUM I was so poorly preserved that virtually every joint in the skeleton was separated and pelvimetry and analysis of bone size were considered unreliable. At autopsy, a shrunken penis established the sex of the mummy. PUM II was thought to be a female, based on equivocal identification of a gynecoid pelvis. The soft tissue detail revealed an elongated opaque area between the shrunken thighs that was originally thought to be extraneous material

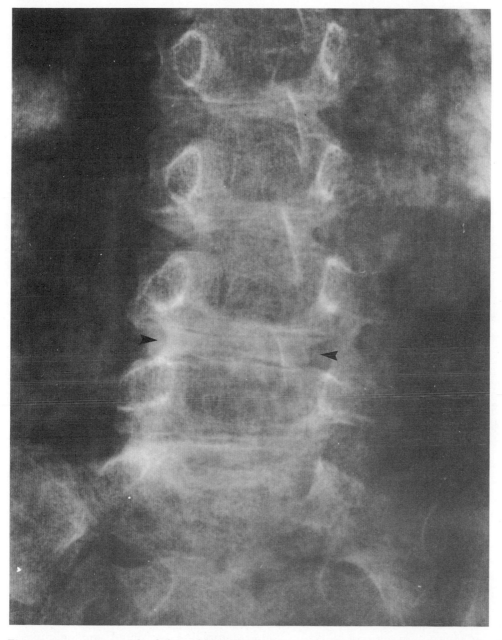

Figure 17.6.   Intervertebral disc of PUM IV, showing a density approaching that of the adjacent vertebral bodies and suggesting calcification. Chemical analysis of the disc did not reveal an inordinately high
calcium content.

between the wrappings because its radiodensity was much greater than that of the adjacent soft tissue. At autopsy, we found that this
density was in fact a 10-cm penis, apparently

uncircumcised, and supported partially erect
by a small wooden slat. The confusion arose
because of the wood and its added density.
ROM I was known from historical data to be a

teenage boy, and radiographic interpretation concurred (see Chapter 5). PUM III was a young female. Pelvimetry and other bone measurements were indeterminate, but pelvic soft tissue detail showed the presence of a uterus and urinary bladder and the sex was established on this basis. Autopsy confirmed these interpretations. PUM IV, a young boy, was so diagnosed not on the basis of bone configuration but by the fact that the penis and scrotum could be seen on the films. In this instance, the density of these structures was similar to that of the other surrounding soft tissue.

EXTREMITIES

The evaluation of the upper extremities before unwrapping has posed some problems because of the fixed positioning of the arms and hands. We have not found any abnormalities in the upper extremity bones. However, any suspicious findings should be reevaluated after autopsy when the arms can be moved or a film cassette can be interposed between the extremity and the body. Often the clavicles and the scapulae have been displaced inward and upward owing to the binding of the arms close to the body by the wrappings. Occasionally, one or several joints have been disarticulated owing to postmortem degeneration.

The lower extremities posed less of a technical problem, though superimposition of the bones on lateral views can be troublesome. PUM II had a distorted foot with the toes severely flexed and turned medially, but this appeared to be an embalming artifact attributable to tight wrapping in this area. The bones and joints of the foot were entirely normal radiographically. The only abnormality we have seen was in the right fibula and tibia of the same mummy. At autopsy, the right leg was noticeably larger than the left and appeared swollen. Radiographically, the right fibula was thickened in the medial aspect of the middle and lower thirds with blurring of the normally sharp cortical contour

(Figure 17.7). There was no apparent medullary component to this lesion. Opposite this area, a similar but very localized zone was noted on the lateral aspect of the tibia at the junction of the middle and lower thirds. The osseous reaction appeared to be periosteal in location in both bones. Microscopic examination showed the lesion to be a nonspecific osteoblastic periostitis. The differential diagnosis includes vascular stasis changes and localized inflammation, but the cause remains uncertain.

Growth arrest lines, or Harris's lines, may be seen as transverse bars radiographically in the metaphysis near the epiphysis in the long bones and have been particularly well seen in the distal femur or tibia (Wells 1967). The lines result from the temporary arrest of growth of the epiphyseal cartilage in children during episodes of acute stress such as malnutrition and acute illnesses. These lines have been present in three of the five mummies examined for this change (Figure 17.7). In some instances, the growth arrest lines have been quite faint, and in the cases where they were not noted, such lines may have been obscured by the cloth covering the mummy. Although the lines are nonspecific indicators, their presence may correlate with subsequent findings from the autopsy. The long bones should be reexamined after unwrapping so that better radiographic study can be achieved.

SOFT TISSUE

Residual soft tissue in the mummies has taken two forms. The first and most common has been the tissue such as skin, subcutaneous fat, and muscle that remained on the body following mummification. These structures have had varying degrees of opacity and could be identified readily. The formation of calcium soaps in subcutaneous tissue owing to postmortem degeneration of the tissue may give irregular density to that tissue, but it has never been great enough to interfere with visualization of underlying bone or soft tis-

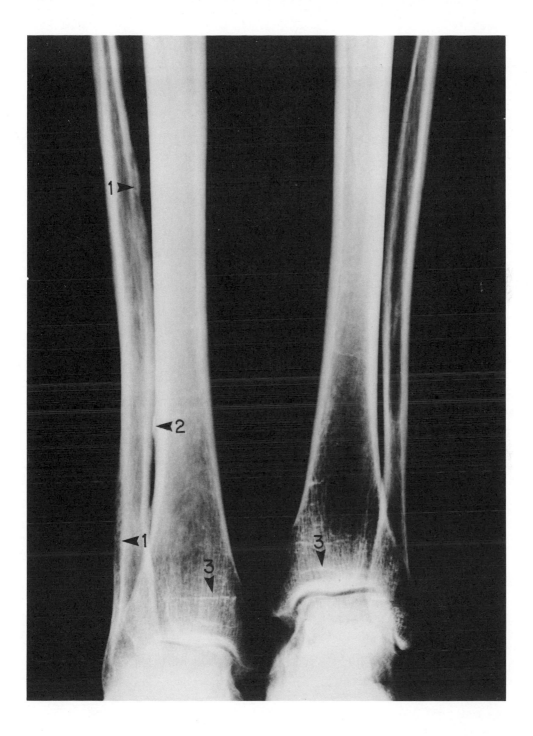

Figure 17.7.   Right fibula of PUM II, showing a wavy irregular contour with loss of sharp cortical margin (1).
A small similar area is seen on the lateral tibia (2). Note the horizontal growth arrest lines present in both
distal tibiae (3).

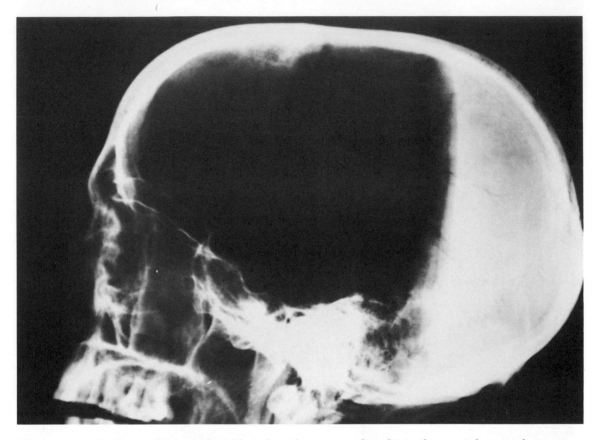

Figure 17.8.   Meniscus effect, attributable to liquid resin introduced into the cranial cavity during embalming, which hardened with the body in the supine position.

sue. Skeletal muscle has been markedly reduced in volume by the loss of water. In the mummies with considerable degenerative changes, air contrast planes have been seen between large muscles; occasionally, severe fissuring was seen as irregular, transversely oriented air contrast defects. If significant dehydration occurred after wrapping, large air spaces may be seen between the wrapping and the shrunken tissues.

Because not all mummies have been eviscerated, some or all of the viscera may be in their normal anatomic locations. Though the size, shape, and density may not conform to living examples, the overall appearance and location have often allowed identification.

Cranial soft tissue has usually been lacking in these mummies, with the exceptions of the skin covering the head, and in some instanc-

es, the presence of the cerebral hemispheres within the cranial cavity. When present, these structures were detected radiographically. The cerebral hemispheres have been present in three (PUM I, PUM IV, ROM I) of the seven skulls we have seen. Oddly enough, in all three the size of the cerebral hemispheres was similar, each measuring about 10 by 5 cm. Because the shrunken brain is loose within the skull, it settles by gravity to the posterior fossa during examination in the supine position. Radiographically, the hemispheres were seen as ill-defined, slightly opaque masses lying over the inner aspect of the occipital bones (Figure 17.3). Another mummy (PUM II) showed a much more dense posterior fossa opacity that had a fluid level effect (Figure 17.8). In our naiveté, we thought that this represented brain. At au-

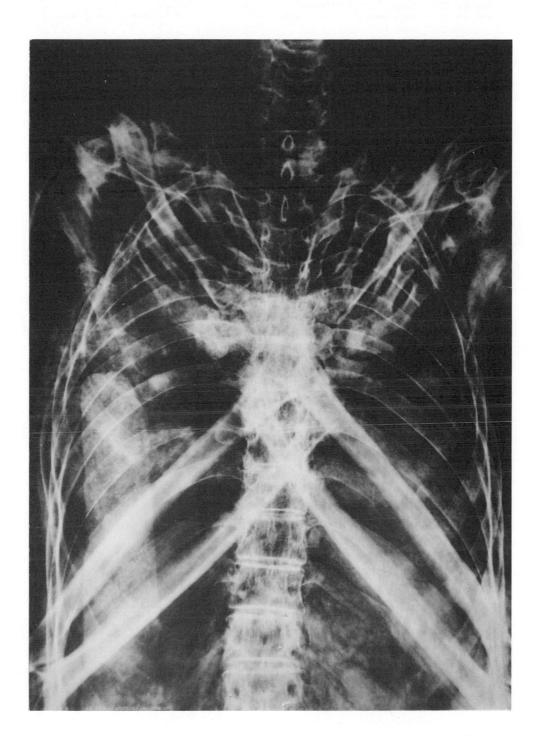

Figure 17.9.   Elongated, opaque cylindrical object in the right hemithorax and right abdominal cavity, which proved to be a visceral package containing the dehydrated organs. Three other similar packages were found in the pelvis.

topsy, we discovered the material to be hardened resin that had been introduced in a liquid state into the cranial cavity through the defect in the left ethmoid sinus and cribriform plate (see Chapter 4). Laminographic examination of the ethmoid sinuses showed the defect in the bony trabeculae. Because the embalming had apparently been performed with the body in the supine position, when the resin solidified it assumed this characteristic pattern. The best example of identification of intact organs was seen in ROM I. No evisceration had been performed at the time of mummification, and all the organs were in situ. Dr. David Rideout was able to identify most of the major organs radiographically. At autopsy, his interpretation was confirmed; his findings are considered in detail in Chapter 5.

A somewhat similar situation was noted following the radiographic examination of PUM III. The pleural cavities were filled with air and the lungs appeared as small, slightly dense, symmetrical paravertebral zones. No intrinsic calcification could be identified. The heart was seen as an ill-defined retrosternal density. The aorta was not seen, and no vascular calcification was noted. There was clear delineation of the thoracic and abdominal cavities by the diaphragm.

Examination of the abdomen revealed a crescent-shaped structure in the right upper quadrant, thought to be shrunken liver. In the pelvis, two central rounded densities could be seen and were identified as urinary bladder and uterus. Within the abdominal cavity there were large irregular densities that contained numerous opacities of varying sizes that approached the density of bone. A sharply outlined but otherwise similar appearing structure protruded past the diaphragm into the thoracic cavity. Although we were tempted to call this intestine, we could not explain all the granular densities. At autopsy, the heart and lungs were present as were remnants of the diaphragm. The right upper quadrant mass was liver, and the pelvic structures were the urinary bladder and uterus. However, the irregular densities in the ab-

dominal cavity were linen wads stuffed into the body during the mummification (Chapter 6). The structure protruding into the chest was of similar material.

Another form of soft tissue found with mummies are the organs that have been removed during mummification and then returned to the body as visceral packages. Two packages were found in DIA I, but examined radiographically later (DIA II). This examination showed that one was very likely soft tissue, and histological examination revealed that it was composed of degenerated but recognizable muscle tissue, perhaps heart. The second package had a peculiar wavy internal configuration with numerous granular densities, similar in type to that seen in the abdomen of PUM III. When sectioned, the packages were found to consist of wads of resin-soaked linen, presumably used to seal the embalmer's incision.

In PUM II, radiographic examination of the chest and abdomen revealed the presence of four elongated, cylindrical masses, one in the right hemithorax and three crisscrossed in the pelvis (Figure 17.9). We thought these were visceral packages, but could not determine radiographically if indeed there was tissue in them. Subsequently, histological examination showed that they contained lungs, spleen, and intestine. Much of the density of the packages was attributed to the several layers of linen impregnated with resin that had been used to wrap the dehydrated organs.

CONCLUSION

As one tool in the investigation of mummies, radiographic examination has proved its worth. In several instances, abnormalities probably would have gone unnoticed without prior radiographic examination. Significant data, both medical and anthropological, have been derived from its use.

REFERENCES

Bourke, J. B. 1967. A review of the paleopathology of the arthritic diseases. In *Diseases in an-*

tiquity, D. Brothwell and A. T. Sandison (eds.), pp. 352–70. Springfield, Ill.: Thomas.

Brothwell, D.; Molleson, T.; Gray, P. H. K.; and Harcourt, R. 1970. The application of x-rays to the study of archeological materials. In *Science in archeology*. D. Brothwell and E. Higgs (eds.), pp. 513–25. New York: Praeger.

Chapman, F. H. 1972. Vertebral osteophytosis in prehistoric populations of central and southern Mexico. *American Journal of Physical Anthropology* 36:31–8.

Gray, P. H. K. 1967. Calcinosis intervertebralis, with specimen reference to similar changes found in mummies of ancient Egypt. In *Diseases in antiquity*, D. Brothwell and A. T. Sandison (eds.), pp. 20–30. Springfield, Ill.: Thomas.

Greulich, W. W., and Pyle, J. I. 1959. *Radiographic atlas of skeletal developmental of the hand and wrist*, 2nd ed. Stanford: Stanford University Press.

Harris, J. E., and Weeks, K. R. 1973. *X-raying the pharaohs*. New York: Scribner's.

Hoerr, N. L.; Pyle, J. I.; and Francis, C. C. 1962. *Radiographic atlas of skeletal development of foot and ankle*. Springfield, Ill.: Thomas.

Pyle, J. I., and Hoerr, N. L. 1955. *Radiographic atlas of the skeletal development of the knee*. Springfield, Ill.: Thomas.

Schine, H. R.; Baensch, W. E.; Friedl, E.; and Uehlinger, E. 1951. *Roentgen diagnostics*. New York: Grune & Stratton.

Wells, C. 1967. A new approach to paleopathology: Harris's lines. In *Diseases in antiquity*, D. Brothwell and A. T. Sandison (eds.), pp. 390–404. Springfield, Ill.: Thomas.

18

Temporal bone studies

JAIME T. BENITEZ
Director, Division of Otoneurology and
the Temporal Bone Research Laboratory
William Beaumont Hospital
Royal Oak, Michigan, U.S.A.

GEORGE E. LYNN
Professor of Audiology
Wayne State University School of Medicine
Detroit, Michigan, U.S.A.

Temporal bones from ancient populations, particularly Egyptian mummies, which were often meticulously embalmed, are usually in a good state of preservation and therefore offer an excellent opportunity for the histopathological investigation of inflammatory, neoplastic, and traumatic ear conditions as well as metabolic disorders involving the temporal bone. Furthermore, with the application of modern techniques, research can be carried out on the haversian systems, associated vascular channels, and osteon activity.

X-RAY STUDIES

X-ray study is the first procedure carried out in the investigation of ancient temporal bones prior to removal of the bones from the skull. Radiographic studies have been enhanced in the past decade with the development of a special technique, polytomography, which allows the survey of tissues in a millimeter-by-millimeter fashion. X-ray polytomography of the temporal bones makes it possible to detect middle ear malformations, otic bony capsule abnormalities, or changes in the internal auditory canal. However, it should be emphasized that certain radiographic findings may be misleading, and final interpretation depends on macroscopic examination and histological studies. For example, with the temporal bones of PUM II, there was a definite increase in density throughout the polytomograms of the left temporal bone as compared with the right, which suggested the possibility of extensive bone disease involving the left specimen. However, macro-scopic examination revealed that the increased density was caused by the presence of large amounts of resin in the air cells of the mastoid and cavities of the middle and inner ears and that there were no anatomical abnormalities in the middle ear and mastoid of these specimens. Furthermore, histological study of undecalcified sections showed no evidence of metabolic bone disease, which could also have caused the unusual X-ray findings. Clearly, the increased radiological density, which might have been attributed to premortem temporal bone disease, was, in fact, the result of a large amount of resin infiltration into the left ear of PUM II during the embalming procedure.

MACROSCOPIC STUDIES WITH ZEISS OPERATING MICROSCOPE

After removal of the temporal bones by means of the block or Schuknecht method (1968), specimens such as that shown in Figure 18.1 are studied with the Zeiss operating microscope, which allows magnification up to 40 times. Observations are recorded in a systematic fashion regarding the preservation or absence of soft tissues in the external ear canal, the surface of the petrous pyramid, and the internal auditory meatus. A detailed description is made of the eardrum or its remnants. The posterior portion of the tegmen tympani is removed with fine rongeur forceps so that the contents of the middle ear cavity may be inspected. The arcuate eminence of the superior surface of the petrous bone is removed with an electric saw to reveal the superior and posterior semicircular canals.

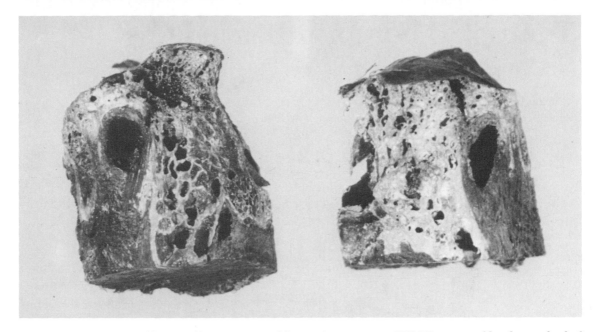

Figure 18.1.   Temporal bones of a 2,000-year-old Egyptian mummy, PUM II, removed by the method of Schuknecht. Specimens include the external auditory canal, middle ear, mastoid process, and the petrous pyramid, which houses the inner ear.

(This portion of the petrous bone is then processed by the undecalcified method.) Color photographs are taken of significant findings of the external canal, eardrum, mastoid process, middle ear cavity, and petrous pyramid. We use a Zeiss Ikon camera (ICAREX 35S) attached to a side viewer that operates with a synchronized flash.

Macroscopic studies by the method described above have been carried out on 12 temporal bones from 6 Egyptian mummies, 19 temporal bones from skeletal remains from 19 Egyptian mummies, a pair of temporal bones from a Nubian skull, 55 temporal bones from the skulls of 41 mummies from Peru, and one set of temporal bones from the skull of an Eskimo from Saint Lawrence Island frozen since her death around A.D. 400. Studies of these specimens have revealed evidence of antemortem disease as well as postmortem abnormalities.

Exostosis of the external auditory canal has been a common finding in material studied by Gregg and Bass (1970) and Hrdlička

(1934), but we have not found a single case in any of our specimens.

Figure 18.2 shows an example of a normal mastoid process from a Peruvian mummy from the Ica culture, A.D. 1200–1400. Pneumatization and preservation of the air cell system is excellent, and there is no evidence that mastoid disease ever occurred. An excellent example of an abnormal mastoid process from ROM II is shown in Figure 18.3. The normally pneumatized bone has been replaced by very dense sclerotic tissue as a sequela of chronic mastoid infection occurring sometime during the individual's lifetime. The opposite mastoid revealed similar findings.

Eardrums are rarely observed in ancient temporal bones because of rapid disintegration after death; however, preservation of eardrums or remnants was found in three Egyptian mummies (PUM II, PUM III, ROM II) and in the Eskimo from Saint Lawrence Island. The right eardrum of PUM II (Lynn and Benitez 1974) showed a small oval perfora-

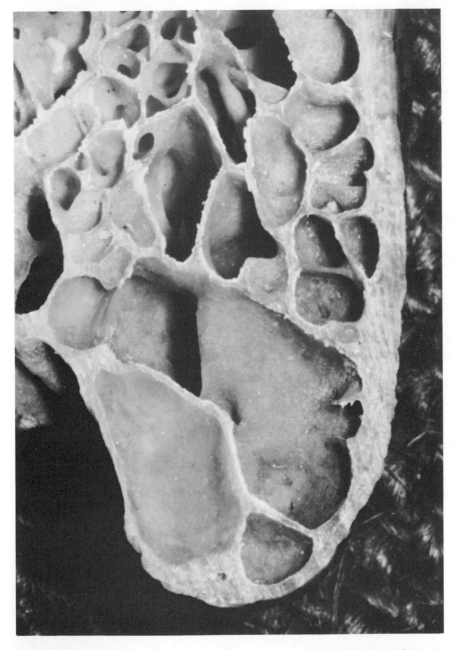

Figure 18.2.   *Left mastoid process of a Peruvian mummy from the Ica culture, A.D. 1200–1400. Pneumatization and preservation of the air cells is excellent, with no evidence of disease.*

tion in the posteroinferior quadrant occupying approximately 10 percent of the pars tensa (Plate 10). The left ear of PUM III revealed a large oval perforation occupying about 40 percent of the posterior portion of the membrane.

These perforations have smooth and well-circumscribed margins suggesting they were caused by middle ear infection occurring during life. The irregular margins of the remnants of the right eardrum of ROM II appear to

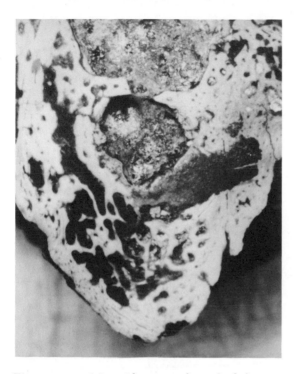

Figure 18.3. Mastoid process from the left temporal bone of a 4,000-year-old Egyptian mummy, ROM II, showing severe sclerosis as a result of chronic mastoid infection.

be postmortem artifacts. However, the well-circumscribed margins of the remnants of the right eardrum of PUM III appear to be the sequela of an inflammatory process.

Intact ossicles found in their normal anatomical position with preservation of the ligaments were seen in the Egyptian mummies PUM II and PUM III. Most of the temporal bone specimens we have studied so far have shown displaced but intact ossicles in the middle ear cavity with no evidence of disease process. Today, fixation of the stapes in the oval window by otosclerosis is a common condition affecting the mobility of the stapes; however, we have not seen this disorder in the 90 temporal bones studied so far.

The internal auditory meatus was occluded with resin in the temporal bones of PUM II. In the left temporal bone of the Eskimo from Saint Lawrence Island, there were remnants of the eighth nerve (Figure 18.4).

In the left external auditory canal of PUM II, a beetle larva measuring about 1.5 by 0.2 mm was found embedded in resin at the medial end close to the rim of the eardrum. It was prepared for examination by scanning electron microscopy (Figure 18.5) and later identified as Staphylinidae, genus probably *Atheta*, by D. M. Anderson, Systematic Entomology Laboratory, U.S. Department of Agriculture.

HISTOLOGICAL STUDIES WITH UNDECALCIFIED SECTIONS

A great advantage of studying undecalcified bone sections is that it provides a superior method for evaluation of bone remodeling, which is a tissue function and can be expected to change with aging and disease. Therefore, in our paleopathological investigations, we routinely remove the upper portion of the petrous bone, which includes the arcuate eminence with the arc of the superior semicircular canal, for such studies. Other portions of the temporal bone can also be removed for these studies without damage to the rest of the specimen that is to be processed by routine histological techniques after decalcification. For histological control on undecalcified sections, a similar specimen is removed from a fresh temporal bone. The specimens are fixed in 80 percent alcohol and processed by the method for undecalcified bone (Frost 1958; Villanueva and Frost 1961; Villanueva et al. 1964).

Light microscope study of sections revealed excellent preservation of the haversian systems in the specimens from the temporal bones of PUM II (Benitez and Lynn 1975). The vascular channels were clearly seen, and, amazingly, they were similar in appearance to those of the specimen from fresh temporal bone that was used as a control. Sections from the temporal bones of PUM II appeared to have a yellowish color that could be detected even after staining (Plate 12). This was attributed to infiltration of resin.

A striking finding in the sections from

Figure 18.4.   Internal auditory canal of the left temporal bone from a 1,600-year-old frozen Eskimo (Saint Lawrence Lady), with remnants of the eighth nerve.

PUM II was the presence of osteocytes in the lacunae with good preservation of the nucleus, which is remarkable for a tissue over 2,000 years old. When specific stains are used, for example, the polychrome bone stains of Villanueva (1964, 1974), osteoid seams can be unequivocally demonstrated and the activity of the osteon may be determined. In the sections from PUM II, we were able to identify osteoid seams. Plate 9 shows an active osteoid seam from the left temporal bone. In the osteon, there were areas of incomplete mineralized bone and areas of resorption (Howship's lacunae); the thickness of the seam is about 5 μ. This is a low-turnover type of bone, but without evidence of metabolic bone disease.

Undecalcified sections from the specimens of the 1,600-year-old frozen Eskimo showed a highly mineralized type of bone, but no evidence of metabolic bone disease. There was good preservation of the vascular channels, and blood cells were identified (Plate 11).

HISTOLOGICAL STUDIES AFTER DECALCIFICATION

The histological preparation of the human temporal bone includes several procedures, which we shall describe with modifications that we believe are convenient for the processing of ancient specimens.

Fixation.   Rehydration of mummified tissues has been recommended for histological studies, in particular for soft tissues (Ruffer 1921). For the temporal bones, we prefer fixation in 10 percent neutral buffered formaldehyde as the first step, which at the same

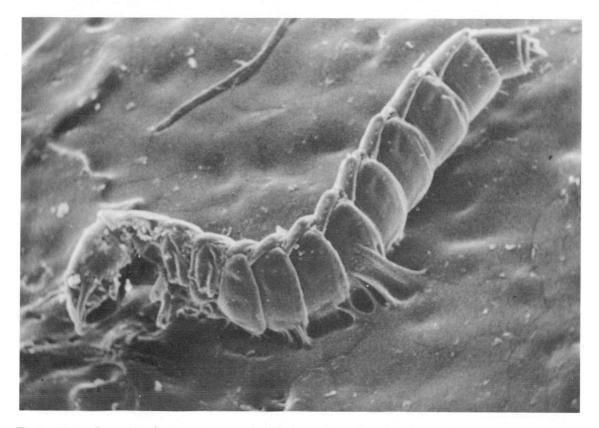

Figure 18.5.   Scanning electron micrograph of the beetle larva found on the eardrum of mummy PUM II. It was identified as Staphylinidae, genus probably Atheta. × 55.

time produces hydration of the specimen. The fixative solution should be prepared at the time of use by mixing 100 ml of 37 to 40 percent formaldehyde solution, 900 ml distilled water, 4 g sodium phosphate (monobasic), and 6.5 g sodium phosphate (anhydrous dibasic). The temporal bones are placed for 2 weeks separately in 500-ml glass jars with screw-on lids containing about 450 ml of the solution. A change of solution is performed on the second day. The jars may be shaken once or twice a day to improve penetration of the fixative. Any changes noted on the anatomical structures of the specimens should be recorded.

Decalcification.   In ancient specimens, the use of strong acid solutions for decalcification should be avoided. We use a solution of ethylenediaminotetraacetic acid (EDTA) at a

pH of 7.2 to 7.8. It is obtained as Versene 100, which is an aqueous solution of the tetrasodium salt of EDTA and can be purchased from Fisher Scientific Company. The 1.0 M Versene stock solution is prepared with 10,500 ml of Versene 100, 7,500 ml of distilled water, and 900 ml of glacial acetic acid. With some modifications, we follow the technique described by Gussen and Donahue (1965). Decalcification is carried out in a 0.66 M Versene solution prepared with 666 ml of Versene stock solution and 333 ml of distilled water. Because distilled water has a pH of approximately 6, the resulting solution will have a pH between 7.2 and 7.8. If the pH is below 7.2, a little Versene stock solution may be added to bring the pH back to between 7.2 and 7.8.

After fixation, each specimen (without

washing in water) is placed in a separate 500 ml jar containing about 450 ml of the decalcifying solution, which is changed twice a week. The period of time required for decalcification is 16 weeks. The end point of decalcification may be more precisely determined by X rays. When decalcification is complete, each temporal bone is placed in a very dilute solution (0.1 to 0.2 M) of Versene, and the solution is changed once every 3 days, three times in all. This dissolves any calcium versenate crystals that may have precipitated within the tissues.

Dehydration.   We modified the dehydration time periods as follows: 40 percent alcohol, 10 minutes; 60 percent, 10 minutes; 80 percent, 10 minutes; 95 percent, 25 minutes; absolute alcohol and ether (1:1), 20 minutes.

Embedding.   Parlodion dissolved in ether–alcohol is used for embedding the specimens. It is similar to celloidin, but dissolves in ether–alcohol more readily, and the parlodion-embedded specimens do not dry as much as the celloidin-embedded specimens, so that a better consistency can be achieved for sectioning. The specimens are placed in a solution of 2 percent parlodion for 1 week. This is followed by a 4 percent solution for 2 weeks, then a 6 percent solution for 3 weeks. Specimens should be oriented so that the opened areas of the bones (epitympanum and superior semicircular canal) are turned upward in an effort to allow air bubbles that form within the bones to rise.

Hardening.   Specimens are placed in 12 percent parlodion solution for 4 weeks, with the superior surface of the petrous bone facing the bottom of the jar. To maintain the specimen in the middle of the jar, it can be suspended with surgical silk (3–0); this helps the orientation of the bone in the desired angle of section. After the 4-week period, the jar top is unscrewed and slipped 0.3 cm back from one side of the jar where it is allowed to rest for about 7 to 8 hours per day, from the beginning to the end of the workday. When the specimen has hardened in this manner for 2 to 3 weeks, the parlodion reaches a concen-

tration that is no longer sticky but is still quite soft and pliable; when this occurs the upper layer of the parlodion around the edge and down the side of the jar is trimmed away with a scalpel blade, and 100 ml of chloroform is poured on top of the block. The top is placed on the jar securely and the specimen left in this manner for 1 week.

Mounting.   When the block has been covered in chloroform for 1 week, the peripheral part of the parlodion is trimmed away with a scalpel blade and the block containing the specimen is removed from the jar. Further trimming is necessary to obtain the approximate shape and orientation desired for cutting. For sectioning on the horizontal plane, the inferior part of the block (which was facing the top of the jar) is cut flat with a scalpel blade in a plane that is parallel to the flat upper portion of the block. Trimming should be performed so as to leave about 1 cm of parlodion around the specimen and about 1.5 cm at the bottom; thus a rectangular block is obtained. While trimming, one should bear in mind that the remainder of the superior semicircular canal is to be positioned in the vertical plane. Other anatomical landmarks used for orientation of the specimen are the internal and external auditory canals. To obtain horizontal sections in the plane of the modiolus of the cochlea, a counterclockwise rotation is given to the specimen so that the posterior surface of the petrous bone is in the vertical plane. If the malleus is present and in its anatomical position, this rotation places the long process also in a vertical plane (Figure 18.6). When the specimen has been trimmed so that this orientation is achieved, the inferior portion of the block is softened in an ether–alcohol solution for about 1 minute. A mounting block is placed in the solution alongside the specimen. The wet portions of the specimen and the mounting block are next dipped in a small amount of 12 percent parlodion, then pressed firmly together and any gaps filled with 12 percent parlodion. The mounted specimen is placed in 150 ml of chloroform for 2 days. This solution is then

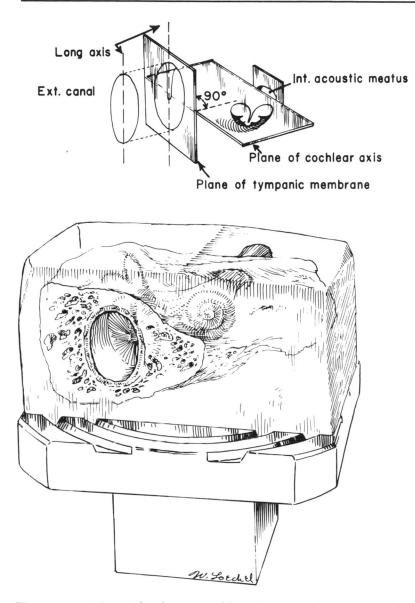

Figure 18.6. Mounted right temporal bone for sectioning in the horizontal plane. Slight counterclockwise rotation has been given to the specimen to bring the cochlea into a horizontal plane. The relationship of the external and internal ear canals is shown above.

replaced with 80 percent alcohol. The specimen may be sectioned on a sliding microtome within 1 hour's time or stored in this solution indefinitely.

Sectioning and Staining. The microtome is set to cut at 20 μ, obtaining about 400 sections from each specimen. All sections are placed in 80 percent alcohol solution on pieces of onionskin paper that are consecutively numbered. Every tenth section is placed in a separate dish for staining. Conventional staining of the temporal bone is usually performed with hematoxylin and eosin. As the cutting proceeds, color photographs are taken of the specimen with the operating microscope at midmodiolar level and at other levels show-

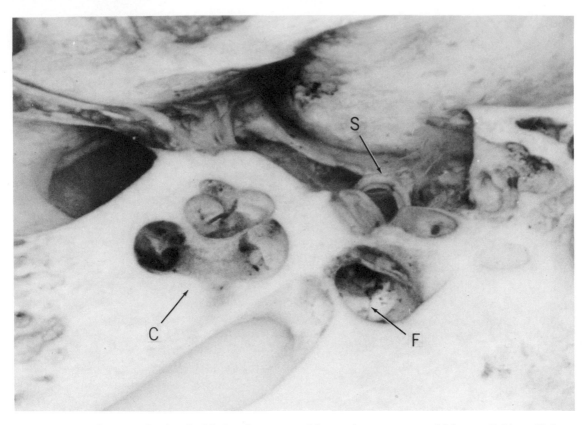

Figure 18.7. Photograph of embedded right temporal bone of a 1,600-year-old frozen Eskimo (Saint Lawrence Lady). Sectioning is at midmodiolus level. There is a fracture line in the vestibule (F) extending through the niche of the oval window. The stapes bone (S) is evulsed. The cochlea (C) and otic capsule around it are normal in appearance.

ing interesting findings. In the right temporal bone of the 1,600-year-old frozen Eskimo, a fracture line was found in the vestibule of the inner ear extending through the niche of the oval window with luxation of the stapes (Figure 18.7). There was histological evidence that this fracture is not an artifact (Plate 13), and it is consistent with fractures of the skull and maxilla on the right side found by Zimmerman and Smith (1975) at autopsy.

Mounting of sections. Sections are mounted on slides in a configuration that centers the important anatomical parts of the temporal bone consistently. The outline of a slide is made on a piece of paper and the desired location of the cochlea marked; the slides are placed over this pattern and the sections aligned with the pattern. All temporal bone sections are mounted in the slide with the left anatomical side of the section placed on the right side of the slide. The posterior anatomical side of the temporal bone section is placed in the anterior portion of the slide.

Comparison of rehydrated and unrehydrated bone. The histological processing of ancient temporal bones offers a challenge to otopathologists and temporal bone technicians. We contemplated the possibility that rehydration could disintegrate tissues essential for the interpretation of certain otopathological conditions, and so we were anxious to carry out a comparative study. The temporal bones from a 4,000-year-old Egyptian mummy skull provided us with this unique

opportunity. As a result of the kind invitation of Patrick Horne, Pathology Department, Banting Institute, Toronto, Canada, a pair of temporal bones was removed by one of us (GEL) from the skull of ROM II following the method of Schuknecht. After removal of the arcuate eminence for study in an undecalcified state and for macroscopic evaluation with photography of both specimens in our Detroit laboratory, the left temporal bone was returned to the Banting Institute for histological study by Horne et al. (1976). They rehydrated the specimen for 1 week in Sandison's modification of Ruffer's solution (1955). Fixation was carried out in 10 percent neutral buffered formaldehyde solution for 2 weeks and decalcification by the method of Gussen and Donahue (1965) with EDTA. The right temporal bone was processed in our laboratory without preliminary rehydration of the specimen. Our first step was fixation in 10 percent neutral buffered formaldehyde solution for 2 weeks, followed by decalcification by the method of Gussen and Donahue (1965) with EDTA. The dehydration time was shortened as described above.

We had an opportunity to study serial sections of both specimens. The general state of structural integrity was similar. Artifactual microfractures of bony tissues produced during decalcification were noted on both embedded specimens, and they were in fact microscopically similar. However, a better preservation of soft tissues was found in the right temporal bone processed without rehydration; a remnant of the basilar membrane was present throughout all turns of the cochlea of the inner ear, whereas in the left specimen, processed with rehydration, a remnant of the basilar membrane was found only in the basal turn. Moreover, in the rehydrated left ear, a good portion of the tympanic membrane disintegrated and histological detail at the level of the perforation was lost, whereas in the unrehydrated right ear, there was excellent preservation of the remnants of the tympanic membrane including the small artifactual postmortem tears (Plate 14).

SUMMARY AND CONCLUSIONS

Techniques have been described for the study of ear disease that may have occurred during the lifetime of ancient people and that may still be preserved in their temporal bones.

The recommended procedures for obtaining maximum information include X rays with polytomography, dissection and macroscopic observation with thorough photographic documentation, and histopathologic studies with undecalcified and decalcified material. The traditional procedure of rehydration for histological studies may be omitted to lessen the amount of tissue disintegration that may occur during preparation of the specimen for sectioning. However, fixation in 10 percent neutral formaldehyde is essential prior to decalcification. Conventional decalcification with trichloroacetic acid cannot be used because it produces disintegration of the tissue. A solution of EDTA is preferred.

With modern laboratory techniques, pre- and postmortem abnormalities involving the mastoid, eardrum, and various structures of the middle and inner ears can be identified in ancient temporal bones, and findings contribute new information about the nature of diseases in ancient cultures.

ACKNOWLEDGMENTS

This research is supported in part by General Research Support Grant RR–05384, Wayne State University School of Medicine (Gel), and a grant from William Beaumont Hospital, Royal Oak (JTB).

REFERENCES

Benitez, J. T., and Lynn, G. E. 1975. Temporal bone studies: findings with undecalcified sections in a 2,600-year-old Egyptian mummy. *Journal of Otology and Laryngology* 89:593–9.

Frost, H. M. 1958. Preparation of thin undecalcified bone sections by rapid method. *Stain Technology* 33:273–6.

Gregg, J. B., and Bass, W. M. 1970. Exostosis in the external auditory canals. *Annals of Otology, Rhinology and Laryngology* 79:834–9.

ssen, R., and Donahue, D. 1965. Decalcification of temporal bones with tetrasodium edetate. *Archives of Otolaryngology* 82:110–14.

Horne, P. D.; MacKay, A.; Jahn, A. F.; and Hawke, M. 1976. Histologic processing and examination of a 4,000-year-old human temporal bone. *Archives of Otolaryngology* 102:713–15.

Hrdlička, A. 1934. Ear exostosis. *Smithsonian Miscellaneous Collection* 93:1–100.

Lynn, G. E., and Benitez, J. T. 1974. Temporal bone preservation in a 2,600-year-old Egyptian mummy. *Science* 183:200–2.

Ruffer, M. A. 1921. *Studies in the Paleopathology of Egypt*. Chicago: University of Chicago Press.

Sandison, A. T. 1955. The histological examination of mummified material. *Stain Technology* 30:277–83.

Schuknecht, H. F. 1968. Temporal bone removal at autopsy. *Archives of Otolaryngolory* 87:33–41.

Villanueva, A. R.; Hattner, R. S.; and Frost, H. M. 1964. A tetrachrome stain for fresh mineralized bone sections useful in the diagnosis of bone disease. *Stain Technology* 39:87–94.

Villanueva, A. R., and Frost, H. M. 1961. Rapid method for obtaining hematoxylin and eosin biopsy sections of bone for table diagnosis. *American Journal of Clinical Pathology* 36:54–9.

Villanueva, A. R. 1974. A bone stain for osteoid seams in fresh unembedded mineralized bone. *Stain Technology* 49:1–8.

Zimmerman, M. R., and Smith, G. S. 1975. A probable case of accidental inhumation of 1,600 years ago. *Bulletin of the New York Academy of Medicine* 51:828–37.

19

Paleobiochemistry

ROBIN A. BARRACO
Associate Professor, Department of Physiology
Wayne State University School of Medicine
Detroit, Michigan, U.S.A.

The most abundant ancient specimens have always been skeletons; consequently, the major investigative effort for paleopathologists has focused on gross observation, morbid anatomy, histology, and to some extent, blood-typing studies. Most discoveries have consisted of lesions of tuberculosis, leprosy, and yaws. However, diseases, particularly infectious diseases, rarely leave detectable traces in bones. The study of surviving soft tissues is required for the comprehensive analysis of ancient diseases. In this regard, the biochemical analysis of ancient tissues may prove a useful tool for the investigation of ancient diseases. Because the chemical composition of an organism provides a record of the interaction of that organism with its environment, biochemical studies, along with archeological and other paleobiological approaches, constitute a coordinated effort to clarify the relationship of some diseases – in particular, infectious diseases – to geography, heredity, and diet and to learn about the subsequent adaptation of people and culture to a disease environment.

This approach should, however, be exercised with caution, as a detectable chemical abnormality will neither explain a di process nor provide specific cause. The abnor pect of the the indiv between (Johnson lar bioch deficien tablish

eases. For example, the particular type of hemoglobin associated with sickle cell anemia is found in some individuals without anemia or evidence of disease. The same reasoning applies to genetics: The realization of genetic potential is not equivalent to having the potential, and abnormalities of genetic origin can be duplicated by nongenetic processes. These considerations emphasize the importance of using collaborative multidisciplinary and contemporary statistical approaches in paleopathology.

This chapter surveys the types of paleobiochemical analysis used to discern a reactive pattern or pathologic process in ancient tissues. The analytic principle in paleobiochemistry is the same as in every other science: first, qualitative, and second, quantitative, analysis.

MACROMOLECULAR CONSTITUENTS IN ANCIENT TISSUES

PROTEINS

At present, attempts are being made to extract, identify protein material in fra s. The prime motivation for gamma globulins and the antibodies oteins lies in the possibility of ain proteins, they may contain, fectious agents, such as hemoglobin, vide genetic markers for certain ological conditions in heterogeneous lations and may suggest something of geographical and ethnic or racial pro- ssion since ancient times.

Although very little information is available about the chemical and physicochemical state of preservation of protein and other macromolecules in mummified tissue, amino acid analysis has been reported for structural proteins of calcified fossils, bone, and hair (Abelson 1957; Wyckoff et al. 1963; Ho 1966; Jope 1967; Miller and Wyckoff 1968; Akiyama and Wyckoff 1970; Gillespie 1970) and of muscle tissue from Egyptian mummies (Lausarot et al. 1972). Further, there is evidence that apparently native proteins persist many years under a variety of preservation conditions. Biologically active proteins have been reported to survive in dried bloodstains exposed to the atmosphere for 10 years (Ducos et al. 1962), in whole blood stored at room temperature in the dried state for up to 40 years (Keilin and Wang 1947; Sensabaugh et al. 1971a,b), in seeds and spores germinating after dormant periods of as long as 2,000 years (Keilin 1959), and in 10,000-year-old samples of Alaska permafrost (Skujins and McLaren 1968). In addition, the long-term stability and antigenicity of material in preserved biological specimens have been demonstrated by blood-typing studies on mummified tissue from Egypt, the Aleutian Islands, the United States, and South America (Boyd and Boyd 1937; Candela 1939; Otten and Flory 1964; Lippold 1971; Stastny 1974).

SALTS

Because natron and other salts were used for chemical mummification by the ancient Egyptian embalmers (Lucas and Harris 1962), the study of the relationship between the degree of preservation of proteins in mummified tissue and tissue cation levels may provide useful archeological information concerning the nature of the mummification process. The cataloging of this kind of information would be useful for characterizing the class or status of the individual and the period and flux of traditions for the ancient society investigated. Also, the characterization of the mummification process for individual paleobiological

specimens may help to standardize paleopathological findings, as the ability to detect diseases varies with the degree of preservation of the ancient specimen.

LIPIDS

The lipid content of ancient tissues is a potentially great source of information in several respects. First, the tendency for lipids to accumulate and be stored within the body in relatively stable form in discrete cellular and subcellular components makes them excellent markers for the integrity of tissues long after death. Second, the qualitative composition of lipid populations in biological material correlates strongly with diet and, to a lesser extent, with more general features of life style, such as exposure to sunlight and extent of physical exertion. Third, lipids provide excellent macromolecular markers for certain pathological conditions, many of which are inherited metabolic defects. Fourth, lipids undergo gradual chemical changes, such as autoxidation of unsaturated fatty acids, which may be time-dependent and therefore of possible value as a biological dating tool.

Lipids and tissue integrity. Lipids have a discrete subcellular distribution. Neutral lipids and the lipid-soluble vitamins are primarily constituents of the cell cytoplasm; phospholipids and sphingolipids are structural components of membranes. The extent to which each type of lipid is recoverable from ancient tissues may reflect (within the limits of obvious chemical modifications) the state of preservation of different subcellular components and organelles.

Lipids, diet, and life style. Because some lipids and lipidic vitamins (A, E, D, and K, principally) are not synthesized in human tissues, the precise nature of these components reflects the dietary availability of them or their precursors. The obvious and best known correlation is that between the amount of polyunsaturated fats ingested and the extent of saturation in the stored fatty acids of triglycerides. A similar correlation

probably exists for the fatty acid components of certain phospholipids and sphingolipids. The presence of tocopherol and many of the isoprenoid derivatives also reflects their availability in the diet. Lipids may also be indicative of the life status and life style of the person from whose tissue they are extracted. For example, the relative distribution of lipoprotein densities in the blood is a function of both age and sex. The amount of various vitamin D precursors might vary in proportion to the amount and intensity of sunlight exposure. (A comparison of tissues from an Egyptian mummy and a frozen Eskimo might be particularly instructive in this regard.) By integrating life status, age, sex, and life style information derived by archeological and other disciplinary approaches with the biochemical information on lipid composition, it should be possible to make some distinctions between the strictly environmental, as opposed to cultural, determinants of the physicochemical characteristics of the ancient tissues under study.

Paleopathology of lipids.   Lipids can provide remarkably reactive signs of disease. Some of these reflect dietary deficiencies. The absence of certain carotene derivatives from some ancient tissues, but not from others, might suggest visual defects in the person from which the former was obtained. Certainly the correlation of lipid composition patterns with other pathological evidence might lead to enlightening conclusions about causative factors in ancient society. In addition, the lipids are peculiarly characterized by susceptibility to certain inborn errors of metabolism that invariably have nonrandom geographical and ethnic distributions (the various gangliosidoses are prominent examples). The possibility exists that evaluation of lipid patterns from ancient tissues will reveal the presence of these defects and suggest something of their geographical and ethnic progression since ancient times.

Lipids as a supplementary dating tool.   The temporal parameters of autooxidation of unsaturated fatty acids have not been studied over the range of time that would be possible with specimens of ancient tissue. Another even less investigated area concerns the degradation of conjugated moieties attached to lipids, such as the carbohydrate portion of sphingolipids, over long periods of time under conditions of mummification. These and similar chemical properties might serve as a supplementary tool for biological dating. Only a detailed analysis of the lipid composition of ancient tissues can verify or reject the feasibility of such a technique.

Overall value of lipid studies.   Lipids can be extracted from ancient tissue samples by a comprehensive protocol to maximize the yield of neutral lipids, phospholipids, sphingolipids, lipidic vitamins, and lipoproteins. They can be fractionated into each of the above categories, and each group can be analyzed for quantitative composition. The end result is a detailed profile of the lipid composition suitable for analysis within the context of information generated by other paleobiologic areas of investigation. As a result, a detailed description of the lipid composition of ancient tissues could provide:

1. Information on the diet of the ancient people and, in conjunction with other types of evidence, the extent to which that diet was based strictly on availability or on cultural and perhaps ritualistic determinants as well

2. Information on the life style and life status of the person in a given ancient society

3. An indication of the state of preservation of different components of ancient tissue – information that will advance understanding of the science of tissue preservation and macromolecular stability in nonliving tissue over long periods

4. Information on the presence and cause of diseases involving lipids

5. Possible information on the geographical and ethnic occurrence of certain heritable maladies in ancient times

6. A basis for evaluating the feasibility of

using lipid chemistry as a biological dating tool

In a more general sense, the overall significance of the study of lipids reflects that of chemical studies on ancient tissues as a whole: namely, the use of standard chemical methodologies to provide information and informed speculation on the biological, cultural, and medical characteristics of the populations and environment of ancient times.

AMINO ACID DATING OF PRESERVED BIOLOGIC SPECIMENS

The standard method for determining the age of most fossils and other organic material is by the carbon-14 technique, which is based on the ratio of carbon 12 to carbon 14 in the specimens. Because carbon 14, a radioactive isotope, decays at a known rate and is not replenished following death of an organism, the proportion of ordinary carbon to carbon 14 slowly increases. However, there are some difficulties with the standard error of the carbon 14 clock (Baxter et al. 1974). Further, the reliability of carbon 14 dating diminishes substantially with samples 50,000 years old or more.

Recent evidence suggests that the amino acid racemization reaction can be used to estimate the age of organic material (Bada and Protsch 1973). Only L-amino acids are usually found in the proteins of living organisms, but over time they undergo slow racemization, producing the nonprotein D-amino acids. Thus, the proportion of D- to L-amino acids in paleobiological material steadily increases with time. However, the racemization reaction is dependent on temperature, and some estimate of the temperature history of a paleobiological specimen must be available for accurate dating. Amino acid dating requires much smaller quantities of material than radiocarbon dating, and because amino acids racemize at different rates, a chosen amino acid can be used for dating such material in a specific age range; this extends the applicability of the technique beyond that of radiocarbon dating.

However, some difficulties persist with the amino acid dating technique. For example, there remains some uncertainty about the effect of pH variation on the racemization rates of amino acids in preserved biologic materials. Further, it is often difficult to obtain a reliable temperature record of the ancient specimen. For more recent specimens (ca. A.D. 500 – B.C. 2000), as is usually the case with mummies, this is even more critical. The temperatures at which most paleobiological specimens are found result in small D/L ratios even for amino acids such as aspartic acid and alanine, which racemize substantially faster than most amino acids. Consequently, it is necessary to detect small amounts of D-amino acid. For this reason, gas chromatography of amino acid mixtures has proved far more sensitive than the traditional ion-exchange chromatographic methods (Barraco 1975a).

Finally, if the relationship between rate of racemization and temperature can be established from a series of rate constants for each amino acid at different temperatures, this would permit calibration of amino acid dating without the use of radiocarbon standard curves, avoiding the incorporation of radiocarbon standard error into the estimation of date by amino acid racemization. This would result in an absolute rate constant for the racemization of individual amino acids at a given temperature. Further, the estimation of age of a given paleobiological specimen could be calculated from approximately 20 data points, representing the estimations of D/L ratios from separate amino acid racemization reactions (not including glycine but including the epimerization of L-isoleucine to the nonprotein amino acid D-alloisoleucine).

TECHNIQUES FOR BIOCHEMICAL ANALYSIS OF ANCIENT TISSUES

In the biochemical studies of six mummies autopsied by our Detroit group (DIA I, PUM I,

PUM II, PUM III, ROM I, Saint Lawrence Eskimo), it became apparent that not only did the nature of the mummification process vary greatly (from the extremely deliberate and meticulous mummification of PUM II to the apparently unintended and natural mummification of the Saint Lawrence Eskimo), but also the degree of preservation of macromolecular constituents extracted from these tissues varied greatly. Consequently, a series of studies was initiated to evaluate not only the persistence and heterogeneity of macromolecular constituents in mummified tissue, but also the relationship between the degree of preservation of these materials in ancient tissues and the nature of the mummification process. Other paleobiological and paleohistological findings from some of these mummies have already been reported (Lewin et al. 1974; Lynn and Benitez 1974; Cockburn et al. 1975; Reyman et al. 1976; Zimmerman 1978; Barraco et al. 1977).

Various types of tissues were used for the biochemical studies. Some samples were taken from the internal aspects of the organ packages (usually located in the abdominal and thoracic cavities of the intentionally preserved bodies) or from the organs in situ (in the case of naturally mummified specimens). However, the following methods and results are intended to apply as general guidelines for the biochemical analysis of mummified tissues. Because ancient relics are difficult to obtain, certain tissues (e.g., spleen, brain) are often not available; also, the preservation condition of some samples is unsatisfactory.

PROTEIN REHYDRATION AND
EXTRACTION PROCEDURES

The tissue samples are cut into small pieces (2 to 3 mm) and placed in various rehydrating mediums. Portions of the rehydrated tissue are used for histology, and the remaining samples are used for biochemical analyses. The methods for rehydration of the mummified tissue were evolved after considerable trial and error. Each tissue requires specific rehydration and extraction conditions depending on its type, its degree of preservation, and the desired macromolecular material, so that standard techniques in protein and lipid chemistry can be used for subsequent analysis. The most satisfactory results for most tissues are obtained when the tissue is rehydrated in 2.5M NaCl (25 ml per gram dry tissue) for 30 hours at 25 °C in a closed vessel. Using this technique, 30 hours later the solution typically appears dark brown, with a sludge at the bottom representing the rehydrated tissue. For unnatronized tissues the rehydrating solution has much less color. The dark color from natronized tissue (i.e., PUM II) is probably attributable to the resins and oils used by the ancient Egyptian embalmers (Lucas and Harris 1962). The total contents of the vessel are then homogenized. The resulting suspension is filtered through cheesecloth and recentrifuged. The supernatant (S_1) is decanted and lyophilized; the pellet (P_1) is stored. Following lyophilization, S_1 is resuspended in 5 mM sodium phosphate buffer, pH7 (10 ml per gram dry tissue).

At this point, two separate procedures are usually employed for the extraction and purification of the protein material (Figure 19.1). Details of these procedures have been reported previously (Barraco et al. 1977). In the first procedure, S_1 is treated with sequential series of solvents to isolate and purify the protein material. In the second procedure, S_1 is ultracentrifuged to form a second supernatant (S_2) and a second pellet (P_2). Further, S_2 is desalted on a Sephadex G–25 column, forming DS_2. DS_2 is subsequently fractionated on Sephadex G–100, forming three major fractions: DS_2–I, DS_2–II, and DS_2–III (Figure 19.3 C). Then polyacrylamide gel electrophoresis (PAGE) is conducted on each fraction (Figure 19.4).

Figure 19.1.   *Procedures for extraction and purification of protein material from mummified tissue.*

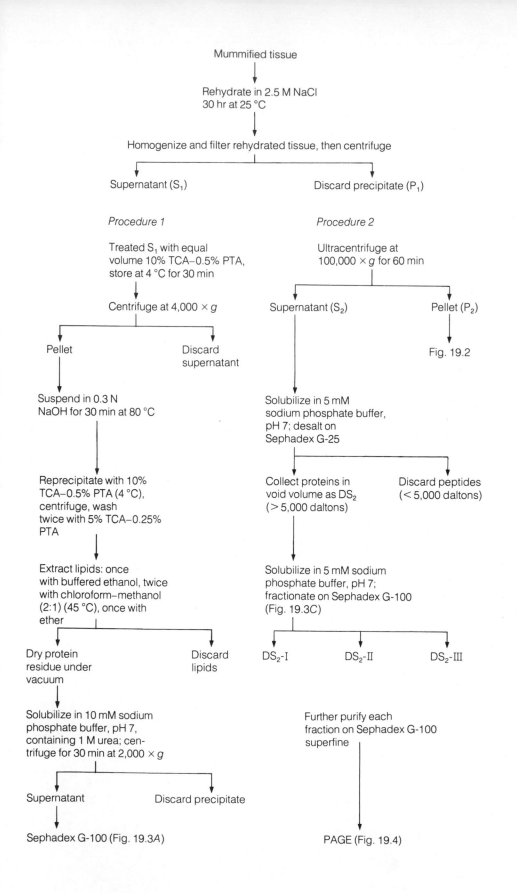

Mummified tissue

↓

Rehydrate in 2.5 M NaCl
30 hr at 25 °C

↓

Homogenize and filter rehydrated tissue, then centrifuge

Supernatant (S₁)   Discard precipitate (P₁)

Procedure 1

Treated S₁ with equal
volume 10% TCA–0.5% PTA,
store at 4 °C for 30 min

↓

Centrifuge at 4,000 × g

Pellet   Discard
supernatant

↓

Suspend in 0.3 N
NaOH for 30 min at 80 °C

↓

Reprecipitate with 10%
TCA–0.5% PTA (4 °C),
centrifuge, wash
twice with 5% TCA–0.25%
PTA

↓

Extract lipids: once
with buffered ethanol, twice
with chloroform–methanol
(2:1) (45 °C), once with
ether

Dry protein   Discard
residue under   lipids
vacuum

↓

Solubilize in 10 mM sodium
phosphate buffer, pH 7,
containing 1 M urea; cen-
trifuge for 30 min at 2,000 × g

Supernatant   Discard precipitate

↓

Sephadex G-100 (Fig. 19.3A)

Procedure 2

Ultracentrifuge at
100,000 × g for 60 min

Supernatant (S₂)   Pellet (P₂)

↓

Fig. 19.2

Solubilize in 5 mM
sodium phosphate buffer,
pH 7; desalt on
Sephadex G-25

Collect proteins in   Discard peptides
void volume as DS₂   (< 5,000 daltons)
(> 5,000 daltons)

↓

Solubilize in 5 mM sodium
phosphate buffer, pH 7;
fractionate on Sephadex G-100
(Fig. 19.3C)

DS₂-I   DS₂-II   DS₂-III

Further purify each
fraction on Sephadex G-100
superfine

↓

PAGE (Fig. 19.4)

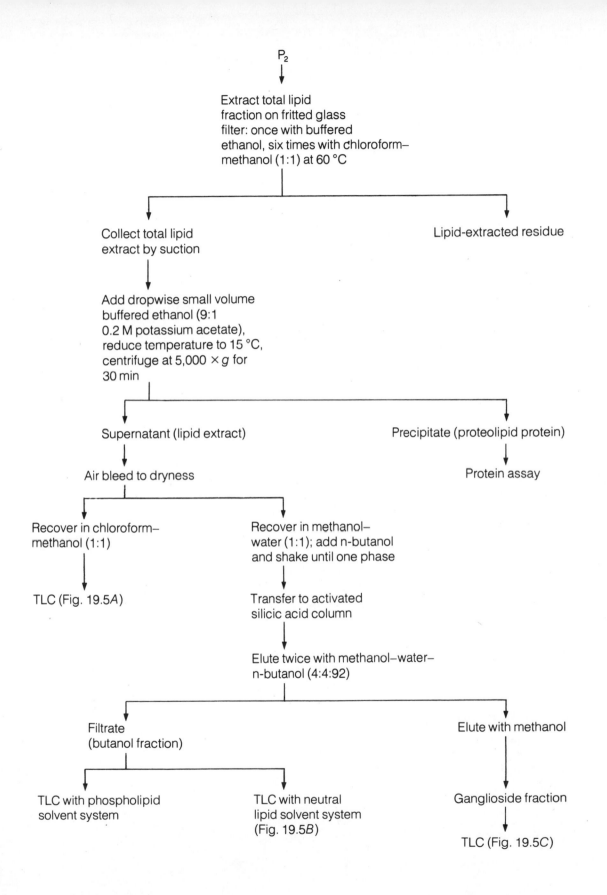

EXTRACTION PROCEDURES AND ANALYTICAL METHODS FOR LIPID MATERIAL

Portions of P_2 are weighed and placed on a medium porosity fritted glass filter previously rinsed with acetone–water. P_2 is treated with a sequence of solvents, and the total lipid extract is fractionated by thin-layer chromatography (TLC) or further partitioned into gangliosidic, neutral lipid, phospholipid, or proteolipid fractions (McCluer et al. 1962), which can be separately analyzed by TLC (Figure 19.5). The plates are sprayed with reagents to visualize neutral lipids, phospholipids, cholesterol and its esters, and gangliosidic sialic acid. The chromatograms are analyzed by densitometry (Canalco, Model G) to standardize mobilities and preliminarily screened by the iodine vapor test. Also, portions of the total lipid extract and the lipid-extracted residue are assayed for protein, neutral hexose, and sialic acid. Details of these procedures have been reported previously (Barraco et al. 1977).

ANALYTICAL TECHNIQUES FOR PROTEIN MATERIAL

Quantitative estimation of protein is done by the method of Lowry et al. (1951) and the biuret assay (Itzhaki and Gill 1964). PAGE is conducted under basic conditions (Davis 1964) and acidic conditions (Brummel and Montgomery 1970). Portions of DS_2 are solubilized in 0.05 M NH_4HCO_3 and rechromatographed on Sephadex G–25 fine (column dimensions 30 by 0.9 cm; eluant 0.5 M NH_4-HCO_3, pH 7.8; flow rate 3.6 ml cm^{-2} hr^{-1}) where the excluded proteins in the void volume (>5,000 daltons) are used for amino acid analysis (Beckman M–121 Micro column with AA Computing Integrator, AA–20 resin) before and after hydrolysis (6 N HCl, 22 hours, 110 °C) by stepwise elution according to a modified procedure of Moore and Stein (1963) (Table 19.1).

DETERMINATION OF TISSUE CATION CONTENT

For the cation studies, tissue samples (2 to 3 mm) are rinsed briefly with methanol to remove surface salt and are lyophilized. Desiccated aliquots are weighed, ashed, and solubilized in a $SrCl_2$–HCl solution for analysis of Na, K, Ca, and Mg by atomic absorption spectrophotometry (Putney and Bianchi 1974) (Table 19.2).

EXPERIMENTAL RESULTS

Most of the following results, particularly the lipid studies, are derived from the analysis of PUM II, the best preserved of the six mummies autopsied. However, the salient features of these studies apply to the entire range of mummies analyzed.

PROTEINS

Depending on the type of tissue used and the degree of preservation of the mummy, the yield of protein in the extracted sample has varied from 5 mg per gram of desiccated mummified tissue for rectus abdominis from ROM I to 30 mg per gram for spleen from PUM II. In fact, all tissue samples analyzed from the series of six mummies have yielded proteinogenous material, except for what appeared to be dried blood located in bone marrow samples from PUM II. The majority of the extracted protein is found in P_2, which consists primarily of membrane material.

Proteins precipitated from S_1 are fractionated on Sephadex G–100 in a urea buffer system (Figure 19.3A and B) and DS_2 is chromatographed on Sephadex G–100 in a nonurea buffer system. The molecular weight distribution in both gel filtration systems is approximately 5,000 to 130,000 daltons. This molecular weight range for the extracted protein is confirmed by PAGE in the presence of sodium dodecyl sulfate (SDS) (Barraco et al.

Figure 19.2. Procedure for desalting S_2 fraction on Sephadex G-25 column.

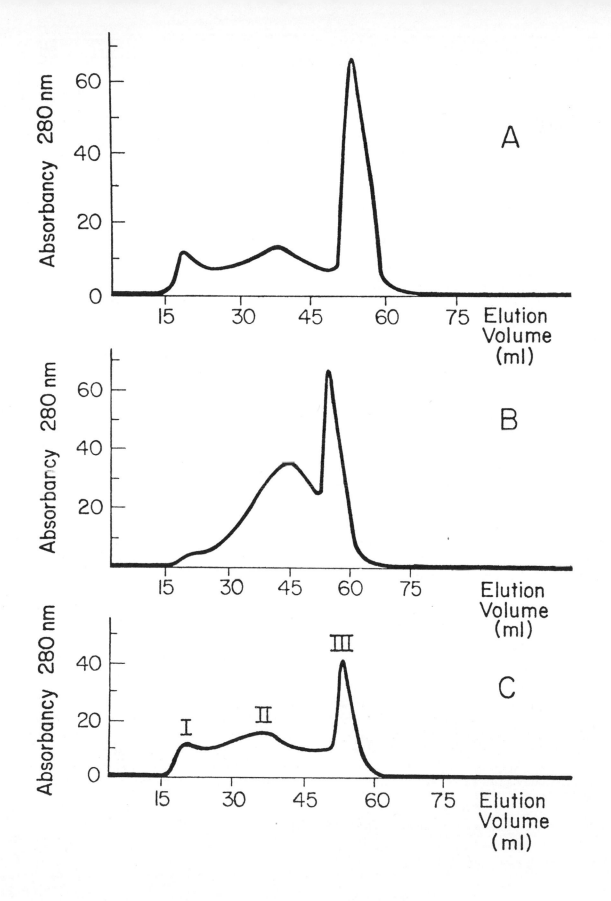

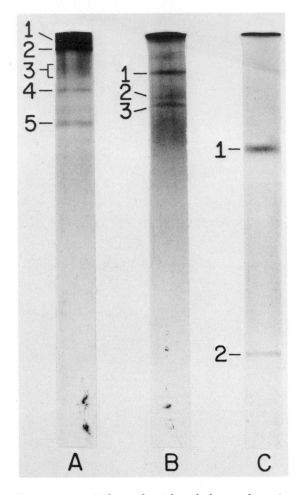

Figure 19.4. *Polyacrylamide gel electrophoresis (PAGE) under basic conditions. Spleen tissue, PUM II; 7.5 percent gels. Electrophoretic run was 1.5 hr in duration. Coomassie brilliant blue was used for staining. All electrophoretic gels were analyzed by densitometry. a. DS₂ -I, bands 1–5. b. DS₂ -II, bands 1–3. c. DS₂ -III, bands 1–2.*

1977). PAGE conducted under basic conditions reveals a significant degree of heterogeneity (Figure 19.4). It is possible to separate and consistently identify 10 compo-

nents from the tissue samples of spleen from PUM II using PAGE with basic conditions. Electrophoresis of rectus abdominis tissue, also from PUM II, under identical conditions results in the separation of fewer components. Using PAGE under acidic conditions, much of the material does not enter the gel, and it has not been possible to produce any major band separation for a variety of tissues from several of the mummies.

LIPIDS

Almost half the dry weight of P_2 is removed by the lipid extraction. A significant amount of protein soluble–chloroform–methanol is detected in the total lipid extract, indicating the presence of intact proteolipid material. The sugar assays indicate the presence of significant quantities of bound neutral hexose in the gangliosidic fraction of the total lipid extract and the lipid-extracted residue. However, significant quantities of bound sialic acid are detected only in the lipid-extracted residue, indicating the persistence of sialic acid in glycoproteins but not gangliosides. Although separation of some gangliosidic material on TLC can be achieved (Figure 19.5), the resolved components do not stain with resorcinol reagent and do not separate into any recognizable pattern, further indicating the degradation of gangliosides. These results held true for P_2 derived from spleen tissue even though splenic tissue is rich in gangliosides and other glycolipids (Weigandt and Bucking 1970).

AMINO ACIDS

Table 19.1 shows the amino acid distribution, as percent of total, for hydrolyzed proteins from rectus abdominis from PUM II (natronized) and ROM I (unnatronized). To rule

Figure 19.3.   a. *Gel chromatography of S₁ from PUM II (rectus abdominis) on Sephadex G-100 (eluant: 10mM sodium phosphate buffer, pH 7, containing 1 M urea). b. Gel chromatography of DS₂ from ROM I (rectus abdominis) on Sephadex G-100 under same conditions as a. c. Gel chromatography of DS₂ from PUM II (splenic tissue) on Sephadex G-100 (eluant: 5 mM sodium phosphate, pH 7). The resulting chromatography produced three fractions DS₂-I, DS₂-II, DS₂III. Each fraction was rechromatographed separately on Sephadex G-100 superfine. Subsequently, each purified fraction was subjected to PAGE.*

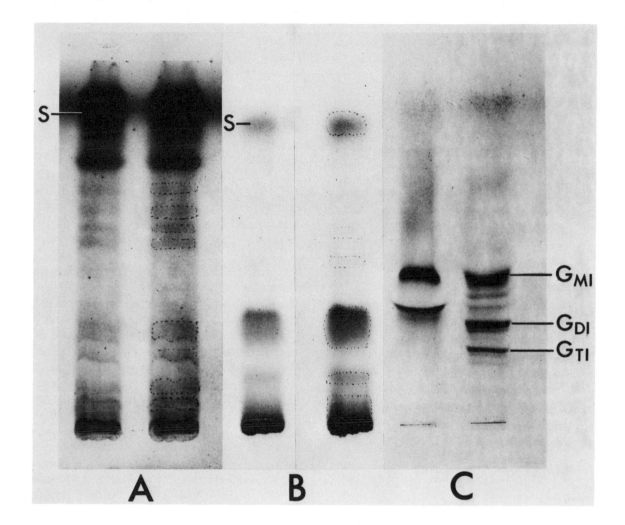

Figure 19.5. Thin-layer chromatography (TLC) of lipid fractions from PUM II. a. TLC of total lipid fraction (not including proteolipids); 20- by 20-cm plate. First solvent system: heptane–ether–methanol–acetic acid (90:20:2:3). Second solvent system: chloroform–methanol–H$_2$O (70:30:4). General detection reagent: sulfuric acid–dichromate. Sterol and sterol ester reagent: sulfuric acid–acetic acid. With the sulfuric acid–dichromate reagent, 11 components were detected, 1 of which stained for sterols (designated S). b. TLC of neutral lipid fraction (butanol filtrate; McCluer et al. 1962), 20- by 20-cm plate. Solvent system: heptane–ether–methanol–acetic acid (90:20:2:3). With the surfuric acid–dichromate reagent, 6 major components were detected, 1 of which stained for sterols (designated S). c. TLC of ganglioside fraction (methanol filtrate; McCluer et al. 1962), 20- by 20-cm plate. Solvent system: n-propanol–H$_2$O–conc. NH$_4$OH (6:2:1). Resorcinol reagent was used for detection of gangliosidic sialic acid: left, spleen gangliosides; right, pigeon brain gangliosides as standards extracted with the same procedure. Major gangliosides designated: G$_{MI}$, monosialoganglioside; G$_{DI}$, disialoganglioside; G$_{TI}$, trisialoganglioside). With the sulfuric acid–dichromate reagent, 2 major spleen gangliosides were detected. However, both components stained negligibly with resorcinol reagent, indicating they are asialogangliosides or other unidentified lipids.

Table 19.1. *Amino acid distribution of hydrolyzed proteins extracted from rectus abdominis of two mummies from the series of six, expressed as percent of total*

Amino acid	PUM II	ROM I
Lysine	3.9	4.1
Histidine	1.2	2.0
Arginine	6.1	3.7
Aspartic acid	11.5	13.1
Threonine	3.6	5.0
Serine	4.7	5.0
Glutamic acid	13.2	18.8
Proline	14.9	4.7
Glycine	12.4	12.0
Alanine	17.1	11.6
Half cystine	—	—
Valine	4.0	4.7
Methionine	—	1.7
Isoleucine	1.2	3.2
Leucine	4.4	4.5
Tyrosine	0.6	3.5
Phenylalanine	1.2	2.4

Dash indicates undetectable at this level of sensitivity.

Table 19.2. *Cation content (micromoles per gram dry weight) of tissue from six mummies and of one fresh autopsy specimen*

Specimen	Na	K	Ca	Mg
Human autopsy	216.9	206.2	9.6	27.5
ROM I	187.0	291.2	7.4	1.1
Eskimo	16.3	3.8	371.2	68.7
PUM I	73.1	122.5	23.4	10.9
PUM II	747.0	57.5	11.5	6.9
DIA I	2,488.0	253.8	74.1	33.8
PUM III	345.0	143.0	5.6	3.4

out contamination by free amino acids, amino acid analysis is also conducted on unhydrolyzed portions of DS_2; however, no free amino acids are detected in the DS_2 fraction. The amino acid analysis for PUM II approximates the amino acid composition of normal mixed tissue proteins except for the absence of half cystine and methionine. However, there appears to be a smaller amount, as percent of total, of basic amino acids, indicating a possible point of preferential degradation of residues in the proteins. This is also evident, particularly with arginine, for ROM I. Further, the high glycine-proline content for PUM II indicates the possible persistence of collagen in the tissues. In fact, electron microscopic analysis of rectus abdominis tissue from PUM II indicates the ubiquitous presence of masses of striated fibrous material resembling collagen (Cockburn et al. 1975). On the other hand, there appears to be little collagen present in the tissue samples from ROM I.

These results can be compared with those of a previous study involving extracted material from an unnatronized Twelfth-Dynasty mummy in which the collagen content was apparently high (Lausarot et al. 1972). In fact, except for the low amount of lysine, which is reported to undergo preferential degradation with time (Lausarot et al. 1972), the percent amino acid composition of extracted protein from ROM I closely approximates that of myoglobin, which may be the major small-molecular-weight protein (ca. 17,000) in skeletal muscle resistant to degradation following desiccation under natural conditions.

CATION STUDIES

In Table 19.2 the various cation levels in tissue samples from the series of six mummies are represented as micromoles of cation per gram of dry weight of tissue. To test the procedures, specimens of freshly autopsied human rectus abdominis are also analyzed by atomic absorption. Table 19.2 shows that ROM I, PUM I, and Saint Lawrence Eskimo were not natronized, as the tissue cation levels approximate closely those of the freshly autopsied specimen. On the other hand, it appears that PUM II and DIA I were natronized (primarily Na salts), and PUM III was partially natronized. Also, the high content of Ca in the muscle of the Saint Lawrence Eskimo indicates the possibility of high lime content in the soil at the burial site.

PRESERVATION OF PROTEINS

One of the major considerations for the paleopathologist is evaluating the degree of preservation of the tissues and the nature of the mummification process. Because it was known that natron and other salts were used for chemical mummification by the ancient Egyptian embalmers (Lucas and Harris 1962), the relationship of Na and other cation levels in the mummified tissue with the molecular weight distribution (a measure of degradation) of extracted protein from the same tissues was investigated. These studies indicate that the degree of preservation (relative distribution of high-molecular-weight protein) of extracted protein varies in relation to the levels of Na in the tissues (natron). For example, compare the molecular-weight-distribution profiles of PUM II and ROM I (Figure 19.3 A and B). In fact, the molecular-weight-distribution profiles for extracted protein from five mummies indicate that natronized tissue undergoes less degradation and the degree of preservation of extracted protein correlates closely with the levels of Na in the tissues (Barraco 1975b, 1978).

CONCLUSIONS

These studies indicate that protein and lipid material can be extracted from mummified tissues and processed using various preparative and analytical techniques. There is evidence that conjugated proteins with carbohydrate and lipid prosthetic groups remain stable and extractable as glycoprotein and proteolipid. Further, it appears that some protein of relatively high molecular weight (ca. 130,000) remains intact, although much of the protein has been degraded with a marked attenuation of heterogeneity. Similar results were reported for protein material extracted from neck tissue of a 4,000 year old Egyptian mummy (Lausarot et al. 1972). In that study, the extracted protein migrated as a single component in a phosphate–SDS PAGE system with a relative mobility corresponding to an apparent molecular weight of 60,000.

In the studies using PAGE under acidic conditions, the extracted protein demonstrates negligible mobility and heterogeneity. Along with the amino acid analyses, these results indicate the possibility that basic amino acid residues in preserved biological specimens undergo preferential degradation. In fact, a net increase in the negative charge of surviving proteins from 8-year-old dried human blood has been reported (Sensabaugh et al. 1971a,b). Also, from the amino acid composition of Egyptian mummy structural protein, it was determined that almost exclusively the basic amino acids undergo chemical modification (Lausarot et al. 1972). This has important implications for recently developed techniques that date organic material using amino acid racemization reactions (Bada and Protsch 1973; Bada et al. 1974).

Greater heterogeneity appears to persist in the lipid than in the protein material. The amount of lipid extractable from S_2 appears negligible. However, from P_2 the total lipids can be fractionated into gangliosidic, neutral lipid, phospholipid, and proteolipid components. The neutral lipid fraction and the total lipid extract demonstrate a persistence of heterogeneity. Using TLC it is possible to identify 6 major components in the neutral lipid fraction and 11 components in the total lipid fraction from the spleen of PUM II. Further, the neutral lipid and total lipid fractions stain positive for cholesterol and sterol esters (Figure 19.5). This indicates the persistence of neutral lipids such as sterols, fatty acids, tocopherols, and triglycerides in mummified tissue. The precise qualitative characteristics of the lipid material can be further evaluated by gas chromatography and other analytical techniques.

These studies also indicate that tissue samples from organ packages appear to be better preserved than tissue from the carcass. This observation is supported not only by histological analysis, but also by the fact that the recovery of extracted proteins (> 5,000 daltons) is greater for these tissues. This may be attributable to a difference in natronization

procedures (Lucas and Harris 1962). The organs, once removed by embalmers, may have been soaked in vessels containing liquid natron, while the remaining carcass, too large for easy submersion in a vessel containing liquid natron, may only have been packed with crystals of natron. Finally, the studies evaluating the relationship between tissue cation levels and preservation of macromolecular tissue components may provide useful archeological information concerning the characterization of the mummification process for different periods and the standardization of paleopathological findings, as the ability to detect disease may vary with the degree of preservation of the ancient specimen.

REFERENCES

Abelson, P. 1957. Some aspects of paleobiochemistry. *Annals of the New York Academy of Sciences* 69:276–85.

Akiyama, M., and Wyckoff, R. W. G. 1970. The total amino acid content of fossil Pecten Shells. *Proceedings of the National Academy of Sciences of the United States* 67:1097–1100.

Bada, J. L., and Protsch, R. 1973. Racemization reaction of aspartic acid and its uses in dating fossil bones. *Proceedings of the National Academy of Sciences of the United States* 70:1331–4.

Bada, J. L.; Schroeder, R. A.; and Carter, G. F. 1974. New evidence for the antiquity of man in North America deduced from aspartic acid racemization. *Science* 184:791–3.

Barraco, R. 1975a. Amino acid dating. *Paleopathology Newsletter* 9:8–9.

– 1975b. Preservation of proteins in mummified tissue. *Paleopathology Newsletter* 11:8–9.

– 1978. Preservation of proteins in mummified tissue. American Journal of Physical Anthropology 48:487–91.

Barraco, R. A.; Reyman, T. A.; and Cockburn, T. A. 1977. Paleobiochemical analysis of an Egyptian mummy. *Journal of Human Evolution* 6:533–46.

Baxter, M. S.; Aitken, M. J.; Clark, R. M.; and Renfrew, C. 1974. Calibration of the radiocarbon time scale in "Matters Arising." *Nature* 249:

Boyd, W. C., and Boyd, L. G. 1937. Blood grouping tests on 300 mummies. *Journal of Immunology* 32:307–19.

Brummel, M. C., and Montgomery, R. 1970. Acrylamide gel electrophoresis of the S-sulfo derivatives of fibrinogen. *Analytical Biochemistry* 33:28–35.

Candela, P. B. 1939. Blood group determinations upon the bones of thirty Aleutian mummies. *American Journal of Physical Anthropology* 24:361–83.

Cockburn, T. A.; Barraco, R. A.; Reyman, T. A.; and Peck, W. H. 1975. Autopsy of an Egyptian mummy. *Science* 187:1155–60.

Davis, B. J. 1964. Polyacrylamide disc gel electrophoresis. *Annals of the New York Academy of Sciences* 121:321–404.

Ducos, J.; Ruffie, J.; and Varsi, M. 1962. Mise en évidence des antigènes Gm^a Gm^b Gm^x dans les taches de sang sec. *Vox Sang* 7:722–31.

Gillespie, M. M. 1970. Mammoth hair: stability of α-keratin structure and constituent proteins. *Science* 170:1100–1.

Ho, T. Y. 1966. The isolation and amino acid composition of the bone collagen in Pleistocene mammals. *Comparative Biochemistry and Physiology* 18:353–8.

Itzhaki, R., and Gill, D. 1964. A microbiuret method. *Analytical Biochemistry* 9:401–5.

Johnson, L. C. 1966. In *Human paleopathology*, S. Jarcho (ed.), pp. 68–81. New Haven: Yale University Press.

Jope, M. 1967. The protein of brachiopod shell. II. Shell protein from fossil articulates: amino acid composition. *Comparative Biochemistry and Physiology* 20:601–5.

Keilin, D. 1959. Problem of anabiosis for latent life: history and current concepts. *Proceedings of the Royal Society*, series B 150:149.

Keilin, D., and Wang, Y. L. 1947. Stability of haemoglobin and of certain endoerythrocytic enzymes in vitro. *Biochemistry Journal* 41:491–500.

Lausarot, P. M.; Ambrosino, C.; Favro, F.; Conti, A.; and Massa, E. R. 1972. Preservation and amino acid composition of Egyptian mummy structure proteins. *Journal of Human Evolution* 1:489–99.

Lewin, P.; Mills, A. J.; Savage, H.; and Vollmer, J. 1974. Nakht, a weaver of Thebes. *Rotunda* 7:15–19.

Lippold, L. K. 1971. The mixed cell agglutination method for typing mummified human tissue. *American Journal of Physical Anthropology* 34:377–83.

Lowry, O. H.; Rosebrough, N. J.; and Farr, A. L. 1951. Protein measurement with the Folin phenol reagent. *Journal of Biological Chemistry* 193:265–75.

Lucas, A., and Harris, J. . 1962. *Ancient Egyptian materials and industries,* pp 270–326. London: Edward Arnold.

Lynn, G. E.; and Benitez, J. T. 1974. Temporal bone preservation in a 2,600-year-old Egyptian mummy. *Science* 183:200–2.

McCluer, R. H.; Coram, E. H.; and Lee, H. S. 1962. A silicic acid absorption method for the determination of ganglioside sialic acid. *Journal of Lipid Research* 3:269–74.

Miller, M. F., and Wyckoff, R. W. G. 1968. Proteins in dinosaur bones. *Proceedings of the National Academy of Sciences of the United States* 60:176–8.

Moore, S., and Stein, W. H. 1963. Chromatographic determination of amino acids by the use of automatic recording equipment. In *Methods in enzymology,* vol. 6, S. Colowick and N. O. Kaplan (eds.), pp. 819–31. New York: Academic Press.

Otten, C. M., and Flory, L. L. 1964. Blood typing of Chilean mummy tissue: a new approach. *American Journal of Physical Anthropology* 21:283–5.

Putney, . W., and Bianchi, C. P. 1974. Site of action of dantrolene in frog sartorius muscle. *Journal of Pharmacology and Experimental Therapeutics* 189:202–12.

Reyman, T. A.; Barraco, R. A.; and Cockburn, T. A. 1976. Histopathological examination of an Egyptian mummy. *Bulletin of the New York Academy of Medicine* 52:506–16.

Sensabaugh, G. F., Jr.; Wilson, A. C.; and Kirk, P. L. 1971a. Protein stability in preserved biological remains. I. Survival of biologically active proteins in an 8-year-old sample of dried blood. *International Journal of Biochemistry* 2:545–57.

1971b. Protein stability in preserved biological remains. II. Modification and aggregation of proteins in an 8-year-old sample of dried blood. *International Journal of Biochemistry* 2:558–68.

Skujins, J. J., and McLaren, A. D. 1968. Persistence of enzymatic activity in stored and geologically preserved soils. *Enzymologia* 34:213–25.

Stastny, P. 1974. HL-A antigens in mummified pre-Columbian tissues. *Science* 183:864–6.

Wiegandt, H., and Bucking, H. W. 1970. Structure and function of ganglioside. *European Journal of Biochemistry* 15 287–91.

Wyckoff, R. W. G.; Wagner, E.; Matter, P.; and Doberenz, A. R. 1963. Collagen in fossil bone. *Proceedings of the National Academy of Sciences of the United States* 50:215–18.

Zimmerman, M. R. 1978. Age determination of an Alaskan mummy: morphologic and biochemical correlation. *Science* 201:11–12.

20

Paleoserology

RAYMOND L. HENRY
Professor of Physiology
Wayne State University School of Medicine
Detroit, Michigan U.S.A.

Genetic determinants are important ancillary factors in the classification of man and his migration history. Antigens of the ABO Landsteiner designation are most often considered because they are distributed throughout body tissue (Landsteiner and Witt 1926). Positive results from typing identify specific antigens present and enable more precise physical characterization of human populations in antiquity. Such characterization can then be compared with present-day populations in the same geographic area. Determination of blood types provides further data for or against hypotheses of spatial and temporal migrations of man. Information about blood group and other antigens from ancient tissues should serve to supplement morphological characterization; it should not be used independently. These data in conjunction with archeological accounts, historical documentation, and ethnographic information can greatly benefit the anthropologist.

Since the initial study of Boyd and Boyd (1933), several attempts have been made to identify the antigen-antibody reactions of ancient tissue. Early investigations were aimed at the tissue antigens, as antibody induction was not satisfactory. However, more recent technology for antibody production seems to be rendering more reliable results. Precipitin tests, agglutination inhibition, mixed-cell agglutination, and antibody induction with precipitin tests have been employed for the serologic study of ancient tissues. Certain pitfalls and inconsistencies in these methods should be recognized to allow yields of useful data from investigative procedures. Anyone attempting serological work should avail himself of the extensive experience of the United States military (Camp et al. 1973). Particular attention should be given to the section on pitfalls encountered in performing blood-grouping tests, in Annex S of the monograph.

Designation of major ABO system blood groups depends on the antigen present on red blood cells or other tissue cells. Ordinarily those specific antigens are identified with specific known antibodies in standard sera developed against known ABO antigens. This procedure is the direct typing system, whether precipitation or agglutination techniques are employed. In some instances, it may be possible to identify antibodies against ABO antigens in the specimen to be tested. These procedures are known as reverse, or indirect, testing. Anti-A agglutinin should be present in specimens of the B group, and anti-B in specimens of the A group. Both anti-A and anti-B agglutinins should be present in group O individuals, and neither is present in group AB persons. Positive results in these tests strongly suggest presence of that particular antibody; however, absence of antibody in ancient serum or tissue still leaves open the question of group identity because the specific antibodies may have deteriorated or been otherwise lost. The ABO system of blood grouping is unique in having naturally occurring isoantibodies always reciprocal to the antigens present on cells of that individual. Isoantibodies have been reported for other blood group systems, such as Rh-Hr, Kell, I, Ss, and Wr[a]; however, such systems are so rare that testing ancient tissues for anything except the ABO system is not war-

ranted. This chapter deals with application of data so far obtained as well as critical evaluation of the techniques employed. Within the discussion, detailed descriptions of methods are presented.

Attention to ABO antigens has gained favor because these antigens are generally distributed throughout body tissues so that bone, hair, skin, muscle, or whatever happens to be available may yield information. Reports concerning negative data should be viewed with caution. Positive data, on the other hand, appear to be more reliable.

Boyd and Boyd (1937) tested 300 mummy specimens and reported a surprising excess of AB types. That the absorptive power of mummified material was present for A and B types was not questioned. However, group A substance was suggested as being perhaps less stable than B type. Nevertheless, this alone could not account for the large number of AB types. Failure of tissue to remove either anti-A or anti-B did not necessarily identify the blood group as O because of the possible disappearance of one or both agglutinogens. The finding of A and B groups in predynastic Egyptian material suggested to the authors that blood groups were not as recent as supposed in 1937, but rather supported postulations of anthropoid ancestors of man.

Brues (1954) discussed selection and genetic drift of the ABO allelomorphs. Early assumptions were made that only the O gene originally existed and that A and B arose and spread at different times, producing the current distribution among living populations. Frequencies of the A gene showed large ranges among populations having reasonably homogeneous morphological features. Investigation of Australian aborigines revealed evidence for vigor of genetic drift under primitive conditions exceeding that in population admixtures. Sufficient data from ancient tissues may substantiate some of these claims. A basic question seems to be whether any selective value can be attached to the ABO blood group genes or to the genotypes formed by them. Distribution of the types in ancient tissues, such as Egyptian mummies, coupled with identification of diseases in these ancient populations, would be of interest.

PRECIPITIN TEST

Precipitin tests depend upon the presence of free antigens of the ABO groups. The precipitate, or interfacial ring, develops if antigens are present to react with antibodies formed in rabbits against laked human blood of known ABO groups. Boyd and Boyd (1937) found that antiserum against human muscle did not prove successful for precipitin tests. These researchers tested 94 Egyptian mummies, with only 4 positive and 1 doubtful result. They found only 2 positive and 7 doubtful cases in 159 American ancient specimens. Some of the positive-reacting specimens also gave nonspecific precipitations with normal serum.

PROCEDURES
These procedures are based on Boyd and Boyd (1937).

1. Extract 0.05 g of material with 0.4 to 0.5 ml saline at 3 °C for 48 hours.
2. Remove supernatant and dilute 1:4 to reduce specific gravity below that of serum.
3. Test against immune rabbit serum prepared by injecting laked human blood of known blood group.
4. An interfacial ring of precipitin marks a positive reaction.

PRECAUTIONS NOTED
Failure of the supernatant from prepared mummified specimens to form a precipitin ring with known antisera to types A or B does not necessarily mean a positive reaction for type O unless precipitation occurs with anti-O serum. Either A or B antigen or both may have deteriorated or be too weak to react. In like

manner, a negative reaction to type O antiserum does not rule out type O for the specimen. Boyd and Boyd (1937) suggested that A may be less stable than B over a long period of time.

MIXED-CELL AGGLUTINATION TEST

Because insufficient blood samples are preserved in ancient mummified tissues, the precipitin tests are not readily adaptable. Coombs et al. (1956) described a method to identify blood group antigens on tissues devoid of blood. This has been termed the mixed-cell agglutination test. The procedure is based on the idea that tissue cell surfaces contain blood group antigens capable of capturing and retaining specific antibodies. Suspensions of isolated tissue cells incubated with erythrocytes of specific A, B, or O will show aggregates induced by interactions of the specific antigen with the specific antibody. The method was successfully used by Otten and Florey (1963) in typing Chilean mummy tissue. Later Lippold (1971) utilized the technique of mixed-cell agglutination to demonstrate the continuity of ABO types between living populations and aboriginal specimens from prehistoric populations of the Aleutian Islands, southwestern United States, Peru, and Chile. In addition to dried tissues, skeletal remains, especially from vertebral bodies, appear to give adequate results with this agglutination test.

Stastny (1974) reported on HL-A antigen profiles in pre-Columbian mummies from 500 to 2,000 years old. Patterns of reactions to the mixed-cell agglutination test were consistent with known relations between HL-A antigens. Distribution was similar in present-day descendants of the ancient populations studied. Allison et al. (1976) found mixed-cell agglutination of fresh or mummified tissue to be equally reliable as agglutination inhibition and induction of antibody production. However, agglutination inhibition would not establish the presence of H(O) antigen.

PROCEDURES
These procedures are based on Otten and Florey (1963).

1. Mummified tissues are shredded and washed gently in saline to remove dust and dirt. Shreds are cut into small pieces, microblended, and incubated at 37 °C for 45 minutes in 2.5 percent trypsin solution to isolate the cells by digestion.

2. Clumps of material are removed, and the free cell suspension is washed three times in physiologic saline by centrifugation and decantation.

3. Washed cells are incubated in the refrigerator at 4 to 5 °C for 1 hour in 5 percent bovine albumin and then washed once again in saline.

4. Cell suspensions are divided into three aliquots. Each is incubated for 2 hours at 4 to 5 °C, one with 2 drops of anti-A antiserum, one with 2 drops of anti-B antiserum, and one with 2 drops of saline extract of *Ulex europaeus* seeds. (The last is prepared by washing 20 to 30 ml fresh seeds in tap water followed by physiologic saline. Seeds are ground in a blender by adding saline until the mixture has the consistency of thick pea soup. The mixture is incubated for 4 to 6 hours at 37 °C and stored in the refrigerator for 24 hours. The mixture is spun at high speed to separate out the clear greenish yellow fluid. A polypore filter may then be used. Sodium azide to 0.01 percent is added to prevent bacterial growth. Specificity will be improved if the mixture is left in the refrigerator for 2 days. The white precipitate that forms can be removed by centrifugation. Titer should be at least 1:32 using pooled group O erythrocytes. Retiter if the reagent is stored frozen.)

5. Red cells of A, B, and O are trypsinized with 1 percent crude trypsin in saline. One part trypsin to nine parts of 2 percent cell suspensions are prepared and incubated 10 minutes at 37 °C, followed by washing the cells until the supernatant shows no agglutination of detector erythrocytes (usually three washings). Each tissue cell aliquot prepared in step 4 is subdivided into three aliquots. Trypsinized A_1 erythrocytes (1 drop) are added to one aliquot of each set, B erythrocytes to the second, and O erythrocytes to the third. The mixtures are incubated at 37 °C for 10 minutes.

6. The nine tubes from step 5 are centrifuged lightly for 1 minute to improve the microscopic detection of red cell aggregates near the bottom. A clearly defined positive will have only one tube of the nine showing aggregates. A or B tissue cells may also demonstrate H specificity in theory. (B erythrocytes with low or no H receptors should be selected as detector cells.)

PRECAUTIONS NOTED

Hair fragments or fragments of hair follicles tend to absorb agglutinins nonspecifically and should be avoided. The inner epidermal layer (stratum granulosum) is more endowed with A, B, and H antigens than other layers of skin. Buccal, lingual, esophageal, or gastric mucosa, or any area of mucus-secreting cells of the respiratory tract, increases success in typing (Szulman 1960, 1962). However, the nonsecretor phenotype may have low levels of mucus-bound antigens. High levels of H antigen could lead to false assignment of blood group O. However, the H antigen is more abundantly distributed in O types and therefore minimizes erroneous assignments.

Negative results may appear if the tissue had undergone sufficient bacterial decomposition. The most common bacteria are of the species *Bacillus*, which act only on H(O) substance (Iseki 1962). However, it is assumed that all antigen types would be equally destroyed so that AB types would not be confused with A or B. Fresh cadaver tissue of known blood groups should be used as controls. All materials used in the test (forceps, scissors, etc.) should be sterilized before each test.

AGGLUTINATION-INHIBITION TEST

Appropriate known red cell types agglutinate according to the unknown antisera tested and deductions are made from the test results. These reactions can be inhibited by first mixing the antisera with specific antigens. If the investigator utilizes known red cell types and known antisera then unknown antigens can be detected by inhibition of the agglutination reaction. Specific antigens will inhibit agglutination by like known antigens only. Testing of tissue cells, even if ancient, depends on the recognition that cells of many tissue types contain surface antigens of the ABO systems. Direct precipitin tests would depend on removal of the antigens from unknown tissue, and the titer endpoints would be difficult to determine. Incubation of tissue with known antisera and removal of the absorbed complexes with tissue material, however, still allows good titers of nonabsorbed antibodies with detectable reduction of the specific antibodies that were absorbed.

PROCEDURE

This procedure is based on Allison et al. (1976).

1. Serial double dilutions of anti-A and anti-B sera are made from 1:2 through 1:2,048. These dilutions are titered with 1 drop of 2 percent saline suspensions of group A cells (in anti-A tubes) and group B cells (in anti-B tubes). The last tubes showing strong agglutination (1+) are taken as endpoints. An eightfold concentration of the titers is used to dilute the original antisera.

2. Tissues to be tested are pulverized by grinding with sand, and 500 mg are di-

vided equally into two tubes. The tissue mixture is moistened with 4 to 5 drops of sterile physiologic saline, and 0.9 ml of diluted antiserum (anti-A or anti-B) is added. Tubes are stoppered and left at 5 °C and 10 °C for 48 hours with shaking at 12-hour intervals.

3. Tubes from step 2 are serofuged until a clear supernatant is obtained. Supernatants are diluted to the original 0.9 ml volume with saline.

4. Serial dilutions from 1:2 to 1:64 are made from the A and B supernatants. One drop of 2 percent group A (or group B for that set) is added to each tube. Tubes are shaken five times by hand.

5. Mild centrifugation assists in reading agglutination. Doubtful agglutinations are read microscopically.

6. If group A, but not group B cells agglutinate, the tissue is considered type A. Conversely, type B tissue is identified if group B but not group A agglutinates. No agglutination in either A or B leads to a designation of type AB tissue; agglutination of both A and B points to type O tissue.

PRECAUTIONS NOTED

Failure of tissue supernatants to absorb A and B agglutinins would lead one to assign the tissue to group O, but this result could also be obtained when A and B antigens are absent owing to destruction as well as by genetic design. The positive finding of A, B, or AB antigens, however, can be taken as valid.

Titer reduction may occur with deterioration of antisera during the test period if proper controls from cadavers of known types are not run simultaneously. Improper refrigeration or bacterial growth could alter results. In addition, nonspecific absorption may occur. False positive agglutination may occur if tissue matter is present in the serofuged supernatant from pulverized cell samples. Even inert materials such as charcoal, kaolin, and benzonite can cause significant antibody

titer reduction (Theime and Otten 1957). Variable particle size of different bone specimens can affect the absorption rate and degree. Fine ground bone quickly loses its antigenicity if not properly sealed in a vacuum container. In addition, a multitude of plant and animal contaminants can result in false inhibition reactions (Salazar Mallen 1951).

Boyd and Boyd (1937) early recognized that antibody serum strength was important for proper results. The absorbing power of mummy tissue is usually very weak, and not all agglutinin can be removed from serum with a high antibody titer. They recommend beginning with a dilution of the antibody serum to 1:8 rather than 1:32, as used by Allison et al. (1976). Caution should be noted with fine powdered bone because it will nonspecifically absorb both anti-A and anti-B antibodies of the dilute serum and the specimen may be thought to be AB. If the sera are too strong, insufficient absorption may occur and the specimen may be incorrectly called type O (Candela 1936).

When both anti-A and anti-B are removed, indicating AB tissue antigens, tissue can be tested for inhibition of diluted anti-chicken serum against chicken red cells. No removal of these antibodies supports immunological specificity. When mummified tissue appears to be group O, the question of whether Schiff's nonsecretor agglutinogens are present becomes important. Boyd and Boyd (1937) tested known nonsecretors and found the proper agglutinogens in muscle tissue.

ANTIBODY INDUCTION WITH PRECIPITIN TESTS

Allison et al. (1976) introduced an antibody-induction test for A, B, and H antigens by immunizing adult New Zealand white rabbits by injecting the antigens mixed with Freund's adjuvant. Highest titers were reached at 3 weeks, and blood group determinations agreed with those based on the agglutination-inhibition technique. The antibody-induction method, however, offers the

advantage of H antibody production and measurement that are in agreement with those obtained by the mixed-cell agglutination technique.

PROCEDURE

Based on Allison et al. (1976).

1. Freund's adjuvant is prepared by grinding 50 mg of heat-killed human tubercle bacilli with 1.0 ml sterile mineral oil. This mixture is stirred in 29 ml of light mineral oil.
2. Mummy tissue (ratio of 100 mg to 2.0 ml sterile 0.9 percent saline) is allowed to soak for 4 days or longer. Adjuvant (0.5 ml per 100 mg tissue in 2.0 ml saline) is added to the tissue suspension and stored at refrigerator temperature for 2 days with periodic shaking.
3. Before injection of rabbits, blood is collected from ear veins and the serum tested to ensure that no initial A or B agglutinins are present. Rabbits are then injected intramuscularly with 0.5 ml of antigen-adjuvant once weekly for 3 weeks. Control animals are injected with 0.5 ml saline plus adjuvant.
4. Rabbits are bled prior to each injection to determine antibody titers. Three sets of eight tubes each are prepared from rabbit serum. Each set is subjected to a twofold serial dilution. A 2 percent solution of human red blood cells of known A, B, or O type is prepared and 1 drop of each type is added to a set of tubes, respectively. All tubes are shaken gently and checked for agglutination macroscopically and microscopically. A threefold rise in titer is considered significant for detection of specific agglutinins.

PRECAUTIONS NOTED

As in other techniques, the antigen for A, B, and H(O) must be present for antibody induction in rabbits. Absence does not imply that the antigen was not there earlier. However, this technique has the advantage of giving a positive result for identification of O type individuals. The relatively long incubation periods of 4 days, 2 days, and so on require careful attention to sterile techniques so that bacterial or other degradation of the desired molecules does not occur. The final step in the procedure is a precipitin-type test. The same precautions for selection of A, B, and O type red cells and reading the degree of precipitation should be recognized.

HL-A ANTIGEN SYSTEM OF TYPING

This technique utilizes a modified cytotoxicity inhibition employing antigens of the HL-A system. The procedure, therefore, is an agglutination-inhibition test. Stastny (1974) found a similarity between prehistoric and present-day American Indian populations in terms of 14 HL-A antigens. The HL-A profile of coastal Peruvian mummies matched well that of American Indians in general. They resembled the Ixils of Guatemala and the Quechuas of the Peruvian highlands with respect to W28 and W27 antigens and the Aymaras of Chile with respect to HL-A5 antigens. This was, admittedly, a small group of mummies tested, but the work does reflect the usefulness of application of agglutination inhibition to typing other than the ABO system.

PROCEDURE

This procedure is based on Stastny (1974).

1. Tissue selected is shredded in a homogenizer and passed through a stainless-steel sieve to obtain a fine powder. Each powder (200 mg) is extracted twice in 2 ml sterile saline with constant stirring at 37 °C for 30 minutes.
2. The two extracts are combined and dialyzed against distilled water at 4 °C. Protein concentration is determined by standard methods.
3. Each sample is freeze-dried and dissolved in fresh rabbit serum to 8 mg per milliliter. Working solutions of 2, 4, and 6 mg per milliliter are made for testing.
4. Antiserums against known HL-A anti-

gens are selected and diluted to the optimal concentration (highest dilution that will still destroy 100 percent of target cells). Target lymphocytes are obtained from typed reference donors. Lymphocytes are prepared by flotation on Ficoll-Hypaque and frozen in small portions with liquid nitrogen.

5. Fluorescein histocompatibility is used for analysis of cytotoxicity.

6. Fluorescein-labeled lymphocytes remain visible if inhibition of cytotoxic antibodies occurs. This inhibition is expressed as a percent of control lymphocytes without antibody. Tissues with at least 50 percent inhibition are considered positive for the presence of that particular antigen.

PRECAUTIONS NOTED

Some variation between duplicate determinations exists so at least two samples should be tested. Controls should consist of antiserum without mummy antigen, antiserum with mummy antigen but without complement, complement alone, and antigen alone. Standard inhibition with lymphocytes known to contain HL-A antigen should also be tested. Skin is probably the best tissue to test, but several different tissues should be run.

Glycoproteins, polysaccharides, lipopolysaccharides, and streptococcal M1 protein have been reported to inhibit antibodies to the HL-A series (Hirata and Terasaki, 1970; Mittal et al. 1973). Contamination may be a factor in some of the tests. Antigens also may have been denatured over time, so negative results should not be considered conclusive.

M-N GROUPS

M-N antigens are rather labile and should be tested for as near to time of blood sample drawing as possible. Matson and Roberts (1949) took careful precautions for refrigeration and transport to preserve the samples for testing. Boyd and Boyd (1934a, 1934b, 1937) were not able to demonstrate M-N antigens in muscle tissue even with fresh or freshly dried

cadaver material. It appears that none of the procedures described above, with possible usefulness to identify other groups, can have any hope of detecting M-N groups from mummified tissue.

REFERENCES

Allison, M. J.; Houssaini, A. A.; Castro, N.; Munizaga, J.; and Pezzia, A. 1976. ABO blood groups in Peruvian mummies: an evaluation of techniques. *American Journal of Physical Anthropology* 44:55–62.

Boyd, W. C.; and Boyd, L. G. 1933. Blood grouping by means of preserved muscle. *Science* 78:578–9.

– 1934a. Group specificity of dried muscle and saliva. *Journal of Immunology* 26:489–94.

– 1934b. An attempt to determine the blood groups of mummies. *Proceedings of the Society of Experimental Biology and Medicine* 31:671–2.

– 1937. Blood grouping tests on 300 mummies. *Journal of Immunology* 32:307–19.

Brues, A. M. 1954. Selection and polymorphism in the ABO blood groups. *American Journal of Physical Anthropology* 12:559–606.

Camp, F. R., Jr.; Conte, N. F.; and Brewer, J. R. 1973. *Military blood banking 1941–1973: a monograph*, pp. 2–24. Fort Knox, Ky.: U.S. Army Medical Research Laboratory.

Candela, P. B. 1936. Blood-group reactions in ancient human skeletons. *American Journal of Physical Anthropology* 21:429–32.

Coombs, R. R. A.; Bedford, D.; and Rouillard, L. M. 1956. A and B blood-group antigens on human epidermal cells demonstrated by mixed agglutination. *Lancet* 270:416–63.

Hirata, A. A.; and Terasaki, P. I. 1970. Cross-reactions between streptococcal M proteins and human transplantation antigens. *Science* 168:1095–6.

Iseki, S. 1962. Blood group specific decomposing enzymes from bacteria. *Bibliotheca Haematologica* 13:215–18.

Landsteiner, K.; and Witt, D. H. 1926. Observations on the human blood groups. *Journal of Immunology* 11:221–47.

Lippold, L. K. 1971. The mixed cell agglutination method for typing mummified human tissue. *American Journal of Physical Anthropology* 34:377–83.

Matson, G. A.; and Roberts, H. J. 1949. Distribution of the blood groups, M-N and Rh types among Eskimos of the Kuskokwin Basin in western Alaska. *American Journal of Physical Anthropology* 7:109–22.

Mittal, K. K.; Terasaki, P. I.; Springer, G. F.; Desai, P. R.; McIntire, F. C.; and Hirata, A. A. 1973. Inhibition of anti-HL-A alloantisera by glycoproteins, polysaccharides and lipopolysaccharides from diverse sources. *Transplant Proceedings* 5:499–506.

Otten, C. M.; and Florey, L. L. 1963. Blood typing of Chilean mummy tissue: a new approach. *American Journal of Physical Anthropology* 21:283–5.

Salazar Mallen, M. 1951. Estudio immunologico de restos oseos antiguos. *Gaceta Médica de México* 81:122–7.

Stastny, P. 1974. HL-A antigens in mummified pre-Columbian tissues. *Science* 183:864–6.

Szulman, A. E. 1960. The historical distribution of blood group substances A and B in man. *Journal of Experimental Medicine* 111:785–800. 1962. The historical distribution of blood group substances in man as disclosed by immunofluorescence. *Journal of Experimental Medicine* 115:977–96.

Theime, F. P.; and Otten, C. M. 1957. The unreliability of blood typing aged bone. *American Journal of Physical Anthropology* 15:387–97.

Index

THE ROUTLEDGE

ATLAS

OF

AFRICAN AMERICAN

HISTORY

Routledge Atlases of American History

Series Editor: Mark C. Carnes

The Routledge Historical Atlas of the American Railroads
John F. Stover

The Routledge Historical Atlas of the American South
Andrew K. Frank

The Routledge Historical Atlas of Women in America
Sandra Opdycke

THE ROUTLEDGE
ATLAS
OF
AFRICAN AMERICAN
HISTORY

JONATHAN EARLE

MARK C. CARNES, SERIES EDITOR

ROUTLEDGE

NEW YORK AND LONDON

Published in 2000 by
Routledge
29 West 35th Street
New York, NY 10001

Published in Great Britain in 2000 by
Routledge
11 New Fetter Lane
London EC4P 4EF

Text copyright © 2000 by Jonathan Earle
Maps and design © 2000 by Arcadia Editions Ltd.

Printed in the United Kingdom on acid-free paper.

10 9 8 7 6 5 4 3 2 1

Library of Congress Cataloging-in-Publication Data

Earle, Jonathan
    The Routledge atlas of African American history / Jonathan Earle.
        p. cm. -- (Routledge atlases of American history)
    Includes bibliographical references and index.
    ISBN 0-415-92136-8 (acid-free paper) -- ISBN 0-415-92142-2 (pbk. : acid-free paper)
    1. Afro-Americans--History. 2. Afro-Americans--History--Maps. I. Title: Atlas of
African American history. II. Title. III. Series.

E185 .E125 2000
973'.0496073--dc21   99-059713

For James Patrick Shenton,

who taught me to love history.

Contents

Foreword

Slavery is the central—arguably the defining—dilemma of the American nation, the first large experiment in democracy. Democracy presumes equality—why else tot up votes to determine a majority?—but from its inception, the American nation countenanced a slave system that defined some human beings as less than human. "That all men are created equal," Thomas Jefferson wrote in the Declaration of Independence, was a "self-evident" truth. Yet he justified owning human beings, degrading them, and exploiting their labor on the grounds that African Americans were "inferior to whites in the endowments both of body and mind."

Slavery and its repercussions are central to this volume, the fourth in the Routledge Atlases of American History. In this atlas Jonathan Earle, a history professor at the University of Kansas, author of *The Undaunted Democracy: Jacksonian Antislavery and Free Soil: 1824–1854*, seeks to show—to demonstrate visually—that slavery was not merely a constellation of ideas, but a cultural institution that left a deep imprint, literally and figuratively, upon the American nation.

Thus the earliest maps in the atlas focus on Africa, where a slave trade antedated the arrival of Portuguese traders in West Africa in the 15th century. The atlas emphasizes the economic forces that caused hundreds of thousands of Africans to be shipped to the Americas, where legal and political arrangements were devised to perpetuate their subjugation.

Earle shows how the slave system in the South matured along the Atlantic seaboard, and then spread westward with the diffusion of cotton cultivation in the early 19th century. This social migration transformed American politics as well, resulting in the fateful stalemate that led to the Civil War. After delineating the slave system in its entirety, Earle looks at the phenomenon close up: from the vantage point of a slave dwelling in one set of maps; from the perspective of a single plantation in another. Earle considers as well the aftermath of slavery, ranging from post-Civil War Reconstruction to Rosa Parks and the Montgomery Bus Boycott nearly a century later.

Slavery was indisputably harsh and degrading; but Earle sides with scholars who contend that slaves were not powerless victims. He examines in gripping detail instances where African Americans defied the slave system: the Stono Revolt in 18th-century South Carolina, Nat Turner's Rebellion in 19th-century Virginia, the flight of thousands of runaway slaves, the exploits of African Americans who fought in the Union Army. Liberation came decades earlier for slaves in the North, where there emerged a distinctive African-American culture, expressed through African-American churches, newspapers, and other unique cultural institutions.

Thus while slavery was central to the evolution of this nation, and to the experiences of African Americans, it did not define those experiences. The history of African Americans is not the history of American slavery: it is something far greater. Earle's special challenge has been to delineate this larger realm of experience.

To that end, he has explored many topics that are commonly overlooked: Marcus Garvey's attempts to repatriate African Americans to Africa, the major role of African-American soldiers in 20th-century wars; the evolution of the Negro Baseball Leagues; the rise of the Nation of Islam; the emergence of the Harlem literary Renaissance; the birth of jazz.

This is a large and complex story: Earle tells it by joining a wealth of information, conveyed through maps, graphs, tables, and pictures, with a concise and authoritative narrative. This atlas will prove indispensable to educators and scholars, and absorbing to those interested in coming to terms with this nation's past, and with its destiny.

Mark C. Carnes
Professor of History
Barnard College, Columbia University

Introduction

> Hither too came the Negro. From the first he was the concrete test
> of that search for Truth, of the strife toward a God, of that body of
> belief which is the essence of true religion. His presence rent and
> tore and tried the souls of men.
> —W. E. B. DuBois, 1924

This small volume strives to tell, with maps, the story of the peoples of African descent in America. The story is a complex one that could easily fill the pages of a dozen atlases, monographs, and textbooks. Instead of attempting to be encyclopedic, then, I tried to focus on elements of African American history that could be enhanced with maps. When executed well and supported with appropriate text, images, and tables, maps can make history more understandable and—in the best circumstances—bring people's stories to life. This is precisely what I have attempted here.

With this in mind, the atlas begins centuries before the arrival of Africans in North America. The first part, "The Roots of Black America," begins by documenting the rich and varied history of African peoples, including the mighty kingdoms of Mali and Songhay. Other maps trace the progression of both the internal African and trans-Atlantic slave trades, which scattered millions of Africans across the globe to toil for others. In the New World, these African immigrants encountered societies in rapid transformation. During the 50 years after the first Africans arrived in Virginia, for example, blacks witnessed their status as unfree laborers (a condition shared by numerous others, including white indentured servants) deteriorate into one of life servitude for themselves and their offspring. Maps illustrate how these mostly African-born blacks developed their own cultures of work, belief, family, and identity. A map of the Stono slave revolt of 1739 shows that slaves did not watch passively as the slave system solidified around them. Stono and the resulting murder of the rebellious slaves starkly demonstrated the growing fear among whites—especially in black-majority colonies like South Carolina—that slaves might rise as one to throw off their chains and seek vengeance. A final map functions as a snapshot of the black population on the eve of the American Revolution, at a time when American-born blacks (or Creoles) had begun to emerge as a dominant force in the community. Unlike slave communities in the Caribbean or in the early years of North American settlement, the slave population in what would soon become the United States began an uninterrupted process of natural increase. This attribute would become extremely significant during the early history of the American republic.

The second part, "Two Communities, Slave and Free," focuses on the years of the new American nation when two black communities—separate but forever linked by heritage, family, and oppression—developed and flourished. Following recent scholarship, the maps attempt to show a people forging a new culture, one rooted squarely in the African past but also incorporating strong American elements. Free blacks, increasingly concentrated in the Northern

states, struggled for rights granted to whites and founded institutions to help stem a rising tide of racism. Maps focus specifically on two of these institutions: the African Methodist Episcopal Church and the black press. Other maps document the abolition of slavery in the North and, in what at first seems like a contradiction, the rise of legal discrimination there. In the South, where slavery became immensely profitable and expanded rapidly into new Western lands, enslaved blacks developed their own religion, traditions, and culture to endure the hardships of thralldom. Four diverse maps illustrate the economy of plantation slavery, and two others focus on slave life, both in the fields and in the quarters. Finally, a detailed map of Southampton County, Virginia, documents the slave rebellion led by Nat Turner. Turner and his band murdered every white person they could find over two nights in 1831, striking a powerful—and bloody—blow for freedom.

"Toward Freedom," the third part, follows African American resistance to slavery through its destruction during the Civil War. One map illustrates how free blacks and their white allies helped build a loosely organized but effective "Underground Railroad" to assist slaves escaping northward, often with the help of "conductors" like the ex-slave Harriet Tubman. Another map details the escape routes taken by several runaway slaves like Frederick Douglass, who became the most famous black orator and author of his day, and William and Ellen Craft, who escaped from Georgia by impersonating a white slaveholder and her trusted servant. Success stories like Douglass's and the Crafts' were rare, however, so the same map also includes the route used to return apprehended fugitive Anthony Burns to slavery and the harrowing journey of the kidnap victim Solomon Northrup, who spent 12 years as a slave before returning home to New York. Maps in this part also detail how attempts to "compromise" over the issue of slavery and its expansion failed—repeatedly—between 1819 and 1857, leading to the Civil War. Once the war started, the escape to Union camps by thousands of runaway slaves (called "contrabands") forced Union leaders to see the possibilities of a war for freedom and forced Abraham Lincoln to issue the Emancipation Proclamation. After the war ended, freed slaves in the South won civil and political rights, and a map illustrates the first elected black officials during Reconstruction. Such victories were fleeting, however, and Reconstruction ended when Union troops ended their occupation in 1877. Soon thereafter,

Born a slave in Maryland, Frederick Douglass became one of the leading figures in African American history. A brilliant speaker and writer, he worked tirelessly for the abolition of slavery, equal justice for blacks, and universal human rights abroad.

facing intimidation, violence, and murder, the first significant wave of African American migrants left the South to found all-black towns in Kansas, Oklahoma, and California.

Part Four, "African Americans Under Arms," temporarily breaks the atlas' chronological organization and traces black participation in the military from the Revolution through recent action in Haiti and the Balkans. Since colonial times, blacks have volunteered for service believing that military action during wartime was a path to greater freedom and opportunity for themselves and their people. At no time was this belief more widely held than during the Civil War, when thousands of African Americans volunteered to combat slavery and the Confederacy by fighting in the Union Army. Frederick Douglass called the decision to use black troops—forbidden before 1863—a "golden opportunity . . . [for blacks to] win for ourselves the gratitude of our country, and the blessings of our posterity through all time." Black valor and courage among soldiers in units like the 54th Massachusetts helped to destroy slavery and the Confederacy, but failed to translate into better treatment for African Americans in the military and in the larger society. This pattern was evident in military actions over the next 130 years: blacks joined up to prove their mettle in battle and to open new paths of opportunity after the shooting stopped; yet time after time black soldiers mustered out only to encounter racism and bitter disappointment. Several maps in this part detail the use of African American soldiers during the era of the segregated armed forces, including the Buffalo Soldiers, the Spanish-American War, and World Wars I and II. After 1948, when President Truman desegregated all U.S. armed forces, blacks served alongside whites both in battle and in peacetime. Additional maps illustrate black soldiers' involvement in conflicts in Korea, Vietnam, and the Persian Gulf and in peacekeeping missions in Somalia and Haiti. Although great strides toward racial equality have been made in the integrated military, African

Blacks lobbied hard for the right to fight against the Confederacy in the Civil War. On July 18, 1863 the 54th Massachusetts Infantry, the first black regiment raised in the North, led a courageous and bloody assault on Fort Wagner, part of the network of defenses surrounding Charleston, SC.

Americans still cite numerous examples of racism and racist behavior as impediments to their advancement.

Part Five, "The Struggle for Equality," takes up where Part Three left off—with the use of terror and murder to intimidate African Americans and trample their constitutional rights. Maps detail the disfranchisement of black voters and the systematic enactment of racist laws (called "Jim Crow" laws) to separate the races. One prominent scholar called the period after Reconstruction "the nadir" of American race relations, yet it was also a time of building and achievement. Maps illustrate the founding of hundreds of black colleges and universities and the stunning success of segregated Negro Leagues baseball. Another map revisits the Great Migration of blacks to Northern and Midwestern cities. The chapter then shifts focus to the civil rights movement and the destruction of Jim Crow, with maps focusing on the Montgomery bus boycott, the rise of the sit-in movement, and the struggle for voting rights. A final map links African decolonization in the 1950s and '60s with the rise of black nationalism, an ideology pioneered in the 19th century and perfected by Marcus Garvey in the 1920s and Malcolm X in the 1960s.

The final part, "The African American Community," again breaks out of the chronological structure. The chapter initially focuses on the formation of a distinct African American culture in the 20th century and its impact on American culture as a whole. A map of the Harlem Renaissance details the block-by-block growth of the "black Mecca" in the 1920s, and demonstrates the cultural influences of authors like Zora Neale Hurston, poets like Langston Hughes, political leaders like Marcus Garvey, and institutions like the Apollo Theater on 125th Street. The literary voices of Harlem had a tremendous impact on the world of letters, just as the "voices" of jazz and blues greats like Louis Armstrong and Robert Johnson profoundly affected the world of music. The black musical tradition also includes the development of genres like rock 'n' roll, soul, and hip hop. Another map attempts to illustrate the diverse origins of African American intellectuals, writers, musicians, and entertainers. The remaining maps in the section detail recent political and demographic changes in the African American community. One focuses on the realignment of black voters from reliable Republicans to die-hard Democrats in the 1920s and 1930s. Another details the urbanization of the black community and spotlights the election of African American mayors. A final map features a snapshot of the African American community as recorded by the 1990 U.S. Census. That study revealed an African American community that continues to be mobile, complex, and influential.

No atlas can present a finished representation of a people, especially one as varied and diverse as Americans of African descent. Experts and amateurs alike will no doubt uncover errors, omissions, and arguments with which they passionately disagree. Yet I hope the picture of the American experience presented here—using maps to tell African Americans' stories—reveals to each interested reader something about a moving and fascinating history.

PART I: THE ROOTS OF BLACK AMERICA

The first object which saluted my eyes when I arrived on the [West African] coast was the sea, and a slave ship which was then riding at anchor and waiting for its cargo . . . I was soon put down under the decks, and there I received such a salutation in my nostrils as I had never experienced in my life: so that I was not able to eat . . . but soon, to my grief, two of the white men offered me eatables, and on my refusing to eat, one of them held me fast and laid me across the windlass, and tied my feet while the other flogged me severely.
—Olaudah Equiano, 1756

In 1619, 20 black people splashed ashore at Jamestown in the British colony of Virginia. A Dutch captain had brought them from Africa and sold them to the English colonists in exchange for water and supplies. The arrival of these Africans was a momentous event in American history, even if it was only briefly noted by Jamestown's leader: human beings from three continents—Africa, Europe, and America—together occupied a ribbon of land on the southeastern coast of North America. Their futures would be intertwined forever after.

Historians cannot answer even the most basic questions about these first black inhabitants of North America. What were their ethnic backgrounds? What was their legal status? How many were men? Women? The Englishman left these important details out of his journal. But one thing is certain: there was more on board that Dutch frigate than human cargo. The people disembarking brought with them the diverse social, linguistic, economic, religious, and political traditions from their African homeland. It was a homeland that was likely the birthplace of humankind—excavations in the Olduvai Gorge in northern Tanzania suggest that hominids began using tools there about 2 million years ago. More than that, it was a homeland that witnessed the emergence and development of vast civilizations, sophisticated cultures, and diverse peoples. Over time, the descendants of those first humans created literally hundreds of separate cultures in Africa. Those cultures varied as widely as the continent's geography. More important, many of these various cultures remained in substantial contact with each other, adding to the diversity. Anthropologists and linguists estimate that modern Africa contains more than 800 linguistic and ethnic groups, each with its own unique culture and religion.

This substantial African past was transported across the Atlantic along with the people aboard that Dutch ship and the thousands that followed. What's more, scholars like Carter G. Woodson, John Blassingame, and Albert Raboteau have demonstrated that this African cultural heritage continues to influence American (not just African American) life to this day. It survives in contemporary American religion; in folk tales told to children; in English words like *banjo*, *yam*, and *canoe*; and in the way we work, play, and congregate socially.

Finally, there was something unique about the Africans who came to America. It was not their status as unfree laborers: thousands of Europeans

crossed the Atlantic as indentured servants and bound laborers, and the labor of Native Americans was similarly coerced in the New World. And as this section points out, slavery and the slave trade existed in Africa before Europeans arrived with their ships in the 15th century. The Africans in the New World were unique because, with very few exceptions, they did not choose to make the trip. The African dispersion was thus forced on unwilling emigrants who, as the 17th century wore on, were legally defined as chattel and denied basic rights. This, as much as their shared African heritage, influenced the formation of African Americans as a people.

For the first hundred years after the arrival of those first Africans at Jamestown, African-born people dominated the emerging slave society in colonial British America. But by the mid-18th century, American-born blacks were forging a new culture, one that combined the African heritage of their ancestors with the New World in which they were born. This process was by no means smooth or easy: at times, as with the Stono rebellion of 1739, it was outright brutal. But by the 1760s these "new" people had laid the foundations of black America.

Slaves at work on a tobacco farm in Virginia. Tobacco production involved indentured servants and slaves in a complex and specialized work process as they planted, cut, cured, stripped, and packaged the leaf for export.

The African Past

The Empires of Africa, 900–1500

African societies developed a large variety of political systems, ranging from loosely structured groups to immense and powerful empires. Anthropologists date the emergence of states in the Nile River valley to about 3500 B.C. Of these ancient societies, Egypt was the most significant. Ancient Egypt's contributions to African, European, and Middle Eastern cultures are well known, especially in such areas as hieroglyphic writing, mathematics, and geography. During the medieval period, the empires of Morocco (controlled by Muslim people called Almoravids), Algiers (controlled by the Almohads), and Egypt (controlled by the Fatimids) dominated the lands of the Mediterranean in both Africa and Europe. Key to their ability to dominate North Africa was their common religion—Islam—which led to lucrative military and trade alliances.

Most African Americans trace their ancestry to West Africa, where numerous powerful and wealthy civilizations rose and fell between the 9th and 16th centuries. The earliest of these was the kingdom of Ghana, located along the grasslands of the Senegal and upper Niger rivers. Records before 900 are sketchy, but by that time Ghana was well known for its fabulous wealth and powerful army. Its economic strength was derived from its vast supplies of gold and participation in an active trans-Saharan trade in items like salt and copper. Ghanaian scholars made advances in mathematics and astronomy, and, at the kingdom's peak in the 11th century, its leaders commanded an army of 20,000 soldiers. In 1076 Muslim Almoravids invaded Ghana, seized the capital, and made it an Islamic city. The conflicts that resulted undermined the kingdom, and it collapsed around 1200.

In its place arose the even more powerful empire of Mali, one of the world's richest and most advanced civilizations. By the early 1300s, Mali extended more than 1,000 miles inland from its border on the Atlantic Ocean, covering much of what is now francophone Africa. Travelers' accounts focus on the kingdom's prosperity, order, sophistication, and Muslim piety. When Mali's king made a pilgrimage to Mecca in 1324, his entourage—which consisted of more than 8,000 retainers, 500 slaves, and 100 camels each carrying 300 pounds of gold—sent Cairo's economy into an inflationary spiral. ("So much gold was current in Cairo that it ruined the value of money.")

In 1468 Mali was conquered by the kingdom of Songhay, which dominated West Africa for more than 300 years. The empire's greatest legacy was the system of universities in Timbuktu, Gao, Walata, and Jenne, which attracted scholars from all over Africa, Europe, and Asia. During this period several smaller West African kingdoms of note rose and fell, including Dahomey, Kanem, Benin, Wolof, Senegambia, and Asante.

Nubia and Ethiopia in East Africa provided vital links with the Mediterranean and the Middle East, and were among the first African nations to convert to Christianity. City-states and smaller nations dotted the east coast of Africa, including Somalia, Tanganyika, Katanga, and Zanzibar. What is

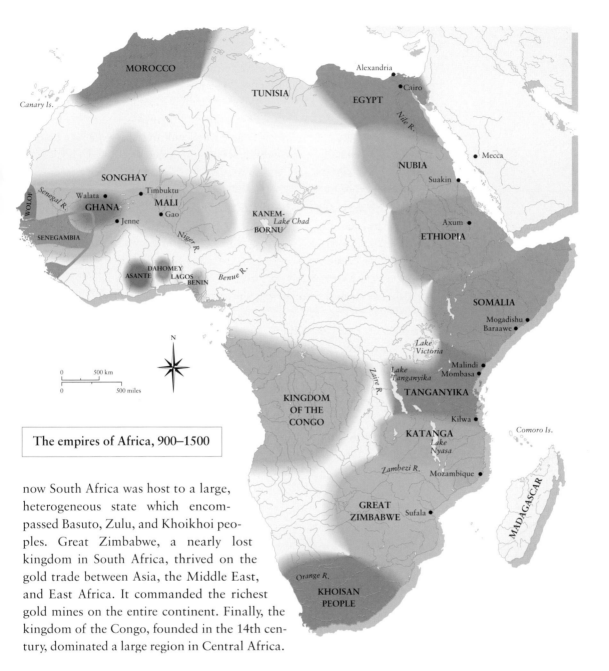

The empires of Africa, 900–1500

now South Africa was host to a large, heterogeneous state which encompassed Basuto, Zulu, and Khoikhoi peoples. Great Zimbabwe, a nearly lost kingdom in South Africa, thrived on the gold trade between Asia, the Middle East, and East Africa. It commanded the richest gold mines on the entire continent. Finally, the kingdom of the Congo, founded in the 14th century, dominated a large region in Central Africa.

The Trans-Saharan Slave Trade, 900–1500

Even before the rise of Islam in the 7th century, North African Berbers used their knowledge of the Sahara Desert's geography to control several lucrative trade routes from northern Africa and southern Europe to West Africa. This elaborate network of trade routes helped initiate the world's first "global" economy by connecting West Africa, the Mediterranean, the Middle East, and

Asia. There was considerable demand in the West African kingdoms of Ghana, Mali, and Songhay for silk, cotton, beads, salt, mirrors, and dates; in return they provided traders with gold, kola nuts, ivory, and slaves. Slaves eventually equaled and then surpassed gold as West Africa's major export.

Slavery thus had a long history on the continent before Europeans arrived in sub-Saharan Africa in the 15th century, but in a form far less brutal than the type the Europeans would impose. Slavery played a prominent part in ancient civilizations like Egypt, Greece, and Rome, where slaves were either captured or purchased to perform menial tasks. Invading Muslims would typically capture African men for military service and personal servants, and women for their harems. It was not uncommon to find black slaves in places like Persia or

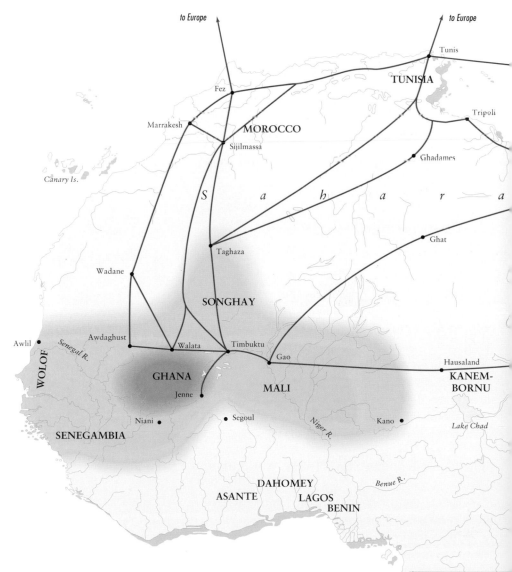

Arabia. As political leaders in West Africa converted to Islam, they often cooperated in the trade of (mostly non-Muslim) slaves.

There is little doubt that this form of African slavery was cruel and exploitative, but it had several differences from the later slave system. First, slavery had no definitive racial basis, and slaves were not isolated as a separate caste. Slaves could be whites from southern Europe, Middle Eastern people, or black Africans. Second, slaves primarily worked as servants for their owners—not as endlessly toiling agricultural laborers whose production of staple crops served to enrich slaveholders. Since there were no large-scale cotton or sugar plantations in the Muslim world, the harshness and permanent exploitation that characterized New World slavery was less apparent. Third, slavery was not always a permanent and inherited status. All of these things changed when Portuguese ships arrived on the coast of West Africa in the 15th century.

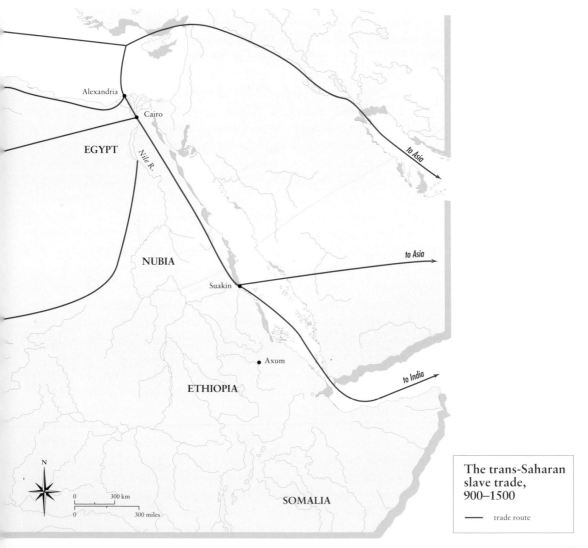

The trans-Saharan
slave trade,
900–1500

——— trade route

The Trans-Atlantic Slave Trade, 1500–1800

Slave Trade in Africa, 1750

After an enslaved captive was purchased by slave traders on the African coast, the purchaser's insignia was branded onto the slave's shoulder, breast, or buttocks.

Right: *Located on the Gold Coast of West Africa, Cape Coast Castle was one of the leading slave-trading forts between the 16th and 19th centuries. Slaves were housed in forts like these—often in cramped, subterranean dungeons— before being loaded onto ships for the dreaded passage across the Atlantic.*

By the end of the 15th century, Europeans had superceded their Arab and African counterparts and established a modern, trans-Atlantic slave trade. A commercial revolution in Europe led to the rise of powerful nation-states like Portugal, Spain, Britain, France, and Holland, as well as new ideas about competition, commodity exploitation, and the accumulation of wealth. The importing and exporting of African slaves became an accepted and profitable part of European commerce.

Portugal was the first European nation to see the economic advantages of the African slave trade. The Portuguese made the slave trade possible by exploiting rivalries among the more than 200 small states and ethnic groups of West and Central Africa. Despite their cultural similarities, West Africans neither viewed themselves as a single people nor shared a universal religion that might have restrained them from selling other Africans into slavery. Christian Europeans who believed that enslaving fellow Christians was immoral had few moral qualms about enslaving pagan or Muslim Africans.

As early as 1502, Portuguese traders were shipping West African slaves to Spanish and Portuguese colonies in the Caribbean and Brazil to work on sugar plantations. As West Indian plantations grew in size and economic importance, the slave trade mushroomed, employing thousands of persons, involving millions of dollars of capital, and beginning the largest forced migration in history. Almost immediately, Europeans (first the Portuguese, then the French, English, Dutch, Swedes, and others) began to build an extensive network of forts along the African coast. The Gold Coast, which became contemporary Ghana, contained more than 50 of these forts along its 300-mile coast on the Gulf of Guinea, including Elmina Castle (the oldest, built by the Portuguese in 1482) and Cape Coast Castle (built by the Swedes in 1653, but later captured by the British). Cape Coast Castle could hold up to 1,500 slaves in its dark, damp dungeons at the peak of the trans-Atlantic slave trade. The forts were constructed with the permission of local rulers, who were paid rent. They also had to be defended against constant assaults from rival Europeans and Africans.

The enslaved Africans lost their liberty in a variety of ways. Wars between African states provided many of the captives who fueled the slave trade (one authority claims that as many as 80 percent of slaves were prisoners of war and came from a state other than that of the seller). Other slaves had been deprived of their civil rights after being convicted of a crime. Still others were kidnap victims, although kidnappers faced severe penalties if caught. Debtors, orphans, and people who lacked kinship ties with other members of their state made up the rest of those sold to slave traders.

African traders usually marched their slaves to the coastal forts in chains, in groups of up to 150 people. Once on the coast, purchasers were allowed to examine the slaves' stature, teeth, limbs, and genitals before deciding whether or not to buy. After purchase the slave's skin was branded with the insignia of

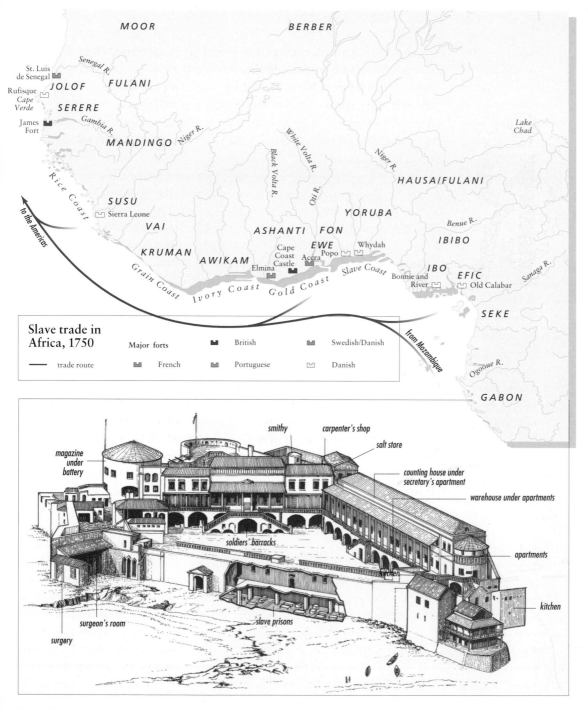

Slave trade in Africa, 1750

Major forts

━━━ trade route

British
French
Portuguese
Swedish/Danish
Danish

MOOR
BERBER
St. Luis de Senegal
JOLOF
Rufisque Cape Verde
FULANI
SERERE
James Fort
MANDINGO
Senegal R.
Gambia R.
Niger R.
White Volta R.
Black Volta R.
Oti R.
Niger R.
Lake Chad
Rice Coast
To the Americas
SUSU
Sierra Leone
VAI
KRUMAN
AWIKAM
Grain Coast
Ivory Coast
Gold Coast
Elmina
Cape Coast Castle
Accra
ASHANTI
FON
EWE
Popo
Whydah
Slave Coast
YORUBA
IBIBO
IBO
Bonnie and River
EFIC
Old Calabar
Benue R.
Samaga R.
HAUSAIFULANI
SEKE
from Mozambique
Ogooue R.
GABON

magazine under battery
smithy
carpenter's shop
salt store
counting house under secretary's apartment
warehouse under apartments
soldiers' barracks
kitchen
apartments
kitchen
surgeon's room
slave prisons
surgery

his or her purchaser. The branded slave then waited, often in subterranean dungeons, for their ships to depart for the Americas. Many died waiting.

The Africans who were brought to the North American mainland (often after spending time in the Caribbean) originated in an area bounded by the

After arriving in the Americas, enslaved Africans were sold at auction. By the mid-18th century slavery was well entrenched in all 13 British colonies in North America.

Senegal River in the north and Angola in the south. Principal ethnic groups inhabiting this area included the Yoruba, Fon, and Ibo peoples from the Bight of Biafra; the Ashanti and other Akan-speaking people from the Gold Coast; the Susu from Sierra Leone; and the Jolof, Serere, Mandingo, and Fulani from the Senegambia. Slaves from the Bakongo, Tio, and Mbundu groups from the Congo-Angola region were also brought to North America.

Destinations of African Slaves

For those who survived the cramped and dangerous confinement on the coast, the nightmare was only beginning. The one-way trip to the Americas was known as the "Middle Passage." Slaves were chained together by twos at their hands and feet, and stacked as tightly as possible in the ship's hold, "like books on a shelf." Since more cargo meant more profit, the slaves were denied standing, lying, or even sitting room. This routine practice of overcrowding led to disastrous outbreaks of disease during the voyage, which lasted from 6 to 16 weeks. Smallpox, dysentery, and various other fevers and fluxes passed quickly from person to person; modern scholars estimate the mortality rate on the ships averaged between 25 and 40 percent during the early years of the slave trade.

Traders learned very quickly to keep slaves securely chained for the duration of the voyage. Given the opportunity, many

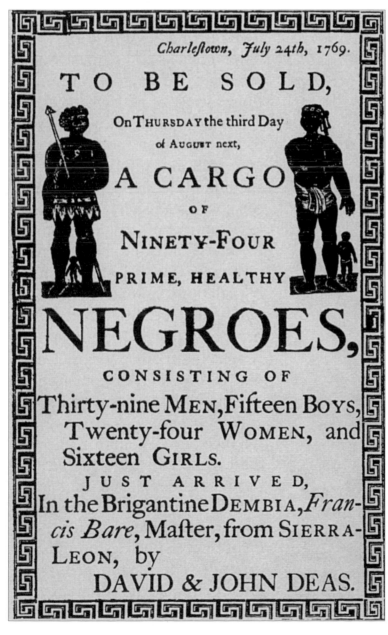

Charlestown, July 24th, 1769.

TO BE SOLD,

On THURSDAY the third Day of AUGUST next,

A CARGO

OF

NINETY-FOUR

PRIME, HEALTHY

NEGROES,

CONSISTING OF

Thirty-nine MEN, Fifteen BOYS, Twenty-four WOMEN, and Sixteen GIRLS.

JUST ARRIVED,

In the Brigantine DEMBIA, *Francis Bare*, Master, from SIERRA-LEON, by

DAVID & JOHN DEAS.

unchained slaves chose suicide over a lifetime of enslavement in the Americas. Moreover violent rebellions by slaves occurred frequently enough to keep fear in the hearts of even experienced traders. As plantations and populations in the Americas grew in size and economic importance, so too did the slave trade. As the map illustrates, most slaves departing Africa did not go to North America: a slave was almost ten times as likely to wind up in the West Indies as he was to land in the United States. After Caribbean islands like Santo Domingo and Cuba, Brazil was the most likely destination.

Between 1619 and 1808, when Congress abolished the slave trade, nearly 400,000 Africans were brought against their will to British North America and the United States. Approximately 8 million more were taken to the sugar and coffee plantations in Brazil and the Caribbean or to the mines of Spanish America.

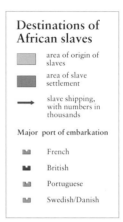

Destinations of African slaves

- area of origin of slaves
- area of slave settlement
- → slave shipping, with numbers in thousands

Major port of embarkation

- French
- British
- Portuguese
- Swedish/Danish

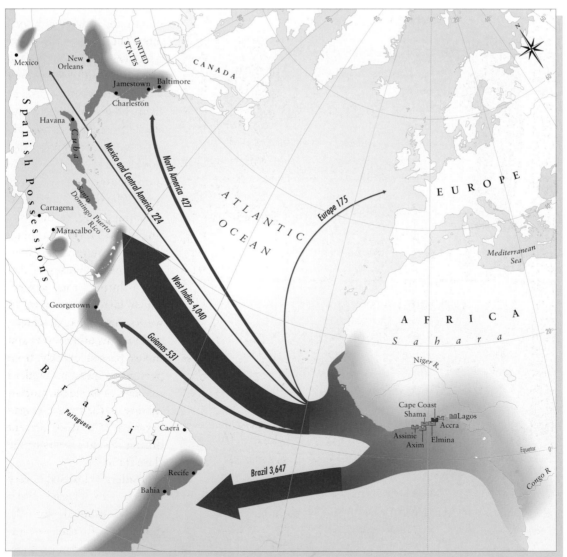

The African Dispersion

An estimated 9 million Africans were taken from their homeland aboard European ships like the one in this wood cut, which dates from the early 1700s, to become forced laborers in the Americas.

Economic development in the New World had far-reaching ramifications for both enslaved Africans and European colonizers. The commercial revolution opened up seemingly boundless markets for New World commodities like tobacco, cotton, rice, and sugar. During the 17th century, European countries and entrepreneurs scrambled to secure colonial possessions, especially in the Caribbean, to produce these profitable staples. Since European diseases and warfare had severely depleted native populations, planters experienced a severe labor shortage. A vast majority of them turned to imported African labor to clear the forests and work the plantations.

Tobacco, the first lucrative staple crop produced in the Americas, quickly glutted the market, and savvy planters began cultivating sugarcane. Sugar and its significant byproducts, molasses and rum, soon became the primary crop produced by slave labor. Slavery exploded in the Caribbean. The result was disastrous for the imported labor force: overcrowding led to severe outbreaks of communicable diseases, death rates soared, and work conditions deteriorated. But it was more profitable for planters to import newer, "fresher" laborers than to care for those already working fields in places like Cuba, Jamaica, and Barbados. With new slaves constantly arriving from Africa, Caribbean planters and overseers developed a system to "break in" newcomers. After being "broken" (in spirit as well as in the habits of freedom) slaves were sometimes re-exported to places like Mexico and North America.

Traditionally, historians have depicted the trans-Atlantic slave trade as a triangle connecting points in western Europe, West Africa, and the Americas. Slave ships first departed from European ports, sailed to Africa, brought their human cargo to the American coast, and returned home. This triangle, however, leaves out the important Caribbean "slave breaking"

Right: Engraving showing a cross section of a slave ship.

link. The slave trade really resembled a lopsided rectangle. Here is how the trade (and the slave economy) worked:

European ships (most west European countries participated in the trade) filled their holds with African slaves in "factories" like Cape Coast Castle or Elmina. In Caribbean ports they replaced most of the slaves in their holds with products like rum and molasses, and sailed for North America. In ports like Charleston and Boston, the remaining slaves and Caribbean products were traded for North American commodities like furs and fish. Traders concluded the rectangle when they unloaded their merchandise back in Lisbon or Liverpool.

The African dispersion and the "triangle trade"

→ European and American slave trade route → Arab and Ottoman slave trade route

Slavery in British North America, 1633–1755

African slave labor was vital to the economic life of British North America. During the 17th century slavery spread into each of the four main regions colonized by England: the Caribbean, the southern, the middle, and the New England colonies. The institution became most entrenched in colonies that exported labor-intensive commodities like sugar, rum, rice, indigo, tobacco, pitch, and turpentine. This meant that the African populations were largest in the Caribbean (where most sugar was produced), the Chesapeake colonies of Maryland and Virginia (which exported tobacco), and the Carolinas (which produced rice, indigo, and small amounts of cotton). In the middle colonies of Pennsylvania, New Jersey, and New York, slaves were used primarily as servants and laborers. New England had the fewest number of slaves in British America because the region was settled by Puritans, who tended to emigrate as families and work the land themselves. This did not, however, keep Puritans from entering and eventually dominating the intercolonial trade in slaves.

The transition to slave labor from labor performed by free people and indentured servants was a complex one, and spanned most of the 17th century. Virginia settlers had enslaved local Indians as early as 1610, but gave up the practice in the face of massive Indian raids and attacks. The switch to African, perpetual, race-based slavery was slow: there were only 1,600 Africans in North America in 1640, with almost a third of them in Dutch New York. During the next four decades slavery was explicitly legalized in Massachusetts (1641), Connecticut (1650), Virginia (1661), Maryland (1663), New York (1665), and South Carolina (1682). Even before it legally recognized slavery in 1663, Maryland lawmakers had mandated slavery as a lifelong condition for Africans and their children; Virginia classified slavery as a lifelong, inheritable, and "racial" status for blacks in 1670. The remaining colonies of British North America legalized slavery in the early 18th century.

With the growth of the institution of slavery in all the British colonies, lawmakers turned their attention to the regulation of slaves' lives. The resulting "slave codes" routinely forbade teaching slaves to read and write; outlawed group gatherings outside of church and contact with free blacks; and required slaves to carry written passes when not on plantation grounds. South Carolina's code, for example, defended the need for legal controls by citing the "barbarous, wild, savage natures [of Africans], rendering them wholly unqualified to be governed by the laws, customs, and practices of this Province." The statute continued: "all negroes, mulatoes, mustizoes or Indians . . . hereafter shall be bought and sold for slaves, are hereby declared slaves; and they, and their children, are hereby made and declared slaves, to all intents and purposes." Life servitude for Africans and their children was becoming a legal reality in British North America.

In Maryland and Virginia, the deteriorating status of black people extended to social relationships as well. Other states quickly followed their lead. In 1664, for example, Maryland passed a law stipulating that a white woman who married a slave would have to serve the slave's master as long as her husband lived.

Maryland's law, in turn, set the pattern for other colonies to penalize black-white marriages.

Not all blacks and slaves lived in the South or performed agricultural labor on rural plantations. Cities and towns on the eastern seaboard contained large African populations as early as the mid-17th century. As cosmopolitan population centers, these cities would always attract free and enslaved blacks and nurture movements and agitation for black rights. Philadelphia, originally settled by anti-slavery Quakers, had the largest free black population in North America. Black Philadelphians helped found the city's schools, literary societies, churches, and fraternal organizations. By 1741, slaves constituted one fifth of the population of New York City, and caused a panic that year when authorities heard that blacks were planning to burn the city and kill all the whites. Calling it the "Great Negro Plot," city leaders arrested 154 blacks and 24 whites, accusing them of conspiracy. Other cities were centers of African colonial life as well: in Baltimore, skilled and unskilled black artisans worked as dockworkers, blacksmiths, coopers, and domestics. Savannah and Charleston quickly became major centers for the importation of slaves from the West Indies and Africa. And the capital of Virginia, Richmond, had the largest percentage of black residents of any city in the 18th century.

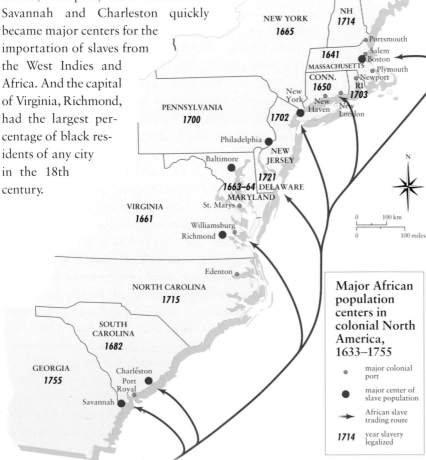

MAINE
(part of Massachusetts)

NEW YORK
1665

NH
1714

Portsmouth

1641
MASSACHUSETTS
CONN.
1650

Salem
Boston
Plymouth
Newport
RI
1703

New York

New Haven

New London

PENNSYLVANIA
1700

1702

Philadelphia

NEW JERSEY

Baltimore

1721

1663–64 DELAWARE
MARYLAND

VIRGINIA
1661

St. Marys

Williamsburg

Richmond

Edenton

NORTH CAROLINA
1715

SOUTH CAROLINA
1682

GEORGIA
1755

Charleston
Port Royal

Savannah

N

0   100 km
0   100 miles

Major African population centers in colonial North America, 1633–1755

- major colonial port
- major center of slave population
→ African slave trading route
1714 year slavery legalized

Blacks in British North America, 1680–1740

Before 1700, black people constituted only a small minority in colonial North America. In 1680, Africans in Virginia (the state with the highest numbers of blacks) composed just 7 percent of the population. Even in South Carolina, which in the 18th century would have a black majority, Africans made up just 17 percent of the population. All this would change in the years after each colony defined the legal status of slaves. With slavery redefined as an inherited, racial status, planters purchased more and more Africans.

What was it like for these Africans? First, it must be underscored that Africans in America, unlike their European counterparts, were forced to immigrate. They were kidnapped and captured from their homes and families, forced to endure the Middle Passage, and sold into a life of toil for others. Second, they had to learn to communicate with whites and even other slaves, many of whom came from different ethnic groups and spoke different languages. Very early on, Africans developed a form of speech that was uniquely their own, combining words, grammatical structures and linguistic traditions. Overall, scholars agree that life for these first African immigrants was full of isolation and alienation.

By 1740, a major demographic shift had occurred. A black population of numerical significance could be found in several colonies, and new arrivals from Africa and the Caribbean would have likely encountered people who shared their religious beliefs, kinship arrangements, language, and (enslaved) status. Also springing into existence by the 1740s was a rapidly expanding native-born population, whose identities spanned two continents and whose experiences laid the groundwork for a new, African American culture. Thus for blacks in British America in the mid-18th century, it became possible to forge a common new culture that retained much of their Africanness. These new immigrants laid the foundations of black America.

The Stono Revolt

At the same time blacks were developing their own cultures of work, belief, family, and identity, slavery became more entrenched in American law and practice. Slaves did not accept these changes passively. On hundreds of plantations and homes in all 13 colonies, slaves learned to work slowly, break tools, steal from their masters, tell stories of resistance, and, in some cases, rebel with violence. In 1739, slaves on the Stono Plantation 20 miles outside Charleston, South Carolina, decided to risk their lives for freedom. The rebellion began when 20 slaves stole guns and ammunition from a nearby store. The armed slaves then headed south for the border with Spanish Florida, a sparsely populated and well-known haven for runaway slaves. En route, they burned and raided several plantations, killing white residents and freeing slaves, swelling their numbers to more than 80. As they reached the Savannah River the slaves were caught and

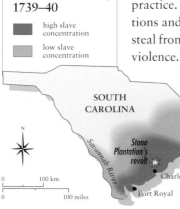

The Stono revolt, 1739–40

■ high slave concentration

■ low slave concentration

SOUTH CAROLINA

N

Savannah River

Stono Plantation's revolt

● Charleston

0 — 100 km

0 — 100 miles

Port Royal

ambushed by the better-armed South Carolina militia, which massacred the runaways. In a chilling act, the militia then beheaded the slaves and lined the road with their heads as a warning to other slaves planning similar revolts.

After the Stono rebellion, one of the most successful slave revolts of the time, South Carolina and the other southern colonies passed harsher slave codes to regulate the movements of both slaves and free blacks. Especially concerned were white South Carolinians, who were out-numbered by blacks two to one. Stono also struck fear about expanding slavery into the hearts of Northern slave owners. This fear of revolt only increased as the colonies grew more and more dependent on slavery and a rising population of enslaved African Americans.

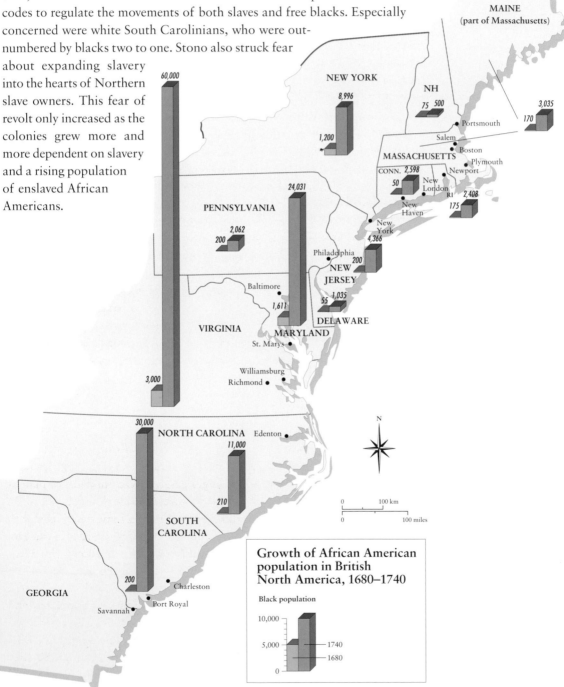

MAINE
(part of Massachusetts)

NEW YORK

60,000

8,996

1,200

NH

75 500

Portsmouth

3,035

170

Salem

Boston

MASSACHUSETTS

Plymouth

CONN. 2,598

Newport

PENNSYLVANIA

24,031

50

New
London

New
Haven

RI 2,408

175

2,062

200

New
York

Philadelphia

4,366

200

NEW
JERSEY

Baltimore

1,035

55

DELAWARE

VIRGINIA

1,611

MARYLAND

St. Marys

Williamsburg

Richmond

3,000

N

30,000

NORTH CAROLINA Edenton

11,000

210

0   100 km

0   100 miles

SOUTH
CAROLINA

GEORGIA

200

Charleston

Port Royal

Savannah

Growth of African American population in British North America, 1680–1740

Black population

10,000

5,000

1740

1680

0

Black Population on the Eve of the Revolution

This watercolor from the 1700s depicts slaves performing a West African dance called the "juba." The banjo fashioned from a gourd and the twisted leather drumsticks are also of African origin, and illustrate the persistence of African cultural heritage in the American colonies.

During the 18th century, two vastly different societies emerged in the American colonies—one white and free and another black and, for the most part, unfree. (Native Americans overwhelmingly lived on the borders of European and black settlement, and within their own nations.) No longer were African Americans isolated on rural plantations: by the time the United States gained its independence the black population had surpassed 500,000. In the middle of the 18th century another watershed occurred as well: for the first time since Africans arrived in North America in 1619, a Creole or locally born black population began to emerge. These Creoles drew strongly on the cultural influences of their African ancestors, but supplemented them with new, hybridized practices and traditions from America, the only home they ever knew. These were the beginnings of the vibrant African American society that exists today.

For the white, free, and dominant society, the mid-18th century was a time of escalating crises with Great Britain. Many colonists questioned the right of the British Parliament to force them to comply with legislation like the Sugar and Stamp Acts, and likened themselves to slaves. It no doubt struck enslaved blacks as ironic when slaveholding patriots like George Washington, Thomas Jefferson, and Patrick Henry demanded liberty and an end to British tyranny. The fact that American independence was in many ways purchased with slave labor, and some of the most eloquent pleas for freedom came from the colonies' largest slaveholders, was labeled the "American paradox" by the historian Edmund Morgan. Yet the revolutionary philosophy that the patriots

honed—culminating in Jefferson's words that all men, being created equal, were endowed with the "unalienable Rights . . . of Life, Liberty, and the pursuit of Happiness"—had powerful implications for blacks and other oppressed minorities. But this battle over the meaning of the Declaration would not take place in the 1770s, or even later in the 18th century; the nation's founding document did not even mention slavery and the slave trade. Jefferson's first draft of the Declaration did contain a factually inaccurate (and anti-slavery) passage blaming George III for the slave trade—but it was excised by a powerful bloc of Southern slaveholders who were perfectly satisfied to make lies out of "self-evident truths."

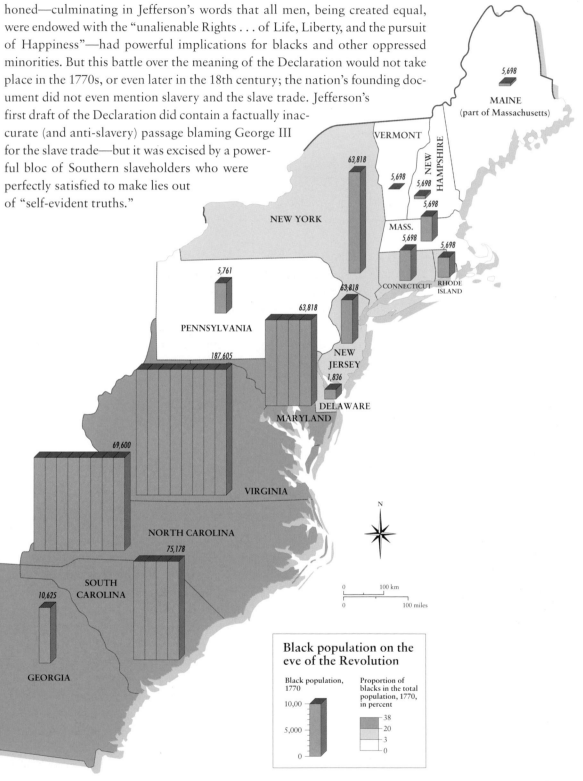

Black population on the eve of the Revolution

Black population, 1770

Proportion of blacks in the total population, 1770, in percent

10,00

5,000

0

38
20
3
0

PART II: TWO COMMUNITIES, SLAVE AND FREE

And about this time I had a vision—and I saw white spirits and black spirits engaged in battle, and the sun was darkened—the thunder rolled in the Heavens, and blood flowed in the streams—and I heard a voice saying, "Such is your luck, such you are called to see, and let it come rough or smooth, you must surely bear it."
—Nat Turner, 1831

A British naval officer painted this watercolor depicting conditions below deck on the slave ship Albanez *in about 1840. Millions of Africans died of disease, dehydration, and suicide during the Middle Passage.*

Slavery was well established in all 13 British colonies by the time a movement for independence began. What's more, slave labor had become the cornerstone of the economic system in every colony south of the Pennsylvania border. Across the South a body of laws called slave codes emerged, codifying the fact that slaves were property, not people. Laws were enacted to protect this property, and to protect whites when their property became "troublesome." That these laws were often enacted by the same men who decried British tyranny is at the center of the "American Paradox." Still the movement for independence from Great Britain was in many ways linked with the problem of slavery, and numerous African Americans attempted to make the Revolution a war for freedom, as both patriots and loyalists. The Massachusetts-born Patriot Prince Hall saw the implications of his own struggle as well as that of his nation—he fought for human freedom as well as political independence. On the other side, thousands of African American slaves saw the British army—which announced

that any slave joining them in arms would be set free—not as an army of oppression, but one of liberation.

After the war was over, the surging revolutionary ideology set in motion a process that historians have termed the "First Emancipation." Within a generation of independence from Great Britain, every state in what was becoming the "North" either abolished chattel slavery or made provisions for gradual emancipation. This process dramatically increased the number of free blacks in the new nation. Free blacks always occupied a precarious place in American society, somewhere between slavery and true freedom. Although not legally property, they were perceived by most whites as inferior. Racist ideology and laws passed by whites denied free blacks their rights and legitimized their treatment as second-class citizens. For example, Congress barred free blacks from serving in the militia in 1792; 18 years later they were banned from delivering the mail. Many of the same states that enacted gradual abolition statutes after the Revolution revoked voting rights from free blacks in the early 19th century. Yet despite rampant discrimination, African Americans in the North intensified their struggle for their own inalienable rights, while also attacking the ongoing system of slavery in the South. In Northern cities, growing communities of free African Americans resisted terrible discrimination to found newspapers, open businesses, and even create their own church.

In the South, slavery was reinvigorated after the Revolution, largely due to the invention of the cotton gin, which made cotton cultivation wildly profitable. Enslaved blacks built an empire based on staple crops in the fertile bottomlands of Alabama, Mississippi, and Georgia. But slaves created more than wealth to be exploited by their white owners. When they left the fields at the end of the day, enslaved blacks tended to their families, their spirituality, and their humanity. They played instruments and danced to a hybridized, African American music. They built strong family bonds in the face of slave law and forced separations. And they fashioned an African American religion that emphasized liberation and a coming day of judgment. "The idea of a revolution in the conditions of the whites and the blacks," said the escaped slave Charles Ball, "is the cornerstone of the religion of the latter."

Slaves used more than religion and family life to resist the inhumanity of slavery. On several occasions, they hatched violent conspiracies and revolts. Denmark Vesey, a free African American from Charleston, planned the city's fiery destruction in 1822. Before the revolt could take place, however, authorities hanged Vesey and 34 other conspirators. Nine years later, the Virginia slave and lay preacher Nat Turner led some 60 slaves in a revolt that left 55 white men, women, and children dead.

During the half century following the Revolution, blacks in slavery and freedom created a culture of their own. They launched assaults on slavery and built institutions. And they prepared themselves for an even greater struggle: to force the United States to live up to the promises and pronouncements in the Declaration of Independence.

The First Emancipation

The War for Independence and the revolutionary ideals of liberty and freedom set in motion forces that forever changed the status of blacks in the North. In 1776, few revolutionaries would have predicted that their arguments against British "tyranny" would also be applied to the tyranny faced by slaves. But that is precisely what happened. Voices rose across the North to contrast the hypocrisy of a war waged for liberty from England on the one hand and the continued existence of human thralldom on the other.

Within a generation, every state in what was becoming the "North" either abolished chattel slavery or made provisions for gradual emancipation. Historians term this the "First Emancipation" (the second would come in the wake of the Civil War). But racism survived the revolutionary era: blacks in the North might have been free, but they were certainly not equal. And in the South, planters were able to stem the anti-slavery tide and build an economy even more dependent on human bondage.

Many black soldiers earned their freedom with their service in the Revolution, and in effect "freed" themselves. Others were emancipated by their state legislatures. The effect of African Americans taking up arms in the war was dramatic: nearly every state that enlisted slaves to serve in the army either freed them immediately or promised manumission at the end of their service.

Residents of the new state of Vermont outlawed slavery when they drafted a constitution in 1777. In some states, emancipation acts were the result of pressure exerted by the slaves themselves. In Massachusetts, where the state's 1780 bill of rights claimed that all people were born "free and equal," an enterprising slave named Elizabeth Freeman sued for freedom and won in 1781, making her surname a reality. Not only did the county court of Great Barrington rule in Freeman's favor—the court also awarded her damages against her former master. Two years later, the chief justice of the Supreme Judicial Court, influenced by Freeman's and other slaves' suits for freedom—declared that the court was "fully of the opinion that perpetual servitude can no longer be tolerated in our government." Hundreds of slaves promptly declared themselves free in Massachusetts and New Hampshire and left their masters.

Legislation was necessary to end slavery in the other Northern states. Pennsylvania, which had a large anti-slavery Quaker population, led the way in 1780 with the first "gradual" emancipation statute in modern history. A state law declared that no black person born after that date could be held in bondage after he or she turned 28. In 1784 Rhode Island and Connecticut followed with their own gradual emancipation statutes. Slaveholders in New York and New Jersey, which had far higher slave populations than the other Northern states, were successful in stalling similar legislation for years. Not until 1799 (New York) and 1804 (New Jersey) did these last two Northern states pass complete gradual emancipation laws. It is important to remember that these gradual emancipation statutes freed no one immediately (adults remained slaves until they were set free or died, and children performed more than two decades of uncompensated labor) and that they were rarely the product of a principled

philosophical stance. More often than not Northern whites responded to a combination of conscience, pragmatism, and their own self-interest.

In 1787 Congress passed the Northwest Ordinance (written by Thomas Jefferson), which outlawed slavery and involuntary servitude in the territories north and west of the Ohio River, thus marking out the future states of Ohio, Michigan, Indiana, Illinois, and Wisconsin as free states. The First Emancipation seemed to be gaining steam in the Upper South as well: Virginia, Maryland, and North Carolina passed legislation after the Revolution making it easier for owners to manumit their slaves. But that was as far as the postwar anti-slavery movement got. In fact, in a reaction against the change sweeping the nation after independence, Southern slaveholders and conservatives joined ranks to insert protections for the institution in the new federal Constitution. Emancipation in the South, they argued, would ruin the planter class and amount to a radical and dangerous social revolution. The First Emancipation was stopped in its tracks.

The First Emancipation

1783 date of emancipation

slavery prohibited by Northwest Ordinance 1787

slavery banned by state constitution

gradual emancipation by statute

slavery allowed by law

Free Blacks in the New Republic

By the beginning of the 19th century there were rapidly growing communities of free blacks in most Northern states. A large majority of free blacks flocked to cities. Philadelphia and New York had long had sizable free black populations, and after the First Emancipation these communities grew dramatically. Free African Americans fled the declining economic opportunities in the countryside in search of a better life in the city.

Urban free black communities were also havens for runaway slaves from the Upper South, where the grip of slavery was growing firmer. In 1800, both New York City's and Philadelphia's populations were more than 10 percent black.

A small minority of the blacks in seaport cities became wealthy entrepreneurs, like the Massachusetts ship builder Paul Cuffe. Others became skilled artisans. Most, however, settled into stable but low-paying jobs as waiters, porters, barbers, coachmen, street vendors, and maids. Numerous sailors, dockworkers, and day laborers at the turn of the century were free blacks.

Free blacks constantly had to struggle with fears of re-enslavement (it was not uncommon for free blacks to be kidnapped and sold into slavery) and the realities of living in a larger society that often despised them. African Americans responded by building strong institutions of their own: black churches, Masonic lodges, schools, and relief societies were created at a dizzying rate around 1800. These institutions joined black dance halls, gambling houses, and saloons as the only racially integrated establishments in Northern cities.

Numerous free blacks made outstanding contributions during the period following the First Emancipation.

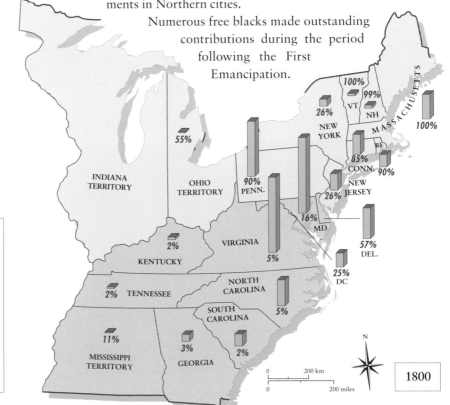

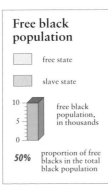

Free black population

- free state
- slave state

free black population, in thousands

50% proportion of free blacks in the total black population

1800

Ex-slave Phillis Wheatley became internationally renowned for her poetry; Benjamin Banneker made important advances in astronomy and surveying; and the gifted author Olaudah Equiano helped found the modern anti-slavery movement with his volume *The Interesting Narrative of the Life of Olaudah Equino, or Gustavus Vassa, the African.*

Much had happened to change these free black communities by 1830. First, the gradual manumission laws passed during the First Emancipation worked: there were only a handful of aging blacks who remained enslaved in the North. Second, with the exception of the Upper South, almost all Southern blacks were slaves. And the slave population had shifted dramatically after 1800, from the Atlantic seaboard states of Maryland, Virginia, the Carolinas, and Georgia to the Gulf states of Florida, Alabama, Mississippi, Louisiana, and Texas.

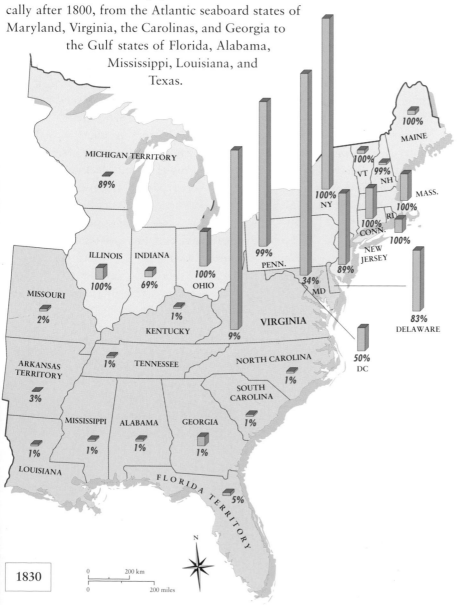

1830

Discrimination in the Antebellum North

Official and unofficial discrimination against free African Americans increased in the early 19th century. At the same time that white men experienced a significant expansion of their rights, especially in terms of voting, free blacks saw their rights eroding in both the North and the South. Many of these erosions were for rights guaranteed in the U.S. Constitution: by 1835, for example, the right of free assembly had been revoked for nearly all of the South's free blacks. They were barred by law from carrying firearms without a license in Virginia, Maryland, and North Carolina. And African Americans in the South were forbidden to hold religious services without the presence of a licensed and "respectable" white minister. Things were little better in the North. Several states followed Pennsylvania's lead in requiring free blacks to work and mandating that their means of support be visible. At the same time, African Americans were almost completely eliminated from the better-paying skilled trades as well as work on the docks, in warehouses, or in the merchant marine. One African American newspaper complained that blacks "have ceased to be hackney coachmen and draymen, and they are now almost displaced as stevedores. They are rapidly losing their places as barbers and servants. Ten families employ white servants now, where one did 20 years ago." A large array of the most menial jobs acquired the contemptuous epithet "nigger work," labor that African Americans performed because it was the only paying work they could get.

Nowhere was discrimination more pronounced in the early republic than in matters of suffrage. For white men, the early 19th century (the "era of the common man") represented a time of expanded suffrage and rights. Free blacks, on the other hand, were routinely barred from the polls. In a bill signed by Thomas Jefferson in 1802, they were excluded from the polls in the newly designated capital of Washington, D.C. Free blacks who had voted for years in Maryland, Tennessee, North Carolina, and Pennsylvania were barred from doing so in 1810, 1834, 1835, and 1838, respectively. New York erected a property qualification of $250 for black voters at the same time it eliminated all such requirements for whites. Between 1819 and 1865, every new state restricted the suffrage to whites. By 1840, the high-water mark for participation by registered (i.e., white male) voters, 93 percent of the free black population of the United States was disfranchised. "An educated colored man in the United States," wrote the abolitionist Frederick Douglass, "unless he has within him the heart of a hero, and is willing to engage in a life-long battle for his rights, as a man, finds few inducements to remain in this country. He is isolated in the land of his birth— debarred by his color from congenial association with whites; he is equally cast out by the ignorance of blacks." Douglass didn't have to mention the even higher barriers erected against black women and uneducated black men.

MICHIGAN TERRITORY

MAINE

VERMONT

NH

NEW YORK

MASSACHUSETTS

CONN. RI

ILLINOIS

INDIANA

OHIO

PENNSYLVANIA

NEW JERSEY

MARYLAND

Washington, DC

DELAWARE

KENTUCKY

VIRGINIA

TENNESSEE

NORTH CAROLINA

SOUTH CAROLINA

MISSISSIPPI

ALABAMA

GEORGIA

FLORIDA

N

0   200 km

0   200 miles

Free black disfranchisement, 1840

Free blacks barred from voting

black suffrage with restrictions

no restrictions

The Free Black Community

The AME Church

African Americans responded to discrimination by creating institutions of their own. The most enduring of these institutions was the African Methodist Episcopal Church, founded in 1790 by the Rev. Richard Allen. Allen saved enough money to purchase his freedom from his Delaware owner in 1777, the same year he experienced a powerful religious conversion. In 1786 he moved to Philadelphia, home to a thriving free black population, and became an itinerant preacher. At first Allen opposed a separate church for blacks, but eventually came to believe that African Americans needed their own organizations to guarantee their autonomy and make their voices heard. Allen, Absalom Jones, and James Forten first established a secular mutual aid society called the Free African Society. Then, in 1794, Allen organized and dedicated the Bethel Church, later known as the Bethel African Methodist Episcopal Church.

Branches of the new AME church quickly appeared in Baltimore, Wilmington, and smaller towns in Pennsylvania and New Jersey. In 1816, the various congregations were officially bound together in a formal organization. Theologically, the members of the church were Methodists, and their numbers grew rapidly. By 1820 there were 4,000 black Methodists in Philadelphia, and nearly half that many in Baltimore. The church even made inroads as far west as Pittsburgh and as far south as Charleston, before the Denmark Vesey slave conspiracy halted the growth of black organizations there. By 1826 the AME Church had almost 8,000 members, and grew to more than 50,000 in 1860 and 500,000 in 1892. The denomination continues to grow today, encompassing more than 3,600 churches worldwide.

Growth of the African Methodist Episcopal Church (AME), 1816–92

1816–26   1827–56   1857–92

African American Newspapers

Free blacks organized far more than religious denominations. Black Masonic lodges attracted hundreds of members, as did black schools, social clubs, lodges, and mutual aid societies. Black restaurants, dance halls, and saloons also dotted free black communities, and were often the only places where blacks and whites met socially. Free African Americans also published scores of newspapers and periodicals to resist discrimination and stake their claim to civil equality. New Yorkers John Russwurm and Samuel Cornish published the first edition of *Freedom's Journal*, the earliest sustained black newspaper, in 1827 (a paper called the *African Intelligencer* had existed briefly in Washington, D.C., in 1820). Committed to civil rights for African Americans, the paper folded after Russwurm emigrated to Liberia. Like many of the paper's successors, it was short-lived, mostly because of financial pressures and a small (often illiterate) readership.

Ten years later Cornish commenced publication of the *Colored American*, which weighed in strongly on the issue of a suitable name for the emerging African American community. "Let us and our friends unite, in baptizing the term 'Colored Americans,' and henceforth let us be written of, preached of, and prayed for as such. It is the true term, and one which is above reproach." Other papers like the *National Watchman*, the *Weekly Anglo-African* and the *Mirror of Liberty* joined in the debate, and gave expression to a growing community. The most successful black journalist of the antebellum period also joined the debate, but Frederick Douglass, editor of the *North Star*, preferred the term "American." Like other editors, Douglass suffered from financial problems and threats of violence, but his paper and its successor endured for more than a decade.

African American newspapers and periodicals, 1820–60

- ● city with 5 or more African American newspapers and/or periodicals
- • city with fewer than 5 African American newspapers and/or periodicals

The Cotton Kingdom

The Antebellum Slave Economy

In 1793 a technological innovation revolutionized Southern agriculture and substantially altered the lives of many African Americans. Great Britain's newly minted textile barons were willing to pay dearly for cotton, but American planters couldn't bear the labor costs required to prepare the crop for export. The problem wasn't the climate (short-staple cotton thrived in the warm, humid climate of the Deep South) or the labor—it was the cotton plant's sticky seeds. Before cotton could be milled, the seeds needed to be removed by hand, a daylong task for a slave cleaning just 1 pound of cotton. By fashioning a machine (or "gin") that combed the seeds from cotton fiber with metal pins on rollers, the Yankee inventor Eli Whitney rejuvenated plantation agriculture in the South. Now a single worker (usually an enslaved African American) could clean 50 pounds of cotton a day.

As a direct result of the cotton gin, slavery spread rapidly south and west. Chesapeake planters, whose land was north of the cotton line, sold their slaves to Alabama, Georgia, and Mississippi, often for huge profits. The movement of

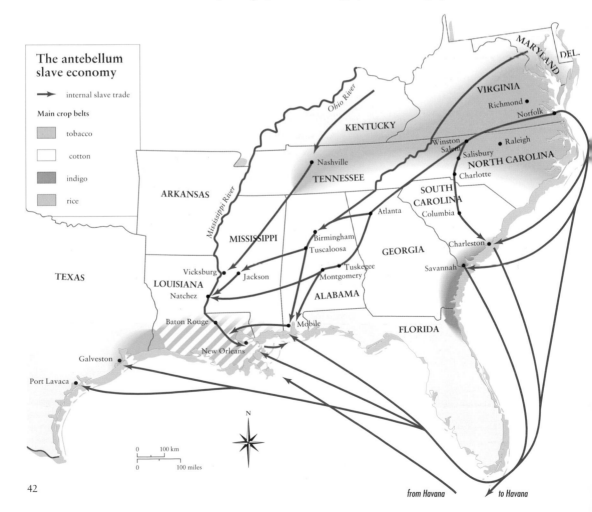

slaves out of the Chesapeake was fast and furious: in 1790, planters in Maryland and Virginia owned 56 percent of all American slaves; by 1860 they owned just 15 percent.

In the years between the invention of the cotton gin and the closing of the slave trade in 1808, more than 250,000 slaves were brought directly to the United States. Seeking profits from the booming demand for cotton, white and black Southerners pushed southeastern Indians and Mexicans in Texas out of their way. In 1850, fully three quarters of all slaves were agricultural laborers: fifty-five percent grew cotton; 10 percent grew rice, hemp, and sugar; and 10 percent grew tobacco. For slaves who were not agricultural laborers, leading occupations were turpentine producer, lumberjack, miner, dock worker, textile worker, and industrial laborer.

Significant Plantations of the Antebellum South

The explosion of slavery into the South and West changed the way enslaved African Americans worked and lived. In the Chesapeake, where agriculture shifted from tobacco production to grain and livestock, more slaves were trained as skilled artisans.

Their work, though complex, required little supervision. Skilled slaves, who were almost always men, were often hired out to work for wages in towns and cities. Women mostly worked under close supervision weeding, hoeing, and maintaining the fields. The large plantations of the 18th century, like Thomas Jefferson's Monticello, largely converted to mixed agriculture by 1800.

In the lowland coastal regions of Georgia and South Carolina, where deadly diseases threatened the lives of both slaves and slave owners, the "task system" prevailed. Slaves were given a daily "task" and allowed to complete it at his or her own pace. This encouraged hard work and took little supervision—and slaves often turned the task system to their own advantage. On many of the region's immense plantations, which included Rose Hill, Ocean Plantation, and Hampton, slaves won the right to cultivate their own "private fields" and sell their products on the open market. The owners of these busy slaves were among the richest Americans, and were therefore tolerant of the slaves' own commercial activity.

In the new cotton lands of the Deep South, however, slaves had very little independence. Cotton cultivation demanded intensive and skilled labor. Slaves on cotton plantations routinely worked in "gangs" from sunup until sundown, under the watchful eye (and whips) of white overseers. Plantations like The Hermitage (owned by President Andrew Jackson) and Melrose in Mississippi reaped tremendous profits from the labor of hundreds of slaves.

This 1862 photograph depicts five generations of a slave family on a South Carolina plantation, illustrating a highly complex network of kin relationships. Despite the horrors of separation and sale, the black family survived slavery.

Slave population distribution

· 2,000 slaves

1830

1860

The Slave Community

Slave Life: Hopeton Plantation, Georgia

According to the 1850 U.S. Census, 1.8 million slaves, or more than half the slaves in the United States, worked on cotton farms or plantations. Thus the cotton plantation was the typical locale for the enslaved African American. On smaller farms, slaves and their owners often worked together in the fields. But the larger plantations were among the most intensely commercialized and specialized farms in the world. Factorylike organization and division of labor was the rule.

The slaves' labor—the exploitation of which provided the economic foundation for the Old South—was the primary concern of the slave owner. The labor force was divided into two distinct groups: house slaves and field hands. Typically under the watchful eye of the slave mistress, house servants cared for children, prepared meals, cleaned the house, yards, and gardens. The field hands were a much larger group. Cotton requires a long growing season and constant attention: slaves worked planting, hoeing, weeding, picking, and preparing the cotton crop for market. Field hands also tended livestock and the large corn and other vegetable fields that fed all the residents of the plantation. Clearing land, burning brush, and maintaining fences and buildings were also the duties of field hands.

Labor on a cotton plantation was almost always performed under the supervision of the owner or overseer. Gangs of slaves were told when to work, when to rest, and when to quit. There was little incentive for the slaves to work hard or develop initiative under this system; in fact, slaves usually worked as slowly as they could without incurring punishment. Whippings were common in an attempt to extract more work from field slaves. Planters justified this practice with racism: African Americans, they reasoned, were a childlike race and required strict punishment.

Hopeton Plantation in Glynn County, Georgia, (see map on page 43) provides an example of the complex microcosm of the plantation. Between 350 and 400 slaves performed various types of labor at Hopeton, from tending the vast rice, cotton, sugarcane, and vegetable fields to manufacturing sugar and

William Henry Brown created this collage depicting a slave harvest crew near Vicksburg, Mississippi, in 1842. Note that during the busy cotton harvest, even children helped in the fields.

serving the planter's family. More than 40 slave cabins and outbuildings stood behind the "big house" occupied by the plantation's manager and part-owner James H. Couper. Slaves worked according to a modified task system under several overseers. According to one observer, both slaves and overseers were punished if work quotas were not met: "as the task of each [slave] is separate, imperfect work can readily be traced to the neglectful worker."

Despite the astonishing variety of crops and products produced at Hopeton (in this respect the plantation was unusual), it was hardly self-sufficient. Necessities such as blankets, cloth, corn, mackerel, needles, pork, salt, shoes, soap, thread, and other items were purchased throughout the year. In 1833 alone Couper had to purchase 3,800 bushels of corn, 29 barrels of prime pork, 53 barrels of mackerel, and 2,710 pounds of beef to feed his African American workforce. As happened on most plantations, slaves received a weekly ration of food. Typically, an adult's share was 3 to 4 pounds of salt pork or bacon, a peck of cornmeal, some sweet potatoes and molasses. Slaves' diet did not vary: this monotonous, though more or less nutritious, regimen prevailed "day to day and week to week." Slaves were also annually issued shoes and clothes, often cheap and shoddy pieces manufactured out of slave-tended cotton in New England.

Hopeton Plantation, Georgia

- ■ owner's residence
- ■ slave quarters
- ▪ service buildings
- woods
- pasture
- idle land

Crops

- cotton
- rice
- corn
- cane
- potatoes
- peas
- barley
- pumpkins

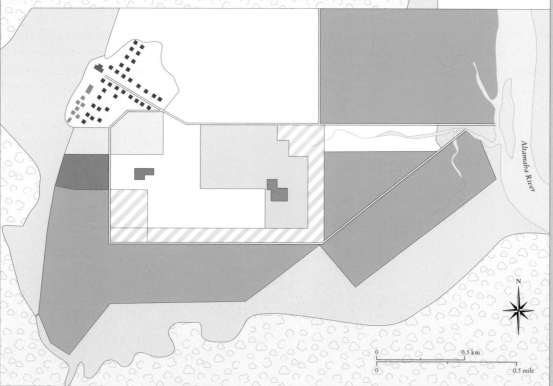

Altamaha River

0   0.5 km

0   0.5 mile

Slave Quarters, Calvitt Plantation, Texas

Plantation living conditions were generally crowded and uncomfortable. Most plantation slaves lived in tiny, rude huts behind the slave owner's "big house." Windows and floors were very rare, as was furniture. Most slaves slept on blankets resting atop corn husks or straw. Rampant overcrowding in the quarters (far more than poor diet or overwork) was the leading threat to slaves' health. Diseases like tuberculosis and dysentery spread rapidly.

This slave cabin on the Calvitt Plantation in Robertson County, Texas, was slightly more comfortable than average. The floorplan is a square of 19 feet, with a dirt floor, fireplace for heat and cooking, and a tiny loft. A front porch provided additional living space. The cabin probably housed an extended slave family, more than eight people in all. It was in the slave quarters, away from the white masters and overseers, that African Americans constructed a community based on family, religion, music, dance, and story telling. From sunup to sundown a slave's time belonged to the master; the rest of the time, however, was his or her own. Most slaves were Christians by the early 19th century, and Christian worship was a vital part of slave culture. Not surprisingly, slave sermons de-emphasized the religion preached by slave owners (obey thy master, work hard, respect property) in favor of an expressive Christianity based on the themes of suffering and redemption. Slaves generally met in secret, often in the woods, to sing spirituals, shout, pray, and dance, away from the watchful eyes of white people.

Family was also central to life in the slave quarters, despite the painful plantation realities of family separation and sexual assault. Enslaved people depended on a wide network of kin relationships to provide emotional support and strength in addition to the everyday tasks of child rearing, food preparation, and care of the

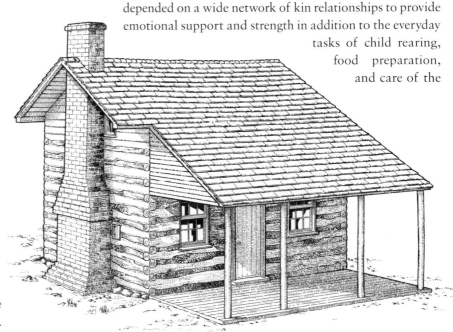

Slave quarters on Calvitt Plantation.

sick or elderly. Owners often encouraged slave marriages because they made life on the plantation more productive and peaceful; yet these families were never sanctioned by law and were always vulnerable. Because a father might live on a neighboring plantation or a mother might be sold away, large, extended kinship groups filled the slave quarters. Numerous adult relatives provided guidance, love, and protection for the children, who often treated aunts, uncles, or grandparents as if they were parents. In the face of great odds, the black family survived slavery.

That is not to say that individual families weren't destroyed by the institution: the historian Herbert Gutman has found that 29 percent of all slave marriages were broken up between 1820 and 1860. Even more families—as many as 2 million by one estimate—experienced the sale of one or more children. The ex-slave Harriet Jacobs recalled a haunting memory of seeing a mother leaving an auction house after each of her seven children had been sold: "I met that mother in the street, and her wild, haggard face lives to-day in my mind. She wrung her hands in anguish, and exclaimed, 'Gone! All gone! Why don't God kill me?'" Despite the constant threat of separation, slave families continued to have children: the right to make and maintain families was one of the most precious and well-guarded privileges slaves had.

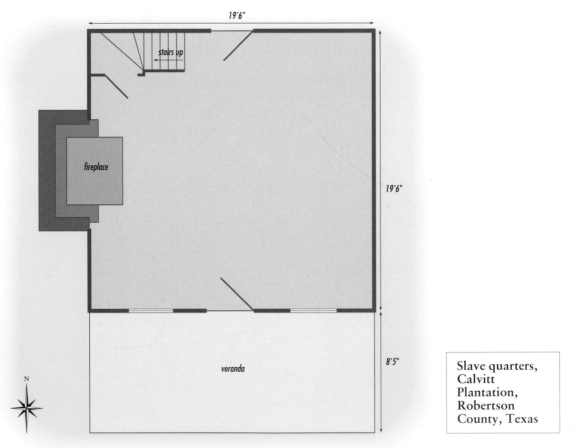

Slave quarters, Calvitt Plantation, Robertson County, Texas

Nat Turner's Rebellion, 1831

Slaves did not endure their servitude passively. They resisted in numerous ways, ranging from the barely noticeable insubordination to episodes of extreme violence. Perhaps the most common way slaves protested the continuous forced labor of plantation life was sabotage: slave hands broke tools, stole food, abused farm animals, and burned down buildings. Other day-to-day acts of resistance included slowing the pace of work, feigning illness, or self-mutilation. Slaves also used their culture to protest their situation. Songs, folk tales, and religion all helped slaves find release from hardship and suffering.

Some slaves ran away, depriving their owners of their labor. Communities of runaway slaves called Maroons existed across remote and swampy areas of the South, especially in Florida. Others, like Frederick Douglass, fought back physically against their owners and overseers.

Rebellion was, of course, the ultimate form of resistance against slavery. Most slaves understood that armed revolt was virtually suicidal. Unlike in the Caribbean, where massive slave revolts took place in Cuba, Jamaica, and most famously, Haiti, rebellion in the American South was difficult to carry out. The white population was large and maintained a near monopoly on firearms; plantations were small and relatively far apart; and panicked slaveholders responded with terror to every rumor of revolt. Still, conspiracies for escape and murder did occur. The Gabriel Prosser conspiracy rocked Richmond in 1800, and in 1822 an ex-slave named Denmark Vesey conspired to seize Charleston's armory, murder the city's white population, and escape by ship to Haiti or Africa. Slaves betrayed the plot at the last minute, and whites responded by hanging Vesey and 35 others after closed-door show trials. Southern whites' escalating fear of slave rebellion was evident by the severity and intensity with which they handled both conspiracies.

In 1829, a free African American named David Walker appealed to Southern slaves to rise up in violent rebellion: "had you not rather be killed than to be a slave to a tyrant, who takes the life of your mother, wife, and dear little children?" Black sailors sewed Walker's tract into their uniforms and spread it across the seaboard South. The next year a Virginia slave and Baptist lay preacher named Nat Turner began experiencing visions of white and black spirits engaged in battle. The son of an African woman and a father who successfully escaped to freedom, Turner learned to read and avidly studied the Bible. Believing himself an instrument of God's wrath, Turner began recruiting other slaves for a massive assault on slavery in the Tidewater.

On August 21, 1831, Turner led a small band of slaves into the home of his master, Joseph Travis, and murdered the entire family. Then, armed with pick-axes, Turner vowed to "carry terror and devastation" throughout the countryside. After attacking whites on neighboring plantations, the growing band hacked Margaret Whitehead to death with swords. The killing spree continued all night and the following day, with at least 60 slaves joining Turner's revolt. On August 23, after killing 55 white men, women, and children, the slaves were pinned down at James Parker's plantation by a group of heavily armed whites.

The next morning U.S. soldiers arrived to put down the rebellion, and most of the slaves were killed or captured. Turner escaped and eluded capture for more than two months, hiding in swamps, caves, and nearby woods.

By the time Virginia authorities apprehended, tried, and hanged Turner, whites had engaged in a retaliatory killing spree of their own. Twenty of Turner's co-conspirators were hanged, and Virginia banished ten slaves from the state. But as many as 120 African Americans, including many innocent victims, lost their lives as a result of the revolt, the bloodiest in U.S. history.

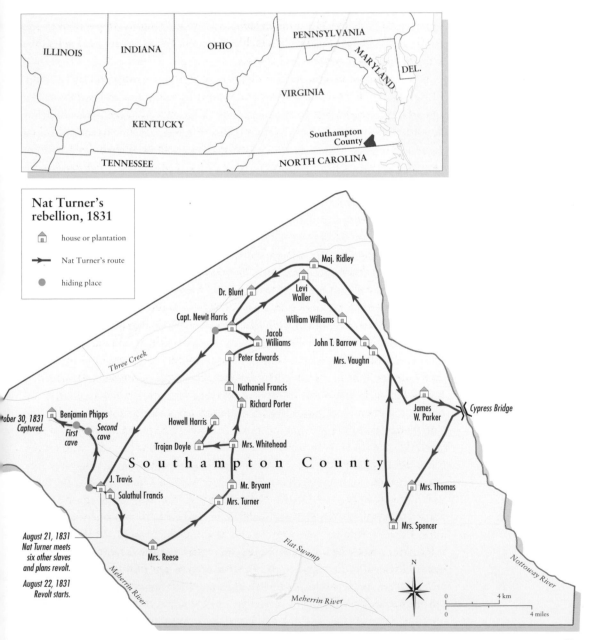

Nat Turner's rebellion, 1831

- 🏠 house or plantation
- ➔ Nat Turner's route
- ● hiding place

PART III: TOWARD FREEDOM

> We are NATIVES of this country, we ask only to be treated as well
> as FOREIGNERS. Not a few of our fathers suffered and bled to
> purchase its independence; we ask only to be treated as well as those
> who fought against it. We have toiled to cultivate it . . . we ask only
> to enjoy the fruits of our labor. Let these moderate requests be
> granted, and we need not go to Africa nor anywhere else to be
> improved and happy.
> —The Rev. Peter Williams, 1822

Slaves resisted their servitude with sabotage, by running away, and, in extraordinary cases, by inciting rebellion. But the assault on slavery was not waged by
slaves alone: beginning in the 17th century, small groups of free blacks and
Quakers began to attack the institution as unjust and against the will of God.
By the middle of the 19th century, black and white abolitionists had fashioned
the greatest reform movement of the age. Through constant agitation, they
kept the slavery issue before the nation, inserting it squarely into the politics of
geographic expansion. And though politicians tried again and again to compromise over the issue of slavery (in 1819, 1850, 1854, and 1857), the actions of
abolitionists, slaveholders, and slaves only exacerbated the crisis between the
free labor North and the slave South. Only through the catastrophe of Civil
War was the destruction of slavery finally accomplished in the United States.

Opponents of slavery often disagreed on tactics and goals. Some abolitionists favored immediate emancipation and the creation of a just, anti-racist society, while most white abolitionists fell short of a belief in racial equality. Many
believed that only by returning ex-slaves to Africa could the United States be
rid of racial problems after emancipation. The anti-slavery movement changed
dramatically between 1829 and 1831 with the publication of an influential
pamphlet by David Walker and a revolutionary newspaper by William Lloyd
Garrison. David Walker's *An Appeal in Four Articles* drips with angry defiance
and repeatedly calls for a violent overthrow of slavery. Walker claimed that
because of slavery, African Americans were the "most wretched, degraded and
abject set of beings that ever lived." And unlike their white oppressors, Walker
said, blacks had the capacity to forgive:

> Treat us like men, and . . . we will love and respect [whites], and pro-
> tect our country. . . . Treat us like men, and there is no danger but
> we will all live in peace and happiness together. For we are not like
> you . . . unforgiving. Treat us like men, and we will be your friends.

Walker, a free black Boston merchant, sewed his *Appeal* into the seams of
the used clothing he sold to sailors. It was read up and down the Atlantic
seaboard and helped African Americans unite in a movement for the immediate abolition of slavery.

William Lloyd Garrison, a white reformer from Massachusetts, thundered in

his newspaper, the *Liberator*, that slavery was a national sin and demanded immediate emancipation. "I am in earnest," he said in his paper's inaugural issue. "I will not equivocate—I will not excuse—I will not retreat a single inch—and I will be heard!" The paper immediately won strong support from African Americans, whose patronage sustained it in its early years. In time the *Liberator* became the quasi-official organ of the American Anti-Slavery Society, founded in 1833 to abolish slavery and improve "the character and condition of the people of color."

Emboldened by the arguments of abolitionists like Walker and Garrison, black and white abolitionists greatly expanded their movement. Abolitionist societies flooded the mail with tracts and petitions. Ex-slaves like Sojourner Truth and Frederick Douglass published their narratives and went on speaking tours. Networks of sympathetic blacks and whites worked with "conductors" like Harriet Tubman to spirit runaway slaves north-

ward on an "underground railroad." And abolitionist journals documented the daily horrors of slavery, the kidnappings of free blacks, and the tragic return of fugitives like Anthony Burns to lives of thralldom.

Born a slave in New York, Isabella Baumfree took the name Sojourner Truth in 1843 and became a prominent itinerant preacher, abolitionist, and feminist. In a famous speech in 1851, Truth attacked both racism and sexism by asking "Ain't I a woman?"

The political crisis came to a head in 1860, with the election of the anti-slavery Republican Abraham Lincoln to the presidency. Lincoln was elected on a platform of halting slavery's expansion without a single electoral vote from the South; Southern states seceded from the Union in response. The ensuing Civil War began as an effort by the North to "restore the Union." But when thousands of runaway slaves escaped behind Union lines, effectively "freeing" themselves, some Northerners began to view the war as a fight for freedom. Military necessity convinced others (including Lincoln) that the only way to win the war was to strike directly at Southern institutions like slavery. Similar sentiments led to the formation of black regiments like the 54th Massachusetts, which suffered heavy casualties fighting to destroy slavery. After the Civil War the nation commenced a bold (and, ultimately, failed) experiment to "reconstruct" the South on the basis of free labor and legal equality for African Americans. Under the protection of the Union Army, African Americans in the South were granted full civil rights, and black men were enfranchised. Several held state and national offices. But when Union troops withdrew in 1877, organizations like the Ku Klux Klan terrorized black voters and enforced racist mores extralegally. Most blacks stayed in the South and attempted to earn a living by sharecropping or tenant farming; others, however, sought a better life elsewhere and began a century of black migration to the North and West.

The Expansion of Slavery, 1819–57

The Missouri Compromise, 1820

Westward migration in the early 19th century inevitably led to tension between residents of slave and free states. Some Northerners resented the stranglehold Virginians seemed to have on the White House and the "three-fifths clause" of the U.S. Constitution, which gave slave states congressional representation based on the white population plus three fifths of the slave population. Both sections eyed each other suspiciously whenever a new state applied for entry into the Union, and were careful to maintain a balance of new slave and free states. The Northwest Ordinance of 1787 barred slavery from the territory that became the states of Ohio, Indiana, Illinois, Michigan, and Wisconsin; and the admission of Kentucky, Tennessee, Alabama, Mississippi, and Louisiana kept the number of free and slave states in balance.

When Missouri applied for admission to the Union as a state in 1819, slavery was already a way of life there. Even before the U.S. purchased Louisiana (including the parts that became Missouri), Spanish and French settlers had owned slaves. By the time the population of Missouri reached the 60,000 people required to apply for statehood, 16 percent of those people were enslaved African Americans. Missouri's application for statehood precipitated an ominous sectional crisis that threatened the unity of the nation.

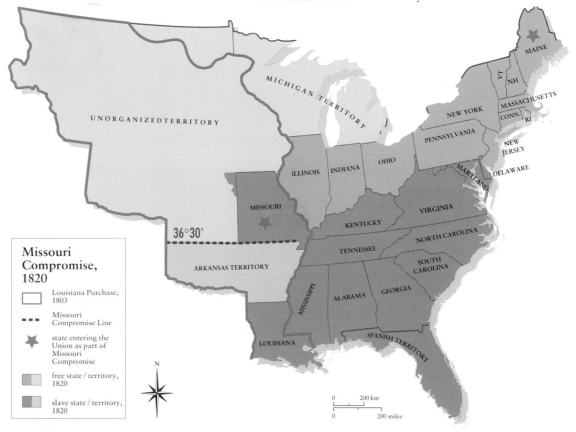

Missouri
Compromise,
1820

- Louisiana Purchase, 1803
- Missouri Compromise Line
- ★ state entering the Union as part of Missouri Compromise
- free state / territory, 1820
- slave state / territory, 1820

36°30'

When Congress received the application, a New York representative added two amendments: one prohibiting the "further introduction of slavery" and another providing for the gradual emancipation of Missouri's 10,000 slaves. Southerners were outraged, and Congress adjourned without approving statehood. The next Congress eked out an uneasy deal known as the Missouri Compromise: Missouri would enter the Union as a slave state while the northernmost counties of Massachusetts became the free state of Maine. This scheme neutralized fears that the South would gain more influence in the U.S. Senate. Next, the South agreed to outlaw slavery north of 36°30' latitude, a line extending west from Missouri's southern border. The compromise opened up the new territory of Arkansas (present-day Oklahoma and Arkansas) to slavery while barring the institution from the remainder of the Louisiana Purchase—territory that would become the states of Kansas, Nebraska, Colorado, Minnesota, Iowa, Montana, Wyoming, and North and South Dakota.

The Missouri Compromise made plain that sectional issues were a political tinderbox. It brought the South's commitment to slavery and the North's resentment of Southern political power into direct confrontation. Thomas Jefferson, in retirement at Monticello, was distraught by the Missouri Compromise: "a geographical line, coinciding with a marked principle, moral and political, once conceived and held up to the angry passions of men, will never be obliterated," he wrote. The dispute over Missouri, "like a fire-bell in the night, awakened and filled me with terror. I considered it at once the knell of the Union."

Compromise of 1850

Thirty years after the Missouri Compromise, the issue of slavery's expansion in the West was as contentious as ever. As a result of the victory in the Mexican-American War, the United States acquired the entire Southwest: territory that would become the states of California, Nevada, Utah, New Mexico, and Arizona. Abolitionists like Frederick Douglass had done much to influence public opinion, and many anti-slavery Northerners wanted to bar slavery from spreading into territory won from Mexico. The political crisis over whether the North or the South would control the settlement of the former Mexican lands almost tore the nation apart in 1850.

Brokered by the aging Senator Henry Clay and newcomer Stephen Douglas, the Compromise of 1850 tried to appease both North and South. First, gold-rich California would be admitted as a free state, while the rest of the Mexican session would be organized without restrictions against slavery. Another provision outlawed the slave trade in the District of Columbia, where African Americans were dragged in chains and families sold apart within shouting distance of the U.S. Capitol. But to appease Southerners, Clay proposed that slavery not be abolished there. Finally, the compromise called for a new, stronger

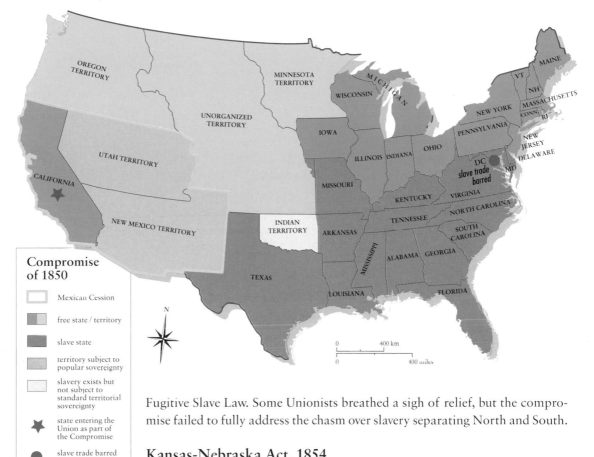

Compromise of 1850

- Mexican Cession
- free state / territory
- slave state
- territory subject to popular sovereignty
- slavery exists but not subject to standard territorial sovereignty
- ★ state entering the Union as part of the Compromise
- ● slave trade barred in nation's capital

Fugitive Slave Law. Some Unionists breathed a sigh of relief, but the compromise failed to fully address the chasm over slavery separating North and South.

Kansas-Nebraska Act, 1854

Just four years after the Compromise of 1850, the battle over slavery's expansion erupted again—this time in the Louisiana Purchase, where the slavery question had supposedly been settled by the Missouri Compromise line at 36°30'. With thousands of settlers retracing Meriwether Lewis and William Clark's path up the Missouri River and entrepreneurs calling for a railroad to connect San Francisco to the east, many Americans favored organizing a territory north of Indian Territory, present-day Oklahoma. Responding quickly, Congress passed a bill in 1853 creating Nebraska Territory, encompassing the area north of Indian Territory all the way to the Canadian border. Southern senators still smarting from the admission of free California in 1850 correctly foresaw that under the Missouri Compromise slavery would be prohibited in each state carved from Nebraska Territory; they desperately wanted to avoid the addition of new free states, which would further dilute their power in Congress.

Senator Stephen Douglas proposed repealing the part of the Missouri Compromise that prohibited slavery north of 36°30', arguing that "popular sovereignty," or the will of the settlers, should determine whether a state was

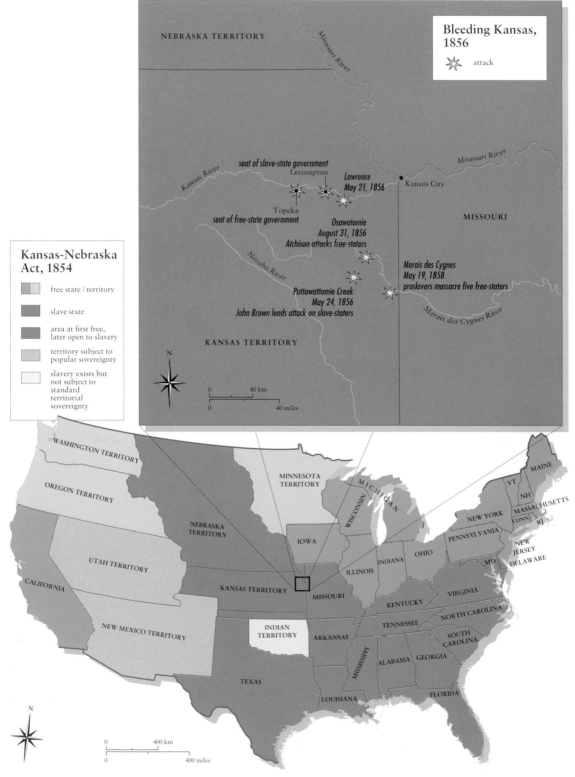

Bleeding Kansas, 1856

✴ attack

NEBRASKA TERRITORY

Missouri River

Kansas River

seat of slave-state government
Lecompton

Lawrence
May 21, 1856

Kansas City

Missouri River

MISSOURI

Topeka
seat of free-state government

Osawatomie
August 31, 1856
Atchison attacks free-staters

Neosho River

Marais des Cygnes
May 19, 1858
proslavers massacre five free-staters

Pottawattomie Creek
May 24, 1856
John Brown leads attack on slave-staters

Marais des Cygnes River

KANSAS TERRITORY

N

0   40 km
0   40 miles

Kansas-Nebraska Act, 1854

free state / territory

slave state

area at first free, later open to slavery

territory subject to popular sovereignty

slavery exists but not subject to standard territorial sovereignty

WASHINGTON TERRITORY

OREGON TERRITORY

MINNESOTA TERRITORY

MICHIGAN

WISCONSIN

MAINE

VT

NH

MASSACHUSETTS

NEBRASKA TERRITORY

IOWA

NEW YORK

CONN.

RI

UTAH TERRITORY

ILLINOIS

INDIANA

OHIO

PENNSYLVANIA

NEW JERSEY

MD

DELAWARE

CALIFORNIA

KANSAS TERRITORY

MISSOURI

KENTUCKY

VIRGINIA

NEW MEXICO TERRITORY

INDIAN TERRITORY

ARKANSAS

TENNESSEE

NORTH CAROLINA

SOUTH CAROLINA

MISSISSIPPI

ALABAMA

GEORGIA

TEXAS

LOUISIANA

FLORIDA

N

0   400 km
0   400 miles

The slaves Dred and Harriet Scott sued for their freedom in 1846, arguing that their travels in the territories of Minnesota and Wisconsin—north of the Missouri Compromise line—made them free people.

free or slave. When a later draft of the bill divided Nebraska in two, creating a new territory of Kansas, most people assumed Kansas would be opened to slavery. People in the North—many of whom viewed the Missouri Compromise as sacrosanct—were outraged. Almost overnight, people who had never publicly opposed slavery's extension became energized against the Kansas-Nebraska Act. Abolitionists, Free Soilers, Northern Whigs, and many Northern Democrats formed "anti-Nebraska" coalitions, which quickly coalesced into the new Republican Party.

Meanwhile, the race to settle Kansas was on between residents of slave and free states. Pro-slavery Missourians and anti-slavery Northerners clashed violently in several battles that became known as "Bleeding Kansas." After pro-slavery "border ruffians" sacked the free-state town of Lawrence, a wild-eyed abolitionist named John Brown retaliated by murdering five pro-slavery Kansans in May 1856. Kansas became, for both the Northerners and the Southerners, a powerful symbol of sectional strife.

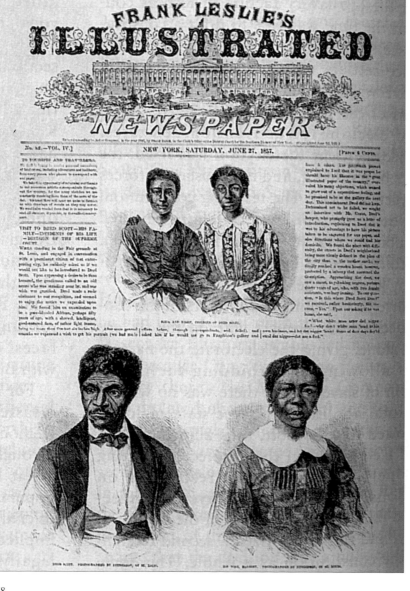

Dred Scott Decision

With "Bleeding Kansas" threatening to draw the entire nation into open warfare, the U.S. Supreme Court decided to settle the question of slavery in the territories once and for all. The Court's justices—five of whom were Southern slaveholders—decided to rule on the case of a slave named Dred Scott in 1857.

Scott's owner, an army doctor from Missouri, had taken his slave with him to military posts in Illinois and Wisconsin Territory (part of which became the state of Minnesota) for several years before returning home. Scott sued for his freedom when the doctor died, claiming that his prolonged stay in territory north of 36°30' made him a free man. Scott's case made it all the way to the Supreme Court as a test case over whether Congress had the power to ban slavery in the territories.

First, the justices, led by the slaveholder Roger B. Taney of Maryland, declared the Missouri Compromise ban unconstitutional. But they didn't stop there. Congress, they ruled, had no power to keep slavery out of any territory, since slaves were private property. Finally, Taney ruled that the case should never have been heard in the first place, since African Americans were not citizens of the United States and therefore could not sue in court. The decision must have come as a surprise to blacks who were legal citizens in several Northern states and therefore citizens of the United States. Most Northerners were shocked: to them, Taney's decision smacked of racist, sectional politics. But the abolitionist editor and ex-slave Frederick Douglass expressed optimism: "We, the abolitionists and colored people, should meet this decision, unlooked for and monstrous as it appears, in a cheerful spirit," he wrote. "This very attempt to blot out forever the hopes of an enslaved people may be one necessary link in the chain of events preparatory to the complete overthrow of the whole slave system."

Dred Scott decision, 1857

- free state
- slave state
- open to slavery by decision
- slavery exists but not subject to standard territorial sovereignty
- → Dred Scott's travel route

Runaways, Kidnappings, and Abolitionists

The Underground Railroad

Far more slaves ran away than engaged in open rebellion. A large majority were captured and returned to their owners. As the 19th century wore on, however, black and white opponents of slavery constructed a large network to assist fugitives in their escapes to freedom. Taken altogether, this network became known as the "Underground Railroad."

Many of the stories about the Underground Railroad are more myth than history. There were few or no stations in the South, where most slaves lived, and runaways had to rely on their own ingenuity to make it to the relative safety of the Northern states. And the network of "stations" never achieved the rates of success claimed by hysterical slave owners (during the 1850s only about 1,000 slaves escaped per year—roughly one quarter of 1 percent of the 4 million enslaved blacks).

Harriet Tubman (far left), a runaway slave from Maryland, personally helped as many as 300 enslaved blacks escape to freedom. She was called the "Moses of her people."

Still, the runaway slave Harriet Tubman made 19 separate trips to the South to spirit 300 slaves to freedom. And Levi Coffin, the reputed "president" of the Underground Railroad, reported receiving up to 100 slaves a year at his house in Cincinnati during its days as one of the railroad's major "stations." Estimates suggest that as many as 3,000 blacks and whites at any one time helped about 75,000 slaves find freedom. Favored routes led through communities in Ohio and Pennsylvania with high populations of Quakers (who were often outspoken abolitionists and volunteered their homes and meeting houses) and through cities with significant free black populations like Philadelphia, New York, and Boston.

In addition to aiding in the escape of slaves and giving countless others hope, the Underground Railroad, myths and all, significantly stepped up tensions between North and South. Southerners deeply resented Northern help given to runaways and demanded strict fugitive slave laws and punishment for those caught violating them. This resentment was far out of proportion to the extent of effectiveness of the Underground Railroad itself. On the other side of the Mason-Dixon line, fugitives risking everything for freedom brought Northerners face to face with the cruelties of slavery. Runaways were real people who needed help, not faraway abstractions. The presence

TEX.

to Mexico

of slave catchers in Northern cities and towns outraged even people who had never opposed slavery in the South or cared for African Americans.

The growth of an abolitionist movement paralleled the existence of the Underground Railroad, dedicated to immediately ending slavery in the United States. Prominent abolitionists like Sojourner Truth, Henry Highland Garnet, and Frederick Douglass raised money to help runaways, printed anti-slavery petitions and newspapers, went on speaking tours, and lobbied politicians to abolish the institution. Their constant pressure—in the face of legal trouble, violence, and even death threats—kept the issue of slavery at the enter of American politics.

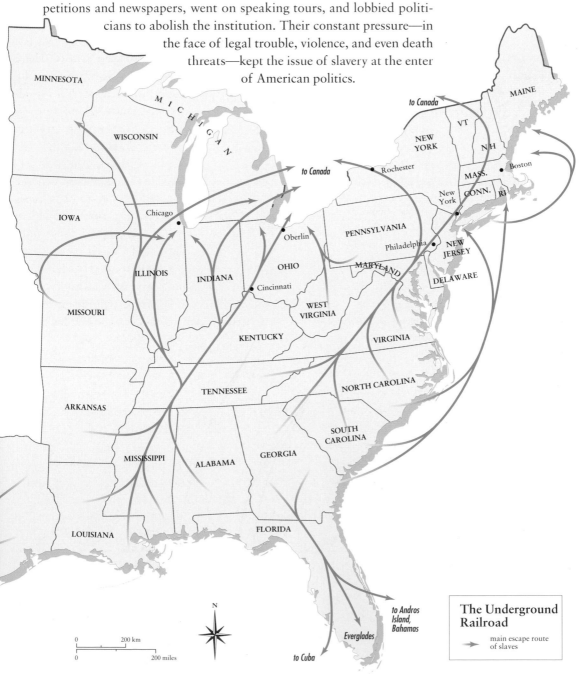

The Underground Railroad

→ main escape route of slaves

61

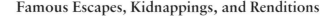

Famous Escapes, Kidnappings, and Renditions

This map illustrates how difficult it was for slaves to escape to freedom from the Deep South. Most of the best-known African American escapees—including Frederick Douglass, Harriet Tubman, Henry Highland Garnet, and Josiah Henson—lived in border states. Henry "Box" Brown was actually shipped from Richmond to Philadelphia in a crate by the Adams Express Company. The harrowing escape of William and Ellen Craft from Georgia was a rare exception. Ellen, whose light skin allowed her to pass as white, dressed as a planter, and William accompanied her as the "planter's" trusty slave. The two boldly boarded a train for Savannah near their plantation in Macon, Georgia, and made their way by boat and rail to Philadelphia. Their escape required several things ordinary slaves did not usually possess: money and an intricate knowledge

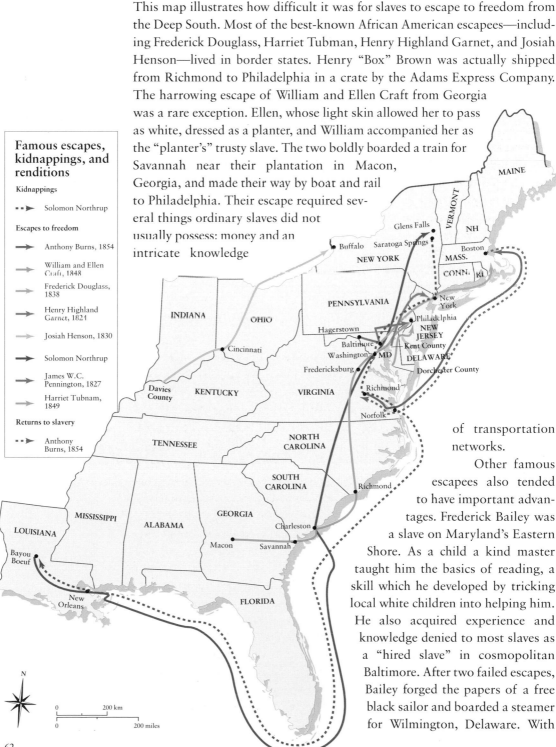

of transportation networks.

Other famous escapees also tended to have important advantages. Frederick Bailey was a slave on Maryland's Eastern Shore. As a child a kind master taught him the basics of reading, a skill which he developed by tricking local white children into helping him. He also acquired experience and knowledge denied to most slaves as a "hired slave" in cosmopolitan Baltimore. After two failed escapes, Bailey forged the papers of a free black sailor and boarded a steamer for Wilmington, Delaware. With

the aid of other black abolitionists he settled in New Bedford, Massachusetts, and he changed his name to Frederick Douglass to avoid easy detection.

Harriet Tubman, known as the "Moses of her people" for helping as many as 300 slaves escape, and Henry Highland Garnet, a famous black abolitionist, were also from the border state of Maryland, as was James W. C. Pennington. Josiah Henson was a Kentucky plantation slave who escaped with his wife and children by walking through Indiana and Ohio

$150 REWARD

RANAWAY from the subscriber, on the night of the 2d instant, a negro man, who calls himself *Henry May*, about 22 years old, 5 feet 6 or 8 inches high, ordinary color, rather chunky built, bushy head, and has it divided mostly on one side, and keeps it very nicely combed; has been raised in the house, and is a first rate dining-room servant, and was in a tavern in Louisville for 18 months. I expect he is now in Louisville trying to make his escape to a free state, (in all probability to Cincinnati, Ohio.) Perhaps he may try to get employment on a steamboat. He is a good cook, and is handy in any capacity as a house servant. Had on when he left, a dark cassinett contec, and dark striped cassinett pantaloons, new—he had other clothing. I will give $50 reward if taken in Louisville; 100 dollars if taken one hundred miles from Louisville in this State, and 150 dollars if taken out of this State, and delivered to me, or secured in any jail so that I can get him again. WILLIAM BURKE

Bardstown, Ky., September 3d, 1838.

before boarding a ship for Buffalo, New York, and freedom in Canada. The Hensons were extremely unusual both for their excellent luck and their escape as an intact family.

Unfortunately, the journey North to freedom was sometimes reversed. Free African American Solomon Northrup of Saratoga Springs, New York, was kidnapped and sold into slavery by two unscrupulous whites in 1841. Northrup was later sold in New Orleans to a cotton planter from remote Bayou Boeuf near the Red River in Louisiana. He remained enslaved for 12 years before he met a kind white carpenter who wrote a letter to Northrup's family and friends in New York. Using an 1840 New York law passed at the urging of the state's free blacks to protect citizens from being sold into slavery, Northrup rejoined his family in 1853. The publication of his harrowing tale, along with Douglass's first autobiography, helped persuade Northerners of the evils of slavery.

Even after a successful escape to freedom, runaways needed to maintain constant vigilance against slave catchers, especially after the passage of the Fugitive Slave Law. The law, part of the Compromise of 1850, required only an affidavit from a slave state court identifying an African American as a runaway to return the person to slavery. Fugitives were forbidden to speak in their own defense. In one spectacular case, a Virginia slave named Anthony Burns escaped to Boston in 1854. Apprehended by a federal marshal, Burns's case became a cause célèbre as angry abolitionists tried to rescue him and block his return to slavery. Thousands of Bostonians somberly lined the streets as Burns was marched by soldiers and marines to the harbor and returned to bondage. Abolitionists purchased his freedom the following year, and Burns died a free man in Canada.

Running away was one of the most common ways slaves rebelled against their condition. Although a large majority of runaways were eventually caught, some fugitives (usually those from the border states) gained freedom in the North or in Canada.

Slavery and the Civil War

Slave "Contraband" Camps, 1861–63

After the early battles of the Civil War, it was clear that the Union would not win a quick victory against the Confederacy. As the war intensified, Union commanders and leaders were faced with a different challenge: everywhere the Northern army advanced, runaway slaves (seeing it as an army of liberation) streamed behind Union lines. Hundreds and then thousands of slaves "voted with their feet" for liberty and left plantations and farms for places occupied by Union soldiers including southeastern Louisiana, Missouri, Tennessee, Virginia, and the coastal Carolinas.

At the beginning of the war, President Lincoln instructed his generals to return the slaves to their masters, insisting the Civil War was about preserving the Union, not freeing the slaves. As the war progressed, however, and the South continued to win battles, Northern military planners began to see the need for "total war": warfare designed to utilize every resource that would bring victory. This included the use of any tactic—even destruction of property or institutions — that would damage the enemy's "will to fight." An attack on slavery, some argued, would strike at the heart of the Confederacy's African American labor force and invest the war with a new moral cause: fighting for freedom.

As early as 1861, runaway slaves forced Union generals to see them as a potential resource. Generals like Benjamin Butler, whose army had penetrated into Virginia, refused to return runaways to their owners and began to use them as laborers and even soldiers. He insisted they were "contraband of war," and should be utilized for the cause.

Lincoln was furious: he

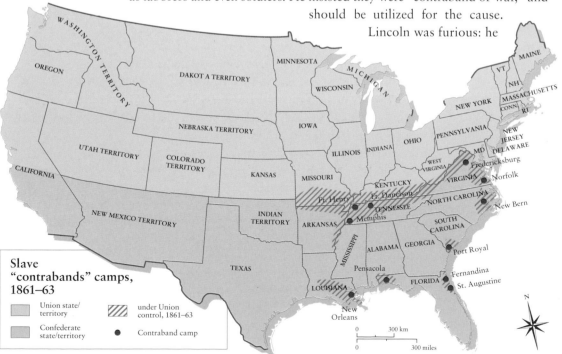

Slave "contrabands" camps, 1861–63

- Union state/territory
- Confederate state/territory
- under Union control, 1861–63
- Contraband camp

was still unwilling to make the war a fight for freedom. Abolitionists fumed. "To fight against slaveholders, without fighting against slavery," said Frederick Douglass, "is a half-hearted business, and paralyzes the hands engaged in it. . . . War for the destruction of liberty must be met with war for the destruction of slavery." But the slaves' action and Butler's response forced the government to compromise. In 1862, Congress passed the Contraband Act, classifying runaway slaves as contraband of war. Instead of returning runaways to their masters, they were placed in internment camps, where they were forbidden to aid the Union Army as soldiers or laborers. Northern religious and social groups led by free blacks were horrified when they visited the contraband camps, and lobbied strenuously for their dismantling.

For the rest of the war slaves who came within Union lines were known as contrabands. Their dramatic act—"freeing" themselves—helped force Lincoln and Congress to transform a war to preserve the Union into one for emancipation.

The Emancipation Proclamation

It was a dire military situation that drove Lincoln to alter his policy on emancipation. Although morally opposed to slavery, Lincoln was faced with the daunting task of keeping the loyal slave states of Missouri, Maryland, and Kentucky in the Union. A general emancipation proclamation would have handed the border states to the secessionists. But pressure from African Americans, his military leaders, and his own political party compelled Lincoln to move toward emancipation.

General Ulysses Grant believed that if the South could not "be whipped in any way other than through a war against slavery, let it come to that." This belief, compounded by Confederate success on the battlefield, forced Lincoln's hand. On July 13, 1862, Lincoln told his secretary of the Navy he intended to issue an emancipation proclamation. "We must free the slaves or ourselves be subdued," he said. "The Administration must set an example, and strike at the heart of the rebellion." The president decided to delay the proclamation until a Union victory in the field, so as not to appear desperate. The proclamation, which would affect the lives of millions of African Americans, languished for two long months in a desk drawer while Lincoln waited for good news from the battlefield.

It came on September 17, when the Battle of Antietam in Maryland ended in a draw. Lincoln retrieved his proclamation and on September 22 proclaimed that if any state (or part of a state) was still in rebellion on January 1, 1863, the slaves there would be "forever free." When the deadline arrived no Southern state had re-entered the Union, and slaves in the border states, of course, were unaffected by the proclamation. Lincoln's edict, therefore, failed to free a single slave. It did, however, set in motion the events that would bring about the legal abolition of slavery for 4 million African Americans. If the Confederacy died as a result of the war, slavery would die along with it.

Radical Reconstruction, 1866–77

African Americans, who constituted a majority in South Carolina, wielded considerable power in the state legislature during Reconstruction. In the state's first legislative session after readmission, blacks held a 2-1 majority in the House.

In his famous address at Gettysburg, Abraham Lincoln said the Union could be reconstructed only by experiencing a "new birth of freedom." For the nation's African Americans, this meant significantly more than the abolition of slavery. Black abolitionists like Frederick Douglass had long argued that freedom meant more than simply "owning" oneself: it meant African Americans had to be treated as equals to whites before the law, and that they should be allowed to vote in order to protect those rights. But many Southerners—joined by significant numbers of Northerners—were unwilling to grant blacks equal rights.

After Lincoln was assassinated in April 1865, his Tennessee-born vice president, Andrew Johnson, assumed the nation's highest office. Johnson quickly proposed to exclude African Americans completely from the Reconstruction process, saying that "white men alone must govern the South." After a delegation led by Douglass confronted Johnson on black suffrage, he told his secretary that "those damned sons of bitches thought they had me in a trap!…[T]hat damned Douglass, he's just like any nigger, and he would sooner cut a white man's throat than not." African Americans and radical Republicans were outraged when the Southern states ratified their new constitutions without enfranchising a single black man. Enfranchising women of any race was not on the agenda of most politicians in either North or South.

Things got worse in the fall of 1865 when some Southern states enacted "Black Codes" that reduced freed people to slavery-like conditions. Congress, controlled by Radical Republicans, took matters into their own hands. They began impeachment proceedings against Johnson and passed the 14th and 15th Amendments, which granted African Americans full citizenship and gave black men the right to vote. Then they divided the South into five military districts, each with a military governor, and forced the states to write new constitutions which included the 13th, 14th, and 15th Amendments.

With Union soldiers occupying the region and black men given the right to vote, large numbers of African Americans were elected to help rewrite the state constitutions. Blacks and their white Unionist and Northern allies organized Union leagues to mobilize new black voters for the Republican Party; by September 1867 there were 735,000 black voters and only 635,000 white voters

in the ten Southern states. The new state constitutions enacted universal male suffrage (putting them ahead of many Northern states), mandated statewide public schools for blacks and whites, and increased the states' responsibility for social welfare.

With African Americans representing almost 80 percent of Southern Republicans, black men were elected to 15–20 percent of the public offices. Fourteen African Americans were elected to the House of Representatives and two to the Senate between 1868 and 1876. African Americans wielded the greatest influence in South Carolina, where they constituted a majority of the population. In the first legislature after the state was readmitted to the Union, there were 87 blacks and 40 whites. Between 1868 and 1896 Louisiana had 133 black legislators, including 38 state senators. Three African Americans—P.B.S. Pinchback, Oscar Dunn, and C. C. Antoine—served as lieutenant governor, and Pinchback was acting governor for 43 days in 1873 when the governor was removed from office. Yet at no time did blacks control a state government.

A large majority of black elected officials were well educated, but racist Democrats chafed under what they called "Negro rule." Venal and racist prop-aganda claimed that all black voters and officeholders were illiterate and incompetent—a lie that was later enshrined in folk memory and history books. The fact remains that both blacks and poor Southern whites benefited greatly from Radical Reconstruction. By 1877, Northern Republicans had grown exhausted from enforcing Reconstruction, and Southern conservatives, acting through terror and propaganda, regained control of the state governments. One by one the Southern states rolled back the gains made by blacks, reinstating poll taxes, literacy tests, and property requirements to disfranchise black citizens. In the decades to come, freed slaves and their descendants suffered second-class citizenship and brutal repression at the hands of Southern whites.

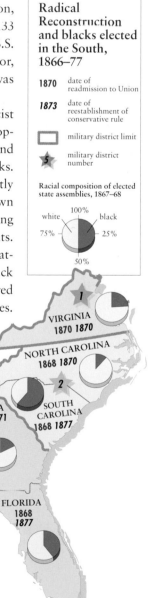

Radical Reconstruction and blacks elected in the South, 1866–77

1870 date of readmission to Union

1873 date of reestablishment of conservative rule

☐ military district limit

⭐**5** military district number

Racial composition of elected state assemblies, 1867–68

white 100% black
75% — — 25%
50%

VIRGINIA
1870 *1870*

NORTH CAROLINA
1868 *1870*

ARKANSAS
1868 *1874*

TENNESSEE 1866 *1869*
(Restored to the Union before Radical Reconstruction)

SOUTH CAROLINA
1868 *1877*

ALABAMA
1868 *1874*

GEORGIA
1870 *1871*

TEXAS
1871 *1873*

MISSISSIPPI
1870 *1876*

LOUISIANA
1868 *1877*

FLORIDA
1868
1877

0   200 km

0   200 miles

"Exodusters"

After federal troops pulled out of the South and Reconstruction collapsed, thousands of Southern blacks decided the time was right to move. With ex-Confederates back in power in most states and hate groups such as the Ku Klux Klan terrorizing the countryside, many African Americans found life in the post-Reconstruction South unbearable. Some, like Henry McNeal Turner, encouraged blacks to leave the country and move to Africa; few African Americans, however, were willing to leave the country that was the only home they knew. Far more significant was the exodus of blacks from the rural South to the North and West after Reconstruction.

Right: Emigrants from the Deep South waiting for a Mississippi River boat in the 1880s. Blacks hoping for a better life in Kansas and the West became known as "Exodusters" after the biblical exodus from Egypt.

Thousands of ex-slaves answered the call sounded by Benjamin "Pap" Singleton, a minister from Tennessee, to abandon the racial prejudice of the South and go West. Singleton issued circulars like "The Advantage of Living in a Free State," encouraging blacks to follow him to Kansas. In a time where their political rights were being trampled, lynchings were on the rise, and economic depressions made sharecropping and tenant farming resemble slavery, some Southern blacks decided they had no future in the South. The largest organized exodus to Kansas began in Nashville in 1880, where more than 60,000 unhappy blacks gathered to begin a journey to a new life. Singleton's followers were quickly dubbed "Exodusters" by newspapers that compared their flight from the South with the Israelites flight from Egyptian slavery.

The movement continued in the 1880s, as black leaders like David Turner in Oklahoma, Edward McCabe in Kansas, and Allen Allensworth in California established and promoted politically independent and economically viable all-black towns. Many of these settlements were victims of white racism, poor planning, and bad luck, but others, like Nicodemus, a town named for an African slave in western Kansas, took root. A flier inviting African Americans to settle in Nicodemus included a song, with the chorus:

> Good time coming, good time coming,
> Long, long time on the way;
> Run and tell Elijah to hurry up Pomp,
> To meet us under the cottonwood tree,
> In the great Solomon Valley
> At the first break of day.

The Exodusters experienced resistance from other blacks, including Booker T. Washington, who urged those weighing flight to "cast your bucket down where you are." The Exodusters were the first mass movement of blacks after the Civil War, but they would be joined in the 20th century by millions of others fleeing the South for a better life in the North, West, and Midwest.

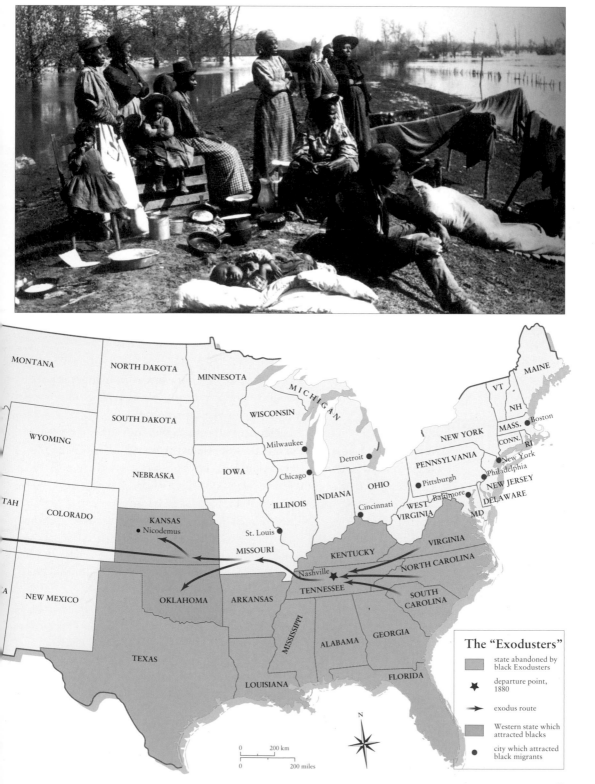

The "Exodusters"

- state abandoned by black Exodusters
- ★ departure point, 1880
- → exodus route
- Western state which attracted blacks
- city which attracted black migrants

PART IV: AFRICAN AMERICANS UNDER ARMS

This is our golden opportunity. Let us accept it, and forever wipe out the dark reproaches unsparingly hurled against us by our enemies. Let us win for ourselves the gratitude of our country, and the best blessings of our posterity through all time.
—Frederick Douglass, 1863

Africans and African Americans have fought bravely in every major military action since colonial times, despite the fact that black soldiers and sailors were forced to battle twin foes: the wartime enemy and withering racism both inside and outside the armed services. Inside the military, blacks were routinely ordered to perform dangerous missions or relegated to auxiliary duties like building roads or burying the dead. They were effectively barred from serving in any military branch besides the army until World War II. And after each war ended blacks repeatedly returned to a larger society that minimized their military contributions and confined them to second class status. Yet with every armed conflict, African Americans clung to a belief that military service in wartime represented a clear path toward greater freedom and opportunity, for the individual as well as the race as a whole.

The 10/th Colored Infantry, photographed at Fort Corcoran, part of the Union defensive perimeter around Washington D.C. during the Civil War.

The Revolutionary War set several precedents for black military service. Africans and African Americans fighting on both sides saw military service as a means by which a slave might win his freedom or a free black improve his standing in the community. Yet, in what would become a pattern in later conflicts, only a relative few blacks were able to take advantage of the war to improve their status. After the war ended, those slaves who sided with the British were either returned to their masters or evacuated to Nova Scotia and Sierra Leone. Northern black soldiers fared slightly better, and their brave service in battles like Bunker Hill and Saratoga helped convince leaders in several states to formally abolish slavery in the years after the Revolution.

Far more significant for African Americans was the Civil War, known to many blacks as the Jubilee War. Seeing the war's potential to become a war for freedom, blacks were among the first to volunteer for service in the Union Army in 1861. But President Lincoln, seeking to mollify the border states that remained in the Union, refused them until early 1863. That year saw the formation of several all-black regiments, although at the time they were paid less than white soldiers and generally served

only as labor battalions and garrison forces. Black troops won the right to fight in combat, however, helping to destroy slavery and the Confederacy and giving the United States a new birth of freedom at the war's conclusion. More than 186,000 African Americans served in the Union Army in the war.

After the Civil War the U.S. government kept several black regiments active, ordering most of them to the West to fight Indians and protect American business interests there. These same Buffalo Soldier units also saw action in the Spanish-American War, in both Cuba and the Philippines. More than a few black soldiers and sailors (and black leaders at home) drew parallels between American imperialism abroad and its treatment of racial minorities at home. Segregated black troops were also a vital part of the U.S. forces that turned the tide for the allies during World War I. Racism was rampant on every level and especially visible in the lily-white officers' corps and harrowing race riots on military bases. Still, all-black units like the Harlem Hell Fighters of the 369th and the Tenth Cavalry fought bravely on the Western Front, halting the German advance and pushing them all the way back to the Rhine in 1918.

World War II struck many African Americans as an ideal opportunity to make gains at home by fighting racism and injustice abroad. The Axis powers included Nazi Germany, where Hitler had come to power preaching racism and anti-Semitism, and Italy, which had invaded the black kingdom of Ethiopia in 1935. Black leaders spoke of a "Double-V Campaign" for twin victories against racial injustice in Europe and in the United States. Approximately 1 million blacks served in the armed services, filling out the ranks in he Marine Corps, Navy, and Coast Guard, which had long been closed off to them. The age of the segregated military came to an end in 1948, when President Harry S Truman issued an executive order integrating all branches of the armed forces. Thus the Korean War (1950–53) was the first armed conflict where African American and white troops fought together side by side.

America's involvement in Vietnam closely paralleled the struggles of the civil rights movement. A larger percentage of the nation's African Americans fought—and died—in the war than the nation's whites, a statistic that added to the racial tumult of the 1960s. When federal troops were called in to quell riots in Detroit in 1967, many saw disturbing parallels between action in Vietnam and in urban America. Blacks continued to volunteer for and serve in the army nonetheless, and many veterans were sorely disappointed at the treatment they received after the United States pulled out. Since Vietnam the United States has deployed its armed forces in several limited engagements including the Persian Gulf War, peacekeeping and humanitarian missions in Haiti and Somalia, and in the Balkans.

Ironically, the U.S. armed forces—the nation's most authoritarian institution—has been more successful than the private sector in implementing integration. African Americans have risen to the military's highest ranks, and occupy positions of authority in numbers that far outpace those in the rest of society. Yet patterns of racism and inequality established hundreds of years ago persist.

Black Patriots and Loyalists

African Americans and the Revolution, 1770–81

The revolutionary spirit following the imperial crisis with Great Britain was not confined to whites. The revolutionary writings of Jefferson, Jean-Jacques Rousseau, and Tom Paine were read and discussed by hundreds of thousands of colonists in the 1760s and '70s, black as well as white, free as well as slave. Many were more than willing to take up arms for the revolutionary cause.

Even before the formal outbreak of hostilities in 1775, blacks were willing to risk their lives to resist British encroachments. A runaway slave named Crispus Attucks, a vehement opponent of British policies including the occupation of Boston in 1770, was one of five protestors gunned down in the Boston Massacre. And as early as the battles of Lexington and Concord, blacks took up arms against British soldiers. Blacks' brave service in these early battles brought a vexing question before the patriot leaders: should blacks be armed to fight against Great Britain? After the Stono rebellion, fear of slave insurrections even in New England led colonists to bar blacks from owning guns and serving in militias. Thus the Committee on Safety ruled that only freemen could be used in the war.

There is ample evidence, however, that the committee's ruling was ignored: enslaved blacks joined their free brethren in the Battle of Bunker Hill, and some slaves were freed explicitly so they could serve in the army. Among those who were commended for service in the war were Prince Hall (a black activist who later established the first black Masonic hall), Cato Tufts, Titus Colburn, and Cuff Hayes.

Still, there was widespread opposition in both North and South to a major black presence in the Continental Army. When George Washington took command in mid-1775, his war council instructed recruiters not to enlist any new black soldiers; an order later that year rejected blacks altogether—dubious thanks for black participation in battles like Bunker Hill.

All this changed in November 1775 when Lord Dunmore, the royal governor of Virginia and a British loyalist, declared "all indentured servants, Negroes, or others . . . free, that they are able and willing to bear arms, they joining his Majesty's troops." Later that month Virginia militiamen defeated a force of almost 800 slaves and 200 redcoats; most of the slaves fell prey to smallpox after the battle. Not surprisingly, thousands of slaves ran away toward British lines to seek the freedom denied them during the previous century and a half (Thomas Jefferson estimated that 30,000 Virginia slaves ran away in 1778 alone). Southern slaves viewed the British troops as an army of emancipation, not invasion.

The next month, due to Dunmore's decree and protests led by Prince Hall, Washington reversed his policy and allowed free blacks to serve in the Continental Army. Washington also cited fears that blacks would seek to fight on the British side if not permitted to serve with the patriots. Approximately 5,000 blacks eventually served in the Continental Army, most of them from the

North. And although there were a few all-black companies (including the Bucks of America in Massachusetts), a large majority of black soldiers fought in groups composed largely of white men. At least some black soldiers participated in nearly every one of the major battles of the Revolution, including Ticonderoga, Princeton, Brandywine, Savannah, and Yorktown.

Resettlement of Black Loyalists after the Revolution

Although 5,000 blacks fought with the patriots in the Revolution, most enslaved African Americans concluded their best chance for liberty lay with the British army. According to historians, almost 10 percent of slaves—more than 50,000—ran away during the war. Of these,

MAINE
(part of Massachusetts)

Fort Ticonderoga

NEW
HAMPSHIRE

Saratoga
Oct. 17, 1777

Bennington
Aug. 16, 1777

Concord and Lexington
April 19, 1775

NEW YORK

MASS.

Boston massacre
1770

Stony Point
July 15–16, 1779

Siege of Boston
1775–76

CONNECTICUT

RHODE
ISLAND

White Plains
Oct. 28, 1776

PENNSYLVANIA

Brooklyn Heights
Aug. 27, 1776

Princeton
Jan. 3, 1777

Monmouth Court House
June 28, 1778

Brandywine
Sept. 11, 1777

Trenton
Dec. 26, 1776

NEW
JERSEY

DELAWARE

MARYLAND

VIRGINIA

Yorktown
Oct. 19, 1781

N

NORTH CAROLINA

0   100 km

0   100 miles

SOUTH
CAROLINA

African Americans and the Revolution, 1770–81

GEORGIA

Savannah
Dec. 29, 1778

New England colonies

Middle colonies

Southern colonies

battle where African Americans were used

approximately 20,000 were evacuated by the British. At least 6,000 loyalist blacks departed from Charleston, 4,000 from Savannah, and more than 3,000 from New York City.

Some runaway loyalists were quickly disappointed with Great Britain: many were treated as contraband and faced resale to new (loyalist) planters; others ran away only to face punishment by their owners when the redcoats failed to rescue them. But, for the most part, reaching British lines meant freedom for runaway slaves. When redcoat troops began their withdrawal after Yorktown, they took blacks with them. If they had served in the British army or navy, they were allowed passage to any part of the British Empire. Some chose to go to Jamaica or London, but most went to Halifax, Nova Scotia, where the British government promised to care for the refugees.

Most Nova Scotians did not welcome the new arrivals, however, and the British treated them with neglect. These people had arrived in Canada with nothing: no possessions, no kinship ties, no familiarity with the climate or landscape. In 1792 Thomas Peters, a black leader in Halifax, traveled to London to plead the refugees' case. As a result of Peters's visit, the British created a new colony for ex-slaves in Sierra Leone, on West Africa's Rice Coast. Not surprisingly, they christened their new home Freetown. Theirs was the first African state founded and controlled by blacks from what is now the United States.

More than 10,000 blacks fought in the Revolutionary War, on both sides of the conflict. Most enslaved blacks saw the British army, which promised to free each slave who volunteered, as one of liberation, not tyranny.

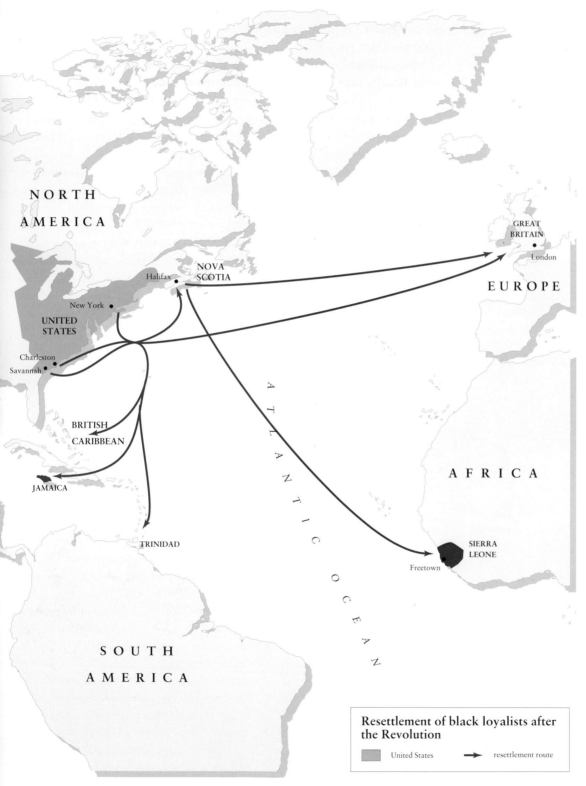

NORTH
AMERICA

GREAT
BRITAIN
•
London

EUROPE

Halifax • NOVA
SCOTIA

New York •

UNITED
STATES

Charleston
Savannah •

BRITISH
CARIBBEAN

AFRICA

JAMAICA

SIERRA
LEONE

Freetown

TRINIDAD

A T L A N T I C   O C E A N

SOUTH
AMERICA

**Resettlement of black loyalists after
the Revolution**

United States   ⟶ resettlement route

Black Men in Blue

African Americans, with an eye to the Civil War's potential to become a war for freedom, were among the first to volunteer to serve in the Union Army in 1861. But a law dating from 1812 barred blacks from service in the U.S. military, and President Lincoln and others still held that the Civil War was a conflict between whites over the fate of the Union. Those opposed to African American enlistment clearly understood that one consequence of blacks' fighting for the Union would be a major step toward racial equality. So, too, did abolitionists like Frederick Douglass. "Once let the black man get upon his person the brass letters, U.S.," he said, "and a musket on his shoulder and bullets in his pocket, and there is no power on earth which can deny that he has earned the right to citizenship." He was right: beginning in 1863, blacks did fight for the Union, helping to destroy slavery and the Confederacy, giving the United States a new birth of freedom after the war was over.

The first African Americans to fight in the Civil War were runaway slaves armed by Union officers in defiance of Lincoln's orders. As early as 1861–62, Union commanders occupying portions of South Carolina and Louisiana began to organize black regiments. But the U.S. government, hoping to strike at the heart of the Confederacy and make up for a lag in white enlistment, didn't lift the ban on African American troops until 1863. When first organized, the new black regiments were paid less than whites, their officers were white, and they served only as labor battalions and garrison forces. Over the next year, however, black soldiers won the right to fight, earn equal pay, and, in rare cases, be led by African American officers.

Black leaders like Frederick Douglass served as recruiting agents, and massive rallies in New York, Boston, and Philadelphia helped spur 186,000 blacks to enlist by the end of the war. Once in combat, black soldiers quickly quashed the racist notion that they wouldn't fight as well as whites. Eight black regiments participated in the assault on Port Hudson in Louisiana, and others bravely defended a Union outpost near Vicksburg called Milliken's Bend. Perhaps most significant was the 54th Massachusetts Infantry, the first black regiment raised in the North. Two of Frederick Douglass's sons were in the regiment, one as sergeant major, and the commanding officer was Robert Gould Shaw, the scion of a prominent abolitionist family. After months of garrisoning forts and supplying white soldiers, the 54th finally won the right to fight on July 18,

Fort Wagner, South Carolina, July 18, 1863

→ attack of 54th Massachusetts

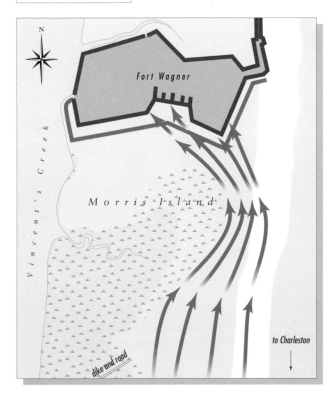

1863. The regiment was assigned the lead position in an assault on Fort Wagner, part of the network of defenses fortifying Charleston, South Carolina. The assault was a bloody one, and the black soldiers were repelled with 50 percent casualties. But every witness saw the 54th Massachusetts fight with tremendous courage. According to the *New York Tribune*, the battle "made Fort Wagner a name to the colored race as Bunker Hill had been for ninety years to the white Yankees."

The Confederate government reacted to the use of African American troops by declaring that they would take no black prisoners. Unlike white troops, blacks in uniform were often massacred as they surrendered, most famously at Fort Pillow in Tennessee. Despite this knowledge, African Americans continued to volunteer for service and fight with astonishing courage in more than 200 battles across the South. More than 38,000 black soldiers lost their lives during the war, a rate of mortality nearly 40 percent higher than that for white troops. Much of this can be attributed to the "take no prisoners" policy practiced by Southerners in dealing with black soldiers.

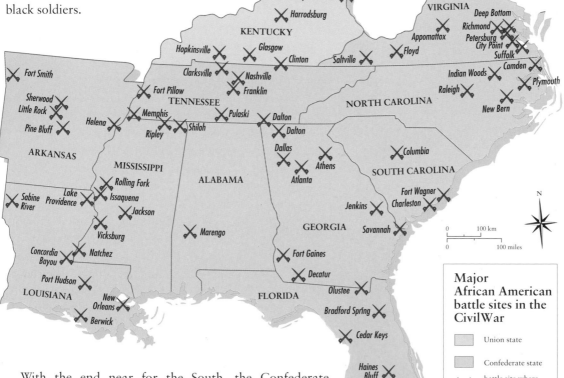

Major African American battle sites in the Civil War

Union state

Confederate state

battle site where African American troops saw action

With the end near for the South, the Confederate Congress authorized the raising of 200,000 black (slave) troops to defend Southern cities and forts. But before any Southern black regiments could be organized, the war ended. The result was clear: African Americans had contributed heavily to the Union victory, making the Civil War a war for freedom.

Buffalo Soldiers and the Indian Wars, 1866–90

Black soldiers proved beyond a shadow of a doubt that they could fight with courage and cunning during the Civil War. After the South surrendered in 1865, the U.S. army quickly focused on protecting white settlement and business interests in the far West. White settlers had followed the transcontinental railroad lines into Indian country, and the region's native inhabitants responded with a series of attacks on towns and rail lines. Those troops not being used to occupy the South, including scores of African American veterans, were quickly diverted to the West.

Blacks had been involved—often against their will—in the struggle between whites and Indians since the first Africans arrived in Virginia in 1619. Knowing they would only inflame hostility in the South, Congress decided to send the best African American army veterans west to fight Plains Indians in 1866. On July 28, Congress authorized the establishment of four permanent black units: two infantry regiments (the U.S. 24th and 25th) and two cavalry regiments (the Ninth and Tenth). Between 1866 and 1880 these four units were engaged almost constantly with recalcitrant Indian tribes. Troops in all four regiments were respectfully nicknamed "Buffalo Soldiers" by their Indian enemies, who saw a resemblance between the hair of the black cavalrymen and the hair of the buffalo, an animal they considered sacred.

The Buffalo Soldiers were involved in countless campaigns against the Sioux, Cheyenne, Crow, Comanche, Arapaho, Navajo, and Apache. They were often the first troops called in to put down rebellions on reservations and face the most fearsome Indian warriors, including Geronimo and Crazy Horse. The regimental officers, who were white, later recalled it was not uncommon for black troops to do virtually all the fighting and sustain nearly all

African American soldiers sent to the West to fight Indians after the Civil War were nicknamed "Buffalo Soldiers" by their adversaries.

the casualties, while white troops received the commendations. The Buffalo Soldiers were also used to police Indian Territory (present-day Oklahoma) until 1901, protecting ranchers and cattle from Indian attacks and protecting Indians from land-hungry white settlers. Many Buffalo Soldiers enlisted because they believed their service would help advance their race in other areas of American society. Yet many were sorely disappointed, despite their distinguished service (17 African Americans received the Congressional Medal of Honor during the Indian wars) and the numerous hardships they experienced compared to their white counterparts.

The Buffalo Soldiers and the Indian wars, 1866–90

- area where Buffalo Soldiers served during the Indian wars
- area policed by Buffalo Soldiers before the 1901 Land Rush
- fort housing black troops
- battle

0   200 km
0   200 miles

Imperial Wars, 1898–1902

The Buffalo Soldiers of the Ninth and Tenth cavalries and the 24th and 25th infantries also saw service in the Spanish-American War, a conflict that established the United States as an imperial power. The black troops joined thousands of others in Cuba, then a Spanish colony, after the USS Maine sank under mysterious circumstances in February 1898.

Called "Smoked Yankees" by the Spanish because of their dark skin, African American soldiers fought at El Caney, Kettle Hill, Las Guásimas, and in the famous battle of San Juan Hill. There, the Ninth Cavalry's "K" Troop achieved fame and respect for saving Teddy Roosevelt and his Rough Riders, who had become bogged down at the foot of San Juan Hill. Amid chaos, black and white troops intermingled in the charge up the hill, which was successful. Militia regiments from Illinois and Kansas, composed of African Americans from the colonel in command to the lowliest private, helped garrison Cuba and the island of Puerto Rico.

Some African Americans questioned whether black citizens, whose own rights were being whittled away by Jim Crow laws, should support a war to bring American-style "freedom" to Cuba and its significant black population. Henry Turner, the senior bishop of the AME church, put it bluntly: "Negroes who are not disloyal to the United States ought to be lynched." Others disagreed, hoping that black participation in the war would spark a new era of comradeship and good feeling among the races at home. As for the black soldiers, most welcomed the prospect of adventure, a steady salary, and an opportunity to prove themselves as warriors.

Black troops also participated in the naval assault of Manila Bay in the Philippines, a Spanish colony in the Pacific. John Jordan, a black gunner's mate, was in charge of the crew that fired the first shot in the short and successful battle. To the Filipino insurgents, however, the Americans who quickly occupied the islands were little different than their Spanish oppressors. Following the short war, members of the 24th and 25th infantries were stationed in the Philippines and Hawaii, where they witnessed (and participated in) some of the atrocities against and mistreatment of the Filipinos who rebelled against American rule. They also experienced poor treatment by white American soldiers. A very

Imperial wars,
1898–1902

→ U.S. territorial expansion, 1867–99

→ territory acquired by military action, 1898

▮ U.S. and under U.S. control, c. 1900

• main source of black volunteers who fought in the wars

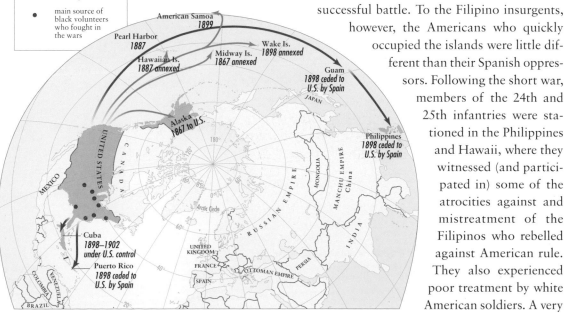

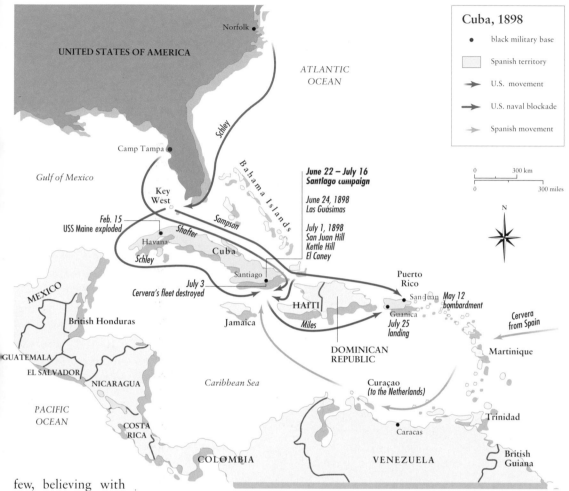

Cuba, 1898

- • black military base
- Spanish territory
- → U.S. movement
- → U.S. naval blockade
- → Spanish movement

UNITED STATES OF AMERICA

ATLANTIC OCEAN

Norfolk

Camp Tampa

Gulf of Mexico

Key West

Feb. 15
USS Maine exploded

Havana

Schley

Shafter

Cuba

Santiago

July 3
Cervera's fleet destroyed

Bahama Islands

Sampson

**June 22 – July 16
Santiago campaign**

June 24, 1898
Las Guásimas

July 1, 1898
San Juan Hill
Kettle Hill
El Caney

Puerto Rico

San Juan   May 12
bombardment

Guanica
July 25
landing

Cervera
from Spain

HAITI

Jamaica

Miles

DOMINICAN
REPUBLIC

Martinique

MEXICO

British Honduras

GUATEMALA

EL SALVADOR

NICARAGUA

PACIFIC
OCEAN

COSTA
RICA

COLOMBIA

Caribbean Sea

Curaçao
(to the Netherlands)

Caracas

VENEZUELA

Trinidad

British
Guiana

0   300 km
0   300 miles

N

few, believing with Bishop Turner that the American occupation constituted an "unholy war of conquest," mutinied and cast their lot with the Filipino rebels. Even those that didn't, like veteran John Calloway of the 24th, recalled being "haunted by the feeling of how wrong morally Americans are in their present affair." Turner's and Calloway's solidarity with oppressed Asians would reverberate across the 20th century, climaxing in the 1960s with the widespread view in the black community that both the Vietnamese and African Americans were victims of white oppression.

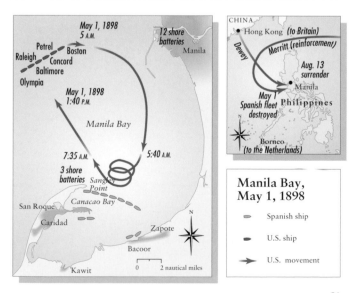

CHINA
Hong Kong (to Britain)

Dewey

Merritt (reinforcement)

Aug. 13
surrender

Manila

May 1
Spanish fleet
destroyed

Philippines

Borneo
(to the Netherlands)

May 1, 1898
5 A.M.

Petrel
Raleigh   Boston
Concord
Baltimore
Olympia

12 shore
batteries

Manila

May 1, 1898
1:40 P.M.

Manila Bay

7.35 A.M.

3 shore
batteries   Sangley
Point

San Roque

Canacao Bay

Caridad

5:40 A.M.

Zapote

N

Bacoor

Kawit

0   2 nautical miles

Manila Bay,
May 1, 1898

- Spanish ship
- U.S. ship
- → U.S. movement

World War I

More than 400,000 African Americans served in the U.S. Army during World War I, despite unprecedented racial oppression inside and outside the armed services. During the years between the Philippine insurrection and the U.S. entry into World War I, black mutinies and riots broke out near bases in Brownsville and Houston. Many black elites hoped that the willingness of African Americans to fight in World War I would somehow break the cycle of white repression and black violence. W.E.B. DuBois, a founder of the NAACP, urged his fellow African Americans "to forget our special grievances and close our ranks shoulder to shoulder with our fellow white citizens and the allied nations that are fighting for democracy."

The all-black 369th Infantry earned the nickname the "Harlem Hell Fighters" for their service in World War I. Their unit was the first to reach the Rhine River after they successfully repelled a German advance.

Almost all the blacks who served in the war did so in the U.S. Army, since they were barred from the Marines and Coast Guard and effectively barred from the Navy. In the army they were forced into service units like the stevedores, the quartermasters, and the pioneer infantrymen, whose responsibilities included cooking, cleaning, digging latrines, burying the dead, and loading supplies. The early days of the war followed a familiar pattern, with white officers commanding all-black units like the Tenth Cavalry and the 805th Pioneer Regiment, and the Harlem Hell Fighters of the 369th. But late in 1917, the government began ROTC programs at several historically black colleges and established a segregated officers training school at Fort Des Moines, Iowa. By the time of the armistice in late 1918, more than 639 black officers led troops during some of the war's most significant campaigns.

In June 1917, black troops landed at the mouth of the Somme River and boarded trains for the Western Front's infamous trenches. They were among the first Americans to arrive, and they fought side by side with soldiers from France and African nations like Senegal and Morocco. In March 1918 Germany launched a major offensive aimed at driving the Allies out of their trenches and then seizing France's English Channel ports. With things looking grim, the all-black 369th, under French command, halted the German advance at Cantigny. The all-black units of the 93rd Division stopped German offensives

at Château-Thierry and Belleau Wood. And in June 1918 the Harlem Hell Fighters of the 369th again proved their mettle by driving the Germans back at the Argonne Forest and the Meuse River. Fittingly, their unit was the first to reach the Rhine River.

The French were keenly aware of the significant contributions made by African Americans on their behalf, and they granted the Croix de Guerre to the First Battalion of the 92nd Division and three of the four regiments of the 93rd Infantry. Henry Johnson and Needham Roberts of the 369th each received individual Croix de Guerre for heroism. When the black veterans arrived home, they expected to be treated the same way they had in France—as honorable soldiers who commanded respect. Indeed, more than 5,000 black soldiers were wounded and 750 killed in the "war to end all wars." Instead, they found whites still viewed them as second-class citizens. The NAACP's *Crisis* spoke for returning black soldiers when it said, "This country of ours . . . is yet a shameful land. It lynches. . . . It disfranchises its own citizens. . . . It encourages ignorance. . . . It steals from us. . . . It insults us. . . . We return. We return from fighting. We return fighting."

World War I

——	German initial advance, 1914
∿∿∿	Paris defensive line, 1914
——	stabilized line, 1914–17
→	German spring offensive, 1918
➤	major U.S. attacks

Armistice line, Nov. 11, 1918

——	held by Belgian army
——	held by British army
——	held by French army
——	held by U.S. army
▨	occupation zones, c. 1919

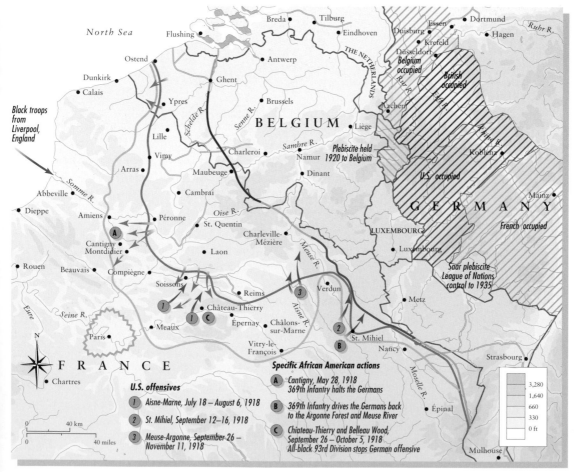

Specific African American actions

U.S. offensives

1 Aisne-Marne, July 18 – August 6, 1918

2 St. Mihiel, September 12–16, 1918

3 Meuse-Argonne, September 26 – November 11, 1918

A Cantigny, May 28, 1918
369th Infantry halts the Germans

B 369th Infantry drives the Germans back to the Argonne Forest and Meuse River

C Chateau-Thierry and Belleau Wood, September 26 – October 5, 1918
All-black 93rd Division stops German offensive

Black troops from Liverpool, England

Plebiscite held 1920 to Belgium

Saar plebiscite League of Nations control to 1935

Belgium occupied

British occupied

U.S. occupied

French occupied

	3,280
	1,640
	660
	330
	0 ft

World War II

Even the most provincial among African Americans became concerned with the rise of European fascism in the 1930s. In Germany, Hitler and the Nazis rose to power on a platform of racism and anti-Semitism. And African Americans strenuously protested the invasion of Ethiopia, a black kingdom in Northeast Africa, by Italian fascists in 1935. When Hitler's invasion of Poland in 1939 triggered the start of World War II in Europe, the neutral United States was far from ready for war. Blacks were caught in a dilemma: their strong desire to combat Nazism was offset by segregation, discrimination, and oppression at home. "Am I a Negro first and then a policeman or soldier second?" asked the sociologist Horace Cayton. "[S]hould I forget in an emergency situation the fact that . . . my first loyalty is to my race?" The dilemma resolved itself when it became clear that the war provided blacks with an explicit opportunity to link their service to improvements in social justice at home. Black leaders like A. Philip Randolph of the Brotherhood of Sleeping Car Porters embraced a "Double-V Campaign," meaning victory for democracy and racial justice both at home and abroad. He predicted that "before the war ends [blacks will] want to see the stuffing knocked out of white supremacy and of empire over subject peoples."

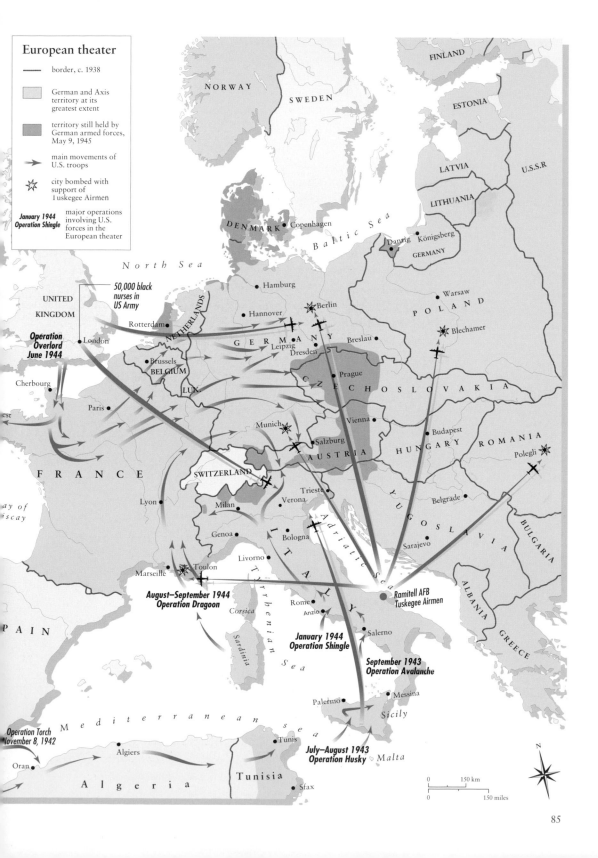

European theater

— border, c. 1938

German and Axis territory at its greatest extent

territory still held by German armed forces, May 9, 1945

→ main movements of U.S. troops

✦ city bombed with support of Tuskegee Airmen

January 1944 Operation Shingle — major operations involving U.S. forces in the European theater

FINLAND

NORWAY

SWEDEN

ESTONIA

LATVIA

U.S.S.R

LITHUANIA

DENMARK • Copenhagen

Baltic Sea

Danzig Königsberg
GERMANY

North Sea

• Hamburg

50,000 black nurses in US Army

• Warsaw

POLAND

UNITED

KINGDOM

• Hannover

Rotterdam•

NETHERLANDS

Berlin

Breslau

• Blechamer

Operation Overlord June 1944

• London

G E R M A N Y

Leipzig

Dresden

•Brussels

BELGIUM

LUX.

Prague

Cherbourg

C Z E C H O S L O V A K I A

est

Paris •

Munich

Vienna •

• Budapest

ay of
scay

FRANCE

Salzburg

A U S T R I A

H U N G A R Y

R O M A N I A

• Polegli

SWITZERLAND

Lyon •

Trieste

Milan •

Verona •

Y U G O S L A V I A

• Belgrade

BULGARIA

Genoa •

I T A L Y

Bologna •

Adriatic Sea

Sarajevo •

ALBANIA

Livorno •

Marseille•

Toulon

Tyrrhenian Sea

August–September 1944 Operation Dragoon

Corsica

Rome •

Anzio•

Ramitell AFB
Tuskegee Airmen

I T A L Y

GREECE

PAIN

January 1944 Operation Shingle

• Salerno

Sardinia

September 1943 Operation Avalanche

Palermo•

• Messina

Sicily

Operation Torch November 8, 1942

M e d i t e r r a n e a n   s e a

Tunis •

July–August 1943 Operation Husky ⬦ Malta

Oran •

Algiers •

Tunisia

A l g e r i a

• Sfax

N

0   150 km

0   150 miles

As the United States began gearing up for war, Randolph used threats of a black march on Washington to force the Roosevelt administration to prohibit discrimination in defense industries and government. Black leaders also pressed for impartial administration of the new draft law, and as a result more than 3 million African Americans registered for service in the armed forces under the Selective Service Act of 1940. By the fall of 1944, when the army was at its peak strength, there were 701,678 blacks in that branch of the service alone (there were also 165,000 in the U.S. Navy, 5,000 in the Coast Guard, and 17,000 in the Marine Corps). These approximately 1 million African Americans had a greater opportunity to serve their country than in any previous war.

Black leaders pressed hard to gain for blacks the opportunity to fly in combat. Only with the greatest reluctance and foot-dragging did the Army Air Force agree to train blacks as pilots and navigators, and early in the war veteran black pilots (even those with combat experience in the Spanish Civil War) were rejected out of hand. In 1940, William Hastie, a black federal judge and dean of Howard University's law school, was appointed as civilian aide to the secretary of war to assist with the large numbers of African Americans in the armed forces. Hastie was constantly frustrated in his attempts to fight segregation and secure equal treatment for black soldiers. In a 1943 article, he explained why he resigned in disgust: "Military men agree that a soldier should be made to feel that he is the best man, in the best unit in the best army in the world. When the Air Command shall direct its policies and practices so as to help rather than hinder the development of such a spirit among its Negro soldiers, it will be on the right road."

Partially in response to Hastie's agitation, the War Department began training African Americans as aviation pilots in Tuskegee, Alabama. As in the rest of the armed services during the war, the pilots were segregated from whites—but by the end of the war, the Tuskegee Airmen gained national recognition for their skilled bombing runs from Ramitell Air Force Base in Italy. In July 1943 First Lieutenant Charles B. Hall shot down a German Focke-Wulf FW 190 over the Mediterranean Sea, and became the first African American to score a verified aerial victory since Eugene Jacques Bullard, fighting for France, had downed a German plane during World War I.

Twenty-two black combat units fought in the European Theater. The 761st Tank Battalion helped turn back the Germans in the Battle of the Bulge, and the 614th Tank Destroyer Battalion earned high praise, with one of its officers, Capt. Charles L. Thomas, receiving the Distinguished Service Cross for heroism in action. In the Pacific, the black 24th

Infantry helped take the New Georgia Islands in May 1942. More than 15,000 black troops helped build the Burma Road, which ran from Burma to China and provided Chinese forces with vital supplies needed to defeat the Japanese. And the black 371st Tank Battalion were the first allied troops to liberate the Buchenwald and Dachau concentration camps in 1944.

The World War II experience was a watershed for African Americans. Jim Crow remained intact, but the ideological bases of white supremacy and colonialism were undermined by the horrors of the Holocaust. Millions of blacks, including large numbers of women, were drawn to jobs in defense-related industries. And once again black soldiers used their brave service abroad to press for justice at home.

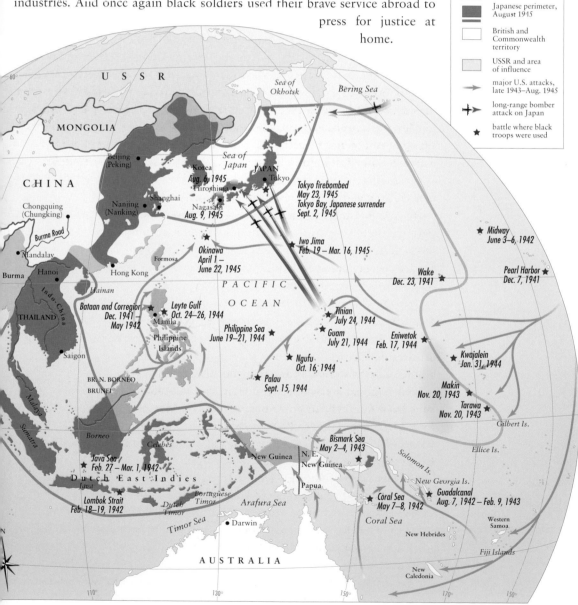

The Cold War: Korea and Vietnam

Although black and white units fought side by side against the Germans in the closing months of World War II, the military's policy was still one of segregation. The long era of segregation in the U.S. Armed Forces came to an immediate end in July 1948 under the express orders of President Harry S Truman. "It is hereby declared," the president ordered, "that there shall be equality of treatment and opportunity for all persons in the armed services without regard to race, color, religion, or national origin." When the president was asked if his advocacy of equal treatment and opportunity in the armed forces foretold the eventual end of segregation in other areas of American society, Truman replied "Yes." He was right: the integration of the military by executive order did presage the end of legal segregation. African Americans took advantage of the executive order to join branches of the military closed to them under segregation, including the Navy, Air Force, and Marines. But blacks still joined the Army in large numbers, even when the battles of the Cold War turned hot. Since 1948, larger percentages of African Americans than whites have served in the armed forces, seeing them as an avenue to better employment and educational opportunities. As a result, black soldiers bore the brunt of the heavy casualties suffered in U.S. involvement in Korea and Vietnam.

When Communist North Korea attacked South Korea in June 1950, one year after Communists triumphed in China, American officials feared all of Asia would follow without an armed response. America's first integrated war, under the flag of the United Nations, began badly: North Korea forced its opponents to retreat to the southeast corner of the Korean peninsula. But in a stunning reversal, U.S. General Douglas MacArthur executed an amphibious landing behind enemy lines at Inchon. With black and white soldiers fighting side by side with startling effectiveness, MacArthur pushed the North Korean troops back to the Chinese border. This was a mistake. Chinese troops entered the war and pushed the U.N. soldiers back across the 1950 border between the two Koreas. With the war at a stalemate, Truman fired MacArthur and initiated peace talks. The armistice line of July 27, 1953, continues to the present day to be one of the most heavily armed places on earth.

There was an even more significant African American involvement in the Vietnam War. As with previous wars, blacks volunteered in large numbers to serve—but even more were conscripted into service by a draft that targeted poor and less educated Americans. Members of the middle and upper classes could more easily evade the draft by entering college or leaving the country, or accept less dangerous assignments in the National Guard. That left many members of the African American community vulnerable to conscription: thirty percent of eligible blacks were drafted, compared to 18 percent of eligible whites. While in Vietnam—which, like Korea, was a war between a Communist North and an anti-Communist South—extremely high numbers of black soldiers were killed and wounded. During 1966–67, for instance, African Americans constituted 11 percent of the total U.S. enlisted personnel in Vietnam; yet black soldiers constituted 22.4 percent of all Army troops killed in action.

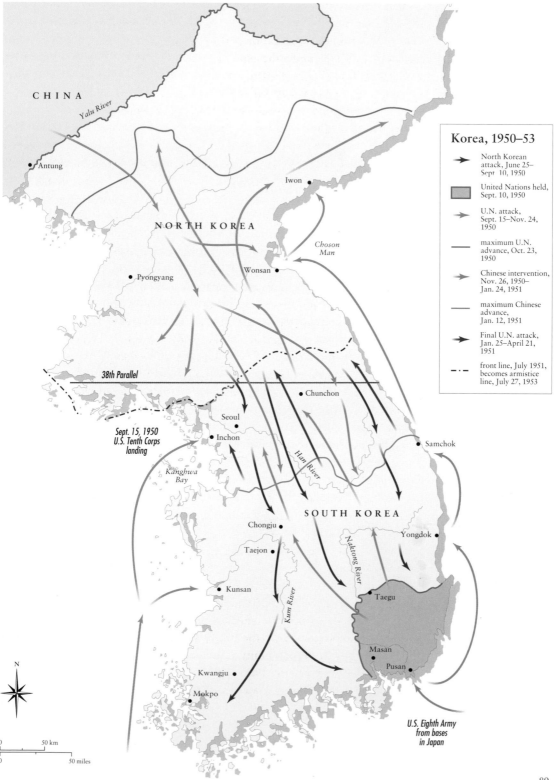

CHINA

Yalu River

Antung

NORTH KOREA

Iwon

Choson
Man

Pyongyang

Wonsan

Korea, 1950–53

→ North Korean
attack, June 25–
Sept. 10, 1950

▨ United Nations held,
Sept. 10, 1950

→ U.N. attack,
Sept. 15–Nov. 24,
1950

— maximum U.N.
advance, Oct. 23,
1950

→ Chinese intervention,
Nov. 26, 1950–
Jan. 24, 1951

— maximum Chinese
advance,
Jan. 12, 1951

→ Final U.N. attack,
Jan. 25–April 21,
1951

–·– front line, July 1951,
becomes armistice
line, July 27, 1953

38th Parallel

Chunchon

Seoul

*Sept. 15, 1950
U.S. Tenth Corps
landing*

Inchon

*Kanghwa
Bay*

Samchok

SOUTH KOREA

Chongju

Yongdok

Taejon

Han River

Naktong River

Kunsan

Kum River

Taegu

Kwangju

Masan

Mokpo

Pusan

N

0   50 km

0   50 miles

*U.S. Eighth Army
from bases
in Japan*

American involvement in Vietnam began in the late 1950s, but escalated rapidly after an August 1964 incident between American and North Vietnamese ships in the Gulf of Tonkin. President Lyndon B. Johnson used the incident to ratchet up U.S. involvement, ultimately raising the number of U.S. troops to an astonishing 535,000. By 1967 the war was costing the United States more than $2 billion a month and severely threatening Johnson's social programs at home. No matter how much the United States bombed, burned, and razed, the North Vietnamese and their southern allies refused to relent. Then in January 1968 the North Vietnamese surprised the United States with a well-orchestrated campaign called the Tet Offensive. Even though the American military claimed victory, Tet convinced many people in the United States that the conflict was un-winnable. Other ironies stood out as well: at the same time as African American soldiers were bearing more than their share of the war in Vietnam, African Americans at home were confronting economic inequalities and racial tension. Between 1965 and 1969 violent riots swept African American neighborhoods in Los Angeles, Newark, and Detroit. Federal troops were called in to quell the Detroit riot, causing many to see uneasy parallels between action in Vietnam and urban America. In 1968, Martin Luther King Jr. publicly condemned America's role in Vietnam, persuading still more blacks and liberal whites to oppose the war.

Johnson's successor, Richard Nixon, campaigned on a "secret plan" to end the war in Vietnam, yet while in office he widened the bombing of North Vietnam and bombed the neighboring countries of Laos and Cambodia. The latter action caused the largest anti-war demonstrations in American history, yet the United States didn't begin pulling out until early 1973. In the spring of 1975, North Vietnamese armies entered Saigon, and America's longest war ended in ignominious defeat. African American veterans, who faced discrimination while in the armed services only to encounter racism and other difficulties at home, felt this defeat especially keenly.

A soldier from the 173rd U.S. Airborne Division calls for medics to aid a wounded comrade in the Vietnam jungle. Blacks served in Vietnam in numbers out of proportion with their percentage in the general population.

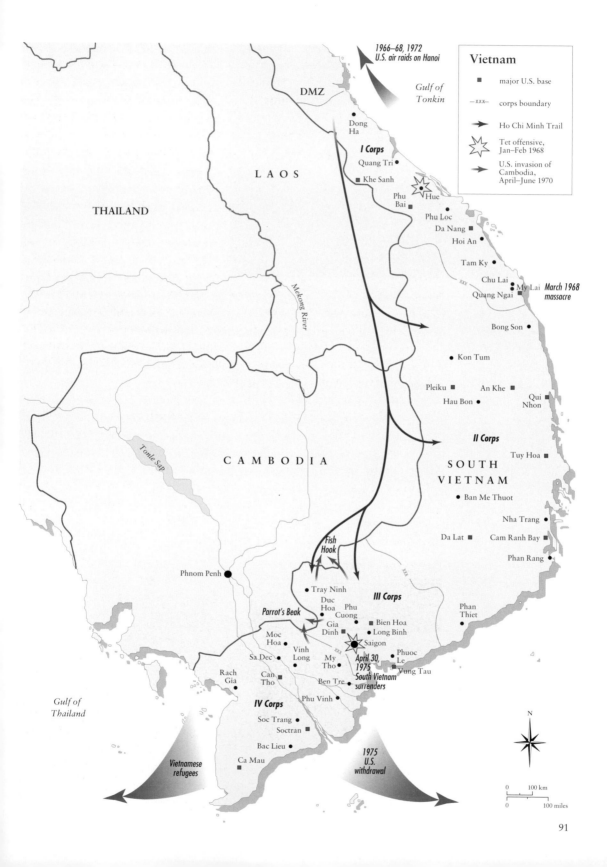

The 1990s: Peacekeeping and the Gulf War

General Colin Powell was the first African American to be named chairman of the Joint Chiefs of Staff, a position he held during the successful war in the Persian Gulf.

The armed forces were especially stung by the "Vietnam syndrome" and tended to resist the use of military force to carry out American objectives abroad. This scenario changed in 1990, when President Bush and Chairman of the Joint Chiefs of Staff General Colin Powell—the first African American to hold the post—sent a massive U.S. force to the Persian Gulf to dislodge Iraqi invaders from Kuwait. Blacks, who made up a little more than 13 percent of the population, accounted for 25 percent of the 500,000 troops sent to the Middle East. Polls taken before Operation Desert Storm began in early 1991 suggested that blacks, dubious of again paying more than their share in a potentially messy war, were less supportive of military action than whites. Yet African Americans expressed ample pride in the strong leadership of General Powell, who emerged as one of the heroes of the Gulf War.

Since the end of the Gulf War, the American military has been involved in several peacekeeping and humanitarian missions, including one in the African nation of Somalia and another in the mostly black Caribbean nation of Haiti. In Somalia, factional infighting left over from the Cold War caused a severe famine in 1991–92. U.S. troops, under the umbrella of a U.N. mission, helped

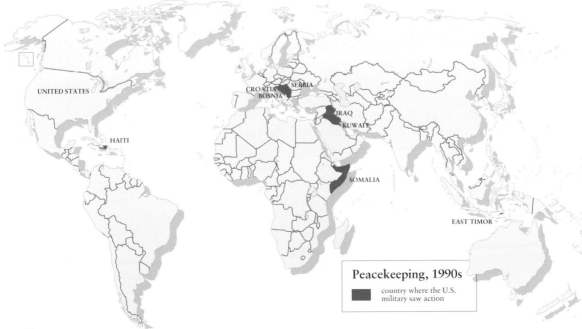

Peacekeeping, 1990s

country where the U.S. military saw action

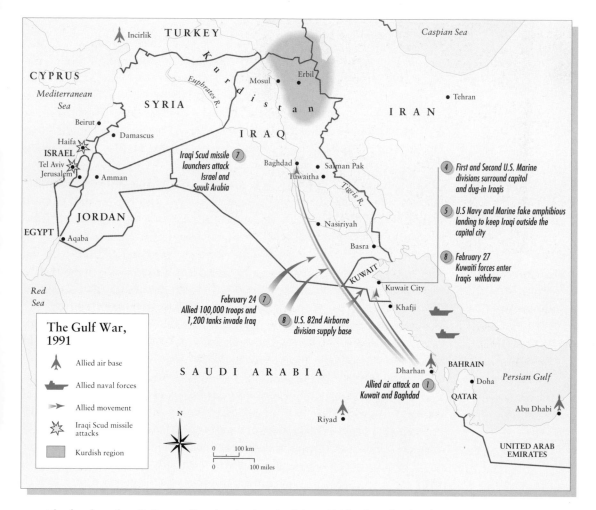

The Gulf War, 1991

Allied air base

Allied naval forces

Allied movement

Iraqi Scud missile attacks

Kurdish region

Iraqi Scud missile ⑦ launchers attack Israel and Saudi Arabia

First and Second U.S. Marine ④ divisions surround capitol and dug-in Iraqis

U.S Navy and Marine fake amphibious ⑤ landing to keep Iraqi outside the capital city

February 27 ⑧ Kuwaiti forces enter Iraqis withdraw

February 24 ⑦ Allied 100,000 troops and 1,200 tanks invade Iraq

U.S. 82nd Airborne ⑧ division supply base

Allied air attack on ① Kuwait and Baghdad

provide food and relief supplies beginning in May 1992. Caught in the middle of the civil unrest, American soldiers suffered several casualties and public humiliation, leading President Clinton to pull the troops out in the spring of 1994.

American interests seemed clearer with regard to Haiti, with whom relations had been rocky since the island's enslaved residents overthrew their masters in 1794. Haiti's first democratically elected leader, a Catholic priest named Jean-Bertrand Aristide, was ousted by a military junta in October 1991. In September 1994, after persistent pressure by the African American community to reinstate Aristide, 3,000 American troops landed in Haiti and the military agreed to step aside. Six months later, with Aristide back in power, the Americans turned the operation over to U.N. forces. More recently, African Americans in the U.S. armed forces have participated in recent peace-building missions in Bosnia-Herzegovina and Kosovo, where bloody spasms of "ethnic cleansing" in the former Yugoslavia brought them face to face with a people subjected to discrimination and genocide.

PART V: THE STRUGGLE FOR EQUALITY

We know through painful experience that freedom is never voluntarily given by the oppressor; it must be demanded by the oppressed. . . . For years now I have heard the word "wait!" It rings in the ear of every Negro with piercing familiarity. This "Wait" has almost always meant "Never."
—Martin Luther King Jr., 1963

Terrorist organizations like the Ku Klux Klan and the White League threatened the legal rights of newly freed blacks. This Northern political cartoon attacks their methods, which effectively denied African Americans access to education and the ballot box.

After the Civil War destroyed the Confederacy and slavery, African American families were reunited and blacks used their newly found mobility to move around the country. Southern blacks were given access to the ballot box, and hundreds of black public officials were elected in the 1870s. Ex-slaves founded scores of churches, schools, and mutual-aid societies. But Frederick Douglass, who had labored so long for the end of slavery, was cautious. "The work," he said, "does not end with the abolition of slavery, but only begins." During Reconstruction, when black freedmen, the federal government, and Northerners held sway in the South, it appeared that a true revolution in race relations might be at hand. Yet the substantial victories won by African Americans in both the South and the North were challenged by Southern whites who resorted to violence, terror, and murder.

Southern blacks remained free after Reconstruction ended, but their freedom was stripped away layer by layer. Strict vagrancy laws again limited black mobility. Poll taxes, literacy tests, and physical intimidation kept blacks away from the ballot box. Jim Crow laws separated blacks from whites and brought racial inequality to every corner of life in the South. Lynching and Ku Klux Klan intimidation became the means through which white Southerners enforced a new racial order. According to the black intellectual W. E. B. DuBois, "the slave went free; stood a brief moment in the sun; then moved back again toward slavery."

The period after Reconstruction has been called "the nadir" by one scholar of American race relations, but it was also a period of transformation and construction among the black community. For example, it was during the era of segregation when a lion's share of the nation's black colleges and universities were founded, schools that would educate generations of African American leaders and professionals (and to this day continue to be among the nation's best). Hundreds of thousands of blacks left the

rural South for jobs and a brighter future during World War I, an exodus known as the Great Migration. And African Americans founded countless organizations to fight for equal rights and justice in both the North and South. In addition to the well-known National Association for the Advancement of Colored People (NAACP) and National Urban League (NUL), blacks founded groups like the National Negro Business League (NNBL), the National Association of Colored Women (NACW), and the United Negro Improvement Association (UNIA). During the 1920s, African Americans flocked to black neighborhoods like New York's Harlem, where the Jamaican immigrant Marcus Garvey refined the ideology of black nationalism. And even though they were barred from official competition with whites, talented black baseball players like Satchel Paige and Josh Gibson battled it out in the Negro Leagues.

African Americans never stopped fighting for equal rights, even in the darkest days of racial violence in the postwar South. They fought in the courts and legislatures to undermine (and hopefully destroy) unjust Jim Crow laws. They joined groups, unions, and political parties that offered an alternative vision of America. And they commenced numerous mass movements to force society to grant the rights already embedded in the U.S. Constitution. The Legal Defense Fund of the NAACP, for example, waged a relentless battle against segregation laws until the U.S. Supreme Court struck them down as unconstitutional in 1954. One year later, after a seamstress named Rosa Parks refused to give up her bus seat for a white man, blacks in Montgomery, Alabama, unleashed a one-year boycott of the bus system which helped integrate public facilities. One of the leaders of the boycott, Dr. Martin Luther King Jr., engaged millions of blacks and whites in a nonviolent struggle for civil rights.

Under the leadership of King and others, the civil rights movement reached its apogee in the 1950s and '60s. A committed Christian inspired by the philosophies of Mohandas Gandhi and Henry David Thoreau, King used non-violence and direct action to inspire the movement for civil rights, broaden its appeal, and legitimate its strategies and demands. "Give us strength to love our enemies and do good to those who despitefully use us and persecute us," King said in 1956. A new generation joined the movement in the 1960s, inspired by King but impatient with the slow pace of integration. In February 1960 four college students from Greensboro, North Carolina, launched the "sit-in" movement to force the desegregation of facilities like lunch counters, restaurants, and bus stations. The Student Non-Violent Coordinating Committee (SNCC) and the Freedom Riders continued this trajectory, aggressively confronting racist whites adamant about maintaining the racial status quo. The violent reaction of the white South to African Americans' claims helped swing the federal government behind the protesters. At President Johnson's urging, Congress enacted legislation in the 1960s to guarantee civil and voting rights for blacks. One hundred years after the end of the Civil War, African Americans won back the rights gained, and then lost, during Reconstruction.

Lynching and the New Racial Order

Lynchings were a gruesome—and all too common—example of extra-legal violence against African Americans. More than 3,000 blacks were lynched in the South between 1880 and 1930.

After federal troops withdrew from the South in 1877, Southern white supremacists filled the power vacuum with terror, violence, and murder. Paramilitary groups like the Ku Klux Klan, founded to terrorize black Republican voters during Reconstruction, used violence to enforce racist social codes as well. African American pedestrians who refused to "step aside" for whites to pass or black sharecroppers who demanded better terms from white landowners were likely to receive a late-night visit from white-hooded Klansmen. White assaults on African Americans, particularly violence precipitated by lynch mobs, became widespread by the 1880s. These murders, usually to avenge a perceived social or sexual transgression, paralleled the passage of Jim Crow segregation laws.

Lynching was not an exclusively Southern phenomenon, and whites as well as blacks fell victim to mob violence. But between 1880 and 1930, 3,220 blacks and 723 whites were lynched in the South, with the greatest numbers taking place in Louisiana, Mississippi, Alabama, Georgia, and Texas. In the West, which was overwhelmingly white, 38 blacks and 447 whites were victims of lynchings during the same period. The worst year for lynching was 1892, when at least 235 African Americans were killed by raging mobs. By the end of the 19th century, lynching had became almost exclusively racially motivated and mostly confined to the Deep South. Lynchings were frequently preceded by a "trial" in which accusers beat confessions for rape, theft, intransigence, or vagrancy out of a bound victim. Accused blacks were often tortured, burned, stabbed, or shot before being hanged.

Black churches and organizations protested vocally against the killings. In the South, courageous black clergymen and civic leaders confronted lynch

mobs, and black newspapers risked almost certain retaliation by publishing anti-lynching articles. A leading crusader against lynching was Ida B. Wells, whose investigative report in the *Memphis Free Speech and Headlight* led racist whites to destroy the paper's headquarters. But Wells could not be silenced: she took her movement for a federal anti-lynching law across the country, arguing that "a national crime requires a national remedy." Wells also sought to disprove the Southern contention that lynch law was necessary to protect white women from black rapists. She found that just one in six lynch mobs even claimed it was avenging the rape of a white woman by an African American man.

Other organizations took up the cause against lynching in the 1930s, including the Association of Southern Women for the Prevention of Lynching. By that time black migration and changes in the regional economy—combined with the successes of the anti-lynching campaigns—had significantly reduced the number of lynchings in the South. In 1953, for the first time, no lynchings were reported in the United States.

Lynchings, by state, 1889–1918

Number of lynchings by state

251 to 386
141 to 250
41 to 140
11 to 40
1 to 10

Proportion of blacks in lynchings, in percent

over 80
51 to 80
under 51

The Spread of Jim Crow

Blacks suffered more than their white counterparts as the Southern economy spiraled downward in the 1890s. Despite this, African Americans became targets for especially virulent white rage during a decade one historian called the "nadir" of the postslavery black experience. Lynchings killed hundreds of blacks per year, and race riots ravaged cities like Atlanta and Wilmington, North Carolina.

Racial discrimination also took on more legal forms. Several state constitutions passed in the South between 1890 and 1900 disfranchised most black voters by using literacy tests, property qualifications, and poll taxes. Whites who were unable to meet the new requirements were often allowed to vote by means of "grandfather clauses." The Supreme Court upheld the disfranchisement clauses in 1898, crippling the Republican Party in the South and handing over state offices to the Democrats for the next 60 years.

In the famous case *Plessy v. Ferguson* (1896), the Supreme Court upheld another kind of law sweeping the South, so-called "Jim Crow" laws. Jim Crow laws (named for a stock character in racist, blackface minstrel shows) mandated regulated racial segregation in public facilities of all kinds, from water fountains to seats on turn-of-the-century streetcars. Homer A. Plessy, a light-skinned black man, was arrested in New Orleans after he refused to ride in a "blacks only" rail car. After a conviction in a Louisiana courtroom, Plessy's case went to the U.S. Supreme Court, which ruled that as long as accommodations for blacks were equal to those of whites, the races could be legally separated. Of course, in the harsh racial realities of the South at the time, facilities were rarely, if ever, equal for blacks and whites. Once blacks were disfranchised and public facilities legally segregated, white supremacy became a reality in the South.

Beginning in Tennessee in 1875, state after state pushed through hundreds of Jim Crow laws, separating blacks and whites on trains, in stations, and on ships. In 1883 (after the Supreme Court declared the 1875 Civil Rights Act unconstitutional) blacks were banned from white schools, theaters, and restaurants. Many African Americans refused to accept the new racial order: in North Carolina, blacks made political alliances with populist whites and initially were able to hold off Jim Crow laws. But in 1898, whites rioted when a black newspaper in Wilmington criticized the tactics of Southern Democrats. The paper's offices were destroyed, as well as several black neighborhoods; North Carolina passed its first Jim Crow law two years later.

Booker T. Washington was the most powerful black leader of the Jim Crow era. In a famous speech in 1895, Washington in effect accepted segregation as a temporary accommodation, in exchange for white support to improve blacks' economic progress, education, and social uplift. In a speech known as the "Atlanta Compromise," Washington told Southern whites, "You can be sure in the future, as in the past, that you and your families will be surrounded by the most patient, faithful, law-abiding, unresentful people that

the world has seen. . . . In all things that are purely social we can be as separate as the fingers, yet one as the hand in all things essential to mutual progress." To Washington's African American critics, such sentiments played right into the hands of white supremacists; to his supporters, Washington was playing the best cards in his hand, hoping to gain acceptance through self-reliance and uplift.

African Americans and their white allies formed The National Association for the Advancement of Colored People (NAACP) in 1909 to end forced segregation, provide equal education for blacks and whites, and completely enfranchise African Americans. W.E.B. DuBois was the only African American officer on the original board, and was placed in charge of research and the NAACP's journal *Crisis*. In its pages DuBois clashed with Booker T. Washington's approach to race relations, urging a legal and popular assault on segregation.

The spread of Jim Crows laws in the South

- Northern city where Jim Crow existed in practice before 1896

1875 legal segregation, with date Jim Crow laws passed

segregation practiced but did not pass Jim Crow laws

✶ major race riot at the start of the Jim Crow era, with date

Black Colleges and Universities

Before 1865 blacks had been systematically denied educational opportunities – in fact, teaching slaves or free blacks to read was a violation of the law in many Southern states. As a result, historians estimate that at least 4 million African Americans emerged from slavery illiterate. The ex-slaves, Northern educators, and members of the black clergy combined efforts to set up schools during Reconstruction, most of them in black churches. According to almost every account of the postwar South, freed slaves' hunger for learning was insatiable. This hunger led to the creation of the first colleges and universities for African Americans—institutions that became vital for educating black leaders and their followers during the age of segregation and beyond.

The earliest black colleges—as well as day, night, religious, and industrial schools—were supervised by the Freedmen's Bureau, a federal agency commissioned by Congress in 1865. Cooperating closely with Northern religious and philanthropic organizations, the Freedmen's Bureau helped found more than 4,000 schools for ex-slaves in the South. The teachers at the first black colleges were mostly white Northerners, although the numbers of black schoolteachers grew steadily. By the time the Freedmen's Bureau halted its educational labors, it had spent more than $5 million schooling almost 250,000 African Americans.

Between 1865 and 1877, nearly 50 black colleges were founded across the South with the help of the Freedman's Bureau. Among them were Howard University (named for the Bureau's leader, Union Gen. O. O. Howard), Fisk University, Atlanta University, Hampton Institute, and St. Augistine's College. Howard, located in the nation's capital, has a long tradition of excellence and is especially known for its law school, which trained generations of civil rights lawyers including Charles Johnson and Thurgood Marshall. Booker T. Washington attended Hampton Institute, a vocational school for blacks run by Samuel Chapman Armstrong. At Hampton, Armstrong stressed physical labor and its capacity to promote honesty and fidelity; Washington drank deeply from Armstrong's well and became convinced that to advance, blacks had to perform "useful" work. When Washington assumed leadership of the Tuskegee Institute in Alabama, he made sure the students performed all the chores on campus (including construction, cleaning, and food preparation) as well as provide community services for local whites. Washington's goals were twofold: he hoped to educate Southern blacks in the vocational trades and also convince Southern whites that the education of blacks was in everyone's best interest. Washington's attempts to accommodate white Southerners angered W. E. B. DuBois, who attended Fisk University in Nashville before receiving his doctorate from Harvard. DuBois believed it was vital to train African Americans as

scholars and leaders. His impeccable scholarship in the fields of history, sociology, and political science paved the way for other black scholars like E. Franklin Frazier, Carter G. Woodson, Kenneth Clark, and Dorothy Porter.

Before the Supreme Court declared segregated education illegal in 1954, black colleges provided African Americans from the South with their only hope for post-secondary education. The schools suffered from chronic shortages of funds, but provided the backbone for several generations of the African American middle class, educating and training a lion's share of the nation's black doctors, lawyers, businessmen, academics, and other professionals. The schools had a large impact on the black struggle for equality as well: many of the leaders of the civil rights movement were educated in all-black, Southern colleges, including Martin Luther King Jr. (Morehouse College), Ralph Abernathy (Alabama State University), and Jesse Jackson (Agricultural and Technical College of North Carolina). Jackson's alma mater also provided a more direct link to the civil rights movement: students at North Carolina Agricultural and Technical (including Jackson) launched the sit-in movement in 1960, which became the largest direct-action protest in U.S. history. Today, almost 50 years after the Supreme Court struck down forced segregation as "separate and unequal," many of the brightest African Americans choose to attend the nation's excellent black colleges. Schools like Spelman College, Lincoln University, and Bethune-Cookman continue to revitalize and pioneer new trends in American higher education.

Towns with historically black colleges and universities

● college or university

The Great Migration

Southern blacks, mired in a cycle of vicious racism and limited economic opportunities, took advantage of a severe labor shortage during World War I to commence a "Great Migration" to Northern and Midwestern cities. Between 1916 and 1920, some 500,000 African Americans left the rural South for the industrial centers of Chicago, New York, Detroit, Cleveland, Philadelphia, St. Louis, and Kansas City. The Great Migration remade the racial landscape of the entire nation.

Like any mass migration of peoples, the African Americans moving north and west responded to both "push" and "pull" factors. Lynchings, segregation, limited educational opportunities, police brutality, and abuse by Southern whites combined with a series of crop failures to "push" blacks northward. "Pull" factors included well-paid industrial jobs and information, spread by the black press, that Northern cities were havens of opportunity for African Americans. Northern cities were hardly free of racism, but World War I caused a sharp decline in European immigration and a skyrocketing demand for industrial goods. Racially biased factory owners were forced to abandon racist hiring practices, and some even recruited in the Black Belt. African American papers like the *Chicago Defender* expounded on the economic opportunities in the North and printed stories

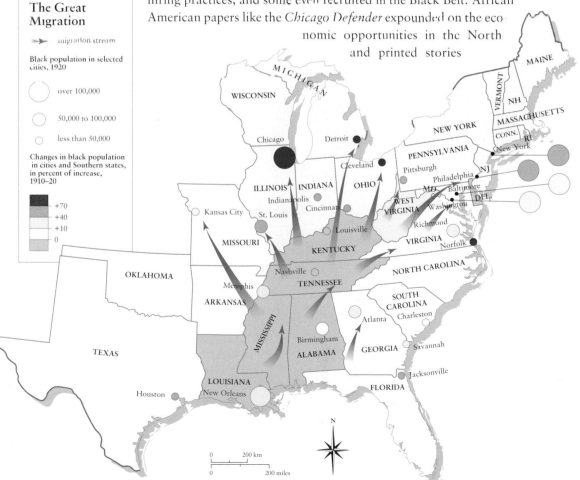

The Great Migration

→ migration stream

Black population in selected cities, 1920

◯ over 100,000

◯ 50,000 to 100,000

◯ less than 50,000

Changes in black population in cities and Southern states, in percent of increase, 1910–20

+70
+40
+10
0

about successful black immigrants. Copies of the paper were passed from hand to hand in the rural South, and people even sought the paper's help in securing jobs. "There is a storm of our people toward the North and especially to your city," read one letter sent to the *Defender* from a group of black Southern women. "Will you please assist us in securing places as we are anxious to come but want jobs before we leave? We want to do any kind of honest labor. Our chance here is so poor."

The Great Migration practically emptied parts of the Southern countryside. Mississippi, Alabama, Georgia, South Carolina, North Carolina, and Virginia each lost more than 100,000 inhabitants between 1910 and 1920, many of them blacks. At the same time, the size of urban African American communities exploded: Detroit's black population, for example, increased by 611 percent over the course of the decade; Cleveland's and Chicago's grew by 307 percent and 148 percent, respectively. These newcomers had a profound impact on the politics, society, and culture of their new homes.

As with the migration of the Exodusters a generation earlier, entire families or communities often made the trek together. And when people did migrate as individuals, they often did so with the aid of family or friends who had journeyed before them. "I know if you come and rent a big house you can get all the roomers you want," wrote one Chicago migrant to friends in the South. "The people are rushing here by the thousands." Even though the newcomers were often disappointed with the weather, the hard work, or the racial animosity of Northern cities, most of them stayed. Dismayed with the condition of life for most Southern migrants, several New York organizations merged in 1910 to form the National Urban League. The league opened branches in most large cities, and sent members to meet newly arrived migrants to direct them to jobs, housing, and educational opportunities.

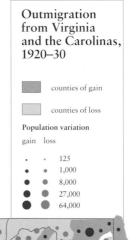

Outmigration from Virginia and the Carolinas, 1920–30

counties of gain

counties of loss

Population variation

gain	loss	
·	·	125
·	·	1,000
●	●	8,000
●	●	27,000
●	●	64,000

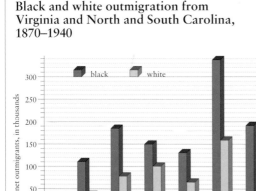

Black and white outmigration from Virginia and North and South Carolina, 1870–1940

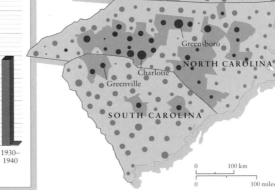

Marcus Garvey and the UNIA

The African Americans who migrated North still faced harsh economic realities and vicious white racism. Most migrants could get only low-paying unskilled jobs, and were often crowded into segregated urban slums. Marcus Garvey was the leader who spoke most directly to the hopes, dreams, and disappointments of the urban black population. Garvey urged African Americans to give up on integration (a central goal of groups like the NAACP) and work to create a separate black nation in Africa, complete with an army to protect it. In the meantime, he stressed a program of self-help and racial pride to help American and Caribbean blacks to achieve economic and cultural independence.

Marcus Garvey was born in Jamaica in 1887 and moved to New York City in 1916. His message of black nationalism stressed that blacks were exploited throughout the world and could never count on whites for help. He also preached a brand of racial pride that reverberated strongly in the black community. In the early 1920s, Garvey's United Negro Improvement Association counted millions of members in 38 states and 41 foreign countries. His newspaper, *The Negro World*, reached 200,000 subscribers. UNIA was one of the most successful all-black organizations in history. Many African Americans closely identified with the dark-skinned Garvey, who insisted that "black" stood for strength and beauty, not weakness and inferiority. He led grandiose parades in Harlem wearing a plumed admiral's hat, and his uniformed followers marched in formation.

Although the New York chapter of the UNIA opened restaurants, groceries, factories, and other examples of racial self-sufficiency, Garvey's ultimate vision was of an independent black nation in Africa. "Wake up Ethiopia! Wake up

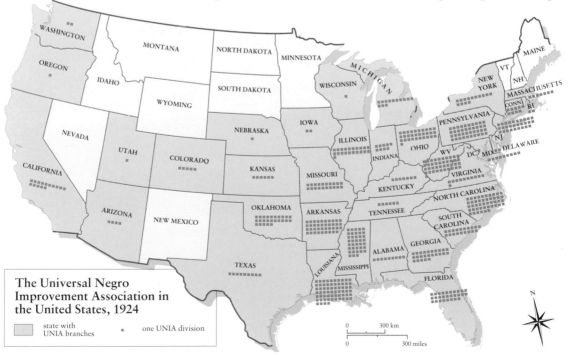

The Universal Negro Improvement Association in the United States, 1924

state with UNIA branches   ▪ one UNIA division

0   300 km
0   300 miles

Africa!" he told his followers. "Let us work toward the one glorious end of a free, redeemed and mighty nation. Let Africa be a bright star among the constellations of nations." Garvey's meteoric success was short-lived, however. He was bitterly attacked by fellow black leaders (W. E. B. DuBois, for example, regarded him as a self-serving demagogue) and the U.S. government. In 1924 Garvey was arrested and charged with mail fraud. He was deported to Jamaica in 1927 after serving one year in prison. Garvey's fantastic rise in the 1920s shows how dubious African Americans were that they would ever be counted as first-class citizens. And his vision of black nationalism endured well beyond his career in the United States.

Marcus Garvey in a black nationalist parade in Harlem. The Jamaica native often appeared in a military-style garb, complete with a plumed admiral's hat and golden epaulets.

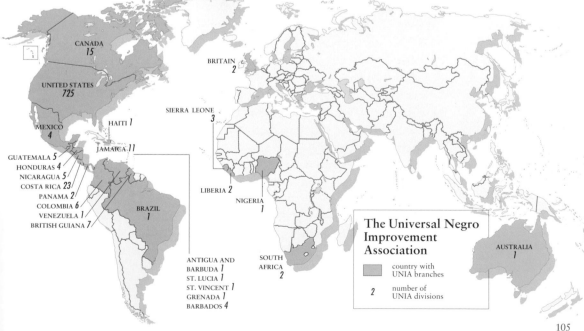

CANADA 15
BRITAIN 2
UNITED STATES 725
SIERRA LEONE 3
MEXICO 4
HAITI 1
JAMAICA 11
GUATEMALA 5
HONDURAS 4
NICARAGUA 5
COSTA RICA 23
PANAMA 2
COLOMBIA 6
VENEZUELA 1
BRITISH GUIANA 7
LIBERIA 2
NIGERIA 1
BRAZIL 1
ANTIGUA AND BARBUDA 1
ST. LUCIA 1
ST. VINCENT 1
GRENADA 1
BARBADOS 4
SOUTH AFRICA 2
AUSTRALIA 1

The Universal Negro Improvement Association
country with UNIA branches
2 number of UNIA divisions

105

The Negro Baseball Leagues

Baseball, variations of which could be found on American playgrounds for more than a century, soared in popularity in the early 20th century. Before 1900, as baseball became a professional sport, a small number of black players appeared on team rosters. But between 1903 and 1946, players with black skin (whether they were African Americans, Cubans, or Latin Americans) were barred from organized baseball. If black ballplayers wanted to play the sport professionally, they had to create their own all-black teams and play in all-black leagues. As a result, some of the best athletes ever to play the game played for teams like the Monarchs, Black Barons, and Cubans, not the Yankees, Cardinals, and White Sox.

In 1920 Andrew "Rube" Foster organized the Negro National League, with teams in Chicago, Dayton, Detroit, Indianapolis, Kansas City, and St. Louis. In the next decade Cincinnati, Pittsburgh, Cleveland, Brooklyn, Washington, and Philadelphia were added to the league. Even though most of the early players hailed from the South, the cities with teams (with the exceptions of Memphis, Nashville, and Birmingham) were in the North or border states. The teams played in existing ballparks, or rented from white major league clubs, but also played games in Cuba, Puerto Rico, and Mexico, where baseball was not organized along racial lines.

The Negro Leagues were filled with ballplayers of immense talent. Josh Gibson, of the Pittsburgh Crawfords, was known as the "black Babe Ruth," but Ruth could have just as easily been called the "white Josh Gibson." Gibson hit more than 70 home runs in 1931 (records were often sketchy in the NNL) and is believed to have hit more than 1,000 during his career (Babe Ruth hit 715; the current major league record, by ex-Negro Leaguer Hank Aaron, is 755). Leroy "Satchel" Paige, who had one of the most memorable wind-ups in baseball history, claimed to have won 2,000 games in his career. In many ways the Negro Leagues were a great success: teams like the Kansas City Monarchs, a perennial powerhouse in the NNL, were extremely profitable, and large (often integrated) crowds turned up to see the annual East-West all-star game.

It was precisely this success that brought on the NNL's demise. The talent pool among black ballplayers was indisputable, and the size and affluence of black populations in major cities was on the rise. But it took the experience of World War II, where black soldiers helped to defeat racist and genocidal regimes, to fully dramatize the gap between America's democratic ideals and its racist reality. In 1947, Jackie Robinson of the Kansas City Monarchs became a Brooklyn Dodger. His immense talent on the field (he hit .297, led the league in stolen bases, helped the Brooklyn Dodgers win the pennant, and was named Rookie of the Year) and coolness in the face of withering racism made him the ideal man to "break the color barrier" in major league baseball. By 1948, Robinson and the Dodgers had been joined by Roy Campanella, and the

Cleveland Indians had hired Larry Doby and the 42-year-old "rookie" Satchel Paige. It took another ten years before all major league teams had at least one black player. In the early 1950s, an astonishing array of talent gave the major leagues a new crop of black stars: Hank Aaron of the Indianapolis Clowns, Ernie Banks of the Kansas City Monarchs, Willie Mays of the Birmingham Black Barons, and Minnie Minoso of the New York Cubans. The integration of major league baseball spelled doom for the NNL; the league folded completely in 1957. When the black community was given a choice between supporting the Negro Leagues or rooting for major league teams with integrated rosters, they overwhelmingly chose the latter. "When Negro players walked in the big league doors," wrote the *Pittsburgh*

Negro League baseball teams, 1920–50

Courier, "Negro baseball walked out." As an unintended result of the integration of major league baseball, many of the greatest players and teams of all time faded, at least temporarily, from memory.

Rosa Parks and the Montgomery Bus Boycott

After the Supreme Court outlawed segregation in the nation's public schools in 1954, the African American struggle for equal rights entered a decisive new phase: non-violent resistance to Jim Crow laws. Organized protest against segregation was not unknown before 1955, but it was dangerous and forced few changes in either law or custom. Rosa Parks, a seamstress from Montgomery, Alabama, offered blacks there an ideal opportunity to overturn the city's ordinance permitting racial segregation on the city's buses. On December 1, 1955, Parks boarded the 5:30 p.m. bus on Dexter Avenue, exhausted after a long day of work at the Montgomery Fair department store. She walked past the "whites only" seats at the front of the bus and took a seat in one of the back rows that were marked for use by whites or blacks. When a group of white men boarded the bus at the Empire Theater stop and couldn't find seats in the front, the driver issued a command common in the Jim Crow South: "Niggers, move back." Parks, who was 42 and a member of the local chapter of the NAACP, refused. Later, she explained that "the only way to let them know I felt I was being mistreated was to do just what I did—resist the order." The driver got off the bus, summoned the police, and Parks was arrested and taken to jail.

Parks' arrest became the basis for a legal challenge to Montgomery's segregation laws, and later a rallying cry for the revitalized Civil Rights Movement. She was an ideal litigant, and local black leaders took full advantage of the opportunity. Within days, the Women's Political Council of Montgomery mimeographed 52,500 leaflets proposing a one-day boycott of the city's buses. That day, most African Americans walked to work and school, or used a make-shift pool of cars and black-owned taxis. The city's bus company lost between 30,000 and 40,000 fares. In the aftermath of the boycott, the new

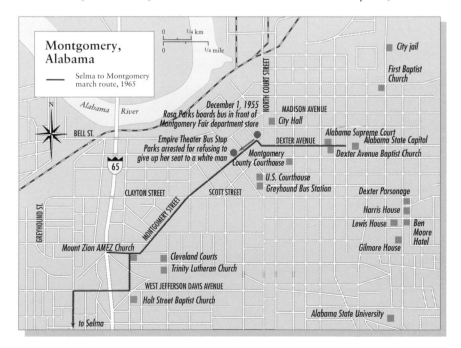

Montgomery Improvement Association elected a new president, Dr. Martin Luther King Jr., the 26-year-old minister of the Dexter Avenue Baptist Church. His first act was to extend the Montgomery Bus Boycott indefinitely. King issued other demands as well: that bus drivers extend courtesy to black passengers; first-come, first-served seating to eliminate the practice of blacks having to give up seats for whites; and the hiring of black bus drivers. When Montgomery's municipal authorities refused to even consider the MIA's modest demands, the MIA intensified its wildly successful boycott, which was 99 percent effective. The brunt of the boycott was borne by Montgomery's black women, many of whom worked as domestics and had to walk long distances to work.

Whites in Montgomery tried to harass the African American community into calling off the boycott. Car pools were ticketed by police on trumped-up charges. MIA leaders received threatening phone calls and indictments for violating Alabama's anti-boycott law. And King's house was firebombed on January 30, 1956. But King, a student of the nonviolent teachings of Gandhi and Thoreau, urged the community to remain passive: "We are not advocating violence. We want to love our enemies. Be good to them. Love them and let them know you love them. I want it to be known that . . . if I am stopped, our work will not stop. For what we are doing is right, what we are doing is just." The boycott finally ended on December 21, 1956, after the Supreme Court declared Alabama's bus segregation laws unconstitutional. That day Parks, King, and thousands of other blacks sat at the front of Montgomery's buses.

The boycott graphically illustrated the economic power of the African American community. But the bus victory was a narrow one, and most facilities in Montgomery remained segregated. In fact, Montgomery's white leaders chose to close the city's parks instead of opening them to blacks. The boycott's leaders were vaulted to national prominence. Martin Luther King followed up the victory in Montgomery by joining other black ministers to form the Southern Christian Leadership Conference (SCLC), which aimed to force integration by means of nonviolent civil disobedience. In 1965, King returned to Alabama for his historic march from Selma to the Capitol in Montgomery to register black voters and air grievances about the slow pace of desegregation.

The Southern Christian Leadership Conference, 1955–65

- Southern state with SCLC activity
- other state with SCLC activity
- • city with SCLC activity

Sit-ins and the Rise of SNCC

On February 1, 1960, four students from North Carolina Agricultural and Technical College in Greensboro launched a radical new phase of the civil rights movement. That day they entered a local Woolworth's dime store, made purchases, then sat down at the lunch counter and ordered coffee. They were refused service because they were black (all seats at the counter were reserved for "whites only"), but they stayed seated until the store closed for the day. The next day they returned with more young people. It was the beginning of the "sit-in" movement, a strategy of peaceful protests against segregation and discrimination, which spread like wildfire across the nation. Young people sat in at white swimming pools, bus stations, hotel lobbies, and restaurants, forcing people to confront the inequalities present in everyday life. "We do not intend to wait placidly for those rights which are legally and morally ours to be meted out to us one at a time," read a newspaper ad purchased by college students in Atlanta.

The demonstrators' courage and moral imperative (and their more aggressive and confrontational tactics) infused new energy into the civil rights movement. Their nonviolent tactics inspired countless others to join the anti-discrimination movement. Soon after the sit-ins began, scores of businesses in the South (including the Woolworth chain's lunch counters) were desegregated, and new, grass-roots organizations began to form. Chief among these were the interracial Congress of Racial Equality (CORE) and the Student Non-Violent Coordinating Committee (SNCC). These new grass-roots activists quickly seized the mantle from the older, more establishment-oriented NAACP and SCLC, and put pressure on the administration of John F. Kennedy to respond to the new demands. It marked a turning point: in 1961 the students forced Kennedy to send federal marshals to the South to protect activists attempting to desegregate buses and waiting rooms; in 1962 and 1963 federal marshals were sent to enforce the integration of several universities in the Deep South.

The SNCC activists prided themselves on their strong ties to local communities. A new generation of African American leaders, including Anne Moody, Julian Bond, John Lewis, Angela Davis, and Stokely Carmichael, emerged from the ranks of SNCC. The protests continued, triggering violent responses from segregationist forces in the South. Civil rights leaders turned to Birmingham, Alabama, in the spring of 1963, calling it the most segregated city in America. Almost nightly, network television broadcast footage of white

police officers using dogs and high-pressure water hoses on young blacks protesting segregation in the city. Later that year four African American girls were killed when white supremacists bombed the Sixteenth Street Baptist Church. Thousands took to the streets in nonviolent protest, and when police officers killed two more black children, the timid Kennedy administration was again forced to act.

Angered by the slow pace of civil rights legislation, some of the young SNCC activists began to attack publicly the tactics of established civil rights groups like the SCLC. Long the shock troops of the civil rights movement, SNCC activists felt that established black leaders were not pushing hard enough for full equality. In 1966 SNCC's new chairman, Stokely Carmichael, insisted that blacks use "black power" to combat the "white power" that held them down. Then SNCC embraced the philosophy of Malcolm X to fight back if attacked, and adopted the slogans and rhetoric of nationalists like the Black Panthers. In 1968, the group officially broke with King and the SCLC and agitated for more immediate changes in American society.

The Student Non-Violent Coordinating Committee, 1960–65

state with major SNCC chapter

★ first major sit-ins organized by SNCC

● other major city with SNCC activity

0   300 km

0   300 miles

Voting Rights

The Rev. Martin Luther King Jr. participating in a march for voting rights in Montgomery, Alabama, in March 1965. When the peaceful marchers were attacked by mounted state troopers, Congress was forced to act.

After the early successes of the sit-in movement, the focus of the civil rights movement shifted from desegregating public facilities to enforcing universal suffrage. The 1964 Civil Rights Act outlawed segregated public facilities and racial discrimination in employment and education, but in many areas of the South, African Americans were still barred from voting. This despite the 15th Amendment, which states that the right to vote shall not be denied on account of "race, color, or previous conditions of servitude." White supremacists utilized a number of tactics to keep Southern blacks disfranchised before 1964, including high poll taxes and literacy tests that kept poor and uneducated African Americans from voting. "Grandfather" clauses enfranchised poor whites. If an African American registered to vote, it usually attracted the attention of the Ku Klux Klan, as well as retribution from employers and local law enforcement.

When a black voting rights activist named Jimmy Lee Jackson was murdered in Alabama in February 1965, Martin Luther King Jr. and other black leaders organized a march from Selma to the state capitol in Montgomery. On March 7, as the protesters attempted to cross the Edmund Pettus Bridge outside Selma, they were attacked by mounted state troopers. As network television cameras rolled, the police beat, whipped, and tear-gassed unarmed and peaceful demonstrators. Two days later, King decided to retreat as a second march

Black voter registration, 1958, 1965

proportion of black voters in nonwhite voting-age population, in percent

1958 1965

was blocked at the bridge. Like his predecessor John F. Kennedy, President Lyndon Johnson was forced to act to aid the demonstrators. On King's third attempt to march to Montgomery, Johnson called the Alabama National Guard into federal service to protect the marchers. More than 50,000 black and white supporters joined the 300 demonstrators as they made their way to the state Capitol.

Johnson had clearly recognized the need for separate federal action to protect the rights of black voters. In an address to Congress and the nation, Johnson quoted the lyrics of a civil rights anthem to make his point: "We intend to fight this battle [for voting rights] where it should be fought—in the courts, in the Congress, and in the hearts of men. And we shall overcome." Within days the president sent Congress his proposals for a voting rights law, and Congress passed it quickly. The Voting Rights Act of 1965 banned literacy tests and other methods to disfranchise blacks, forced suspect counties to obtain Justice Department approval when they changed election procedures, and authorized the attorney general of the United States to send federal examiners to register black voters when he concluded that local registrars were not doing their job.

The Voting Rights Act did not immediately alter voting patterns in the South, but its eventual impact was immense. Federal studies proved that most incidents of disfranchisement occurred in remote rural areas where African Americans made up a majority of the population. In 1967 the attorney general sent federal agents into 62 counties (mostly in rural Alabama, Mississippi, and Louisiana) to register black voters. By 1971, 62 percent of eligible blacks had been registered in the South, compared to 20 percent in 1960. As Colin Palmer, a prominent historian of the African American experience, wrote, "the vote could now be used as a weapon to force other changes upon a reluctant larger society."

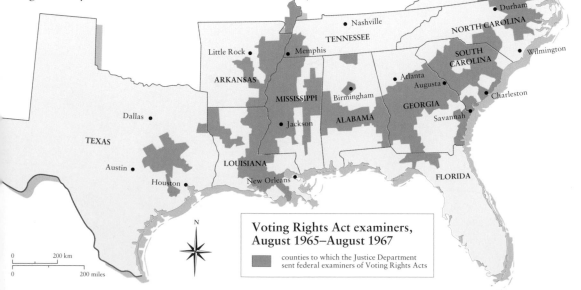

Voting Rights Act examiners, August 1965–August 1967

counties to which the Justice Department sent federal examiners of Voting Rights Acts

African Decolonization and Black Nationalism

At the height of the age of European imperialism, ministers of several countries met in Berlin in the winter of 1884–85 to divvy up the African continent. By the turn of the century, the great powers of Europe had effectively divided Africa for their individual exploitation. The basin of the Congo in Central Africa was "awarded" to King Leopold of Belgium, while England received a huge swath of East Africa stretching from Cairo to Cape Town. France acquired vast portions of North and West Africa including the kingdoms of Algeria, Mali and the Ivory Coast. Germany, Italy, Portugal, and Spain also received colonial possessions.

But the nationalism that led the European countries to become imperial powers also took root on the African continent, especially after World War I. At the insistence of Woodrow Wilson, national identity was used to redraw the boundaries of Europe at the Versailles Peace Conference, and was at the center of the vision for the League of Nations. African people who had been colonial "subjects" for years began to petition and even fight for independence. African Americans, who linked the colonial dependency of Africans with their own subjugation to whites, were central to the movement of decolonization in Africa. Marcus Garvey's Universal Negro Improvement Association transcended national boundaries and helped make "black nationalism" a worldwide ideology. "I do not believe," said the African American scholar W. E. B. DuBois, "that the descendants of Africans are going to be received as American citizens so long as the peoples of Africa are kept by white civilization in semi-slavery, serfdom, and economic exploitation."

The idea that blacks the world over needed to unite to combat racism and colonialism became the basis of a new "Pan-African" movement. DuBois called the first Pan-African Congress in 1919; others convened in 1921, 1923, and 1927. The Ghanian nationalist Kwame Nkrumah recalled how his own movement for independence was inspired by the struggle of African Americans. Nkrumah's movement was among the first in postwar Africa to achieve its goal—it won independence from Great Britain in 1957. Ghana's

Local children play on a toppled statue of the Portuguese founder of Nova Lisboa, in Angola. Angola declared its independence from Portugal in 1975.

success was quickly followed by Nigeria and most of francophone Africa (including Mali, Ivory Coast, Niger, Chad, and Mauritania) in 1960. African Americans were present in large numbers for the ceremonies marking the independence of Kenya, Sierra Leone, and Uganda, and they celebrated when white-run Rhodesia became Zimbabwe in 1980. Andrew Young, an American civil rights leader who became the U.S. ambassador to the United Nations in 1977, said "Africa's emergence from centuries of colonialism made us as African Americans feel part of a world movement for the liberation and self-determination of a subjugated peoples."

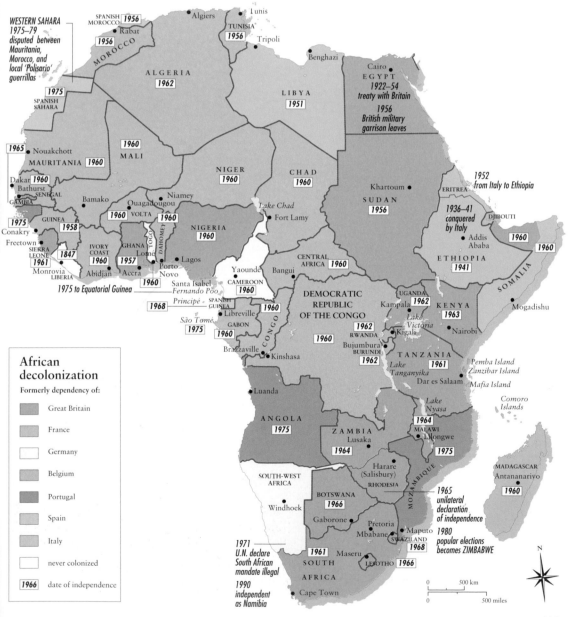

African decolonization

Formerly dependency of:

- Great Britain
- France
- Germany
- Belgium
- Portugal
- Spain
- Italy
- never colonized

1966 date of independence

WESTERN SAHARA 1975–79 disputed between Mauritania, Morocco, and local 'Polisario' guerrillas

115

The Nation of Islam

Blending the methods and arguments of Marcus Garvey with Islamic religion, the Black Muslims became one of the most enduring black nationalist groups in American history. The Hon. Wallace D. Fard founded the Nation of Islam in Detroit in 1929. In its early years, the group grew slowly and won few adherents. The Black Muslims' mission and message changed substantially in 1950, when Elijah Poole, who renamed himself Elijah Muhammad, took control. Muhammad moved the organization to Chicago and opened temples in New York, Detroit, and Los Angeles. Like Garvey, Muhammad preached that blacks should be proud of their African heritage—but he also claimed that whites were devils and inherently evil. This message reverberated powerfully at a time when black nationalism was spreading throughout Africa, creating new black nations, and white segregationists were holding the line in the U.S. South. The Black Muslims rejected all names that implied a connection with slavery and white America and sought economic and physical separation from the white community. This set them dramatically apart from the integrationist strategies of civil rights leaders like Martin Luther King Jr.

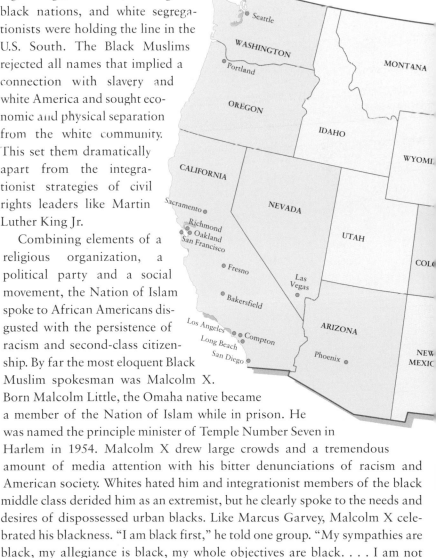

Malcolm X emerged in the early 1960s as the chief spokesperson for the Nation of Islam. His fiery pronunciations of black nationalism appealed to large numbers of urban African Americans.

Combining elements of a religious organization, a political party and a social movement, the Nation of Islam spoke to African Americans disgusted with the persistence of racism and second-class citizenship. By far the most eloquent Black Muslim spokesman was Malcolm X. Born Malcolm Little, the Omaha native became a member of the Nation of Islam while in prison. He was named the principle minister of Temple Number Seven in Harlem in 1954. Malcolm X drew large crowds and a tremendous amount of media attention with his bitter denunciations of racism and American society. Whites hated him and integrationist members of the black middle class derided him as an extremist, but he clearly spoke to the needs and desires of dispossessed urban blacks. Like Marcus Garvey, Malcolm X celebrated his blackness. "I am black first," he told one group. "My sympathies are black, my allegiance is black, my whole objectives are black. . . . I am not

interested in being an American, because America has never been interested in me." In stark contrast to King's strategies of nonviolence, Malcolm X asserted that blacks needed to resist their condition "by any means necessary." After a pilgrimage to Mecca, Malcolm X moderated some of his views and distanced himself from the Nation of Islam. He was assassinated in early 1965.

Louis Farrakhan, Malcolm X's successor as spokesman for the Nation of Islam, has at the same time marginalized the group and led it toward the mainstream. Farrakhan's separatism and open anti-Semitism (he called Judaism a "gutter religion") has fueled substantial protests against him. He achieved a notable success, however, in October 1995 as leader of the Million Man March in Washington, D.C. To a large crowd (though less than the touted million) of African American men, Farrakhan offered a chance to "atone" for brushes with the law, abusive relationships, or fathering illegitimate children. The protest was peaceful and, for many, inspirational.

The Nation of Islam

- center of importance for the Nation of Islam
- city with mosque(s) by 1997

state with mosque(s)

PART VI: THE AFRICAN AMERICAN COMMUNITY

> We want to be Americans, full-fledged Americans, with all the rights of other American citizens. But is that all? Do we want simply to be Americans? Once in a while through all of us there flashes some clairvoyance, some clear idea, of what America really is. We who are dark can see America in a way that white Americans cannot. And seeing our country thus, are we satisfied with its present goals and ideals?
> —W. E. B. DuBois, October 1926

The black intellectual W.E.B. DuBois wrote these words in the NAACP's influential newspaper *The Crisis* during the heyday of what became known as the "Harlem Renaissance." In the editorial, entitled "Criteria of Negro Art," he wrote of "a new desire to create . . . a new will to be" within the African American community. According to DuBois, this new realization of itself as both quintessentially American and also separate from the dominant, white society, was what allowed the African American artistic community to flourish after World War I. The emergence of African American arts, letters, music, and political activism burst forth on the streets of upper Manhattan in the 1920s and '30s and spread to other locales as well. The black people who made their homes in Northern cities did not find a paradise of racial equality and riches, but they did (along with their kin who remained in the South) make communities that changed the intellectual, political, and cultural life of the nation.

In Harlem, poets like Langston Hughes and Countee Cullen wrote of a black experience that was, at its core, an American experience as well. "I, too, sing America," exclaimed Hughes in one famous poem, echoing (and building upon) the art of Walt Whitman, who took as his muse the masses of New York City a century before. New Yorkers of every color flocked uptown to hear a revolutionary and evolving black musical form called jazz—imported from New Orleans in the second decade of the 20th century—at a nightclub named for the crop African Americans toiled over in slavery and freedom: the Cotton Club. Black orators stood on street corners, extolling the alternative visions of political parties like the Socialists, or the black nationalism of Marcus Garvey's United Negro Improvement Association.

What does the Harlem Renaissance tell us about the African American community in the 20th century? First, that a mostly Southern and rural people was in the process of becoming an urban one. Second, that voices too often repressed and ignored in the South were being heard more loudly than ever before. And finally, that black people had something unique and powerful to say about what it means to be an American. This final part of the atlas does not share the chronological or thematic organization of its predecessors, skipping across the 20th century and mapping snapshots of African American culture, politics, and demographics. The map of Harlem in the 1920s shows how,

A march by the National Association for the Advancement of Colored People (NAACP) to end legalized segregation and racism winds its way through Harlem. The formation of the NAACP in 1909 marked the beginning of the modern civil rights movement.

block by block, an urban neighborhood became a center for the intellectual and cultural life of the United States. Another map shows the diverse geography of black musical contributions, from the Delta blues to hip hop. A third chronicles a profound political shift in the African American community from the Republican to the Democratic Party. Blacks did not abandon the party of Abraham Lincoln to become one of the Democrats' key constituencies overnight, but voting trends underwent a significant change in the 1920s and '30s. Finally, the section concludes with an illustration of black political clout in America's cities and an examination of the African American community as reflected in the 1990 U.S. Census. The picture it portrays is of a community making continual gains and at the same time confronting persistent inequalities and racism. How else to explain the notable increases in black median earnings and educational levels that parallel a poverty and infant mortality rate almost three times that of white Americans? The African American struggle to survive, endure, and prosper in this country continues even as we enter a new millennium.

Since the first Africans arrived on the shores of Jamestown, black people have been a central part of the struggle for freedom in the Americas. It is in this aspect that African Americans' contributions transcend their relatively small numbers and minority status. On one side, in the words of the historian John Hope Franklin, blacks have been "constant reminders of the imperfection of social order and the immorality of its human relationships." On the other, African Americans have used their unique perspective to point out the weaknesses and the strengths, the horrors and the beauty of the American experience.

The Harlem Renaissance

African Americans continued to leave the rural South in unprecedented numbers in the 1920s. And despite the race riots of 1919, most continued to flock to industrial cities in the North. Already established black enclaves in cities like Chicago and New York beckoned to Southern African Americans and even Afro-Caribbeans like Marcus Garvey. Sociologists and demographers began to write about the "black metropolis"—a city within a city containing a complex melange of workers, artists, entertainers, intellectuals, and businesspeople. Nowhere was this mixture more culturally significant than in Harlem, a black neighborhood in upper Manhattan. There black writers, entertainers, and artists created a movement that forever changed the arts in America.

The black historian and writer James Weldon Johnson called Harlem the "Negro capital" of the United States: "[Harlem] is the Mecca for the sightseer, the pleasure-seeker, the curious, the adventurous, the enterprising, the ambitious, and the talented of the entire Negro world." Americans of all races flocked to the Cotton Club on Lenox Avenue to listen to jazz. Authors like Jean Toomer and poets like Countee Cullen helped insert the concept of race into the center of American literature. Harlem also attracted Langston Hughes, who recalled turning down an invitation to see Europe in the 1920s "because I wanted to see Harlem. . . . More than Paris or the Shakespeare country, or Berlin, or the Alps, I wanted to see . . . the greatest Negro city in the world."

Harlem's luxurious brownstones and apartment buildings were built around the turn of the century, before the neighborhood became predominantly African American. Perhaps the most elegant street in Harlem was Edgecombe Street on Sugar Hill, which became the home of elite Harlem society. Nearby was the Dunbar (named for the black poet Paul Lawrence Dunbar), a large garden apartment complex that was home to Harlem luminaries such as the poet Countee Cullen, the intellectual W.E.B. DuBois, the union leader A. Philip Randolph, and the actor/activist Paul Robeson. Langston Hughes settled in southern Harlem, in a brownstone facing Mt. Morris Park, now Marcus Garvey Memorial Park. Other Harlem landmarks include the Apollo Theater, where numerous black entertainers including Ella Fitzgerald began their careers; the Cotton Club, where Duke Ellington's Orchestra was the house band; the 369th Regiment Armory, home of the famed black regiment; and the Abyssinian Baptist Church, home of the largest black Protestant congregation in the United States.

The Harlem Renaissance displayed significant longevity, stretching into the darkest days of the Great Depression. Zora Neale Hurston, a brilliant intellectual and anthropologist, began collecting American and Caribbean folklore in the late 1920s. She wrote short stories, scholarly articles, and novels including *Moses, Man of the Mountain*, *Their Eyes Were Watching God*, and *Dust Tracks on the Road* between 1931 and 1943. The confidence and creativity of the Harlem Renaissance helped to rejuvenate entire ranges of American art in the 20th century. It also provided inspiration for generations of black and white artists, writers, and activists to come.

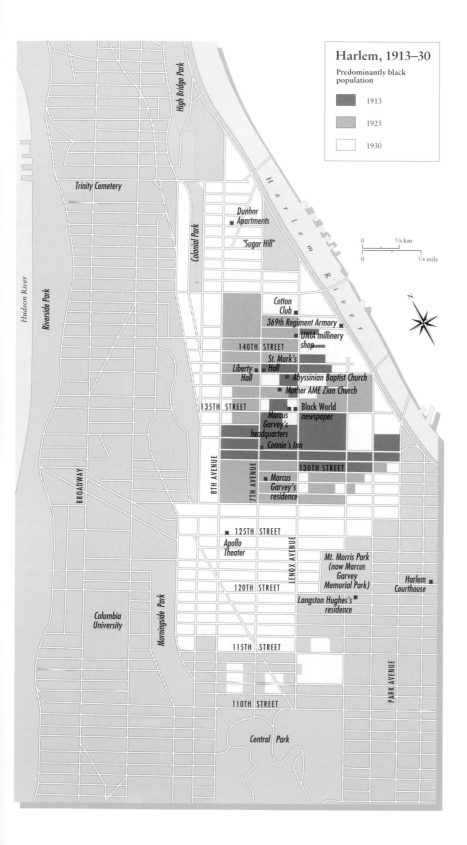

Harlem, 1913–30

Predominantly black population

■ 1913

■ 1925

□ 1930

Trinity Cemetery

High Bridge Park

Harlem River

Hudson River

Riverside Park

Colonial Park

Dunbar Apartments

"Sugar Hill"

0   1/4 km

0   1/4 mile

Cotton Club

369th Regiment Armory

UNIA millinery shop

140TH STREET

St. Mark's Hall

Liberty Hall

Abyssinian Baptist Church

Mother AME Zion Church

135TH STREET

Black World newspaper

Marcus Garvey's headquarters

Connie's Inn

8TH AVENUE

7TH AVENUE

130TH STREET

Marcus Garvey's residence

BROADWAY

125TH STREET

Apollo Theater

LENOX AVENUE

Mt. Morris Park (now Marcus Garvey Memorial Park)

Harlem Courthouse

120TH STREET

Columbia University

Morningside Park

Langston Hughes's residence

115TH STREET

PARK AVENUE

110TH STREET

Central Park

African American Musical Traditions

African Americans and whites flocked to the Cotton Club in Harlem to hear jazz music. The nightclub—home to Duke Ellington's band—epitomized uptown elegance during the Harlem Renaissance.

The African American musical tradition stretches back to songs and laments brought from Africa on the Middle Passage. Work songs, "shouts," and spirituals were always part of the lives of African Americans. In the 20th century, black Americans fashioned entirely new and significant musical traditions onto the older genres, including jazz, blues, and rock 'n' roll. With the advent of new technologies like radio and sound recording, African American music spread across the globe.

In the early 1900s, blacks from the Mississippi Delta and various other regions in the South developed a distinctive musical style called the blues. Ragtime, a distinctive new musical form that composers like Scott Joplin fashioned out of classically European and American techniques, was popular at the same time. The streets, clubs, bars, and brothels of New Orleans became a breeding ground for a new, American art form called jazz, which incorporated influences from both the blues and ragtime. Black and white audiences found the new sound irresistible. Jazz followed the Great Migration to places like Kansas City, Chicago, and New York, where the music's spontaneity, its overt sexuality ("jazz" was well-known as a slang term for sex), and its contagious rhythms and melodies packed cabarets and nightclubs during what became known as the "Jazz Age." Many early jazz greats, including Louis Armstrong, Bessie Smith, and Jelly Roll Morton influenced the next generation of performers to expand the music into a new, modern art form. Duke Ellington emerged in the 1920s as one of the greatest composers of all time, perfectly encapsulating jazz music's innovative creativity.

The blues also survived on its own, and artists like Ma Rainey, Robert Johnson, and Tampa Red (Hudson Woodbridge) each made contributions to the music's unique communication of loneliness and longing. Jazz underwent significant transformations in the 1930s, with "swing," and the 1940s, when Charlie Parker, Dizzy Gillespie, and Thelonious Monk helped invent "bebop." In the 1950s, rock 'n' roll burst on the scene as an amalgam of blues and country traditions. Little Richard and James Brown inspired both blacks and whites, and Ray Charles helped fashion a new genre called soul. Soul eventually became associated with Detroit, where the black-owned "Motown" label recorded blockbuster artists including the Supremes, Smokey Robinson, and the Jackson 5. In the late 1970s, young African American disc jockeys from poor urban neighborhoods combined elements of funk, disco and rock with older black, oral traditions to create rap. Rap's anti-establishment lyrics (often directed against white police officers) angered many politicians but became wildly popular with young people of all races. It is only the latest example of the African American musical tradition and its continued ability to affect global popular culture.

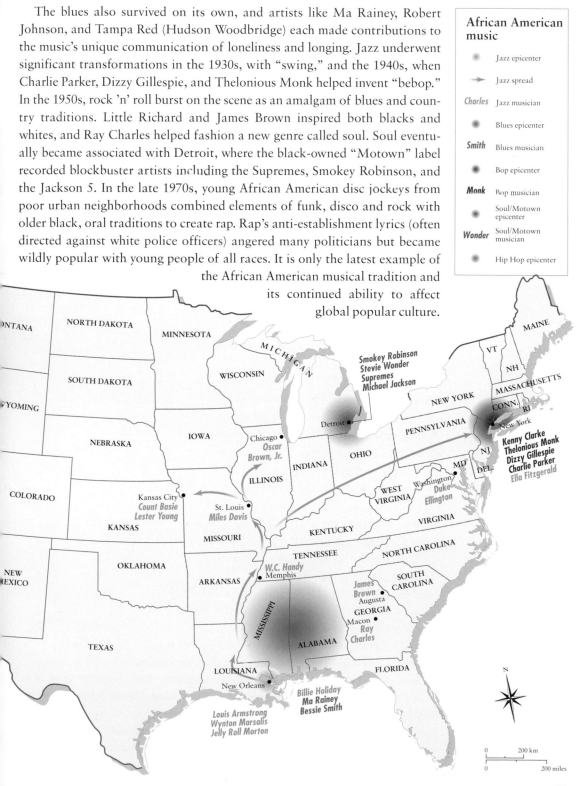

African American music

- Jazz epicenter
- Jazz spread
- *Charles*   Jazz musician
- Blues epicenter
- *Smith*   Blues musician
- Bop epicenter
- *Monk*   Bop musician
- Soul/Motown epicenter
- *Wonder*   Soul/Motown musician
- Hip Hop epicenter

African American Culture

African Americans' cultural heritage is both diverse and complex, drawing from a deep well of influences including West Africa, the Caribbean, the American South, and the urban environment. Perhaps this is why, as scholars like Henry Louis Gates Jr. have argued, blacks have had such a dramatic effect on the wider culture of the United States as a whole. On the one hand separate and different, African Americans have repeatedly (and profoundly) used their uniqueness to speak to human conditions and emotions that are shared by everyone. As an example, Gates cites the way the African American author Toni Morrison (who won the Nobel Prize for Literature in 1993) takes "the blackness of the culture for granted, as a springboard" to write about the larger American and, indeed, human, experience.

Morrison has used her home state of Ohio, which shares a border with the former slave state of Kentucky, as a setting for her fiction, most memorably in *Beloved*, a ghost story about life after slavery. Many significant black authors have come from Southern states, including Zora Neale Hurston and James Weldon Johnson (Florida), Richard Wright (Mississippi), and Alice Walker (Georgia). Ralph Ellison (Oklahoma), Gwendolyn Brooks (Kansas), and Lorraine Hansberry (Illinois), all of whom deal honestly and dramatically with the topic of race in their fiction, hail from the Midwest. The "black Mecca" of Harlem produced many significant black authors, including James Baldwin, Toni Cade Bambara, and Paul Marshall.

African American contributions to American music, from gospel to jazz to rock 'n' roll, cannot be overstated. Duke Ellington, perhaps America's greatest composer, was born in Washington, D.C., in 1899. Jazz great and bebop legend

Charlie Parker, and jazz and rythm 'n' blues singer Big Joe Turner both came from post-Great Migration Kansas City, Missouri. Paul Robeson, a multitalented actor, athlete, orator, and baritone, came from a small but well-established black community in Princeton, New Jersey. The Mississippi Delta region produced Bluesmen B. B. King (Indianola), Muddy Waters (Rolling Fork), Mississippi John Hurt (Teoc), and Robert Johnson (Hazelhurst). Other blues singers hailed from a wide swath across the South, ranging from Chattanooga, Tennessee (Bessie Smith), Columbus, Georgia (Ma Rainey), Albany, Georgia (Ray Charles), and tiny Shiloh, Louisiana (Leadbelly). African Americans have also made substantial contributions to American culture in the visual arts (painter Jacob Lawrence and photographer Gordon Parks), dance (choreographer Alvin Ailey), and film (Melvin Van Peebles and Spike Lee).

African American artists

author

musician

painter

dancer

Black Political Realignment, 1928–40

Between the Civil War and the election of 1928, an overwhelming majority of African American voters were Republicans. This loyalty was completely understandable: from its inception in 1854, the Republican Party was anti-slavery. The Democratic Party, on the other hand, was closely identified with slaveholders and secession. The first Republican President, Abraham Lincoln, had signed the Emancipation Proclamation, and his allies and successors passed and (for a time) enforced the 13th, 14th, and 15th amendments. During Reconstruction, newly enfranchised black voters in the South helped the Republican Party achieve national political hegemony that lasted until at least 1880.

The South remained overwhelmingly Democratic after Reconstruction, and the party solidified its dominance by disfranchising African Americans and passing Jim Crow laws. When they could, blacks voted against Southern racism and the Democratic Party by voting Republican, which they did in every major local, state, and national election for more than 60 years. Not that a single 20th-century Republican president had even remotely approached Lincoln's legacy. At first Theodore Roosevelt, who publicly opposed lynching, appointed blacks to federal offices, and sought the advice of Booker T. Washington, seemed likely to fill Lincoln's shoes. But Roosevelt retreated significantly in his second term, and African American voters began to feel abandoned by the "party of Lincoln."

By the late 1920s, the national Republican Party hit upon a new strategy: to use its conservative principles to woo white Southern voters. This "lily-white" strategy, combined with the economic misery of the Great Depression, began a process by which African American voters switched their party allegiances. In 1934, Arthur W. Mitchell switched from Republican to Democrat and won election to the House of Representatives from Illinois. Mitchell was the first black Democrat elected to Congress. New Deal Democrats in Northern cities—and especially in New York and Chicago—began reaching out to black voters and promising relief from the Great Depression. Like most other Americans, black voters supported the policies of Democratic President Franklin Delano Roosevelt.

African Americans especially responded to Roosevelt's "Black Cabinet," an unofficial group of prominent blacks who advised the president on race relations, and to the first lady Eleanor Roosevelt, who publicly associated with blacks and was an outspoken proponent of racial equality. Many older blacks continued to vote Republican, but in New York City more than 81 percent of black voters cast Democratic ballots in 1936—an increase of almost 50 percent from 1928. New Deal programs like the Works Progress Administration (WPA) and the National Recovery Administration (NRA) helped African Americans learn to read and get back on their feet financially. The WPA also sent oral historians to interview aging ex-slaves about their lives before and during the Civil War.

Southern blacks were less likely to support the party of Roosevelt, because to them, the Democratic Party was and had always been the party of

segregation and white supremacy. Today, however, a large majority of both Northern and Southern blacks routinely cast votes for Democratic candidates while, in another major political shift, many of the former "Dixiecrats" have become Republicans.

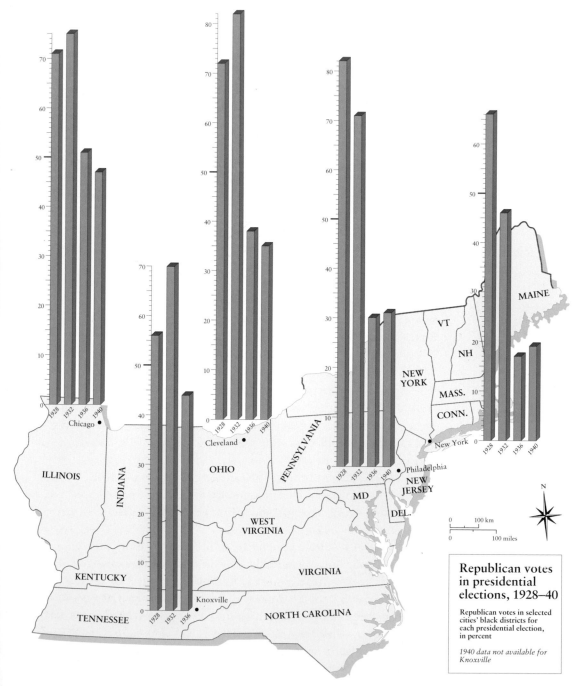

Republican votes in presidential elections, 1928–40

Republican votes in selected cities' black districts for each presidential election, in percent

1940 data not available for Knoxville

127

Urbanization and Black Mayors

After World War II, blacks continued to leave the rural South in large numbers, with a large majority moving to cities in the North and West. Such cities as Chicago, Cleveland, New York, Detroit, and Philadelphia experienced explosions in black residents. But blacks moved to Southern cities, too. New Orleans, Atlanta, Charleston, Montgomery, Birmingham, and Baltimore each received huge influxes of African Americans in the postwar years. Increasingly concentrated in cities, in black neighborhoods called ghettos, African Americans contributed a growing share of the electorate in states like New York, Illinois, Ohio, Pennsylvania, and Michigan. No longer could politicians for local and national office ignore the black population at election time. Although black voters formed a small proportion of the electorate as a whole and often faced intimidation, harassment, and even violence at the polls, they often had an impact on the electoral outcome in several cities. In some localities, blacks banded together to enhance their political clout. The Atlanta Negro Voters League, for example, conscientiously represented the black constituency for decades. By 1954, blacks had been elected to office in 11 municipalities in the South.

Mass migration to urban areas and this growing awareness of the power of the ballot box propelled hundreds of African Americans into elective office after 1954, especially at the citywide level. The first black mayor was elected in 1885 with the incorporation of Princeville, North Carolina, the oldest black incorporated town in the country. Since then hundreds of blacks have been elected mayors in the United States, in majority-black and majority-white towns alike. Between 1963 and 1973 blacks served as mayors of large cities such as Los Angeles, Newark, Cleveland, and Gary, Indiana, as well as dozens of small towns including Tuskegee, Alabama, and Fayette, Arkansas. The decade of the 1970s saw the elections of African American mayors to large Southern cities like New Orleans and Atlanta. In the 1980s, African Americans won mayoral races in other significant cities: Harold Washington in Chicago, Sharon Pratt Kelley in Washington, D.C., and David Dinkins in New York.

By 1990, 314 African Americans were mayors of their cities, nearly twice the number in 1980 and eight times as many as in 1970. In the 1990s African American mayors of the nation's largest cities either stepped down or were defeated for re-election. But in 1999, there were black mayors in cities and towns as varied as Houston, Texas (Lee Brown); San Francisco (Willie Brown); Pasadena, California (Chris Adams); Trenton, New Jersey (Douglas Bradley); Detroit, Michigan (Dennis Archer), and Minneapolis, Minnesota (Sharon Campbell).

Black elected officials, 1970–90

	members of Congress	state senators	state representatives	county officials	mayors	city council members	school board members	total black elected officials	percentage of change
1970	10	31	137	92	48	552	362	1,469	-
1975	18	53	223	305	135	1,237	894	3,503	138.5
1980	17	70	247	451	182	1,809	1,149	4,912	40.2
1985	20	90	302	611	286	2,189	1,368	6,056	23.3
1990	26	108	340	810	314	2,972	1,561	6,131	1.2

Cities with black mayor, 1999

City size, in inhabitants

● over 1 million

● 500,000 to 1 million

● 100,000 to 500,000

· 50,000 to 100,000

African American Population, 1990

According to the 1990 U.S. Census, there were 29,986,000 African Americans in the United States. Blacks thus made up 12 percent of the total population of 248,710,000.

The historical legacy of slavery continues to affect where African Americans live at the end of the 20th century. The states with the highest percentages of African Americans are all former slaveholding states: Mississippi (35%), South Carolina (30%), Louisiana (29%), Georgia (27%), Alabama (26%), Maryland (23%), and North Carolina (22%). Yet the Great Migration of the 20th century has also placed an indelible African American mark on the country's urban landscape. The largest populations of African Americans are in New York (2.1 million), Chicago (1.4 million), Detroit (800,000), Philadelphia (700,000), and Los Angeles (600,000). Large cities (those with a population of 300,000 or higher) with the greatest percentages of blacks are still mostly in the South: Washington, D.C (70%), Atlanta (67%), New Orleans, (55%) and Memphis (40%), with the exception of Detroit (65%).

Although African Americans have made tremendous strides in the struggle for equality in the United States, demographic evidence suggests blacks still lag behind in significant areas. Income is one: nowhere in the United States is the median African American family income ($18,098) equal to that of the white family ($30,853). The greatest disparities exist in the Midwest and Northeast. Twenty-nine percent of black families live below the poverty line while only 11 percent of white families do. The rate of infant mortality for African Americans is more than twice as high as for whites, and blacks attain 11.5 years of education compared to 12.5 years for whites. Despite the centuries-old struggle for America to live up to the words in the Declaration of Independence—that all men are created equal and endowed by their creator with the inalienable rights of life, liberty, and the pursuit of happiness—it is clear that true equality, be it racial, economic, or gender-based, has proved an elusive goal.

A multicultural crowd gathers on the streets of Brooklyn, New York for a recent West India Day parade, highlighting the diversity of the cities black population.

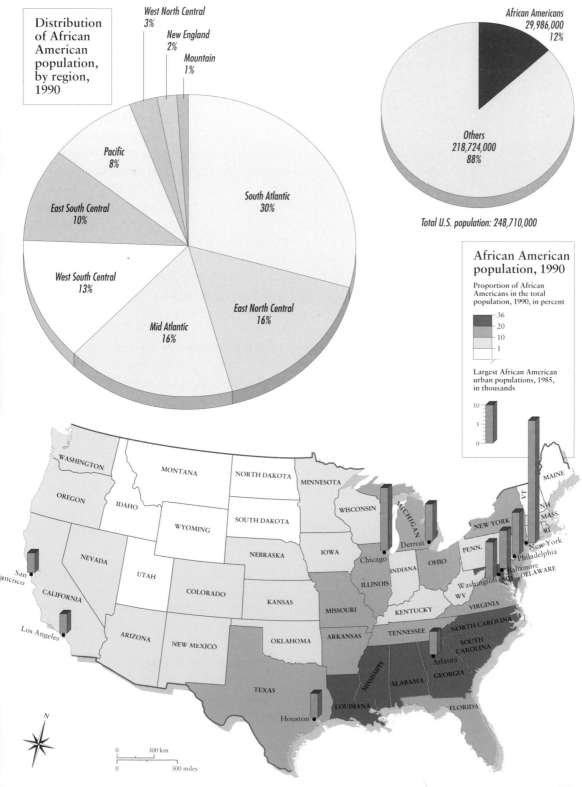

Distribution of African American population, by region, 1990

West North Central 3%

New England 2%

Mountain 1%

Pacific 8%

East South Central 10%

West South Central 13%

South Atlantic 30%

East North Central 16%

Mid Atlantic 16%

African Americans 29,986,000 12%

Others 218,724,000 88%

Total U.S. population: 248,710,000

African American population, 1990

Proportion of African Americans in the total population, 1990, in percent

36
20
10
1

Largest African American urban populations, 1985, in thousands

10
5
0

Chronology

3200 B.C.   Egypt founded in Nile River Valley in Northeast Africa.

A.D. 610–733   Islamic faith spreads across Northern Africa.

700–900   North African Berbers establish Trans-Saharan slave trade.

900–1200   Ghanaian Empire reaches its peak.

1200–1450   Empire of Mali reaches its peak.

1450–1800   Songhay Empire reaches its peak.

1494   Treaty of Tordesillas divides New World between Portugal and Spain.

1502   Portuguese ships begin transporting West African slaves to New World.

1619   Dutch merchant delivers 20 Africans to Jamestown colony in Virginia.

1641   Massachusetts becomes first British North American colony to legalize slavery.

1661   Maryland mandates slavery as a life-long condition for Africans and their children.

1739   Stono Rebellion near Charleston, South Carolina; 80 slaves and 20 whites killed.

1741   Great Negro Plot detected in New York City.

1770   Crispus Attucks killed by British soldiers in Boston Massacre.

1770   Phillis Wheatley publishes her first poem, "On the Death of the Reverend George Whitefield."

1775   British General Lord Dunmore offers to free slaves willing to join British Army.

1775–83   More than 5,000 free blacks serve in Continental Army; at least that many runaway slaves and free blacks fight with the British.

1775   First abolition society organized by Philadelphia Quakers.

1776   33 enslaved Africans die in a revolt at Cape Coast Castle slave fort.

1777   Vermont bans slavery in Constitution; first state to do so.

1780   Massachusetts judge rules that state's bill of rights outlaws slavery.

1787   Northwest Ordinance bans slavery in territory north and west of the Ohio River.

1787   Slavery written into U.S. Constitution.

1789   Olaudah Equiano publishes *The Interesting Narrative of Olaudah Equiano*.

1790   The Rev. Richard Allen founds the African Methodist Episcopal (A.M.E.) Church in Philadelphia.

1791   Haitian Revolution begins with a slave revolt.

1791   Astronomer Benjamin Banneker issues the first edition of his almanac.

1793   Congress enacts the first Fugitive Slave Law, requiring free states to return runaways.

1793   Inventor Eli Whitney builds cotton gin; makes cotton a profitable crop and reinvigorates slavery.

1794   Toussaint L'Ouverture overthrows French rule in Haiti, proclaims a republic.

1800   Virginia militia thwarts Gabriel Prosser's slave conspiracy.

1808	Foreign slave trade officially ended by U.S. Constitution.
1820	Missouri Compromise establishes 36°30' parallel as dividing line between slave and free states.
1821	New York erects $250 property qualification for black voters at the same time it eliminates such requirements for whites.
1822	Denmark Vesey slave conspiracy foiled in Charleston, S.C.
1827	John Russworm and Samuel Cornish publish first edition of *Freedom's Journal* in New York.
1829	David Walker's radical *Appeal . . . to the Colored Citizens of the World* widely circulated along eastern seaboard.
1831	Abolitionist William Lloyd Garrison publishes first issue of the *Liberator*.
1831	Nat Turner leads bloodiest slave revolt in U.S. history in Southampton County, Virginia, killing 55 whites; Turner and at least that many killed in retaliation.
1838	Frederick Douglass escapes from slavery and begins career as an abolitionist writer and lecturer.
1841	Upstate New Yorker Solomon Northrup kidnapped and sold into slavery.
1846	"Wilmot Proviso" attempts to ban slavery in territory acquired as a result of Mexican War.
1848	Ellen Craft poses as a sick slaveholder as she and her husband escape from slavery. Anti-slavery Free Soil Party founded.
1850	Compromise of 1850 admits California as a free state, bans the slave trade in Washington, D.C., and enacts a strict Fugitive Slave Law.
1853	Kidnap victim Solomon Northrup rejoins his family in New York.
1854	Kansas-Nebraska Act opens vast new territory to slavery and leads to formation of anti-slavery Republican Party.
1854	Fugitive slave Anthony Burns returned to slavery as thousands in Boston line the streets in protest.
1857	In Dred Scott Decision, U.S. Supreme Court declares slaves are not citizens and rules Missouri Compromise unconstitutional.
1859	John Brown leads raid on federal arsenal at Harper's Ferry in Virginia.
1861	Southern states form Confederate States of America. Civil War begins.
1862	Congress passes Contraband Act, classifying runaway slaves as contraband of war.
1863	Lincoln issues Emancipation Proclamation, declaring slaves in rebellious states to be "forever free." Union Army's ban on black soldiers lifted; 186,000 African Americans enlist during final two years of the Civil War. Black soldiers of the 54th Massachusetts fight bravely in assault on Ft. Wagner.
1865	Lincoln assassinated. Southern states enact "Black Codes." Congress passes 13th Amendment forever outlawing slavery. Reconstruction begins.

1866	Congress passes 14th Amendment granting citizenship to African Americans. Ku Klux Klan formed in Tennessee. Congress authorizes establishment of four permanent black units to fight Indians in the West. Fisk University founded as Fisk Free Colored School in Nashville.
1869	Congress passes 15th Amendment enfranchising black men.
1870–71	Federal Ku Klux Klan Acts attempt to protect black voters from white terror.
1875	Tennessee becomes first state to institute Jim Crow law.
1877	Federal troops withdraw from the South; Reconstruction ends.
1880	60,000 "Exodusters" leave Nashville for Kansas.
1880s	Southern states institute Jim Crow laws.
1884	European nations convene in Berlin and divide Africa into colonies.
1890	Mississippi begins using literacy tests to disfranchise black voters.
1892	A record 235 African Americans killed by lynch mobs.
1896	In *Plessy v. Ferguson*, U.S. Supreme Court upholds Jim Crow laws as constitutional.
1909	African Americans and their white allies form the National Association for the Advancement of Colored People (NAACP) to end forced segregation.
1910	Several organizations merge to form the National Urban League (NUL) to direct black immigrants to jobs, housing, and educational opportunities.
1912	Self-proclaimed "inventor of jazz," Ferdinand "Jelly Roll" Morton, publishes his first song, "The Jelly Roll Blues."
1916	Great Migration begins; 500,000 leave rural South by 1920.
1917	Jamaican immigrant Marcus Garvey founds the Universal Negro Improvement Association (UNIA).
1917–19	More than 400,000 African Americans serve in U.S. Army during World War I.
1917	James Weldon Johnson publishes *Fifty Years and Other Poems*; marks beginning of the Harlem Renaissance.
1919	W.E.B. DuBois convenes first Pan-African Congress.
1920	Negro National Baseball League founded.
1926	Langston Hughes, the poet laureate of the Harlem Renaissance, publishes *Weary Blues*.
1929	Nation of Islam (Black Muslims) formed in Detroit.
1932	African Americans begin to switch allegiance from the Republican to the Democratic Party.
1937	Zora Neale Hurston publishes *Their Eyes Were Watching God*.
1940	The War Department begins training black pilots at Tuskegee, Alabama.
1944	U.S. Army reaches peak strength during World War II; 701,678 African Americans in service in U.S. Army.
1945–47	Jazz greats Thelonious Monk, Charlie Parker, and Dizzy Gillespie pioneer "bebop" jazz at Minton's Play House, an after-hours club in Harlem.

1947	Jackie Robinson joins the Brooklyn Dodgers, breaking the color barrier in major league baseball.
1948	President Harry S Truman integrates the U.S. armed forces.
1954	U.S. Supreme Court declares segregation unconstitutional in *Brown v. Board of Education of Topeka, Kansas*.
1955	Montgomery bus boycott begins when Rosa Parks refuses to give up her seat to a white man. The one-year boycott brings Dr. Martin Luther King Jr. to national prominence.
1957	President Eisenhower sends troops to enforce school desegregation in Little Rock, Arkansas. Southern Christian Leadership Conference (SCLC) formed. Several African Americans on hand as Ghana wins independence from Great Britain.
1959	Berry Gordy founds Motown Records in Detroit.
1960	Four students from North Carolina A&T College launch the sit-in movement at the Greensboro, N.C., Woolworth's lunch counter.
1961	Freedom Riders protest segregated buses in Deep South.
1962	James Meredith integrates the University of Mississippi.
1963	Civil rights protests explode in Birmingham, Alabama. More than 250,000 Americans participate in March on Washington.
1964	Congress passes Civil Rights Act.
1965	King leads protesters on a march from Selma to Montgomery to press for voting rights. Congress passes Voting Rights Act. Malcolm X assassinated in New York.
1966	Student Non-violent Coordinating Committee (SNCC) splits with SCLC; adopts "black power" slogan. Black Panther Party organized.
1967	Thurgood Marshall named associate justice of U.S. Supreme Court.
1968	Martin Luther King assassinated in Memphis.
1972	New York Congresswoman Shirley Chisholm runs for the Democratic presidential nomination.
1984	The Rev. Jesse Jackson runs for the Democratic presidential nomination and garners significant black and white support.
1988	Jackson again competes for the Democratic nomination and wins 6.6 million votes.
1989	L. Douglas Wilder wins gubernatorial election in Virginia, becoming first African American elected chief executive of a state.
1992	Carol Moseley Braun becomes the first black woman elected to the U.S. Senate.
1993	Toni Morrison wins the Nobel Prize for Literature
1995	Nation of Islam leader Louis Farrakhan leads Million Man March in Washington, D.C.
1997	President Bill Clinton announces his Initiative on Race, and appoints an advisory board chaired by the historian John Hope Franklin to report on the state of race relations in the United States.

Further Reading

Narrative Overviews and Reference Works

Bennett, Lerone, *Before the Mayflower: A History of Black America*, Penguin, 1993.

Blassingame, John, *The Slave Community: Plantation Life in the Antebellum South*, Oxford University Press, 1979.

Franklin, John Hope, and Alfred A. Moss, Jr., *From Slavery to Freedom: A History of African Americans* (7th ed.), McGraw-Hill, 1994.

Harding, Vincent, *There Is a River:The Black Struggle for Freedom in America*, Harcourt Brace, 1981.

Jones, Jacqueline, *Labor of Love, Labor of Sorrow: Black Women, Work and Family, From Slavery to the Present*, Vintage, 1986.

Palmer, Colin A., *Passageways: An Interpretive History of Black America* (2 vols.), Holt, Rinehart & Winston, 1998.

Part I: The Roots of Black America

Blackburn, Robin, *The Making of New World Slavery: From the Baroque to the Modern, 1492–1800*, Verso, 1998.

Brown, Kathleen M., *Good Wives, Nasty Wenches, and Anxious Patriarchs: Gender, Race, and Power in Colonial Virginia*, University of North Carolina Press, 1996.

Curtin, Philip D., *The Atlantic Slave Trade: A Census*, University of Wisconsin Press, 1969.

Jordan, Winthrop D., *White Over Black: American Attitudes Towards the Negro, 1550–1812*, University of North Carolina Press, 1968.

Kolchin, Peter, *American Slavery, 1619–1877*, Hill and Wang, 1993.

Morgan, Edmund, *American Slavery/American Freedom*, W.W. Norton and Co., 1975.

Thomas, Hugh, *The Slave Trade: The Story of the Atlantic Slave Trade, 1440–1870*, Simon and Schuster, 1997.

Thornton, John, *Africa and Africans in the Making of the Modern World, 1400–1800* (2d ed.), Cambridge University Press, 1992.

Wood, Peter, *Black Majority: Negroes in Colonial South Carolina from 1670 through the Stono Rebellion*, W.W. Norton and Co., 1974.

Part II: Two Communities, Slave and Free

Aptheker, Herbert, *American Negro Slave Revolts* (6th ed.), International Publishers, 1993.

Berlin, Ira, *Many Thousands Gone: The First Two Centuries of Slavery in North America*, Belknap Press, 1998.

Campbell, James T., *Songs of Zion: The African Methodist Episcopal Church in the United States and South Africa*, University of North Carolina Press, 1995.

Fox-Genovese, Elizabeth, *Within the Plantation Household: Black and White Women of the Old South*, University of North Carolina Press, 1988.

Genovese, Eugene, *Roll, Jordan, Roll: The World the Slaves Made*, Vintage Books, 1972.

Gutman, Herbert, *The Black Family in Slavery and Freedom, 1750–1825*, Random House, 1976.

Horton, James O., *Free People of Color: Inside the African American Community*, 1993

Litwack, Leon F., *North of Slavery: The Negro in the Free States, 1790–1860*, University of Chicago Press, 1961.

Raboteau, Albert J., *Slave Religion: The Invisible Institution in the Antebellum South*, Oxford University Press, 1980.

Voelz, Peter M., *Slave and Soldier: The Military Impact of Blacks in the Colonial Americas*, Garland, 1993.

Part III: Toward Freedom

DuBois, W. E. B., *Black Reconstruction in America*, Maxwell Macmillan International, 1992.

Fehrenbacher, Don E., *Slavery, Law, and Politics: The Dred Scott Case in Historical Perspective*, Oxford University Press, 1981.

Filler, Louis, *Crusade Against Slavery, 1830–1860*, Reference Publications, 1986.

Foner, Eric, *Reconstruction: America's Unfinished Revolution, 1863–1877*, Harper and Row, 1988.

Freehling, William W., *The Road to Disunion: Secessionists at Bay, 1776–1854*, Oxford University Press, 1990.

Gara, Larry, *The Liberty Line: The Legend of the Underground Railroad*, University of Kentucky Press, 1961.

Levine, Lawrence, *Black Culture and Black Consciousness: Afro-American Folk Thought from Slavery to Freedom*, Oxford University Press, 1978.

McPherson, James M., *Battle Cry of Freedom: The Civil War Era*, Oxford University Press, 1988.

Painter, Nell I., *Exodusters: Black Migration to Kansas After Reconstruction*, Alfred A. Knopf, 1977

Quarles, Benjamin, *Black Abolitionists*, Da Capo Press, 1991.

Stampp, Kenneth M., *The Peculiar Institution: Slavery in the Ante-Bellum South*, Vintage Books, 1989.

Part IV: African Americans Under Arms

Aptheker, Herbert, *The Negro in the Civil War*, International Publishers, 1962.

Berlin, Ira, Joseph P. Reidy, and Leslie S. Rowland, eds., *The Black Military Experience*, Cambridge University Press, 1982.

Francis, Charles E., *The Tuskegee Airmen: The Men Who Changed A Nation* (3rd ed.), Branden Publishing, 1993.

McGregor, Morris J., *Integration of the Armed Forces, 1940–1965*, Center for Military History, 1981.

Mullin, Robert W., *Blacks in America's Wars: The Shift in Attitudes from the Revolutionary War to Vietnam*, Monad Press, 1974.

Nalty, Bernard C., *Strength for the Fight: A History of Black Americans in the Military*, Free Press, 1986.

Williams, Charles H., *Negro Soldiers in World War I: The Human Side*, AMS Press, 1970.

Part V: The Struggle for Equality

Branch, Taylor, *Parting the Waters: America in the King Years, 1954–63*, Simon and Schuster, 1988.

Carmichael, Stokely, and Charles Hamilton, *Black Power: The Politics of Liberation in America*, Penguin, 1969.

Carson, Clayborne, *In Struggle: SNCC and the Black Awakening of the 1960s*, Harvard University Press, 1995.

Chafe, William H., *Civilities and Civil Rights: Greensboro, North Carolina, and the Black*

Struggle for Freedom, Oxford University Press, 1981.

Cronon, E. David, *Black Moses: The Story of Marcus Garvey and the Universal Negro Improvement Association* (2d. ed.), University of Wisconsin Press, 1981.

DuBois, W. E. B., *The Souls of Black Folk*, Essays and Sketches (1903), Penguin, 1996.

Gaines, Kevin K., *Uplifting the Race: Black Leadership, Politics, and Culture in the Twentieth Century*, University of North Carolina Press, 1996.

Garrow, David J., *Bearing the Cross: Martin Luther King, Jr., and the Southern Christian Leadership Conference*, Vintage, 1993.

Haley, Alex, *The Autobiography of Malcolm X*, Ballantine Books, 1965.

Harrison, Alferdteen, ed., *Black Exodus: The Great Migration from the American South*, University Press of Mississippi, 1991.

Lawson, Steven F., *Black Ballots: Voting Rights in the South, 1944–1969*, Columbia University Press, 1976.

Lewis, David Levering, *The Race to Fashoda: European Colonialism and African Resistance to the Scramble for Africa*, Bloomsbury, 1988.

_____, *W. E. B. DuBois: Biography of a Race*, Henry Holt, 1994.

Peterson, Robert, *Only the Ball Was White: A History of Legendary Black Players and All-Black Professional Teams*, Oxford University Press, 1992.

Swain, Carol M., *Black Faces, Black Interests: The Representation of African Americans in Congress*, Harvard University Press, 1995.

Trotter, Joe William, ed., *The Great Migration in Historical Perspective: The Making of an Industrial Proletariat, 1915–45*, Indiana

University Press, 1991.

Woodward, C. Vann, *Origins of the New South*, Louisiana State University Press, 1951.

_____, *The Strange Career of Jim Crow*, Oxford University Press, 1982.

Part VI: The African American Community

Bruce, Dickson, *Black American Writing from the Nadir: The Evolution of a Literary Tradition, 1877–1915*, Louisiana State University Press, 1989.

Huggins, Nathan I., *Harlem Renaissance*, Oxford University Press, 1973.

Lewis, David Levering, *When Harlem Was in Vogue*, Oxford University Press, 1989.

Raboteau, Albert, *A Fire in the Bones: Reflections on African American Religious History*, Beacon Press, 1995.

Sitkoff, Harvard, *A New Deal for Blacks: The Emergence of Civil Rights as an Issue*, Oxford University Press, 1981.

Southern, Eileen, *The Music of Black Americans: A History* (3rd ed.), Norton, 1997.

Stuckey, Sterling, *Going Through the Storm: The Influence of African American Art in History*, Oxford University Press, 1994.

Weiss, Nancy J., *Farewell to the Party of Lincoln: Black Politics in the Age of FDR*, Princeton University Press, 1983.

Woodson, Carter G., *The History of the Negro Church* (3rd ed.), Associated Publishers, 1992.

Index

Acknowledgments

The author wishes to thank the University of Kansas, the Huntington Library, the National Endowment for the Humanities and the American Council of Learned Societies for supporting this project, and Leslie Tuttle, Andrew Frank, and Eric T. L. Love for their suggestions on how to make it better.

Pictures are reproduced by permission of, or have been provided by the following:

American Antiquarian Society: 22
Arcadia Editions Limited: 21, 48
Archive Pictures Inc.: 105
Associated Press Worldwide: 92
Bettmann Archive: 66, 116, 119
Brown Brothers: 96
Brown University: 74
Culver Pictures: 84
e.t. archives: 24, 32
Frank Leslie's Illustrated Newspaper archive: 58
Historic New Orleans Collection: 46
Hulton Getty: 114
Kansas State Historical Society: 69
Library Company of Philadelphia: 44
Library of Congress: 70
Link Picture Library, London: 130
Montana Historical Society: 78
National Archives: 11, 12, 82
Peter Newarks American Pictures: 53, 63, 112, 139
Rockefeller Folk Art Museum: 30
U.S. Army Archives: 90
Private collections

Design: Malcolm Swanston, Elsa Gibert

Cartography: Elsa Gibert, Malcolm Swanston

Drawing p. 21, 48: Peter A. B. Smith.